D0903175

Synapses

From the Howard Hughes Medical Institute

Synapses

Edited by

W. Maxwell Cowan

Howard Hughes Medical Institute
Chevy Chase, Maryland

Thomas C. Südhof

Howard Hughes Medical Institute
Center for Basic Neuroscience
and Department of Molecular Genetics
The University of Texas Southwestern Medical School
Dallas, Texas

Charles F. Stevens

Howard Hughes Medical Institute
The Salk Institute for Biological Studies
La Jolla, California

with the assistance of

Kevin Davies

Howard Hughes Medical Institute

The Johns Hopkins University Press
Baltimore and London

© 2001 The Johns Hopkins University Press
All rights reserved. Published 2001
Printed in the United States of America on acid-free paper
9 8 7 6 5 4 3 2 1

The Johns Hopkins University Press
2715 North Charles Street
Baltimore, Maryland 21218-4363
www.press.jhu.edu

Library of Congress Cataloging-in-Publication Data

Synapses / edited by W. Maxwell Cowan, Thomas C. Südhof, Charles F. Stevens.
     p.   cm.
   Includes bibliographical references and index.
   ISBN 0-8018-6498-4 (alk. paper)
   1. Synapses 2. Neural transmission. I. Cowan, W. Maxwell. II. Südhof,
Thomas C. III. Stevens, Charles F., 1934– .
  QP364 .S945   2000
  612.8—dc21   00-042399

A catalog record for this book is available from the British Library.

Contents

Preface

The introduction of the term *synapse* for the sites at which axons make functional contacts with their target cells marked the beginning of a new era in the study of the nervous system. Sherrington, who coined the term, had become convinced from the work of several neuroanatomists, but principally Ramón y Cajal, that the earlier notion that neuronal processes are in direct "protoplasmic" continuity was untenable. Not only was it difficult to conceive how specific neural functions could be executed if the central nervous system was an elaborate syncitium, but there was also compelling—albeit indirect—evidence from developmental, pathological, and various neuroanatomical studies that argued strongly for the view that neurons are morphologically distinct entities whose processes are contiguous with those of other cells but structurally separated from them. Although the acrimonious debate between those who had espoused the "reticular theory" of neuronal interactions and those who were identified with the "neuron theory" continued well into the twentieth century, for all but the most committed reticularists (like Camillo Golgi and his followers) the issue was effectively settled by 1897, when Sherrington was moved to write that, "As far as our present knowledge goes, we are led to think that the tip of a twig of the [axonal] arborescence is not continuous with but merely in contact with the substance of the dendrite or cell body on which it impinges. Such a special connection of one nerve cell with another might be called '*synapsis*.'" And more than 50 years later he was to say of Cajal's great contribution: "The so-called nerve networks with unfixed direction of travel he swept away. The nerve circuits are valved, he said, and he was able to point out where the valves lie—namely where one nerve cell contacts the next one."

To Cajal's identification of synapses as the mediators of nerve cell interactions, Sherrington was to add a major contribution of his own—the discovery that *inhibition* is as important as *excitation* in determining coordinated neural activity—or, to use his lucent phrase, the "integrative action of the nervous system."

Once the morphological issue of how nerve cells interact had been resolved, attention naturally turned toward understanding the mechanism of synaptic transmission: was it electrical or was it chemical? The fact that there was already considerable evidence to suggest that the transmission of nerve impulses was electrical led most physiologists to espouse the view that transmission at synapses was probably also electrical. But some physiologists and most pharmacologists were convinced—and argued, at times with great vehemence—that it must be chemical. The debate between the two camps—later referred to as the "soup versus spark" controversy—was to continue for more than half a century. It was only resolved in the late 1950s with the recognition that, although transmission at most central and peripheral synapses is mediated by chemical transmitters, at some sites it is clearly electrical, and, at a few, it is both chemical *and* electrical.

The fascinating history of this debate and how it was finally settled is described at some length in the first chapter of this volume. Here we need only mention that, as in many of the long-lasting controversies in biology, substantial progress was made only when the alternative hypotheses were sufficiently clearly articulated as to be amenable to verification (or, more importantly, falsification), when techniques of adequate "resolving power" had been developed, and when appropriate model systems had been identified.

From among the many examples that could be cited, the following may serve to make these points. Langley's use of nicotine as a chemical probe to analyze the functional organization of the autonomic nervous system led him to postulate the existence of "receptive substances" or neurotransmitter receptors, as we would now refer to them. Loewi's simple yet ingenious experiment of transferring the perfusate from the heart of one frog whose vagus nerve was being stimulated to that of another animal provided the first unequivocal evidence of a chemical mediator—his "Vagusstoff," later identified as acetylcholine. The use of the dorsal muscle of the leech as a biological assay for acetylcholine permitted Feldberg to establish that acetylcholine was released at autonomic preganglionic synapses and later, with Dale and Vogt, to pinpoint its release at the neuromuscular junction. Close arterial injection of acetylcholine (a technique perfected by Brown, Dale, and Feldberg) settled once and for all that transmission at the neuromuscular junction is cholinergic. The isolation of single nerve and muscle fibers by Kuffler permitted the first really critical analysis of the endplate potential, but this approach was soon overtaken by the use of intracellular recording with micropipettes of the type introduced by Ling and Gerard. And this, in turn, led to the discovery by Katz and his colleagues of miniature endplate potentials and in short order the formulation of the "vesicle hypothesis."

The development, at about the same time, of methods for fixing, sectioning, and staining tissues for electron microscopy made it possible not only to establish unequivocally that neurons are morphologically distinct but also to clarify the various cellular elements that make up synapses—the presynaptic process with its complement of vesicles, the synaptic cleft, and the postsynaptic specializations. Intracellular recordings from spinal motoneurons led Eccles to disprove his own most carefully crafted electrical hypothesis for synaptic inhibition and, shortly thereafter, to discover presynaptic inhibition. Using the techniques developed for cell fractionation, Whittaker and his colleagues were able to isolate first synaptosomes and later virtually pure populations of synaptic vesicles. And, finally, there was Fatt's thoughtful prediction that if the anatomical relationships were favorable transmission could be electrical, which was soon followed by Furshpan and Potter's discovery that this was indeed correct.

Once the basic mechanisms of synaptic transmission had been established, the question arose: is acetylcholine the only neurotransmitter or are there others that remain to be discovered? In the 1970s and early 1980s, this issue was promptly answered: a variety of different transmitter substances was uncovered, beginning with noradrenaline and thereafter including several biogenic amines, a number of excitatory and inhibitory amino acids, a host of neuropeptides, and, most recently, certain gases. For most of these transmitters the relevant receptors were in time identified, their genes were cloned, and the way in which the transmitter actions are terminated was discovered. The cloning of the genes for the various receptors coincided with the emergence of molecular neurobiology as an important subdiscipline within neuroscience, and it coincidentally marked the beginning of molecular and genetic approaches to clinical neurology and psychiatry.

The cloning and sequencing of the genes that encode the many proteins involved in each aspect of synaptic transmission continue to be among the central activities in the field and appropriately form the major part of this volume. Among these proteins are the channels that permit the influx of Ca^{2+} into the axon terminal, the kinases that activate the vesicle release mechanism, the proteins that compose the vesicles themselves (including the transporters involved in neurotransmitter loading and the fusion machinery), the proteins responsible for postrelease endocytosis, those that serve to link the pre- and postsynaptic processes, the various receptors associated with the postsynaptic membrane and the proteins that interact with the receptors to bring about their localization and mediate their signaling, and finally the enzymes and transmitter reuptake mechanisms that bring to an end transmitter action.

Although the molecular biology of synapses had its origins in the late 1970s, it was only in the 1980s that it moved to center stage, and it has been responsible for the greatest progress in the past 15 years. However, other unresolved issues still command attention, including, perhaps most importantly, synaptogenesis and synaptic plasticity. Much has been learned in the past two decades about the general development of synapses and the factors responsible for their stabilization or elimination. And, ever since Bliss and Lømo discovered what they termed "long-term potentiation" in the hippocampus in 1973, behavioral neuroscientists and neurophysiologists have considered the elucidation of the mechanisms responsible for long-lasting changes in synaptic strength as critical for our understanding of the cellular basis of learning and memory.

The impetus for the present volume, in which most of these topics are reviewed, came from a workshop on synapses held at the Howard Hughes Medical Institute in June 1999. Since the workshop had brought together more than 40 of the leading figures in the field, whose work collectively covered nearly every topic of current interest to synaptologists, it seemed to us appropriate to extend this effort to a wider audience in the form of a monograph—one that would not be simply the proceedings of the workshop but rather a collection of authoritative reviews covering most aspects of the subject. Fortunately several participants agreed to write chapters in their areas of special expertise. Only one constraint was imposed upon the authors: they were asked to present a balanced view of the current state of our knowledge and not just a summary of their own work. Since it was expected that each chapter would stand on its own, it was understood that there would be some degree of overlap among them. Where the overlap was extensive, we attempted to reduce it while, at the same time, allowing the authors' distinctive voices and viewpoints to come through with as little attenuation as possible.

We are especially grateful to our colleague Dr. Kevin Davies, science editor at the Howard Hughes Medical Institute, who not only handled most of the logistics for the workshop but also kept reminding the authors of the impending deadlines for the chapters. Kevin has assisted us in immeasurable ways throughout the editorial process, and the appearance of his name on the title page only hints at the extent of his contribution to the volume. Finally our thanks go to the Johns Hopkins University Press, and especially its director, Jim Jordan, for their willingness to publish the book and for the care they have taken to produce such a splendid volume.

W.M.C.
T.C.S.
C.F.S.

Contributors

Thomas M. Bartol
Howard Hughes Medical Institute
Computational Neurobiology
  Laboratory
The Salk Institute for Biological Studies
10010 North Torrey Pines Road
La Jolla, California 92037

Mark F. Bear
Howard Hughes Medical Institute
Department of Neuroscience
Brown University, Box 1953
Providence, Rhode Island 02912

Lennart Brodin
Nobel Institute for Neurophysiology
Department of Neuroscience
Karolinska Institute
S-171 77 Stockholm, Sweden

W. Maxwell Cowan
Howard Hughes Medical Institute
4000 Jones Bridge Road
Chevy Chase, Maryland 20815

Ann Marie Craig
Department of Anatomy and
  Neurobiology
Washington University School
  of Medicine
660 South Euclid Avenue
St. Louis, Missouri 63110

Pietro De Camilli
Howard Hughes Medical Institute
Department of Cell Biology
Yale University School of Medicine
295 Congress Avenue
New Haven, Connecticut 06510

Christopher D. Ferris
Department of Medicine
Division of Gastroenterology
Medical Center North
Vanderbilt University Medical Center
Nashville, Tennessee 37232

Michael E. Greenberg
Division of Neuroscience
Children's Hospital
and Department of Neurobiology
Harvard Medical School
Boston, Massachusetts 02115

Paul D. Grimwood
Department of Neuroscience
University of Edinburgh
Crichton Street
Edinburgh EH8 9LE, United Kingdom

Volker Haucke
Howard Hughes Medical Institute
Department of Cell Biology
Yale University School of Medicine
295 Congress Avenue
New Haven, Connecticut 06510

Eric R. Kandel
Howard Hughes Medical Institute
Center for Neurobiology and Behavior
Columbia University College of
  Physicians and Surgeons
New York, New York 10032

Edward S. Lein
Howard Hughes Medical Institute
Vision Center Laboratory
The Salk Institute for Biological
  Studies
10010 North Torrey Pines Road
La Jolla, California 92037

Jeff W. Lichtman
Department of Anatomy and
    Neurobiology
Washington University School of
    Medicine
660 South Euclid Avenue
St. Louis, Missouri 63110

David J. Linden
Department of Neuroscience
Johns Hopkins University School
    of Medicine
725 North Wolfe Street
Baltimore, Maryland 21205

Robert C. Malenka
Department of Psychiatry
Stanford University School
    of Medicine
1201 Welch Road
Palo Alto, California 94304

Stephen J. Martin
Department of Neuroscience
University of Edinburgh
Crichton Street
Edinburgh EH8 9LE, United Kingdom

Richard G. M. Morris
Department of Neuroscience
University of Edinburgh
Crichton Street
Edinburgh EH8 9LE, United Kingdom

Enrico Mugnaini
Institute for Neuroscience
Northwestern University
320 East Superior Street
Chicago, Illinois 60611

Wade G. Regehr
Department of Neurobiology
Harvard Medical School
220 Longwood Avenue
Boston, Massachusetts 02115

Edwin E. Salpeter
Departments of Physics and
    Astronomy
Cornell University
Ithaca, New York 14853

Miriam M. Salpeter
Department of Neurobiology and
    Behavior
Cornell University
Ithaca, New York 14853

Richard H. Scheller
Howard Hughes Medical Institute
Department of Molecular and Cellular
    Physiology
Stanford University School of Medicine
Palo Alto, California 94305

Terrence J. Sejnowski
Howard Hughes Medical Institute
Computational Neurobiology
    Laboratory
The Salk Institute for Biological
    Studies
10010 North Torrey Pines Road
La Jolla, California 92037

Carla J. Shatz
Department of Neurobiology
Harvard Medical School
220 Longwood Avenue
Boston, Massachusetts 02115

Morgan H.-T. Sheng
Howard Hughes Medical Institute
Department of Neurobiology
Massachusetts General Hospital
and Harvard Medical School
Boston, Massachusetts 02114

Oleg Shupliakov
Nobel Institute for Neurophysiology
Department of Neuroscience
Karolinska Institute
S-171 77 Stockholm, Sweden

Steven A. Siegelbaum
Howard Hughes Medical Institute
Center for Neurobiology
Department of Pharmacology
Columbia University
New York, New York 10032

Vladimir I. Slepnev
Howard Hughes Medical Institute
Department of Cell Biology
Yale University School of Medicine
295 Congress Avenue
New Haven, Connecticut 06510

Solomon H. Snyder
Departments of Neuroscience,
    Pharmacology and Molecular
    Sciences, and Psychiatry and
    Behavioral Sciences
The Johns Hopkins University School
    of Medicine
Baltimore, Maryland 21205

Charles F. Stevens
Howard Hughes Medical Institute
The Salk Institute for Biological Studies
10010 North Torrey Pines Road
La Jolla, California 92037

Joel R. Stiles
Biomedical Applications Group
Pittsburgh Supercomputing Center
Pittsburgh, Pennsylvania 15213

Thomas C. Südhof
Howard Hughes Medical Institute
Center for Basic Neuroscience
and Department of Molecular
    Genetics
University of Texas Southwestern
    Medical School
Dallas, Texas 75235

Kohji Takei
Department of Biochemistry
University of Okayama School of
    Medicine
Okayamashi, Okayama 1700-8558,
    Japan

Edward B. Ziff
Howard Hughes Medical Institute
Department of Biochemistry
New York University Medical Center
New York, New York 10016

Synapses

1 A Brief History of Synapses and Synaptic Transmission

W. Maxwell Cowan and Eric R. Kandel

1

his chapter provides a historical account of the development of our current ideas about the structure and function of synapses. Many of the developments that led to our understanding of synapses and of synaptic transmission occurred between the late 1870s and the mid-1970s and revolved around the resolution of two major controversies. The first of these concerned the morphology of the neuron and more specifically the question: Are individual neurons discrete cells or part of a large syncytium? The second controversy concerned the physiology of the synapse and in particular the question: Is synaptic transmission electrical or chemical? In both cases the controversies arose because the techniques available at the time did not have sufficient analytic power to address the questions that were being asked. However, in each case, the resolution of the disputes revealed new features about the synapse.

In introducing this volume on the modern status of our understanding of the synapse and of synaptic transmission we begin at the beginning and trace the origin and evolution of these controversies, highlighting the methodological improvements that led to their resolution. In an appendix we provide a chronology of the major discoveries that paved the way for the work of the past two decades, together with the names of the investigators who made them.

Galvani, Volta, and Animal Electricity

Although the term *synapse* was not introduced until 1897, the history of what we now refer to as *synaptic transmission* extends back at least until the middle of the nineteenth century and, in one sense, as far back as the end of the eighteenth century and Luigi Galvani's discovery of "animal electricity."[1] In his great treatise *De viribus Electricitatis in Moto Musculari: Commentarius* of 1791, Galvani summarized his experiments on the contractions induced in limb muscles when he inserted one end of a metal hook into the medulla of a frog and attached the other end to an iron railing. These experiments, Galvani wrote, "worked no little wonderment within us and began to give rise to a suspicion that electricity was inherent in the animal itself. . . . [The muscular contractions] were increased by the flow, so to speak, of a very fine fluid from the nerves to the muscles, which we notice took place during the phenomenon, in the same way as the electric fluid is set free in a Leyden jar."

Galvani's contemporary and rival Alessandro Volta was later to challenge this interpretation, but his immediate reaction was one of admiration: "Signor Galvani['s] . . . brilliant discoveries . . . mark a new era in the annals of physics and medicine. The existence of a real and inherent animal electricity . . . is preserved and continues in the dissected limbs so long as some vitality is there, the play and movement of which takes place primarily between nerve and muscles" (Volta, 1792; cited in Stevens, 1971).

Volta's later claim that Galvani's frogs were simply serving as a sensitive galvanometer, reacting to the currents set up by the contact between the different metals in the hook and the railing, was soon shown to be correct. Nevertheless, the activation of the nerve-muscle synapse and the contraction of the limb musculature that followed the electrical stimulation of the brain (and the consequent activation of the synapses between the motor nerve fibers and the muscles) may rightly be regarded as the first experimental demonstration of synaptic transmission.

Bernard, Curare, and the Early Analysis of Synaptic Transmission

More direct evidence bearing on the transmission of activity from nerves to muscles came almost 90 years later, in experiments by Claude Bernard that still have a surprisingly modern ring (Bernard, 1878). Bernard's primary objective was to determine whether a muscle could be caused to contract, independent of its nerve supply. Taking advantage of the recently introduced South American Indian arrow poison curare, Bernard isolated a nerve-muscle preparation and found that, whereas an electrical stimulus to the nerve was ineffective after administering curare, a contraction could still be obtained if the stimulus was applied directly to the muscle. He carried this study one step further by preparing a frog with a ligature that interrupted the blood supply to the lower part of the body but did not interfere with the innervation of the hind limbs. When curare was now introduced above the level of the ligature, a paralysis developed that affected only the upper parts of the body, including the forelimbs. He now found that pinching the skin above the ligature did not produce movements in the upper parts of the body but nevertheless caused normal reflex movements of the hind limbs. From this experiment Bernard drew two conclusions: first, that curare does not cause a loss of sensation, and second, that the effect of the drug must be ascribed to a specific poisoning of the motor nerve or its link to the muscle because, as he had earlier shown, the muscle could still be excited directly even in the presence of curare. Since curare also did not affect the motor nerve in its more central course—from the spinal cord to the level of the ligature—Bernard concluded that the poison acts only on the most distal part of the motor nerve, probably where it makes contact with the muscle.

That a distinct process—*synaptic transmission*—was interposed between nerve and muscle was first recognized by Willy Kühne, Helmholtz's successor in Heidelberg, and by Wilhelm Krause of Göttingen. In the early 1860s Kühne and Krause provided the first good descriptions of the neuromuscular junction (Kühne, 1862; Krause, 1863) and showed a clear separation between the nerve endings and the skeletal muscle fibers. Independently they suggested that a nerve throws a muscle into contraction by means of its "currents of action" (reviewed in Kühne,

1888). The suggestion that neuromuscular transmission was essentially electrical was subjected to a rigorous analysis a few years later by Emil du Bois-Reymond, who is rightly regarded as the father of electrophysiology for his discoveries of the resting potential (which he determined by measuring the demarcation potential) and of the action potential (which he realized was due to current flow). In his two-volume work on the physiology of nerve and muscle, du Bois-Reymond (1877) raised two objections to the electrical transmission hypothesis. The first was that if current flow were responsible for synaptic transmission, it would almost certainly activate adjoining muscle fibers in addition to those innervated directly. Second, if the current were small, its cathodal (excitatory) effect would probably be counteracted by the anodal (inhibitory) current set up in the immediately adjoining area (see Grundfest, 1975, for discussion). Summarizing his views, du Bois-Reymond provided the first proposal that synaptic transmission could be mediated chemically: "Of known natural processes that might pass on excitation, only two are, in my opinion, worth talking about: either there exists at the boundary of the contractile substance a stimulatory secretion in the form of a thin layer of ammonia, lactic acid, or some other powerful stimulatory substance; or the phenomenon is electrical in nature" (1877, 2:700; cited in Davenport, 1991).[2]

We shall return later to the vexing question of electrical versus chemical transmission. But before doing so, we must consider the even more controversial and fundamental issue that dominated the thinking of neuroanatomists and physiologists between about 1870 and 1920—namely, the nature of the contacts between nerve cells. This controversy revolved around the question of whether the nervous system is composed of independent cellular units whose processes contact (but are physically separated from) other cells, or whether the nervous system is a complex syncytium consisting of a network of interlacing fibers that are not physically separate from one another but in direct cytoplasmic continuity. The debate over these two opposing views, generally referred to as the *neuron theory* (or doctrine) and the *reticular theory*, was more prolonged and decidedly more acrimonious than any other in the history of neuroscience (Shepherd, 1991).

Neuron Theory and Reticular Theory: The Remarkable Contributions of Cajal

Rather than conceiving of the processes of nerve cells as being strictly separated from each other by their surrounding surface membranes, as did the advocates of the neuron theory, the reticularists saw axonal and dendritic processes as being continuous with the processes of other cells. In retrospect, it seems difficult to understand how, at the end of the nineteenth century—more than 40 years after Schleiden and Schwann had formulated the cell theory (the idea that cells are the structural and functional units of all living tissues and organs [see Schwann, 1839])—neuroanatomists would still be questioning whether this theory applied to the nervous system. Yet, until well into the twentieth century, some neuroanatomists continued to question whether the nervous system was made up of morphologically discrete cells. In fact, between 1870 and 1920, the reticular hypothesis had the support of some of the leading figures in the field, including Held (the discoverer of "end feet" and terminal calyces), Apáthy, Dogiel, and especially Camillo Golgi, the great Italian neurohistologist still honored for his discoveries of the chrome silver method and the cytoplasmic organelle that bears his name (Cajal, 1954).

The reticularist view, which challenged both the cell theory in general and the neuron doctrine in particular, dates from Gerlach's discovery in 1872 of a fine network of fibers in sections of the spinal cord, cerebral cortex, and cerebellum, stained with carmine and gold chloride. Gerlach interpreted this network of fibers as being formed by the anastomosis of neuronal processes. This view was reinforced in 1897 by Apáthy's studies on the leech nervous system, from which he proposed that the neurofibrils within nerve processes are continuous from one cell to the next and serve as conductors, much like electrical wires, for the flow of current from one cell to another (Peters et al., 1976).

The persistence of the reticularist view, despite the strong evidence advanced against it, is one of the more remarkable episodes in the early history of neuroscience (see van der Loos, 1967; Clarke and Jacyna, 1987; Shepherd, 1991; and Jacobson, 1993, for reviews), but it is easily accounted for by the inability of the anatomical methods available at the time to resolve cell membranes. As Cajal stated, "To settle the question [of contiguity versus continuity] definitely, it was necessary to demonstrate clearly, precisely, and indisputably the final ramifications of the nerve fibers, which no one had seen, and to determine which parts of the cells made the imagined contacts" (Cajal, 1937).

Since the visualization of cell membranes was well beyond the resolution of the light microscope, this morphological issue could not be definitively settled until electron microscopy was applied to neural tissue in the early 1950s (Bodian, 1966). It is therefore to the great credit of several neuroanatomists working at the turn of the century that, by the late 1890s, all but the most die-hard reticularists were convinced of the morphological discreteness of individual neurons. In the absence of direct microscopic evidence, Wilhelm His, August Forel, van Gehuchten, Waldeyer, Retzius, and especially Santiago Ramón y Cajal were able to adduce several independent lines of evidence for the neuron doctrine.

Waldeyer (1891) is usually credited with having formulated the neuron theory and with having clearly stated that neurons are developmentally, structurally, functionally, and pathologically discrete. However, as Cajal so trenchantly points out, although Waldeyer "supported [the theory] with the prestige of his authority, [he] did not contribute a single personal observation. He limited himself to a short brilliant exposition

[1891] of the objective proofs, adduced by His, Kölliker, Retzius, van Gehuchten and myself, and he invented the fortunate term *neuron*" (Cajal, 1954).[3]

It is important to acknowledge two contributions to the neuron theory that predated Cajal's work. These derived from the studies of His and Forel, who respectively provided developmental and pathological evidence for the individuality of neurons. His provided two types of evidence based on his histological studies of the developing spinal cord in humans and other vertebrates. First, he found that at early stages the wall of the neural tube consists of a single layer (now called the *ventricular zone*), which he described as consisting of two cell types: germinal cells, which he considered to be the progenitors of neurons, and spongioblasts, which he assumed were the precursors of the ependymal and glial elements. We now know that His's interpretation of the cellular composition of the ventricular zone was incorrect—his germinal cells are simply neuroepithelial cells that are in, or have just passed through, the M-phase of the cell cycle, whereas his spongioblasts are cells in G1, S, and G2 (Sauer, 1935). (However, his claim that the cells remained distinct throughout their phase of migration remains correct.) Second, and more important, His recognized that nerve processes are direct outgrowths of young neurons and that they end freely, without fusing with the processes of other cells. With remarkable self-confidence he wrote in 1886, "I consider it as an established principle that each nerve fiber emerges as an outgrowth from a single cell. This is its genetic, trophic and functional center. All other connections of fibers are either indirect or secondary" (cited in Hamburger, 1980).

August Forel's contribution to the neuron/reticular theory debate was based both on the earlier experimental findings of Gudden (1870) and on his own observations of Golgi-stained preparations. Gudden had noted that when a nerve is severed, the resulting neuronal atrophy (or retrograde degeneration, as we now call it) is confined to the relevant cell group and does not spread to involve neighboring populations of neurons, as might be predicted if the cells were physically continuous.[4]

A surprising aspect of Cajal's many contributions to this issue was that he began his studies completely unaware of the earlier contributions of His and Forel. The success of his work, which led him to become the foremost advocate of the neuron theory, derived in large part from his application of the chrome silver impregnation method or "reazione nera" that had been introduced by Golgi in 1873. This method, which was still widely used until recently, offered two advantages. First, the method stains, in an apparently random manner, only about 1% of the cells in any particular region of the brain or spinal cord. This makes it possible to study the morphology of individual nerve cells in isolation from their neighbors. But the method has an additional advantage: the neurons that are stained are often impregnated throughout their entire extent, so that one can clearly visualize cell bodies, axons, axon collaterals, the full dendritic arbor, and, in developing brains, axonal and dendritic growth cones. As Cajal discovered, the method works especially well in embryonic and immature brains, and because the developing nervous system is considerably less complex than the mature brain, it is much easier to study individual neurons against the background of unstained cells. He stated:

> Since the full grown forest turns out to be impenetrable and indefinable, why not revert to the study of the young wood, in the nursery stage, as we might say? . . . If the stage of development is well chosen . . . the nerve cells, which are still relatively small, stand out complete in each section; the terminal ramifications of the axis cylinder are depicted with the utmost clearness and perfectly free; the pericellular nests, that is the interneuronal articulations, appear simple, gradually acquiring intricacy and extension. (Cajal, 1937:324–325)

By examining in detail nerve cells and their contacts in histological sections of almost every brain region, Cajal was able to describe not only differences between various types of nerve cells but also the great variety of axonal endings found in the central nervous system (CNS). This led him inexorably to conclude that the axon terminals of neurons end freely upon the surfaces of other cells and that at the sites of interaction they are not continuous with their cellular targets and therefore not part of a diffuse network.

We are fortunate in having Cajal's own reminiscences about how he arrived at this view in his delightful—if at times somewhat exuberant—*Recollections of My Life.*[5] In May 1888, he published his first critical observations on the termination of the axons of the stellate cells of the cerebellum:

> [The axons of the stellate neurons] take up a direction transverse to the cerebellar convolution, describing an arc and giving off numerous collateral branches characterized by progressive thickening. Finally both the end of the main fibre and its numerous descending processes break up into terminal fingers or tufts applied closely to the bodies of the cells of Purkinje, about which they form . . . complicated nests or baskets. (*Recollections,* 330; see also Fig. 1.1)

In the same paper he drew attention to the termination in the cerebellar cortex of the mossy fiber afferents, which "exhibit both at [their] ultimate ending and in [their] collateral branches, bunches or rosettes of short tuberous appendages ending freely" (*Recollections,* 331).

Later he was able to show that the rosettes (or "excrescences" as he referred to them) articulate directly with the clawlike ends of the granule cell dendrites. In August of the same year he made two further observations. First he identified the axons of the granule cells, which he named the parallel fibers. Then came the finding that he described in his *Recollections* as "the most beautiful [discovery] which fate vouchsafed to me in that fertile epoch, [that] formed the final proof of the transmission of *nerve impulses by contact*" (*Recollections,* 332; his emphasis). This

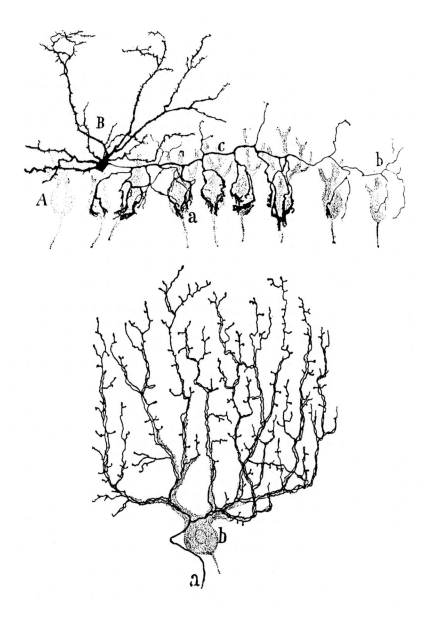

Figure 1.1. Two types of nerve ending that convinced Cajal that axons are not in continuity with their targets. The top panel depicts a cerebellar stellate neuron stained by the Golgi method, showing the basketlike arrangement of its axon collaterals that surround the bodies of the Purkinje cells. The bottom panel shows the pattern of termination of a climbing fiber along the dendrites of a Purkinje cell. Reproduced from Cajal (1995) by permission of Oxford University Press.

was his discovery of the climbing fibers to the cerebellum, which, in characteristic fashion, he described as follows:

> These robust conductors arise in the centres of the pontine region of the brain; invade the white core of the cerebellar lamella; cross the granule layer without branching; afterwards attain the level of the cells of Purkinje; and finally run over the bodies and principal outgrowths of these elements, to which they adapt themselves closely. When they reach the level of the first branches of the dendritic trunks of the Purkinje cells, they break up into twining parallel networks which ascend along the protoplasmic branches, to the contours of which they apply themselves like ivy or lianas to the trunks of trees. (*Recollections*, 332; see also Cajal, 1911)

And, if further evidence were needed that central axons ended freely, he extended his observations to the retina of birds, demonstrating that centrifugal fibers (which arise in the brain) "cross the internal plexiform zone [of the retina] and end by a free varicose arborization among the spongioblasts [i.e., amacrine cells]" and that "in the internal plexiform zone . . . the dendrites of the ganglion cells come into relation with the axons and collaterals of the bipolar cells by contact and not through a diffuse net" (*Recollections*, 339, 340).

We cite these observations at some length to illustrate not only Cajal's justifiable pride in his discoveries but also his pugnacious advocacy of the neuron theory. The defense of this theory was to be a theme to which he would return again and again, and indeed it was the subject of his last major publication, his monograph *¿Neuronismo o Reticularismo?*, published shortly after his death.[6] In this monograph he summarized the extensive body of evidence, both descriptive and experimental, that he had accumulated over the years in support of the neuron theory and listed the many varieties of axosomatic and axodendritic synaptic contacts he and others had observed in different parts of the nervous system.

Until the late 1880s, Cajal's work (which was published in Spanish, in a journal of limited circulation that he had founded) was not widely known or appreciated. But in 1889 he attended a meeting of the German Anatomical Society in Berlin and there attracted the attention of Kölliker, at that time the doyen of European anatomists. Kölliker encouraged Cajal to have his work translated into French or German. As Cajal later recalled, "Although he [Kölliker] had been a supporter of the reticular theory, he abandoned it completely, adapting himself with the flexibility of a young man to the new conceptions of contact and of the morphological independence of the neurons. In his friendliness for me, he carried his goodwill so far as to learn Spanish in order to read my earliest communications" (*Recollections*, 358).

In addition to providing the critical evidence for the neuron doctrine, Cajal also outlined two other rules that governed the functioning of nerve cells. First, he restated the *principle of dynamic polarization* that had been originally articulated by van Gehuchten. According to this principle signaling within a neuron flows in a single, predictable direction, from the dendrites and cell body that receive inputs from other neurons to the axon and from there to the presynaptic terminals, which contact yet other neurons or effector cells. Second, he outlined the *principle of connectional specificity,* according to which nerve cells do not connect indiscriminately with one another or form random networks. Rather, each cell communicates only with certain postsynaptic targets, but not with others, and always at special points of synaptic contact. Taken together, the principles of dynamic polarization and connectional specificity form the cellular basis for the modern connectionist approach to the brain.

Sherrington and the Integrative Action of Synaptic Transmission

Although he was helped by Kölliker's support, in the long run Cajal's greatest advocate proved to be not a fellow anatomist but the physiologist Charles Sherrington, who coined the term *synapse.* In Part III of the seventh edition of Michael Foster's *Textbook of Physiology,* published in 1897, Sherrington wrote:

> So far as our present knowledge goes we are led to think that the tip of a twig of the [axonal] arborescence is not continuous with but merely in contact with the substance of the dendrite or cell body on which it impinges. Such a special connection of one nerve cell with another might be called "*synapsis.*" . . . Each synapsis offers an opportunity for a change in the character of nervous impulses, such that the impulse as it passes over from the terminal arborescence of an axon into the dendrite of another cell, starts in that dendrite an impulse having characters different from its own. (929, 969)

Again we are fortunate to have Sherrington's account of the origin of the term *synapse,* from a letter he wrote to John Fulton:

> You enquire about the introduction of the term "synapse"; it happened thus. M. Foster had asked me to get on with the Nervous System part [Part III] of a new edition of his "Text of Physiol." for him. I had begun it, and had not got far with it before I felt the need of some name to call the junction between nerve-cell and nerve-cell (because the place of junction now entered physiology as carrying functional importance). I wrote him of my difficulty, and my wish to introduce a specific name. I suggested using "syndesm." . . . He consulted his Trinity friend Verrall, the Euripidean scholar, about it, and Verrall suggested "synapse" (from the Greek "clasp") and as that yields a better adjective form, it was adopted for the book. (Fulton, 1938)

Sherrington was quick to acknowledge his indebtedness to Cajal, whom he had first met in 1885 when, as a young physician/pathologist interested in infectious disease, he had visited Spain to investigate an outbreak of cholera. We have no record of what transpired at their meeting, but from all we know of both men we can be sure that they discussed more than the weather and the differences between Spanish and English academic institutions on issues governing academic tenure. What we do know is that, a few years later, Sherrington successfully persuaded the Royal Society and his mentor Michael Foster (who by then was serving as secretary of the society) to invite Cajal to give the Croonian Lecture for 1894. Sherrington also arranged to have Cajal stay at his home during his visit to London. As Cajal later recalled, "a letter from Sherrington finally decided me [to agree to give the lecture]. . . . [He] claimed generously, as a neurologist, the right to have me stay in his home. At the present time, my host, who was then young [he was, in fact, 36] can be regarded as the leading physiologist in England." In addition to hosting his illustrious guest, Sherrington assisted him in preparing slides and in coloring drawings for his lecture, which was focused largely on his "discoveries concerning the morphology and connections of the nerve cells in the spinal cord, the ganglia, the cerebellum, the retina, the olfactory bulb, etc." (*Recollections*, 419–420).

Writing several years later, Sherrington amusingly described how, during his brief stay at their home, Cajal had succeeded in converting their guest bedroom into a temporary laboratory. He was especially impressed by Cajal's ability to glean so much about the function of nerve cells from looking at their structure, even in the poorest microscope preparation. Cajal's artistic skill enabled him to capture rapidly the essence of an observation, and to describe it with an effective use of anthropomorphisms. As Sherrington later wrote,

> A trait very noticeable in [Cajal] was that in describing what the microscope showed he spoke habitually as though it were a living scene. This was perhaps the more striking because not only were his preparations all dead and fixed, but they were to appearance roughly made and rudely treated—no cover-glass and as many as half a dozen tiny scraps of tissue set in one large blob of balsam and left to dry, the curved and sometimes slightly wrinkled surface of the balsam creating a difficulty for microphotography. . . . Such scanty illustrations as he vouchsafed for the preparations he demonstrated were a few slight, rapid sketches of points taken here and there—depicted, however, by a master's hand.
>
> The intense anthropomorphism of his descriptions of what the preparations showed was at first startling to accept. He treated the microscopic scene as though it were alive. . . . A nerve-cell by its emergent fibre "groped to find another"! Listening to him I asked myself how far this capacity for anthropomorphizing might not contribute to his success as an investigator. I never met anyone else in whom it was so marked.

And, in a more serious vein, Sherrington summarized Cajal's great contribution to synaptic transmission—the subject of this chapter—as follows:

> He solved at a stroke the great question of the direction of nerve currents in their travel through the brain and spinal cord. He showed, for instance, that each nerve path is always a line of one-way traffic only, and that the direction of that traffic is at all times irreversibly the same. The so-called nerve networks with unfixed direction of travel [the reticular theory] he swept away. The nerve-circuits are valved, he said, and he was able to point out where the valves lie—namely where one nerve cell meets the next one. (Sherrington, 1949)

Needless to say, Sherrington's contribution to this topic was not limited to coining the term *synapse* and boosting Cajal's reputation (as if this was necessary). His own work, much of which was summarized in the ten Silliman lectures he gave at Yale University—later published under the title *Integrative Action of the Nervous System*[7] and in the collection entitled *Selected Writings of Sir Charles Sherrington* edited by Denny-Brown (1979)—stands with Cajal's *Histologie du Système Nerveux de l'Homme et des Vertébrés* (1909, 1911) as one of the unquestioned foundation pillars of modern neuroscience.

Something of the flavor of Sherrington's unusual style, and also of the clarity of his thinking, is found in the following quotation from the first Silliman lecture:

> If there exists any surface or separation at the nexus between neurone and neurone, much of what is characteristic of the conduction exhibited by the reflex-arc might be more easily explainable. . . . It seems therefore likely that the nexus between neurone and neurone in the reflex-arc, at least in the spinal arc of the vertebrate, involves a surface of separation between neurone and neurone; and this as a transverse membrane across the conductor must be an important element in intercellular conduction. The characters distinguishing reflex-arc conduction from nerve-trunk conduction may therefore be largely due to intercellular barriers, delicate transverse membranes, in the former.
>
> In view, therefore, of the probable importance physiologically of this mode of nexus between neurone and neurone, it is convenient to have a term for it. The term introduced has been synapse. (Sherrington, 1906)

Sherrington's work on spinal reflexes not only convinced him that the reticular theory was untenable but also led him to delineate several defining features of synaptic transmission. First, he was struck by the valvelike, unidirectional flow of information across a synapse made by different afferent inputs to the spinal motoneurons that form the final common pathway to the muscles. The second feature was what we now call the *synaptic delay*—a measurable delay at the site of interaction beyond that attributable to the conduction time in the afferent fibers. The methods available to Sherrington were too insensitive to measure the delay at individual synapses, but, as we shall see later, others have since provided such measurements, ranging from less than 0.5 msec to 1.3 msec (Eccles, 1964). The third feature was his conclusion that the interaction between the afferent input and the motoneurons must involve many synapses acting in concert. Central to this notion was the idea that each afferent fiber has only a small effect on a given motoneuron and that many afferent fibers need to sum, both spatially and temporally, to cause the motoneuron to discharge. Finally, Sherrington made the fundamental discovery that not all synaptic actions are excitatory: there are also inhibitory inputs, and, from a functional point of view, these are at least as important as those that lead to excitation.

The recognition of inhibition as an independent and active process was one of Sherrington's greatest accomplishments. It is all the more remarkable when one considers that neither Cajal nor any of his neuroanatomical contemporaries considered the possibility of inhibition in any of their writings. In fact—and this is especially surprising—they continued to ignore the role of inhibition and to make no reference to it long after it had been generally recognized by physiologists.

The concept of central inhibition derived largely from Sherrington's studies on what he termed the "principle of reciprocal innervation." This principle—which, as Adrian (1957) remarked, "was the clue to the whole system of traffic control in the spinal cord and throughout the central pathways"—emerged from Sherrington's experiments on flexor and extensor reflexes. These experiments demonstrated that reflex excitation of motoneurons that activate one group of muscles (e.g., the extensor muscles of a limb) is always accompanied by the inhibition of the motoneurons that innervate the antagonistic group of muscles (in this case, the limb flexors; see Creed et al., 1932, and Fig. 1.2). Writing in 1908, Sherrington addressed the issue of the interaction of excitatory and inhibitory actions on a group of motoneurons in these terms:

> It seems clear that the reflex effect of concurrent stimulation of [an] excitatory afferent nerve with [an] inhibitory afferent nerve on the vastocrureus nerve-muscle preparation is an algebraic summation of the effects obtainable from the two nerves singly. . . . One inference allowable from this is that . . . the two afferent arcs employed act in opposite direction at one and the same point of application in the excitable apparatus. . . . As to the common locus of operation, the point of collision of the antagonistic influences, it seems permissible to suppose either that it lies at a synapse . . . or that it lies in the substance of the "central" portion of a neurone. The net change which results there when the two areas are stimulated concurrently is an algebraic sum of the *plus* and *minus* effects producible separately by stimulating singly the two antagonistic nerves. (Cited in Eccles, 1964)

Sherrington was to remain interested in the nature of central inhibition throughout his career and chose it as the topic of his 1932 Nobel

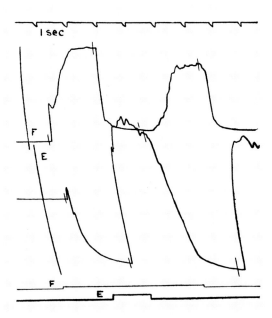

Figure 1.2. An example of reciprocal innervation. Record F is of the contraction of a knee flexor (the semitendinosus) and record E of the contraction of a knee extensor (vasto-crureus) in a decerebrate cat. The upper signal line (F) along the bottom marks the duration of the faradic stimulation of the ipsilateral peroneal nerve, while the lower signal line (E) marks the stimulation of the contralateral peroneal nerve. Note that when the flexor muscle contracts the extensor is inhibited; conversely, when the extensor is thrown into contraction the flexor is inhibited. The lower myograph recording is shifted slightly to the right, for the sake of clarity; in reality the two recordings were synchronous. Reproduced from Denny-Brown (1979) by permission of the Rockefeller Medical Library, Institute of Neurology, London.

Prize address, which he entitled "Inhibition as a Coordinative Factor." He stated with remarkable foresight:

> It is still early to venture any definite view of the intimate nature of "central Inhibition" . . . the suggestion is made that it consists in the temporary stabilization of the surface-membrane which excitation would break down. As tested against a standard excitation the inhibitory stabilization is found to present various degrees of stability. The inhibitory stabilization of the membrane might be pictured as a heightening of the "resting" polarization, somewhat on the lines of an electrotonus. Unlike the excitation-depolarization it would not travel; and, in fact, the inhibitory state does not travel.

Sherrington was, of course, not the first scientist to consider inhibition as an important physiological mechanism. The Weber brothers had described the slowing of the heart on stimulation of the vagus nerve as early as 1845. Earlier, Sir Charles Bell (1834) had spoken of a "nervous bond" that caused muscles to "conspire in relaxation as well as to combine in contraction," while Fridrich Goltz (in whose laboratory in Strasbourg Sherrington had begun his studies of the nervous system) had demonstrated that strong cutaneous stimuli could inhibit movements in spinal dogs (Goltz and Freusberg, 1874). However, Sherrington not only saw it as an independent physiological process but also put it into the context of synaptic transmission and the integrative action of reflexes. To this extent his contribution went well beyond that of his predecessors, and, in addition, he provided the first quantitative assessment of the magnitude and time course of the inhibitory process. Most important, he correctly placed the locus of inhibition, not at the periphery (as had previously been assumed) but in the spinal cord and more particularly at the level of the synapses upon the motoneurons. In doing so he set the stage for the study of the *mechanisms* whereby synaptic transmission occurs.

The Stimulation of Certain Forms of Synaptic Transmission by Known Chemicals

While Sherrington was busy studying spinal reflexes and the role of inhibition, his Cambridge colleague (and fellow student of Michael Foster) J. Newport Langley was providing the first conclusive evidence that synaptic transmission may occur by chemical means, by investigating transmission through the peripheral autonomic ganglia. According to Davenport (1991), Langley's interest in this subject was first aroused when he received from Professor Liversidge of Sydney a sample of pituri, an alkaloid extracted from the leaves of an Australian plant. With the help of one of his students, Langley quickly established that pituri's physiological actions were identical to those of nicotine, which he then began to use as a chemical probe to study the autonomic nervous system. Among his first observations was that nicotine, applied to the superior cervical ganglion of an anesthetized cat, caused retraction of the nictitating membrane, dilatation of the pupil, and piloerection on the treated side. He subsequently used this approach to analyze the distribution of each nerve root that contributed to the sympathetic nervous system—work that he summarized in his important monograph *The Autonomic Nervous System,* published in 1921.

Initially Langley was inclined to accept the then-current view that the action of drugs such as nicotine, curare, and atropine was on nerve conduction, and that the preganglionic axons and the ganglion cells were continuous: "In the earlier accounts by Dickenson and myself upon the action of nicotine . . . we spoke of it as first stimulating and then paralyzing the nerve-cells of the ganglia since at that time we held the common view that the axis cylinder of a nerve-fibre which excited a nerve-cell was continuous with it." However, he changed his mind when he discovered that the application of nicotine to the ganglion produced "effects like those produced by brief stimulation of its preganglionic fibers" even after a ganglion had been completely denervated by cutting the preganglionic fibers and allowing them time to degenerate (cited in Davenport, 1991). It followed that the action of nicotine must be directly on the ganglion cells. Langley came to the same conclusion when he cut the nerves to the leg muscles of a chicken and showed (again after the nerve fibers had degenerated) that injecting nicotine still caused the muscles to contract, and that this contraction could be blocked by curare. Reviewing these findings, he concluded that

> In all cells two constituents at least must be distinguished, (1) substances concerned with carrying out the chief functions of the cells, such as contraction, secretion, the formation of special metabolic products, and (2) receptive substances especially liable to change and capable of setting the chief substance in action. Further, that nicotine, curari . . . as well as the effective material of internal secretions produce their effects by combining with the receptive substance, and not by an action on axon-endings if these are present, nor by a direct action on the chief substance. (Langley, 1905)[8]

As neither nicotine nor curare prevented contraction on direct stimulation of the muscle (as Bernard had earlier reported), he concluded that the drugs must exert their (antagonistic) effects on "the receptive substance," and "this seems in its turn to require that the nervous impulse should not pass from nerve to muscle by an electric discharge, but by the secretion of a special substance at the end of the nerve, a theory suggested in the first instance by du Bois Reymond" (Langley, 1906).

Langley went on to test his ideas about receptive substances on the neuromuscular synapse. He soaked the tip of a sewing thread in nicotine and carefully touched it to the surface of the muscle fibers so as to stimulate only a small area of the muscle membrane. He found that in the innervated muscle he could produce contraction only when he applied nicotine directly to the region near the nerve endings, a region we now call the *endplate*. This experiment provided the first evidence for the localization of nicotinic acetylcholine (ACh) receptors to the subsynaptic region. Langley subsequently found that, several days after denervation, receptor sensitivity had spread throughout the muscle membrane (Langley, 1907).

In addition to his recognition of molecular receptors (as we now know them)—for which he may, with justification, be considered the "father of neuropharmacology"—Langley's earlier work on the regeneration of preganglionic fibers to the superior cervical ganglion provided the first convincing evidence for synaptic specificity and for what would later be known as the *chemoaffinity hypothesis*. Briefly, he showed that following regeneration of the preganglionic afferents, the end-organ responses to stimulating different thoracic roots (e.g., dilatation of the pupil on stimulating T_2 and constriction of the blood vessels to the ears and piloerection on stimulating T_4) were restored to their normal pattern, from which he concluded that there must be a special (chemical) relationship between the different preganglionic fibers and the related subsets of ganglion cells (see Purves and Lichtman, 1985, for discussion).

A short while before Langley published his studies on the effects of nicotine on sympathetic ganglia and striated muscle, Oliver and Schäffer reported that intravenous injections of a glycerol extract of the adrenal gland produce a marked increase in arterial blood pressure (Oliver and Schäfer, 1895). At about the same time, a Czech scientist, Szymonowicz, made essentially the same observation but found, in addition, that if the vagus nerve was cut the extracts of the adrenal medulla also produced a marked increase in heart rate (Szymonowicz, 1896; cited in Davenport, 1991). Although it was to be some years before this work was extended to the postganglionic sympathetic outflow, these findings were most easily accounted for by postulating that the actions of the adrenal extracts were mediated by a second class of Langley's postulated "receptive substances."

Shortly after "adrenalin" was purified and its chemical structure identified by the Japanese scientist Jokichi Takamine (Davenport, 1982), one of Langley's students, T. R. Elliott, examined its effects on the peripheral target tissues innervated by the sympathetic nervous system. In short order, Elliott was able to show that the effects of adrenaline, whether excitatory or inhibitory, closely paralleled those observed on stimulating the relevant postganglionic fibers. And (again echoing Langley), Elliott found that degeneration of the target tissues' sympathetic input had no effect on (or, on occasion, even accentuated) adrenaline's effect on the target tissues. This led Elliott to conclude that impulses in sympathetic nerves lead to the release of a small quantity of adrenaline or a related compound. In a paper presented to the British Physiological Society, he wrote that "Adrenaline might . . . be the chemical stimulant liberated on each occasion when the impulse arrives at the periphery" (Elliott, 1904).

Prompted by this success, Elliott turned his attention to striated muscle. However, apart from demonstrating that it was not stimulated by adrenaline, he made little progress and regretfully concluded that he had "tried in vain to discover an active substance in the muscle plates of striped muscle" (cited in Davenport, 1991).

Inspired by Elliott's findings on adrenaline, Walter E. Dixon turned his attention to the mediator of the known inhibition of the heart on stimulation of the vagus nerve. To study this phenomenon he removed the heart of a dog (while it was inhibited by stimulating the vagus) and rapidly made an extract of the heart tissue, which he applied to a beating frog heart. Although the frog heart was clearly inhibited by some factor in the extract, Dixon was never able to purify the active component. But his findings led him to postulate that parasympathetic stimulation acts through the release of a substance whose effects mimic those of the natural alkaloid muscarine (Dixon, 1906).

Ironically, at a meeting of the British Medical Association in Toronto where Dixon reported his findings, Reid Hunt of the U.S. Public Health and Marine Hospital Service (the precursor to the National Institutes of Health) described his efforts with his colleague René de M. Taveau to isolate agents that slowed the heart and lowered blood pressure. They had found no fewer than 19 derivatives of choline that lowered blood pressure, of which ACh was by far the most potent (Hunt and Taveau, 1906). Yet neither Dixon nor Hunt (nor anyone else who was present) seems to have made the connection between the two sets of observations, and it was not until several years later that ACh was positively identified as the transmitter involved.

Dale, Loewi, and Feldberg: Establishing Acetylcholine as a Chemical Transmitter

At this point another towering figure appeared on the stage of the synaptic transmission saga: Henry Hallett Dale, another of Langley's students and a close friend of Elliott.[9] Working at the Wellcome Physiological Research Laboratories, Dale began to analyze a large number of sympathomimetic amines that had been isolated by his colleague George Barger. He tested not only their effects on blood pressure but also their actions on several organs and tissues (the pupil of the eye, the nictitating membrane, the salivary and lachrymal glands, the bladder, and the nonpregnant uterus). In this way, Dale identified αl-aminoethyl catechol (later known as noradrenaline) as a particularly potent vasopressor that, unlike αl-adrenaline, did not inhibit contractions of the nonpregnant uterus. Concurrently he began a long series of studies on the actions of ergot, another naturally occurring alkaloid whose effects on the pregnant uterus had been known for almost 2000 years (Gilman et al., 1990). Again he turned to one of his colleagues, Arthur Ewins, to isolate components from ergot mixtures, and he began to look for components with pharmacological properties similar to those of muscarine (which he originally thought might be in the ergot preparations). Although they failed to find muscarine in any of the isolates, they discovered what later proved to be ACh. They found that each preparation, which had a muscarinelike action in their standard assay (loops of rabbit intestine), was associated with choline. And the fact that the active factor was alkaline-sensitive suggested to Dale that it might be ACh. In 1914 he described his experiences with various choline esters:

> The question of a possible physiological significance, in the resemblance between the action of choline esters and the effects of certain divisions of the involuntary nervous system, is one of great interest, but one for the discussion of which little evidence is available. Acetylcholine is, of all the substances examined, the one whose action is most suggestive in this direction. The fact that its action surpasses even that of

adrenine, both in intensity and evanescence, when considered in conjunction with the fact that each of these two bases reproduces those effects of involuntary nerves which are absent from the action of the other, so that the two actions are in many directions at once complementary and antagonistic, gives plenty of scope for speculation. (Dale, 1914)

Later, looking back on this time, he remarked:

Such was the position in 1914. Two substances were known, with actions very suggestively reproducing those of the two main divisions of the autonomic system; both, for different reasons, were very unstable in the body, and their actions were in consequence of a fleeting character; and one of them was already known to occur as a natural hormone. These properties would fit them very precisely to act as mediators of the effects of autonomic impulses to effector cells, if there were any acceptable evidence of their liberation at the nerve endings. The actors were named, and the parts allotted; a preliminary hint of the plot had, indeed, been given ten years earlier, and almost forgotten; but only direct and unequivocal evidence could ring up the curtain, and this was not to come till 1921. (Dale, 1938)

What happened in 1921, of course, was Otto Loewi's great discovery of "Vagusstoff." The story of this discovery and of the dreams that led him to perform the critical experiment have become part of the folklore of neuropharmacology. In his own words:

The night before Easter Sunday of that year I awoke, turned on the light, and jotted down a few notes on a tiny slip of paper. Then I fell asleep again; it occurred to me at six o'clock in the morning that during the night I had written down something most important, but I was unable to decipher the scrawl. The next night, at three o'clock, the idea returned. It was the design of an experiment to determine whether or not the hypothesis of chemical transmission that I had uttered seventeen years ago was correct. I got up immediately, went to the laboratory, and performed a simple experiment on a frog heart according to the nocturnal design.

Like most remembered dreams, Loewi's account is distorted. It misstated the chronology (the experiment was performed in 1921, not 1920 as he first reported) while, at the same time, exaggerating the most dramatic event: "[I went to the laboratory at three o'clock in the morning] and at five o'clock the chemical transmission of [the] nervous impulse was proved" (Loewi, 1953).

A more prosaic account would simply recall that, using a readily available technique, Loewi carried out 14 experiments on two species of frogs and on four toads over a period of several days. The experiment involved isolating two hearts and perfusing each through a glass cannula with Ringer's solution. After the vagus nerve to one of the hearts was stimulated and its well-known inhibitory effect observed, the fluid from that heart was transferred (probably using a glass pipette) to the second heart, where it promptly caused a decrease in the strength and frequency of beating.

In many respects Loewi was fortunate in both the timing of the experiments and his choice of preparation. The frog vagus contains both stimulatory and inhibitory fibers, but in winter the inhibitory fibers predominate (his experiments were carried out in February or March). And the cholinesterase content of the frog heart is low (compared with that of the mammalian heart), so that the released transmitter remained active (at the low temperature of the unheated laboratory) long enough for its effects on the second heart to be observable. In an important control experiment showing that the inhibitory action of Vagusstoff could be completely blocked by the prior administration of atropine, Loewi was able to rule out an alternative possible explanation for his findings—namely that they were due to the release of potassium, as Howell and Duke (1908) had suggested.

Davenport (1991) has discussed at some length the inconsistencies in Loewi's account and the difficulties encountered by others who attempted to replicate his experiments. However, the issue was finally resolved in 1933, when Wilhelm Feldberg and Otto Krayer conclusively demonstrated (with appropriate controls) that stimulating the vagus nerve of a dog released an ACh-like substance into the coronary sinus. This caused the dorsal muscle of a leech to contract and the blood pressure of a cat to fall, provided that the ACh-degrading enzyme, cholinesterase, was blocked by the prior administration of eserine (Feldberg and Krayer, 1933).

Feldberg was to dominate the early thinking on ACh and on chemical synaptic transmission in the period 1930–1950 much as Bernard Katz was later to dominate the field from 1950 to 1970. Feldberg's experiments with Krayer were only the first of a long series of studies he carried out on the role of ACh as a neurotransmitter (Feldberg, 1950, 1977; Bisset and Bliss, 1997). As a medical student in Berlin he had begun working during his vacations at the Physiological Institute of the university and had become fascinated by Langley's book on the autonomic nervous system. His mentor, Schilf, accordingly arranged for him to work with Langley at Cambridge, but unfortunately Langley died within six months of Feldberg's arrival. Yet the two years Feldberg spent at Cambridge had a lasting effect on his career: "I read and re-read all of Dale's papers," he wrote. In 1927 he returned to the Physiological Institute in Berlin but was summarily dismissed from his position in the university in 1933, at the outset of the Nazi purge of Jewish academics.[10] Hearing of his plight, Dale invited Feldberg to join him at the National Institute for Medical Research in London, and it was here that he carried out much of his later work on ACh as a transmitter.

While still in Berlin, Feldberg had undertaken a series of experiments to clarify what was known as the Vulpian-Heidenhain paradox—contraction of the muscles of the tongue on stimulating the parasympathetic outflow in the lingual nerve, after interruption of the hypoglossal nerve. Dale had suggested that this might be due to the release of ACh from the endings of the parasympathetic fibers to the tongue. To prove this, Feldberg (1933) succeeded in collecting fluid from the cannulated lingual vein and passing it over his favorite assay, the dorsal muscle of the Hungarian leech, which Führer (1917; cited in Davenport, 1991) had shown to be exquisitely sensitive to ACh.

Shortly after moving to London, Feldberg succeeded (with Gaddum) in perfusing the superior cervical ganglion with eserinized Locke's solution, and he was able to show that stimulation of the preganglionic fibers (which caused retraction of the nictitating membrane) resulted in the release of ACh into the perfusate (Feldberg and Gaddum, 1934). Over the next 15 years he carried out many other studies on the role of ACh (see Bisset and Bliss, 1997, for references). These included investigations of the mechanism of transmission by the gastric vagus (with Dale), by the preganglionic fibers to the adrenal medulla (with Mintz and Tsudzimura), and by the postganglionic sympathetic fibers to the sweat glands (again with Dale).

These several actions all belonged to the category called *muscarinic* by Dale, because they were simulated by muscarine and shared several additional features: (1) there was a long delay between the electrical stimulation of the nerve and the onset of the response of the innervated organ; and (2) the response itself was long-lasting and often persisted well after nerve stimulation had come to an end, with only a very gradual return of the organ or tissue to its baseline level of activity. One needed only to look at the results of experiments in which these autonomic nerves were stimulated to become convinced that by far the best general theory that could account for all these slow actions is one that allows a more or less labile substance to be interposed between the nerve endings and the effector cells, be they smooth muscle or glands.

Following the Feldberg-Krayer experiment of 1933, and the subsequent experiments of Feldberg, little doubt remained in the minds of almost everyone—not only the pharmacologists such as Loewi, Dale, and Feldberg, but even the neurophysiologists such as Eccles, Lorente de Nó, and Erlanger—that muscarinic actions were mediated by chemical transmitters and specifically by ACh. Indeed, the elegance of the Feldberg experiments made it seem that muscarinic actions were the very prototype of chemical synaptic actions. However, in addition to the muscarinic actions of ACh, Dale had described a second class of cholinergic actions, which he termed *nicotinic* because they could be elicited when muscarinic actions had been blocked by atropine. Nicotinic actions were found in the adrenal medulla, in the preganglionic neurons of the sympathetic and parasympathetic ganglia, in skeletal muscle, and in the electroplaques of electric fish. But, in contrast to the general acceptance of muscarinic actions, in 1935 it seemed unlikely to almost everyone, but particularly to neurophysiologists, that nicotinic actions could be mediated by a chemical process. Unlike muscarinic actions, which were very slow, nicotinic actions were fast, and their rapidity did not seem consistent with a chemical mechanism.

There were, of course, some preliminary clues that even nicotinic effects might be chemically mediated. We have mentioned the experiments of Claude Bernard and others that had shown that curare could block neuromuscular transmission. It was subsequently found that curare could also block transmission in autonomic ganglia. Finally, it was known that ACh also caused contraction of isolated striated muscles of frogs and toads and that, as Langley had observed, nicotine first excited and then blocked skeletal muscle and sympathetic postganglionic cells. But these pharmacological actions of nicotine and ACh on sympathetic ganglion cells and on skeletal muscles seemed for the longest time without physiological significance, and it was not until Feldberg turned his attention to the issue of fast synaptic transmission that the tide began to turn.

The first critical step was taken by Feldberg and Mintz as early as 1933 when they found that nicotinic transmission to the adrenal medulla was cholinergic. As the chromaffin cells of the adrenal medulla are homologous to sympathetic ganglion cells and are innervated by preganglionic sympathetic fibers, it seemed likely that preganglionic sympathetic fibers that synapse in sympathetic ganglia would also be cholinergic. As noted previously, in 1934 Feldberg and Gaddum, using a perfusion method, showed beyond question that on preganglionic stimulation ACh is liberated from the superior cervical ganglion, as it is from the adrenal medulla. And with Dale, Feldberg showed that ACh was the mediator of vagus effects on the stomach (Dale and Feldberg, 1934). From here it was only one step further to examine skeletal muscle. This junction was particularly important for the doubting Thomases, because transmission at the neuromuscular junction had long been considered the most critical test of the chemical hypothesis.

In 1936 Dale, Feldberg, and Marthe Vogt studied the effects of stimulation of the hypoglossal nerve (the motor nerve to the tongue) in cats and dogs in which the parasympathetic outflow in the chorda tympani had previously been severed, and also the effect of stimulation of ventral roots on skeletal muscles in the hind leg, after section of the sympathetic chain. These experiments showed unequivocally that stimulation of the motor fibers (but not sensory or sympathetic fibers) resulted in the release of ACh. Such release did not occur on stimulating denervated muscles or muscles treated with tubocurarine. Also, in 1936, G. L. Brown, Dale, and Feldberg were able to induce a muscle twitch, similar to that seen after nerve stimulation, by injecting ACh directly into the artery supplying the gastrocnemius muscles of cats and dogs close to its entry into the muscles. They further showed that in the presence of eserine a single electrical stimulus to the sciatic nerve elicited a brief tetanus (rather than a simple twitch). The response to tetanic stimulation, on the other hand, was depressed under these conditions because of the accumulation of ACh at the neuromuscular junction.

To summarize, by 1936 Feldberg and his colleagues had found that the motor nerve impulse releases ACh at the neuromuscular junction, and that when it is induced by close intraarterial injection, it causes a brief muscle twitch or, following a single electric shock to the nerve, it gives rise to a short tetanic contraction if the breakdown of ACh is prevented by the prior administration of an anticholinesterase drug. Quantitatively, the amounts of ACh liberated by a nerve impulse and the amount injected are of different orders of magnitude, but this is to be expected because the experimental application is not directly to the active site. Since a blocking dose of curare does not affect the liberation of ACh, but does block its ability to cause muscle contraction, the conclusion that ACh is critical for the neural excitation of muscle fibers was inescapable.

Finally, Feldberg crossed the English Channel in 1939, just before World War II broke out, to collaborate in Arcachon, France, with Albert Fessard on the electric organs of the electric ray, *Torpedo*. They found that the nerves to the electroplaques released ACh and that, on injecting the transmitter into the solution perfusing the electric organ, there was an electrical discharge that could be markedly potentiated with eserine (Feldberg and Fessard, 1942). To the British pharmacologists the situation was now crystal clear. Peripheral transmission was obviously chemical, and it seemed very likely that this would be proved to be true also of central transmission (Feldberg, 1945). The only question remaining was: are ACh and adrenaline (or noradrenaline) the sole transmitters or are there others yet to be discovered?

The Soup versus Spark Controversy

Feldberg's pioneering work had convinced many electrophysiologists, especially Bernard Katz, who, like Feldberg, had emigrated from Germany and was working in A. V. Hill's department at University College, London. However, some physiologists still remained skeptical and favored electrical transmission at sites of nicotinic actions. The short latency and the brevity of the postulated transmitter actions (lasting, at most, just a few milliseconds) suggested that transmission was simply too fast to be chemical. Thus began the hotly debated argument between the two factions—facetiously referred to as the "soup versus spark controversy"—that was to govern thinking about synaptic transmission from 1936 until the early 1950s.

The finding that ACh was released by nerve stimulation at sites of nicotinic action and the speed of the synaptic actions were reconciled by John Eccles, one of Sherrington's last students and certainly his most productive. Eccles argued that there are two components to transmission at nicotinic synapses: (1) an initial, fast excitatory action mediated electrically by the presynaptic action currents; and (2) a prolonged residual action mediated by chemical transmitters such as ACh. Although Eccles was the most forceful advocate of this view and persistently presented it in a number of reviews (Eccles, 1936, 1937, 1946, 1949), he was not alone in this belief. For example, Monnier and Bacq (1935) had shown that the response of smooth muscle to stimulation of the relevant nerves exhibited an initial fast and a later slow phase. In the United States, the electrical hypothesis found strong support among the group of electrophysiologists whom Ralph Gerard referred to as the "axonologists."[11] As documented in the Symposium on the Synapse held in 1939, their consensus view was that the current generated by the presynaptic axon flows into the postsynaptic cell, where it excites an impulse, much as one active segment of an axon excites the next (Fig. 1.3). In a word, the process of cell-to-cell transmission was simply an extension, to the synapse, of Alan Hodgkin's local circuit model for conduction of the nerve impulse (Hodgkin, 1937a,b).

However, two discoveries made it necessary to reconsider these views of electrical transmission. The first came from a quantitative analysis of the amount of current that a presynaptic neuron could inject into a postsynaptic cell, and the second was the discovery of the endplate (and, later, other synaptic) potentials.

Current Flow between Contiguous Axons

In retrospect it is surprising that not one of the proponents of the electrical theory seems to have bothered to ask: is the current from a presynaptic axon quantitatively adequate to excite a postsynaptic cell? The first attempt to estimate this current came in 1940 from studies of *ephapses*, artificial synapses constructed by closely approximating two axons to one another. Some of these studies appeared at first to provide support for electrical transmission by showing that there is an excellent correspondence between the effects predicted by the local circuit theory of nerve conduction and the effects observed in the neighboring axons. It was therefore concluded that the effects are caused by the electrical current flow across the ephapses (Arvanitaki, 1942; Eccles, 1946).

However, Bernard Katz and O. H. Schmitt (1940) pointed out that the penetrating current acting on the resting fiber at an ephapse is virtually a mirror image of the current in the active fiber. As a result the active fiber produces a triphasic excitability change in the inactive fiber—depression, followed by excitation, and then depression. In addition, the excitatory action produced by one fiber in the other is normally much too weak to initiate an impulse in the inactive fiber. For example, in Katz and Schmitt's experiments the maximum excitatory effect never exceeded 20% of the threshold required to initiate an action potential in the second fiber. It followed from this finding that if electrical excitation were to be adequate for synaptic transmission, special conditions would have to prevail at synaptic contacts. In fact, in Arvanitaki's

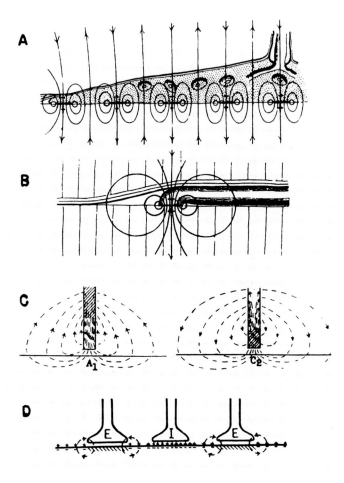

Figure 1.3. Some models of electrical transmission.

(A) Du Bois-Reymond's "modified discharge" hypothesis for the neuromuscular junction.

(B) Du Bois-Reymond's diagram of current flow at the "point of contact" between the motor axon and the muscle fiber.

(C) Eccles's 1946 model for electrical transmission between neurons, showing the pattern of current flow as the nerve impulse approaches the terminal, causing first a hyperpolarization due to inward current flow (A_1) and then, when it reaches the terminal, a depolarization and excitation (C_2).

(D) Brooks and Eccles's (1949) hypothesis for electrical transmission at excitatory (E) and inhibitory (I) synapses. In I the terminal was postulated to be that of a nonspiking Golgi II interneuron, which would produce an anodal focus at its point of contact with the target cell.

Reproduced from Grundfest (1959) by permission of the American Physiological Society.

experiments this was achieved by adjusting the Ca^2 concentration of the bathing solution, thereby increasing excitability at the ephapse; under these conditions impulses were initiated across the experimental ephapse, but only when the axons were chelated.

The experiments by Katz and Schmitt made Eccles realize that electrical synaptic transmission could not function as at an ephapse, and in particular that it could not occur at any arbitrary point where a presynaptic fiber happened to contact a postsynaptic cell. Rather, for electric transmission to occur there would have to be some form of membrane specialization where special conditions for current flow would prevail. He therefore modified his earlier view and postulated that the postsynaptic membrane at the synaptic region had what he termed "the special property of electroreception."[12]

Discovery of the Synaptic Potential

Not only did the discovery in 1938 of the endplate potential present a second serious challenge to the electrical transmission hypothesis, but its later analysis by Stephen Kuffler and Katz was to provide the foundation of our current views of chemical synaptic transmission.

In 1938 Göpfert and Schaefer discovered that the action potential in the presynaptic fiber does not lead directly to the initiation of an action potential in the postsynaptic muscle fiber, a finding that was confirmed the following year by Eccles and O'Connor (1939). Both groups observed that the action potential in the muscle cell did not arise directly out of the baseline but was preceded by a smaller and slower transitional potential. This slower potential is normally lost in the much larger action potential, but it can be unmasked by large doses of curare. In the presence of curare, it was evident that the nerve impulse in the presynaptic axon sets up a local depolarization at the muscle endplate. This local depolarization, which soon became known as the *endplate potential* (EPP), seemed to act like an electronic potential produced by a subthreshold current (Eccles et al., 1941). Equally important, Kuffler was soon to show that there is an irreducible delay, a *synaptic delay* (as first suggested by Sherrington), between the action potential in the axon terminals and the start of the endplate potential.

Similar local, graded potentials were soon demonstrated at other synapses—in the cat sympathetic ganglia, the squid stellate ganglion, and spinal motoneurons in frogs and cats. In each case excitation of the presynaptic axons was found to give rise, with a measurable delay, to a slow depolarization of the postsynaptic cell: these local depolarizations were appropriately termed *synaptic potentials*. Thus by the late 1940s it was generally agreed (1) that synaptic potentials probably occur at all sites of synaptic transmission; (2) that they provide an essential functional link between the action potential in the presynaptic terminal and that in the postsynaptic cell; and (3) that the properties of the synaptic potential are distinctly different from those of either the pre- or the

Figure 1.4. Three of the major contributors to synaptic transmission. Photographed at an international scientific meeting in the 1960s are, from left to right, Stephen Kuffler, Bernard Katz, and John Eccles. (Another photograph, taken when they worked together in Sydney in the 1940s, has been reproduced frequently elsewhere.)

postsynaptic spike in that they are much slower than action potentials and are graded rather than all-or-none.

In his continuing rearguard action against the growing evidence for chemical transmission, Eccles interpreted the synaptic potential as a reflection of the special property of electroreception. He argued that the postsynaptic subjunctional membrane is specialized to give only local responses of graded intensity without the sudden all-or-nothing breakdown of resistance that occurs with the initiation of an impulse. According to this view, the presynaptic impulse sets up an electric current that exerts a diphasic effect on the junctional region of the postsynaptic cell, first an inhibitory (or anodal) focus followed by an excitatory (cathodal) focus with an inhibitory (anodal) surround (Fig. 1.4). The excitatory focus would, Eccles believed, set up a brief and intense local response at the synaptic region that would spread electrotonically over the postsynaptic cell membrane. On reaching a certain threshold, the depolarization of the extrajunctional membrane would finally set up a propagated impulse. The subsynaptic specialization of the postsynaptic membrane was thus seen as an amplifier of the small electrical currents that flowed from the presynaptic axons, acting until the synaptic potential was of sufficient amplitude to trigger an action potential in the postsynaptic cell.

The Experiments of Kuffler and of Fatt and Katz Turn the Tide toward Chemical Transmission

Despite Eccles's progressively more ingenious explanations, by the late 1940s the tide was clearly turning against the electrical hypothesis for transmission at peripheral synapses, as Eccles himself later acknowledged. In his 1964 monograph *The Physiology of Synapses*, he wrote that "In the Paris symposium of 1949 there was fairly general agreement that both neuro-muscular and ganglionic transmission were mediated by ACh, particularly as Kuffler . . . reported most convincing experiments against the electrical hypothesis. *However, there was still fairly general agreement that central synaptic transmission was likely to be electrical*" (emphasis added).

The specific experiments of Kuffler to which Eccles alluded were directed toward three key issues: (1) the synaptic delay; (2) the consequence for the EPP of altering the configuration of the action potential in the presynaptic terminals; and (3) the effects of subthreshold stimulation of the nerve terminals.

The earlier measurements of what was referred to as the *synaptic delay* or *neuromuscular delay* had not been precise because they determined only the latency between the action potential in the presynaptic fibers and that of the muscle or postsynaptic ganglion cell, rather than that between the action potential in the axon terminals and the onset of the EPP or synaptic potential. Kuffler was able to address this issue critically for the first time by carefully dissecting single nerve–muscle fiber preparations, which Katz considered "a brilliant technical feat [that] immediately and deservedly put [Kuffler] 'on the map'" (Katz, 1982).

By stimulating within 0.5 mm of the electrode used to record the EPP, Kuffler found that, in frogs at 20°C, the synaptic delay is on the order of 0.8–0.9 msec. This delay was not appreciably reduced even when the stimulating electrode was as close as 50 μm to the endplate region. If the entire delay were attributable to conduction in the presynaptic terminals, this finding would have implied that the presynaptic action potential was slowed by a factor of about 300 from its prior velocity in the distal part of the nerve, which seemed unlikely. Moreover, since the duration of the EPP is long compared to that of the preceding action potential, it would be necessary to assume that current flow in the presynaptic terminals lasts at least 4–5 msec. This could only occur if the action potential was followed by a depolarizing afterpotential of long duration, and only if the postulated potential was important in depolarizing the terminals and effecting the release of the transmitter. To test this idea, Kuffler exposed the nerve terminals to veratrin, an alkaloid that enhances depolarizing afterpotentials, and found that even though the depolarizing afterpotential was greatly increased, it had no effect on the amplitude of the EPP as recorded from the muscle. Finally, when a subthreshold depolarizing current pulse was applied within 0.5 mm of the terminals, Kuffler found that it had no appreciable effect on the transjunctional potential changes. In a word, the severe attenuation of current spread from the nerve to the muscle effectively ruled out the possibility that transmission at the endplate could be electrical (Kuffler, 1942a,b).

The final blow to the theory of electrical transmission at the neuro-muscular junction was delivered in the early 1950s by Bernard Katz and his colleagues in the department of biophysics at University College, London, in a series of experiments that elevated the analysis of synaptic transmission to an entirely new level. Katz was another of that distin-guished group of scientists who had been compelled to leave Germany in the 1930s, and for a short period in the early 1940s he had been as-sociated with Eccles and Kuffler in Sydney, Australia. Katz had received a thorough grounding in biophysics in Germany, and as early as 1939 he had published an important monograph, *Electrical Excitation of Nerve*. On returning to England from Australia, he first collaborated with Alan Hodgkin on the study which established that the rising phase and over-shoot of the action potential is due to a sudden increase in sodium per-meability (Hodgkin and Katz, 1949). He then joined Hodgkin and Hux-ley in their initial experiments to test the Na^+ hypothesis by carrying out voltage clamp experiments to analyze Na^+ and K^+ currents during and immediately following the action potential (Hodgkin et al., 1952). But it is for his seminal series of studies with José del Castillo, Paul Fatt, and Ricardo Miledi on synaptic transmission that he is perhaps best known. Indeed, it was for this work that he shared the Nobel Prize for Medicine or Physiology in 1970 with Julius Axelrod and Ulf von Euler.

In the first set of studies with Fatt, Katz extended the ionic hypothe-sis to synaptic transmission by providing a critical analysis of the ions that flow during the synaptic potential. For these experiments Fatt and Katz used sharp-tipped intracellular recording microelectrodes of the type developed by Ling and Gerard (1949) and used earlier to analyze ion fluxes in muscle fibers by Nastuk and Hodgkin (1950). Intracellular microelectrode recording enabled Fatt and Katz to circumvent many of the technical difficulties involved in dissecting single nerve–muscle fiber preparations and the uncertainties associated with extracellular meas-urements due to the shunting effects of interstitial fluid (Fatt and Katz, 1951, 1952). Also by using curare, they were able to reduce the ampli-tude of the EPP below the threshold for action potential initiation and in this way to study the EPP in isolation.

Fatt and Katz found that the EPP produced in the muscle cell by the action of the motor nerve was largest when they placed the recording intracellular electrode precisely at the endplate. As the electrode was moved progressively farther away from the endplate region, the ampli-tude of the EPP decreased systematically (Fig. 1.5). From these findings they concluded that the EPP is generated by inward current that is con-fined to the endplate and spreads passively away along the muscle fiber from the region of the endplate. They further found that the synaptic potential at the endplate rises rapidly but decays more slowly. They at-

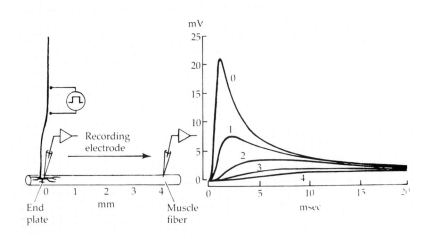

Figure 1.5. Decay of synaptic potentials with distance from the endplate region of a muscle fiber. Records taken by an intracellular electrode at distances of 1, 2, 3, and 4 mm from the endplate show a progressive decrease in size and slowing of rise time of the synaptic potential. Reproduced from Kuffler and Nicholls (1976), after Fatt and Katz (1951), by permission of Sinauer Associates, Inc.

tributed the rapid rise to the sudden release of ACh into the synaptic cleft by the action potential in the presynaptic terminal. Once released, the ACh would diffuse rapidly to the receptors on the surface of the muscle fiber. However, not all the released ACh reaches the postsynaptic receptors, because two processes act to remove it from the cleft: some simply diffuses away out of the synaptic cleft and some is hydrolyzed by the enzyme acetylcholinesterase, which is localized in the intervening basal membrane. Not surprisingly, after treatment with a cholinesterase inhibitor, the EPP is greatly prolonged, and the charge transfer through the endplate can be increased by as much as 50-fold.

Fatt and Katz also examined the mechanism underlying the EPP and suggested that it involved an increase in conductance that was non-selective for Na^+, K^+, and Cl^- and served, as it were, to short-circuit the resting membrane potential. (Functionally it was equivalent to placing a fixed leak resistance across the membrane.) As a result, there was a direct relationship between the value of the resting membrane potential and the amplitude of the EPP. When they examined the reversal potential (i.e., the membrane potential at which the EPP is nullified), they found it to be 14 mV. At more positive potentials, the normally depolarizing EPP reversed its sign and became a hyperpolarizing response. The fact that the values of the postsynaptic membrane potential determined the amplitude, and even the sign, of the synaptic potential indicated that the "battery" responsible for the endplate current must be located in the postsynaptic membrane. This finding effectively excluded a presynaptic source for the current, as had been predicted by the "spark hypothesis," but was exactly what was to be expected if transmission were chemical.

Since the value of the reversal potential, 14 mV, did not match the equilibrium potential of Na^+, K^+, or Cl^-, the EPP could not be attributed to an increased conductance of any *single* ion species (as is the case for Na^+ influx during an action potential). This reversal potential was instead consistent with a simultaneous increase in the conductance of several ion species.

On the basis of these findings, Fatt and Katz proposed that the release of ACh produced a drastic change in the membrane at the endplate—in effect a short circuit—so that it became transiently permeable to all the major ions: Na^+, K^+, and Cl^-. However, they could not distinguish such a generalized increase in conductance to all ions from a more selective but simultaneous increase in two-ion species such as Na^+ and K^+, or Na^+ and Cl^-. These alternatives were tested several years later by A. Takeuchi and N. Takeuchi (1960) using a voltage clamp technique. They found that in curarized preparations there was an almost linear relationship between the membrane potential and the endplate current, as had earlier been observed by Fatt and Katz. But when the Takeuchis changed the ions in the bathing solution they found that in response to the release of ACh, the endplate became selectively permeable to Na^+ and K^+, but not to Cl^-. The finding of a selective increase in cation permeability, rather than a general permeability breakdown, has subsequently proven to be common at excitatory synapses.

After the appearance of the Fatt and Katz paper, Eccles (1964) rapidly accepted this model for synaptic transmission at the endplate and other peripheral synapses: "It would seem probable," he wrote, "that like the endplate transmitter, the synaptic transmitter [at other peripheral sites] would cause its intense depolarizing action by a large nonselective increase in the ionic permeability of the subsynaptic membrane." He was not, however, ready to accept a similar mechanism for central synaptic transmission.

Eccles's Discovery of Synaptic Inhibition Ends the Soup versus Spark Controversy

As the evidence against electrical transmission at peripheral synapses became incontrovertible, Eccles retreated to the CNS, where he thought that the evidence for electrical transmission was still compelling. In his last major review on this controversy in 1949, entitled "A Review and Restatement of the Electrical Hypothesis of Synaptic Excitatory and Inhibitory Action," written after Kuffler's experiments of 1942, he wrote: "In view of the exclusion of the electrical hypothesis from the neuromuscular junction and the uncertainty of its application to synaptic transmission in ganglia, where acetylcholine transmission also is operative, it would seem expedient to restrict it in the first instance to monosynaptic transmission through the spinal cord, where chemical transmission by acetylcholine seems highly improbable, and where the experimental investigation has been more rigorous than elsewhere in the [central] nervous system."

A few years before Eccles wrote this review, he had become friendly with the Viennese philosopher Karl Popper, and he was soon to be much influenced by Popper's way of thinking (Eccles, 1975). Popper argued that since a scientific hypothesis can never be proven—it can only be falsified—the strength of a scientific theory is directly related to the precision with which it is formulated so as to allow it to be falsified by experiment. "The criterion of the scientific status," Popper wrote, "is its falsifiability or refutability" (Eccles, 1975). The falsification of a theory, he stressed, should not be viewed as an embarrassment. On the contrary, it is evidence of the rigor and precision of the hypothesis: to specify a hypothesis so precisely as to allow it to be falsified is the highest goal of science. Popper therefore convinced Eccles to continue to define the electrical hypothesis as rigorously as possible, and this Eccles proceeded to do, not only for excitation but also for synaptic inhibition. In the event, the critical falsification that finally led Eccles to abandon his theories of electrical transmission in the CNS came not from his studies of excitatory synaptic actions but from his discovery of the mechanism underlying synaptic inhibition. As he was to write on a later occasion, "I had been encouraged by Karl Popper to make my hypothesis as precise as possible, so that it would call for experimental attack and falsification. It turned out that it was I who succeeded in this falsification" (Eccles, 1975).

Popper's influence is particularly evident in Eccles's models of synaptic inhibition. Synaptic inhibition had posed enormous difficulties for the proponents of electrical transmission, and over a number of years several imaginative hypotheses had been put forward to account for central inhibition by electrical means. These included (1) the view that impulses in inhibitory fibers blocked the excitatory impulses in the presynaptic terminals, presumably by some hyperpolarizing action; (2) the notion that the inhibitory effect is exerted on the postsynaptic cell because the terminals of the inhibitory fibers end at some special spatial location on the cell, for example around the site of impulse initiation at the axon hillock; and (3) the hypothesis that there are specific inhibitory synapses at which the presynaptic impulses exert an electrical effect that is the functional inverse of excitation.

But the most elegant of the hypotheses for electrical inhibition was proposed in 1947 by Eccles and Chandler Brooks (Brooks and Eccles, 1947). They postulated that inhibitory inputs to the spinal cord ended on short axon cells (like Golgi's type II cells), which formed close electrical contacts with the motoneurons. They further hypothesized that the Golgi cells were non-impulse-generating neurons and that the afferents that impinged upon them would not readily excite the cells because of their high threshold. An incoming volley of impulses would, however,

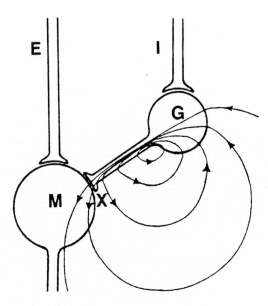

Figure 1.6. Eccles's postulated mechanism for electrical inhibition. The mechanism involves a Golgi II (nonspiking) neuron (G) interposed between the inhibitory input (I) and the motoneuron (M). The inhibitory input subliminally excites the Golgi cell and generates the pattern of current flow indicated by the arrows, which produces an anelectrotonic focus on the motoneuron. Reproduced from Eccles (1982), with permission from the *Annual Review of Neuroscience,* Volume 5, ©1982 by Annual Reviews; http://www.AnnualReviews.org.

set up a synaptic potential in the Golgi cells. Although this synaptic potential would be too weak to initiate an impulse, it would give rise to an inward current that would increase the conductance of the Golgi cells at the site at which the inhibitory fibers ended. Depolarization at this site would then give rise to an outward current flow throughout the rest of the neuron. Because the Golgi cell is small and inactive, and because of the close apposition of its axon terminals to the motoneuron, this outward current flow, it was argued, would penetrate the membrane of the postsynaptic cell and produce an inward current flow at a localized region of the motoneuron membrane. It followed from this hypothesis that synaptic inhibition mediated by the Golgi cell's axon would be diphasic in character—a combined inhibitory-excitatory action (see also Eccles, 1949, and Fig. 1.6).

In the early 1950s Eccles and his colleagues began to apply to motoneurons the same intracellular recording methods used by Fatt and Katz for their studies of transmission at the neuromuscular junction. The initial interest of Eccles's group was to determine if the properties that Hodgkin and Huxley had reported for the giant axons of invertebrates were shared by vertebrate motoneurons (Brock et al., 1951;

Eccles, 1953). But from the beginning they were also interested in the mechanism of excitation and inhibition in the CNS (Brock et al., 1952). When they examined the effects of stimulating inhibitory inputs to the motoneurons they found, to Eccles's surprise, that synaptic inhibition caused a hyperpolarization of the motoneuron without any associated depolarization (Fig. 1.7). This led him to abandon, without reservation, the electrical transmission hypothesis for inhibition that he had so recently espoused. As he now wrote, "The potential change observed is directly opposite to that predicted by the Golgi-cell hypothesis which is thereby falsified" (Brock et al., 1952; see Fig. 1.8). And in the discussion of their 1952 paper he went on to write off electrical excitation in equally strong terms: "Since the experimental evidence has falsified the

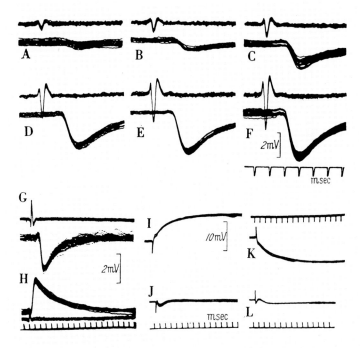

Figure 1.7. Inhibitory postsynaptic potentials. It was IPSPs such as this one that led Eccles to abandon his electrical hypothesis for synaptic transmission. The lower records give intracellular responses of a biceps-semitendinosus motoneuron following stimulation of a quadriceps volley of progressively increasing size, as shown by the upper records, which are recorded from the L_6 dorsal root by a surface electrode (downward deflections signal negativity). All records are formed by the superposition of about 40 faint traces. *G* shows IPSPs similarly generated in another biceps-semitendinosus motoneuron, the monosynaptic EPSPs of this motoneuron being seen in *H. I–L* show changes in potential produced by an applied rectangular pulse of 12×10^{-9} A in the depolarizing and hyperpolarizing directions, *I* and *K* being intracellular and *J* and *L* extracellular. Reproduced from Eccles (1964) by permission.

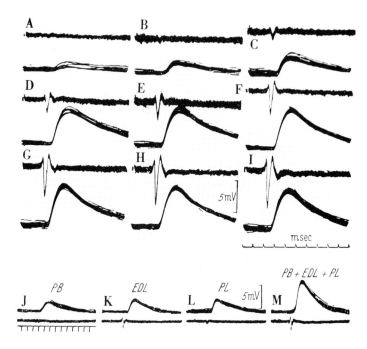

Figure 1.8. Monosynaptic excitatory postsynaptic potentials recorded intracellularly in motoneurons. Each record is formed by the superposition of about 25 faint traces. In *A–I,* the EPSP is generated in a medial gastrocnemius motoneuron by an afferent volley from the medial gastrocnemius nerve of progressively increasing size, as indicated by the spike potentials in the upper records from the dorsal roots. The EPSP attained its maximum in *F,* where the afferent volley was probably maximal for group Ia fibers. In *J–M,* EPSPs are similarly recorded in a peroneus longus motoneuron in response to maximum group Ia volleys from the nerves to peroneus brevis, extensor digitorum longus, and peroneus longus, and by all three volleys together, as indicated by the symbols. Reproduced from Eccles (1964) by permission.

Golgi-cell hypothesis of inhibition and left the chemical transmitter hypothesis as the only likely explanation, it suggests further that the excitatory synaptic action is also mediated by a chemical transmitter" (Brock et al., 1952).

Eccles and his colleagues suggested that the hyperpolarization was inhibitory because it moved the membrane potential from its resting level so that subsequent *excitatory postsynaptic potentials* (EPSPs) acted from a more negative baseline. However, subsequent studies by Fatt and Katz on inhibition in the crayfish showed that inhibitory synaptic actions could occur without a change in membrane potential, simply because of the short-circuit or shunting action of inhibition. Again, as was the case with excitation, the ions responsible for inhibition were at first not clear. Fatt and Katz postulated a nonspecific increase in small ions, specifically Cl⁻ and K⁺. Later, more detailed studies of a variety of inhibitory actions were to establish that in no case does inhibition involve the simultaneous increase in membrane permeability to more than one ion species. Transmission is due to either chloride permeability or potassium permeability being selectively and independently turned on (Fatt and Katz, 1953).

In reviewing the long struggle leading to the acceptance of chemical transmission in the CNS, Dale (1954) wrote, with more than a little smugness:

> Eccles and his team conclude that this positive variation in the motor horn cell could only be due to the release of a chemical agent from the endings of the afferent fibre making synaptic contacts with its surface, and that, if synaptic inhibition was thus chemically transmitted, synaptic excitation was unlikely to be transmitted by an essentially different process, though the transmitter might probably be a different one. By obvious analogy, it was to be supposed that some chemical agent or other would be effective at all central synapses, and that being accepted, Eccles was naturally ready to take cholinergic transmission in the ganglion in his stride. A remarkable conversion indeed! One is reminded, almost inevitably, of Saul on his way to Damascus, when the sudden light shone and the scales fell from his eyes.

However, although the soup and spark controversy was effectively resolved in 1952, the nature of the excitatory and inhibitory transmitters involved would not be discovered for some years.

The Surprising Discovery of Electrical Transmission

Having been converted to chemical transmission by the discovery of the hyperpolarizing nature of synaptic inhibition, Eccles celebrated the falsification of the electrical hypothesis that he had so vigorously championed by converting wholeheartedly to the chemical hypothesis for synaptic transmission, arguing with equal enthusiasm and vigor for its universality. It was at this point, in October 1954, that Paul Fatt, Katz's collaborator, wrote a masterly review of junctional transmission in which he took a farsighted view of synaptic transmission and presciently pointed out that it was premature to conclude that chemical transmission is in fact universal. He concluded his review as follows:

> Although there is every indication that chemical transmission occurs across those junctions which have been discussed in this review and which are most familiar to the physiologist, *it is probable that electrical transmission occurs at certain other junctions*. The geometry of the junction is decidedly unfavorable for electrical transmission at the junctions which have been mentioned. The pre-junctional structure, which according to the electrical hypothesis would generate the electric current

for transmission, is usually much smaller than the post-junctional structure, which would have its excitatory state altered by those currents. Conditions are much more favorable for transmission in the reverse direction, and, in fact, the excitation of pre-junctional motor nerve fibers by the action currents of post-junctional muscle fibers has been observed to occur in mammalian muscle. This argument does not hold when one considers the synapses between giant nerve fibers where the pre- and post-junctional structures have usually about the same dimensions. In this case some electrical interaction will be expected, its intensity depending on how closely the fibers approach each other. One possible arrangement, which may be envisaged to give a high degree of interaction, is for the two fibers to be actually touching and for the membrane in contact to have a low electric resistance compared with that in neighboring parts of the fibers. The synapse would then serve to direct current between the interior of the two fibers, while active membrane changes would occur in neighboring regions. The ultimate development of such a system would be the elimination of the contacting membrane to secure greatest efficiency of transmission, should other factors permit this. This view of electrical transmission has been taken because it is possible to observe in certain giant fiber preparations a protoplasmic continuity existing between, what in an earlier stage of phylogenetic development must have been independent, synapsing nerve cells. The available evidence, however, does not indicate that a single mechanism of transmission operates at all giant fiber synapses. . . .

A case in which there can be little doubt that electrical transmission operates is in the nervous system of the crayfish, where successive giant nerve cells, each extending along one segment of the thoracic or abdominal region, butt upon each other to form the lateral giant nerve fibers. . . . Transmission takes place in either direction so that an impulse initiated at any level travels over the whole chain of segmental nerve cells, both cranially and caudally. . . . A more perplexing case is the synapses in the crayfish ganglion between the central giant nerve fibers and the motor nerve fibers. The fact that this synapse is polarized to transmit impulses only in the direction from giant fiber to motor fiber cannot be taken as an indication of nonelectrical transmission, since the geometrical arrangement is the reverse of that ordinarily obtaining at synapses: the pre-junctional structure is here larger than the post-junctional structure. (Fatt, 1954)

Three years later, the correctness of Fatt's view was convincingly demonstrated by Edwin Furshpan and David Potter, who analyzed synaptic transmission between the presynaptic giant axon and the postsynaptic motor axon in the crayfish nerve cord and found it to be electrical (Fig. 1.9). The several tests for electrical transmission that Kuffler had carried out at the neuromuscular junction—where he had failed to find evidence for current flowing from the pre- to the postsynaptic cell—turned out positive at this electrical synapse. The latency was extremely short, and even small electrical currents flowed from the pre- to the postsynaptic cell. As Furshpan and Potter (1957) wrote, "It is difficult to

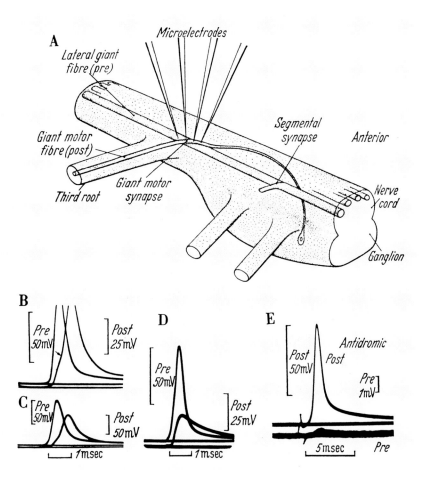

Figure 1.9. The first convincing evidence for electrical transmission.

(A) Semidiagrammatic drawing of a portion of a crayfish abdominal nerve cord, containing one ganglion. The course of a motor giant axon is shown from its cell body in the ventral part of the ganglion until it leaves the third ganglionic root on the opposite side of the cord. Only its junction with the lateral giant pre-fiber is shown, but its synapses with the two medial giant fibers are located just centrally, where the fibers cross the motor axon. A septal synapse between two segments of the lateral giant fiber is also shown.

(B–D) Orthodromic nerve impulse transmission at the giant synapse shown in A with simultaneous intracellular recording from pre- and postfibers, the pre-fiber potential being recorded in the upper traces. B and C were recorded from the same synapse at different amplifications, the postspike origin being indicated by the arrow in B. In E the upper trace is the postfiber antidromic spike potential, which produces a negligible potential in the prefiber (lower trace). Note the separate potential scales for pre- and postfiber records in B–E.

Reproduced from Eccles (1964), after Furshpan and Potter (1957), by permission.

assign a value to the response between the pre-spike and the synaptic response. Both potentials seem to arise about the same time but at different rates . . . even subthreshold electric currents passed through one of the internal electrodes can produce appreciable changes in the membrane potential recorded with another electrode on the opposite side of the synapse."

Even more astonishing, several years later Furshpan and Furukawa came up with another surprise—the demonstration that inhibition could occur by electrical means and, in fact, by a mechanism somewhat analogous to that postulated by Eccles several years before (Furshpan and Furukawa, 1964). At the initial segment of the Mauthner cell axon an impulse in the presynaptic fiber generates a positive field in the surrounding extracellular space. This extracellular positivity hyperpolarizes the membrane of the initial segment (which is the point of the lowest threshold for excitation in the Mauthner cell) and causes effective inhibition. As they point out, this is because at any one time the membrane potential is simply the difference between the extracellular and intracellular voltage; if the extracellular voltage becomes more positive, the voltage difference across the membrane of the initial segment would consequently be increased.

Over several years Michael Bennett and his collaborators extended the analysis of electrical synaptic transmission in several important directions (see Bennett, 1966, 1972, for reviews; see also Fig. 1.10). Among

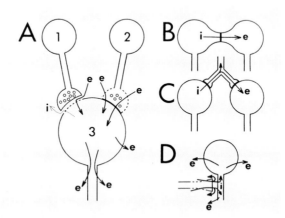

Figure 1.10. Patterns of current flow in a conventional chemical synapse (A₁) and in an electrical synapse (A₂). The broken lines indicate the areas active in generating postsynaptic potentials.
   (B) A dendrodendritic electronic contact.
   (C) Two axosomatic electronic contacts that can synchronize cell firing.
   (D) Electrically mediated inhibition found at a Mauthner cell axon hillock.
   Reproduced from Bennett (1972) by permission.

other things, they discovered that in most cases electrical synapses could pass current in either direction, that is, they were not rectifying (their junctional resistance was constant). This led Bennett to think of them as electrotonic synapses by analogy with electrotonic spread along a core conductor. With Pappas and Nakajima, Bennett (Bennett et al., 1967a–d) carried out a series of electrophysiological and fine structural analyses that enabled them to correlate the biophysical properties of coupling with the occurrence of close membrane appositions, which subsequently came to be known as *gap junctions*.

These studies, culminating in a detailed comparison of electrical and chemical transmission, allowed Bennett to challenge the confidence of those who held that all central synaptic transmission was chemical. Bennett demonstrated that many of the interesting properties that were supposedly the exclusive purview of chemical transmission were also to be found in electrical synapses, and he pointed out that the speed of electrotonic transmission and the reciprocal action of these synapses imparted specific advantages. In particular, Bennett demonstrated that electrotonic synapses are most commonly found in rapidly activated circuits—since they transmit without the delay incurred by the complexities of chemical transmission and in synchronously active ensembles of neurons, in which reciprocity as well as speed is important.

For the sake of completeness we should mention in this context the discovery by Martin and Pilar (1963, 1964) that in the chick ciliary ganglion transmission is both electrical and chemical. Although the preganglionic axons terminate by forming large calyces that embrace much of the surface of ganglion cells, forming extensive, characteristic chemical synapses (De Lorenzo, 1960), there are also more restricted foci of close membrane apposition between the axons and the ganglion cells. De Lorenzo (1966) originally described these as tight junctions, but we now know they are, in fact, typical gap junctions.

Before leaving this topic, we should note also that the discovery of synapses that are based on gap junctions was quickly seized upon by a number of "latent reticularists" who saw in it a modern-day challenge to the neuron doctrine. This view was strengthened when it was shown that although a narrow intercellular cleft exists at gap junctions, this gap is filled with an array of junctional channels that serve to connect the cytoplasm of the two related cells, permitting current to flow freely from one neuron to the other. The gap junction channels are in fact large enough to permit the flow of small organic metabolites between the connected neurons. To this extent, the electronic coupling of cells that are united by such junctions can be regarded as evidence of intercellular continuity and can be considered as an interesting exception to the neuron doctrine as Cajal and others had initially conceived of it. But it is important to appreciate that not only are electrical synapses relatively uncommon in the mammalian CNS, they also provide a form of cell-cell interaction quite different from that conceived of by the reticularists. Each neuron is bounded by its own cell membrane, each has its own nucleus and array of organelles and, although they may be connected by specialized channels, they function as independent entities in every other respect. Viewed as a whole, the nervous system is unequivocally not a network of cytoplasmically continuous cells.

In summary, by the late 1960s the field had come full circle: both chemical and electrical mechanisms for transmission had been shown to exist, and both mechanisms were known to display a variety of subtypes. Moreover, not only do both mechanisms exist, but some models for synaptic action that had been jettisoned during the soup versus spark controversy had been resurrected during the subsequent détente.

The Quantal Nature of Transmitter Release

With the discovery that transmission at the vertebrate neuromuscular junction and at most synapses in the CNS is chemical in nature, and with the specification of the ionic mechanisms for generating excitatory and inhibitory postsynaptic potentials, attention next turned to the mechanisms whereby the chemical transmitter is released. Here, again, the field was opened up by Katz.

During the course of their experiments, Fatt and Katz (1951) had made a remarkable chance observation: when they recorded from the endplate region of a muscle fiber, there were often small, spontaneous depolarizing potentials even in the complete absence of presynaptic stimulation (Fig. 1.11). These spontaneous potentials were about 0.5–1.0 mV in amplitude and resembled miniature versions of the EPP in their time course and in their response to various drugs. Thus drugs that enhance the action of ACh, such as inhibitors of acetylcholinesterase, prolonged the spontaneous potentials much as they prolonged the EPP, whereas agents that block the ACh receptors, such as curare, also abolished the miniature EPPs. Moreover, as was the case with the EPP, the miniature potentials were recorded only at the endplate, at the point of contact between the nerve and muscle. Fatt and Katz (1952) therefore called these miniature potentials *spontaneous miniature EPPs* (mEPPs).

Fatt and Katz next found that the mEPP frequency could be increased by depolarizing the presynaptic terminal and that the mEPPs disappeared after the presynaptic axons were cut and the motor nerves had degenerated, only to reappear when the muscle was reinnervated. Together these manipulations established that the mEPPs derive from the presynaptic terminals. They also found that removal of Na^+ from the bathing solution abolished both the EPP and the mEPPs, but that reducing the external Ca^{2+} reduced only the size of the EPP but had no effect on the size of the mEPPs.

In 1954 del Castillo and Katz showed that with sufficiently low Ca^{2+} levels the size of the EPP, normally about 70 mV, became no larger than the size of the mEPPs (0.5–1.0 mV). Under these circumstances, succes-

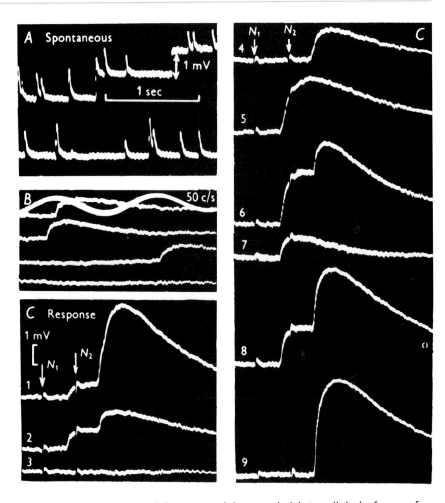

Figure 1.11. Miniature endplate potentials recorded intracellularly from a frog skeletal muscle. The muscle was bathed in a solution containing 0.45 mM Ca^{2+} and 6 mM Mg^{2+}. Records *A* and *B* are of spontaneous mEPPs; record *C* shows the responses to paired nerve impulses with failures to the first impulse (N_1) in records C_4 and C_9, and to the second (N_2) in C_4 and C_9. Reproduced from del Castillo and Katz (1954a) by permission of the Physiological Society.

sive impulses in the motor nerve to the muscle fiber evoked, in a random fashion, EPPs that varied in a stepwise manner so that each EPP was an integral multiple (0, 1, 2, 3, or more) of the mEPP. From this they concluded that the normal EPP also was constituted of an integral number of miniature units and that the effect of lowering the calcium concentration was to reduce the EPP *"in definite quanta,* as though it blocks individual nerve terminals, or active patches within them, in an all-or-none manner" (emphasis added). Thus was born the *quantal hypothesis,* which was to dominate thinking about synaptic transmission for the next three decades.

Quanta: Multimolecular Packets of Transmitter Released from Vesicles by Exocytosis

Following a suggestion by Alan Hodgkin, Fatt and Katz (1952) initially attributed the mEPPs to active localized patches of membrane that generate potentials in fine branches of individual nerve terminals, causing the all-or-none release of a certain number of molecules of transmitter: "We must therefore think of a local mechanism by which acetylcholine is released at random moments in fairly large quantities; and that the most plausible explanation is the occurrence of excitation at individual nerve terminals, evoked by spontaneous fluctuations of their membrane potential . . . but it does not lead to a propagated impulse if the affected area is too small." They were inclined to attribute this spontaneous excitation at the nerve terminals to "electrical noise" generated across the membrane by random fluctuations of the resting potential as a result of thermal agitation of ions within the membrane. This noise, they argued, can become sufficiently large at a small structure occasionally to exceed the threshold for the release of transmitter.

In 1954 Katz abandoned this idea because he and del Castillo had found that the spontaneous release of mEPPs still occurred when all electrical activity had been blocked by the application of an isotonic solution of potassium sulfate. Moreover, extracellular recordings from nerve terminals at branch points showed that the fluctuation of ACh release occurs, even though the action potential invaded the nerve terminals without failure and retained a constant amplitude throughout.

Del Castillo and Katz considered and rejected a second possible basis for the mEPPs: a membrane shutter mechanism. Suppose, they argued, there are specialized areas in the terminal axon membrane that act as ACh gates. In the absence of an action potential these gates would usually be closed. However, the gates could reach a degree of instability when they open for a brief period, during which time a small amount of ACh would be released from the interior of the nerve ending. An action potential would greatly increase the likelihood of an ACh gate opening. A mechanism of this kind is feasible, but it did not seem very attractive, for one would have to explain why such a membrane-controlled flip-flop process leads to a quantal amount of ACh release that is identical for the spontaneous mEPPs and for evoked release, despite the fact that the potential difference across the membrane changes from the resting level to the peak of spike activity during evoked release. Any alteration in the gating action, they argued, would necessarily be reflected in a change in quantal size during the nerve impulse, but this is ruled out by a large body of experimental evidence (Katz and Miledi, 1965).

Finally del Castillo and Katz concluded that the most straightforward explanation for the constancy of quantal amplitude is that the transmitter is released from the axon terminal in discrete multimolecular packets, which they called *quanta*.[13] The relative constancy of the packets of ACh suggested to them that the size of the quanta is controlled not by the rapidly changing properties of the membrane, but by some cellular process that is not disturbed from the outside. They further proposed that transmitter is stored in preformed submicroscopic packages inside the cell, from which it can be released at the cell surface in an all-or-none fashion (del Castillo and Katz, 1954a–d). *Synaptic vesicles*, which were discovered at about the same time by Palay and Palade (1955) and by de Robertis and Bennett (1955), seemed to provide just the right structural counterpart. The vesicles were of fairly uniform size and were found at the right place within the nerve terminal, whereas no other structures could be seen nearby that would meet the requirements of the quantal hypothesis. In addition, the notion that the membrane-bound vesicles contained small packets of transmitter was consistent with Feldberg's finding that most of the ACh stored in the nervous system is protected or bound within subcellular organelles, to which it remains attached even during processes of homogenization and high-speed centrifugation.

Del Castillo and Katz further argued that the vesicle could actively accumulate the transmitter substance and maintain it at a much higher concentration than exists in the surrounding axoplasm. Moreover, when packaged within vesicles, the transmitter is separated from its post-synaptic target by two membranes: the membrane surrounding the vesicle itself and the plasma membrane surrounding the axon terminal. For the transmitter to be released so that the entire "quantum" reaches the receptors in the postsynaptic membrane more or less synchronously and at a sufficiently high concentration, del Castillo and Katz assumed that the transmitter is released by an exocytotic process in which the vesicle membrane fuses with the presynaptic membrane and thereby discharges its contents into the synaptic cleft (Katz and Miledi, 1965).

In a series of papers del Castillo and Katz explored the statistical nature of quantal synaptic transmission and developed the modern view of transmitter release (del Castillo and Katz, 1954a–d). According to this view, ACh is released in quanta—made up of multimolecular packets. The release of the quanta is probabilistic. It occurs spontaneously even in the complete absence of action potentials, at a rate of about one quantum released per second per endplate. An action potential transiently increases the probability that quanta of transmitter will be released, so that the normal EPP is generated by the release of, on average, about 150 quanta in less than 1 msec, with each quantum contributing about 0.5 mV to the EPP. The exact number of quanta released by any given nerve impulse fluctuates in a random fashion that can only be described in statistical terms. Formally del Castillo and Katz (1954b) expressed this as follows: a nerve terminal contains a large number (n) of quanta, each released in response to an action potential with a probability p:

The average "quantum content" of the e.p.p. depends on the probability of response of the individual units and this varies with the external Ca and Mg concentration. . . . If one accepts the present results as showing that the miniature e.p.p. is the basic unit of response, then the effect of Ca must be to raise the quantum content of the e.p.p. either by increasing the size of the population n or its probability of responding p.

This line of investigation led to what has come to be known as the Ca^{2+} *hypothesis* (Katz and Miledi, 1965). This hypothesis emerged from a remarkable series of studies of the frog neuromuscular junction and the giant synapse of the squid stellate ganglion by Katz and Ricardo Miledi, who found that the depolarization following an action potential in the nerve terminals opens Ca^{2+} channels and increases the conductance to Ca^{2+}. The entry of Ca^{2+} into the terminal leads, after various delays, to the release of transmitter. They next showed that neither Na^+ entry nor the K^+ efflux associated with the action potential is required for normal transmitter release. Indeed, the only role of the action potential is to depolarize the terminals and thus open the Ca^{2+} channels. Thus when Na^+ and K^+ channels were blocked by tetrodotoxin and tetraethylammonium, respectively, graded depolarizations of the terminals could activate a graded Ca^{2+} influx, which, in turn, results in the graded release of transmitter.

The finding that depolarization of the terminals by the action potential serves to open voltage-dependent Ca^{2+} channels was later confirmed by Rodolfo Llinas and his colleagues, who also found that the synaptic delay—the time from the onset of the action potential in the presynaptic terminals to the onset of the postsynaptic potential—is due in large part to the time required for the Ca^{2+} channels to open. Because the voltage-dependent Ca^{2+} channels are located very close to the transmitter release sites, they can act to trigger transmitter release within as little as 0.2 msec. It has been estimated that the resultant influx of Ca^{2+} produces localized concentrations of up to 200–300 µM in microdomains within the presynaptic terminal near the release sites. Such local increases in Ca^{2+} concentration greatly enhance the probability of vesicle fusion and transmitter release (Llinas et al., 1972).

In most nerve cells there are at least three (and probably more) classes of voltage-sensitive Ca^{2+} channels. One class (the L-type channel) is characterized by a slow rate of inactivation, so that it remains open during a prolonged depolarization of the membrane. The other two classes (N-type and P-type) inactivate more rapidly, and the available evidence suggests that it is the influx of Ca^{2+} through these latter channels that contributes most directly to transmitter release.

The Ultrastructure of the Synapse Visualized in the Electron Microscope

As noted previously, at the same time as Katz's electrophysiological studies were being carried out, cell biologists and neuroanatomists were beginning to use the electron microscope (EM) to study neural tissue. The EM had been developed in Germany in the 1930s, but its application to biological material was delayed until after World War II and until appropriate methods had been developed for fixing, embedding, and sectioning tissues. Thanks largely to the efforts of George Palade and Keith Porter at the Rockefeller University, most of these difficulties had been overcome by the early 1950s. The first high-quality EM images of neural tissue were published in the mid-1950s (de Robertis and Bennett, 1955; Palay and Palade, 1955; Palay, 1956), including the first observations on the fine structure of neurons and their processes and (most important in the present context) the first descriptions of synapses.

From a historical prospective, these observations were of great significance. By directly visualizing the structural discontinuity of the pre- and postsynaptic elements—a process that was possible only with the increased resolution afforded by the EM—they provided the final unequivocal evidence for the neuron doctrine. In addition, they clarified definitively the characteristics of each of the three elements of the synapse: the pre- and postsynaptic elements and the intervening synaptic cleft. Subsequent EM observations extended these initial observations and led to the discovery of new types of synapses that had not been anticipated in the classical literature, providing a new (albeit tentative) basis for the classification of functional types of synapses on the basis of their fine structure (Gray, 1959; Pappas and Waxman, 1972; Peters et al., 1976).

The Presynaptic Components and the Process of Exocytosis

It became evident from an examination of the presynaptic components of the synapse that they contain many (in some instances, hundreds of) vesicular organelles ranging in size from about 20 to 150 nm in diameter. Palay (1967) aptly likened them to "chocolates [coming] in a variety of shapes and size, and . . . stuffed with different kinds of fillings."

As we have seen, Katz and del Castillo immediately recognized that these might be the organelles that store the quanta of transmitter. Subsequent work on the transmitter content of cholinergic vesicles has provided strong supporting evidence for this correlation. And the finding that other transmitters (such as glutamate and glycine) are also released in quantal fashion has established that this is a general feature of all chemically transmitting synapses (Fig. 1.12).

The most common vesicular forms are small (20–40 nm diameter) round or spherical vesicles with clear (i.e., electron-lucent) centers. They are found at the neuromuscular junction, in autonomic ganglia and several other peripheral synapses, and throughout the CNS. Somewhat larger vesicles, about 40–60 nm in diameter with electron-dense centers,

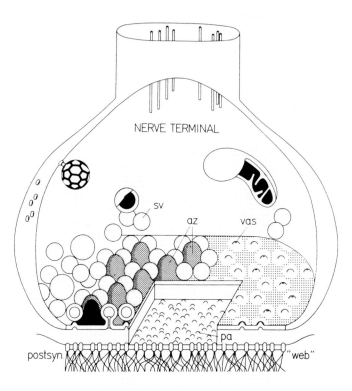

Figure 1.12. Schematic drawing of a "typical" chemically transmitting synapse in the CNS. The illustration is reconstructed from transmission EM and freeze-fracture preparations. az, Active zones; pa, particle aggregate in the postsynaptic membrane; postsyn. "web," postsynaptic density; sv, synaptic vesicle; vas, vesicle attachment site. Reproduced from Akert et al. (1975) by permission of Lippincott, Williams & Wilkins.

are commonly found at sites of aminergic transmission in the CNS and in the peripheral sympathetic system. A third type of synaptic vesicle, characterized by its larger size (80–100 nm diameter) and again possessing a central, dense core, is fairly ubiquitous, but usually occurs in small numbers and always associated with small, clear-centered vesicles. For many years the significance of this third type of vesicle remained uncertain, but they are now thought to be associated with various synaptically released peptides, such as the calcitonin gene related peptide found at the neuromuscular junction and elsewhere. Finally, there is a fourth class of very large vesicles (120–150 nm diameter), characteristic of neurosecretory nerve endings such as those in the neurohypophysis, that contain the peptides oxytocin and vasopressin. These very large vesicles are again commonly found in association with many more small, clear vesicles whose functional significance in neurosecretory terminals is still unknown.

Following the introduction of aldehyde fixation for electron microscopy, several workers observed that in some synapses the small, clear vesicles assumed a flattened or ellipsoidal form. Uchizono (1965) seems to have been the first investigator to have suggested that such flattened (or F-type) vesicles might be associated with the presence of an inhibitory transmitter, having observed them in the terminals of Purkinje cell axons and axon collaterals, as well as in the axons of other known inhibitory neurons in the cerebellum. The physical basis of this vesicle flattening has been shown to be artifactual, in the sense that it is associated with the high osmolarity of the fixing solution. However, its occurrence has proved to be a useful indicator of inhibitory synapses in some (but by no means all) regions of the CNS (e.g., Walberg, 1965; Bodian, 1966).

Del Castillo and Katz postulated that synaptic vesicles discharge their contents by fusing with the presynaptic membrane in the process known as *exocytosis*. This point proved difficult to investigate, even in the EM using conventionally fixed tissue, because the chances of finding a vesicle in the act of opening are relatively small. A thin section through a terminal at the neuromuscular junction of a frog, for example, shows only 1/40,000th of the total presynaptic membrane. As a result, in the 1970s investigators began to apply freeze-fracture techniques to this problem. Heuser and his colleagues (Heuser and Reese, 1973; Heuser et al., 1975) used this technique in an attempt to demonstrate that one vesicle undergoes exocytosis for each quantum of transmitter release. Statistical analysis of the spatial distribution of discharge sites along the active zone showed that individual vesicles fuse with the plasma membrane independently of each other. These results were consistent with the physiological studies indicating that quanta of transmitter are released independently. These freeze-fracture studies therefore provided indirect evidence that synaptic vesicles store the transmitter and that exocytosis is the mechanism by which transmitter is released into the synaptic cleft (see Heuser, 1977, for review).

An alternative approach to the study of synaptic vesicles was pioneered by Viktor Whittaker, who, as early as 1959, had reported the isolation (by homogenization and differential centrifugation of brain tissue) of particles that bound ACh. Later, with George Gray, he was able to show that his fractionation procedure produced *synaptosomes*—pinched-off axon terminals containing synaptic vesicles and mitochondria, attached to postsynaptic densities (Gray and Whittaker, 1962). Two years later Whittaker and his colleagues had further refined the fractionation procedure and obtained preparations of isolated synaptic vesicles that proved to be enriched for ACh (Whittaker et al., 1964). The development of methods for the preparation of relatively pure populations of synaptic vesicles paved the way for the later molecular studies on the characterization of vesicle membrane proteins and the mechanism of exocytosis that are dealt with elsewhere in this volume (see Südhof and Scheller, this volume).

Among the other components of the presynaptic process—such as mitochondria, occasional smooth endoplasmic membranes and cisternae—only three call for special mention here. These are coated vesicles, the presynaptic membrane density, and intermediate filaments. The significance of the *coated vesicles* (we now know that the coat is formed by a meshwork of clathrin) at synapses was not generally appreciated until the freeze-fracture and horseradish peroxidase uptake studies of Heuser and Reese (1973). These experiments showed convincingly that, following the fusion of synaptic vesicles with the presynaptic membrane, the vesicle membrane retains its identity and is endocytotically returned into the presynaptic process, where it can be recycled.

As Fig. 1.13 (which is taken from their work) indicates, the recycling involves several steps: (1) the assembly of a clathrin coat at the site of the invagination of the membrane, usually just beyond the presynaptic membrane density; (2) the movement of the coated vesicle into the presynaptic process; (3) the loss of the clathrin coat; (4) the fusion of the returned vesicle with a membranous cisterna (where it was thought to be reconstituted as a synaptic vesicle); and (5) the recharging of the vesicle with transmitter (now known to be effected through the action of specific

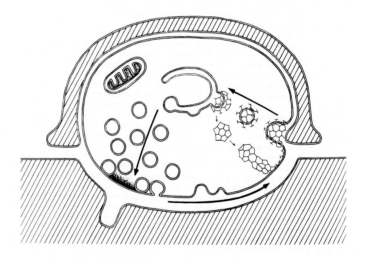

Figure 1.13. Heuser and Reese's synaptic vesicle membrane recycling hypothesis. Based on their studies of the frog neuromuscular junction, they proposed that synaptic vesicles discharge their content of transmitter as they coalesce with the plasma membrane at specific regions adjacent to the muscle. Equal amounts of membrane are then retrieved when coated vesicles pinch off from regions of the plasma membrane adjacent to the Schwann sheath. Finally the coated vesicles lose their coats and coalesce to form cisternae, which accumulate in regions of vesicle depletion and slowly give rise to new synaptic vesicles. Reproduced from Heuser and Reese (1973) by copyright permission of the Rockefeller University Press.

neurotransmitter transporters). More recent work suggests that the fourth step in this process may not always occur and that the vesicles can be re-filled with transmitter shortly after losing their clathrin coats.

These findings on the recycling of synaptic vesicles have gone a long way toward clarifying their origin, an issue that had been debated for some time (see Peters et al., 1976, for discussion). Among the many suggestions originally put forward were that they arose (1) from tubular components of the smooth endoplasmic reticulum; (2) from complex multivesicular bodies; or (3) from microtubules. But the most widely held view was that they were transported from the cell body to the axon terminals along microtubules. Since it was known that much intracellular protein trafficking is mediated by vesicles, and that in axons most of the proteins destined for axon terminals are conveyed by fast axonal transport involving microtubules, this view had much to commend it. Moreover, it had been clearly established for the large neurosecretory vesicles in the neurohypophysis that they are assembled within the cell body and transported down the axons that make up the supraoptico-hypophysial tract (Palay, 1957). However, the current consensus is that most of the smaller vesicles are assembled locally within presynaptic processes from components that either are recycled or were previously transported from the cell soma.

The second component of the presynaptic process that merits comment is the *presynaptic density,* the specialized region of the membrane directly opposed to the postsynaptic element. In most EM preparations, but especially those stained with phosphotungstic acid or bismuth iodide, this portion of the membrane appears to be thicker or denser than others. This appearance is actually due to the presence of a submembranous meshwork of electron-dense material that in some cases has the appearance of a series of pyramidal projections extending into the presynaptic process. En face views of such projections suggest that they may form a regular gridlike arrangement, which Konrad Akert and his colleagues have termed the *presynaptic vesicular grid* (Akert et al., 1972). Their notion is that the spaces between the presynaptic projections are sites at which synaptic vesicles align themselves prior to fusing with the presynaptic membrane. From this hypothesis has emerged the notion that there are specific docking sites for vesicles and specific sites (or *synaptic pores*) where vesicle fusion and transmitter release occur—sites that Couteaux and Pecot-Dechavassine (1970) have termed the *active zones.* In support of this idea is the fact that in nearly every synapse examined there is a small cluster of vesicles closely associated with the presynaptic density (Birks et al., 1960). These vesicles are thought to contain the readily releasable pool of synaptic transmitter, and infrequently Ω-like membrane infoldings can be seen at the active zone, an appearance suggestive of vesicles that had been fixed immediately after fusing with the presynaptic membrane.

In some, but by no means all, presynaptic processes bundles of *intermediate filaments* (or "neurofilaments," as they used to be called) are evident. These are especially prominent when they are aggregated around clusters of mitochondria[14] or other organelles; this arrangement is thought to account for the ringlike boutons seen in reduced silver preparations, apparently as the result of the deposition of metallic silver on the filament bundles. The presence of such clusters of intermediate filaments is of special interest following axonal injury, when they can become a particularly prominent feature of the degenerating axon terminals (Guillery, 1970).

The Synaptic Cleft

The finding that at all chemical synapses the pre- and postsynaptic processes are bounded by distinct membranes and separated from each other by a clearly defined space—the *synaptic cleft*—was, as we have pointed out, the final vindication of the neuron hypothesis. But the synaptic cleft is of interest in its own right (see Südhof, this volume). In most chemical synapses it is somewhat wider than the usual intercellular spaces, being between 20 and 30 nm in width; however, it is not simply a free space. In EM preparations it can be seen to contain filamentous or dense material that spans the interval between the surrounding membranes and is thought to account for the firm attachment of presynaptic processes to the postsynaptic membrane in synaptosomal preparations (Gray and Whittaker, 1962). In preparations stained with ethanolic phosphotungstic acid the material in the cleft often appears as a distinct intercellular plaque that from cytochemical studies appears to consist of a variety of glycoproteins and glycolipids, similar to the glycocalyx that surrounds most cells (Peters et al., 1976).

The neuromuscular junction is distinctive in this regard, in that there is a well-defined basement membrane (or *basal lamina*) interposed between the longitudinally arranged axon terminals and the muscle membrane (or *sarcolemma*). At the endplate the sarcolemma is marked by a series of deep transverse folds into which the basal lamina extends. The presynaptic densities, and an associated cluster of synaptic vesicles, are aligned opposite the openings of the sarcolemmal folds (Birks et al., 1960), a region that is now known to be densely packed with ACh receptors.

The Postsynaptic Density

The region of the postsynaptic membrane directly opposed to the presynaptic process is marked by the presence on its cytoplasmic face of a zone of electron-dense material. In general this is more prominent than the presynaptic density, and in some synapses it is associated with filamentous material that extends for a short distance into the subjacent cytoplasm. The width of the *postsynaptic density* varies considerably. In many synapses—especially those found on the somata of neurons or on dendritic shafts—it is not much greater than that of the presynaptic density, leading to the suggestion that such "symmetric synapses" constitute a separate class, distinct from those in which the postsynaptic densities are appreciably thicker and hence appear distinctly "asymmetric." Such asymmetric synapses are especially prominent on dendritic spines. The distinction between the two classes of synapses is, however, not absolute, and considerable variation in postsynaptic densities can be found (Colonnier, 1968).

George Gray (1959) was the first electron microscopist to draw attention to the differences between synapses on the basis of their postsynaptic densities. From his studies of the cerebral cortex he suggested that they fall into two classes: type I, with pronounced postsynaptic densities and a somewhat wider synaptic cleft; and type II (corresponding to *symmetric synapses* in later terminology), in which the postsynaptic densities were much less prominent. Not surprisingly, this suggestion was promptly taken up by physiologists, who identified Gray's type I synapses as excitatory and his type II as inhibitory. As with the appearance of the synaptic vesicles, this correlation appears to hold true for many, but not all, regions of the CNS.

Further analysis of the nature of the postsynaptic density had to await the development of techniques for its isolation and chemical characterization, including the characterization of the postsynaptic receptor molecules. Since this research is the subject of later chapters (see especially Sheng, this volume), we will not discuss it further. However, it is worth mentioning here that once molecular probes for receptors (such as the ACh receptor) had been developed, it came as something of a surprise that their density was so high, amounting to as many as 10,000–20,000 receptors/μm^2 at the neuromuscular junction.

Varieties of Synapses

As more and more neural tissues were examined under the EM, it became evident that synapses come in many varieties and that the prototypical synapses of the type considered previously, though common (especially in the CNS), are but one among a host of different forms. A complete account of all the different forms is beyond the scope of this chapter; suffice it to say that variations in each of the principal components—the pre- and postsynaptic elements and the synaptic cleft—have been observed at one or more sites (Fig. 1.14).

We have already mentioned the presence of the basal lamina between the pre- and postsynaptic membranes at the neuromuscular junction. In retinal photoreceptors and at the squid giant synapse, there are prominent ribbonlike structures within the presynaptic processes (Sjostrand, 1958; Dowling and Boycott, 1968; Martin and Miledi, 1972). At spine synapses in the hippocampus and cerebral cortex there is often a

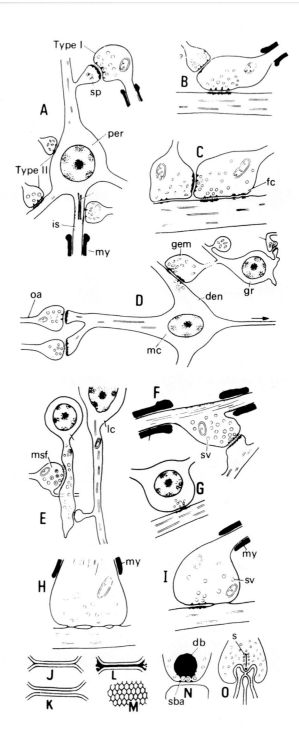

distinct "spine apparatus" consisting of two or more membrane-bound sacs or cisternae separated by plaques of electron-dense material (Hamilyn, 1962). At many synapses in these same regions clusters of polyribosomes, which are now thought to be involved in local protein synthesis, are seen at the base of dendritic spines (Steward and Levy, 1982). In many other regions (including the thalamus, the retina, and the olfactory bulb) both the pre- and postsynaptic elements have the morphological features of dendrites that display "reciprocal synapses," in which one process is presynaptic to another at one point and postsynaptic to that process at an adjoining site (Rall et al., 1966; Price and Powell, 1970; Famiglietti and Peters, 1972). And, most interestingly, in many regions axoaxonic synapses have been observed, with the postsynaptic element being either the axon hillock or the terminal portion of a second axon (see Peters et al., 1976, for a detailed account).

Presynaptic Inhibition

The finding of synapses upon axon terminals in the spinal cord (Gray, 1962, 1963) is of particular historical interest since it provided morphological evidence in support of Eccles's view of the mechanism of presynaptic inhibition. The initial observation of a reduction in the amplitude of an EPSP elicited in a motoneuron by stimulating one afferent when a second afferent is activated (that itself has no effect on the resting potential of the motoneuron) was first made by Frank and Fuortes (1957). They proposed that this was a form of presynaptic inhibition. Later Frank (1959) suggested that the depression of the EPSP, without other detectable changes in the motoneuron, could also be brought about if the terminals of the relevant afferents end on the distal dendrites of the

Figure 1.14. (opposite) Various types of synapses seen in the electron microscope.

(A) Axosomatic, axodendritic, and axoaxonic contacts on a cortical pyramidal cell.

(B) Axoaxonal contacts of the type thought to be involved in presynaptic inhibition of the spinal cord.

(C–D) Serial synapses seen in the thalamus (C) and the olfactory bulb.

(E) Synapses between amacrine (i.e., axonless) cells.

(F) En passant synapses at a node of Ranvier.

(G) A somatodendritic synapse.

(H) An electronic contact in the brain of a fish.

(I) A combined electrical and chemically transmitting synapse.

(J–M) Various forms of gap junction.

(O) A photoreceptor synapse in the retina, showing the typical presynaptic ribbon.

Reproduced from Gray (1974) by permission of Oxford University Press.

motoneurons, at a distance too remote to be detected by an electrode within the cell body, and act remotely to shunt the EPSP. He accordingly termed the phenomenon "remote inhibition." In a series of papers published between 1961 and 1962, Eccles reexamined this issue and provided convincing evidence that the observed inhibition is due to a direct action upon the terminals of the primary afferents. As such, it should appropriately be termed *presynaptic inhibition* (Eccles et al., 1961, 1962; Fig. 1.15).[15] This interpretation also served to account for several earlier observations, such as the *dorsal root potential*—an activity-induced depolarization of the dorsal root fibers, studied by Barron and Matthews (1938) and others, that in its time course paralleled presynaptic inhibition. At the time Eccles first proposed it, there was no evidence for the postulated axoaxonal endings, but Eccles was undeterred by this fact. In a seminar at Oxford in 1961, he confidently predicted that such synapses would soon be found "because the anatomists are good boys and always find what they are told to look for."

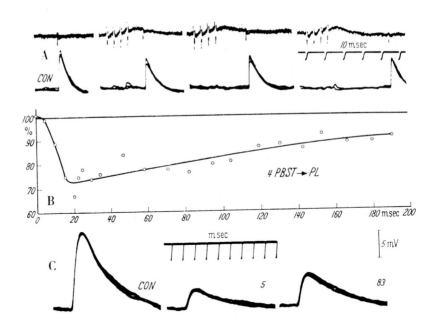

Figure 1.15. Depression of a monosynaptic EPSP by presynaptic inhibition.
    (A) The control EPSP (CON) in a plantaris motoneuron is seen to be depressed by four group I conditioning volleys in the nerve to the knee flexors and posterior biceps plus semitendinosus (PBST). The timing of the conditioning and testing afferent volleys is shown in the upper traces (positivity upward in both traces).
    (B) The time course of the EPSP depression (expressed as a percentage of control) is shown for the series illustrated in A.
    (C) The control EPSP (CON) of another experiment is seen to be greatly depressed at both 5 and 83 msec after a conditioning tetanus of 22 group I volleys.
    Reproduced from Eccles (1964), after Eccles et al. (1961), by permission.

While these observations on presynaptic inhibition in the mammalian spinal cord were being made, Dudel and Kuffler were studying a closely related phenomenon in the crayfish nerve-muscle system (Dudel and Kuffler, 1961; Dudel, 1962). Here there are two independent innervations of the muscle fiber mediated by a single excitatory and a single inhibitory axon. The essential finding in their work was that when an impulse in an inhibitory fiber preceded an impulse in the excitatory fiber to the same muscle, there was a marked depression of the evoked EPSP. Although there was more than one possible explanation for this phenomenon, Dudel and Kuffler clearly showed the inhibitory impulse acted on the terminals of the excitatory axon to depress the release of transmitter by the excitatory impulse (in addition, of course, to its direct action on the muscle fiber). The inhibitory impulse accomplished this inhibition by reducing the number of quanta released from the terminals of the excitatory nerve.

By contrast, the size of the individual quanta—a measure of receptor sensitivity—was unchanged. Moreover, from the timing and other features of the inhibitory response, Dudel and Kuffler concluded that the observed inhibition was chemically mediated and that the transmitter involved was probably γ-aminobutyric acid (GABA), the same transmitter that the inhibitory axon released directly onto the muscle fiber.

The Search for Neurotransmitters

In an influential review published in 1958, Paton set out five criteria that must be satisfied before a substance can be considered a neurotransmitter (Paton, 1958; see also McLennan, 1963): (1) the enzymes involved in the synthesis of the substance must be present within the presynaptic neurons; (2) the substance must be released from the axon terminals when the presynaptic fibers are stimulated; (3) the action of the substance when applied to the postsynaptic cells must accurately mimic that seen during normal synaptic transmission; (4) a mechanism must be present at the site of the synapses to terminate the action of the putative transmitter; and (5) the effect of drugs (whether agonists or antagonists) on the postsynaptic cells must be the same when the putative transmitter substance is applied to the synapse (usually by microiontophoresis). To this list we would now add a sixth criterion, namely, that the postsynaptic cells must bear the appropriate receptors for the substance.[16]

At the time Paton wrote, only ACh and noradrenaline came close to satisfying these criteria, and until the early 1950s there was considerable skepticism among physiologists that even these substances could be regarded as transmitters in the vertebrate CNS. All this was to change during the next two decades as evidence began to accumulate for a variety of transmitter substances, ranging from simple amino acids such as glutamate and GABA to various biogenic amines such as dopamine, norepinephrine, and serotonin, and, somewhat later, a host of different neuropeptides. It is impossible within the scope of this chapter to give anything like a full account of the discovery of all the currently recognized transmitters, but some are of particular historical significance and should be mentioned briefly. But before considering these examples, reference should be made to another organizing principle, commonly referred to as *Dale's law*.

Dale (1935), with his usual insight and prescience, had concluded that a neuron would release the same transmitter substance from all its synaptic terminals.[17] Over time this simple and clear statement gave rise to two mistaken notions: first, that the action of a neuron must be the same at all its postsynaptic targets; and second, that a neuron can release only one transmitter. In the 1960s, experiments in *Aplysia* clearly established that a single (cholinergic) neuron could have an excitatory action on one target neuron and an inhibitory action on another (Tauc and Gerschenfeld, 1961; Kandel et al., 1967; Kandel, 1968). And in the 1970s, when appropriate cytochemical markers for different transmitters became available, Hökfelt and his colleagues provided convincing evidence for the co-release of transmitters—usually one or more neuropeptides in association with a so-called "conventional transmitter" (e.g., Lundberg et al., 1979).

Acetylcholine

As we have noted, by the late 1940s it was generally accepted (even by Eccles) that ACh is an excitatory transmitter at several sites in the peripheral nervous system (PNS)—including all autonomic ganglia, parasympathetic postganglionic targets, some sympathetic effector cells, and the neuromuscular junction—and also that it functions to inhibit activity at other sites, such as the heart. However, there was still considerable resistance to the notion that it might also serve as a transmitter within the CNS (Eccles, 1949). This despite the fact that Feldberg and his colleagues had provided rather strong evidence that ACh is a central transmitter (see Feldberg, 1945, 1950, for reviews). The principal evidence for this role of ACh derived from a study that Feldberg had carried out with Marthe Vogt, in which they had described the distribution of the enzyme choline acetylase within the brain and spinal cord, where it appeared to be restricted to certain cranial nerve nuclei and the anterior horn of the spinal cord (Feldberg and Vogt, 1948). Later, with Harris and Lin, Feldberg found that the levels of choline acetylase were lowest in the sensory pathways and also low in the motor cortex. This finding led them to suggest that there might, in many systems, be an alternation between noncholinergic and cholinergic neurons. In the motor system, for example, the so-called "upper motor neurons" in the cortex would be noncholinergic while the lower spinal and cranial motoneurons that they contact are cholinergic (Feldberg et al., 1951). So, as far as Feldberg was concerned, by 1950 "the theory of acetylcholine as [a] central transmitter [was] all but settled" (Feldberg, 1950).

Shortly after Eccles had become convinced that central transmission is chemical, he and his colleagues addressed the mechanism responsible for the recurrent inhibition first described by Renshaw (1946). Since this inhibition is due to collateral branches of the axons of motoneurons and is mediated by the repetitive firing of a population of small interneurons near the ventral margin of the anterior horn (which Eccles had termed *Renshaw cells;* see Fig. 1.16), Eccles thought that the motoneuron axon

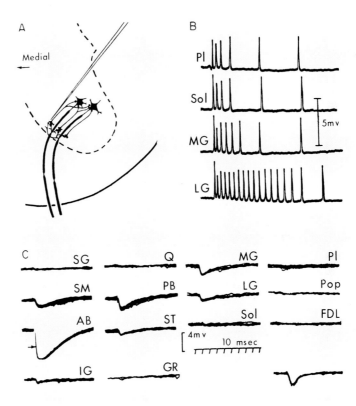

Figure 1.16. Responses of Renshaw cells involved in recurrent inhibition. Renshaw cells are cholinergic interneurons located near the margin of the anterior horn (A).

(B) Recordings from a Renshaw cell that fires repetitively to single volleys in the motor fibers to four different muscles.

(C) Intracellular recording reveals that IPSPs of various sizes are produced in an anterior biceps motoneuron by single volleys in the motor fibers supplying eight different muscles of the same hind limb. AB, anterior biceps; FDL, flexor digitorum longus; GR, gracilis; IG, inferior gluteal; LG, lateral gastrocnemius; MG, medial gastrocnemius; PB, posterior biceps; Pl, plantaris; Pop, popliteus; Q, quadriceps; SG, superior gluteal; SM, semimembranosus; Sol, soleus; ST, semitendinosus.

Reproduced from Eccles (1967), after Eccles et al. (1961), by copyright permission of the Rockefeller University Press.

collateral–Renshaw cell synapse would be a good candidate site at which to test Dale's principle. Since motoneurons release ACh at the neuro-muscular junction, their axon collaterals should (if Dale's principle is correct) form cholinergic synapses upon Renshaw cells. In the mid-1950s, Eccles collaborated with his daughter Rosamond, Paul Fatt, and K. Koketsu to demonstrate this point convincingly. They recorded from Renshaw cells and found that local administration of ACh and nicotine increased their firing. Moreover, administration of eserine, an inhibitor of acetylcholinesterase, greatly increased and prolonged the discharge of the Renshaw cells in response to single shocks to the ventral root (Eccles et al., 1954, 1956).

Later studies by others soon demonstrated (1) the presence of cholin-ergic neurons in several regions of the brain, including the so-called basal nucleus of the forebrain that provides the cholinergic input to the cerebral cortex and hippocampus; (2) the release of ACh in different brain regions after appropriate stimulation of the relevant afferent path-ways; (3) the presence of receptors for ACh at sites of termination of the cholinergic fibers; and (4) the release of ACh on electrical or chem-ical stimulation of brain slices. Since these studies have been discussed extensively elsewhere (Waser, 1975), they need not be considered fur-ther here.

Noradrenaline (Norepinephrine)

Dale, who had introduced the term *cholinergic,* used the term *adrenergic* to describe the postganglionic nerves that release an adrenaline-like sub-stance at their terminals. The great interest in adrenaline as a hormone that had been isolated from adrenal extracts in the early part of the century, and the similarities between the actions of adrenaline and those that followed stimulation of postganglionic sympathetic nerves, led to the erroneous assumption that the transmitter liberated by sympathetic nerve endings was in fact adrenaline. However, as early as 1910, Barger and Dale had sounded a cautionary note when they wrote that "the ac-tion of some of the other bases, particularly the amino acid and amino-ethyl-bases of the catechol group [noradrenaline] corresponds more closely with that of sympathetic nerves than does that of adrenaline."

Nevertheless the idea that adrenaline was the sympathetic transmit-ter persisted until the 1940s, in large part because of the report by Cannon and Lassak (1939) that certain organs seem to contain adrena-line in their sympathetic nerves[18] and the finding of Gaddum and Kwiatkowski (1939) that postganglionic stimulation of the nerves to the rabbit ear released what appeared to be adrenaline. It was only in 1946 —when Ulf von Euler succeeded in showing that noradrenaline, not adrenaline, was the principal compound isolated from mammalian sympathetic nerves—that the pharmacological community came around to accepting noradrenaline as the transmitter. Soon thereafter all of Paton's criteria were met as, in rapid succession, the mechanisms of synthesis, storage, release, and inactivation of noradrenaline at noradrenergic nerve terminals were elucidated. For example, in 1956 von Euler and Hillarp found noradrenaline storage particles in the sympathetic nerve trunk. In 1957 Brown and Gillespie showed that stimulation of the sympathetic nerves to the spleen resulted in the release of noradrenaline into the perfusing fluid. In 1961 de Robertis and Pellegrino De Iraldi described characteristic large, synaptic vesicles with electron-dense cores in sympathetic nerve terminals. And finally, in 1961, Hertting and Axelrod showed that labeled norepinephrine, which is taken up by sympathetic nerves, is released when the nerves are stimulated (see Iversen, 1967, for review).

As early as 1955 Eranko had observed that formaldehyde condensation can cause catecholamines to fluoresce. In 1962 Falck and Hillarp discovered that by freeze-drying and using gaseous formaldehyde they could prevent catecholamine diffusion within tissues, thereby increasing the sensitivity of the method so that it was possible to visualize catecholaminergic neurons and their nerve terminals in histological sections when viewed under a fluorescence microscope (Falck, 1962; Falck et al., 1962). Although all parts of the catechol-containing neurons could be visualized, the strongest fluorescence (and, by inference, the highest concentration of the amine) was found in the nerve terminal and in axonal varicosities. Later work using this method and immuno-histochemistry for the enzyme dopamine β-hydroxylase (Swanson and Hartman, 1975) was able to show that most of the noradrenergic neurons in the brain have their cell bodies in a small nucleus of the brainstem, the locus coeruleus, and in two other cell groups in the lower pons and medulla. The locus coeruleus is a remarkable structure; it contains only a few thousand neurons, yet it gives rise to axons that extend over considerable distances to innervate neurons throughout much of the brain (Bloom, 1977; Moore and Bloom, 1979).

GABA and Glycine: The Search for Inhibitory Transmitters

In 1950 Eugene Roberts and Jorge Awapara independently discovered GABA in the brain and determined the mechanism of its biosynthesis from glutamic acid (Awapara et al., 1950; Roberts and Frankel, 1950). Nevertheless its significance remained unclear for some years. However, as early as 1953 Florey had identified in crude brain extracts a fraction, which he termed factor I, that had a powerful inhibitory effect on the slowly adapting neuron of the crayfish abdominal stretch receptor organ (Florey, 1953). Four years later, with Blazemore and Elliott, he succeeded in purifying the inhibitory factor and identified it as GABA (Blazemore et al., 1957). The following year Kuffler and Edwards (1958) demonstrated that GABA could accurately mimic the action of the crayfish inhibitory neuron, and in 1963 Kravitz and his colleagues found that the concentration of GABA in the inhibitory neuron far exceeded that in the adjoining sensory neuron (Kravitz et al., 1963). Conclusive evidence that inhibition at this site is mediated by GABA was provided by the observation that GABA is released on stimulating the inhibitory nerve in the lobster (Otsuka et al., 1966).

Not long after these studies in crustaceans, several workers demonstrated that GABA played a similar role in the mammalian CNS. In 1966 Krnjevic and Schwartz found that the microiontophoresis of GABA into the cerebral cortex could mimic the action of the local inhibitory neurons. Using much the same approach, Obata and his colleagues showed that this is true also of the action of Purkinje cells on neurons in Deiter's nucleus (Obata et al., 1967). Furthermore, GABA was released into the fourth ventricle on stimulating the axons of Purkinje cells (Obata and Takeda, 1969). Finally, in 1974, Roberts and his colleagues succeeded in raising antibodies against glutamic acid decarboxylase, the key enzyme in the synthesis of GABA, and showed immunocytochemically that they labeled many of the known inhibitory neurons in the brain (Roberts et al., 1976). McGeer et al. (1975) achieved essentially the same result by examining, in autoradiographs, the uptake of ^{3}H-GABA by inhibitory neurons and its transport to their axon terminals.

Glycine

The other known ionotropic inhibitory transmitter in the vertebrate CNS is the amino acid glycine. Unlike GABA, glycine (as a neurotransmitter) is confined to the pons, medulla, and spinal cord, where it is found mainly in interneurons (and their axon terminals) that mediate the inhibition of motoneurons, Renshaw cells, and some of the large neurons of the reticular system (Aprison and Werman, 1965; Aprison et al., 1970). In the period covered by this review much less had been done on glycine than on GABA. However, the finding that iontophoresing glycine onto the spinal motoneurons closely mimicked the naturally occurring inhibition induced by stimulation of group Ia afferents from an antagonist muscle was generally accepted as evidence that glycine is *the* inhibitory transmitter in the lower brainstem and spinal cord (see Aprison et al., 1974, for review).

Glutamate

The surprising discovery that the amino acid glutamate is the major excitatory transmitter in the brain had its origins in the laboratory of David Curtis, one of Eccles's students. In the late 1950s Jeffrey Watkins, working with Curtis and Phillis, found that glutamate and aspartate (and a series of more than 100 analogues) had strong excitatory actions when iontophoresed into the vicinity of spinal neurons in vivo or when added to the bathing solution of isolated spinal cord preparations in vitro. However, when it became clear that glutamate has similar excitatory actions on virtually every neuron in the nervous system, the balance of opinion swung against the view that glutamate might be a neurotransmitter. How, it was asked, could a molecule that participates in a highly specific signaling mechanism have such widespread and general effects?

In the 1960s and 1970s several different lines of evidence were adduced in an attempt to answer this question. For example, in 1971 Solomon Snyder and colleagues demonstrated high-affinity uptake of glutamate and aspartic acids into a distinctive population of synaptosomes from the brains of rats (Wofsey et al., 1971). In 1967 Aprison and his colleagues had shown that L-glutamate is more concentrated in dorsal than in ventral spinal roots and is found in higher concentrations in the dorsal medulla than in its ventral half. This finding led to the suggestion that glutamate might be the transmitter in primary sensory afferents (Graham et al., 1967). And in the late 1970s, Storm-Mathisen showed that interruption of the perforant path to the hippocampus resulted in a reduction of more than 50% in the uptake of glutamate by the dentate gyrus. Moreover, uptake of ^{3}H-glutamate by mossy fiber and other axon terminals could be readily demonstrated autoradiographically (Storm-Mathisen, 1977; Storm-Mathisen and Iversen, 1979).

Convincing though this type of evidence was to some investigators, many more remained skeptical until Watkins synthesized a number of structurally related analogues of glutamate and aspartate and set out to study the structure-activity relationships of these putative excitatory amino acid receptors. The most notable compound he produced was N-methyl-D-aspartate (NMDA). By comparing the excitatory potency of NMDA with that of other analogues (such as kainic acid), Watkins proposed that there must be multiple receptors for glutamate, one of which he named the NMDA receptor.

In 1981 Watkins and Evans published a highly influential review that suggested that glutamate receptors could be divided into two broad categories: NMDA and non-NMDA receptors (the latter including kainate and α-amino-3-hydroxy-5-methyl-4-isoxazolepropionate [AMPA] receptors). The NMDA receptors had a number of interesting characteristics. In particular, they were pharmacologically distinct, being blocked by phosphono-substituted amino acid derivatives and also, surprisingly, by Mg^{2+} ions (Watkins and Evans, 1981). The discovery of the blockade of the NMDA receptor by Mg^{2+} was both unique and puzzling. Through the subsequent work of Mark Mayer, Gary Westbook, and Philipe Ascher, it became clear that Mg^{2+} plugged the NMDA channel in a voltage-dependent manner and that, unlike the non-NMDA channels, the NMDA channel was permeable to Ca^{2+}.

In addition to synthesizing agonists, Watkins also synthesized a number of antagonists of the glutamate receptors. His first success was with the NMDA antagonist 2-amino-5-phosphonovalerate. Watkins used these NMDA antagonists to provide the first direct evidence that NMDA receptors are involved in synaptic transmission in the CNS. Others showed that NMDA receptors played key roles in synaptic plasticity (including long-term potentiation) and in neuropathology (including epilepsy and neuronal cell death; see Bear and Linden, this volume).

It would be difficult to exaggerate the impact of the discovery of the NMDA receptor. Its unique combination of properties allows this receptor to participate in many of the fundamental mechanisms in the brain, of which the following are but a few examples. In 1949 Donald Hebb proposed a theory of associative memory based on the idea that if a presynaptic neuron excites its postsynaptic partner sufficiently strongly, so that it fires an action potential, the synapse would be strengthened. For many years this seemed to be a theory in search of a mechanism, but the discovery of the NMDA receptor provided just the mechanism needed. A large body of evidence now exists supporting the hypothesis that NMDA receptors are critically involved in synaptic plasticity in the hippocampus and elsewhere through an essentially Hebbian mechanism. The essence of the hypothesis is that NMDA receptors usually do not participate in normal synaptic transmission in the hippocampus because, at the resting membrane potential, the channel mouth is blocked by Mg^{2+}. However, when the postsynaptic neuron is sufficiently depolarized by a level of activity through the non-NMDA receptors, the Mg^{2+} blockade of the NMDA receptor channel is relieved. This allows the second defining characteristic of the NMDA channel, its permeability to Ca^{2+}, to come into play. The resulting Ca^{2+} influx triggers a biochemical cascade that ends with the strengthening of transmission at the synapse. Thus the NMDA receptor underlies a highly specific, associative form of synaptic plasticity. This is not all: the same mechanisms allow the NMDA receptor to play a fundamental role in the development of wiring specificity in the nervous system. For example, in the development of ocular dominance columns in the optic tectum of the frog, the NMDA receptor seems to act as a "coincidence detector" that enables neighboring ganglion cells to capture and maintain synaptic contacts with neighboring tectal neurons. Another example of the importance of NMDA receptors comes from their role in excitotoxicity. It is well known that the ischemia caused by a stroke causes many neurons to die and can result in debilitating brain damage. It now appears that the NMDA receptor plays a major role in causing the superadded death of cells outside the immediate zone of ischemic necrosis. The release of glutamate from the oxygen-deprived neurons massively activates NMDA receptors, causing a huge Ca^{2+} influx into nearby neurons, and this Ca^{2+} influx is sufficient to trigger the events that lead to cell death or apoptosis.

Dopamine

In a short but prescient note written in 1939, Herman Blaschko delineated the biosynthetic pathway that leads from the amino acid tyrosine to adrenaline and noradrenaline, in which dopamine is a critical intermediate (Blaschko, 1939, 1942). Until the 1950s this was thought to be dopamine's only role; however, it was during that decade that Arvid

Carlsson noted the marked differences in the regional distribution of dopamine and noradrenaline, both in peripheral tissues and in the mammalian CNS. This led him to suggest that dopamine might act as a transmitter in its own right and have a role quite independent of its function as a precursor to noradrenaline (Carlsson 1959; see Carlsson, 1987, for review).

The introduction of the Falck fluorescence method for mapping the distribution of central aminergic neuronal groups and their projections permitted Dahlstrom and Fuxe in 1964 to delineate, for the first time, the location of dopamine-containing neurons in the brainstem and to show their rostral projections to the hypothalamus, the limbic cortex, and the striatum (caudate nucleus and putamen). The projection to the striatum from the pars compacta of the substantia nigra is of particular interest, since the loss of cells in the substantia nigra had long been recognized as the principal pathological finding in Parkinson's disease. The suggestion by Oleb Hornykiewicz that a loss of dopamine from the striatum might underlie the extrapyramidal motor signs of the disorder, and that some of its clinical features might be relieved by the administration of L-DOPA, is one of the great success stories of clinical neurology (see Hornykiewicz, 1973, for review). Of the other two projections, that to the meso-limbic and neocortex would later be proposed to be important in schizophrenia, and the projection from the tubero-infundibular region of the hypothalamus proved to be critically involved in the regulation of pituitary function. For a general account of the physiological roles played by dopamine and of the various pharmacologically recognizable dopamine receptor subtypes, reference should be made to the review by Gingrich and Caron (1993).

Serotonin

Serotonin, whose role in brain function only began to be understood in the 1980s, was initially isolated from blood platelets. But as early as 1953, Betty Twarog, using a clam heart bioassay, had discovered a high concentration of serotonin in the brain (Twarog and Page, 1953). The following year Amin, Crawford, and Gaddum (1954) were able to show that serotonin was particularly concentrated in the limbic system and hypothalamus. These findings assumed new significance when it was shown the action of serotonin was antagonized by lysergic acid diethylamide (LSD), which had been known for some time to induce mental states reminiscent of those seen in schizophrenia (Woolley and Shaw, 1954).

In the 1960s the distribution of serotonin-containing neurons was mapped by Dahlstrom and Fuxe, using the Falck method. The majority of the cells were found to be confined to two of the raphe nuclei of the brainstem, and their axons could be readily traced rostrally to the hypothalamus and the limbic cortex, and caudally to the spinal cord, where they may act to inhibit the transmission of pain sensibility (Dahlstrom and Fuxe, 1964).

Neuropeptides

Since the discovery of the first neuropeptide—substance P—by von Euler and Gaddum in 1931, a large and continuously expanding number of neuronally active peptides have been identified in both vertebrates and invertebrates. As many of the precursor molecules or prohormones are known to give rise (either by alternative mRNA splicing or peptide cleavage) to two or more biologically active peptides, it is difficult to predict what the total number of such peptide transmitters (or, more correctly in some cases, neuromodulators) is likely to be. Moreover, many peptides that were originally isolated from other tissues, such as the skin and the gastrointestinal tract, have later been shown to be present in particular classes of neurons in the CNS and PNS, and to give rise to distinct neural projections. Some of these peptides now have well-documented physiological and behavioral roles, and many more neurally active peptides of this kind will probably be discovered. Although the primary role of many of these peptides is well known—as in the case of the neurohypophysial hormones vasopressin and oxytocin and some of the hypothalamic releasing hormones—their other functions, as putative neurotransmitters or neuromodulators within the CNS, remain to be determined. And only in a few cases, such as luteinizing hormone releasing hormone (LHRH)—which is responsible for the slow potential changes seen in sympathetic ganglia (Jan and Jan, 1982)—has convincing evidence been adduced about their actions at identified synaptic sites.

As it is impossible in the space available to review the discovery of all the known neuropeptides, we shall limit ourselves to just one—substance P (SP). Here we shall mention only the major historical events that led to its recognition as an important mediator of pain sensibility (see McGeer et al., 1978, for review).

When von Euler and Gaddum first identified the peptide in 1931, they noted that although it resembled ACh in its action on smooth muscle, its effects were not abolished by atropine. It was more than 20 years later before Pernow (1953) showed that it was present in several regions of the brain, including the thalamus, hypothalamus, and basal ganglia, and, interestingly, also in the dorsal roots. This latter observation led Lembeck (1953; cited in McGeer et al., 1978) to propose that SP might be a neurotransmitter. Again, almost 20 years passed before Susan Leeman and her colleagues, while trying to isolate the corticotrophin-releasing factor, discovered a substance that promoted salivary secretion and found that it too was not blocked by atropine. On further study the substance proved to be an undecapeptide that had all the properties of von Euler and Gaddum's SP (Chang and Leeman, 1970). That the peptide had the properties of a neurotransmitter, including its release from neurons in a Ca^{2+}-dependent manner, was subsequently demonstrated by Iversen et al. in 1976. And, at about the same time, Hökfelt and his colleagues, using antibodies against the peptide, were able to map its distribution within the sensory ganglia and in the CNS (Hökfelt et al., 1975).

The history of the discovery of several of the other neuropeptides is at least as interesting as that of SP, and in some cases the discoveries were extremely controversial at the time they were made (Wade, 1981). But here it will suffice simply to mention a number of features that most of the known neuropeptides have in common. First, they are synthesized in the neuronal bodies, packaged into large dense-core vesicles (where they may undergo further processing), and axonally transported along microtubules to their sites of release either at en passant contacts or at axon terminals. Second, at their release sites they are nearly always associated with one of the more conventional neurotransmitters in small, clear vesicles and are co-released with the conventional transmitter. Third, compared with the action of the ionotropic transmitters, their action is slow and long lasting. Fourth, the known receptors for peptide transmitters are of the seven-transmembrane-domain, G-protein-coupled variety, whose intracellular actions on the target cells are mediated by second messengers. Fifth, the relevant receptors may be located at some distance from the release site, and the peptide often has autocrine and paracrine effects. And, finally, unlike conventional transmitters, which are usually present in synaptic vesicles at concentrations in the 100 mM range and whose affinity for the associated receptors is on the order of 100 μM to 1 mM, neuropeptides are present at concentrations of 2–10 mM at most, and they bind to their receptors with affinities in the nanomolar to low micromolar range.

Synaptic Receptors Coupled to Second Messenger Pathways

By the 1970s, it was clear that virtually all the conventional, small-molecule transmitters—ACh, GABA, glutamate, norepinephrine, dopamine, serotonin—not only activate ionotropic receptors and ligand-gated channels to produce rapid synaptic potentials that last for only milliseconds but also interact with a second, even larger class of seven-transmembrane-domain metabotropic receptors that produce slow synaptic responses that can persist for seconds or minutes. Metabotropic receptors consist of a receptor molecule that is coupled to its effector molecule by a nucleotide-binding G protein. G proteins couple the receptors to secondary effectors—such as cAMP, cGMP, diacylglycerol, and metabolites of arachidonic acid—that can activate channels directly. More commonly, however, these second messengers activate a protein kinase that regulates channel function by phosphorylating the channel protein or an associated regulatory protein. This family of receptors is remarkably large, and its members serve not only as receptors for small molecule and peptide transmitters, but also as the sensory receptors for vision and olfaction.

The study of slow synaptic potentials mediated by second messengers has added three new features to our understanding of chemical transmission. Two of these are particularly important. First, in addition to their action on ion channels, transmitters that act on metabotropic receptors can (by means of their action through second messengers) modify proteins other than the channels, thereby activating a coordinated molecular response within the postsynaptic cell. Second, the second messengers that are activated by these receptors can translocate to the cell nucleus and modify transcriptional regulatory proteins; in this way they are able to regulate gene expression rather directly. Thus second messengers can both produce a covalent modification of preexisting proteins and regulate the synthesis of new proteins. This latter class of synaptic action can lead to long-lasting structural changes at synapses. Finally, fast synaptic transmission is used primarily to mediate behavior; by contrast, slow synaptic actions are often used to modulate behavior.

The Plastic Properties of Synapses

It is perhaps fitting that we should conclude this chapter by returning to an issue that was first clearly articulated by Cajal. Knowing that in most regions of the mammalian brain no additional neurons are generated in postembryonic development, and knowing that the patterns of connectivity that are laid down during development are, of necessity, highly specific, Cajal pondered two fundamental questions: (1) How can the brain acquire new information, in the process usually referred to as learning? (2) How can such information be retained in the form of memory?

In his 1894 Croonian Lecture to the Royal Society, to which we referred earlier, Cajal proposed a possible solution to these problems:

> These observations . . . have suggested to us an hypothesis which will enable us to understand . . . intelligence acquired by good mental training, the inheritance of intelligence . . . and even the creation of . . . artistic ability. . . .
>
> Mental training cannot better the organization of the brain by adding to the number of cells; we know that nervous elements have lost the property of multiplication past embryonic life; but it is possible to imagine that mental exercise facilitates a greater development of the protoplasmic apparatus and of the nervous collaterals in the part of the brain in use. In this way, pre-existing connections between groups of cells could be reinforced by multiplication of the terminal branches of protoplasmic processes and nervous collaterals. But the pre-existing connections could also be reinforced by the formation of new collaterals and protoplasmic expansions. (Cajal, 1894)

No better hypothesis was forthcoming until Donald Hebb and Jerzy Konorski proposed that the strength or effectiveness of specific synapses may be changed as a result of activity:

> The application of a stimulus . . . leads to changes of a two-fold kind in the nervous system. . . . The first property by virtue of which nerve cells

react to the incoming impulses with certain cycles of changes we call *excitability*, and that changes arising in the centers because of this property we should call *changes due to excitability*. The second property, by virtue of which certain permanent functional transformations arise in particular systems of neurons, as a result of appropriate stimuli or combinations, we shall call *plasticity*, and the corresponding changes *plastic changes*. (Konorski, 1948; his emphasis)

And, to provide a specific neuronal basis for such changes, Hebb noted that "When an axon of cell A . . . excite[s] cell B and repeatedly and persistently takes part in firing it, some growth process or metabolic change takes place in one or both cells so that A's efficiency as one of the cells firing B is increased" (Hebb, 1949).

The idea that learning might produce plastic alterations in synaptic strength, and that the persistence of these changes would give rise to memory storage, was first systematically tested in invertebrates, where studies of synaptic transmission in the neural circuit responsible for the gill-withdrawal reflex in the marine snail *Aplysia* showed that simple forms of learning—habituation, sensitization, and classical conditioning—produce changes in synaptic strength that can persist for one or more days and that parallel the time course of the memory process (Kandel, 1976). This functional plasticity hypothesis was dramatically extended to the mammalian brain by Bliss and Lømo (1973), who found that high-frequency tetani applied to the perforant pathway in the hippocampus— a structure known to be critically involved in memory storage—could produce alterations in synaptic strength, which they termed long-term potentiation (LTP). In brain slices LTP lasts for hours, and in the intact animal, for days.

Since these and several of the other topics we have touched upon are dealt with in detail in the succeeding chapters of this volume, we may end by once again quoting Cajal:

> Functional theories based on the localization of different cortical areas, no matter how good, fail completely to explain mechanisms underlying cognitive activity, which is almost certainly accompanied by molecular changes in neurons, as well as by very complex changes in relationships between neurons. Therefore, to understand cognitive activity, it will be necessary to understand these molecular and connectional changes, not to mention the exact histology of each cortical area and all of their pathways. However, this is still not enough; we also need to understand the properties of neural impulses: What energy transformations are required for their initiation, spread, and involvement in the phenomena accompanying perception and thought, namely consciousness, volition, and emotion?
>
> Our knowledge is far from complete. While waiting for chemistry, cell biology, and histology to help achieve this goal, which will take a very long time, we must be content with hypotheses that occasionally lead to the discovery of a useful observation or formulate a more precise concept. (Cajal, 1995:721–722)

Appendix

A Chronology of the Major Events in the Study of Synapses and Synaptic Transmission

1791   **Galvani** observes the contraction of muscles in the hind limbs of frogs when a metal hook is inserted into the medulla and then attached to an iron railing. He claims to have discovered "animal electricity."

1793   **Volta** recognizes that the source of the current in Galvani's experiments was the interaction of two unlike metals, not the animal itself.

1862–   **Kühne** and **Krause** independently describe the structure of the neuro-
1863   muscular junction and suggest that transmission from nerve to muscle is an electrical process.

1877   **Du Bois-Reymond** calls into question the notion that transmission between nerve and muscle is electrical and suggests that it may be mediated by the release of a chemical substance.

1878   **Bernard** experiments with curare and concludes that it acts to block transmission at or near the neuromuscular junction while having no effect on nerve conduction.

1886   **His** provides strong evidence from his developmental studies for the structural independence of neurons and for the outgrowth of their processes from the cell body.

1887   **Forel** shows that the interruption of neuronal projections leads to the atrophy of only the injured neurons and does not spread to other neuronal populations, thus providing further evidence for neuronal independence.

1888   **Cajal** observes the termination of the axons of the stellate cells of the cerebellum in pericellular "baskets" around Purkinje cells and launches a long series of studies on the mode of axon terminations, in support of the neuron theory.

1891   **Waldeyer** formulates the neuron theory and introduces the term *neuron*.

1897   **Sherrington** coins the term *synapse* for the site at which an axon terminal or collateral makes a functional contact with another cell.

         Elliott concludes that adrenaline (epinephrine) is the transmitter released by sympathetic postganglionic fibers.

1905   Based on his experiments on transmission through the superior cervical ganglion, **Langley** suggests that it is mediated by "receptive" substances on the ganglion cells. This is the first clear statement of the concept of receptors for neurotransmitters.

1906   Publication of **Sherrington's** *Integrative Action of the Nervous System,* which summarizes a vast body of experimental work on spinal and other reflexes, including observations on the role of inhibition as an active physiological mechanism.

1909–   Publication of the definitive French translation of **Cajal's** great work
1911   *Textura del Sistema Nervioso del Hombre y de los Vertebrados* under the title *Histologie du Système Nerveux de l'Homme et des Vertébrés.*

1914   From studies of the actions of various choline esters, **Dale** concludes that acetylcholine is probably the transmitter released at preganglionic synapses and by most parasympathetic postganglionic fibers.

1921   **Loewi** discovers "Vagusstoff" and later shows that it is probably acetylcholine.

1931   **Von Euler** and **Gaddum** discover a depressor substance in various tissues; they name it substance P.

1933   **Feldberg** and **Krayer** repeat Loewi's experiment in dogs and provide clear evidence for the release of acetylcholine on stimulating the vagus nerve. This is the first of what would be a long series of studies that Feldberg was to carry out on acetylcholine as a transmitter in the autonomic nervous system, at the neuromuscular junction, and in the central nervous system (CNS).

1934   Publication of **Cajal's** last work, ¿*Neuronismo o Reticularismo?*, later translated into English as *Neuron Theory or Reticular Theory*, which finally lays to rest the mistaken notion that the nervous system is a syncytium.

1935   **Dale** formulates the principle that neurons release the same transmitter at all their axon terminals.

1936   **Dale, Feldberg,** and **Vogt** demonstrate that acetylcholine is released by motor fibers at the neuromuscular junction.

1938   **Göpfert** and **Schaefer** discover what would later be known as the endplate potential (EPP). Their findings are confirmed by **Eccles** and **O'Connor** (1939).

1939   **Blaschko** identifies dopamine as an intermediate in the biosynthesis of adrenaline and noradrenaline.

1942   **Kuffler** develops the single nerve–muscle fiber preparation and provides the first elementary analysis of the endplate potential.

       Arvanitaki describes the properties of artificially constructed synapses, called *ephapses*.

1946   **Von Euler** finds that noradrenaline (not adrenaline) is the transmitter released at most postganglionic sympathetic terminals.

1947   **Brooks** and **Eccles** put forward a rigorously defined electrical hypothesis for both central excitation and inhibition, the latter calling for the interposition of a nonspiking Golgi II cell between the afferent input and the target neuron.

1948   **Konorski** predicts that associative learning may be due to long-term changes in neuronal excitability and defines such changes as "plastic."

1949   **Lloyd** discovers posttetanic potentiation.

       Hebb postulates that the near-coincident firing of a presynaptic afferent fiber and its target postsynaptic cell may lead to a strengthening of the synaptic connection. Such synapses become known as "Hebbian."

1950   **Roberts** and **Awapara** independently discover γ-aminobutyric acid (GABA), which is later shown to be the principal inhibitory transmitter in the brain.

1951   **Fatt** and **Katz** extend the analysis of the EPP using intracellular recording. They observe among other things spontaneous discharges, later known as miniature EPPs (mEPPs).

Brock, Coombs, and Eccles report the first intracellular recordings from motoneurons in the spinal cord. Their later work (1952) leads to the identification of excitatory and inhibitory postsynaptic potentials (EPSPs and IPSPs, respectively). The finding that inhibition is marked by a simple hyperpolarization causes Eccles to abandon his earlier electrical hypothesis and enthusiastically embrace the view that central synaptic transmission is chemical.

1953   **Twarog** and **Page** find that serotonin is present in high concentration in the brain.

Pernow finds that substance P is present in several regions of the brain, spinal cord, and dorsal root ganglia.

1954   **Del Castillo** and **Katz** describe mEPPs in detail and formulate the quantal hypothesis for transmitter release.

Fatt, in an extensive review of junctional transmission, predicts that at certain sites (especially where the pre- and postjunctional elements are about the same size) transmission may be found to be electrical.

1955   **Palay** and **Palade** as well as **De Robertis** and **Bennett** identify synaptic vesicles in electron microscopic (EM) preparations and relate these to the quantal release hypothesis.

Eranko finds that formaldehyde condensation causes biological amines to fluoresce. In 1962 **Falck** and **Hillarp** develop this as a method for identifying such neurons and their projections in the CNS.

1956   **Eccles** and colleagues establish that transmission at the motoneuron axon collateral–Renshaw cell synapse is cholinergic.

1957   **Frank** and **Fuortes** observe a reduction in the amplitude of EPSPs evoked by stimulation of muscle afferents on stimulation of other spinal inputs. Frank later refers to this as "remote inhibition."

Brazmore, Elliott, and Florey purify the inhibitory factor I that Florey had found in brain extracts and show that it is GABA.

1958   **Kuffler** and **Edwards** provide convincing evidence that GABA is the transmitter in the crayfish inhibitory neuron.

Paton sets out five criteria that must be satisfied before any substance can be considered a neurotransmitter.

1959   **Furshpan** and **Potter** discover and provide a detailed analysis of rectifying electrical synapses in the abdominal nerve cord of the crayfish. **DeLorenzo**'s EM studies show that the intercellular gap is considerably narrowed at these sites; such contacts are later identified as *gap junctions*.

Curtis, Phillis, and Watkins demonstrate that acidic amino acids excite neurons in the spinal cord.

Based on the density of postsynaptic membranes, **Gray** identifies type I and type II synapses, and it is suggested that these are excitatory and inhibitory respectively.

Whittaker identifies particles from homogenized and centrifuged brain tissue that bind acetylcholine.

Carlsson suggests that dopamine may function as a neurotransmitter in its own right.

1961   **Eccles, Eccles,** and **Magni** identify the phenomenon reported by Frank and Fuortes as *presynaptic inhibition* and predict that it is due to endings on the terminals of the excitatory inputs.

Dudel and **Kuffler** find presynaptic inhibition in the crayfish.

Tauc and **Gerschenfeld** show in *Aplysia* that acetylcholine can be excitatory at some neurons and inhibitory at others.

1962   **Gray** and **Whittaker** isolate pinched-off presynaptic terminals with attached postsynaptic densities, which they term *synaptosomes.*

Gray observes axoaxonic synapses in EM studies of the spinal cord and suggests that they may be responsible for presynaptic inhibition, as postulated by Eccles.

1963   **Martin** and **Pilar** show that transmission at single synapses in the chick ciliary ganglion can be both electrical and chemical.

Kravitz and colleagues provide strong evidence that GABA is the transmitter released by the crayfish inhibitory neuron.

1964   **Whittaker** and colleagues succeed in isolating fairly pure populations of synaptic vesicles.

Furukawa and **Furshpan** find that inhibition at the Mauthner cell axon hillock is electrical.

Dahlstrom and **Fuxe** identify cells in the raphe nuclei of the brainstem as the source of serotonin projections to the forebrain and spinal cord.

1965   **Uchizono** observes flattened synaptic vesicles in certain cerebellar synapses and concludes that their presence indicates that the presynaptic fibers are inhibitory. The symmetry of the pre- and postsynaptic specializations at these and in other synapses leads **Colonnier** (1968) to introduce the term *symmetric synapses.*

Aprison and **Werman** provide the first evidence that glycine is an inhibitory transmitter in the spinal cord and brainstem.

1965,   **Katz** and **Miledi** demonstrate the critical role of Ca^{2+} entry into
1967   the axon terminals for synaptic vesicle release at the neuromuscular junction.

1966   **Krnjevic** and **Schwartz** show that microiontophoresis of GABA mimics the action of cortical inhibitory neurons.

Rall and colleagues identify "reciprocal synapses" in the olfactory bulb.

1967   **Kandel, Frazier,** and **Coggeshall** demonstrate that different branches of an identified cholinergic interneuron in *Aplysia* can be excitatory at some synapses and inhibitory at others.

Bennett, Pappas, and **Nakajima,** in a series of four papers, provide a detailed account of the ultrastructural appearance and functional characteristics of electrical synapses in the brains of various fish.

Obata and colleagues establish that GABA is the transmitter at sites of termination of cerebellar Purkinje cell axons.

1969   **Katz** and **Miledi** study synaptic transmission in the giant synapse in the squid stellate ganglion, which permits simultaneous intracellular recording from both the pre- and postsynaptic processes.

1970   **Couteaux** and **Pecot-Dechavassine** introduce the term *active zones* for the sites on the presynaptic membrane where vesicle release occurs.

Kandel and his colleagues find that habituation and sensitization, two simple forms of learning, produce alterations in the strength of specific synaptic connections between sensory and motor neurons mediating the gill-withdrawal reflex and that the persistence of these changes contributes to short-term memory storage in *Aplysia*.

Chang and **Leeman** purify an undecapeptide from hypothalamic tissue and identify it as having the properties of substance P.

1972   **Llinas** and his colleagues confirm the findings of Katz and Miledi on the squid stellate ganglion. They also discover that most of the synaptic delay is attributable to the time required for the opening of the Ca^{2+} channels and document the high local concentration of Ca^{2+} near the sites of transmitter release.

1973   **Heuser** and **Reese** provide functional and EM evidence for the recycling of synaptic vesicles.

Hornykiewicz summarizes the evidence that the motor disabilities of Parkinson's disease associated with the death of cells in the substantia nigra are due to the loss of dopamine within the striatum. He proposes the use of L-DOPA to treat the disorder.

Bliss and **Lømo** demonstrate long-lasting changes in synaptic transmission in the dentate gyrus of rabbits following brief tetanic stimulation of the perforant path (one of the major afferent inputs to the dentate). They name this phenomenon *long-term potentiation*.

1976   **Iversen** and colleagues show that substance P is released in a Ca^{2+}-dependent manner from neurons.

1977   **Storm-Mathisen** establishes that glutamate is the excitatory transmitter in the hippocampus by showing that its levels are markedly reduced after interruption of the perforant path and (with **Iversen** in 1979) that ^{3}H-glutamate is taken up by excitatory terminals.

1979   **Lundberg, Hökfelt,** and colleagues demonstrate immunohistochemically that a peptide and a conventional neurotransmitter can coexist in the same presynaptic process and predict that the co-release of such transmitters may be fairly common.

1981   **Watkins** provides evidence for three different types of ionotropic glutamate receptors based on their binding of N-methyl-D-aspartate (NMDA), kainic acid, or α-amino-3-hydroxy-5-methyl-4-isoxazolepropionate (AMPA). Considerable attention is later paid to NMDA receptors as mediating long-term changes in neurons.

Notes

1. For a succinct account of Galvani's experiments, their antecedents, and the reception they received, see Clarke and Jacyna (1987).

2. As Krnjevic (1974) has pointed out, since du Bois-Reymond wrote at a time when it was generally believed that axons were in direct continuity with the cells they innervated, it is perhaps misleading to suggest that he conceived of chemical transmission in the same way as it is now understood.

3. The reference is to the 1954 translation into English of Ramón y Cajal's last monograph, *¿Neuronismo o Reticularismo?*, which appeared shortly after his death in 1934.

4. Later work showed that in some situations degenerative changes extend to other cell populations. Indeed, in some of Gudden's experiments (which involved lesions of the cerebral cortex in young rabbits), he reported an atrophy of the mammillary body, which we now know to be secondary to the retrograde degeneration in the anterior thalamic nuclei (see Cowan, 1970, for review). But at the time Forel wrote (1887), his interpretation of Gudden's finding was widely considered a significant ancillary line of evidence for the "trophic independence" of neurons.

5. Originally published in Madrid as *Recuerdos De Mi Vida* between 1901 and 1917. References here to *Recollections* are from the English translation by E. Horne Craigie with the assistance of Juan Cano, first published as Volume 8 of *Memoirs of the American Philosophical Society* in 1937. The translation was reissued by MIT Press in 1966 and published in paperback in 1989.

6. This work was translated in part into French and German quite soon after the publication of the original Spanish version but did not appear in English until 1954 (Ramón y Cajal, 1954).

7. The work was first published by Yale University Press in 1906. It was reissued by Cambridge University Press in 1947, on the occasion of the International Congress of Physiology held at Cambridge, and in 1961 as a paperback. Like Darwin's *Origin of Species*, *Integrative Action* is distinguished for being more frequently cited than read.

8. As Davenport (1991) has pointed out, Langley used the rather cumbersome phrase "receptive substances" rather than *receptors* (which soon became the accepted term) because "receptor" at the time was widely used for sensory receptors in skin, muscles, and the special senses.

9. The list of Langley's students reads like a veritable Who's Who of British physiology, pharmacology, and biophysics in the first half of the twentieth century. It includes three Nobel laureates—A. V. Hill, Edgar Adrian, and Henry Dale—as well as such other giants as Keith Lucas, Joseph Barcroft, and T. R. Elliott.

10. We cannot resist pointing out that many of the critical experiments on synaptic transmission derived from the work of physiologists, pharmacologists, and chemists who left Germany between 1933 and 1937. Among them were Herman Blaschko, Edith Bulbring, Bernard Katz, Otto Krayer, David Nachmanson, Marthe Vogt, and, of course, Wilhelm Feldberg.

11. In addition to Gerard himself, the group included George Bishop, Detlev Bronk, Halowell Davis, Joseph Erlanger, Alexander Forbes, Herbert Gasser, and "Iron Wire" Lillie, as well as various occasional visitors, such as Ragnar Granit, A. Monnier, and William Rushton (Rushton, 1975).

12. In this section we are dealing with electrical transmission as a general synaptic mechanism; we shall consider later some of the special sites at which true electrical synapses occur (see pp. 38–42).

13. Responding to criticisms about his use of the term *quantal*, Katz later wrote:

> Controversies about words, like arguments about priority, are dominated by emotion, and I well remember W. Feldberg's dictum, namely that there is a type of scientist who, if given the choice, would rather use his colleague's toothbrush than his terminology! My colleagues and I were looking for an adjective which would adequately describe the important property of evoked transmitter release, namely, that it occurs in standard "packets" of large multi-molecular size which are identical with the spontaneously occurring units, and whose size is independent of the event (e.g., impulse, local potential change, chemical or osmotic stimuli) which causes the release. We chose the term "quantal" for this purpose, which seems entirely proper and unobjectionable to me. I have, nevertheless, found myself challenged on two grounds: (a) for supposedly basking in the reflected glory of quantum physics, and (b) for applying the term "quantum" to something which is not constant in size, but subject both to random variation and to experimental change. Objection (a) is, of course, impossible to disprove, and to protest would be in vain. All I would say is that I take my authority for the use of the words from an ordinary dictionary (the entry "quantum" in the *Concise Oxford English Dictionary* may serve), and not from books on quantum physics. This may not satisfy the objectors, but I will take that risk rather than discard an adjective which is singularly apt in describing a whole set of characteristic features. (1969:41)

14. The high density of mitochondria in presynaptic processes is a reflection of their high metabolic activity. It also formed the basis of a quasi-selective staining method for synapses.

15. During the 1950s and 1960s, Eccles's laboratory had become a mecca for neurophysiologists from around the world. A partial list (in alphabetical order) of his collaborators during this extraordinary period includes the following: Anderson, Araki, C. McC. Brooks, V. B. Brooks, Coombs, Curtis, Downman, R. Eccles, Fatt, Hubbard, Iggo, Ito, Kostyuk, Krnjevic, Landgren, Liley, Lundberg, McIntyre, Magni, Malcolm, Miledi, Oscarson, Phillis, Rall, Sears, Schmidt, Watkins, and Willis.

16. The recent discovery that carbon monoxide and nitric oxide are released from active neurons and can have both local and more widely distributed effects has challenged the uniqueness of these six criteria.

17. Useful as Dale's principle has been, it is worth noting that it was formulated half a century before it was known that neurons could contain more than one transmitter. It is now known that in some cases different neuropeptides derived from a common prohormone can be targeted to different processes of a cell (Sossin et al., 1990).

18. The question of whether the transmitter released by postganglionic sympathetic fibers is adrenaline or noradrenaline was for a time unnecessarily complicated by Cannon's suggestion that the transmitter (which he termed sympathin) existed in two forms—one excitatory (sympathin E) and another inhibitory (sympathin I). In 1933 he wrote: "Sympathin is defined as the chemical mediator of sympathetic impulses, ME or MI, which in the (effector) cell induces the typical response, contraction or relaxation, and which, escaping from the cell into the blood stream, induces effects elsewhere in organs innervated by the sympathetic" (Cannon and Rosenblueth, 1933; cited by Davenport, 1991).

References

Adrian, E. D. (1957). The analysis of the nervous system. Sherrington Memorial Lecture. *Proc. R. Soc. Med.* 50:993–998.

Akert, K., Pfenninger, K., Sandri, C., Moor, J. (1972). Freeze etching and cytochemistry of vesicles and membrane complexes in synapses of the central nervous system. In *Structure and Function of Synapses,* Pappas, G. D., Purpura, D. P., eds. New York: Raven Press. pp. 67–86.

Akert, K., Peper, K., and Sandri, C. (1975). Structural organization of motor end plate and central synapses. In *Cholinergic Mechanisms,* Waser, P. G., ed. New York: Raven Press.

Amin, A. H., Crawford, T. B., Gaddum, J. H. (1954). The distribution of substance P and 5-hydroxytryptamine in the central nervous system of the dog. *J. Physiol.* 126:596–618.

Apáthy, S. (1897). Das leitende Element des Nervensystems und seine topographischen Beziehungen zu den Zellen. *Mittheil. Zool. Stat. Neapal.* 12:495–748.

Aprison, M. H., Werman, R. (1965). The distribution of glycine in cat spinal cord and roots. *Life Sci.* 4:2075–2083.

Aprison, M. H., Davidoff, R. A., Werman, R. (1970). Glycine: Its metabolic and possible roles in nervous tissue. In *Handbook of Neurochemistry,* Vol. 3, Lajtha, A., ed. New York: Plenum Press.

Aprison, M. H., Tachiki, K. H., Smith, J. E., Lane, J. D., McBride, W. J. (1974). In *Advances in Biochemical Pharmacology,* Costa, E., Gessa, L., Sandler, M., eds. New York: Raven Press.

Arvanitaki, A. (1942). Effects evoked in an axon by the electrical activity of a contiguous one. *J. Neurophysiol.* 5:89–108.

Awapara, J., Landau, A. J., Fuerst, R., Seale, B. (1950). Free γ-aminobutyric acid in brain. *J. Biol. Chem.* 187:35–39.

Barger, G., Dale, H. H. (1910). Chemical structure and sympathomimetic action of amines. *J. Physiol.* 41:19–59.

Barron, D. H., Matthews, B. H. C. (1938). The interpretation of potential changes in the spinal cord. *J. Physiol.* 92:276–321.

Bell, C. (1834). On the functions of some parts of the brain, and on the relations between the brain and nerves of motion and sensation. *Philos. Trans. R. Soc. Part 1:* 471–483.

Bennett, M. V. L. (1966). Physiology of electrotonic junctions. *Ann. N. Y. Acad. Sci.* 137:509–539.

Bennett, M. V. L. (1972). A comparison of electrically and chemically mediated transmission. In *Structure and Function of Synapses,* Pappas, G. D., Purpura, D. P., eds. New York: Raven Press. pp. 221–256.

Bennett, M. V. L., Nakajima, Y., Pappas, G. D. (1967a). Physiology and ultrastructure of electrotonic junctions. I. Supramedullary neurons. *J. Neurophysiol.* 30:161–179.

Bennett, M. V. L, Pappas, G. D., Aljure, E., Nakajima, Y. (1967b). Physiology and ultrastructure of electrotonic junctions. II. Spinal and medullary electromotor nuclei in *mormyrid* fish. *J. Neurophysiol.* 30:180–208.

Bennett, M. V. L., Nakajima, Y., Pappas, G. D. (1967c). Physiology and ultrastructure of electrotonic junctions. III. Giant electromotor neurons of *Malapterus electricus. J. Neurophysiol.* 30:209–235.

Bennett, M. V. L., Pappas, G. D., Giménez, M., Nakajima, Y. (1967d). Physiology and ultrastructure of electrotonic junctions. IV. Medullary electromotor neurons in *gymnotid* fish. *J. Neurophysiol.* 30:236–300.

Bernard, C. (1878). Le curare. In *La Science Expérimentale*. Paris: Baillière. pp. 237–315.

Birks, R. I., Huxley, H. E., Katz, B. (1960). The fine structure of the neuro-muscular junction of the frog. *J. Physiol*. 150:134–144.

Bisset, G. W., Bliss, T. V. P. (1997). Wilhelm Siegmund Feldberg CBE. *Biogr. Mem. Fellows R. Soc*. 43:143–170.

Blaschko, H. (1939). The specific action of L-Dopa decarboxylase. *J. Physiol*. 96:50–51.

Blaschko, H. (1942). The activity of L-Dopa decarboxylase. *J. Physiol*. 101:337–349.

Blazemore, A. W., Elliott, K. A. C., Florey, E. (1957). Isolation of factor I. *J. Neurochem*. 1:334–339.

Bliss, T. V. P., Lømo, T. (1973). Long lasting potentiation of synaptic transmission in the dentate area of the anaesthetized rabbit following stimulation of the perforant path. *J. Physiol*. 232:331–356.

Bloom, F. E. (1977). Central noradrenergic systems: physiology and pharmacology. In *Psychopharmacology—A 20 Year Progress Report*, Lipton, M. E., Killam, K. C., Di Mascio, A., eds. New York: Plenum Press. pp. 131–141.

Bodian, D. (1966). Electron microscopy: two major synaptic types on spinal motoneurons. *Science* 151:1093–1094.

Brock, L. G., Coombs, J. S., Eccles, J. C. (1951). Action potentials of moto-neurones with intracellular electrode. *Proc. Univ. Otago Med. Sch*. 29:14–15.

Brock, L. G., Coombs, J. S., Eccles, J. C. (1952). The recording of potentials from motoneurones with an intracellular electrode. *J. Physiol*. 117:431–460.

Brooks, C. McC., Eccles, J. C. (1947). An electrical hypothesis of central inhibition. *Nature* 159:760–764.

Brown, G. L., Gillespie, J. S. (1957). The output of sympathetic transmitter from the spleen of the cat. *J. Physiol*. 138:81–102.

Brown, G. L., Dale, H. H., Feldberg, W. (1936). Reactions of the normal mammalian muscle to acetylcholine and eserine. *J. Physiol*. 87:394–424.

Cajal, S. Ramón y. (1894). The Croonian Lecture: La fine structure des centres nerveux. *Proc. R. Soc. London Ser. B* 55:444–467.

Cajal, S. Ramón y. (1909, 1911). *Histologie du Système Nerveux de l'Homme et des Vertébrés*, 2 vols. Madrid: Consejo Superior de Investigaciones Cientificas. [Reprinted 1955. English translation by Swanson, N., Swanson, L. W. (1995) published as *Histology of the Nervous System*, 2 vols. New York: Oxford University Press.]

Cajal, S. Ramón y. (1937). *Recollections of My Life*. Translated by E. H. Craigie and J. Cano. *Am. Philos. Soc. Mem*. 8.

Cajal, S. Ramón y. (1954). *Neuron Theory or Reticular Theory: Objective Evidence of the Anatomical Unity of Nerve Cells*. Translated by M. U. Purkiss and C. A. Fox. Madrid: Consejo Superior de Investigaciones Cientificas.

Cannon, W. B., Lassak, K. (1939). Evidence for adrenaline in adrenergic neurons. *Am. J. Physiol*. 125:765–777.

Cannon, W. B., Rosenblueth, A. (1933). Sympathin E and I. *Am. J. Physiol*. 104: 574–577.

Carlsson, A. (1959). Occurrence, distribution and physiological role of catecholamines in the nervous system. *Pharmacol. Rev*. 11:300–304.

Carlsson, A. (1987). Perspectives on the discovery of central monaminergic neurotransmission. *Annu. Rev. Neurosci*. 10:19–40.

Chang, M. M., Leeman, S. E. (1970). Isolation of a sialogic peptide from bovine hypothalamus tissue and its characterization as substance P. *J. Biol. Chem*. 245:3784–3790.

Clarke, E., Jacyna, L. S. (1987). *Nineteenth-Century Origins of Neuroscientific Concepts.* Berkeley: University of California Press.

Colonnier, M. (1968). Synaptic patterns on different cell types in the different laminae of the cat visual cortex. An electron microscope study. *Brain Res.* 9:268–287.

Couteaux, R., Pecot-Dechavassine M. (1970). Synaptic vesicles and pouches at the level of "active zones" of the neuromuscular junction. *C. R. Hebd. Seances Acad. Sci. D. Sci. Nat.* 217:2346–2349.

Cowan, W. M. (1970). Anterograde and retrograde transneuronal degeneration in the central and peripheral nervous system. In *Contemporary Research Methods in Neuroanatomy,* Nauta, W. J. H., Ebbesson, S. O. E., eds. New York: Springer-Verlag. pp. 217–249.

Creed, R. S., Denny-Brown, D., Eccles, J. C., Lidde II, E. G.T., and Sherrington, C. S. (1932). *Reflex Activity of the Spinal Cord.* Oxford: Clarendon Press.

Dahlstrom, A., Fuxe, K. (1964). Evidence for the existence of monoamine-containing neurons in the central nervous system. *Acta Physiol. Scand.* 62 (Suppl. 232).

Dale, H. H. (1914). The action of certain esters of choline and their relation to muscarine. *J. Pharmacol. Exp. Ther.* 6:147–190.

Dale, H. H. (1935). Pharmacology and nerve endings. *Proc. R. Soc. Med.* 28:319–332.

Dale, H. H. (1938). Acetylcholine as a chemical transmitter substance of the effects of nerve impulses. (The William Henry Welch Lectures of 1937). *J. Mt. Sinai Hosp.* 4:401–429.

Dale, H. H. (1954). The beginnings and the prospects of neurohumoral transmission. *Pharm. Rev.* 6:7–13.

Dale, H. H., Feldberg, W. (1934). The chemical transmitter of vagus effects to the stomach. *J. Physiol.* 81:320–334.

Dale, H. H., Feldberg, W., Vogt, M. (1936). Release of acetylcholine at voluntary motor nerve endings. *J. Physiol.* 86:353–380.

Davenport, H. W. (1982). Epinephrin(e). *Physiologist* 25:76–82.

Davenport, H. W. (1991). Early history of the concept of chemical transmission of the nerve impulse. *Physiologist* 34:129–142.

Del Castillo, J., Katz, B. (1954a). Quantal components of the end-plate potential. *J. Physiol.* 124:560–573.

Del Castillo, J., Katz, B. (1954b). Statistical factors involved in neuromuscular facilitation and depression. *J. Physiol.* 124:574–585.

Del Castillo, J., Katz, B. (1954c). Changes in end-plate activity produced by presynaptic polarization. *J. Physiol.* 124:586–604.

Del Castillo, J., Katz, B. (1954d). The membrane change produced by the neuromuscular transmitter. *J. Physiol.* 125:546–565.

De Lorenzo, A. J. D. (1960). The fine structure of synapses in the ciliary ganglion of the chick. *J. Biophys. Biochem. Cytol.* 7:31–36.

De Lorenzo, A. J. D. (1966). Electron microscopy: tight junctions in the synapses of the chick ciliary ganglion. *Science* 152:76–78.

Denny-Brown, D., ed. (1979). *Selected Writings of Sir Charles Sherrington.* London: Hamish Hamilton.

De Robertis, E., Bennett, H. S. (1955). Some features of the submicroscopic morphology of synapses in frog and earthworm. *J. Biophys. Biochem. Cytol.* 1:47–58.

De Robertis, E., Pellegrino De Iraldi, A. (1961). Pleurivesicular secretory processes and nerve endings in the pineal gland of the rat. *J. Biophys. Biochem. Cytol.* 10:361–372.

Dixon, W. E. (1906). Vagus inhibition. *Br. Med. J.* 2:1807.

Dowling, J. E., Boycott, B. B. (1968). Organization of the primate retina: electron microscopy. *Proc. R. Soc. London Ser. B* 166:80–111.

Du Bois-Reymond, E. (1877). *Gesammelte Abhandlungen zur Allgemeinen Muskel- und Nervenphysik.* 2 vols. Leipzig: von Veit Verlag.

Dudel, J. (1962). Effect of inhibition on the presynaptic nerve terminal in the neuromuscular junction of the crayfish. *Nature* 193:587–588.

Dudel, J., Kuffler, S. W. (1961). Presynaptic inhibition at the crayfish neuromuscular junction. *J. Physiol.* 155:543–562.

Eccles, J. C. (1936). Synaptic and neuromuscular transmission. *Ergeb. Physiol.* 38: 339–444.

Eccles, J. C. (1937). Synaptic and neuromuscular transmission. *Physiol. Rev.* 17: 538–555.

Eccles, J. C. (1946). An electrical hypothesis of synaptic and neuromuscular transmission. *Ann. N. Y. Acad. Sci.* 47:429–455.

Eccles, J. C. (1949). A review and restatement of the electrical hypothesis of synaptic excitatory and inhibitory action. *Arch. Sci. Physiol.* 3:567–584.

Eccles, J. C. (1953). *The Neurophysiological Basis of Mind. The Principles of Neurophysiology.* Oxford: Clarendon Press.

Eccles, J. C. (1964). *The Physiology of Synapses.* New York: Academic Press.

Eccles, J. C. (1967). Postsynaptic inhibition in the central nervous system. In *The Neurosciences,* Quarton, G. C., Melnechuk, T., Schmitt, F. O., eds. New York: Rockefeller University Press. pp. 408–427.

Eccles, J. C. (1975). Under the spell of the synapse. In *The Neurosciences: Paths of Discovery,* Worden, F. G., Swazey, J. P., Adelman, G., eds. Cambridge, Mass.: MIT Press. pp. 157–179.

Eccles, J. C. (1982). The synapse: from electrical to chemical transmission. *Annu. Rev. Neurosci.* 5:325–339.

Eccles, J. C., O'Connor, W. J. (1939). Responses which nerve impulses evoke in mammalian striated muscles. *J. Physiol.* 97:44–102.

Eccles, J. C., Katz, B., Kuffler, S. W. (1941). Nature of the "endplate potential" in curarized muscle. *J. Neurophysiol.* 4:362–387.

Eccles, J. C., Fatt, P., Koketsu, K. (1954). Cholinergic and inhibitory synapses in a pathway from motor-axon collaterals to motoneurons. *J. Physiol.* 216:524–562.

Eccles, J. C., Eccles, R. M., Fatt, P. (1956). Pharmacological investigations on a central synapse operated by acetylcholine. *J. Physiol.* 131:154–169.

Eccles, J. C., Eccles, R. M., Magni, F. (1961). Central inhibitory action attributable to presynaptic depolarization produced by muscle afferent volleys. *J. Physiol.* 159:147–166.

Eccles, J. C., Schmidt, R. F., Willis, W. D. (1962). Presynaptic inhibition of the spinal monosynaptic reflex pathway. *J. Physiol.* 161:282–297.

Elliott, T. R. (1904). On the action of adrenalin. *J. Physiol.* 31:xx–xxi. [Reprinted in Hall et al. (1974).]

Eranko, O. (1955). Histochemistry of noradrenaline in the adrenal medulla of rats and mice. *Endocrinology* 57:363–368.

Falck, B. (1962). Observations on the possibilities of the cellular localization of monoamines by a fluorescence method. *Acta Physiol. Scand.* 56(Suppl.)197: 1–25.

Falck, B., Hillarp, N. A., Thieme, G., Torp, A. (1962). Fluoresence of catechol amines and related compounds condensed with formaldehyde. *J. Histochem. Cytochem.* 10:348–354.

Famiglietti, E. V., Peters, A. (1972). The synaptic glomerulus and the intrinsic neuron in the dorsal lateral geniculate nucleus of the cat. *J. Comp. Neurol.* 144:285–334.

Fatt, P. (1954). Biophysics of junctional transmission. *Physiol. Rev.* 34:674–710.

Fatt, P., Katz, B. (1951). An analysis of the end-plate potential recorded with an intra-cellular electrode. *J. Physiol.* 115:320–370.

Fatt, P., Katz, B. (1952). Spontaneous subthreshold activity at motor nerve endings. *J. Physiol.* 117:109–128.

Fatt, P., Katz, B. (1953). The effect of inhibitory nerve impulses on a crustacean muscle fibre. *J. Physiol.* 121:374–389.

Feldberg, W. (1933). Die Empfindlichkeit der Zungenmuskulatur und der Zungerfässe des Hundes auf Lingualisreizung und auf Acetylcholin. *Pflügers Arch. Gesamte Physiol.* 232:75–87.

Feldberg, W. (1945). Present views on the mode of action of acetylcholine in the central nervous system. *Physiol. Rev.* 25:596–642.

Feldberg, W. (1950). The role of acetylcholine in the central nervous system. *Br. Med. Bull.* 6:312–321.

Feldberg, W. (1977). The early history of synaptic and neuromuscular transmission by acetylcholine: reminiscences of an eye witness. In *The Pursuit of Nature*. Cambridge: Cambridge University Press. pp. 65–83.

Feldberg, W., Fessard, A. (1942). The cholinergic nature of the nerves to the electric organ of the torpedo (*Torpedo marmorata*). *J. Physiol.* 101:200–216.

Feldberg, W., Gaddum, J. H. (1934). The chemical transmitter at synapses in a sympathetic ganglion. *J. Physiol.* 81:305–319.

Feldberg, W., Krayer, O. (1933). Das Auftreten eines azetylcholinartigen Stoffes im Jerzenenblut von Warmblütern bei Reizung der Nervi Vagi. *Arch. Exp. Pathol. Pharmakol.* 172:170–193.

Feldberg, W., Mintz, B. (1933). Das Auftreten eines acetylcholinartigen Stoffes im Nebennierenvenenblut bei Reizung der Nervi splanchnici. *Pflügers Arch. Gesamte Physiol.* 233:657–682.

Feldberg, W., Vogt, M. (1948). Acetylcholine synthesis in different regions of the central nervous system. *J. Physiol.* 107:372–381.

Feldberg, W., Harris, G. W., Lin, R. C. Y. (1951). Observations on the presence of cholinergic and non-cholinergic neurones in the central nervous system. *J. Phyiol.* 112:400–404.

Florey, E. (1953). Über einen nervösen Hemmungsfaktor in Gehirn und Rückenmark. *Naturwissenschaften* 40:295–296.

Forel, A. (1887). Einige hirnanatomische Betrachtungen und Ergebnisse. *Arch. Psychol.* (Berlin) 18:162–198.

Foster, M. (1897). *A Textbook of Physiology,* 7th ed., Part III. London: Macmillan.

Frank, K. (1959). Basic mechanisms of synaptic transmission in the central nervous system. *IRE Trans. Med. Electron.* 6:85–88.

Frank, K., Fuortes, M. (1957). Presynaptic and postsynaptic inhibition of monosynaptic reflexes. *Fed. Proc.* 16:39–40.

Fühner, H. (1917). Die chemische Erregbarkeitssteigerung glatter Muskulatur. *Naunyn-Schmiedebergs Arch. Exp. Pathol. Pharmakol.* 62:51–80.

Fulton, J. F. (1938). *Physiology of the Nervous System.* London: Oxford University Press.

Furshpan, E. J., Potter, D. D. (1957). Mechanism of nerve impulse transmission at a crayfish synapse. *Nature* 180:342–343.

Furukawa, T., Furshpan, E. J. (1964). Two inhibitory mechanisms in the Mauthner neurons of goldfish. *J. Neurophysiol.* 26:140–176.

Gaddum, J. H., Kwiatkowski, H. (1939). Properties of the substance liberated by adrenergic nerves in the rabbit's ear. *J. Physiol.* 96:385–391.

Galvani, L. (1791). *De viribus Electricitatis in Moto Musculari: Commentarius.* Bologna: Ex Typographia Instituti Scientarium.

Gilman, A. G., Rall, W., Nies, A. S., Taylor, P. (1990). *Goodman and Gilman's The Pharmacological Basis of Therapeutics,* 8th ed. New York: Pergamon.

Gingrich, J. A., Caron, M. G. (1993). Recent advances in the molecular biology of dopamine receptors. *Annu. Rev. Neurosci.* 16:299–321.

Golgi, C. (1873). Sull sostanza grigia del cervello. *Gazz. Med. Lombarda* 6:224–246.

Goltz, F., Freusberg, A. (1874). Über die Funktionen des Ledenmarks des Hundes. *Pflügers Arch. Ges. Physiol.* 8:460–486.

Göpfert, H., Schaefer, H. (1938). Über den direkt und inderekt errgten Aktionsstrom und die Funktion der motorischen Endplatte. *Pflügers Arch. Ges. Physiol.* 239:597–619.

Graham, L. T., Shank, R. P., Werman, R., Aprison, M. H. (1967). Distribution of some synaptic transmitter suspects in cat spinal cord: glutamic acid, aspartic acid, γ-aminobutyric acid, glycine and glutamine. *J. Neurochem.* 14:465–472.

Gray, E. G. (1959). Axo-somatic and axo-dendritic synapses of the cerebral cortex: an electron microscope study. *J. Anat.* 93:420–433.

Gray, E. G. (1962). A morphological basis for presynaptic inhibition? *Nature* 193:82–83.

Gray, E. G. (1963). Electron microscopy of presynaptic organelles of the spinal cord. *J. Anat.* 97:101–106.

Gray, E. G. (1974). Synaptic morphology with special reference to microneurons. In *Essays on the Nervous System,* Bellairs, R., Gray, E. G., eds. Oxford: Clarendon Press. pp. 155–178.

Gray, E. G., Whittaker, V. P. (1962). The isolation of nerve endings from brain: an electron microscopic study of cell fragments derived by homogenization and centrifugation. *J. Anat.* 96:79–88.

Grundfest, H. (1959). Synaptic and ephaptic transmission. In *Handbook of Physiology,* Section I: *Neurophysiology.* Washington, D.C.: American Physiological Society. pp. 147–197.

Grundfest, H. (1975). History of the synapse as a morphological and functional structure. In *Golgi Centennial Symposium: Perspectives in Neurobiology,* Santini, M., ed. New York: Raven Press. pp. 39–50.

Gudden, B. (1870). Experimentaluntersuchungen über das peripherische und centrale Nervensystem. *Arch. Psychiatr. Nervenkr.* 2:693–723.

Guillery, R. W. (1970). Light- and electron-microscopical studies of normal and degenerating axons. In *Contemporary Research Methods in Neuroanatomy,* Nauta, W. J. H., Ebbesson, S. O. E., eds. New York: Springer-Verlag. pp. 77–104.

Hall, Z. W., Hildebrand, J. G., Kravitz, E. A. (1974). *Chemistry of Synaptic Transmission.* Newton, Mass.: Chiron Press.

Hamburger, V. (1980). S. Ramón y Cajal, R. G. Harrison, and the beginnings of neuroembryology. *Perspect. Biol. Med.* 23:600–616.

Hamilyn, L. H. (1962). The fine structure of the mossy fiber ending in the hippocampus of the rabbit. *J. Anat.* 96:112–120.

Hebb, D. O. (1949). *The Organization of Behavior: A Neuropsychological Theory.* New York: John Wiley and Sons.

Hertting, G., Axelrod, J. (1961). Fate of tritiated noradrenaline at sympathetic nerve endings. *Nature* 192:172–173.

Heuser, J. E. (1977). Synaptic vesicle exocytosis revealed in quick-frozen frog neuromuscular junctions treated with 4-amino-pyridine and given a single electric shock. In *Approaches to the Cell Biology of Neurons,* Cowan, W. M., Ferrendalli, J. A., eds. Washington, D.C.: Society for Neuroscience. pp. 215–239.

Heuser, J. E., Reese, T. S. (1973). Evidence for recycling of synaptic vesicle membrane during transmitter release at the frog neuromuscular junction. *J. Cell Biol.* 57:315–344.

Heuser, J. E., Reese, T. S., Landis, D. M. D. (1975). Functional changes in frog neuromuscular junction studied with freeze fracture. *J. Neurocytol.* 3:109–131.

His, W. (1886). Zur Geschichte des menschlichen Rückenmarks und der Nervenwurzeln. *Abh. Kgl. Sächs. Ges. Wiss.* 13:147–209, 447–513.

Hodgkin, A. L. (1937a). Evidence for electrical transmission in nerve. Part I. *J. Physiol.* 90:183–210.

Hodgkin, A. L. (1937b). Evidence for electrical transmission in nerve. Part II. *J. Physiol.* 90:211–232.

Hodgkin, A. L., Katz, B. (1949). The effect of sodium ions on the electrical activity of the giant axon of the squid. *J. Physiol.* 108:37–77.

Hodgkin, A. L., Huxley, A. F., Katz, B. (1952). Measurement of current-voltage relations in the membrane of the giant axon of *Loligo. J. Physiol.* 116:424–448.

Hökfelt, T., Kellerth, J. O., Nilsson, G., Pernow, B. (1975). Substance P localization in the central nervous system and in some primary sensory neurons. *Science* 190:889–890.

Hornykiewicz, O. (1973). Dopamine in the basal ganglia: its role and therapeutic implications (including the clinical use of ʟ-Dopa). *Br. Med. Bull.* 29:172–178.

Howell, W. H., Duke, W. W. (1908). The effect of vagus inhibition on the output of potassium from the heart. *Am. J. Physiol.* 21:51–63.

Hunt, R., Taveau, R. D. (1906). On the physiological action of certain cholin derivatives and new methods for detecting cholin. *Br. Med. J.* 2:1788–1791.

Iversen, L. L. (1967). *The Uptake and Storage of Noradrenalin in Sympathetic Nerves.* London: Cambridge University Press.

Iversen, L. L., Jessell, T., Kanazawa, I. (1976). Release and metabolism of substance P in rat hypothalamus. *Nature* 264:81–83.

Jacobson, M. (1993). *Foundations of Neuroscience.* New York: Plenum Press.

Jan, L. Y., Jan, Y. N. (1982). Peptidergic transmission in sympathetic ganglia of the frog. *J. Physiol.* 327:219–246.

Kandel, E. R. (1968). Dale's principle and the functional specificity of neurons. In *Psychopharmacology: A Review of Progress,* Efron, E. F., ed. U.S. Public Health Service Publication 1936. Washington, D.C.: Government Printing Office. pp. 1957–1967.

Kandel, E. R. (1976). *Cellular Basis of Behavior: An Introduction to Behavioral Neurobiology.* San Francisco: W. H. Freeman.

Kandel, E. R., Frazier, W. T., Coggeshall, R. E. (1967). Opposite synaptic actions mediated by different branches of an identifiable neuron in *Aplysia. Science* 155:346–349.

Katz, B. (1939). *Electrical Excitation of Nerve.* London: Oxford University Press.

Katz, B. (1969). The release of neural transmitter substances. In *The Xth Sherrington Lecture.*Springfield, Ill.: Charles C Thomas.

Katz, B. (1982). Stephen William Kuffler: 24 August 1913–11 October 1980. *Biogr. Mem. Fellows R. Soc.* 28:225–259.

Katz, B., Miledi, R. (1965). The quantal release of transmitter substances. In *Studies in Physiology,* Curtis, D. R., McIntyre, A. K., eds. New York: Springer-Verlag.

Katz, B., Schmitt, O. H. (1940). Electrical interaction between two adjacent nerve fibres. *J. Physiol.* 97:471–488.

Konorski, J. (1948). *Conditioned Reflexes and Neuron Organization.* Cambridge: Cambridge University Press.

Krause, W. (1863). Über die Endigung der Muskelnerven. *Z. Rat. Med.* 18:136–160.

Kravitz, E. A., Kuffler, S. W., Potter, D. D. (1963). Gamma-aminobutyric acid and other blocking compounds in *Crustacea.* III. Their relative concentrations in separated motor and inhibitory axons. *J. Neurophysiol.* 26:739–751.

Kravitz, E. A., Potter, D. D. (1965). A further study of the distribution of gamma-aminobutyric acid between excitatory and inhibitory axons of the lobster. *J. Neurochem.* 12:323–328.

Krnjevic, K. (1974). Chemical nature of synaptic transmission in vertebrates. *Physiol. Rev.* 54:418–540.

Krnjevic, K., Schwartz, S. (1966). Is γ-aminobutyric acid an inhibitory transmitter? *Nature* 211:1372–1374.

Kuffler, S. W. (1942a). Electrical potential changes at an isolated nerve-muscle junction. *J. Neurophysiol.* 5:211–230.

Kuffler, S. W. (1942b). Further study on transmission in an isolated nerve-muscle fiber preparation. *J. Neurophysiol.* 5:309–322.

Kuffler, S. W., Edwards, C. (1958). Mechanism of gamma-aminobutyric acid (GABA) action and its relation to synaptic inhibition. *J. Neurophysiol.* 21:589–610.

Kuffler, S. W., Nicholls, J. G. (1976). *From Neuron to Brain.* Sunderland, Mass.: Sinauer.

Kühne, W. (1862). *Über die peripherischen Endorgane der motorischen Nerven.* Leipzig: Engelman.

Kühne, W. (1888). On the origin and causation of vital movement. *Proc. R. Soc. London Ser. B* 44:427–448.

Langley, J. N. (1905). On the reaction of cells and of nerve endings to certain poisons, chiefly as regards the reactions of striated muscle to nicotine and curari. *J. Physiol.* 33:374–413.

Langley, J. N. (1906). On nerve endings and on special excitable substances in cells. *Proc. R. Soc. London Ser. B* 78:170–194.

Langley, J. N. (1907). On the contraction of muscle, chiefly in relation to the presence of "receptive" substances. I. *J. Physiol.* 36:347–384.

Langley, J. N. (1921). *The Autonomic Nervous System.* Cambridge: Heffer.

Lembeck, F. (1953). Zur Frage der zentralen Übertragung afferenter Impulse. *Arch. Exp. Pathol. Pharmakol.* 219:197–213.

Ling, G., Gerard, R. W. (1949). The normal membrane potential of frog sartorius fibers. *J. Cell. Comp. Physiol.* 34:383–396.

Llinas, R., Blinks, J. R., Nicholson, C. (1972). Calcium transient in presynaptic terminal of squid giant synapse: detection with aequorin. *Science* 176:1127–1129.

Loewi, O. (1921). Über humorale Übertragbarkeit der Herznerven-wirkung. *Pflügers Arch.* 189:239–242.

Loewi, O. (1953). *From the Workshop of Discoveries.* Lawrence: University of Kansas Press.

Lundberg, J. M., Hökfelt, T., Schultzberg, M., Uvnäs-Wallenstein, K., Kohler, C., Said, S. I. (1979). Occurrence of vasoactive intestinal polypeptide (VIP)–like immunoreactivity in certain cholinergic neurons of the cat. Evidence from combined immunohistochemistry and acetylcholinesterase staining. *J. Neurosci.* 4:539–559.

McGeer, P. L., Hattori, T., McGeer, E. G. (1975). Chemical and radioautographic analysis of γ-aminobutyric acid transport in Purkinje cells of the cerebellum. *Exp. Neurol.* 47:26–41.

McGeer, P. L., Eccles, J. C., McGeer, E. G. (1978). *Molecular Neurobiology of the Mammalian Brain.* New York: Plenum Press.

McLennan, H. (1963). *Synaptic Transmission.* Philadelphia: W. B. Saunders.

Martin, A. R., Miledi, R. (1972). A presynaptic complex in the giant synapse of the squid. *J. Neurocytol.* 4:121–129.

Martin, A. R., Pilar, G. (1963). Dual mode of synaptic transmission in the avian ciliary ganglion. *J. Physiol.* 168:443–463.

Martin, A. R., Pilar, G. (1964). An analysis of electrical coupling at synapses in the avian ciliary ganglion. *J. Physiol.* 171:454–475.

Monnier, A. M., Bacq, Z. M. (1935). Recherches sur la physiologie de la pharmacologie du système nerveux autonome. XVI. Dualité du mécanisme de la transmission neuromusculaire de l'excitation chez le muscle lisse. *Arch. Int. Physiol.* 40:485–510.

Moore, R. Y., Bloom, F. E. (1979). Central catecholamine neuron systems: anatomy and physiology of the norepinephrine and epinephrine systems. *Annu. Rev. Neurosci.* 2:113–168.

Nastuk, W. L, Hodgkin, A. L. (1950). The electrical activity of single muscle fibres. *J. Cell. Comp. Physiol.* 35:39–73.

Obata, K., Takeda, K. (1969). Release of GABA into the fourth ventricle induced by stimulation of the cat cerebellum. *J. Neurochem.* 16:1043–1047.

Obata, K., Ito, M., Och, R., Sato, N. (1967). Pharmacological properties of the postsynaptic inhibition of Purkinje cell axons and the action of γ-aminobutyric acid on Deiter's neurons. *Exp. Brain Res.* 4:43–57.

Oliver, G., Schäffer, E. A. (1895). The physiological effects of extracts of the suprarenal capsule. *J. Physiol.* 18:230–276.

Otsuka, M., Iversen, L. L., Hall, Z. W., Kravitz, E. A. (1966). Release of gamma-aminobutyric acid from inhibitory nerves of the lobster. *Proc. Natl. Acad. Sci. USA* 56:1110–1115.

Palay, S. L. (1956). Synapses in the central nervous system. *J. Biophys. Biochem. Cytol.* 2(Suppl.):193–202.

Palay, S. L. (1957). The fine structure of the neurohypophysis. In *Progress in Neurobiology*, Vol. 2, Waelsch, H., ed. New York: Hoeber. pp. 31–44.

Palay, S. L. (1967). Principles of cellular organization in the nervous system. In *The Neurosciences*, Quarton, G. C., Melnechuk, T., Schmitt, F. O., eds. New York: Rockefeller University Press. pp. 24–31.

Palay, S. L., Palade, G. (1955). The fine structure of neurons. *J. Biophys. Biochem. Cytol.* 1:69–88.

Pappas, G. D., Waxman, S. G. (1972). Synaptic fine structure—morphological correlates of chemical and electrotonic transmission. In *Structure and Function of Synapses*, Pappas, G. D., Purpura, D. P., eds. New York: Raven Press. pp. 1–43.

Paton, W. D. M. (1958). Central and synaptic transmission in the nervous system (pharmacological aspects). *Annu. Rev. Physiol.* 20:431–470.

Pernow, B. (1953). Studies on substance P. Purification, occurrence and biological actions. *Acta Physiol.* 29(Suppl.)105:1–90.

Peters, A., Palay, S. L., Webster, H. de F. (1976). *The Fine Structure of the Nervous System: The Neurons and Supporting Cells*, 2nd ed. Philadelphia: W. B. Saunders.

Price, J. L., Powell, T. P. S. (1970). The synaptology of the granule cells of the olfactory bulb. *J. Cell Sci.* 7:125–156.

Purves, D., Lichtman, J. W. (1985). *Principles of Neural Development.* Sunderland, Mass.: Sinauer.

Rall, W., Shepherd, G. M., Reese, T. S., Brightman, M. W. (1966). Dendrodendritic synaptic pathway for inhibition in the olfactory bulb. *Exp. Neurol.* 14:44–56.

Renshaw, B. (1946). Central effects of centripetal impulses in axons of spinal ventral roots. *J. Neurophysiol.* 9:191–204.

Roberts, E., Frankel, S. (1950). γ-Aminobutyric acid in the brain: its formation from glutamic acid. *J. Biol. Chem.* 187:55–63.

Roberts, E., Chase, T. N., Tower, D. B., eds. (1976). *GABA in Nervous System Function*. New York: Raven Press.

Rushton, W. A. H. (1975). From nerves to eyes. In *The Neurosciences: Paths of Discovery*, Worden, F. G., Swazey, J. P., Adelman, G., eds. Cambridge, Mass.: MIT Press. pp. 277–292.

Sauer, F. C. (1935). Mitosis in the neural tube. *J. Comp. Neurol.* 62:377–405.

Schwann, T. (1839). *Mikroskopische Untersuchungen über die Uebereinstimmung in der Struktur und dem Wachsthum der Thiere und Pflanzen*. Berlin: G. E. Reimet. [English translation by H. Smith (1845); reprinted 1969.]

Shepherd, G. M. (1991). *Foundations of the Neuron Doctrine*. New York: Oxford University Press.

Sherrington, C. S. (1897). The central nervous system. In *A Textbook of Physiology*, 7th ed., Part III, Foster, M., ed. London: Macmillan.

Sherrington, C. S. (1906). *Integrative Action of the Nervous System*. New Haven, Conn.: Yale University Press.

Sherrington, C. S. (1932). Inhibition as a Coordinative Factor. Nobel Lecture. Stockholm: P. A. Norstedt.

Sherrington, C. S. (1949). A memoir of Dr. Cajal. In *Explorer of the Human Brain: The Life of Santiago Ramón y Cajal*, Cannon, D. F., ed. New York: H. Schuman.

Sjostrand, F. S. (1958). Ultrastructure of retinal rod synapses of the guinea pig eye as revealed by three-dimensional reconstructions from serial sections. *J. Ultrastruct. Res.* 2:122–170.

Sossin, W. S., Sweet, C. A., Scheller, R. H. (1990). Dale's hypothesis revised: different neuropeptides derived from a common prohormone are targeted to different processes. *Proc. Natl. Acad. Sci. USA* 87:4845–4848.

Stevens, L. A. (1971). *Explorers of the Brain*. New York: Alfred A. Knopf.

Steward, O., Levy, W. B. (1982). Preferential localization of polyribosomes under the base of dendritic spines in granule cells of the dentate gyrus. *J. Neurosci.* 2:284–291.

Storm-Mathisen, J. (1977). Glutamic acid and excitatory nerve endings. Reduction of glutamic acid uptake after axotomy. *Brain Res.* 120:379–386.

Storm-Mathisen, J., Iversen, L. L. (1979). Uptake of [^{3}H]glutamic acid in excitatory nerve endings: light and electron microscopic observations in the hippocampal formation of the rat. *Neuroscience* 4:1237–1253.

Swanson, L. W., Hartman, B. K. (1975). The central adrenergic system: an immunofluorescence study of the localization of cell bodies and their efferent connections in the rat, utilizing dopamine-β-hydroxylase as a marker. *J. Comp. Neurol.* 163:467–500.

Szymonowicz, L. (1896). Die Funktion der Nebenniere. *Pflügers Arch.* 64:97–164.

Takeuchi, A., Takeuchi, N. (1960). On the permeability of the end-plate membrane during the action of transmitter. *J. Physiol.* 154:52–67.

Tauc, L., Gerschenfeld, H. M. (1961). Cholinergic transmission mechanisms for both excitation and inhibition in molluscan central synapses. *Nature* 192: 366–367.

Twarog, B. M., Page, I. H. (1953). Serotonin content of some mammalian tissues and urine and a method for its determination. *Am. J. Physiol.* 175:157–161.

Uchizono, K. (1965). Characteristics of excitatory and inhibitory synapses in the central nervous system of the cat. *Nature* 207:642–643.

Van der Loos, H. (1967). The history of the neuron. In *The Neuron*, Hyden, H., ed. Amsterdam: Elsevier. pp. 1–47.

Volta, A. (1792). *Memoria prima sull-electricità*. Pavia: Giorancale Fisico-Medico, L. Brugnatell.

Von Euler, U. S. (1946). A specific sympathomimetic ergone in adrenergic nerve fibres (sympathin) and its relation to adrenaline and nor-adrenaline. *Acta Physiol. Scand.* 12:73–97.

Von Euler, U. S., Gaddum, J. H. (1931). An unidentified depressor substance in certain tissue extracts. *J. Physiol.* 72:74–87.

Von Euler, U. S., Hillarp, N. A. (1956). Evidence for the presence of noradrenaline in submicroscopic structures of adrenergic axons. *Nature* 177:44–45.

Wade, N. (1981). *The Nobel Duel.* New York: Doubleday.

Walberg, F. (1965). A special type of synaptic vesicles in boutons in the inferior olive. *J. Ultrastruct. Res.* 12:237A.

Waldeyer-Hartz, H. W. G. von. (1891). Über einige neuere Forschungen im Gebiete der Anatomie des Centralnervensystems. *Dtsch. Med. Wochenschr.* 17: 1213–1218; 1244–1246; 1267–1270; 1287–1289.

Waser, P. G., ed. (1975). *Cholinergic Mechanisms.* New York: Raven Press.

Watkins, J. C., Evans, R. H. (1981). Excitatory amino acid transmitters. *Annu. Rev. Pharmacol. Toxicol.* 21:165–204.

Whittaker, V. P., Michelson, I. A., Kirkland, R. J. A. (1964). The separation of synaptic vesicles from disrupted nerve ending particles (synaptosomes). *Biochem. J.* 90:293–303.

Wofsey, A. R., Kuhar, M. J., Snyder, S. H. (1971). A unique synaptosomal fraction which accumulates glutamic and aspartic acids in brain tissue. *Proc. Natl. Acad. USA* 68:1102–1106.

Woolley, D. W., Shaw, E. (1954). A biochemical and pharmacological suggestion about certain mental disorders. *Science* 119:587–588.

2

The Structure of Synapses

Pietro De Camilli, Volker Haucke, Kohji Takei, and Enrico Mugnaini

ynapses are specialized intercellular junctions between neurons or between neurons and other excitable cells where signals are propagated from one cell to another with high spatial precision and speed. Synapses are defined as electrical or chemical depending upon whether transmission occurs via direct propagation of the electrical stimulus in the presynaptic process or via a chemical intermediate. Electrical synapses are gap junctions between neurons and have all the typical structural features of such junctions. They commonly allow bidirectional propagation of the signal, and they play a role in synchronizing neuronal activity (Bennett, 2000). This chapter, like others in this book, will focus on chemical synapses, sites of discontinuity of the neuronal network where propagation of the signal is highly regulated. At chemical synapses, referred to henceforth simply as *synapses,* the presynaptic electrical signal is converted into a secretory response, leading to the release of a chemical intermediate into the synaptic cleft. This chemical message is then reconverted postsynaptically into an electrical signal. Thus chemical synapses are fundamentally asymmetric, although some retrograde, feedback signaling does occur. In spite of their great morphological variability, all synapses share some common structural characteristics (Palay, 1958; Pappas and Waxman, 1972; Heuser and Reese, 1977; Peters and Palay, 1996), which are designed to optimize transmission as well as the speed and the spatial precision of the process.

General Structure of Synapses

The basic feature of a synapse is a close apposition of specialized regions of the plasma membranes of the two participating cells to form the synaptic interface. On the presynaptic side a cluster of neurotransmitter-filled synaptic vesicles is associated with the presynaptic plasma membrane. On the postsynaptic membrane an accumulation of neurotransmitter receptors is marked by a thickening of the membrane and by the presence of a submembranous electron-dense scaffold (Figs. 2.1 and 2.2).

In the most typical case, the presynaptic compartment is localized in the outpocketing of an axonal branch and the postsynaptic compartment is located at the surface of the cell body or of a dendrite (Fig. 2.2). This organization reflects the general functional polarity of the neuron, according to which dendrites are the primary afferent elements and axons the primary efferent elements. However, there are exceptions to this general rule. Among these special cases are axonless olfactory bulb and retinal neurons that possess only short neurites that serve both effector (axonal) and receptive (dendritic) functions. Some olfactory bulb and thalamic neurons are provided with a standard axon, but their cell bodies and dendrites, which typically receive synaptic inputs, are also capable of acting as presynaptic elements (Peters et al., 1991). The nomenclature of synapses defines the primary direction of signal transmission (for example, axodendritic, axosomatic, dendrodendritic) and

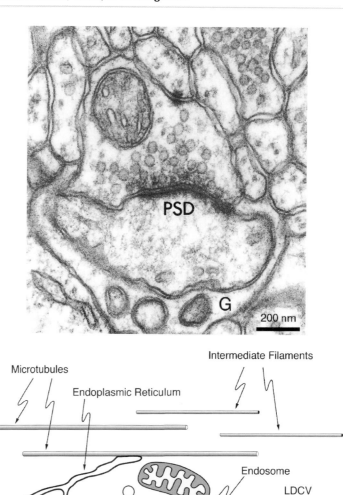

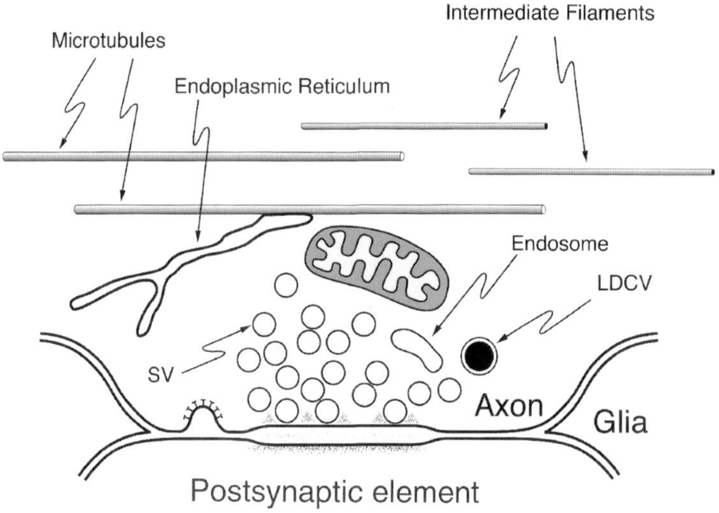

Figure 2.1. *(top) Ultrastructure of the synapse.* Electron micrograph of a synapse between a parallel fiber axon terminal and a Purkinje cell dendritic spine in the rat cerebellar cortex. Docked vesicles surrounded by a dense matrix are present at the presynaptic plasma membrane. Other vesicles are clustered behind these "front row" vesicles. A thick membrane undercoat, the postsynaptic density (PSD), is present postsynaptically, in precise register with the specialized portion of the presynaptic plasma membrane. Note that in the postsynaptic compartment a dendritic spine is completely ensheathed by a glial cell process (G).

Figure 2.2. *(bottom) Schematic illustration of the major organelles present in the presynaptic compartment.* This compartment represents an outpocketing of the axon, as is the case for the majority of synapses in the CNS. LDCV, large dense-core vesicle; SV, synaptic vesicle.

is based on the notion that such direction is indicated morphologically by the clustering of synaptic vesicles presynaptically.

Morphological studies of central nervous system (CNS) synapses had led to their classification into two major groups, type I and type II (Gray, 1959a) or asymmetric and symmetric (Colonnier, 1968), respectively, suggesting a corresponding functional dichotomy. The basis of this classification is ultrastructural features seen after conventional fixation for electron microscopy, such as the shape of the vesicles (generally spherical in type I synapses and often ovoid or flattened in type II synapses), the width of the synaptic cleft, and the thickness of the submembranous scaffolds (both greater in type I synapses). This classification clearly represents an oversimplification. However, there is evidence that at most sites it reflects a difference in the morphology of excitatory (type I, primarily glutamatergic) and inhibitory (type II, primarily γ-aminobutyric acid [GABA]– or glycinergic) synapses. Recent progress in the biochemical characterization of these two types of synapses has revealed compositional differences (primarily postsynaptically) that may underlie this structural dichotomy.

Presynaptic Compartment

The great distance of nerve terminals from the cell body poses special problems for the supply of newly synthesized components to the nerve terminal. The bulk of protein synthesis occurs in the perikaryal-dendritic region, where the early stations of the secretory pathway are localized (rough endoplasmic reticulum and Golgi complex). Although electrical signals can travel rapidly between the cell body and the axon terminals, the time required for macromolecules and organelles to reach the periphery of the axon can be much longer. Simple diffusion cannot guarantee the delivery of material to the axonal periphery, nor efficient retrograde flow for the recycling and degradation of materials. Anterograde and retrograde axonal flow is therefore assisted by a large number of molecular motors, which use microtubules (Hirokawa et al., 1998) or actin (Kuznetsov et al., 1992) as tracks. Even the fast microtubular motors, however, may take hours (days in the case of very long axons) to deliver their membrane-bound cargo to the synapses. Furthermore, transport of cytosolic proteins, including many proteins of the cytoskeletal matrix of nerve terminals, occurs at a very low rate, resulting in a transit time of up to several weeks for very long axons (Lasek et al., 1984).

To cope with these unique characteristics, nerve terminal proteins have a prolonged half-life in the axon, suggesting interesting differences in the mechanisms that control protein turnover in the axon and in the cell body. Furthermore, to compensate for the slow resupply of membranous organelles, the nerve terminal uses a specialized secretory apparatus based on synaptic vesicles. Synaptic vesicles can rapidly reform by local recycling (see the chapter on synaptic vesicle endocytosis by De Camilli et al.) and be refilled locally with their neurotransmitter content without involving the protein-synthesizing machinery confined to the cell body. In fact, local recycling may also represent the main (or only) biosynthetic pathway for these organelles (Hannah et al., 1999) because newly synthesized synaptic vesicle proteins are transported to nerve terminals in precursor membranes rather than as mature synaptic vesicles (Tsukita and Ishikawa, 1980; Yonekawa et al., 1998; see also the chapter on synaptic vesicle endocytosis by De Camilli et al.). Nerve terminals may also secrete peptide neurotransmitters via large dense-core vesicles, which are organelles of the classical regulated secretory pathway (Burgess and Kelly, 1987; De Camilli and Jahn, 1990; Thureson-Klein and Klein, 1990). However, large dense-core vesicles are only co-players in synaptic transmission (Fig. 2.3). The properties of typical synaptic vesicles and of other major organelles of the presynaptic compartment are now summarized.

Synaptic Vesicles

As soon as they were discovered in the early days of electron microscopy (Palay and Palade, 1955), synaptic vesicles were linked to chemical transmission (Katz, 1962). They have served ever since as one of the morphological hallmarks of chemical synapses (Pappas and Purpura, 1972). Synaptic vesicles are small, electron-lucent vesicles with a size range of 35–50 nm that store nonpeptide neurotransmitters, such as acetylcholine, glutamate, GABA, and glycine (Jahn et al., 1990). Within a given nerve terminal they are of uniform size, although their diameter and shape may vary slightly in different classes of synapses (Gray, 1959a; Akert et al., 1972). In inhibitory synapses they can appear ovoid or flattened in conventional electron microscopy owing to an artifact associated with the osmolarity of the fixative (Peters et al., 1991). Small dense-core vesicles of aminergic neurons may be a subclass of these vesicles. They derive their name from the electron-dense core that is visible in their lumen after certain fixations, which probably represents a precipitate induced by the presence of amines (Hökfelt, 1968; Bloom et al., 1970). Vesicles that are similar in size and molecular composition to the synaptic vesicles (sometimes referred to as synaptic-like microvesicles) are present in neuroendocrine cells such as chromaffin cells and cells of the endocrine pancreas. They may be involved in a form of local, paracrine signaling (Fig. 2.3C). Thus the secretory apparatus of the synapse may also be present in a more rudimentary form in endocrine cells (De Camilli and Jahn, 1990; Thomas-Reetz and De Camilli, 1994; Hannah et al., 1999).

Synaptic vesicles can be easily obtained from brain tissue in high yield and purity, owing to their abundance in the CNS, homogeneous size, and buoyant density. They have therefore been thoroughly characterized (Fig. 2.4). An inventory of their membrane proteins has been

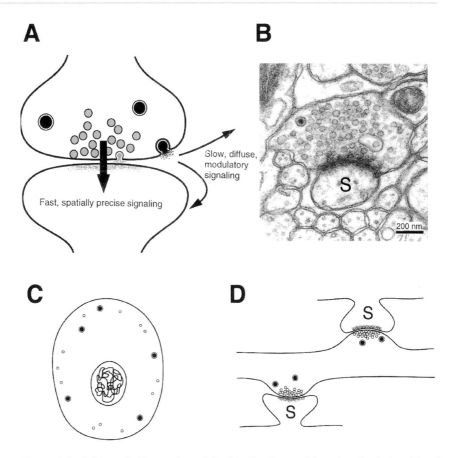

A

Slow, diffuse, modulatory signaling

Fast, spatially precise signaling

B

S

200 nm

C

D

S

S

Figure 2.3. Schematic illustration of the localization and functional relationship of synaptic vesicles (small circles) and large dense-core vesicles (large filled circles) at synapses and of corresponding organelles in endocrine cells.

(A) Secretion of nonpeptide neurotransmitters via synaptic vesicles mediates fast transmission, whereas peptide secretion from large dense-core vesicles is associated with a slower, more diffuse form of signaling. Note that nonpeptide neurotransmitter secreted from synaptic vesicles may also contribute to modulatory signaling via effects on metabotropic receptors not clustered at the synaptic cleft (not shown).

(B) Electron micrograph of a synapse onto a dendritic spine (S) in the cerebellar cortex, demonstrating the presence of a large dense-core vesicle.

(C–D) Large dense-core vesicles of axons *(D)* correspond to peptide-containing secretory granules of endocrine cells *(C)*; small vesicles biochemically similar to synaptic vesicles (synaptic-like microvesicles) are also present in endocrine cells. These organelles have been used for studies on the biogenesis of synaptic vesicles. S, dendritic spine.

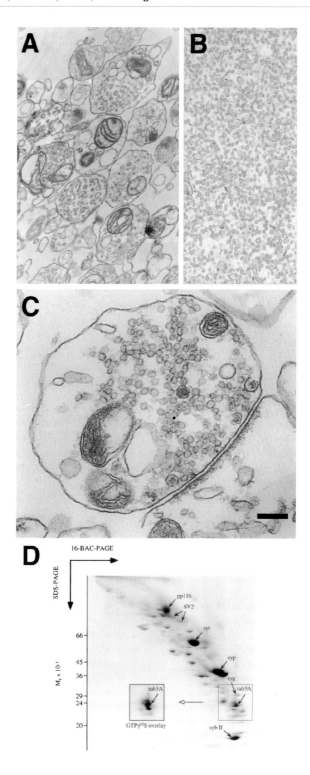

compiled and the genes for several of them have been cloned. They can be divided into two main groups: those with a known or putative role in ion and neurotransmitter transport (e.g., the proton pump, neurotransmitter transporters, SV2) and those with a role in vesicle trafficking (e.g., synaptobrevin/Vamp, synaptotagmin, the Rab3-rabphilin complex) (Jahn and Südhof, 1994; Calakos and Scheller, 1996; Südhof and Scheller, this volume). Information on this second class of proteins has greatly advanced our knowledge of the molecular mechanisms involved not only in synaptic vesicle cycling but also, more generally, in membrane trafficking.

The presynaptic vesicle cluster can consist of a few dozen to several hundred vesicles (Harris and Sultan, 1995) at most synapses in the CNS (e.g., Figs. 2.1 and 2.3B), but it may contain up to thousands of vesicles at specialized synapses. In some very large terminals (e.g., those of the mossy fiber endings in the hippocampus and cerebellum) a single large cluster of synaptic vesicles is connected to multiple, independent synaptic junctions with distinct postsynaptic elements (Fig. 2.5) (Peters et al., 1991). The cluster of vesicles is very compact at some synapses and more dispersed at others. Interactions underlying vesicle clustering may vary with the distance from the presynaptic membrane. At each synapse a few vesicles are in physical contact with the presynaptic plasma membrane at release sites (so-called docked vesicles), also called "active zones" (Harris and Sultan, 1995) (Figs. 2.1 and 2.6). Some of these vesicles may be frozen at an intermediate fusion stage, ready to undergo full fusion upon Ca^{2+} stimulation. They may correspond to partially engaged fusion complexes (SNARE complexes) between the vesicle membrane and the plasma membrane (Rothman and Söllner, 1997; Sutton et al., 1998). However, morphologically docked vesicles can also be seen, and actually in increased numbers (Hunt et al., 1994; Neale et al., 1999;

Figure 2.4. *(opposite) Subcellular fractionation of the components of the synapse.*

(A) Electron micrograph of a synaptosomal fraction from rat brain. Synaptosomes correspond to pinched-off nerve terminals with attached portions of the postsynaptic membrane, which are generated by a mild homogenization of brain tissue.

(B) Purified rat synaptic vesicle fraction obtained by the further fractionation of synaptosomes. Reproduced with permission from Huttner et al. (1983).

(C) Higher-magnification view of a synaptosome, demonstrating that the physical connection between the pre- and postsynaptic plasma membrane has been preserved by the fractionation. Scale bar = 500 nm in *A,* 400 nm in *B,* and 200 nm in *C.*

(D) Two-dimensional separation of synaptic vesicle proteins by 16-BAC/ SDS-polyacrylamide gel electrophoresis. Proteins were stained with Coomassie blue and subsequently identified by immunoblotting. pp116, 116-kDa subunit of the vacuolar proton pump; syb II, synaptobrevin II; syp, synaptophysin; syt, synaptotagmin.

Reproduced with permission from Hartinger et al. (1996).

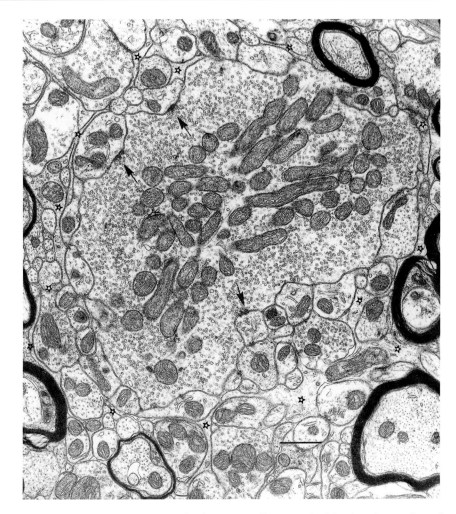

Figure 2.5. Electron micrograph of a mossy fiber terminal in the glomerulus of the cerebellum. This giant terminal forms multiple synaptic contacts with distinct postsynaptic elements. It contains numerous mitochondria and a mass of synaptic vesicles. Astrocytic lamellae (stars) ensheath most of the glomerular periphery, but they do not abut the synapses. Only a minority of the synaptic vesicles are clustered at synapses (arrows), which are marked by the presence of postsynaptic densities. The bulk of the synaptic vesicles represent a common reserve pool for all the synapses. Scale bar = 1 μm.

Fig. 2.7), after clostridial toxin–mediated disruption of SNARE complex formation. (Clostridial toxins are powerful irreversible inhibitors of synaptic vesicle exocytosis [Jahn and Niemann, 1994; Pellizzari et al., 1999].) This finding suggests that docked vesicles may represent multiple stages of engagement in the fusion process, including early stages that are independent of SNAREs.

Another subpopulation of vesicles is tightly bound to a specialized cytoskeletal scaffold that is anchored to the presynaptic membrane and

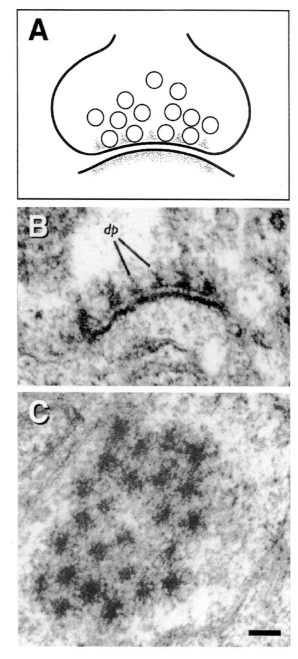

Figure 2.6. Presynaptic grid at central synapses. The grid is represented by an array of electron-dense projections that lie along the presynaptic membrane at the active zone. Docked vesicles lie within the mesh of this matrix.

(A) Schematic drawing of the presynaptic grid (gray punctate triangles) with interspersed synaptic vesicles.

(B–C) Electron micrographs of cat spinal cord synapses cut perpendicularly (*B*) or tangentially (*C*) and showing regularly arranged dense projections (dp) in pre-synaptic processes. The electron density of the grid in these images was enhanced by the use of 1% phosphotungstic acid during specimen preparation. (Vesicle membranes were not preserved by the fixative.) Scale bar = 100 nm. Panels B and C are reproduced from Gray (1963) with permission of Cambridge University Press.

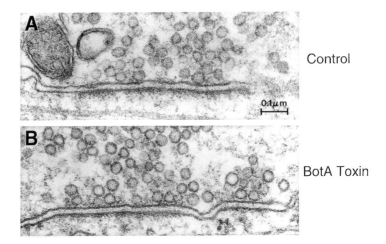

Figure 2.7. Docked synaptic vesicles are present in nerve terminals after blocking of synaptic vesicle exocytosis by the action of Botulinum neurotoxin A. Synaptic junctions of cultured spinal cord neurons are shown before (*A*) and after (*B*) application of the toxin. The number of docked vesicles is increased in the toxin-treated synapse. Reproduced with permission from Neale et al. (1999) by copyright permission of the Rockefeller University Press.

has an electron-dense appearance after conventional fixation. This scaffold varies in morphology at different synapses. At central synapses it often forms a presynaptic grid in whose cavities docked synaptic vesicles are localized (Gray, 1959a,b; Bloom and Aghajanian, 1966; Akert et al., 1972) (Fig. 2.6). In specialized synapses of sensory organs (collectively called ribbon synapses) it is organized into very dense masses, which form ribbons or spherical dense bodies (see Fig. 2.15) (Hama and Saito, 1977; Rao-Mirotznik et al., 1995; Lenzi et al., 1999). In *Drosophila* neuromuscular junctions it forms T-bars (Koenig et al., 1993; Fig. 2.8A,B) and in *Caenorhabditis elegans* it forms a dense mass, often at the center of a bipartite postsynaptic site (Fig. 2.8C; White et al., 1986; Jorgensen and Nonet, 1995). The main function of this dense matrix appears to be the concentration of synaptic vesicles in close proximity to release sites,

Figure 2.8. (opposite) Synapses in Drosophila *and* C. elegans.

(A) Electron micrograph of a resting *Drosophila* neuromuscular junction. Two presynaptic dense bodies are visible (white arrows). Reproduced with permission from Koenig et al. (1993).

(B) High-power view of a presynaptic dense body at a *Drosophila* neuromuscular junction. Reproduced with permission from Koenig et al. (1998). Copyright 1998 National Academy of Sciences, U.S.A.

(C) Electron micrograph of a synapse in *C. elegans*. A dense matrix marks the presynaptic active zone. Unpublished micrograph by E. Salcini, L. Daniell, and P. De Camilli.

Scale bar = 100 nm in *B* and 200 nm in *A* and *C*.

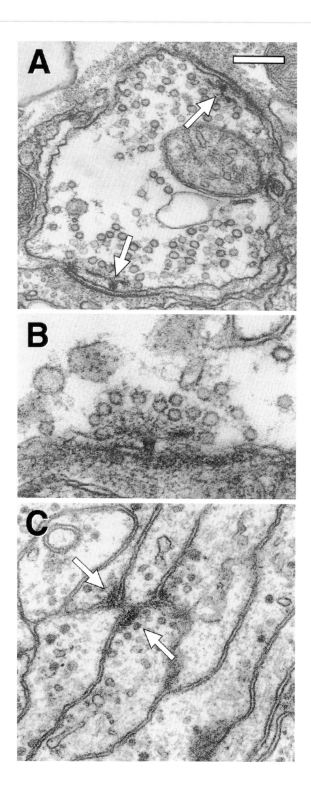

so that new vesicles can rapidly replace docked vesicles as they undergo exocytosis. The molecular composition of this matrix remains elusive, but some protein components, which share blocks of homology, have recently been identified. They include RIM (Wang et al., 1997, 2000) and two very large proteins, aczonin/piccolo (Cases-Langhoff et al., 1996; Wang et al., 1999; Fenster et al., 2000) and bassoon (tom Dieck et al., 1998). The properties of these molecules are consistent with those of scaffold and adaptor proteins. RIM binds the synaptic vesicle protein Rab3 in its GTP-bound state and may therefore mediate a Rab3-dependent attachment of the vesicle to this scaffold (Wang et al., 1997). Piccolo was shown to interact with the protein Pra1, which in turn binds to both Rab3 and the vesicle protein synaptobrevin/VAMP (Fenster et al., 2000). RIM, bassoon, and piccolo were localized both to the ribbon of photoreceptors and to active zones of conventional synapses (Wang et al., 1997; Brandstatter et al.,1999; Fenster et al., 2000), confirming that in spite of their different morphologies these cytoskeletal scaffolds share at least some molecular properties.

The majority of the vesicles at most synapses are connected to active zones by a more labile link. After lysis of the nerve terminal, this peripheral pool is lost more readily from the synapse than the pool of vesicles adjacent to the active zones. Actin fibers were hypothesized to function as an anchoring scaffold for these vesicles (Hirokawa et al., 1989), and synapsin, which is a vesicle and actin binding protein (De Camilli et al., 1990), was proposed to play a role in the anchoring of vesicles to actin (Greengard et al., 1993). Consistent with its putative role in vesicle clustering, acute disruption of synapsin function by antisynapsin antibody injection at a giant synapse has been found to lead to dissociation and dispersal of the peripheral pool of vesicles without affecting the vesicles adjoining the active zones (Pieribone et al., 1995). Furthermore, this pool of vesicles is smaller in synapsin knockouts than in control mice (Rosahl et al., 1995; Li et al., 1995). Recent studies have shown that synapsin has the structural properties of an ATP-utilizing enzyme, although its precise enzymatic function is not yet known (Esser et al., 1998). It is possible therefore that the main function of synapsin in the regulation of vesicle clustering may involve its enzymatic activity. Strikingly, the loss of the peripheral pool of vesicles produced by synapsin disruption does not affect release of neurotransmitter induced by low-frequency stimulation. These vesicles represent a reserve pool that is critically required only during high-frequency impulse bursts (Pieribone et al., 1995; Brodin et al., 1997).

The vesicle clusters have very plastic properties. It has been shown that two clusters can merge into one, and that when they do so they do not merge progressively but rather collapse rapidly into a single mass (Betz et al., 1992). However, individual vesicles seem to have restricted mobility within the cluster, as shown by experiments involving fluorescence recovery after photobleaching (Henkel et al., 1996; Kraszewski et al., 1996). The boundaries of a photobleached spot within a fluorescent vesicle cluster do not change during sustained nerve terminal stimulation, indicating that the translocation of vesicles to the release site is not correlated with their detachment from a restraining matrix (Cochilla et al., 1999). An interesting possibility is that actin-myosin–based motility may underlie this translocation. Protein phosphorylation is likely to play an important role in the regulation of molecular interactions within the cluster because the protein phosphatase inhibitor okadaic acid induces vesicle dispersion (Betz and Henkel, 1994), whereas the protein kinase inhibitor staurosporine has the opposite effect (Henkel et al., 1996; Kraszewski et al., 1996).

Studies of hippocampal neurons in culture have shown that during synaptogenesis the assembly of synaptic vesicles at presynaptic sites results in the recruitment of preassembled vesicle clusters (Kraszewski et al., 1996; see also Ahmari et al., 2000). Developing axons contain functionally competent, neurotransmitter-filled synaptic vesicles from their earliest stages. These form small mobile clusters enriched in the distal axon, where they undergo a constitutive exo- and endocytosis (Fig. 2.9A–C, E). Such vesicles are clearly distinct from the vesicular carriers that deliver new membrane components to the growth cone, where axonal elongation takes place (Craig et al., 1995; Kraszewski et al., 1996; Zakharenko et al., 1999; see Fig. 2.9B). As a growing axon contacts an appropriate target cell, these preassembled groups of vesicles, which also include other presynaptic components, rapidly coalesce at sites of contact (this site will become the presynaptic site) until the mature size and morphology of the presynaptic compartment are reached (Fletcher et al., 1994; Kraszewski et al., 1996; Ahmari et al., 2000; see Craig and Lichtman, this volume). Concomitantly synaptic vesicle clusters disappear from the nonsynaptic portions of the axons (Fig. 2.9D,F). Critical factors are likely to limit the maximal size of a synapse because variability in the synaptic interface between two cells is achieved by variability in the number of synaptic contacts (which have roughly similar size), rather than by the variability of the size of individual synapses.

The elucidation of the mechanisms underlying the function and maintenance of the presynaptic cluster is clearly an important priority in the field of synaptic transmission. It will also be important to establish the structural and molecular basis of the functionally distinct pools of vesicles defined by biophysical studies of synaptic vesicle exocytosis (Rosenmund and Stevens, 1996; Stevens and Wesseling, 1999). Whereas the readily releasable pool is likely to coincide with the population of docked vesicles, it is unclear whether a subpopulation of vesicles located close to the plasma membrane may recycle more efficiently than vesicles at more distant sites. Genetic studies in simple organisms are likely to provide important insights into the ontogenesis of the vesicle cluster. Recently alterations in the organization of the active zone and of the underlying presynaptic vesicle cluster have been reported in *C. elegans* with mutations in a member of the liprin family (Zhen and Jin, 1999). Liprins bind tyrosine phosphatase receptors of the LAR (for leukocyte

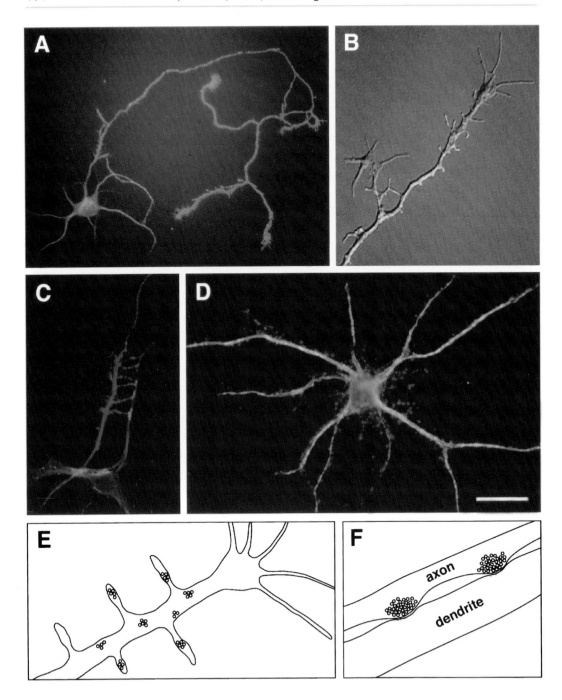

antigen–related) family, suggesting a role for tyrosine phosphorylation in the ontogenesis of presynaptic active zones. Considering the fundamental homology of synapses and other types of intercellular junctions, some of the mechanisms implicated in the organization of submembranous scaffolds at other junctions may also play a role at the synapse.

Large Dense-Core Vesicles

Large dense-core vesicles (70–200 nm diameter) usually contain a cocktail of neuroactive peptides, referred to as peptide neurotransmitters or neurohormones, and they may also contain amines (Hökfelt et al., 1980; Thureson-Klein and Klein, 1990; Stjarne, 1999). These secretory granules, which have the same properties as the secretory granules of endocrine cells, can only be assembled in the somatodendritic compartment of the neuron, and are transported to the nerve terminals as mature organelles. They cannot be regenerated at the cell periphery by recycling. Their number in nerve terminals is highly variable but usually orders of magnitude lower than the number of synaptic vesicles (Fig. 2.3). Although most (and possibly all) neurons have this type of secretory pathway, large dense-core vesicles are abundant only in specialized neuronal subpopulations, including many neurons of the hypothalamus (Navone et al., 1989). Large dense-core vesicles are not selectively concentrated in proximity to the presynaptic plasma membrane and indeed are often

Figure 2.9. (opposite) Development of synapses in hippocampal neurons in cultures. The distribution of synaptic vesicles is visualized by immunolabeling for the synaptic vesicle marker synaptotagmin.

(A, C, D) Living neurons were incubated with antibodies directed against the lumenal portion of the synaptic vesicle protein synaptotagmin, which are internalized in an activity-dependent manner (Kraszewski et al., 1996). They were subsequently fixed and double stained for either the internalized antibodies (red) or the dendritic marker MAP-2 (green).

(B) Distal axon from a living neuron incubated with CY3-conjugated antibodies to the lumenal domain of synaptotagmin and visualized by differential interference contrast microscopy and immunofluorescence. Reproduced with permission from Kraszewski et al. (1996). The distal axon is enriched in recycling synaptic vesicles from the earlier stages of its differentiation (A). In these immature, isolated axons, synaptic vesicles are organized in highly mobile microclusters, which move into and out of lateral filopodia. This is shown at higher magnification in B.

Upon contact with a postsynaptic cell, vesicle microclusters coalesce at contact sites (C) and eventually form large, stable presynaptic clusters. Eventually synaptic vesicles are no longer visible in the nonsynaptic regions of the axon (D). Calibration bar = 25 μm in A and D, 10 μm in B and C.

(E–F) Schematic illustration of the distribution of synaptic vesicles in a isolated developing axon (E) (such as the one shown in B) and at mature synapses (F) (such as the ones shown in D).

excluded from the synaptic vesicle cluster (Navone et al., 1984). Their exocytosis does not occur at active zones (Thureson-Klein and Klein, 1990), is preferentially triggered by trains of action potentials, and is a relatively rare event compared with synaptic vesicle exocytosis (Hökfelt et al., 1986; Bruns and Jahn, 1995)—as might be expected considering their critical dependence on the cell body for their formation. Their se-cretory product may act on receptors at a significant distance from the release sites, and their receptors are not ligand-gated channels (iono-tropic receptors) but usually slower-acting receptors (e.g., G protein–coupled receptors) such as those that transduce the signals of peptide hormones (Fuxe and Agnati, 1991). In conclusion, secretion from these organelles does not mediate synaptic transmission directly but partici-pates in a slower, endocrinelike mode of signaling and probably plays a modulatory role in synaptic transmission.

Endocytotic Organelles

Nerve terminals contain a variety of endocytotic intermediates respon-sible for the recycling of membrane delivered to the cell surface from the exocytosis of synaptic vesicles. These recycling intermediates are often infrequent, and they are virtually absent in most resting nerve terminals but very abundant after strong stimulation (Ceccarelli et al., 1973; Heuser and Reese, 1973; Shupliakov and Brodin, 2000). Like any other periph-eral compartment of the cell, nerve terminals also contain vesicles in-volved in the continued renewal and recycling of plasma membrane components and in the constitutive and regulated internalization of re-ceptors, including those for peptide neurohormones and neurotrophins. This recycling traffic partially intersects with the recycling of synaptic vesicles, which is by far the predominant form of traffic in the pre-synaptic compartment (see the chapter on synaptic vesicle endocytosis by De Camilli et al.). For this reason, nonspecific markers of endocytosis, such as horseradish peroxidase or FM1-43, are often used to follow the dynamics of synaptic vesicle membranes (Ceccarelli et al., 1973; Heuser and Reese, 1973; Cochilla et al., 1999). However, it is important to con-sider that individual endocytotic events may also represent traffic that is independent of synaptic vesicles.

Endocytotic intermediates are represented by clathrin-coated pits and vesicles and by pleiomorphic membrane compartments, such as vac-uoles, cisternae, and tubulovesicular elements (Heuser and Reese, 1973; Fig. 2.10B). Endocytotic elements are typically excluded from active zones and synaptic vesicle clusters, indicating a segregation of exocytotic from endocytotic sites (see the chapter on synaptic vesicle endocytosis by De Camilli et al.). Large endocytotic compartments represent a functionally heterogeneous population. Some of them are likely to represent the classical sorting endosomes downstream of clathrin-coated vesicles (Mell-man, 1995). Others—in particular the vacuoles that rapidly accumulate after massive stimulation of secretion—may be generated by bulk endo-

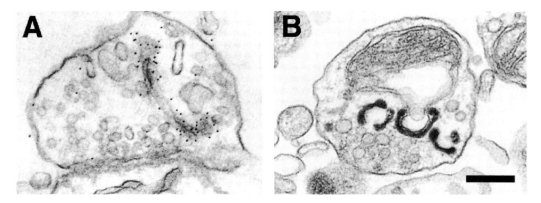

Figure 2.10. *Smooth pleiomorphic tubules and cisternae visible in nerve terminals by electron microscopy are either endocytotic organelles or endoplasmic reticulum and can be differentiated only by cytochemical markers.* (A) *Purkinje cell nerve terminal labeled by immunogold for the InsP$_3$ receptor, a resident protein of the endoplasmic reticulum (black dots). Reproduced with permission from Takei et al. (1992).* (B) *Rat synaptosome fixed after a short incubation in depolarizing conditions (high K$^+$) and in the presence of extracellular horseradish peroxidase. Presence of the peroxidase reaction product in membranous organelles defines them as endocytotic compartments. Scale bar = 100 nm.*

cytosis from the plasma membrane (Miller and Heuser, 1984; Takei et al., 1996; see also the chapter on synaptic vesicle endocytosis by De Camilli et al.). The membranes of these vesicles may be functionally equivalent to the plasma membrane and accordingly would serve as donors for the budding of clathrin-coated vesicles (Miller and Heuser, 1984; Takei et al., 1996). Small multivesicular bodies (late endosomes) may also be present at synapses. They derive from the maturation of early endosomes and are organelles destined to join retrograde axoplasmic flow (Tsukita and Ishikawa, 1980) for delivery to lysosomes in the perikaryal-dendritic region of the cell.

Endoplasmic Reticulum

Some of the membrane profiles present in the nerve terminal represent elements of the smooth endoplasmic reticulum (ER), an interconnected system of tubules and cisternae that extends throughout the cells from dendrites to nerve terminals. Among its many roles, an important one is its ability to function as a regulated store of intracellular Ca^{2+} (Pozzan et al., 1994). A variety of presynaptic receptors can induce Ca^{2+} release from the ER via the generation of InsP$_3$, which acts on InsP$_3$-gated Ca^{2+} channels in its membrane. The resulting elevations in cytosolic Ca^{2+} do not have a significant effect on synaptic vesicle exocytosis, which is triggered selectively by a depolarization-dependent Ca^{2+} influx (Smith and Augustine, 1988; Berridge, 1993). However, Ca^{2+} release from intracellular stores may regulate a variety of presynaptic functions, including the efficiency of the depolarization-secretion coupling and the release of large dense-core vesicles (Nicholls, 1998). The similarity in structure of elements of the ER and endosomes makes it difficult to distinguish them morphologically. However, they can be easily differentiated by cytochemical markers. For example, the ER is immunoreactive for ER-resident components, including the InsP$_3$ receptor (Takei et al., 1994; Fig. 2.10A), whereas endosomes can be labeled by extracellular endocytotic tracers (Fig. 2.10B).

Mitochondria

Mitochondria are frequently observed in the vicinity of the synaptic vesicle clusters, in agreement with the ATP requirement of several steps of the vesicle cycle. Known ATP-dependent stages in the cycle include vesicle priming for exocytosis (Eberhard et al., 1990; Robinson and Martin, 1998), disassembly of the synaptic SNARE complex (Söllner et al., 1993), the endocytotic reaction (see the chapter on synaptic vesicle endocytosis by De Camilli et al.), clathrin-coat uncoating (Ungewickell et al., 1995), and the uptake of neurotransmitters into the vesicles (Jahn et al., 1990). Many other steps are likely to require ATP. For example, ATP is needed for reversible changes in membrane lipids that correlate with exo- and endocytosis (Hawthorne and Pickard, 1979; Cremona et al., 1999), for the remodeling of the actin cytoskeleton, for protein phosphorylation reactions that control the vesicle cycle (Greengard et al., 1993), and for efficient buffering of cytosolic Ca^{2+}. In addition, ATP is used for the generation of GTP. Several GTPases—including dynamin (Koenig and Ikeda 1989; Takei et al., 1995), Rab proteins, and other small GTPases (Novick and Zerial, 1997)—have a key role at specific stages of synaptic vesicle release and reuptake. There is evidence that in neurons, as in other cells, mitochondria may be structurally and functionally connected by intermitochondrial contacts (Bakeeva et al., 1983, 1985; Amchenkova et al., 1988). The organization of presynaptic mitochondria differs depending on the temporal pattern of transmitter release at a given synapse. The number of mitochondria is considerably higher in tonic-firing than in phasic-firing synapses, most likely reflecting a higher requirement for ATP in the former (Fig. 2.11). Moreover, the close vicinity of mitochondria in tonic synapses to each other and to the active zone makes it conceivable that they may function as a single electrically coupled unit (Brodin et al., 1999).

Cytoskeletal Elements

Microtubules and intermediate filaments represent the main structural scaffold of the axons. Microtubules also function as the tracks for the intense traffic of organelles from the cell body to axon terminals and vice versa. They are therefore involved in delivering synaptic vesicle precursor membranes to the synapse (Hirokawa et al., 1998). In optimally fixed

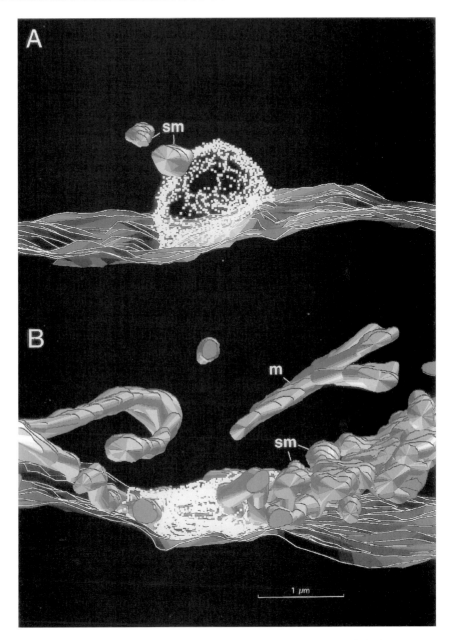

Figure 2.11. Distribution of synaptic mitochondria at synapses of the lamprey as revealed by three-dimensional reconstruction of serial sections. (A) Mitochondria in the vicinity of a reticulospinal synapse (see also Fig. 2.16), a phasic synapse with a large reserve pool of synaptic vesicles and a low capacity for vesicle recycling. *(B)* Mitochondria in the vicinity of a dorsal column synapse, a tonic synapse with a small reserve pool of synaptic vesicles and a high capacity for vesicle recycling. White dots delineate the outer surface of synaptic vesicle clusters. White lines correspond to single ultrathin sections. m, Elongated mitochondrion; sm, synaptic mitochondria. Reproduced with permission from Brodin et al. (1999).

synapses, however, both microtubules and intermediate filaments are generally excluded from the presynaptic vesicle cluster. This cluster often lies in an outpocketing of the axon, at one side of its main core, that contains microtubules and intermediate filaments. Microtubules, therefore, are unlikely to play a direct role in the exo- and endocytotic cycle of synaptic vesicles (see Fig. 2.2). One study reported the presence of microtubules in direct apposition to the presynaptic grid of dense material that underlies the active zone (Gray, 1983). The significance of these striking images, which differ from most other published images of synapses, remains unclear. They may be partially explained by rapid changes occurring at the time of fixation, but the consensus is that microtubules do not reach the presynaptic terminal density.

By contrast, there is strong evidence for an important role of the actin-based cytoskeleton in vesicle dynamics. Ultrastructural studies based on quick-freezing, deep-etching techniques have suggested the presence of actin filaments among vesicles (Hirokawa et al., 1989). Furthermore, a dynamic and regulated meshwork of actin may be present at the periphery of the vesicle cluster (Gustaffson et al., 1998). Evidence for the existence of a pool of polymerized actin has come from electron microscopic analysis of the presynaptic compartment of the giant reticulospinal synapse of the lamprey. At this synapse the area surrounding the vesicle cluster is free of other organelles, thus allowing optimal visualization of the cytoskeleton. The peripheral actin meshwork is clearly visible after microinjection of reagents that stabilize or induce the formation of actin filaments (Gustaffson et al., 1998; Gad et al., 1999). The site where this actin pool is localized coincides with identified "endocytotic hot spots." It is therefore of interest that a variety of studies have implicated actin in the endocytotic reaction (see the chapter on synaptic vesicle endocytosis by De Camilli et al.). A honeycomb-like network, which surrounds the synaptic vesicle clusters and is enriched in endocytotic proteins, was recently described at *Drosophila* neuromuscular junctions (Roos and Kelly, 1999).

The Presynaptic Plasma Membrane

In most central synapses, the synaptic area of the presynaptic membrane is structurally defined by its parallel and close apposition to the postsynaptic membrane and by the presence of presynaptic densities interspersed between the docked vesicles on its cytoplasmic side. In small synapses the synaptic interface has roughly the shape of a disc. In larger synapses this interface is often interrupted by discontinuities (perforated or segmented synapses) as if there was a critical size beyond which an uninterrupted synaptic junction cannot grow. In the case of perforated synapses the synaptic interfaces may have the shapes of rings (single perforations) or of complex honeycomb structures (Andres, 1975). Clathrin-coated pits are frequently observed at synaptic perforations both presynaptically and postsynaptically, suggesting that these areas play an important role in membrane turnover.

Based on its many functions, the presynaptic membrane is thought to contain several specialized proteins in addition to the housekeeping proteins common to all plasma membranes. However, only a few of these specialized proteins have been characterized. Two presynaptic membrane proteins are the target membrane SNAREs (t-SNAREs) syntaxin and SNAP-25, which have a role in synaptic vesicle exocytosis. They act as the partners of the synaptic vesicle SNARE (v-SNARE) synaptobrevin/VAMP in the formation of the exocytotic fusion complex (Rothman and Söllner, 1997). Neither syntaxin nor SNAP-25, however, is selectively concentrated in the presynaptic membrane (Garcia et al., 1995), suggesting that other factors present either in the membrane or in close proximity to it are responsible for the selective exocytosis of synaptic vesicles at the synapse.

Voltage-dependent Ca^{2+} channels, responsible for the depolarization-secretion coupling, are also crucial elements at sites of exocytosis (Robitaille et al., 1990). A direct connection between plasma membrane Ca^{2+} channels and docked vesicles is suggested by the reported interaction of these channels with two synaptic vesicle proteins, synaptotagmin (Charvin et al., 1997; Sheng et al., 1997) and cysteine string proteins (Mastrogiacomo et al., 1994; Leveque et al., 1998). Such a link would allow ready-to-fuse vesicles to be exposed to transient, but very high, peaks of cytosolic Ca^{2+} following depolarization-dependent channel opening (Smith and Augustine, 1988). In freeze-fracture views the presynaptic membrane is characterized by the presence of peculiar arrays of large intramembranous particles. The shape of these arrays varies in different presynaptic membranes, but it generally reflects the distribution of docked synaptic vesicles (Heuser et al., 1974; Peper et al., 1974; Ceccarelli et al., 1979). It has therefore been proposed that these particles represent Ca^{2+} channels (Robitaille et al., 1990).

Based on the selective presynaptic action of α-latrotoxin, a component of black widow spider venom, it seemed likely that its receptor would be a unique component of the presynaptic membrane (Valtorta et al., 1984). Subsequent studies on α-latrotoxin-binding sites have led to the identification of a family of transmembrane cell adhesion proteins—the neurexins—that have multiple localizations in the nervous system (Ushkaryov et al., 1992; Südhof, this volume). A G-protein-coupled receptor for α-latrotoxin has also been identified, although its precise localization in the brain has not yet been determined (Lang et al., 1998; Ichtchenko et al., 1999). Some members of the neurexin family bind through their extracellular domains to neuroligins, transmembrane proteins of postsynaptic membranes (Ichtchenko et al., 1996; Song et al., 1999). Furthermore, the cytoplasmic domains of both the neurexins and the neuroligins bind PDZ domains of the membrane-associated guanylate kinase (MAGUK) family of proteins—CASK and PSD-95, respectively (Butz et al., 1998).

MAGUK family members are modular proteins that play a key role in the organization of submembranous scaffolds at a variety of intercellular junctions. These findings are therefore consistent with the other similarities in the molecular interactions that underlie synaptic and other types of intercellular junctions, with the notable difference that synaptic contacts are asymmetric in nature. CASK binds presynaptic cytosolic proteins that participate in neurosecretion (Butz et al., 1998). PSD-95 binds postsynaptic ionotropic receptors (Sheng and Kim, 1996; Sheng, this volume). Thus the neurexin-neuroligin pair may function as a major organizer of the synaptic junction (see Südhof, this volume).

Besides the neurexins, two other classes of cell adhesion molecules have been implicated in the specific adhesion function at synapses—the syndecans and the cadherins. Syndecans are transmembrane proteoglycans that bind the PDZ domain of CASK (Hsueh and Sheng, 1999). Cadherins participate in homophilic interactions through their extracellular domains and are anchored to the cytoskeleton via a direct link to the catenins at their cytoplasmic side. They may supply, at least in part, "the molecular code necessary for the point-to-point specificity required for central nervous system synapses" (Tanaka et al., 2000). Different classes of cadherins have been detected at synapses (Shapiro and Colman 1999); some of these have been shown to be localized at the synaptic junction proper, and their adhesive properties have been found to be regulated by the functional activity of the synapse (Tanaka et al., 2000). Other cadherins are localized at the outer margins of the synapse (Uchida et al., 1996), where they may contribute to the organization of endocytotic zones.

Finally the presynaptic plasma membrane contains a variety of receptors involved in presynaptic regulation. These include receptors for both peptide neurohormones and nonpeptide neurotransmitters released by synaptic vesicles. Through these receptors, neurotransmitters produce autocrine effects and mediate heterosynaptic modulation (Nakanishi et al., 1997; Vogt and Nicholl, 1999).

Postsynaptic Compartment

The postsynaptic compartment is represented by a patch of plasma membrane containing a packed array of neurotransmitter receptors and by an underlying dense matrix, the postsynaptic density (PSD) (Figs. 2.1 and 2.2). In freeze-fracture preparations, the postsynaptic plasma membrane associated with the PSD has a characteristic appearance owing to the presence of a compact array of large intramembranous particles that probably correspond to the receptors. Within this patch, the receptors may be arranged heterogeneously. For example, at glutamatergic synapses two classes of receptors, usually referred to as N-methyl-D-aspartate (NMDA) and α-amino-3-hydroxy-5-methyl-4-isoxazolepropionic acid (AMPA) receptors, are localized in the central and peripheral regions, respectively (Kharazia and Weinberg, 1997). The postsynaptic membrane also includes other proteins needed for the formation and maintenance of the synaptic junction, such as the partners of the cell adhesion molecules of the presynaptic membrane discussed earlier. These include the neuroligins, syndecans, cadherins, and (at least at some synapses) densin-180 (Apperson et al., 1996).

The PSD is a scaffold of modular proteins that differs in molecular composition and morphology at excitatory and inhibitory synapses. Some of its major components include Cam Kinase II, PSD-95, PSD-93, SAP-102, GKAP, CRIPT, SynGAP, Homer, and gephyrin (Ziff, 1997; Sheng, this volume). The differential expression of these components at excitatory and inhibitory synapses must account, at least in part, for the differences in the ultrastructural appearance of the postsynaptic densities in the two types of synapses. These scaffolding proteins mediate receptor clustering, regulate receptor function, control receptor internalization and turnover, link the postsynaptic membrane to the cytoskeleton, have signaling roles, and coordinate electrical responses induced by neurotransmitter gating of ion channels with longer-lasting cellular responses.

Dendritic Spines

In many cortical synapses, the postsynaptic site is localized to a small outpocketing of the postsynaptic cell called a spine (Gray, 1959b; Spacek, 1985; Fig. 2.12). The vast majority of such spines are located on peripheral dendrites, but certain neurons also form spines on their proximal dendrites and even the cell body. The term *spines* is derived from the thorny appearance that they confer to the dendritic surface in low-power views of metal-impregnated or dye-filled dendrites. Dendritic spines are the locus of the vast majority of excitatory synapses in the CNS, although some spines can form synapses with inhibitory axons.

Most spines are shorter than 2 μm in length and have an enlarged head about 1 μm in diameter and a relatively thin stalk or neck (Harris and Kater, 1994). Spines differ in shape and size, but neurons of the same type are usually characterized by the same dendritic pattern of spines. Their presence greatly enhances the dendritic surface area. In addition they compartmentalize the dendritic arbor chemically and electrically. The narrow neck limits the diffusion of ions and chemical messengers, thus playing a role in the induction of input-specific plasticity. There is evidence that the diameter of the neck is subject to regulation (Segal et al., 2000), thus representing an important site for the modulation of the flow of signals from the spine to the dendritic shaft (Harris and Kater, 1994).

Spines are provided with a prominent actin-myosin-based cytoskeleton (Matus, 1999). This may explain the intense spine motility recently demonstrated by video microscopy (Fischer et al., 1998; van Rossum and Hanisch, 1999). Several actin-binding proteins are present or enriched in spines, including spinophillin-neurabin (Nakanishi et al., 1997; Burnett

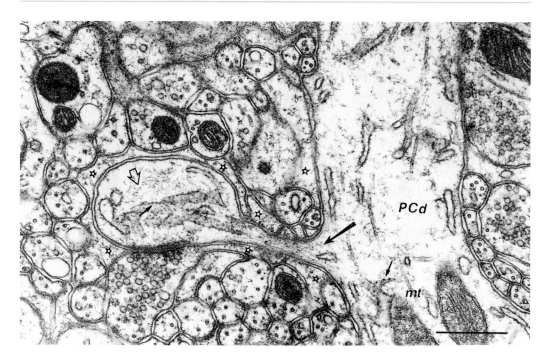

Figure 2.12. Morphology of a dendritic spine on a Purkinje cell dendrite (PCd) in the cerebellum. The spine receives a synapse from a parallel fiber axon. Actin forms a lattice in the spine head (open arrow) and parallel microfilaments in the spine neck (thick arrow). Small arrows point to dense spots on the surface of the ER that do not represent ribosomes but the large cytoplasmic domains of InsP$_3$ receptors. These Ca^{2+} channels are extremely abundant in Purkinje cells. Microtubules (mt) are confined to the shaft of the dendrite. Side branches of the astrocytic Bergmann fibers (stars) surround the synaptic profiles, with the exception of the synaptic apposition. Scale bar = 0.5 μm.

et al., 1998; Satoh et al., 1998), α-actinin (Ziff, 1997), synaptopodin (Mundel et al., 1997), and drebrin (Hayashi and Shirao, 1999). The actin cytoskeleton is directly anchored to the postsynaptic receptor cluster and regulates receptor function (Rosenmund and Westbrook, 1993; Allison et al., 1998; Krupp et al., 1999). Conversely receptor function regulates actin dynamics as well as the shape of the spine (Buonomano and Merzenich, 1998; Matus, 1999; Kins et al., 2000). Spines do not contain neurofilaments and microtubules, although components of the PSD have been shown to interact biochemically with tubulin and microtubule-associated proteins (MAPs) (Kins et al., 2000).

Spines contain at least two classes of intracellular membranous organelles. The predominant membranes are elements of the smooth ER. In pyramidal neurons of the cerebral cortex and medium-sized spiny neurons in the striatum, this reticulum forms a peculiar structure termed the spine apparatus (Gray, 1959b; Spacek, 1985). One of the important functions of the smooth ER in spines is to contribute to the local regulation of Ca^{2+} signaling and homeostasis via Ca^{2+} channels (Fig. 2.12) and Ca^{2+} transporters in its membrane (Takei et al., 1992; Connor et al., 1994). Other membranous organelles of spines—seen only occasionally in thin sections because of their very low abundance—are exocytotic and endocytotic vesicles that participate in the turnover of integral membrane proteins of the spine plasma membrane. Recent evidence suggests an important role for the clathrin-dependent endocytosis of receptors in the regulation of postsynaptic sensitivity, thus providing these organelles with a new critical significance (Lüscher et al., 1999; Shi et al., 1999; Man et al., 2000). Polyribosomes are often found in the main dendritic shaft near the entry of the spine neck, raising the possibility that the synthesis of spine proteins may be locally regulated by patterns of synaptic activity (Kiebler and Desgroseillers, 2000).

The Synaptic Cleft

The link between the pre- and the postsynaptic membrane is very tight, and the synaptic junction cannot be disrupted without the use of chaotropic agents. The intercellular space—the so-called synaptic cleft—is occupied by a regular array of material of moderate electron density. At central synapses the precise molecular nature of this material is not known, but it must comprise the extracellular domains of many of the pre- and postsynaptic membranes. At cholinergic synapses, it also includes the acetylcholine-degrading enzyme acetylcholinesterase (AchE). The only synapse for which some of the molecular components of the cleft have been identified and characterized is the neuromuscular junction, where the cleft is considerably wider than at central synapses and is occupied by a specialized basal lamina, as discussed subsequently.

Interactions occurring in the synaptic cleft play a critical role in synaptic function beyond their role in bringing the pre- and postsynaptic membrane into precise register. They are needed for the formation of the pre- and postsynaptic specializations and therefore represent the starting point for signaling cascades that lead to synapse formation and stabilization (Craig, 1998; Sanes and Lichtman, 1999). Most likely they also mediate at least some aspects of retrograde signaling, through which the activation of postsynaptic receptors may modulate the efficiency of the presynaptic compartment (Kandel et al., 2000). Finally it is through these interactions that cognate receptors are selected for axons that secrete a given neurotransmitter.

Glia

The intercellular space in the general vicinity of synaptic profiles is marked by the presence of glial cell processes that more or less completely enwrap the related axonal and dendritic processes, excluding of course the apposed pre- and postsynaptic surfaces. Central synapses, with the exception of cholinergic synapses, lack extracellular enzymes for neurotransmitter inactivation, and the rapid reuptake of other transmitters from the extracellular space is therefore particularly important. One of the functions of the surrounding glial elements is to contribute to the rapid clearance of neurotransmitter from the synaptic cleft via neurotransmitter transporters with fast kinetics. This function of glial cells is crucial not only to terminate the postsynaptic response but also to partially shield synapses from each other and from other nearby neuronal processes and thereby to enhance spatial specificity of signaling. In the synaptic glomeruli of the cerebellum (Fig. 2.5), the thalamus, and other centers, the astrocytic glial profiles are situated at the periphery of the arrays—an arrangement that may facilitate crosstalk between different synapses as a result of transmitter spillover (Wall and Usowicz, 1997; Rossi and Hamann, 1998). Glial cell elements also contribute to ionic homeostasis in the synaptic cleft and may have additional regulatory functions.

Model Synapses

Studies of synaptic structure and function have been greatly helped by the availability of model systems that have unique functional properties or are particularly suitable to experimental manipulations.

Neuromuscular Junction

Before the introduction of primary cultures of CNS neurons (Banker and Cowan, 1977), peripheral synapses had been the preferred model for in vitro studies of synaptic function, and among them the motor endplate (in particular, the frog neuromuscular junction) had been the most widely used model (Katz, 1966). Several features made the neuromuscular junction an especially favorable synapse for physiological studies and for systematic correlative analyses of structure and function. These include the experimental accessibility of the junction, its viability after dissection, the suitability of the postsynaptic compartment for intracellular recordings, and the fact that in adult animals each muscle fiber is innervated by a single motor axon.

The cholinergic axons that innervate a muscle fiber usually split into two or more branches that, in mammals, form a rosette (Fig. 2.13A,C), and in amphibians run parallel to the long axis of the muscle fibers (Fig. 2.14A,B). These terminal axon branches form multiple synaptic contacts with the muscle cell, each of which can be seen as corresponding to a varicosity on CNS axons (Fig. 2.13). In amphibian motor endplates, such contacts are represented by stripes perpendicular to the main axis of the fiber (Fig. 2.14C). Freeze-fracture views of the presynaptic plasma membrane underlying these stripes reveal the presence of

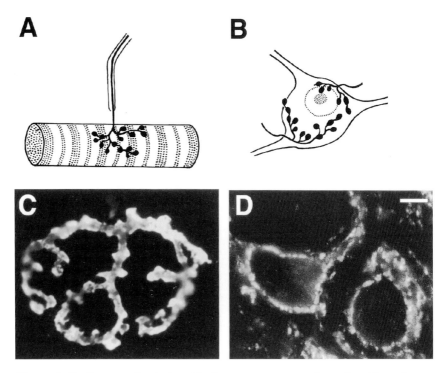

Figure 2.13. Structural relationship between presynaptic varicosities of a rat neuromuscular junction (A–C) and of axosomatic synapses in the rat central nervous system (B–D). Schematic representations (A–B) and immunofluorescence images (C–D) of presynaptic varicosities. Varicosities in C and D were revealed by immunofluorescence for the presynaptic marker synapsin. Scale bar = 5 µm. A and B are partially adapted from Kuffler et al. (1984).

a double row of intramembranous particles (Couteaux and Pecot-Dechavassine, 1970). They are in precise register with a double row of docked vesicles in the underlying cytoplasm (Heuser et al., 1974) and most likely represent Ca^{2+} channels (Robitaille et al., 1990; Fig. 2.14D,E). Accordingly they define sites where exocytosis sites can be seen in freeze-fracture preparations (Ceccarelli et al., 1979; Heuser et al., 1979; Torri-Tarelli et al., 1985; Fig. 2.14E). A dense presynaptic matrix is also located at these stripes, in the space between two rows of vesicles, as clearly seen in conventional thin sections (Fig. 2.14F). Large dense-core vesicles are present, but only in very low number. They represent about 1% of the total vesicle population and have been found to contain calcitonin-gene-related peptide in addition to other peptides (Matteoli et al., 1988).

The postsynaptic membrane of the motor endplate is not flat; rather it is interrupted by deep infoldings that are in precise register with presynaptic active zones and mirror their geometry (see Stiles et al., this volume). The postsynaptic ionotropic receptors at this synapse (nicotinic acetylcholine receptors) are concentrated on the edges of the infoldings

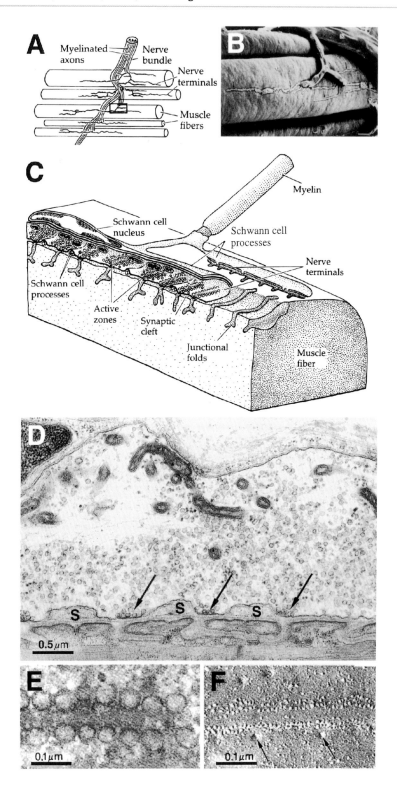

and therefore in direct proximity to the release sites (Heuser et al., 1974). The synaptic cleft is relatively wide and is occupied by a matrix that represents a specialized form of basal lamina. Among other proteins it contains AChE (Rotundo et al., 1998), agrin (McMahan, 1990; Rupp et al., 1991; Gautam et al., 1996), a unique laminin isoform (Martin et al. 1995), and a variety of other proteins that mediate the structural and functional connections between the pre- and postsynaptic compartments. AChE, which inactivates acetylcholine by cleaving it into acetate and choline, is directly associated with the basal lamina via a posttranslationally added collagenlike tail that links its tetrameric catalytic subunit to a heparan sulfate proteoglycan (Silman and Futerman, 1987; Rotundo et al., 1998). Inhibitors of this enzyme are potent neurotoxins and have been used as nerve gases, insecticides, and drugs for the treatment of certain neurological disorders.

In recent years simple model organisms in which mechanisms of synaptic function can be genetically dissected have come to center stage. Both classical and reverse genetics is being increasingly applied to the study of the nervous system in *Drosophila* and *C. elegans*. Even in these organisms, neuromuscular junctions represent optimal synapses for the analysis of synaptic function because of their large size, peculiar morphology, or both; their role in a function that can be easily assessed (motor behavior); and their suitability for electrophysiological recording.

Figure 2.14. (opposite) Morphology of the frog neuromuscular junction.

(A–B) Schematic illustration and scanning electron microscopy image of several muscle fibers and their innervation.

(C) Three-dimensional sketch of a portion of a synaptic contact area. Synaptic vesicles are clustered in the nerve terminal in special regions opposite the opening of the postsynaptic folds. These zones are the sites of neurotransmitter release. Such zones and the corresponding synaptic folds are arranged in stripes perpendicular to the main axis of the muscle fiber. Processes of Schwann cells, also perpendicular to the muscle axis, penetrate the region between the nerve terminal and the postsynaptic membrane, separating active zones.

(D) Electron micrograph of a longitudinal section of the junction, showing many of the features illustrated in *C,* including active zones (arrows) and Schwann cell processes (S). Scale bar = 0.5 μm.

(E) Cytoplasmic side of an active zones cut en face shows the presence of synaptic vesicles aligned in two rows at the outer edge of an electron-dense stripe. Scale bar = 0.1 μm.

(F) Freeze-fracture view of the cytoplasmic half (P face) of the presynaptic membrane of an active zone. The two rows of membrane particles are in precise register with the two rows of vesicles in the underlying cytoplasm. Small pits (arrows) represent exocytotic opening of synaptic vesicles in the process of fusion. Scale bar = 0.1 μm.

Panels *A–C* are reproduced with permission from Kuffler et al. (1984); *D* is courtesy of Dr. B. Ceccarelli; *E* is reproduced from Couteaux and Pecot-Dechavassine (1970); *F* is reproduced from Heuser et al. (1974), with permission from Elsevier Science.

Knowledge of the structure of these synapses is therefore becoming important. As in vertebrate motor endplates, the motor axons of *Drosophila* split into several short branches at the surface of the muscle, and these branches generate rows of synaptic boutons. These terminals use glutamate as the primary neurotransmitter, but subsets of neurons also secrete octopamine and peptides as co-transmitters (Gramates and Budnik, 1999). Based on their morphology (size, branching pattern, infolding of the postsynaptic membrane, ratio of small to large dense-core vesicles), these nerve terminals can be divided into three main types. The morphology of the active zone of type I nerve terminals has been extensively studied. It is characterized by T-bars, that is, presynaptic dense bodies, so called because of their appearance in cross sections as seen by transmission electron microscopy (Fig. 2.8A,B). Synaptic vesicles are located above and to the sides of the dense body, as well as in the cytoplasm surrounding it. The dense body may function in the delivery of vesicles to the release sites and may play a function similar to that of the presynaptic grid at vertebrate synapses and of the synaptic ribbon at certain sensory synapses. Typically only a small percentage of all sections through the active zone have a vesicle docked under the dense body at a position compatible with rapid release (Koenig et al., 1993).

Ribbon Synapses

A very specialized type of synapse, the so-called ribbon synapse, is found in the receptor cells of certain sensory organs. These are the cells that either mediate signal transduction (e.g., the retinal photoreceptors and hair cells of the inner ear) or are directly downstream to the primary receptors (as in the bipolar cells of the retina). They do not have a bona fide axon and do not generate action potentials. Instead their membrane potential undergoes graded changes (receptor potentials) that are proportional to the intensity of the sensory stimulus and tonically regulate the rate of synaptic vesicle exocytosis (Dowling, 1987). In photoreceptors and the bipolar cells of the retina, release sites are localized at the end of a short process and are marked by the presence of an electron-dense presynaptic specialization. This specialization appears as a bar perpendicular to the membrane in cross section, but as a ribbon (hence the name) in three-dimensional reconstructions (Rao-Mirotznik et al., 1995; Fig. 2.15). Synaptic vesicles fill the space surrounding the membrane, and a pool of vesicles appears to be directly anchored to the ribbon. In the vestibular and cochlear hair cells of the inner ear, release sites are localized at the basolateral surface of the cell body and are marked by the presence of electron-dense spheres (dense bodies), which are thought to be chemically and functionally similar to the ribbons of retinal synapses—hence the general name *ribbon synapses* is often used for all these synapses. The dense bodies are decorated by docked synaptic vesicles and are surrounded by a halo of undocked vesicles. They are directly anchored to the cell membrane, and the contact sites in the

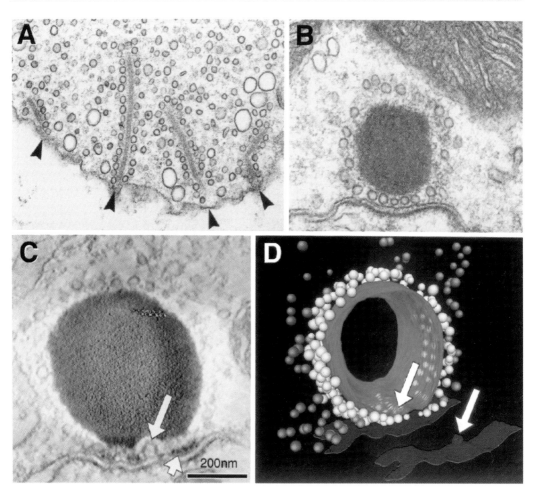

Figure 2.15. Ultrastructural features of ribbon synapses of the retina (A) and vestibular hair cells (C–D).

(A) Electron micrograph of the release area (pedicle) of an isolated salamander retinal rod cell, demonstrating the presence of ribbons (electron-dense rods in the section; arrowheads) with attached synaptic vesicles. Reproduced with permission from Townes-Anderson et al. (1985).

(B) Thin section of a synaptic dense body in a saccular hair cell of the goldfish. Reproduced with permission from Hama and Saito (1977).

(C–D) Electron tomography of the synaptic dense body of a frog saccular hair cell. *C* shows an electron micrograph and *D* a three-dimensional reconstruction. The long white arrows point to the same exocytotic image shown in *D,* in the whole reconstruction as well as in isolation. The dense body is shown in blue, dense body–associated vesicles in yellow, surrounding vesicles in green, and the regions of the plasma membrane that form the synaptic cleft in red. Panel D reproduced with permission from Lenzi et al. (1999).

plasma membrane are characterized by peculiar patches of intra-membranous particles, most likely Ca^{2+} channels (Hama and Saito, 1977; Lenzi et al., 1999). Ribbon synapses, which function tonically, may sustain very high rates of vesicle exocytosis. The presynaptic ribbon and dense bodies may facilitate the rapid replacement of fused vesicles at release sites.

Giant Synapses

Synapses of giant axons are particularly useful experimental models for the study of synaptic transmission because of their suitability for the introduction of microinjection or recording pipettes into the presynaptic processes. Two such synapses, which have been extensively used for functional and structural studies, are the giant synapse of the squid and the synapses of the giant reticulospinal axon of the lamprey.

The giant synapse of the squid represents the contact between a very large axon terminal and a very large postsynaptic element. However, the true synaptic interface is represented by a large number of small individual synaptic patches, which are very similar to conventional synapses. At each of these sites the postsynaptic compartment forms an outpocketing that contacts the presynaptic axon (Llinas, 1999).

The giant axon of the lamprey is a very large, rectilinear axon that forms numerous en passant synapses with postsynaptic cells along its course (Fig. 2.16). The large number and tight clustering of synaptic vesicles at each of the presynaptic sites, as well as other peculiar structural features of these synapses (phasic properties with the virtual absence

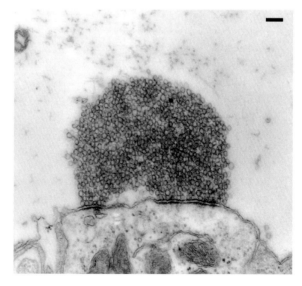

Figure 2.16. Electron micrograph of a reticulospinal synapse. This synapse is characterized by an extremely large synaptic vesicle cluster. Scale bar = 200 nm. Courtesy of Drs. Oleg Shupliakov and Lennart Brodin, Karolinska Institute, Stockholm.

of endocytotic intermediates at rest; absence of other organelles in the proximity of the vesicle cluster; lack of boundaries to the dispersion of vesicles upon experimental disruption of the clustering mechanisms) have made them a powerful model for studying the cell biology of synaptic vesicle recycling (Shupliakov and Brodin, 2000).

Structure-Function Relationships

The description of synaptic morphology summarized in this chapter makes it clear that the structural features of synapses are optimally designed to maximize spatial precision and rapidity of signal transmission. Spatial precision is achieved by the selective exocytosis of synaptic vesicles at active zones, by the localization of these sites in close proximity to clusters of ionotropic receptors on the postsynaptic membrane, and by the presence of mechanisms that prevent the lateral diffusion of neurotransmitters. Speed is achieved presynaptically by a very fast stimulus-secretion coupling that capitalizes on a pool of vesicles already docked at fusion sites. As these vesicles undergo exocytosis, they are rapidly replaced by vesicles from a reserve pool. This large reserve pool of vesicles allows reliable function even during sustained stimulation at very high frequencies (Brodin et al. 1997). Postsynaptically speed is achieved by fast chemoelectrical transmission mediated by ionotropic receptors.

Chemical messengers implicated in synaptic transmission are characterized by low affinity and fast "on-off" rates, a property that allows their selective and rapidly reversible effects only at sites where they are generated. For example, the synaptic vesicle fusion machinery responds only to the high (100 μM range), but very transient, rise in cytosolic Ca^{2+} that occurs at or near the mouth of Ca^{2+} channels. These localized Ca^{2+} transients are rapidly reversed by channel closure, diffusion, cytosolic Ca^{2+} buffering systems, and active extrusion of Ca^{2+} from the cytosol (Smith and Augustine, 1988). Likewise ionotropic neurotransmitter receptors localized on the postsynaptic membranes are optimally activated by the high, but very transient, concentration of neurotransmitter achieved in the synaptic cleft after each secretory event. This concentration decays very rapidly, not only through diffusion and reuptake mechanisms in presynaptic processes and glia but also, in some cases (e.g., acetylcholine), by extracellular catabolism. The transient and local action of chemical messengers at the synapse is responsible for the rapid reversibility of synaptic responses that enables synapses to fire at high frequency. Furthermore, the steep temporal and spatial decay of the concentration of extracellular neurotransmitters makes it possible for synapses to use only very few "fast" neurotransmitters and yet to act independently of one another despite the dense packing of synapses in the CNS. High-frequency firing may result in a tonic overall increase of neurotransmitter beyond the synaptic cleft, resulting in the activation of metabotropic receptors. These receptors have higher affinities for neurotransmitters than ionotropic receptors and are generally not concentrated at postsynaptic densities. The slow, modulatory action of metabotropic receptors cooperates with modulatory signals elicited by secreted neuropeptides.

Finally the tight physical connection between pre- and postsynaptic compartments mediates an important form of bidirectional signaling that is likely to play a key role not only in the development and maintenance of the synapse but also in its structural and functional plasticity (Tanaka et al., 2000).

Acknowledgments

We thank Henry Tan for help with the figures and Laurie Daniell for providing the electron micrographs shown in Figs 2.1, 2.3B, 2.8C, and 2.10B.

References

Ahmari, S. E., Buchanan, J., Smith, S. J. (2000). Assembly of presynaptic active zones from cytoplasmic transport packets. *Nature Neurosci.* 3:445–451.

Akert, K., Pfenninger, K., Sandri, C., Moor, H. (1972). Freeze etching and cytochemistry of vesicles and membrane complexes in synapses of the central nervous system. In *Structure and function of synapses,* Pappas, G. D., Purpura, D. P., eds. New York: Raven Press. pp. 67–86.

Allison, D. W., Gelfand, V. I., Spector, I., Craig, A. M. (1998). Role of actin in anchoring postsynaptic receptors in cultured hippocampal neurons: differential attachment of NMDA versus AMPA receptors. *J. Neurosci.* 18:2423–2436.

Amchenkova, A. A., Bakeeva, L. E., Chentsov, Y. S., Skulachev, V. P., Zorov, D. B. (1988). Coupling membranes as energy-transmitting cables. I. Filamentous mitochondria in fibroblasts and mitochondrial clusters in cardiomyocytes. *J. Cell Biol.* 107:481–495.

Andres, K. (1975). Morphological criteria for the differentiation of synapses in vertebrates. *J. Neurol. Transm.* 12:1–37.

Apperson, M. L., Moon, I. S., Kennedy, M. B. (1996). Characterization of densin-180, a new brain-specific synaptic protein of the O-sialoglycoprotein family. *J. Neurosci.* 16:6839–6852.

Bakeeva, L. E., Chentsov, Y. S., Skulachev, V. P. (1983). Intermitochondrial contacts in myocardiocytes. *J. Mol. Cell Cardiol.* 15:413–420.

Bakeeva, L. E., Kirillova, G. P., Kolesnikova, O. V., Konoshenko, G. I., Mokhova, E. N. (1985). Effect of palmitate on energy coupling in lymphocyte mitochondria. *Biokhimiia* 50:774–781.

Banker, G. A., Cowan, W. M. (1977). Rat hippocampal neurons in dispersed cell culture. *Brain Res.* 126:397–442.

Bennett, M. V. (2000). Seeing is relieving: electrical synapses between visualized neurons. *Nature Neurosci.* 3:7–9.

Berridge, M. J. (1993). Inositol trisphosphate and calcium signalling. *Nature* 361: 315–325.

Betz, W. J., Henkel, A. W. (1994). Okadaic acid disrupts clusters of synaptic vesicles in frog motor nerve terminals. *J. Cell Biol.* 124:843–854.

Betz, W. J., Bewick, G. S., Ridge, R. M. (1992). Intracellular movements of fluorescently labeled synaptic vesicles in frog motor nerve terminals during nerve stimulation. *Neuron* 9:805–813.

Bloom, F. E., Aghajanian, G. K. (1966). Cytochemistry of synapses: selective staining for electron microscopy. *Science* 154:1575–1577.

Bloom, F. E., Iversen, L. L., Schmitt, F. O. (1970). Macromolecules in synaptic function. *Neurosci. Res. Program Bull.* 8:325–455.

Brandstatter, J. H., Fletcher, E. L., Garner, C. C., Gundelfinger, E. D., Wassle, H. (1999). Differential expression of the presynaptic cytomatrix protein bassoon among ribbon synapses in the mammalian retina. *Eur. J. Neurosci.* 11:3683–3693.

Brodin, L., Low, P., Gad, H., Gustafsson, J., Pieribone, V. A., Shupliakov, O. (1997). Sustained neurotransmitter release: new molecular clues. *Eur. J. Neurosci.* 9:2503–2511.

Brodin, L., Bakeeva, L., Shupliakov, O. (1999). Presynaptic mitochondria and the temporal pattern of neurotransmitter release. *Philos. Trans. R. Soc. London Ser. B.* 354:365–372.

Bruns, D., Jahn, R. (1995). Real-time measurement of transmitter release from single synaptic vesicles. *Nature* 377:62–65.

Buonomano, D. V., Merzenich, M. M. (1998). Cortical plasticity: from synapses to maps. *Annu. Rev. Neurosci.* 21:149–186.

Burgess, T. L., Kelly, R. B. (1987). Constitutive and regulated secretion of proteins. *Annu. Rev. Cell Biol.* 3:243–293.

Burnett, P. E., Blackshaw, S., Lai, M. M., Qureshi, I. A., Burnett, A. F., Sabatini, D. M., Snyder, S. H. (1998). Neurabin is a synaptic protein linking p70 S6 kinase and the neuronal cytoskeleton. *Proc. Natl. Acad. Sci. USA* 95:8351–8356.

Butz, S., Okamoto, M., Südhof, T. C. (1998). A tripartite protein complex with the potential to couple synaptic vesicle exocytosis to cell adhesion in brain. *Cell* 94:773–782.

Calakos, N., Scheller, R. H. (1996). Synaptic vesicle biogenesis, docking, and fusion: a molecular description. *Physiol. Rev.* 76:1–29.

Cases-Langhoff, C., Voss, B., Garner, A. M., Appeltauer, U., Takei, K., Kindler, S., Veh, R. W., De Camilli, P., Gundelfinger, E. D., Garner, C. C. (1996). Piccolo, a novel 420 kDa protein associated with the presynaptic cytomatrix. *Eur. J. Cell Biol.* 69:214–223.

Ceccarelli, B., Hurlbut, W. P., Mauro, A. (1973). Turnover of transmitter and synaptic vesicles at the frog neuromuscular junction. *J. Cell Biol.* 57:499–524.

Ceccarelli, B., Grohovaz, F., Hurlbut, W. P. (1979). Freeze-fracture studies of frog neuromuscular junctions during intense release of neurotransmitter. I. Effects of black widow spider venom and Ca^{2+}-free solutions on the structure of the active zone. *J. Cell Biol.* 81:163–177.

Charvin, N., L'Eveque, C., Walker, D., Berton, F., Raymond, C., Kataoka, M., Shoji-Kasai, Y., Takahashi, M., De Waard, M., Seagar, M. J. (1997). Direct interaction of the calcium sensor protein synaptotagmin I with a cytoplasmic domain of the alpha1A subunit of the P/Q-type calcium channel. *EMBO J.* 16:4591–4596.

Cochilla, A. J., Angleson, J. K., Betz, W. J. (1999). Monitoring secretory membrane with FM1-43 fluorescence. *Annu. Rev. Neurosci.* 22:1–10.

Collonnier, M. (1968). Synaptic patterns on different cell types in the different laminae of the cat visual cortex. *Brain Res.* 9:268–287.

Connor, J. A., Miller, L. D., Petrozzino, J., Muller, W. (1994). Calcium signaling in dendritic spines of hippocampal neurons. *J. Neurobiol.* 25:234–242.

Couteaux, R., Pecot-Dechavassine, M. (1970). Synaptic vesicles and pouches at the level of "active zones" of the neuromuscular junction. *C. R. Acad. Sci. Ser. D* 271:2346–2349.

Craig, A. M. (1998). Activity and synaptic receptor targeting: the long view. *Neuron* 21:459–462.

Craig, A. M., Wyborski, R. J., Banker, G. (1995). Preferential addition of newly synthesized membrane protein at axonal growth cones. *Nature* 375:592–594.

Cremona, O., Di Paolo, G., Wenk, M. R., Luthi, A., Kim, W. T., Takei, K., Daniell, L., Nemoto, Y., Shears, S. B., Flavell, R. A., McCormick, D. A., De Camilli, P. (1999). Essential role of phosphoinositide metabolism in synaptic vesicle recycling. *Cell* 99:179–188.

De Camilli, P., Jahn, R. (1990). Pathways to regulated exocytosis in neurons. *Annu. Rev. Physiol.* 52:625–645.

De Camilli, P., Benfenati, F., Valtorta, F., Greengard, P. (1990). The synapsins. *Annu. Rev. Cell Biol.* 6:433–460.

Dowling, J. E., Boycott, B. B. (1966). Organization of the primate retina: electron microscopy. *Proc. R. Soc. London Ser. B* 166:80–111.

Eberhard, D. A., Cooper, C. L., Low, M. G., Holz, R. W. (1990). Evidence that the inositol phospholipids are necessary for exocytosis. Loss of inositol phospholipids and inhibition of secretion in permeabilized cells caused by a bacterial phospholipase C and removal of ATP. *Biochem. J.* 268:15–25.

Esser, L., Wang, C. R., Hosaka, M., Smagula, C. S., Südhof, T. C., Deisenhofer, J. (1998). Synapsin I is structurally similar to ATP-utilizing enzymes. *EMBO J.* 17:977–984.

Fenster, S., Chung, W., Cases-Langhoff, C., Zhai, R., Voss, B., Garner, A. M., Kaempf, U., Kindler, S., Gundelfinger, E. D., Garner, C. C. (2000). The novel presynaptic zinc finger protein piccolo is a structural homolog of bassoon and interacts with PRA1 in nerve terminals. *Neuron* 25:203–214.

Fischer, M., Kaech, S., Knutti, D., Matus, A. (1998). Rapid actin-based plasticity in dendritic spines. *Neuron* 20:847–854.

Fletcher, T. L., De Camilli, P., Banker, G. (1994). Synaptogenesis in hippocampal cultures: evidence indicating that axons and dendrites become competent to form synapses at different stages of neuronal development. *J. Neurosci.* 14:6695–6706.

Fuxe, K., Agnati, L. F. (1991). *Volume Transmission in the Brain,* Vol. 1. New York: Raven Press.

Gad, H., Shupliakov, O., Low, P., Kjaerluff, O., Ringstad, N., De Camilli, P., Brodin, L. (1999). Perturbation of synaptojanin function impairs synaptic vesicle endocytosis at a living synapse. In *Annual Meeting of the Society for Neuroscience,* Vol. 25. Washington, D.C.: Society for Neuroscience. p. 1744.

Garcia, E. P., McPherson, P. S., Chilcote, T. J., Takei, K., De Camilli, P. (1995). rbSec1A and B colocalize with syntaxin 1 and SNAP-25 throughout the axon, but are not in a stable complex with syntaxin. *J. Cell Biol.* 129:105–120.

Gautam, M., Noakes, P. G., Moscoso, L., Rupp, F., Scheller, R. H., Merlie, J. P., Sanes, J. R. (1996). Defective neuromuscular synaptogenesis in agrin-deficient mutant mice. *Cell* 85:525–535.

Gramates, S. L., Budnik, V. (1999). Assembly and maturation of the Drosophila larval neuromuscular junction. *Int. Rev. Neurobiol.* 43:93–117.

Gray, E. (1959a). Axo-somatic and axo-dendritic synapses of the cerebral cortex: an electron microscopic study. *J. Anat.* 93:420–432.

Gray, E. G. (1959b). Electron microscopy of synaptic contacts on spines of dendrites of the cerebellar cortex. *Nature* 183:1592–1593.

Gray, E. G. (1963). Electron microscopy of presynaptic organelles of the spinal cord. *J. Anat.* 97:101–106.

Gray, E. G. (1983). Neurotransmitter release mechanisms and microtubules. *Proc. R. Soc. London Ser. B* 218:253–258.

Greengard, P., Valtorta, F., Czernik, A. J., Benfenati, F. (1993). Synaptic vesicle phosphoproteins and regulation of synaptic function. *Science* 259:780–785.

Gustaffson, J., Shupliakov, O., Takei, K., Low, P., De Camilli, P., Brodin, L. (1998). GTPgS induces an actin matrix associated with coated endocytic intermediates in presynaptic regions. In *Annual Meeting of the Society for Neuroscience*, Vol. 24. Washington, D.C.: Society for Neuroscience. p. 823.

Hama, K., Saito, K. (1977). Fine structure of the afferent synapse of the hair cells in the saccular macula of the goldfish, with special reference to the anastomosing tubules. *J. Neurocytol.* 6:361–373.

Hannah, M. J., Schmidt, A. A., Huttner, W. B. (1999). Synaptic vesicle biogenesis. *Annu. Rev. Cell Dev. Biol.* 15:733–798.

Harris, K. M., Kater, S. B. (1994). Dendritic spines: cellular specializations imparting both stability and flexibility to synaptic function. *Annu. Rev. Neurosci.* 17:341–371.

Harris, K. M., Sultan, P. (1995). Variation in the number, location and size of synaptic vesicles provides an anatomical basis for the nonuniform probability of release at hippocampal CA1 synapses. *Neuropharmacology* 34:1387–1395.

Hartinger, J., Stenius, K., Hogemann, D., Jahn, R. (1996). 16-BAC/SDS-PAGE: a two-dimensional gel electrophoresis system suitable for the separation of integral membrane proteins. *Anal. Biochem.* 240:126–133.

Hawthorne, J. N., Pickard, M. R. (1979). Phospholipids in synaptic function. *J. Neurochem.* 32:5–14.

Hayashi, K., Shirao, T. (1999). Change in the shape of dendritic spines caused by overexpression of drebrin in cultured cortical neurons. *J. Neurosci.* 19: 3918–3925.

Henkel, A. W., Simpson, L. L., Ridge, R. M., Betz, W. J. (1996). Synaptic vesicle movements monitored by fluorescence recovery after photobleaching in nerve terminals stained with FM1-43. *J. Neurosci.* 16:3960–3967.

Heuser, J. E., Reese, T. S. (1973). Evidence for recycling of synaptic vesicle membrane during transmitter release at the frog neuromuscular junction. *J. Cell Biol.* 57:315–344.

Heuser, J. E., Reese, T. (1977). Structure of the synapse. In *Handbook of Physiology*, Vol. 1, Kandel, E. R., ed. Bethesda, Md.: American Physiology Society. pp. 261–294.

Heuser, J. E., Reese, T. S., Landis, D. M. (1974). Functional changes in frog neuromuscular junctions studied with freeze-fracture. *J. Neurocytol.* 3:109–131.

Heuser, J. E., Reese, T. S., Dennis, M. J., Jan, Y., Jan, L., Evans, L. (1979). Synaptic vesicle exocytosis captured by quick freezing and correlated with quantal transmitter release. *J. Cell Biol.* 81:275–300.

Hirokawa, N., Sobue, K., Kanda, K., Harada, A., Yorifuji, H. (1989). The cytoskeletal architecture of the presynaptic terminal and molecular structure of synapsin 1. *J. Cell Biol.* 108:111–126.

Hirokawa, N., Noda, Y., Okada, Y. (1998). Kinesin and dynein superfamily proteins in organelle transport and cell division. *Curr. Opin. Cell Biol.* 10:60–73.

Hökfelt, T. (1968). In vitro studies on central and peripheral monoamine neurons at the ultrastructural level. *Z. Zellforsch. Mikrosk. Anat.* 91:1–74.

Hökfelt, T., Johansson, O., Ljungdahl, A., Lundberg, J. M., Schultzberg, M. (1980). Peptidergic neurones. *Nature* 284:515–521.

Hökfelt, T., Fried, G., Hansen, S., Holets, V., Lundberg, J. M., Skirboll, L. (1986). Neurons with multiple messengers—distribution and possible functional significance. *Prog. Brain Res.* 65:115–137.

Hsueh, Y. P., Sheng, M. (1999). Regulated expression and subcellular localization of syndecan heparan sulfate proteoglycans and the syndecan-binding protein CASK/LIN-2 during rat brain development. *J. Neurosci.* 19:7415–7425.

Hunt, J. M., Bommert, K., Charlton, M. P., Kistner, A., Habermann, E., Augustine, G. J., Betz, H. (1994). A post-docking role for synaptobrevin in synaptic vesicle fusion. *Neuron* 12:1269–1279.

Huttner, W. B., Schiebler, W., Greengard, P., De Camilli, P. (1963). Synapsin I (protein I), a nerve terminal–specific phosphoprotein. III. Its association with synaptic vesicles studied in a highly purified synaptic vesicle preparation. *J. Cell Biol.* 96:1374–1388.

Ichtchenko, K., Nguyen, T., Südhof, T. C. (1996). Structures, alternative splicing, and neurexin binding of multiple neuroligins. *J. Biol. Chem.* 271:2676–2682.

Ichtchenko, K., Bittner, M. A., Krasnoperov, V., Little, A. R., Chepurny, O., Holz, R. W., Petrenko, A. G. (1999). A novel ubiquitously expressed alpha-latrotoxin receptor is a member of the CIRL family of G-protein-coupled receptors. *J. Biol. Chem.* 274:5491–5498.

Jahn, R., Niemann, H. (1994). Molecular mechanisms of clostridial neurotoxins. *Ann. N.Y. Acad. Sci.* 733:245–255.

Jahn, R., Südhof, T. C. (1994). Synaptic vesicles and exocytosis. *Annu. Rev. Neurosci.* 17:219–246.

Jahn, R., Hell, J., Maycox, P. R. (1990). Synaptic vesicles: key organelles involved in neurotransmission. *J. Physiol. (Paris)* 84:128–133.

Jorgensen, E. M., Nonet, M. (1995). Neuromuscular junctions in the nematode *C. elegans. Sem. Dev. Biol.* 6:207–220.

Kandel, E. R., Schwartz, J. H., Jessell, T. M. (2000). *Principles of Neural Science,* 4th ed. New York: McGraw-Hill.

Katz, B. (1962). The transmission of impulses from nerve to muscle and the subcellular unit of synaptic action. *Proc. R. Soc. London Ser. B* 155:455–477.

Katz, B. (1966). *Nerve, Muscle and Synapse.* New York: McGraw-Hill.

Kharazia, V. N., Weinberg, R. J. (1997). Tangential synaptic distribution of NMDA and AMPA receptors in rat neocortex. *Neurosci. Lett.* 238:41–44.

Kiebler, M. A., Desgroseillers, L. (2000). Molecular insights into mRNA transport and local translation in the mammalian nervous system. *Neuron* 25:19–28.

Kins, S., Betz, H., Kirsch, J. (2000). Collybistin, a newly identified brain-specific GEF, induces submembrane clustering of gephyrin. *Nature Neurosci.* 3:22–29.

Koenig, J. H., Ikeda, K. (1989). Disappearance and reformation of synaptic vesicle membrane upon transmitter release observed under reversible blockage of membrane retrieval. *J. Neurosci.* 9:3844–3860.

Koenig, J. H., Yamaoka, K., Ikeda, K. (1993). Calcium-induced translocation of synaptic vesicles to the active site. *J. Neurosci.* 13:2313–2322.

Koenig, J. H., Yamaoka, K., Ikeda, K. (1998). Omega images at the active zone may be endocytotic rather than exocytotic: implications for the vesicle hypothesis of transmitter release. *Proc. Natl. Acad. Sci. USA* 95:12,677–12,682.

Kraszewski, K., Daniell, L., Mundigl, O., De Camilli, P. (1996). Mobility of synaptic vesicles in nerve endings monitored by recovery from photobleaching of synaptic vesicle–associated fluorescence. *J. Neurosci.* 16:5905–5913.

Krupp, J. J., Vissel, B., Thomas, C. G., Heinemann, S. F., Westbrook, G. L. (1999). Interactions of calmodulin and alpha-actinin with the NR1 subunit modulate Ca^{2+}-dependent inactivation of NMDA receptors. *J. Neurosci.* 19:1165–1178.

Kuffler, S. W., Nicholls, J. G., Martin, A. R. (1984). *From Neuron to Brain,* 2nd ed. Sunderland, Mass.: Sinauer Associates.

Kuznetsov, S. A., Langford, G. M., Weiss, D. G. (1992). Actin-dependent organelle movement in squid axoplasm. *Nature* 356:722–725.

Lang, J., Ushkaryov, Y., Grasso, A., Wollheim, C. B. (1998). Ca^{2+}-independent insulin exocytosis induced by alpha-latrotoxin requires latrophilin, a G protein–coupled receptor. *EMBO J.* 17:648–657.

Lasek, R. J., Garner, J. A., Brady, S. T. (1984). Axonal transport of the cytoplasmic matrix. *J. Cell Biol.* 99:212s–221s.

Lenzi, D., Runyeon, J. W., Crum, J., Ellisman, M. H., Roberts, W. M. (1999). Synaptic vesicle populations in saccular hair cells reconstructed by electron tomography. *J. Neurosci.* 19:119–132.

Leveque, C., Pupier, S., Marqueze, B., Geslin, L., Kataoka, M., Takahashi, M., De Waard, M., Seagar, M. (1998). Interaction of cysteine string proteins with the alpha1A subunit of the P/Q-type calcium channel. *J. Biol. Chem.* 273:13,488–13,492.

Li, L., Chin, L. S., Shupliakov, O., Brodin, L., Sihra, T. S., Hvalby, O., Jensen, V., Zheng, D., McNamara, J. O., Greengard, P., Andersen, P. (1995). Impairment of synaptic vesicle clustering and of synaptic transmission, and increased seizure propensity, in synapsin I–deficient mice. *Proc. Natl. Acad. Sci. USA* 92: 9235–9239.

Llinas, R. R. (1999). *The Squid Synapse: A Model for Chemical Transmission.* Oxford: Oxford University Press.

Lüscher, C., Xia, H., Beattie, E. C., Carroll, R. C., von Zastrow, M., Malenka, R. C., and Nicholl, R. A. (1999). Role of AMPA receptor cycling in synaptic transmission and plasticity. *Neuron* 24:649–658.

Man, H. Y., Liu, J. W., Ju, W. H., Ahmadian, G., Liu, L., Becker, L. E., Sheng, M., Wang, Y. T. (2000). Regulation of AMPA receptor–mediated synaptic transmission by clathrin-dependent receptor internalization. *Neuron* 25:649–662.

McMahon, U. J. (1990). The origin hypothesis. *Cold Spring Harbor Symp. Quant. Biol.* 55:407–418.

Martin, P. T., Ettinger, A. J., Sanes, J. R. (1995). A synaptic localization domain in the synaptic cleft protein laminin beta 2 (s-laminin). *Science* 269:413–416.

Mastrogiacomo, A., Parsons, S. M., Zampighi, G. A., Jenden, D. J., Umbach, J. A., Gundersen, C. B. (1994). Cysteine string proteins: a potential link between synaptic vesicles and presynaptic Ca^{2+} channels. *Science* 263:981–982.

Matteoli, M., Haimann, C., Torri-Tarelli, F., Polak, J. M., Ceccarelli, B., De Camilli, P. (1988). Differential effect of alpha-latrotoxin on exocytosis from small synaptic vesicles and from large dense-core vesicles containing calcitonin gene–related peptide at the frog neuromuscular junction. *Proc. Natl. Acad. Sci. USA* 85:7366–7370.

Matus, A. (1999). Postsynaptic actin and neuronal plasticity. *Curr. Opin. Neurobiol.* 9:561–565.

Mellman, I. (1995). Molecular sorting of membrane proteins in polarized and nonpolarized cells. *Cold Spring Harbor Symp. Quant. Biol.* 60:745–752.

Miller, T. M., Heuser, J. E. (1984). Endocytosis of synaptic vesicle membrane at the frog neuromuscular junction. *J. Cell Biol.* 98:685–698.

Mundel, P., Heid, H. W., Mundel, T. M., Kruger, M., Reiser, J., Kriz, W. (1997). Synaptopodin: an actin-associated protein in telencephalic dendrites and renal podocytes. *J. Cell Biol.* 139:193–204.

Nakanishi, H., Obaishi, H., Satoh, A., Wada, M., Mandai, K., Satoh, K., Nishioka, H., Matsuura, Y., Mizoguchi, A., Takai, Y. (1997). Neurabin: a novel neural tissue–specific actin filament–binding protein involved in neurite formation. *J. Cell Biol.* 139:951–961.

Navone, F., Greengard, P., De Camilli, P. (1984). Synapsin I in nerve terminals: selective association with small synaptic vesicles. *Science* 226:1209–1211.

Navone, F., Di Gioia, G., Jahn, R., Browning, M., Greengard, P., De Camilli, P. (1989). Microvesicles of the neurohypophysis are biochemically related to small synaptic vesicles of presynaptic nerve terminals. *J. Cell Biol.* 109:3425–3433.

Neale, E. A., Bowers, L. M., Jia, M., Bateman, K. E., Williamson, L. C. (1999). Botulinum neurotoxin A blocks synaptic vesicle exocytosis but not endocytosis at the nerve terminal. *J. Cell Biol.* 147:1249–1260.

Nicholls, D. G. (1998). Presynaptic modulation of glutamate release. *Prog. Brain Res.* 116:15–22.

Novick, P., Zerial, M. (1997). The diversity of Rab proteins in vesicle transport. *Curr. Opin. Cell Biol.* 9:496–504.

Palay, S. L. (1958). The morphology of synapses in the central nervous system. *Exp. Cell Res. Suppl.* 5:275–293.

Palay, S. L., Palade, G. E. (1955). The fine structure of neurons. *J. Biophys. Biochem. Cytol.* 1:69–88.

Pappas, G. D., Purpura, D. P. (1972). *Structure and Function of Synapses*. New York: Raven Press.

Pappas, G. D., Waxman, S. G. (1972). Synaptic fine structure–morphological correlates of chemical and electrotonic transmission. In *Structure and Function of Synapses*, Vol. 1, Pappas, G. D., Purpura, D. P., eds. New York: Raven Press. pp. 1–44.

Pellizzari, R., Rossetto, O., Schiavo, G., Montecucco, C. (1999). Tetanus and botulinum neurotoxins: mechanism of action and therapeutic uses. *Phil. Trans. R. Soc. London Ser. B* 354:259–268.

Peper, K., Dreyer, F., Sandri, C., Akert, K., Moor, H. (1974). Structure and ultrastructure of the frog motor endplate. A freeze-etching study. *Cell Tissue Res.* 149:437–455.

Peters, A., Palay, S. L. (1996). The morphology of synapses. *J. Neurocytol.* 25:687–700.

Peters, A., Palay, S. L., Webster, F. H. (1991). *The Fine Structure of the Nervous System.* New York: Oxford University Press.

Pieribone, V. A., Shupliakov, O., Brodin, L., Hilfiker-Rothenfluh, S., Czernik, A. J., Greengard, P. (1995). Distinct pools of synaptic vesicles in neurotransmitter release. *Nature* 375:493–497.

Pozzan, T., Rizzuto, R., Volpe, P., Meldolesi, J. (1994). Molecular and cellular physiology of intracellular calcium stores. *Physiol. Rev.* 74:595–636.

Rao-Mirotznik, R., Harkins, A. B., Buchsbaum, G., Sterling, P. (1995). Mammalian rod terminal: architecture of a binary synapse. *Neuron* 14:561–569.

Robinson, L. J., Martin, T. F. (1998). Docking and fusion in neurosecretion. *Curr. Opin. Cell Biol.* 10:483–492.

Robitaille, R., Adler, E. M., Charlton, M. P. (1990). Strategic location of calcium channels at transmitter release sites of frog neuromuscular synapses. *Neuron* 5:773–779.

Roos, J., Kelly, R. B. (1999). The endocytic machinery in nerve terminals surrounds sites of exocytosis. *Curr. Biol.* 9:1411–1414.

Rosahl, T. W., Spillane, D., Missler, M., Herz, J., Selig, D. K., Wolff, J. R., Hammer, R. E., Malenka, R. C., Südhof, T. C. (1995). Essential functions of synapsins I and II in synaptic vesicle regulation. *Nature* 375:488–493.

Rosenmund, C., Stevens, C. F. (1996). Definition of the readily releasable pool of vesicles at hippocampal synapses. *Neuron* 16:1197–1207.

Rosenmund, C., Westbrook, G. L. (1993). Calcium-induced actin depolymerization reduces NMDA channel activity. *Neuron* 10:805–814.

Rossi, D. J., Hamann, M. (1998). Spillover-mediated transmission at inhibitory synapses promoted by high affinity alpha6 subunit $GABA_A$ receptors and glomerular geometry. *Neuron* 20:783–795.

Rothman, J. E., Söllner, T. H. (1997). Throttles and dampers: controlling the engine of membrane fusion. *Science* 276:1212–1213.

Rotundo, R. L., Rossi, S. G., Peng, H. B. (1998). Targeting acetylcholinesterase molecules to the neuromuscular synapse. *J. Physiol. (Paris)* 92:195–198.

Rupp, F., Payan, D. G., Magill-Solc, C., Cowan, D. M., Scheller, R. H. (1991). Structure and expression of a rat agrin. *Neuron* 6:811–823.

Rupp, F., Hoch, W., Campanelli, J. T., Kreiner, T., Scheller, R. H. (1992). Agrin and the organization of the neuromuscular junction. *Curr. Opin. Neurobiol.* 2:88–93.

Sanes, J. R., Lichtman, J. W. (1999). Development of the vertebrate neuromuscular junction. *Annu. Rev. Neurosci.* 22:389–442.

Satoh, A., Nakanishi, H., Obaishi, H., Wada, M., Takahashi, K., Satoh, K., Hirao, K., Nishioka, H., Hata, Y., Mizoguchi, A., Takai, Y. (1998). Neurabin-II/spinophilin. An actin filament–binding protein with one PDZ domain localized at cadherin-based cell-cell adhesion sites. *J. Biol. Chem.* 273:3470–3475.

Segal, I., Korkotian, I., Murphy, D. D. (2000). Dendritic spine formation and pruning: common cellular mechanisms? *Trends Neurosci.* 23:53–57.

Shapiro, L., Colman, D. R. (1999). The diversity of cadherins and implications for a synaptic adhesive code in the CNS. *Neuron* 23:427–430.

Sheng, M., Kim, E. (1996). Ion channel–associated proteins. *Curr. Opin. Neurobiol.* 6:602–608.

Sheng, Z. H., Yokoyama, C. T., Catterall, W. A. (1997). Interaction of the synprint site of N-type Ca^{2+} channels with the C2B domain of synaptotagmin I. *Proc. Natl. Acad. Sci. USA* 94:5405–5410.

Shi, S. H., Hayashi, Y., Petralia, R. S., Zaman, S. H., Wenthold, R. J., Svoboda, K., Malinow, R. (1999). Rapid spine delivery and redistribution of AMPA receptors after synaptic NMDA receptor activation. *Science* 284:1811-1816.

Shupliakov, O., Brodin, L. (2000). A model glutamate synapse. I. The lamprey giant reticulospinal axon. In *Handbook of Chemical Neuroanatomy*, Storm-Mathisen, J., Ottersen, O. P., eds. Amsterdam: Elsevier Science. In press.

Silman, I., Futerman, A. H. (1987). Modes of attachment of acetylcholinesterase to the surface membrane. *Eur. J. Biochem.* 170:11–22.

Smith, S. J., Augustine, G. J. (1988). Calcium ions, active zones and synaptic transmitter release. *Trends Neurosci.* 11:458–464.

Söllner, T. Whiteheart, S. W., Brunner, M., Erdjument-Bromage, H., Geromanos, S., Tempst, P., Rothman, J. E. (1993). SNAP receptors implicated in vesicle targeting and fusion. *Nature* 362:318–324.

Song, J. Y., Ichtchenko, K., Südhof, T. C., Brose, N. (1999). Neuroligin 1 is a postsynaptic cell-adhesion molecule of excitatory synapses. *Proc. Natl. Acad. Sci. USA* 96:1100–1105.

Spacek, J. (1985). Three-dimensional analysis of dendritic spines. II. Spine apparatus and other cytoplasmic components. *Anat. Embryol. (Berlin)* 171:235–243.

Stevens, C. F., Wesseling, J. F. (1999). Identification of a novel process limiting the rate of synaptic vesicle cycling at hippocampal synapses. *Neuron* 24:1017–1028.

Stjarne, L. (1999). Catecholaminergic neurotransmission: flagship of all neurobiology. *Acta Physiol. Scand.* 166:251–259.

Sutton, R. B., Fasshauer, D., Jahn, R., Brunger, A. T. (1998). Crystal structure of a SNARE complex involved in synaptic exocytosis at 2.4 Å resolution. *Nature* 395:347–353.

Takei, K., Stukenbrok, H., Metcalf, A., Mignery, G. A., Südhof, T. C., Volpe, P., De Camilli, P. (1992). Ca^{2+} stores in Purkinje neurons: endoplasmic reticulum subcompartments demonstrated by the heterogeneous distribution of the InsP3 receptor, Ca^{2+}-ATPase, and calsequestrin. *J. Neurosci.* 12:489–505.

Takei, K., Mignery, G. A., Mugnaini, E., Südhof, T. C., De Camilli, P. (1994). Inositol 1,4,5-trisphosphate receptor causes formation of ER cisternal stacks in transfected fibroblasts and in cerebellar Purkinje cells. *Neuron* 12:327–342.

Takei, K., McPherson, P. S., Schmid, S. L., De Camilli, P. (1995). Tubular membrane invaginations coated by dynamin rings are induced by GTP-gamma S in nerve terminals. *Nature* 374:186–190.

Takei, K., Mundigl, O., Daniell, L., De Camilli, P. (1996). The synaptic vesicle cycle: a single vesicle budding step involving clathrin and dynamin. *J. Cell Biol.* 133:1237–1250.

Tanaka, H., Shan, W., Phillips, G. R., Kirsten Arndt, K., Ozlem Bozdagi, O., Shapiro, L., Huntley, G. W., Benson, D. L., Colman, D. R. (2000). Molecular modification of N-cadherin in response to synaptic activity. *Neuron* 25:93–107.

Thomas-Reetz, A. C., De Camilli, P. (1994). A role for synaptic vesicles in nonneuronal cells: clues from pancreatic beta cells and from chromaffin cells. *FASEB J.* 8:209–216.

Thureson-Klein, A. K., Klein, R. L. (1990). Exocytosis from neuronal large dense-cored vesicles. *Int. Rev. Cytol.* 121:67–126.

Tom Dieck, S., Sanmarti-Vila, L., Langnaese, K., Richter, K., Kindler, S., Soyke, A., Wex, H., Smalla, K. H., Kampf, U., Franzer, J. T., Stumm, M., Garner, C. C., Gundelfinger, E. D. (1998). Bassoon, a novel zinc-finger CAG/glutamine-repeat protein selectively localized at the active zone of presynaptic nerve terminals. *J. Cell Biol.* 142:499–509.

Torri-Tarelli, F., Grohovaz, F., Fesce, R., Ceccarelli, B. (1985). Temporal coincidence between synaptic vesicle fusion and quantal secretion of acetylcholine. *J. Cell Biol.* 101:1386–1399.

Townes-Anderson, E., MacLeish, P. R., Raviola, E. (1985). Rod cells dissociated from mature salamander retina: ultrastructure and uptake of horseradish peroxidase. *J. Cell Biol.* 100:175–188.

Tsukita, S., Ishikawa, H. (1980). The movement of membranous organelles in axons. Electron microscopic identification of anterogradely and retrogradely transported organelles. *J. Cell Biol.* 84:513–530.

Uchida, N., Honjo, Y., Johnson, K. R., Wheelock, M. J., Takeichi, M. (1996). The catenin/cadherin adhesion system is localized in synaptic junctions bordering transmitter release zones. *J. Cell Biol.* 135:767–779.

Ungewickell, E., Ungewickell, H., Holstein, S. E., Lindner, R., Prasad, K., Barouch, W., Martin, B., Greene, L. E., Eisenberg, E. (1995). Role of auxilin in uncoating clathrin-coated vesicles. *Nature* 378:632–635.

Ushkaryov, Y. A., Petrenko, A. G., Geppert, M., Südhof, T. C. (1992). Neurexins: synaptic cell surface proteins related to the alpha-latrotoxin receptor and laminin. *Science* 257:50–56.

Valtorta, F., Madeddu, L., Meldolesi, J., Ceccarelli, B. (1984). Specific localization of the alpha-latrotoxin receptor in the nerve terminal plasma membrane. *J. Cell Biol.* 99:124–132.

Van Rossum, D., Hanisch, U. K. (1999). Cytoskeletal dynamics in dendritic spines: direct modulation by glutamate receptors? *Trends Neurosci.* 22:290–295.

Vogt, K. E., Nicholl, R. A. (1999). Glutamate and gamma-aminobutyric acid mediate a heterosynaptic depression at mossy fibers in the hippocampus. *Proc. Natl. Acad. Sci. USA* 96:1118–1122.

Wall, M. J., Usowicz, M. M. (1997). Development of action potential–dependent and independent spontaneous GABA_A receptor–mediated currents in granule cells of postnatal rat cerebellum. *Eur. J. Neurosci.* 9:533–548.

Wang, X., Kibschull, M., Laue, M. M., Lichte, B., Petrasch-Parwez, E., Kilimann, M. W. (1999). Aczonin, a 550-kD putative scaffolding protein of presynaptic active zones, shares homology regions with Rim and Bassoon and binds profilin. *J. Cell Biol.* 147:151–162.

Wang, Y., Okamoto, M., Schmitz, F., Hofmann, K., Südhof, T. C. (1997). Rim is a putative Rab3 effector in regulating synaptic-vesicle fusion. *Nature* 388:593–598.

Wang, Y., Sugita, S., Südhof, T. C. (2000). The RIM/NIM family of neuronal C2 domain proteins: interactions with Rab3 and a new class of neuronal SH3-domain proteins. *J. Biol. Chem.* (in press).

White, J. G., Southgate, E., Thomson, J. N., Brenner, S. (1986). The structure of the nervous system of the nematode *Caenorhabditis elegans*. *Phil. Trans. R. Soc. London Ser. B* 314:1–340.

Yonekawa, Y., Harada, A., Okada, Y., Funakoshi, T., Kanai, Y., Takei, Y., Terada, S., Noda, T., Hirokawa, N. (1998). Defect in synaptic vesicle precursor transport and neuronal cell death in KIF1A motor protein–deficient mice. *J. Cell Biol.* 141:431–441.

Zakharenko, S., Chang, S., O'Donoghue, M., Popov, S. V. (1999). Neurotransmitter secretion along growing nerve processes: comparison with synaptic vesicle exocytosis. *J. Cell Biol.* 144:507–518.

Zhen, M., Jin, Y. (1999). The liprin protein SYD-2 regulates the differentiation of presynaptic termini in *C. elegans*. *Nature* 401:371–375.

Ziff, E. B. (1997). Enlightening the postsynaptic density. *Neuron* 19:1163–1174.

3

Physiology of Synaptic Transmission and Short-Term Plasticity

Wade G. Regehr and Charles F. Stevens

The landmark work of Katz and his colleagues on the neuromuscular junction (NMJ) has framed all succeeding discussion of synaptic function (Katz, 1969). Since Katz's formulation of the quantal theory of synaptic transmission, a central problem of synaptic physiology has been to identify the morphological and molecular correlates of the three key variables that characterize quantal neurotransmitter release: the number of release sites (N), the probability of a quantal release (p), and the size of the quantal response (q). This chapter describes the current state of knowledge relating to Katz's three variables and reviews short-term synaptic plasticity in light of these basic mechanisms.

Basic Mechanisms of Neurotransmission

Basic Properties of Different Types of Synapses

Synapses display a wide range of properties. Some cells are connected by a single synaptic contact site, whereas others are connected by many, and the number of active zones per contact varies from one to hundreds. At some synapses a presynaptic action potential reliably triggers vesicle fusion at a single synaptic contact, whereas at others an action potential rarely triggers vesicle fusion ($p < 0.1$). For some synapses each quantum produces a large depolarization in a cell (q is large); at others the depolarization produced by quanta is difficult to resolve (q is small). These differences in N, p, and q influence the manner in which synapses are studied and, in part, determine the short-term plasticity exhibited by synapses. With such diversity in the behavior of different synapses, it is worthwhile to compare the basic features of the commonly studied ones.

Our basic understanding of synaptic transmission is derived primarily from classic studies on the frog NMJ (Katz, 1969). At the frog NMJ N and p are both large under natural circumstances (that is, at physiological calcium and magnesium concentrations), and individual quanta are readily resolved.

The squid giant synapse has been central for experiments requiring presynaptic recording (Bullock and Hagiwara, 1957; Llinás et al., 1981). This synapse is part of the escape response circuit; when stimulated, it triggers the ejection of water through the mantle of the squid, which propels the squid away from danger. The presynaptic terminal is exceptionally large—about 50 μm in diameter and 700 μm long—and it is possible to place a number of electrodes directly into this presynaptic terminal and simultaneously to monitor the voltage of the postsynaptic terminal. At this synapse N and p are both very large to evoke reliably the release of sufficient neurotransmitter for a rapid escape. Because the postsynaptic cell is so large, individual quantal events are difficult to resolve at this synapse.

Many types of synapses in the mammalian brain are also commonly studied. Some of these synapses, such as the calyx of Held, or the climbing fiber—Purkinje cell synapse, are similar to the frog NMJ and the squid giant synapse in that they appear to be specialized to fire their post-synaptic targets rapidly and reliably. These synapses are also high-N and -p synapses. At the other extreme, many cells receive a single synaptic contact from other cells (Gulyas, 1993), such as CA3 and CA1 pyramidal cells in the hippocampus and granule cells—Purkinje cells in the cerebellum. The probability of release can be quite variable from one such synaptic contact to the next.

Experimental Conditions

The experimental conditions under which synapses are studied can have a profound impact on synaptic behavior. Synapses are usually investigated in brain slice and tissue culture, rarely in vivo. These preparations have the advantage that a variety of optical and electrophysiological techniques can be readily applied, an advantage that is particularly evident for cultured cells. Such preparations have, however, a potential drawback: there is always the possibility that the properties of the synapses differ in important ways from those exhibited in the intact animal. This is particularly true for cultured cells, but it can also be a factor for synapses in brain slices. In vitro studies must always therefore be complemented with information on in vivo behavior.

Another important potential problem is that many in vitro studies are carried out at room temperature in the presence of nonphysiological concentrations of Ca^{2+} and Mg^{2+}. Synaptic transmission is sensitive to temperature for many reasons: the presynaptic waveform, the properties of the Ca^{2+} channels, presynaptic Ca^{2+} regulation, and the processes driving short-term plasticity are all temperature dependent. In addition, the extracellular concentrations of Ca^{2+} have a significant impact on many aspects of synaptic transmission. Although the measured concentrations in cerebrospinal fluid are about 1.5 mM for Ca^{2+} (Manthei et al., 1973), external Ca^{2+} concentration is often set at 2–4 mM to make synaptic currents easier to resolve. Elevated Mg^{2+} concentrations are also often used to minimize recurrent excitation. However, elevating the external Ca^{2+} concentration increases the probability of release (magnesium does the opposite) and alters the plasticity exhibited by a synapse. Thus synaptic properties are crucially dependent upon divalent ion levels, temperature, and other experimental conditions.

Stimulating Presynaptic Cells

An obvious requirement for the study of synaptic transmission is that the experimenter be able to stimulate the release of neurotransmitter. One of the most common ways of doing this is to pass pulses of current through an extracellular electrode. This can bring nearby axons to threshold and trigger an action potential that in turn provides depolarization to open Ca^{2+} channels in the presynaptic bouton. A more direct approach is to evoke an action potential in the presynaptic cell with an intracellular electrode in the cell body, the axon, or the synaptic bouton.

A number of additional methods that do not rely on presynaptic action potentials are used to evoke transmitter release. Voltage clamping of presynaptic terminals provides more precise control over presynaptic Ca^{2+} influx, but it can only be applied to a limited number of preparations (Llinás et al., 1981; Stanley, 1989; Forsythe, 1994; Vyshedskiy and Lin, 1997). Another approach is to induce release by exposing presynaptic boutons to high-osmolarity solutions, a method that produces release through a Ca^{2+}-independent mechanism (Rosenmund and Stevens, 1996). This method has been useful for studying readily releasable pools and individual synapses in cultured cells. Finally, optical techniques provide a powerful means of evoking synaptic transmission. After loading cells with photolabile calcium chelators, exposure to ultra-violet light can trigger release. This technique has the advantage of elevating Ca^{2+} to known levels throughout the terminal (Zucker, 1993). Photostimulation by local release of glutamate from a caged compound is also a very effective method for achieving local stimulation via neurotransmitter receptors (Callaway and Katz, 1993).

Detecting Neurotransmitter Release

Once the presynaptic cell has been stimulated to fire an action potential the next task is to detect the released neurotransmitter. A variety of detection methods have been employed. Most commonly, the electrical response of the postsynaptic cell is recorded. As neurotransmitter activates postsynaptic receptors, ligand-gated channels open. The resulting current flow can be detected with an extracellular electrode. Such extracellular recordings are widely used and have made an important contribution to our understanding of synapses. For example, extracellular recording is the most common way of quantifying synaptic responses in studies of the CA3 to CA1 synapse in hippocampal slices. Such recordings have been so widely used primarily because they can be remarkably stable and easy to implement. In these experiments the extracellular signal is the population response of hundreds or thousands of activated CA1 pyramidal cells. Although extracellular methods most often record population responses, in some circumstances one can record extracellularly from individual presynaptic boutons and detect individual quantal events.

Intracellular recording methods can have significant advantages over extracellular methods: individual cells can be studied with intracellular recording, single quantal events can be detected, and both the contents of the cell and its potential can be controlled. With such intracellular methods, synaptic currents can be measured in whole-cell voltage clamp. In general, voltage-clamp recordings of synaptic properties are best suited to small cells, whereas many synapses of interest occur on large cells that are not well suited to voltage clamping. Postsynaptic depolarization can be measured in current-clamp mode, although such recordings can be affected by voltage-gated ion channels in the postsynaptic cell and can be significantly distorted when the potential of the postsynaptic cell approaches the reversal potential of the ligand-gated channel.

A number of approaches have been used to bypass the postsynaptic cell and to detect released neurotransmitter in other ways. Presynaptic vesicles can be loaded with FM dyes, which fluoresce brightly within vesicles (Betz et al., 1992; Cochilla et al., 1999). Fusion of a vesicle labeled with FM dyes leads to a loss of fluorescence that can be used to quantify the release of neurotransmitter. Until recently, FM dyes were used primarily in the study of the NMJ and cultured neurons. Recent advances may help to extend the use of FM dyes to brain slices and intact preparations where background staining has been a problem (Kay et al., 1999; Pyle et al., 1999).

Measurements of changes in capacitance have been important tools in studying fusion in endocrine cells, but more recently have been used to examine exocytosis in neurons (Parsons et al., 1994; von Gersdorff and Matthews, 1994). When exocytosis occurs the surface membrane of the cell increases slightly—by the area of the number of vesicles incorporated—and these area increases can sometimes be measured as an increase in capacitance (Neher and Marty, 1982). This method is restricted to those synapses for which the presynaptic terminal can be patch clamped, and care must be taken in interpreting the results because concomitant endocytosis can confound exocytosis (von Gersdorff and Matthews, 1999).

Electrochemical methods can also provide a powerful means of detecting secreted molecules. This approach is based on oxidation or reduction of transmitter molecules, with the consequent flow of electrons across the electrode surface, and has been used to great effect to study secretion from chromaffin cells and mast cells. However, a number of transmitters, such as glutamate, GABA, and acetylecholine, cannot be detected readily by this technique, because they are not easily oxidized or reduced. For transmitters such as dopamine, norepinephrine, and serotonin, electrochemical detection is a powerful means of mesuring transmitter release.

Constant innovations are providing new ways for monitoring neurotransmitter release. For example, in brain slices glutamate release can be detected by measuring currents associated with glutamate transporters (Bergles and Jahr, 1997; Diamond et al., 1998). Another new method, evanescent-wave microscopy, has allowed visualization of the movements of single vesicles in chromaffin cells. These two examples illustrate the types of innovative ways of studying release that will benefit future studies of synaptic transmission.

Population Studies versus Studies of Single Synapses

Most studies of synaptic transmission are based on recordings from populations of synapses. In such studies synaptic currents reflect the aggregate response from a population of synapses which have a variety of different properties. Such population studies form the basis for most of what we know regarding synaptic function and synaptic plasticity.

Studies of synapse populations can sometimes provide information on the properties of single synapses. One clever approach is the MK-801 blocking method (Hessler et al., 1993; Rosenmund et al., 1993). With this method, the NMDA component of synaptic transmission is recorded for a population of synapses in the presence of the open-channel blocker MK-801, and the decline in response amplitude is recorded with repeated stimulation. Because the blocking of NMDA receptor channels is essentially irreversible (on the time scale and under the conditions of these experiments), the rate at which the response declines depends on the release probability of the synapses. The decline in response amplitude is rapid at first (indicating that some synapses have a high release probability) and then slows (showing that other synapses have a lower release probability). Although one cannot infer the distribution of release probabilities with this method (Huang and Stevens, 1997), it can be used to reveal that a population of synapses has heterogeneous properties and can show that the shape of this distribution did or did not change with some manipulation.

Just as studies of single channels give information that was inaccessible when populations of channels were investigated, so can the study of single synapses provide special insights into synaptic transmission. With technical advances, particularly improvements in imaging methods, investigation of individual synapses is becoming increasingly common. What is a single synapse? An adult frog muscle fiber typically receives synaptic input from only a single axon, so one might suppose that single synapses had been the subject of classical investigations of synaptic transmission at the NMJ. Structurally, however, an NMJ is made up of many—typically more than 100—identifiable units organized around separate and distinct active zones (Steinbach and Stevens, 1976). If the smallest synaptic unit is identified as "a synapse," then the NMJ is actually composed of over 100 synapses made by a single axon. Most central excitatory synapses possess only a single active zone, so the smallest synaptic unit that can be investigated is a single active zone with its docked vesicles and corresponding postsynaptic density. In this chapter we shall adopt the convention of identifying a single synapse with a single active zone.

Methods for Studying Single Synapses

Various methods have been used to investigate single synapses. All of them have been revealing, but none is ideal. In this section we describe some of the methods that have been used to study single synapses and evaluate their advantages and disadvantages.

Imaging Methods

In the past several decades powerful new imaging methods have been developed that permit visualization of individual synapses in culture and in slices; soon they will be extended to the living brain (Svoboda et al., 1997). Confocal microscopy (Wijnaendts Van Resandt et al., 1985), only fifteen years old, has now become a standard tool, with many types of instruments commercially available, and the two-photon laser scanning fluorescence microscope (Denk et al., 1990), only ten years old, is also rapidly becoming a common tool. Complementing these new microscopy techniques are a range of optical indicators of synaptic function, such as fluorescent dyes to follow synaptic vesicle trafficking (Cochilla et al., 1999), an impressive array of Ca^{2+} concentration indicators (Takahashi et al., 1999), and the green fluorescent protein and related molecules, which can serve as both a structural tag and an indicator (Tsien, 1998).

Synapses are small (close to the resolution of the light microscope), and so their visualization in the living state is a challenge. The confocal scanning microscope offers great advantages for imaging single synapses in that the resolution is excellent and the depth of field is very small, so that closely packed structures can be resolved. The microscope separately illuminates each point in the specimen under study and then collects the light from that point only where it should appear in the image; this method rejects out-of-focus light and yields resolution near or even beyond the theoretical minimum set by the diffraction limit associated with the aperture of the microscope's objective. Because only light from the image point is collected, the depth of focus is very narrow, which means that the microscope effectively sections the specimen optically. Combined with computer image processing, the confocal scanning microscope reconstructs imaged objects in three dimensions with a resolution of less than 1 μm—a requirement for the resolution of single synapses. The main disadvantage of the confocal microscope, apart from its high cost, is the relatively long time needed for the scanner to form an image. Thus the temporal resolution of the imaging system can be unsatisfactory.

The confocal microscope can also be used to excite fluorescent indicators in experiments of the sort described later in this chapter. Although this can be very effective, the intense laser beam used for imaging can bleach the indicator dyes or cause photodynamic damage to the living tissue. A very effective way around the problem is a microscope that excites fluophors with photon pairs of two different wavelengths, so the probability that two of the right photons arrive at the correct place is the square of the individual light intensities at that point. Because of the nonlinear addition (the square of intensities at two wavelengths), the energy of the light is very effectively focused to a single point. Two-photon microscopy has the same optical advantages of confocal microscopy, but it greatly decreases the bleaching and photo-dynamic damage that occurs with the confocal microscope. The disadvantages of two-photon microscopy are the same as those of confocal microscopy.

Most optical studies of single synapses employ a fluorescent indicator dye to follow membrane trafficking or to measure calcium concentration, and single synapses can sometimes be best studied by simply equipping a good microscope with a very efficient method for collecting light, a cooled charge coupled device camera of the sort used by astronomers. For an appropriate specimen—neurons grown in culture, for example—this efficient imaging method, together with computer programs for manipulating the images, can be as effective as the more expensive option of confocal microscopy for answering certain questions.

One of the most widely used methods for optically monitoring function at single synapses is to image vesicle trafficking with charged dyes such as FM1-43 that fluoresce when they enter the membrane (Cochilla et al., 1999). These dyes have a hydrophilic head and a lipid-soluble tail, so that they partition between the bathing solution and the membrane but cannot (because of the charged head) cross the membrane. When endocytosis occurs, the dyes are taken up so that the fluorescence intensity is a measure of the quantity of neurotransmitter (number of vesicles) that was released. Furthermore, stained vesicles can release their dye on exocytosis and cause fluorescence to decrease. Thus both neurotransmitter release and synaptic membrane retrieval can be followed optically. Employed with care, this method permits the resolution of dye movement into or out of a single vesicle (Murthy et al., 1997; Ryan et al., 1997; Murthy and Stevens, 1998).

Another imaging method that permits study of single synapses is the detection of local changes in Ca^{2+} concentration, either pre- or post-synaptically (Malinow et al., 1994; Murphy et al., 1994, 1995) through the use of calcium indicator dyes developed by Roger Tsien and colleagues (Takahashi et al., 1999). The Ca^{2+} influx associated with the presynaptic action potential may be detected, as can the Ca^{2+} concentration increases that occur postsynaptically when N-methyl-D-aspartate (NMDA) receptor channels are activated. Both of these classes of events produce Ca^{2+} concentration transients that can be monitored and are sufficiently well localized to report on the behavior of a single synapse. These imaging studies in culture often require the use of autaptic circuits—neurons isolated in culture in which the source of all the cell's synapses is the cell itself—so that one can know for sure the origin and destination of all synapses studied (Furshpan et al., 1976; Segal and Furshpan, 1990; Bekkers and Stevens, 1991).

Minimal Stimulation

The most direct way to study a single synapse is to stimulate only one and record its response; the method of minimal stimulation offers an approximation to this approach. The underlying principle is straightforward (Raastad et al., 1992; Raastad, 1995; Raastad and Lipowski, 1996). For those situations in which most axons make only a single synapse onto a given postsynaptic neuron (e.g., Schaffer collateral synapses on CA1 pyramidal cells in the hippocampus or parallel fiber synapses onto Purkinje cells in the cerebellum), the idea is to decrease the stimulus intensity until only a single relevant axon is stimulated; a "relevant" axon is one that makes a single synapse onto the neuron from which a recording is being made. This method was developed by Raastad et al. (1992), who used it to show that individual synapses in hippocampus are very unreliable (in the sense that they release neurotransmitter with low probability) and also that hippocampal neurons receive synaptic inputs with diverse individual properties (Raastad and Lipowski, 1996).

The criterion for minimal stimulation originally was that the average size of the response remain constant as the stimulus intensity is increased. Generally, if multiple axons are being stimulated the response will grow in size as the stimulus is made more intense and more synapses are recruited; Raastad's criterion thus eliminates many cases in which multiple synapses are being activated. Unfortunately, however, one can never be sure that just one synapse is being activated (Raastad, 1995), even when additional criteria—such as requiring that the postsynaptic response have an invariant latency and shape (Stevens and Wang, 1995; Dobrunz and Stevens, 1997)—are used. For this reason great care must be used in interpreting results from minimal stimulation experiments. The rule is that one cannot draw a firm conclusion based on a result that might have been an artifact of having multiple synapses when only one was believed to be present. For example, Raastad et al. (1992) concluded that at least some hippocampal synapses release with low probability; this is a firm conclusion because mistaking multiple synapses for a single one would not decrease the apparent release probability. If one were to find in a minimal stimulation experiment that synapses appeared to release with high probability, this would not be a firm conclusion because having many synapses when the experimenter thought only one was present would artifactually make the release probability appear to be higher than it actually is. The minimal stimulation technique can, if used with care, yield accurate results, as has been demonstrated by confirming the conclusions of these experiments with alternative techniques.

Cell Pairs in Slices

Because a neuron often makes only a single synaptic connection with its target cells, another similar method for investigating single synapses is to study cell pairs (Thomson et al., 1993). In this method one cell is stimulated and the response of the target cell is recorded. Insofar as each axon makes only one synapse on its target, this method provides a reliable way to study single synapses. The difficulty with this approach is that the probability of finding a connected pair of cells is often low, and, when it is not, the connection is often made by multiple synapses. When the source cell is stimulated with an electrode in the hippocampus, the results have been the same as those obtained by minimal stimulation (Stevens and Wang, 1994; Bolshakov and Siegelbaum, 1995). An interesting variant of this technique uses focal release of caged glutamate to stimulate the source cell (Callaway and Katz, 1993; Katz and Dalva, 1994); for the most part this approach has so far been used to elucidate the pattern of connections in neocortex.

Cell Pairs in Culture

Whereas neurons in the brain often make only one or a few synaptic connections with each other, neurons in culture typically make multiple synapses. To study single synapses in culture, then, three general approaches are possible. The first is to isolate a single synapse by pharmacologically blocking synaptic transmission at all synapses but one. Such isolation can be accomplished by increasing the extracellular concentration of Mg^{2+} and lowering that of Ca^{2+} throughout the bath and then carrying out local superfusion with a solution that supports synaptic transmission in an area with one or a few synapses (Bekkers and Stevens, 1995). This approach requires some method for identifying synapses (see "Imaging Methods") that are sufficiently distant from their neighbors that they can be selectively studied in this way.

A second set of methods employs either stimulation of a single synapse (or a few synapses) locally or the recording of activity from isolated synapses. Local application of a hypertonic solution causes local release and has been used, together with histochemical methods for identifying single synapses, to study their properties (Bekkers et al., 1990). An alternative method uses focal application of a hyperkalemic solution (together with blockade of synaptic transmission at other synapses) to depolarize axons locally and produce local transmitter release (Liu and Tsien, 1995). Yet another variant of local stimulation is to place a patch electrode over the bouton and stimulate it electrically (Forti et al., 1997), although this method is technically difficult to employ because boutons are so infrequently placed in a way that permits it.

Finally one can record electrical activity and simultaneously use calcium indicator imaging; the coincidence of a pre- or postsynaptic calcium transient and a spontaneous synaptic current can then be used to "tag" the current (Umemiya et al., 1999).

An alternative to stimulating a single synapse is to stimulate many and record the behavior of one. This can be accomplished through the use of the imaging methods discussed previously.

The Quantum (q): A Vesicle

The discovery of synaptic vesicles led immediately to the proposal that Katz's quantum corresponds to the exocytosis of a single vesicle (Katz, 1969). This view remains the standard one, but direct proof is still lacking and various authors have proposed at one time or another a variety of alternatives. One proposal holds that a quantum consists of the simultaneous release by multiple vesicles. Although the question of subquanta—the notion that a quantum is the concerted exocytosis of multiple vesicles (Kriebel and Gross, 1974; Kriebel, 1988)—has not been popular recently and the evidence seems to run against the idea (Frerking et al., 1997), this question has still not been answered definitively.

Edwards et al. (1990) have proposed that the quantal response represents properties of the postsynaptic membrane rather than the release mechanism. According to these authors, different synapses have a different number of postsynaptic "modules," so that the size of the postsynaptic response is determined by how many modules are present. The modules have a constant size and thus define the quantum. According to this view, the postsynaptic receptors are saturated by the neurotransmitter contained in a single vesicle, so that the size of the response reflects the number of modules present rather than the quantity of neurotransmitter released.

The identification of the quantum with the release of a single vesicle is still accepted by most workers, although alternatives of the sort just described have not been finally excluded for all synapses.

Quantal Properties

If one records a series of responses from a population of synapses, the size of the response is found to vary from one stimulus to the next. Katz and co-workers (Katz, 1969) found that the size of the postsynaptic response was quantized—that is, the peak amplitude of the postsynaptic current increased in steps of a specific size, the quantum—and that the statistical properties of the variability could be accounted for by a theory that assumed N release sites are present and that a quantum of size q is released independently at each site with a probability p. Thus the amplitude of postsynaptic responses varies according to a binomial distribution. Fitting the data to such a binomial distribution and extracting the best-fitting values of N, q, and p is called "doing a quantal analysis."

The quantal size was originally estimated from the amplitude of spontaneous miniature postsynaptic currents (mPSCs), identified as single-vesicle exocytotic events, which have an approximately constant size at the NMJ. At central synapses, however, spontaneous miniature synaptic currents (single-vesicle events) vary greatly in amplitude from one to the next (Bekkers and Stevens, 1989), even at single synapses (Bekkers et al., 1990; Liu and Tsien, 1995; Forti et al., 1997). Because the variability in quantal amplitude is generally large enough to obscure the quantal steps in the amplitude of postsynaptic responses (see, e.g., Bekkers and Stevens, 1989, 1995), the quantal nature of central synaptic transmission is not immediately apparent, and the applicability of quantal theory to central synapses has been less compelling than it is for the NMJ.

Although the precise cause of this quantal variability is unknown, several authors have proposed that it is simply a reflection of the variation in synaptic vesicle size or filling (Bekkers et al., 1990; Frerking et al., 1995). The most compelling case for the quantal variability arising presynaptically comes from studies of inhibitory synapses made by retinal bipolar cells in culture, in which two distinct postsynaptic currents (in two different cells) arising from the same quantal release can be recorded simultaneously (Frerking et al., 1995). These triadic synapses show that the quantal postsynaptic responses fluctuate in amplitude for both cells, but that the size of the quantal response in the two cells is highly correlated. Because the fluctuations in quantal size occur in parallel in different cells, the variability must reflect different amounts of neurotransmitter released from vesicle to vesicle.

Several authors have reported that the amplitude distribution of mPSCs exhibits quantal peaks; that is, the quanta are themselves quantized and come in a number of sizes that are integral multiples of some elementary unit. Although most laboratories have found that quanta are rather variable in amplitude at central synapses, the existence of quanta with very sharply defined amplitudes have been inferred to exist (Larkman et al., 1991, 1997).

In fact, the standard deviation of these amplitude distributions is less than the standard deviation in the recording noise. These very precise quanta were originally detected in experiments in which synapses were stimulated rapidly until depletion occurred, and the postsynaptic response became very small. Epochs of responses that revealed the amplitude histograms with very precisely spaced peaks were then selected. These peaks have been proposed (as described previously) to reflect a modular construction of synapses in which sensitivity of the postsynaptic membrane is expressed in multiples of a smallest sensitivity unit (Edwards et al., 1990). Because such peaked histograms do not always appear, and because no method has been devised to control the phenomenon experimentally, it has been difficult to eliminate statistical artifacts completely. Most laboratories have not found this substructure in the amplitude histograms of mPSCs, and the notion of a postsynaptic quantal structure for synapses has not attracted much attention in recent years.

Postsynaptic Saturation

An important though unresolved issue is whether neurotransmitter released by a single vesicle saturates postsynaptic receptors. Several workers have proposed that this is the case (Edwards et al., 1990; Clements, 1996;

see also Frerking and Wilson, 1996). This conclusion is based on three main pieces of evidence. First, calculations of neurotransmitter concentration in the synaptic cleft and considerations of postsynaptic channel properties have predicted that the postsynaptic receptors should be saturated by a single quantum (Harris and Sultan, 1995). Second, the peaks in the amplitude histogram of mPSCs and evoked PSPs (discussed previously) have been interpreted as meaning that the postsynaptic receptors must be saturated. Finally, brief applications of glutamate to membrane patches, at concentrations believed to be achieved in the synaptic cleft, have revealed that the receptors are saturated (Tong and Jahr, 1994).

Other evidence calls this conclusion of saturation of postsynaptic receptors into question (Liu et al., 1999). A recent study used a novel approach to provide evidence that this same conclusion holds for the NMDA receptor component of excitatory synaptic currents (Mainen et al., 1999). The authors in this study measured the amplitude of the postsynaptic response by imaging calcium concentration in single spines in hippocampal slices (CA1 region) and stimulated the presynaptic axon with a pair of pulses spaced 10 msec apart. They found that the total increase in Ca^{2+} concentration postsynaptically was greater for the response pairs than with single stimuli, an observation that is inconsistent with saturation.

The issue of postsynaptic saturation is central to understanding why quantal size is so variable at central synapses. Several authors have found that quantal size is quite variable even at single synapses (Bekkers et al., 1990; Liu and Tsien, 1995; Forti et al., 1997), so the phenomenon is not entirely a reflection of different postsynaptic sensitivities at different synapses. The source of quantal size fluctuations must be either pre- or postsynaptic. If it is presynaptic—that is, if fluctuations have their source in the quantity of transmitter released from one vesicle to the next—then the postsynaptic receptors cannot be saturated; saturated receptors would give a response size that is independent of the quantity of agonist present in a vesicle. If the quantal variability has a postsynaptic origin, then it must reside in moment-to-moment fluctuations in the postsynaptic sensitivity. As noted earlier, the size of the response to the same quantum of inhibitory transmitter in separate retinal bipolar cells is highly correlated, so that the quantal size variability here must be presynaptic, implying that the postsynaptic receptors at this inhibitory synapse are not saturated (Frerking et al., 1995).

The source of quantal variability at individual synapses and postsynaptic receptor saturation remain matters of some debate. An important factor could be the variability of the size of the postsynaptic densities observed at different synapses. For example, at hippocampal CA3-CA1 synapses there is a large difference in the size of the postsynaptic densities at different synapses (Harris et al., 1992). Similarly, at inhibitory synapses in the cerebellum there is considerable variability in the synaptic current produced by individual quanta at different release sites (Auger and Marty, 1997). Such pronounced variability in the postsynaptic properties of individual synapses could contribute to differences in the extent of receptor saturation at different synapses.

Endocytosis

Implicit in the original identification of a vesicular exocytosis with quantal release was the notion that the vesicular membrane would have to be recovered to keep the surface area of the nerve terminal from expanding without limit. The question of vesicle recycling is discussed in detail in the chapter on synaptic vesicle endocytosis by De Camilli et al., but several key facts are important for interpreting experiments described here. At some central synapses, exocytosis normally exposes the vesicle interior to the extracellular medium for about 20 sec (Ryan et al., 1996), long enough for high-affinity lipophilic dyes that fluoresce to equilibrate between the extracellular medium and the inward-facing vesicular lipids. Using charged styryl dyes such as FM1-43 (Cochilla et al., 1999) that will not pass through lipid membranes, investigators have followed synaptic vesicular trafficking. They find that vesicle membrane is internalized with a time constant of about 20 sec and that it takes close to a minute for a vesicle to be reclaimed and readied for another round of exo- and endocytosis. The cycle time for synaptic vesicles is functionally important because, together with the number of vesicles present in the nerve terminal, it sets the upper limit for the maximum sustained average release rate at a synapse. For example, if the average cycle time for a vesicle is 1 min, and if the total number of vesicles present in the terminal is, say, 60, the maximum average sustained release rate would be one release per second. One of the important outstanding problems is to understand how vesicle trafficking is regulated to suit the synapse for its role in information transmission.

Katz's Release Site: An Active Zone

The number of release sites (N) was originally thought to correspond to the number of vesicles, or perhaps to the number of specific membrane sites that could support exocytosis. Several early workers noted, however, that the number of release sites determined by quantal analysis was very similar to the number of synapses (Zucker, 1973; Redman, 1990; Korn and Faber, 1991). This observation led to the notion that some mechanism exists to restrict the number of vesicle fusions per nerve impulse to a single one per terminal, so that a single release site is a single active zone. Some synaptic connections possess a large number of active zones, but in many cases each synapse has only a single active zone (e.g., Schikorski and Stevens, 1997). This notion of "one vesicle release per synapse," then, was effectively equivalent to the idea that some mechanism restricts the number of exocytotic events at an active zone per action potential to a single one.

The weakness of this early work was that it relied on quantal analysis, and some of the assumptions necessary to use quantal analysis are known to be incorrect. For example, quantal analysis assumes that all the synapses in the population under study have the same release probability. However, all available data indicate that members of a population of synapses, even those made by a single axon, have very different release probabilities (Robitaille and Tremblay, 1987, 1991; Murthy et al., 1997). Since one cannot be sure whether a particular conclusion based on quantal analysis is correct or simply an artifact due to an incorrect assumption, the one-synapse–one-vesicle hypothesis was not widely accepted. This issue has been reinvestigated more recently with methods that do not rely on quantal analysis. These newer experiments have confirmed and extended the original notion: a single active zone can normally release only a single vesicle per nerve impulse arrival; the dead time for release persists for about 10 msec (Stevens and Wang, 1995); and the requirement for at most a single vesicle release has other manifestations, such as the relation between the amount of paired-pulse facilitation and a synapse's release probability (Dobrunz and Stevens, 1997).

However, several authors have questioned the one-active-zone–one-vesicle rule. Tong and Jahr (1994) found evidence for multivesicular release in cultured hippocampal neurons, and Auger et al. (1998) have reported that inhibitory synapses in the cerebellum can also release multiple vesicles. At present it is unclear whether the "rule" is not obeyed at some synapses or the instances of multivesicular release are occurring at synapses with multiple active zones. Excitatory synapses with multiple active zones are more common for hippocampal cells in culture than in vivo, and inhibitory synapses frequently have multiple active zones (see Schikorski and Stevens, 1997).

The mechanism of the proposed one-active-zone–one-vesicle release is obscure. After an exocytotic event, the inhibition of remaining vesicles at an active zone happens too rapidly to involve biochemical processes, so presumably some mechanical effect is responsible. Such an explanation makes sense in that one might expect that the energy barrier for vesicle fusion—itself a mechanical process—could depend on other mechanical processes, such as increased stress in the active zone membrane from a vesicle that has fused.

Release Probability (p): Regulated by the Size of the Readily Releasable Pool

Synapses vary greatly in their release probabilities (p) (Hessler et al., 1993; Rosenmund et al., 1993; Allen and Stevens, 1994), even in the case of neighboring synapses made by the same presynaptic axon (Murthy et al.,

1997). Release probability averages about 0.3 for synapses in hippocampal cultures and in slices, but it is broadly distributed, with a maximum near 0.15 and a pronounced skew to low release probabilities. Changes in release probability constitute one of the most common ways in which synaptic strength is regulated, as described in detail in this section.

A typical hippocampal synapse has around 5–10 vesicles docked at the active zone (Harris and Sultan, 1995; Schikorski and Stevens, 1997)—the docked vesicle pool. Although at most only one of the vesicles can be released by the arrival of a nerve impulse, the current evidence is that all or most of these vesicles can be released and that, in fact, Katz's release probability (p) depends on the number of vesicles available to be released. Evidence for this proposition is summarized in this section.

Several authors have identified the docked vesicle pool with the physiologically defined readily releasable vesicle pool—those vesicles that are immediately available to be released (Zucker, 1973). The size of the readily releasable pool is estimated—actually defined—by stimulating the synapse rapidly and counting the number of quanta that can be released before the supply of immediately releasable quanta is exhausted. When a synapses is stimulated rapidly in this way, the size of the post-synaptic response declines to a small steady-state level, and the response size recovers its initial value after a number of seconds (Birks and McIntosh, 1961; Elmqvist and Quastel, 1965; Thies, 1965; Magleby, 1987; Zucker, 1989; Thomson and West, 1993; Liu and Tsien, 1995; Stevens and Tsujimoto, 1995; Rosenmund and Stevens, 1996; Abbott et al., 1997). Since the classical era of synaptic physiology, this decline in response size has been attributed to depletion in the available stores of neurotransmitter. The recovery of response amplitude is interpreted as the refilling of the store of vesicles (usually identified as the docked vesicles); the phenomena relating to depression are discussed in greater detail subsequently.

Implicit in the interpretation of these depletion and recovery experiments is the notion that release probability decreases as the pool of available vesicles is depleted; that is, release probability depends on the size of the available store of vesicles. It was not possible to check this interpretation when populations of synapses were studied, but more recent experiments studying single active zones have determined the quantitative relationship between release probability and the size of the pool of releasable vesicles (Dobrunz and Stevens, 1997; Murthy et al., 1997).

In summary, then, all three of Katz's variables—the quantum, the release site, and the release probability—have morphological counterparts. The quantum is, as originally proposed, the exocytosis of a single synaptic vesicle; the release site corresponds to the single active zone; and the morphological correlate of release probability is the number of docked vesicles.

Short-Term Synaptic Plasticity

During the normal operation of the brain, many neuron types fire at high frequencies and often in bursts, sometimes with long silent intervals interposed. In response to such activity patterns, synapses can undergo profound changes in strength. Neither the mechanisms nor the precise consequences of this dynamic regulation of synaptic strength are well understood, but the synapses clearly are processing as well as transmitting information by applying a complex time-varying filter to the incoming spike train. Here we discuss some advances in understanding how such mechanisms combine to control synaptic strength during complex activity patterns. Synaptic strength can be increased or decreased more than 10-fold by use-dependent mechanisms. These forms of short-term plasticity can differ markedly for different types of synapses (Dittman et al., 2000), as illustrated by two types of excitatory connections onto cerebellar Purkinje cells. In these experiments presynaptic fibers were activated with an extracellular electrode and synaptic currents were recorded by whole-cell voltage clamp. The presynaptic fibers were stimulated with a pattern of activity that is similar to that experienced by many neurons in vivo in that there is a relatively high rate of activity, and the interpulse interval is variable. The synapses between inferior olivary neurons and Purkinje cells, the climbing fiber synapses, are depressed with repetitive activation (Fig. 3.1A, middle). During pauses in stimulation the climbing fiber response recovers partially, and the extent of recovery is greater as the length of the pauses increases. The synapses between granule cells and Purkinje cells, the parallel fiber synapses, have a very different response to the same activity pattern (Fig. 3.1A, bottom). Synaptic strength increases with high-frequency activation and begins to return to basal levels during lulls in presynaptic firing. For both types of synapses there is considerable variability in the amplitude of the synaptic responses during the train, which reflects the preceding activity.

Such variable properties of short-term plasticity are observed throughout the nervous system; they have also been described in olfactory cortex (Bower and Haberly, 1986), hippocampus (Berzhanskaya et al., 1998), and neocortex (Thomson, 1997; Galarreta and Hestrin, 1998; Reyes et al., 1998). Indeed Markram et al. (1998) found that terminals made by the same axon onto different target cells can exhibit quite different plasticity properties. Recent work by Gupta et al. (2000) has begun to systematize short-term plasticity differences for inhibitory terminals, and Schikorski and Stevens (1999) have offered an explanation for different expression of facilitation by olfactory cortical synapses in terms of the fine structure of the boutons.

Another aspect of synaptic plasticity on these time scales is that responses for a given population of synapses can be remarkably stereotyped. Activation of parallel fibers with the same pattern of activity produced similar responses in four successive trials (Fig. 3.1B). In these experiments

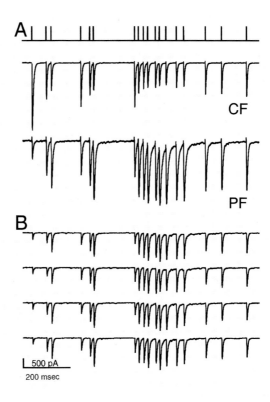

Figure 3.1. Variable-amplitude synaptic currents evoked by irregular stimulation.
(A) Synaptic currents evoked by irregular stimulation (top) of the climbing fiber synapse (CF) and the parallel fiber synapse (PF). These synaptic currents are averages of four trials.
(B) Individual trials show the reproducibility of the response to parallel fiber activation. Both CF and PF synaptic currents were measured in a whole-cell voltage clamp preparation of a Purkinje cell.
Figure courtesy Anatol Kreitzer.

a large enough number of release sites was activated to make the stochastic variations relatively small.

Again, the stereotypy in short-term plasticity is also seen at hippocampal synapses transmitting a natural spike train previously recorded from the hippocampus of a behaving animal. Although synaptic strength varies in a complex way throughout the spike train, repetitions of the same pattern presented to a population of synapses give a nearly identical pattern of responses on each presentation of the spike train. For putative single synapses, the stereotypy is limited only by the probabilistic functioning of the synapse (Dobrunz and Stevens, 1999).

These examples raise a number of fundamental questions about synapses. How is it that within a given trial the synaptic currents show such large variations in strength, yet each of the trials is so stereotyped?

What is it about different synapses that makes them respond in such a different manner to the same activity pattern? Are the short-term plasticity mechanisms adapted to the temporal characteristics of the spike trains they need to filter? What mechanisms control synaptic strength on these time scales?

In trying to understand how such stereotyped responses are produced, and why different types of synapses can exhibit such varied behavior, it is helpful to consider the individual mechanisms that can influence synaptic strength on these time scales. Many steps are involved in generating a postsynaptic response: (1) presynaptic fibers are activated at a remote site, (2) the action potential reaches the presynaptic terminal, (3) the depolarization opens voltage-gated calcium channels, (4) Ca^{2+} enters the presynaptic bouton, (5) Ca^{2+} triggers vesicle fusion, and (6) neurotransmitter diffuses across the synaptic cleft, where it binds to receptors and produces a response in the postsynaptic cell. During a train presynaptic Ca^{2+} builds and the effective number of release sites is altered. We consider each mechanism in turn, discuss when it might contribute, and evaluate the magnitude of its contribution.

Threshold for Fiber Activation and Spike Invasion

One way in which synaptic strength can be altered is through modified fiber excitability at the point of stimulus application or by changes in the extent of presynaptic spike invasion over the axon arbor. In many experiments presynaptic fibers are stimulated with an extracellular electrode, and the threshold for fiber activation can change during a train, leading to a change in the number of activated fibers. In such experiments the number of activated fibers is typically monitored with an extracellular electrode to ensure that no changes occur in the effective stimulus. Although most workers find that the axon threshold remains quite constant, McNaughton et al. (1994) reported that very-long-lasting changes in fiber excitability occur following tetanic stimulation; however, this work has not yet been repeated by other laboratories.

In some experiments cell pairs are studied, and the presynaptic cell is stimulated with an intracellular electrode. The resulting action potential can be detected directly, and failures in action potential generation can be excluded. But even when one is sure that action potentials are being produced by each stimulus, the possibility remains that spike conduction may fail and the action potential may not reliably invade all parts of the presynaptic arbor (Parnas, 1972; Hatt and Smith, 1976; Grossman et al., 1979; Gu and Huang, 1991; Bielefeldt and Jackson, 1993; Debanne et al., 1997). Many studies of branch point failures have been carried out in invertebrate preparations, but their extrapolation to particular mammalian preparations is of uncertain validity. Mackenzie et al. (1996) used a combination of imaging and electrophysiology to determine the rate at which branch point failures occur in hippocampal neurons maintained in culture, and they found that essentially all action potentials propagated throughout the axonal arbor. Based on recordings from pairs of neocortical pyramidal neurons in slices, Williams and Stuart (1999) have argued that each action potential in a high-frequency burst propagates throughout the axonal arbor and can lead to transmitter release onto the target cell.

Changes in Calcium Entry during Trains

For all synapses the probability of release is extremely sensitive to changes in calcium entry, with the relationship between calcium influx and release described (in the low Ca^{2+} concentration limit) by a power law relationship:

$$EPSC = constant \times (Ca^{2+} \; influx)^{n},$$

where EPSC is the excitatory postsynaptic current and n is typically between 2 and 4, which means that doubling the Ca^{2+} influx will increase the synaptic response 4- to 16-fold (Dodge and Rahamimoff, 1967; Goda and Stevens, 1994; Wu and Saggau, 1994; Mintz et al., 1995; Borst and Sakmann, 1996; Takahashi et al., 1996; Reid et al., 1998). Thus even relatively small changes in Ca^{2+} entry could contribute to use-dependent plasticity during trains.

Such alterations in calcium entry could arise in a number of ways. In some neurons the properties of calcium channels vary in a use-dependent way, so that calcium entry per spike is not constant throughout the train. Several studies have demonstrated that calcium influx can either increase or decrease during trains of action potentials (Borst and Sakmann, 1998; Cuttle et al., 1998; Forsythe et al., 1998; Patil et al., 1998). Such changes in calcium entry during trains can be quite large, and different types of calcium channels can respond differentially to the same pattern of presynaptic activity. In many cases such changes in calcium entry do not, however, appear to be the dominant form of plasticity during a train. For example, at the calyces of Held, the Ca^{2+} influx can be enhanced during high-frequency trains to 150% of control values while the EPSC is depressed to less than 25% of control (Cuttle et al., 1998; Forsythe et al., 1998).

Another mechanism that may lead to changes in Ca^{2+} influx through a spike train is use-dependent changes in action potential waveforms, which can affect Ca^{2+} channel activation and Ca^{2+} influx. Synaptic transmission is extremely sensitive to changes in presynaptic waveforms, with a 20% increase in spike width doubling synaptic strength at some synapses. Although the waveform of the action potential at the small presynaptic terminals in the brain is inaccessible, the presynaptic waveform can be recorded at a few special synapses where the presynaptic bouton is large enough to permit intracellular or patch recording, such as the caliciform synapses (Martin and Pilar, 1964; Borst et al., 1995) and the squid giant synapse (Bullock and Hagiwara, 1957). Changes in presynaptic waveform can also be detected during trains of action potentials when presynaptic fibers run in well-defined tracts, such as the parallel fiber synapses onto Purkinje cells. Modified spike waveforms can be detected in these cases by using either extracellular recording techniques (Katz and Miledi, 1965) or optical detection of the presynaptic waveform with voltage-sensitive dyes (Salzberg, 1989; Sabatini and Regehr, 1996).

Changes in synaptic strength can also result through the action of neuromodulators such as adenosine, γ-aminobutyric acid, glutamate, or other agents that act through metabotropic receptors to modulate presynaptic Ca^{2+} and K^+ channels (Stefani et al., 1996; Takahashi et al., 1996; Scanziani et al., 1997; Wang and Lambert, 2000). Generally such processes have slow kinetics, and changes in synaptic strength lag behind the changes in firing frequency by hundreds of milliseconds. In many cases pharmacological agents are available to test the contributions of such transmitters or modulators to changes in synaptic strength.

Another way in which Ca^{2+} influx can change during trains is that synaptic activation can relieve G protein—mediated blocking of voltage-gated Ca^{2+} channels (Bean, 1989; Park and Dunlap, 1998; Tosetti et al., 1999; Brody and Yue, 2000). It has been shown previously that Ca^{2+} channels can be inhibited by G protein activation and that large prolonged depolarization can relieve this inhibition. Many synapses in vivo are likely to be tonically inhibited to some extent.

Postsynaptic Mechanisms

The response to postsynaptic receptors can be altered by insertion or removal of receptors (O'Brien et al., 1998; Lissin et al., 1999; Lüscher et al., 1999; Shi et al., 1999) or modifications of the response properties of existing receptors by, for example, phosphorylation or desensitization (Trussell and Fischbach, 1989; Greengard et al., 1991; Trussell et al., 1993; Liao et al., 1995). On the relatively short time scales dealt with here, it is unlikely that insertion or removal of receptors from the membrane contributes to changes in synaptic strength.

Facilitation of Synaptic Transmission

At many synapses, when two action potentials depolarize a presynaptic bouton in rapid succession, the second action potential releases more neurotransmitter than the first (Magleby, 1987; Zucker, 1989). This short-term enhancement of release that persists for tens to hundreds of milliseconds following a conditioning pulse is known as facilitation (or frequently as paired-pulse facilitation). Facilitation can increase synaptic strength 10-fold at some synapses and may play an important role in the synaptic filtering of information contained in the temporal pattern of spike trains. Figure 3.2A shows an example of paired-pulse facilitation at the synapse between granule cells and Purkinje cells. The inset shows an

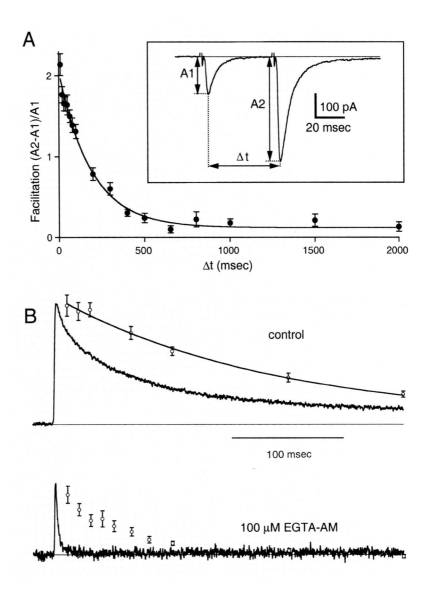

Figure 3.2. *Paired-pulse facilitation driven by residual Ca²⁺.* The synapse between
parallel fibers and Purkinje cells was stimulated twice in rapid succession and
synaptic currents were recorded.

(A) The amplitude of the EPSC evoked by the second stimulus (A2) was
larger than that evoked by the first (A1), and the amplitude of paired pulse facilitation, (A2–A1)/A1, was largest when the interpulse interval (Δt) was brief.

(B) Comparison of the time course of paired pulse facilitation and presynaptic residual calcium at the parallel fiber—Purkinje cell synapse under control conditions (top) and after loading the presynaptic fibers with the Ca²⁺ chelator EGTA
(bottom). Adapted from Atluri and Regehr (1996).

example of synaptic currents evoked by two stimulus pulses of identical magnitude. The second stimulus evokes an EPSC (A2) that is about three times larger than that evoked by the first stimulus (A1). In this example the EPSC is facilitated maximally when the separation between the stimuli is 10 to 20 msec and the duration of facilitation is approximated by an exponential with a time constant of 200 msec. Many synapses display facilitation of similar amplitude and duration.

Facilitation has long been believed to reflect a transient increase in the probability of transmitter release (Magleby, 1987; Zucker, 1989). Quantal analysis at a number of synapses has found increased release probability in response to the second stimulus pulse, and studies of individual synapses have directly confirmed this increase in release probability and described the relationship between the degree of facilitation and the initial release probability (Stevens and Wang, 1995; Debanne et al., 1996; Dobrunz and Stevens, 1997; Murthy et al., 1997; Jiang et al., 2000). Even neighboring synapses made by the same axon onto the same target neuron can exhibit large differences in facilitation. This increase in release probability is not due to increased calcium entry into the presynaptic terminal, nor is it the result of an increase in the number of fibers that are excited. Rather it is thought that residual presynaptic Ca^{2+} contributes to the increased probability of release that is responsible for the observed facilitation (Katz and Miledi, 1968).

To appreciate the manner in which presynaptic Ca^{2+} can produce long-lasting changes in synaptic strength, it is helpful to review some of the basic properties of presynaptic Ca^{2+} signaling. Resting Ca^{2+} concentrations within a presynaptic terminal are about 50–100 nM and external Ca^{2+} levels are 1–2 mM, so there is an impressive differential between internal and external Ca^{2+} concentrations. When an action potential invades a presynaptic bouton, it opens Ca^{2+} channels, giving rise to a locally high Ca^{2+} signal (10–100 µM) near open channels (Ca_{local}) (Chad and Eckert, 1984; Fogelson and Zucker, 1985; Simon and Llinas, 1985; Roberts et al., 1990; Matthews, 1996). As Ca^{2+} equilibrates throughout the bouton, a process that requires about 1 msec, a much smaller (<1 µM), longer-lasting residual Ca^{2+} signal persists for tens of milliseconds to seconds (Ca_{res}) (Delaney et al., 1989; Regehr and Atluri, 1995; Feller et al., 1996; Helmchen et al., 1997; Ravin et al., 1997). Ca_{local} is difficult to measure, but Ca_{res} is readily measured with fluorescent Ca^{2+} indicators.

Ca_{res} has long been known to contribute to facilitation at the neuromuscular junction and the squid giant synapse (Katz and Miledi, 1968; Magleby, 1987; Bittner, 1989; Zucker, 1989; Liu and Stanley, 1995), but there are many different models for how Ca_{res} acts. According to the simplest form of the residual calcium hypothesis, Ca^{2+} entering during the test stimuli sums with Ca_{res} remaining from the conditioning stimulus and enhances release by acting at the low-affinity Ca^{2+}-binding sites involved in triggering release. Another possibility is that Ca_{res} acts at highly sensitive Ca^{2+}-binding sites distinct from those involved in triggering release (Atluri and Regehr, 1996).

An example of an experiment that supports this second model of facilitation is shown in Fig. 3.2B. A comparison of the time course of facilitation and the time course of presynaptic residual Ca^{2+} concentration indicates that facilitation persists longer than the time in which Ca^{2+} concentration remains elevated (Fig. 3.2B, top). It is possible to test the involvement of residual Ca^{2+} at this synapse further by introducing ethylene glycol-bis(β-aminoethyl ether)-N,N,N',N'-tetraacetic acid (EGTA) into the presynaptic terminal. This slow Ca^{2+} chelator has little effect on Ca_{local} but greatly speeds the decay of Ca_{res} (Fig. 3.2B, bottom). In addition the amplitude and duration of facilitation are reduced. Experiments such as this suggest that at the parallel fiber synapse there is a slow, high-affinity Ca^{2+}-binding site involved in facilitation. At the granule cell–Purkinje cell synapse, the time course and amplitude of facilitation are determined through a combination of Ca_{res} dynamics and the properties of a Ca^{2+}-driven reaction.

Depression

Many synapses exhibit decreased efficacy with repeated use, an effect that may persist for seconds to minutes (Liley and North, 1953; Del Castillo and Katz, 1954; Thies, 1965; Betz, 1970; Kusano and Landau, 1975; Thomson and West, 1993; Varela et al., 1997). Such synaptic depression was first described nearly 60 years ago (Eccles et al., 1941; Feng, 1941).

Figure 3.3 shows an example of paired-pulse depression at the synapse between cerebellar climbing fibers and Purkinje cells. Activation of the climbing fiber in rapid succession evokes a response that is reduced in amplitude (Fig. 3.3A). The EPSC is most depressed at short interpulse intervals and the synaptic efficacy recovers as the interpulse interval is lengthened. This is illustrated by a plot of the ratio of the amplitudes of the second and first EPSCs (Fig. 3.3B). If A1 is the amplitude of the first stimulus and A2 is the amplitude produced by the second stimulus then the relation

$$PPD = 1 - A2/A1$$

provides a convenient measure of paired-pulse depression (Fig. 3.3C). Very often the recovery from depression following one or a small number of stimuli can be approximated with a single exponential of several hundred milliseconds to several seconds, as shown for the climbing fiber response in Fig. 3.3C.

There have been many attempts to model depression and recovery from depression. One approach is based on activity-dependent "depletion" of available release sites (Liley and North, 1953; Takeuchi, 1958; Elmqvist and Quastel, 1965; Thies, 1965; Betz, 1970). The basic idea is

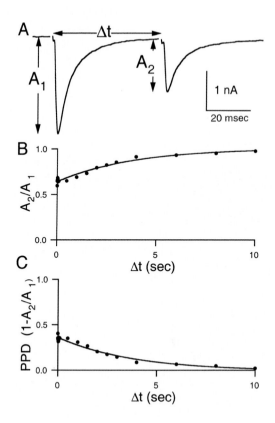

Figure 3.3. *Paired pulse depression.* The climbing fiber—Purkinje cell synapse was activated twice in rapid succession and synaptic currents were recorded. *(A)* The EPSC evoked by the second of two closely spaced stimuli (A_2) was smaller than that evoked by the first stimulus (A_1). The extent of synaptic depression was more pronounced at short interpulse intervals, as illustrated by the time dependence of the A_2/A_1 ratio *(B)*, and by plotting PPD ($1-A_2/A_1$) *(C)*. Experiments were conducted in 1 mM external Ca^{2+}. Adapted from Dittman and Regehr (1998).

that release is proportional to the product of the number of release sites and the probability of release:

$$A1 = Npq,$$

where A1 is the amplitude of the response to the first stimulus, N is the number of release sites, p is the release probability, and q is the size of the quantum. Immediately after the first stimulus the number of release sites that remain effective is $N2 = N–Np$, and

$$A2 = N(1 - p)pq.$$

A2 is the amplitude of the second response, assuming that no facilitation is present so that the release probability is again p for the second stimulus. Often this assumption will not be valid; it then follows that the initial magnitude of depression is given by the initial probability of release:

$$PPD = 1 - A2/A1 = p.$$

It is then assumed that recovery of depression is a single exponential. This provides a good description of data such as those shown in Fig. 3.3, with $p = 0.35$.

Such approaches to modeling depression fail in several ways. This type of model predicts that presynaptic activity at rates faster than the time required to recover from depression would deplete available release sites, thereby making synapses extremely ineffective during high-frequency trains. Although depression occurs during trains, synapses are generally about ten times more effective during high-frequency trains than would be predicted by such a depletion model. This observation gives rise to the hypothesis that recovery from depression might be more rapid during stimulus trains (Kusano and Landau, 1975; Byrne, 1982).

Recent studies have provided new insight into depression and recovery from depression. For example, at the climbing fiber synapse, although recovery from depression follows a single exponential in 1 mM external Ca^{2+} (Fig. 3.4, top), in 4 mM Ca^{2+} a rapid phase of recovery is

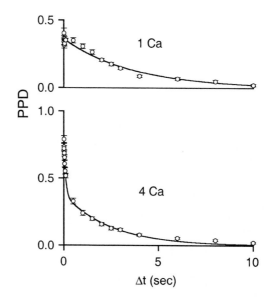

Figure 3.4. *Ca^{2+}-dependent recovery from depression at the climbing fiber—Purkinje cell synapse.* PPD curves were recorded in the presence of external Ca^{2+} concentrations of 1 mM (1 Ca) and 4 mM (4 Ca). Recovery from depression is approximated with a single exponential with a time constant of 3.6 sec in 1 Ca, and by a double exponential with time constants of 90 msec and 2.5 sec in 4 Ca. Adapted from Dittman and Regehr (1998).

apparent. This suggests the hypothesis that high levels of residual calcium accelerate recovery from depression. To test the involvement of residual Ca^{2+}, EGTA was introduced into the presynaptic terminal to accelerate decay of residual Ca^{2+}. This did not affect the initial probability of release but eliminated rapid recovery from depression (Dittman and Regehr, 1998).

Ca_{res} accelerates recovery from depression at the climbing fiber—Purkinje cell synapse (Dittman and Regehr, 1998), an effect that is also observed at hippocampal synapses (Stevens and Wesseling, 1998) and at the calyces of Held (Wang and Kaczmarek, 1998). This Ca_{res}-dependent recovery plays a major role in determining synaptic strength during periods of repetitive activation when Ca_{res} is high. The mechanisms responsible for depression and recovery from depression differ for various types of synapses. They include inhibition or neuromodulation of Ca^{2+} channels through metabotropic receptors, depletion of docked vesicles, desensitization of postsynaptic receptors, and other as yet incompletely identified mechanisms (Trussell et al., 1993; Mennerick and Zorumski, 1995; Stevens and Tsujimoto, 1995; Cummings et al., 1996; Otis et al., 1996; Rosenmund and Stevens, 1996; Takahashi et al., 1996; von Gersdorff and Matthews, 1997; Dittman and Regehr, 1998; Forsythe et al., 1998; Hashimoto and Kano, 1998; Bellingham and Walmsley, 1999).

Control of Synaptic Strength during Spike Trains

Because natural spike trains contain such a wide range of interspike intervals, it is inescapable that multiple forms of short-term plasticity will appear in a synapse's response to them. A variety of models has been developed to account for important features of synaptic behavior during trains of impulses (Magleby, 1987; Sen et al., 1996; Abbott et al., 1997; Lisman, 1997; O'Donovan and Rinzel, 1997; Tsodyks and Markram, 1997; Varela et al., 1997). Although these diverse models have provided important insights into synaptic function, they often do not adequately account for synaptic behavior during high-frequency trains or during highly variable activity patterns as might be observed in vivo. Studies of depression, facilitation, and delayed release suggest that to succeed in describing the behavior of synapses during trains, the Ca^{2+} dependence of the underlying processes must be considered. Here we present an example to illustrate the manner in which multiple forms of plasticity interact to control synaptic strength.

One synapse that has been considered in this regard is the granule cell–Purkinje cell synapse. This synapse has the advantage of being a system in which it is relatively straightforward to monitor presynaptic Ca^{2+}, which is a vital determinant of release probability, facilitation, and recovery from depression. Many of the other individual mechanisms described previously were examined for their potential contributions to synaptic plasticity. Changes in fiber excitability and changes in Ca^{2+}

influx were found to be small, and they could not account for the large changes in synaptic strength observed at this synapse. It was known that facilitation owing to residual calcium is prominent at this synapse, and this must be included. In addition, during stimulus trains depression must also be taken into account. This leads to a general consideration of how to account for transmission at synapses that display both facilitation and depression. One approach to this problem is summarized in Fig. 3.5. According to this model, presynaptic depression controls the number of release-ready sites (D) and facilitation controls the probability of release (F). Synaptic strength is proportional to the product of these two terms. This model is closely linked to the data, is mechanistically realistic, and contains few free parameters.

A simplistic way of thinking about this model is to consider the parameter F as being proportional to the probability of release (P) and the parameter D as corresponding to the number of release-competent sites (see Dittman et al. 2000 for a more detailed discussion). During a train, the residual Ca^{2+} level in the presynaptic terminal increases. This has the effect of increasing the probability of release through the Ca^{2+}-driven process for facilitation (see Fig. 3.2). The behavior of depression is based on the properties of depression at the climbing fiber—Purkinje cell synapse. This is a refractory depression model in which sites that release neurotransmitter are not immediately available to release neurotransmitter. The sites then return to a release-ready state with a rate that is accelerated by elevations in residual calcium.

Facilitation and depression are coupled in two ways. First, they are both driven by residual calcium; second, as F increases there is a large fractional decrease in D. The probability of release is simply the product of these two quantities. In the example shown, facilitation increases rapidly during a train and then saturates. Depression decreases and then reaches a steady-state level, which is determined by the rate of recovery from depression. The net effect is to increase transiently the EPSC, followed by a steady-state response that barely differs from the initial EPSC.

Simulations based on this model permit the dissection of the contributions of facilitation and Ca^{2+}-dependent recovery from depression to different aspects of synaptic transmission (Fig. 3.6). Four model synapses starting with the same initial probability of release show very different transient and steady-state behaviors. In the absence of facilitation (Fac) and Ca^{2+}-dependent recovery from depression (CDR), the synapse gradually becomes less effective during a train (Fig. 3.6A). With Fac and no CDR there is a transient enhancement, but without the rapid recovery from depression provided by CDR the synapse is not very effective at steady state (Fig. 3.6B). When CDR is present but facilitation is absent, the synapse shows only a slight reduction in synaptic strength (Fig. 3.6C). Such a synapse shows a lack of use-dependent plasticity. Finally, when facilitation and CDR are both present there is a rapid elevation in the EPSC amplitude and then a slight reduction (Fig. 3.6D).

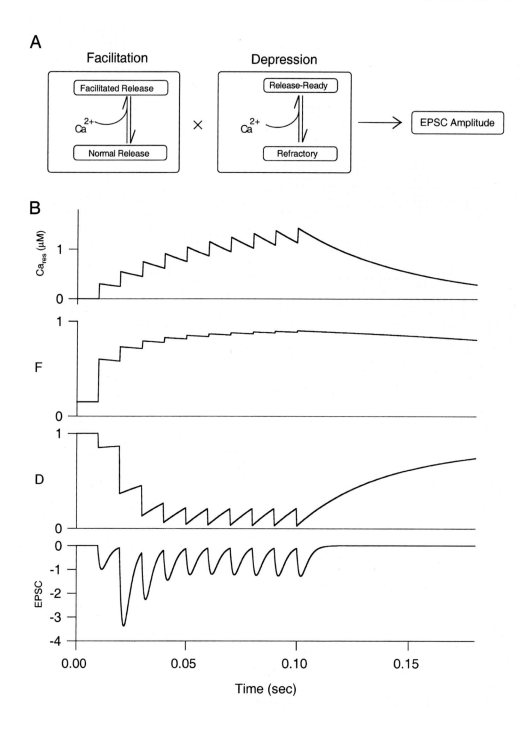

At steady state the synapse remains effective. The steady-state behavior as a function of frequency reflects the presence or absence of facilitation and calcium-dependent recovery from depression (Fig. 3.6E).

Despite the great many assumptions made to keep this model simple, it does a remarkably good job of describing the behavior of the climbing fiber and parallel fiber synapses (Dittman et al., 2000). Such models will benefit greatly from a better understanding of the mechanisms that contribute to short-term plasticity. As we learn more about calcium cooperativity, the kinetics of facilitation and recovery from depression, and about the interactions between these processes, the accuracy of this model will undoubtedly improve but at the expense of its simplicity.

A variety of other known mechanisms that have not been included in the model will have to be taken into account at particular synapses. The addition of augmentation and posttetanic potentiation (Magleby, 1987) may also need to be considered, particularly if we wish to model the behavior of synapses for sustained high-frequency stimulus trains. The inclusion of additional forms of depression will also be required, because the model is restricted to a single mechanism whereas multiple mechanisms are known to contribute at other synapses. For example, changes in the size of the readily releasable pool play an important role at other synapses and under some patterns of stimulation (Birks and McIntosh, 1961; Magleby, 1987; Zucker, 1989; Liu and Stanley, 1995; Liu and Tsien, 1995; Stevens and Wesseling, 1999). During prolonged periods of presynaptic activity, additional forms of depression must also be included (Dittman and Regehr, 1998; Galarreta and Hestrin, 1998; Silver et al., 1998). In addition many mechanisms for synaptic plasticity, such as changes in Ca^{2+} influx, are not considered in the model, although they may be important at some synapses (Dobrunz and Stevens, 1997). The model also does not take into account the stochastic nature of release and the variability in the initial release probability for different sites. Despite these simplifications, the model describes remarkably well the behavior of three very different types of synapses under a variety of stimulus conditions.

Figure 3.5. (opposite) FD model for Ca^{2+} dependence of short-term plasticity.

(A) Release is the product of parameters F and D. Facilitation (F) is increased by elevated levels of residual Ca^{2+}. Recovery from refractory depression (D) is accelerated by increases in residual calcium Ca_{res}.

(B) During a train the residual Ca^{2+} levels increase, which in turn increases F. Each stimulus results in refractory depression, which in turn decreases D. During the train, elevations of residual Ca^{2+} in turn accelerate recovery from depression. The EPSC amplitude was determined from the product of F and D.

Adapted from Dittman et al. (2000), in which the model and parameters are described.

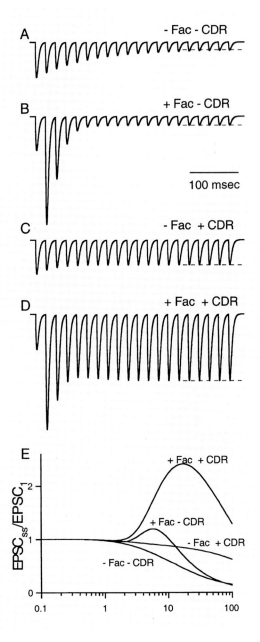

Figure 3.6. Effects of facilitation (Fac) and Ca^{2+}-dependent recovery from depression (CDR) on synaptic properties. Simulation of 20 EPSCs evoked with 50-Hz stimulation using FD as described in Dittman et al. (2000), with the only difference in the four simulations being whether or not facilitation and Ca^{2+}-dependent recovery from depression were included.

(A) No facilitation or CDR.

(B) Facilitation only and no CDR.

(C) CDR only and no facilitation.

(D) Both facilitation and CDR.

(E) Steady-state EPSC versus frequency curves. For A and C, F was held constant at F_1. For A and B, the recovery rate was held constant at k_o.

Adapted from Dittman et al. (2000).

Conclusion

Considerable progress has been made in understanding fundamental properties of transmission at synapses in the central nervous system and in characterizing the mechanisms responsible for various aspects of short-term synaptic plasticity. In coming years it is reasonable to expect that a great deal more will be learned about the molecular determinants of these mechanisms and the role these forms of plasticity play in the filtering and transmission of information in the central nervous system.

References

Abbott, L. F., Varela, J. A., Sen, K., Nelson, S. B. (1997). Synaptic depression and cortical gain control. *Science* 275:220–224.

Allen, C., Stevens, C. F. (1994). An evaluation of causes for unreliability of synaptic transmission. *Proc. Natl. Acad. Sci. USA* 91:10,380–10,383.

Atluri, P. P., Regehr, W. G. (1996). Determinants of the time course of facilitation at the granule cell to Purkinje cell synapse. *J. Neurosci.* 16:5661–5671.

Auger, C., Marty, A. (1997). Heterogeneity of functional synaptic parameters among single release sites. *Neuron* 19:139–150.

Auger, C., Kondo, S., Marty, A. (1998). Multivesicular release at single functional synaptic sites in cerebellar stellate and basket cells. *J. Neurosci.* 18:4532–4547.

Bean, B. P. (1989). Neurotransmitter inhibition of neuronal calcium currents by changes in channel voltage dependence. *Nature* 340:153–156.

Bekkers, J. M., Stevens, C. F. (1989). NMDA and non-NMDA receptors are co-localized at individual excitatory synapses in cultured rat hippocampus. *Nature* 341:230–233.

Bekkers, J. M., Stevens, C. F. (1991). Excitatory and inhibitory autaptic currents in isolated hippocampal neurons maintained in cell culture. *Proc. Natl. Acad. Sci. USA* 88:7834–7838.

Bekkers, J. M., Stevens, C. F. (1995). Quantal analysis of EPSCs recorded from small numbers of synapses in hippocampal cultures. *J. Neurophysiol.* 73:1145–1156.

Bekkers, J. M., Richerson, G. B., Stevens, C. F. (1990). Origin of variability in quantal size in cultured hippocampal neurons and hippocampal slices. *Proc. Natl. Acad. Sci. USA* 87:5359–5362.

Bellingham, M. C., Walmsley, B. (1999). A novel presynaptic inhibitory mechanism underlies paired pulse depression at a fast central synapse. *Neuron* 23:159–170.

Bergles, D. E., Jahr, C. E. (1997). Synaptic activation of glutamate transporters in hippocampal astrocytes. *Neuron* 19:1297–1308.

Berzhanskaya, J., Urban, N. N., Barrionuevo, G. (1998). Electrophysiological and pharmacological characterization of the direct perforant path input to hippocampal area CA3. *J. Neurophysiol.* 79:2111–2118.

Betz, W. J. (1970). Depression of transmitter release at the neuromuscular junction of the frog. *J. Physiol.* 206:629–644.

Betz, W. J., Mao, F., Bewick, W. J. (1992). Activity dependent staining and destaining of living motor nerve terminals. *J. Neurosci.* 12:363–375.

Bielefeldt, K., Jackson, M. B. (1993). A calcium-activated potassium channel causes frequency-dependent action-potential failures in a mammalian nerve terminal. *J. Neurophysiol.* 70:284–298.

Birks, R. I., McIntosh, F. C. (1961). Acetylcholine metabolism of a sympathetic ganglion. *Can. J. Biochem. Physiol.* 39:787–827.

Bittner, G. D. (1989). Synaptic plasticity at the crayfish opener neuromuscular preparation. *J. Neurobiol.* 20:386–408.

Bolshakov, V. Y., Siegelbaum, S. A. (1995). Regulation of hippocampal transmitter release during development and long-term potentiation. *Science* 269:1730–1734.

Borst, J. G., Sakmann, B. (1996). Calcium influx and transmitter release in a fast CNS synapse. *Nature* 383:431–434.

Borst, J. G., Sakmann, B. (1998). Facilitation of presynaptic calcium currents in the rat brainstem. *J. Physiol.* 513:149–155.

Borst, J. G., Helmchen, F., Sakmann, B. (1995). Pre- and postsynaptic whole-cell recordings in the medial nucleus of the trapezoid body of the rat. *J. Physiol.* 489:825–840.

Bower, J. M., Haberly, L. B. (1986). Facilitating and nonfacilitating synapses on pyramidal cells: a correlation between physiology and morphology. *Proc. Natl. Acad. Sci. USA* 83:1115–1119.

Brody, D. L., Yue, D. T. (2000). Relief of G-protein inhibition of calcium channels and short-term synaptic facilitation in cultured hippocampal neurons. *J. Neurosci.* 20:889–898.

Bullock, T., Hagiwara, S. (1957) Intracellular recording from the giant synapse of squid. *J. Gen. Physiol.* 40:565–577.

Byrne, J. H. (1982). Analysis of synaptic depression contributing to habituation of gill-withdrawal reflex in *Aplysia californica. J. Neurophysiol.* 48:431–438.

Callaway, E. M., Katz, L. C. (1993). Photostimulation using caged glutamate reveals functional circuitry in living brain slices. *Proc. Natl. Acad. Sci. USA* 90: 7661–7665.

Chad, J. E., Eckert, R. (1984). Calcium domains associated with individual channels can account for anomalous voltage relations of Ca-dependent responses. *Biophys. J.* 45:993–999.

Clements, J. D. (1996). Transmitter timecourse in the synaptic cleft: its role in central synaptic function. *Trends Neurosci.* 19:163–171.

Cochilla, A. J., Angleson, J. K., Betz, W. J. (1999). Monitoring secretory membrane with FM1-43 fluorescence. *Annu. Rev. Neurosci.* 22:1–10.

Cummings, D. D., Wilcox, K. S., Dichter, M. A. (1996). Calcium-dependent paired-pulse facilitation of miniature EPSC frequency accompanies depression of EPSCs at hippocampal synapses in culture. *J. Neurosci.* 16:5312–5323.

Cuttle, M. F., Tsujimoto, T., Forsythe, I. D., Takahashi, T. (1998). Facilitation of the presynaptic calcium current at an auditory synapse in rat brainstem. *J. Physiol.* 512:723–729.

Debanne, D., Guerineau, N. C., Gahwiler, B. H., Thompson, S. M. (1996). Paired-pulse facilitation and depression at unitary synapses in rat hippocampus: quantal fluctuation affects subsequent release. *J. Physiol.* 491:163–176.

Debanne, D., Guerineau, N. C., Gahwiler, B. H., Thompson, S. M. (1997). Action-potential propagation gated by an axonal I(A)-like K+ conductance in hippocampus. *Nature* 389:286–289. [Erratum appears in *Nature* (1997) 390:536.]

Delaney, K. R., Zucker, R. S., Tank, D. W. (1989). Calcium in motor nerve terminals associated with posttetanic potentiation. *J. Neurosci.* 9:3558–3567.

Del Castillo, J., Katz, B. (1954). Quantal components of the end-plate potential. *J. Physiol.* 124:560–573.

Denk, W., Strickler, J. H., Webb, W. W. (1990). Two-photon laser scanning fluorescence microscopy. *Science* 248:73–76.

Diamond, J. S., Bergles, D. E., Jahr, C. E. (1998). Glutamate release monitored with astrocyte transporter currents during LTP. *Neuron* 21:425–433.

Dittman, J. S., Regehr, W. G. (1998). Calcium dependence and recovery kinetics of presynaptic depression at the climbing fiber to Purkinje cell synapse. *J. Neurosci.* 18:6147–6162.

Dittman, J. S., Kreitzer, A. C., Regehr, W. G. (2000). Interplay between facilitation, depression, and residual calcium at three presynaptic terminals. *J. Neurosci.* 20:1374–1385.

Dobrunz, L. E., Stevens, C. F. (1997). Heterogeneity of release probability, facilitation, and depletion at central synapses. *Neuron* 18:995–1008.

Dobrunz, L. E., Stevens, C. F. (1999). Response of hippocampal synapses to natural stimulation patterns. *Neuron* 22:157–166.

Dodge, F. A., Jr., Rahamimoff, R. (1967). Co-operative action of calcium ions in transmitter release at the neuromuscular junction. *J. Physiol.* 193:419–432.

Eccles, J. C., Katz, B., Kuffler, S. W. (1941). Nature of the "endplate potential" in curarized muscle. *J. Physiol.* 124:574–585.

Edwards, F. A., Konnerth, A., Sakmann, B. (1990). Quantal analysis of inhibitory synaptic transmission in the dentate gyrus of rat hippocampal slices: a patch-clamp study. *J. Physiol.* 177:109–128.

Elmqvist, D., Quastel, D. M. J. (1965). A quantitative study of end-plate potentials in isolated human muscle. *J. Physiol.* 178:505–529.

Feller, M. B., Delaney, K. R., Tank, D. W. (1996). Presynaptic calcium dynamics at the frog retinotectal synapse. *J. Neurophysiol.* 76:381–400.

Feng, T. P. (1941). Studies on the neuromuscular junction. *Chin. J. Physiol.* 16: 341–372.

Fogelson, A. L., Zucker, R. S. (1985). Presynaptic calcium diffusion from various arrays of single channels. Implications for transmitter release and synaptic facilitation. *Biophys. J.* 48:1003–1017.

Forsythe, I. D. (1994). Direct patch recording from identified presynaptic terminals mediating glutamatergic EPSCs in the rat CNS, in vitro. *J. Physiol.* 479:381–387.

Forsythe, I. D., Tsujimoto, T., Barnes-Davies, M., Cuttle, M. F., Takahashi, T. (1998). Inactivation of presynaptic calcium current contributes to synaptic depression at a fast central synapse. *Neuron* 20:797–807.

Forti, L., Bossi, M., Bergamaschi, A., Villa, A., Malgaroli, A. (1997). Loose-patch recordings of single quanta at individual hippocampal synapses. *Nature* 388: 874–878.

Frerking, M., Wilson, M. (1996). Saturation of postsynaptic receptors at central synapses? *Curr. Opin. Neurobiol.* 6:395–403.

Frerking, M., Borges, S., Wilson, M. (1995). Variation in GABA mini amplitude is the consequence of variation in transmitter concentration. *Neuron* 15:885–895.

Frerking, M., Borges, S., Wilson, M. (1997). Are some minis multiquantal? *J. Neurophysiol.* 78:1293–1304.

Furshpan, E. J., MacLeish, P. R., O'Lague, P. H., Potter, D. D. (1976). Chemical transmission between rat sympathetic neurons and cardiac myocytes developing in microcultures: evidence for cholinergic, adrenergic, and dual-function neurons. *Proc. Natl. Acad. Sci. USA* 73:4225–4229.

Galarreta, M., Hestrin, S. (1998). Frequency-dependent synaptic depression and the balance of excitation and inhibition in the neocortex. *Nature Neurosci.* 1:587–594.

Goda, Y., Stevens, C. F. (1994). Two components of transmitter release at a central synapse. *Proc. Natl. Acad. Sci. USA* 91:12,942–12,946.

Greengard, P., Jen, J., Nairn, A. C., Stevens, C. F. (1991). Enhancement of the glutamate response by cAMP-dependent protein kinase in hippocampal neurons. *Science* 253:1135–1138.

Grossman, Y., Parnas, I., Spira, M. E. (1979). Differential conduction block in branches of a bifurcating axon. *J. Physiol.* 295:283–305.

Gu, Y., Huang, L.-Y. M. (1991). Block of kainate receptor channels by Ca^{2+} in isolated spinal trigeminal neurons of rat. *Neuron* 6:777–784.

Gulyas, A. I., Miles, R., Sik, A., Toth, K., Tamamaki, N., Freund, T. F. (1993). Hippocampal pyramidal cells excite inhibitory neurons through a single release site. *Nature* 366:683–687.

Gupta, A., Wang, Y., Markram, H. (2000). Organizing principles for a diversity of GABAergic interneurons and synapses in the neocortex. *Science* 287:273–278.

Harris, K. M., Sultan, P. (1995). Variation in the number, location and size of synaptic vesicles provides an anatomical basis for the nonuniform probability of release at hippocampal CA1 synapses. *Neuropharmacology* 34:1387–1395.

Harris, K. M., Jensen, F. E., Tsao, B. (1992). Three-dimensional structure of dendritic spines and synapses in rat hippocampus (CA1) at postnatal day 15 and adult ages: implications for the maturation of synaptic physiology and long-term potentiation. *J. Neurosci.* 12:2685–2705.

Hashimoto, K., Kano, M. (1998). Presynaptic origin of paired-pulse depression at climbing fibre–Purkinje cell synapses in the rat cerebellum. *J. Physiol.* 506:391–405.

Hatt, H., Smith, D. O. (1976). Synaptic depression related to presynaptic axon conduction block. *J. Physiol.* 259:367–393.

Helmchen, F., Borst, J. G. G., Sakmann, B. (1997). Calcium dynamics associated with a single action potential in a CNS presynaptic terminal. *Biophys. J.* 72:1458–1471.

Hessler, N. A., Shirke, A. M., Malinow, R. (1993). The probability of transmitter release at a mammalian central synapse. *Nature* 366:569–572.

Huang, E. P., Stevens, C. F. (1997). Estimating the distribution of synaptic reliabilities. *J. Neurophysiol.* 78:2870–2880.

Jiang, L., Sun, S., Nedergaard, M., Kang, J. (2000). Paired-pulse modulation at individual GABAergic synapses in rat hippocampus. *J. Physiol.* 523:425–439.

Katz, B. (1969). *The Release of Neural Transmitter Substances.* Liverpool: Liverpool University Press.

Katz, B., Miledi, R. (1965). The measurement of synaptic delay, and the time course of acetylcholine release at the neuromuscular junction. *Proc. R. Soc. London Ser. B* 161:483–495.

Katz, B., Miledi, R. (1968). The role of calcium in neuromuscular facilitation. *J. Physiol.* 195:481–492.

Katz, L. C., Dalva, M. B. (1994). Scanning laser photostimulation: a new approach for analyzing brain circuits. *J. Neurosci. Meth.* 54:205–218.

Kay, A. R., Alfonso, A., Alford, S., Cline, H. T., Holgado, A. M., Sakmann, B., Bergles, D. E., Jahr, C. E. (1999). Synaptic activation of glutamate transporters in hippocampal astrocytes. *Neuron* 19:1297–1308.

Korn, H., Faber, D. S. (1991). Quantal analysis and synaptic efficacy in the CNS. *Trends Neurosci.* 14:439–445.

Kriebel, M. E. (1988). The neuromuscular junction. In *Handbook of Experimental Pharmacology*, Whittaker, V. P., ed. Berlin: Springer-Verlag. pp. 537–566.

Kriebel, M. E., Gross, C. E. (1974). Multimodal distribution of frog miniature endplate potentials in adult, denervated and tadpole leg muscle. *J. Gen. Physiol.* 64:85–103.

Kusano, K., Landau, E. M. (1975). Depression and recovery of transmission at the squid giant synapse. *J. Physiol.* 245:13–32.

Larkman, A., Stratford, K., Jack, J. (1991). Quantal analysis of excitatory synaptic action and depression in hippocampal slices. *Nature* 350:344–347.

Larkman, A. U., Jack, J. J., Stratford, K. J. (1997). Quantal analysis of excitatory synapses in rat hippocampal CA1 in vitro during low-frequency depression. *J. Physiol.* 505:457–471.

Liao, D., Hessler, N. A., Malinow, R. (1995). Activation of postsynaptically silent synapses during pairing-induced LTP in CA1 region of hippocampal slice. *Nature* 375:400–404.

Liley, A. W., North, K. A. K. (1953). An electrical investigation of effects of repetitive stimulation on mammalian neuromuscular junction. *J. Neurophysiol.* 16:509–527.

Lisman, J. E. (1997). Bursts as a unit of neural information: making unreliable synapses reliable. *Trends Neurosci.* 20:38–43.

Lissin, D. V., Carroll, R. C., Nicoll, R. A., Malenka, R. C., von Zastrow, M. (1999). Rapid, activation-induced redistribution of ionotropic glutamate receptors in cultured hippocampal neurons. *J. Neurosci.* 19:1263–1272. [Erratum appears in *J. Neurosci.* (1999) 19:3275.]

Liu, G., Tsien, R. W. (1995). Properties of synaptic transmission at single hippocampal synaptic boutons. *Nature* 375:404–408.

Liu, G., Choi, S., Tsien, R. W. (1999). Variability of neurotransmitter concentration and nonsaturation of postsynaptic AMPA receptors at synapses in hippocampal cultures and slices. *Neuron* 22:395–409.

Liu, Y., Stanley, E. F. (1995). Calcium binding sites of the transmitter release mechanism: clues from short-term facilitation. *J. Physiol.* 89:163–166.

Llinás, R., Steinberg, I. Z., Walton, K. (1981). Relationship between presynaptic calcium current and postsynaptic potential in squid giant synapse. *Biophys. J.* 33:323–352.

Lüscher, C., Xia, H., Beattie, E. C., Carroll, R. C., von Zastrow, M., Malenka, R. C., Nicoll, R. A. (1999). Role of AMPA receptor cycling in synaptic transmission and plasticity. *Neuron* 24:649–658.

Mackenzie, P. J., Umemiya, M., Murphy, T. H. (1996). Ca^{2+} imaging of CNS axons in culture indicates reliable coupling between single action potentials and distal functional release sites. *Neuron* 16:783–795.

Magleby, K. L. (1987). Short-term changes in synaptic efficacy. In *Synaptic Function*, Edelman, G. M., Gall, V. E., Cowan, W. M., eds. New York: John Wiley & Sons. pp. 21–56.

Mainen, Z. F., Malinow, R., Svoboda, K. (1999). Synaptic calcium transients in single spines indicate that NMDA receptors are not saturated. *Nature* 399:151–155.

Malinow, R., Otmakhov, N., Blum, K. I., Lisman, J. (1994). Visualizing hippocampal synaptic function by optical detection of Ca^{2+} entry through the N-methyl-D-aspartate channel. *Proc. Natl. Acad. Sci. USA* 91:8170–8174.

Manthei, R. C., Wright, D. C., Kenny, A. D. (1973). Altered CSF constituents and retrograde amnesia in rats: a biochemical approach. *Physiol. Behav.* 10:517–521.

Markram, H., Wang, Y., Tsodyks, M. (1998). Differential signaling via the same axon of neocortical pyramidal neurons. *Proc. Natl. Acad. Sci. USA* 95:5323–5328.

Martin, A. R., Pilar, G. (1964). Quantal components of the synaptic potential in the ciliary ganglion of the chick. *J. Physiol.* 175:1–16.

Matthews, G. (1996). Neurotransmitter release. *Annu. Rev. Neurosci.* 18:219–233.

McNaughton, B. L., Shen, J., Rao, G., Foster, T. C., Barnes, C. A. (1994). Persistent increase of hippocampal presynaptic axon excitability after repetitive electrical stimulation: dependence on N-methyl-D-aspartate receptor activity, nitric-oxide synthase, and temperature. *Proc. Natl. Acad. Sci. USA* 91:4830–4834.

Mennerick, S., Zorumski, C. F. (1995). Paired-pulse modulation of fast excitatory synaptic currents in microcultures of rat hippocampal neurons. *J. Physiol.* 488:85–101.

Mintz, I. M., Sabatini, B. L., Regehr, W. G. (1995). Calcium control of transmitter release at a cerebellar synapse. *Neuron* 15:675–688.

Murphy, T. H., Baraban, J. M., Wier, W. G., Blatter, L. A. (1994). Visualization of quantal synaptic transmission by dendritic calcium imaging. *Science* 263:529–532.

Murphy, T. H., Baraban, J. M., Wier, W. G. (1995). Mapping miniature synaptic currents to single synapses using calcium imaging reveals heterogeneity in postsynaptic output. *Neuron* 15:159–168.

Murthy, V. N., Stevens, C. F. (1998). Synaptic vesicles retain their identity through the endocytic cycle. *Nature* 392:497–501.

Murthy, V. N., Sejnowski, T. J., Stevens, C. F. (1997). Heterogeneous release properties of visualized individual hippocampal synapses. *Neuron* 18:599–612.

Neher, E., Marty, A. (1982). Discrete changes of cell membrane capacitance observed under conditions of enhanced secretion in bovine adrenal chromaffin cells. *Proc. Natl. Acad. Sci. USA* 79:6712–6716.

O'Brien, R. J., Kamboj, S., Ehlers, M. D., Rosen, K. R., Fischbach, G. D., Huganir, R. L. (1998). Activity-dependent modulation of synaptic AMPA receptor accumulation. *Neuron* 21:1067–1078.

O'Donovan, M. J., Rinzel, J. (1997). Synaptic depression: a dynamic regulator of synaptic communication with varied functional roles. *Trends Neurosci.* 20:431–433.

Otis, T., Zhang, S., Trussell, L. O. (1996). Direct measurement of AMPA receptor desensitization induced by glutamatergic synaptic transmission. *J. Neurosci.* 16:7496–7504.

Park, D., Dunlap, K. (1998). Dynamic regulation of calcium influx by G-proteins, action potential waveform, and neuronal firing frequency. *J. Neurosci.* 18: 6757–6766.

Parnas, I. (1972). Differential block at high frequency of branches of a single axon innervating two muscles. *J. Neurophysiol.* 35:903–914.

Parsons, T. D., Lenzi, D., Almers, W., Roberts, W. M. (1994). Calcium-triggered exocytosis and endocytosis in an isolated presynaptic cell: capacitance measurements in saccular hair cells. *Neuron* 134:875–883.

Patil, P. G., Brody, D. L., Yue, D. T. (1998). Preferential closed-state inactivation of neuronal calcium channels. *Neuron* 20:1027–1038.

Pyle, J. L., Kavalali, E. T., Choi, S., Tsien, R. W. (1999). Visualization of synaptic activity in hippocampal slices with FM1-43 enabled by fluorescence quenching. *Neuron* 24:803–808.

Raastad, M. (1995). Extracellular activation of unitary excitatory synapses between hippocampal CA3 and CA1 pyramidal cells. *Eur. J. Neurosci.* 7:1882–1888.

Raastad, M., Lipowski, R. (1996). Diversity of postsynaptic amplitude and failure probability of unitary excitatory synapses between CA3 and CA1 cells in the rat hippocampus. *Eur. J. Neurosci.* 8:1265–1274.

Raastad, M., Storm, J. F., Andersen, P. (1992). Putative single quantum and single fibre excitatory postsynaptic currents show similar amplitude range and variability in rat hippocampal slices. *Eur. J. Neurosci.* 4:113–117.

Ravin, R., Spira, M. E., Parnas, H., Parnas, I. (1997). Simultaneous measurement of intracellular Ca^{2+} and asynchronous transmitter release from the same crayfish bouton. *J. Physiol.* 501(2):251–262.

Redman, S. (1990). Quantal analysis of synaptic potentials in neurons of the central nervous system. *Physiol. Rev.* 70:165–198.

Regehr, W. G., Atluri, P. P. (1995). Calcium transients in cerebellar granule cell presynaptic terminals. *Biophys J.* 68:2156–2170.

Reid, C. A., Bekkers, J. M., Clements, J. D. (1998). N- and P/Q-type Ca^{2+} channels mediate transmitter release with a similar cooperativity at rat hippocampal autapses. *J. Neurosci.* 18:2849–2855.

Reyes, A., Lujan, R., Rozov, A., Burnashev, N., Somogyi, P., Sakmann, B. (1998). Target-cell-specific facilitation and depression in neocortical circuits. *Nature Neurosci.* 1:279–285.

Roberts, W. M., Jacobs, R. A., Hudspeth, A. J. (1990). Colocalization of ion channels involved in frequency selectivity and synaptic transmission at presynaptic active zones of hair cells. *J. Neurosci.* 10:3664–3684.

Robitaille, R., Tremblay, J. P. (1987). Non-uniform release at the frog neuromuscular junction: evidence of morphological and physiological plasticity. *Brain Res. Rev.* 12:95–116.

Robitaille, R., Tremblay, J. P. (1991). Non-uniform responses to Ca^{2+} along the frog neuromuscular junction: effects on the probability of spontaneous and evoked transmitter release. *Neuroscience* 40:571–585.

Rosenmund, C., Stevens, C. F. (1996). Definition of the readily releasable pool of vesicles at hippocampal synapses. *Neuron* 16:1197–1207.

Rosenmund, C., Clements, J. D., Westbrook, G. L. (1993). Nonuniform probability of glutamate release at a hippocampal synapse. *Science* 262:754–757.

Ryan, T. A., Smith, S. J., Reuter, H. (1996). The timing of synaptic vesicle endocytosis. *Proc. Natl. Acad. Sci. USA* 93:5567–5571.

Ryan, T. A., Reuter, H., Smith, S. J. (1997). Optical detection of quantal presynaptic membrane turnover. *Nature* 388:478–482.

Sabatini, B. L., Regehr, W. G. (1996). Timing of neurotransmission at fast synapses in the mammalian brain. *Nature* 384:170–172.

Salzberg, B. M. (1989). Optical recording of voltage changes in nerve terminals and in fine neuronal processes. *Annu. Rev. Physiol.* 51:507–526.

Scanziani, M., Salin, P. A., Vogt, K. E., Malenka, R. C., Nicoll, R. A. (1997). Use-dependent increases in glutamate concentration activate presynaptic metabotropic glutamate receptors. *Nature* 385:630–634.

Schikorski, T., Stevens, C. F. (1997). Quantitative ultrastructural analysis of hippocampal excitatory synapses. *J. Neurosci.* 17:5858–5867.

Schikorski, T., Stevens, C. F. (1999). Quantitative fine-structural analysis of olfactory cortical synapses. *Proc. Natl. Acad. Sci. USA* 96:4107–4112.

Segal, M. M., Furshpan, E. J. (1990). Epileptiform activity in microcultures containing small numbers of hippocampal neurons. *J. Neurophysiol.* 64:1390–1399.

Sen, K., Jorge-Rivera, J. C., Marder, E., Abbott, L. F. (1996). Decoding synapses. *J. Neurosci.* 16:6307–6318.

Shi, S. H., Hayashi, Y., Petralia, R. S., Zaman, S. H., Wenthold, R. J., Svoboda, K., Malinow, R. (1999). Rapid spine delivery and redistribution of AMPA receptors after synaptic NMDA receptor activation. *Science* 284:1811–1816.

Silver, R. A., Momiyama, A., Cull-Candy, S. G. (1998). Locus of frequency-dependent depression identified with multiple-probability fluctuation analysis at rat climbing fibre–Purkinje cell synapses. *J. Physiol.* 510:881–902.

Simon, S. M., Llinas, R. R. (1985). Compartmentalization of the submembrane calcium activity during calcium influx and its significance in transmitter release. *Biophys. J.* 48:485–498.

Stanley, E. F. (1989). Calcium currents in a vertebrate presynaptic nerve terminal: the chick ciliary ganglion calyx. *Brain Res.* 505:341–345.

Stefani, A., Pisani, A., Mercuri, N. B., Calabresi, P. (1996). The modulation of calcium currents by the activation of mGluRs: functional implications. *Mol. Neurobiol.* 13:81–95.

Steinbach, J. H., Stevens, C. F. (1976). Neuromuscular transmission. In *Handbook of Frog Neurobiology*, Llinas, R., Precht, W., eds. Berlin: Springer-Verlag. pp. 33–92.

Stevens, C. F., Tsujimoto, T. (1995). Estimates for the pool size of releasable quanta at a single central synapse and for the time required to refill a pool. *Proc. Natl. Acad. Sci. USA* 92:846–849.

Stevens, C. F., Wang, Y. (1994). Changes in reliability of synaptic function as a mechanism for plasticity. *Nature* 371:704–707.

Stevens, C. F., Wang, Y. (1995). Facilitation and depression at single central synapses. *Neuron* 14:795–802.

Stevens, C. F., Wesseling, J. F. (1998). Activity-dependent modulation of the rate at which synaptic vesicles become available to undergo exocytosis. *Neuron* 21: 415–424.

Stevens, C. F., Wesseling, J. F. (1999). Augmentation is a potentiation of the exocytotic process. *Neuron* 22:139–146.

Svoboda, K., Denk, W., Kleinfeld, D., Tank, D. (1997). *In vivo* dendritic calcium dynamics in neocortical pyramidal neurons. *Nature* 385:161–165.

Takahashi, A., Camacho, P., Lechleiter, J. D., Herman, B. (1999). Measurement of intracellular calcium. *Physiol. Rev.* 79:1089–1125.

Takahashi, T., Forsythe, I. D., Tsujimoto, T., Barnes-Davies, M., Onodera, K. (1996). Presynaptic calcium current modulation by a metabotropic glutamate receptor. *Science* 274:594–597.

Takeuchi, A. (1958). The long-lasting depression in neuromuscular transmission of frog. *Jpn. J. Physiol.* 8:102–113.

Thies, R. E. (1965). Neuromuscular depression and the apparent depletion of transmitter in mammalian muscle. *J. Neurophysiol.* 28:427–442.

Thomson, A. M. (1997). Activity-dependent properties of synaptic transmission at two classes of connections made by rat neocortical pyramidal axons *in vitro*. *J. Physiol.* 502:131–147.

Thomson, A. M., West, D. C. (1993). Fluctuations in pyramid-pyramid excitatory postsynaptic potentials modified by presynaptic firing pattern and postsynaptic membrane potential using paired intracellular recordings in rat neocortex. *Neuroscience* 54:329–346.

Thomson, A. M., Deuchars, J., West, D. C. (1993). Large, deep layer pyramid-pyramid single axon EPSPs in slices of rat motor cortex display paired pulse and frequency-dependent depression, mediated presynaptically and self-facilitation, mediated postsynaptically. *J. Neurophysiol.* 70:2354–2369.

Tong, G., Jahr, C. E. (1994). Block of glutamate transporters potentiates postsynaptic excitation. *Neuron* 13:1195–1203.

Tosetti, P., Taglietti, V., Toselli, M. (1999). Action-potential-like depolarizations relieve opioid inhibition of N-type Ca^{2+} channels in NG108-15 cells. *Pflugers Arch.* 437:441–448.

Trussell, L. O., Fischbach, G. D. (1989). Glutamate receptor desensitization and its role in synaptic transmission. *Neuron* 3:209–218.

Trussell, L. O., Zhang, S., Raman, I. M. (1993). Desensitization of AMPA receptors upon multiquantal neurotransmitter release. *Neuron* 10:1185–1196.

Tsien, R. Y. (1998). The green fluorescent protein. *Annu. Rev. Biochem.* 67:509–544.

Tsodyks, M. V., Markram, H. (1997). The neural code between neocortical pyramidal neurons depends on neurotransmitter release probability. *Proc. Natl. Acad. Sci. USA* 94:719–723.

Umemiya, M., Senda, M., Murphy, T. H. (1999). Behaviour of NMDA and AMPA receptor-mediated miniature EPSCs at rat cortical neuron synapses identified by calcium imaging. *J. Physiol.* 521:113–122.

Varela, J. A., Sen, K., Gibson, J., Fost, J., Abbott, L. F., Nelson, S. B. (1997). A quantitative description of short-term plasticity at excitatory synapses in layer 2/3 of rat primary visual cortex. *J. Neurosci.* 17:7926–7940.

Von Gersdorff, H., Matthews, G. (1994). Dynamics of synaptic vesicle fusion and membrane retrieval in synaptic terminals. *Nature* 367:735–739.

Von Gersdorff, H., Matthews, G. (1997). Depletion and replenishment of vesicle pools at a ribbon-type synaptic terminal. *J. Neurosci.* 17:1919–1927.

Von Gersdorff, H., Matthews, G. (1999). Electrophysiology of synaptic vesicle cycling. *Annu. Rev. Physiol.* 61:725–752.

Vyshedskiy, A., Lin. J. W. (1997). Activation and detection of facilitation as studied by presynaptic voltage control at the inhibitor of the crayfish opener muscle. *J. Neurophysiol.* 77:2300–2315.

Wang, L. Y., Kaczmarek, L. (1998). High-frequency firing helps replenish the readily releasable pool of synaptic vesicles. *Nature* 394:384–388.

Wang, X. Y., Lambert, N. A. (2000). GABA(B) receptors couple to potassium and calcium channels on identified lateral perforant pathway projection neurons. *J. Neurophysiol.* 83:1073–1078.

Wijnaendts Van Resandt, R. W., Marsman, H. J. B., Kaplan, R., Davoust, J., Stelzer, E. H. K., Stricker, R. (1985). Optical fluorescence microscopy in three dimensions: microtomoscopy. *J. Microsc.* 138:29–34.

Williams, S. R., Stuart, G. J. (1999). Mechanisms and consequences of action potential burst firing in rat neocortical pyramidal neurons. *J. Physiol.* 521:467–482.

Wu, L. G., Saggau, P. (1994). Presynaptic calcium is increased during normal synaptic transmission and paired-pulse facilitation, but not in long-term potentiation in area CA1 of hippocampus. *J. Neurosci.* 14:645–654.

Zucker, R. S. (1973). Changes in the statistics of transmitter release during facilitation. *J. Physiol.* 229:787–810.

Zucker, R. S. (1989). Short-term synaptic plasticity. *Annu. Rev. Neurosci.* 12:13–31.

Zucker, R. S. (1993). The calcium concentration clamp: spikes and reversible pulses using the photolabile chelator DM-nitrophen. *Cell Calcium* 14:87–100.

4

Mechanism and Regulation of Neurotransmitter Release

Thomas C. Südhof and Richard H. Scheller

euronal information processing requires exquisitely specific and rapid signaling mechanisms that must be flexible and easily modified. Neurons communicate primarily via the process of chemical synaptic transmission at a specialized site, the synapse. The biochemical mechanisms underlying chemical synaptic transmission include a built-in plasticity to accommodate the storage and retrieval of past activities. A central goal of contemporary neuroscience is to understand the cell biology and biochemistry of the synapse, because this endeavor will help us explain who we are, how we became who we are, and how we function moment by moment.

Synaptic signaling is mediated by a variety of chemical neurotransmitters that carry a signal from the presynaptic to the postsynaptic neuron. Independent of the neurotransmitter type, the same fundamental mechanisms are used at all synapses: Neurotransmitters are stored in synaptic vesicles in presynaptic nerve terminals and are released when excitation of a presynaptic terminal triggers vesicle exocytosis. Released neurotransmitters then stimulate postsynaptic receptors to complete synaptic transmission.

Katz and co-workers developed the first cellular picture of the mechanism of transmitter release during the 1950s. Electrophysiological studies of the frog neuromuscular junction demonstrated that neurotransmitters are released in discrete quanta (Fatt and Katz, 1952). With the advent of electron microscopy, abundant membrane-bound organelles were observed in the presynaptic nerve terminal, leading to the discovery of synaptic vesicles. These observations resulted in the vesicle hypothesis, which states that a vesicle contains a single quantum of transmitter (Katz, 1969). This hypothesis was validated when synaptic vesicles were purified and shown to contain neurotransmitters (Whittaker, 1968). More sophisticated morphological techniques, including the examination of rapidly frozen stimulated nerve terminals, resulted in a clearer picture of the flow of membranes in the presynaptic nerve terminal (Heuser and Reese, 1973).

A key feature of neurotransmitter release is that it involves fusion of the membrane of synaptic vesicles with the presynaptic plasma membrane, a process that is tightly regulated by calcium. After exocytosis, vesicles are endocytosed and recycle. As a result, synaptic vesicles undergo a cycle of membrane traffic composed of exocytosis, endocytosis, and recycling (Fig. 4.1). More recent studies have further delineated the various stages of the synaptic vesicle cycle, including attachment (also called docking), prefusion (priming), triggering, recycling, and reloading of the vesicles with transmitter. In this chapter, we present a current view of the biochemistry of neurotransmitter release from presynaptic nerve terminals, and we attempt to correlate the physiological and morphological properties of presynaptic nerve terminals with the underlying molecular events.

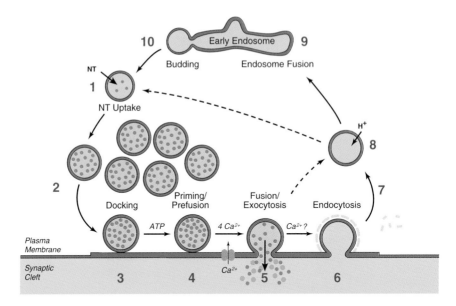

Figure 4.1. The synaptic vesicle cycle. The trafficking of synaptic vesicles in the nerve terminal is conceptually divided into 10 stages: (1) Synaptic vesicles are filled with neurotransmitters (NT) by active transport. (2) The vesicles are translocated to the active zone and (3) attach to the active zone of the presynaptic plasma membrane (docking). (4) The vesicles undergo a prefusion reaction (probably involving several substeps) that primes them for calcium-triggered neurotransmitter release. (5) Calcium influx through voltage-gated channels triggers fusion pore expansion and neurotransmitter release in <0.1 msec. (6) Empty synaptic vesicles are coated by clathrin and associated proteins prior to endocytosis. (7) Coated synaptic vesicles bud from the plasma membrane and shed their clathrin coat. (8) Empty vesicles acidify via proton pump activity and return to the interior of the nerve terminal. (9) Synaptic vesicles fuse with early endosomes as an intermediate sorting compartment to eliminate aged or missorted proteins. (10) Synaptic vesicles are freshly generated by budding from endosomes. Synaptic vesicles can probably bypass the steps in the cycle at two points (indicated by broken arrows): First, steps 6 and 7 (coating of the fused vesicles with clathrin and clathrin-dependent endocytosis) may be replaced by a rapid endocytosis process ("kiss and run"). Second, although some synaptic vesicles recycle via endosomes (steps 9 and 10), the endosomal intermediate is probably not obligatory for recycling, and some vesicles go directly from step 8 to step 1. Adapted from Südhof (1995).

The Synaptic Vesicle Cycle

The synaptic vesicle cycle can be envisioned to start with the uptake of neurotransmitters into synaptic vesicles by an energy-dependent transport activity that requires ATP (step 1 in Fig. 4.1). Specialized transporter proteins in the synaptic vesicle membrane mediate this process. Of central importance here is the vacuolar proton pump, which acidifies the vesicle interior, resulting in an electrochemical gradient across the synaptic vesicle membrane. This gradient provides the energy for the actual neurotransmitter uptake process. Uptake is mediated by specific transport proteins that are specialized for the various transmitters. There appear to be separate transport proteins for glutamate, ATP, and acetylcholine, while γ-aminobutyric acid and glycine share the same transporter, as do all catecholamines (Usdin et al., 1995; Fykse and Fonnum, 1996; Liu and Edwards, 1997; McIntire et al., 1997). Synaptic vesicles filled with neurotransmitters move to the active zone of the presynaptic plasma membrane in a translocation process that probably occurs by diffusion, although the participation of molecular motors has not been excluded (step 2). At the active zone, the vesicles become attached to the plasma membrane (step 3). Attachment (or docking) of the vesicles involves a specific interaction between the vesicles and the active zone; the vesicles do not attach to any other part of the presynaptic plasma membrane except for the active zone. Attached vesicles undergo an ATP-dependent prefusion reaction(s) that primes them for calcium-dependent release and may involve a partial fusion process (step 4). Calcium then triggers the completion of fusion in a rapid reaction that can occur in less than 0.1 msec (Sabatini and Regehr, 1999). Calcium triggering of release involves the binding of multiple calcium ions at a synaptic calcium-binding site (step 5).

After exocytosis, synaptic vesicles are rapidly retrieved by endocytosis, most likely via coated pits (step 6). Coated synaptic vesicles shed the coats (step 7) and recycle to the interior of the synaptic nerve terminal (step 8). The empty vesicles either refill immediately with neurotransmitters to reinitiate the cycle (step 1) or pass through an endosomal intermediate (steps 8 and 9) as a sorting station before being redirected into the recycling vesicle population. Synaptic vesicles go through the whole cycle in approximately 60 sec (Betz and Bewick, 1992). Vesicle attachment and prefusion (steps 3 and 4) may require approximately 10–20 msec, calcium-triggered fusion (step 5) less than 1 msec, and endocytosis (step 6) only a few seconds. Thus most of the vesicle cycle consists of neurotransmitter uptake and vesicle recycling.

In accordance with their specialized functions, presynaptic nerve terminals are characterized by three morphological hallmarks (Peters et al., 1991). The most prominent feature of synaptic terminals is their cluster of some 200–500 synaptic vesicles that are situated next to the active zone. Given that the mammalian brain has on the order of 10^{14}–10^{15} synapses, synaptic vesicles are extremely abundant. Second, at the point of synaptic contact, the presynaptic plasma membrane is thickened into an active zone to which several synaptic vesicles are attached. The active zones occupy an area of 5–20 μm^2. A variable yet relatively small number of synaptic vesicles is attached to the active zone in a nerve terminal, usually 5–10. Consistent with the fact that most of the vesicle cycle is spent on recycling and not at the active zone, the majority of synaptic vesicles are located in the cytosol of the nerve terminal. Finally, at the active zone, presynaptic nerve terminals are connected with the postsynaptic neuron by undefined material spanning the synaptic cleft. This material presumably represents cell adhesion molecules that function in the recognition and precise alignment of pre- and postsynaptic parts of the synapse (see Südhof, this volume).

In eukaryotic cells, membrane trafficking is universal. At any given time, all cells continuously execute many different, simultaneous trafficking reactions. The same fundamental process governs all intracellular membrane traffic: transport vesicles bud from the membrane of origin, move toward and attach to the target organelle, and fuse with the target membrane. Different kinds of membrane traffic in a cell are distinguished by the specific organelles involved, the time course of the reactions, and the regulatory mechanisms controlling them. Cells keep the many different types of simultaneous membrane traffic separate and successfully avoid "traffic jams." Thus each membrane trafficking reaction is well regulated and has identifying features that distinguish it from others. The fact that the various membrane trafficking processes are fundamentally similar, but individually regulated and targeted, suggests that all membrane trafficking processes share the same basic mechanisms but differ in their regulatory components. The synaptic vesicle cycle has the same characteristics as other types of intracellular membrane traffic. It involves budding of transport vesicles from the plasma membrane and endosomes (steps 7 and 10 in Fig. 4.1), transfer of the vesicles from the originating to the target organelle (steps 2 and 8), and attachment and fusion of the transport vesicles to the target membrane (steps 3 and 9). Thus progress in understanding the mechanisms of neurotransmitter release has provided insight into the general mechanisms of membrane traffic that apply to all cells (Bennett and Scheller, 1993).

At the same time, neurotransmitter release has unique regulatory mechanisms. It is exquisitely targeted and regulated to an extent beyond any other trafficking pathway; consequently knowledge of the specific mechanisms that confer upon synapses their unusual properties will be important for understanding synaptic transmission. If synaptic vesicles do not recycle via the endosomal intermediate (Fig. 4.1), the whole cycle is composed of only a single translocation, attachment-fusion, and budding reaction, making it similar to a single step in the secretory pathway. The local autonomous recycling of synaptic vesicles in nerve terminals enables the presynaptic terminals to sustain repeated rounds of neurotransmitter secretion, independent of the cell body to which they belong. Since at least some synaptic terminals are separated from their corresponding neuronal cell body by more than 1 m (e.g., synaptic terminals of spinal cord motoneurons), the functional autonomy of the vesicle cycle is essential for the functioning of the nervous system.

Synaptic Vesicles: Molecular Anatomy of an Organelle

A fruitful approach to understanding the molecular mechanisms of the synaptic vesicle cycle and neurotransmitter release was the characterization of the proteins associated with synaptic vesicles (Südhof, 1995). Synaptic vesicles are homogeneous in size and density, making it relatively easy to isolate large amounts of pure vesicles by a combination of density-gradient and size-fractionation techniques. The characterization of the proteins localized to synaptic vesicles has opened the door to genetic, physiological, and biochemical studies of their functions (Fernandez-Chacon and Südhof, 1999). Through this set of diverse approaches, we now have elaborate and testable biochemical models of vesicle attachment and fusion at the nerve terminal.

As judged by electron microscopy, synaptic vesicles are electron-lucent organelles that are about 35 nm in diameter and may be slightly larger under native conditions. Because they are relatively small, synaptic vesicles can accommodate only a limited number of lipids and proteins. Calculations suggest that each synaptic vesicle is composed of approximately 10,000 molecules of phospholipids and of proteins with a combined approximate molecular weight of $5–10 \times 10^3$ kDa (Jahn and Südhof, 1993). Since an average protein has a molecular weight of around 50 kDa, synaptic vesicles contain approximately 200 protein molecules. The only known function of synaptic vesicles is their role in neurotransmitter release; thus they are also functionally relatively simple. In principle, the limited number of functions and proteins associated with synaptic vesicles should make it possible to obtain a complete molecular dissection of these organelles. Indeed, as a result of extensive biochemical studies, synaptic vesicles are possibly the best described of all organelles. We now know the structures of most vesicle proteins; we have deduced the functions of some of these proteins, and we have testable ideas about the functions of others.

Synaptic vesicle proteins can be conceptually divided into two functional classes: transport proteins, which mediate the uptake of neurotransmitters and other components into synaptic vesicles, and trafficking proteins that execute the intracellular traffic of the vesicles. The most prominent transport protein of synaptic vesicles is the proton pump, which generates the electrochemical transmembrane gradient that fuels neurotransmitter uptake. The synaptic vesicle proton pump is similar to other vacuolar proton pumps. It is composed of at least 13 subunits, with a total size of some 0.8×10^3 kDa (see Forgac, 1999, for review). A single copy of the proton pump per vesicle is probably sufficient to drive neurotransmitter uptake; because of its large size, a single copy would account for approximately 10% of the total vesicle protein. One of the determinants of the transmitter type that is used by a particular synapse is the class of neurotransmitter transporter present in the vesicle. For example, as glutamate is a universal component of the cytosol, any synaptic vesicle that transports glutamate will make the corresponding nerve terminal glutamatergic. In addition to the proton pump and neurotransmitter transporters, synaptic vesicles contain ancillary transport proteins that mediate zinc transport, chloride flux, and possibly other transport activities; most of these have not yet been identified.

There are currently nine families of synaptic vesicle proteins that appear to be involved in membrane traffic (Fig. 4.2). It is possible that additional proteins of synaptic vesicles remain to be identified, but it is unlikely that many proteins are missing. In addition to the known proteins, many other proteins have been reported to be associated with synaptic vesicles (e.g., calcium-calmodulin–dependent protein kinase II, actin, SNAPIN, and dynein). However, these proteins have only rarely been found in highly purified vesicles, and, when present, they were of such low abundance in the purified vesicles that they cannot be stoichiometric components of all vesicles. Although we cannot currently

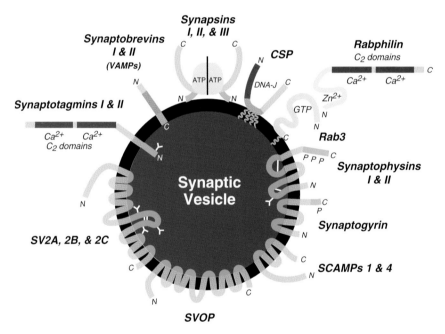

Figure 4.2. *Structures of putative trafficking proteins of synaptic vesicles.* Shown are the structures of the major transmembrane trafficking proteins of synaptic vesicles, defined as proteins with a likely function in the vesicle cycle but not in neurotransmitter uptake. Only proteins that are present on all vesicles and are tightly associated with synaptic vesicles are pictured. Note that synapsins, rab3's, and rabphilin are reversibly bound to the vesicles and dissociate from them in a regulated manner at the synapse (Stahl et al., 1996; Hosaka et al., 1999). Most of the proteins (with the exception of CSP) are present on synaptic vesicles in several isoforms that are encoded by multiple genes; only one isoform for each protein is shown. Adapted from Südhof (1995).

exclude the possibility that some of these proteins transiently associate with a subset of vesicles, it is unlikely that they are requisite components of all synaptic vesicles.

As illustrated in Fig. 4.2, synaptic vesicle proteins share no common structural theme but consist of all types of membrane proteins. The proteins include peripheral membrane proteins (e.g., synapsins and rabphilin) and proteins that are attached to the vesicles by a posttranslational lipid modification (CSP, rab proteins) in addition to transmembrane proteins with single (synaptotagmins, VAMP/synaptobrevins) or multiple (synaptophysins, synaptogyrins, SV2s, SCAMPs) transmembrane regions. Interestingly, synaptic vesicles contain three families of proteins with 4 transmembrane regions arranged in the same topology (synaptophysins, synaptogyrins, SCAMPs) and two families of proteins with 12 transmembrane regions (SV2s and SVOP) but only a single family each of type I and type II transmembrane proteins (synaptotagmin and VAMP/synaptobrevin, respectively).

All synaptic vesicle trafficking proteins except for CSP are members of gene families composed of closely related isoforms that are on synaptic vesicles. In many cases, these proteins or more distant relatives are also found outside synapses. A typical example is synaptophysin, which is expressed in two isoforms in brain, synaptophysin I and synaptoporin/synaptophysin II (Knaus et al., 1990; Fykse et al., 1993), and at least two isoforms in nonneuronal tissues, pantophysin and mitsugumin29 (Haass et al., 1996; Takeshima et al., 1998). Other synaptic vesicle proteins that are present in two or more isoforms on the vesicles are synaptogyrins, synaptotagmins, VAMP/synaptobrevins, SV2s, rab3's, synapsins, and SCAMPs. All of these proteins except for synapsins and SV2s have nonneuronal isoforms. The synaptic vesicle isoforms of these proteins are differentially expressed in the brain in a pattern that does not correlate with either the neurotransmitter type or the known functions of the different brain regions. No protein families exhibit the same differential distribution. For example, the synaptophysins are generally co-expressed in all brain areas (Fykse et al., 1993), whereas synaptotagmins I and II are almost always expressed in different neurons (Geppert et al., 1991). SV2A and SV2B are largely co-expressed, with SV2A as the dominant isoform present almost everywhere (Bajjalieh et al., 1994), whereas SV2C is selectively expressed at high levels only in a subgroup of neurons of the basal forebrain (Janz and Südhof, 1999).

Some synaptic vesicle protein families are more distantly related to each other. Synaptogyrins and synaptophysins represent distinct families of proteins that exhibit a low but significant sequence similarity, as does SVOP with SV2s (Janz and Südhof, 1998; Janz et al., 1999b). At least synaptophysin and synaptogyrin are partly functionally redundant (Janz et al., 1999c). In addition to containing nonneuronal isoforms, many synaptic vesicle proteins are also expressed in nonneuronal cells. Synaptophysin is present on endocrine granules, CSP is located on both endocrine and exocrine granules (Lowe et al., 1988; Chamberlain and

Burgoyne, 1996), and VAMP/synaptobrevin II appears to be ubiquitously expressed in all cells tested. However, for none of these proteins
has a nonneuronal function been demonstrated, and it is possible that
their nonneuronal expression could be nonfunctional.

Finally, the evolutionary history of the different vesicle proteins is
quite distinct. Some proteins are evolutionarily conserved to a high degree (e.g., VAMP/synaptobrevin, synaptotagmin). For other proteins
(e.g., synapsins, synaptophysin) only distantly related homologues can
be found in invertebrates, and some proteins (e.g., SV2) seem to be absent from invertebrates. These differences in evolutionary conservation
indicate that some proteins may be components of the basic machinery
for release, whereas others may have more peripheral regulatory functions that are important in complex nervous systems but not in simpler
organisms.

Taken together, synaptic vesicle proteins are characterized by a great
variety of structures and topologies (Fig. 4.2). The only characteristic
that all vesicle proteins share is that they lack a cleaved signal peptide,
usually the most common mechanism of membrane protein insertion.
Although most transmembrane proteins of synaptic vesicles probably
include internal signal peptides, the VAMP/synaptobrevins contain a
C-terminal membrane anchor. As a result, protein synthesis is completed
before the translocation machinery would recognize the hydrophobic
sequence. Thus, as expected, the transmembrane regions of these proteins are probably inserted into lipid bilayers posttranslationally, independent of the translocation apparatus (Kutay et al., 1995). In spite of
this unconventional membrane insertion, the site of the initial membrane association of VAMP/synaptobrevins appears to be the endoplasmic reticulum; the proteins then flow through the secretory pathway
in a normal fashion.

Although most vesicle proteins are highly concentrated in presynaptic nerve terminals, lower levels of these molecules can be found throughout the cell. In immature neurons or cells in culture, a considerable
concentration of the vesicle proteins is found in the Golgi apparatus.
Following exit from the trans-Golgi network, synaptic vesicle membrane
proteins have been proposed to follow constitutive vesicular trafficking
pathways to the plasma membrane. From the plasma membrane,
the proteins may be internalized and sorted from other proteins within
the endosomal system (Johnston et al., 1989; Linstedt et al., 1992).
Vesicle proteins travel via fast anterograde axonal transport to synaptic
sites, assisted by a diverse array of motor proteins (see Foletti et al., 1999,
for review). The final assembly of a fully mature synaptic vesicle may not
occur until further rounds of exo- and endocytosis at or close to the
synapse (Matteoli et al., 1992), but the exact mechanisms involved are
obscure.

Despite a large body of work, the functions of most synaptic vesicle
proteins remain unknown. For several proteins, a point of action has
been identified, but it is unclear what exactly these proteins do at that

point. For example, CSP contains a DNA-J domain that interacts with HSC70 and is essential in *Drosophila* for normal calcium-triggered release (Buchner and Gundersen, 1997; Ranjan et al., 1998). It seems likely that CSP functions as a component of a chaperone for a target protein complex, but this hypothetical target protein complex has not yet been identified. Similarly, synapsins are ATP-binding proteins that are structurally closely related to ATP-dependent synthases (Esser et al., 1998; Hosaka and Südhof, 1998a,b). Although it has been shown in mouse knockouts that synapsins are essential for the normal regulation of neurotransmitter release (Rosahl et al., 1995), an enzymatic activity has yet to be identified for these abundant and enigmatic proteins. A further complicating factor is that the relative simplicity of the protein composition of synaptic vesicles suggests that many vesicle proteins must have multiple functions. The synaptic vesicle cycle contains many independent steps (Fig. 4.1), all of which are likely to involve protein-protein interactions. These protein-protein interactions must in the end target synaptic vesicle proteins. Since there are only 9–11 different families of putative trafficking proteins (some of which, such as SV2, may turn out to be specialized transport proteins, as indicated by the phenotype of the SV2 knockout mice suggesting a role in regulating calcium levels in the terminals [Janz et al., 1999a]), many proteins must have multiple functions in order to account for all the steps in the vesicle cycle. The best-characterized example of this are synaptotagmins I and II, which are thought to function both in triggering exocytosis (Geppert et al., 1994b) and in nucleating AP-2 for endocytosis (Zhang et al., 1994).

Characteristics of Synaptic Vesicle Exocytosis

Arguably the most important step in the synaptic vesicle cycle is vesicle exocytosis, which mediates neurotransmitter release (Fig. 4.1). Although the synaptic vesicle cycle is generally similar to other trafficking pathways, the final step in synaptic exocytosis is more tightly regulated, especially by calcium influx, and more localized than any other known membrane fusion event. Synaptic exocytosis is triggered when an action potential invades the nerve terminal and opens voltage-gated calcium channels, allowing calcium to flow into the nerve terminal. Calcium triggers exocytosis of docked vesicles at the active zone; since vesicles are attached only to the active zone, exocytosis is tightly restricted to the synaptic junction by this mechanism.

Membrane fusion during exocytosis requires a merging of the inner and outer leaflets of the two phospholipid bilayers. Calcium triggers neurotransmitter release in less than 0.1 msec (Sabatini and Regehr, 1999), which is faster than many enzyme reactions, let alone a cascade of chemical reactions. This suggests that calcium only induces completion of a fusion reaction that was largely performed before calcium came into play. Based on these considerations, we conceptually divide the exocytotic reaction in the synaptic vesicle cycle into three steps (steps 3–6 in Fig. 4.1): synaptic vesicle attachment, which docks the vesicles at the active zone; prefusion, which primes the vesicles for calcium-triggered exocytosis (and which may consist of a partial fusion reaction); and the actual calcium-triggered step. Existence of a prefusion step preceding the point of calcium action is suggested not only by the speed of the calcium action but also by the finding that synaptic vesicle exocytosis can be non-physiologically elicited by hypertonic sucrose in the absence of calcium (Rosenmund and Stevens, 1996). Thus calcium is not required for fusion as such, but rather for an added regulatory event.

Synaptic vesicle attachment occurs only at the active zone. This implies a molecular recognition process between the active zone and the synaptic vesicle, the nature of which is unclear. However, this recognition step is ultimately responsible for the tight targeting of exocytosis to the active zone. The prefusion step that follows vesicle attachment may itself consist of several separate steps. In endocrine cells, the prefusion reaction—often referred to in this context as priming (note that in vacuole membrane fusion "priming" is a different step, preceding membrane attachment; Ungermann et al., 1998)—involves a series of steps whose rate can also be calcium-regulated (Neher, 1998). In addition to calcium influx, synaptic vesicle exocytosis can also be stimulated by two calcium-independent procedures, treatment with hypertonic sucrose and a spider toxin called α-latrotoxin.

Several proteins have been identified as essential for calcium-triggered exocytosis and for calcium-independent exocytosis. These proteins are the so-called SNAREs—VAMP/synaptobrevin, syntaxin, and SNAP-25—and the syntaxin-binding protein munc18a/nsec1. In addition, munc13-1 is essential for normal prefusion in glutamatergic nerve terminals (Augustin et al., 1999). Munc13-1 is a large active zone protein that binds phorbol esters and contains multiple C_2 domains, which in other proteins (such as synaptotagmins) function as calcium-binding modules (Brose et al., 1995). Interestingly, munc13-1 is required for normal calcium-triggered exocytosis and for exocytosis induced by hypertonic sucrose but not for transmitter release evoked by α-latrotoxin. This suggests that α-latrotoxin acts at a step preceding the point of action of hypertonic sucrose and that munc13-1 functions downstream of the SNARE complexes. These results provide support for the notion that the prefusion step includes multiple separate reactions. It is unclear what actually happens during prefusion to set up the calcium-responsive state, and what the endpoint of the process is that ultimately constitutes the substrate for calcium action. One possibility is that prefusion results in a hemifusion intermediate that involves fusion of only one of the two leaflets of the phospholipid bilayers (Südhof, 1995). Another hypothesis is that fusion is completed at the end of prefusion but that fusion pore expansion is blocked until it is triggered by calcium.

Triggering of neurotransmitter release by calcium exhibits a steep calcium concentration dependence, which suggests that, in order to induce fusion, at least three to four calcium ions must act simultaneously (and possibly cooperatively) at the same site (Dodge and Rahamimoff, 1967). Very high local calcium concentrations (>100 mM) are required for triggering exocytosis (Llinas et al., 1992). Thus calcium probably triggers release by binding to one or several synaptic proteins that contain multiple calcium-binding sites of low affinity. At any given time, 5–10 vesicles are attached to most active zones. All of the attached vesicles are apparently "ready" for release, since they can all be stimulated by hypertonic sucrose to undergo exocytosis (Rosenmund and Stevens, 1996). Nevertheless, a calcium signal during an action potential does not always trigger exocytosis. At most synapses, release is observed with a relatively low probability—about once for every 5–10 calcium signals. Furthermore, when calcium is successful in stimulating exocytosis at an active zone, it usually triggers only the fusion of a single vesicle, although multiple vesicles are ready to be released at the active zone.

These unique characteristics of synaptic exocytosis—its low probability and the selection of a single vesicle for release—make presynaptic terminals "reliably unreliable." Since every action potential normally leads to a fairly uniform flooding of the active zone with calcium, there must be negative regulatory elements involved in addition to the positive regulatory elements. In other words, there must be mechanisms that inhibit fusion just as there are mechanisms to trigger fusion, in order to account for the restriction of fusion to one vesicle for every 5–10 "ready" vesicles and every 5–10 calcium signals. As discussed later in this chapter, at present our ideas about the molecular mechanisms involved in this regulation are quite rudimentary.

In terms of overall synaptic signaling in the central nervous system, the unreliability of synaptic exocytosis is advantageous. The low release probability gives the synapse considerable leeway for regulation. By changing release probability at individual synapses, the properties of synaptic networks can be carefully modulated. In addition, the triggering of only one out of several "ready" vesicles makes it possible to elicit repetitive release events in rapid succession at a rate that would be too fast for replenishment. Synaptic exocytosis is subject to a great deal of plasticity. Previous activity and the reception of neurotransmitter signals, neuropeptide signals, or both from other synaptic terminals can dramatically change the pattern of exocytosis. These modulation processes are referred to as use-dependent homosynaptic plasticity and heterosynaptic plasticity, respectively (see Malenka and Siegelbaum, this volume).

Physiologically, one important form of presynaptic plasticity consists of the various types of short-term synaptic plasticity, such as paired pulse facilitation or posttetanic potentiation. These forms of plasticity result in short-duration (millisecond to minute) changes in the release probability of a synaptic terminal; they depend strictly on the previous activity of the synapse. The associated changes in release are very large, resulting in severalfold increases (or decreases) in synaptic transmission.

Mechanistically, it is known that virtually all forms of short-term synaptic plasticity are evoked by calcium, but it is unclear how the various forms are generated (Zucker, 1989). In addition to short-term synaptic plasticity, synapses can be modified by long-term changes in synaptic strength. These are referred to as long-term potentiation (LTP) and long-term depression (LTD), which are long-lasting (>3 h) increases and decreases, respectively, in synaptic strength. The most thoroughly investigated (and controversial) form is N-methyl-D-aspartate (NMDA) receptor–dependent LTP in the CA1 region of the hippocampus. NMDA-receptor-dependent LTP is induced postsynaptically; the participation of presynaptic mechanisms in this form of LTP is unclear (see Malenka and Siegelbaum, this volume). However, a second form of LTP in the hippocampus, LTP of mossy fiber synapses in the CA3 region, is clearly dependent on presynaptic activity. In mossy fiber LTP, the synaptic terminals experience a major, long-lasting change in release probability, suggesting that presynaptic mechanisms can also trigger changes that last for hours.

Molecular Basis for Synaptic Membrane Fusion: The SNARE Cycle and munc18a/nsec1

The early dissection of synaptic vesicle proteins led to the characterization of a variety of molecules critical for vesicle function. One of these proteins, called VAMP (Trimble et al., 1988) in electric rays or synaptobrevin in vertebrates (Südhof et al., 1989a), has an intriguing structure that is now well understood in functional terms. VAMP/synaptobrevin contains a hydrophobic carboxyl terminus that is anchored in the membrane, possibly (at least in some isoforms) completely spanning the lipid bilayer. At the amino terminus, VAMP/synaptobrevin has a 30-residue proline-rich sequence that is poorly conserved between species. The central 70 amino acids of VAMP/synaptobrevin, between the proline-rich sequence and the membrane anchor, are predicted to form an α-helical structure with a hydrophobic face centered on an arginine residue.

VAMP/synaptobrevin forms a complex with two proteins that are largely localized to the plasma membrane, SNAP-25 and syntaxin (Söllner et al., 1993a). SNAP-25, or synaptosomal protein of 25 kDa, was discovered as an abundant protein in the nerve terminal, and for years after its discovery its function was not known (Oyler et al., 1989). SNAP-25 is anchored to the membrane through multiple palmitoylated cysteines situated in an alternatively spliced domain near the middle of the protein. Syntaxin was initially discovered serendipitously using a monoclonal antibody and named HPC-1 (Barnstable et al., 1985; Inoue et al., 1992). This protein also languished without a known function until it was identified as a component of a complex with the synaptic vesicle protein synaptotagmin (Bennett et al., 1992; Yoshida et al., 1992). Like VAMP/synaptobrevin, syntaxin is anchored to the membrane via a

C-terminal hydrophobic stretch of 22 amino acids. Syntaxin, SNAP-25, and VAMP/synaptobrevin were the founding members of a large family of membrane fusion proteins referred to as SNAREs.

Direct evidence that the complex of syntaxin, SNAP-25, and VAMP/synaptobrevin is involved in membrane fusion was first obtained when the mechanisms of action of botulinum and tetanus toxins were elucidated (see Südhof et al., 1993, and Pellizzari et al., 1999, for reviews). These toxins are site-specific proteases that are taken up into the cytosol of nerve terminals. In the terminals, the proteolytic activity of these toxins blocks synaptic vesicle exocytosis. Because the toxins act catalytically, they are among the most toxic substances known. Indeed, a single molecule of a toxin is sufficient to poison a whole nerve terminal. Fascinatingly, although exocytosis is blocked by the toxins, no other major changes are observed in the nerve terminal. In particular, no significant structural changes can be detected in the synapse after toxin treatment. Both toxins act as proteases with very high substrate specificity: tetanus toxin and botulinum toxins B, D, F, and G cleave only VAMP/synaptobrevin; botulinum toxins A and E cleave only SNAP-25; and botulinum toxin C1 cleaves both syntaxin and SNAP-25 (Link et al., 1992; Schiavo et al., 1992, 1993, 1994; Blasi et al., 1993a,b; Binz et al., 1994). The fact that the proteolytic cleavage of these SNARE proteins selectively blocks exocytosis proves that these three proteins perform a restricted function in membrane fusion, but not in vesicle attachment or the maintenance of active zones (Südhof et al., 1993). The toxins inhibit not only calcium-triggered release but also calcium-independent release evoked by hypertonic sucrose or α-latrotoxin, indicating that the SNAREs act prior to, or in concert with, the calcium-dependent step (Chen et al., 1999).

The heterotrimeric complex of SNAP-25, VAMP/synaptobrevin, and syntaxin is often referred to as the core complex, and it has a number of interesting properties. The three-dimensional structure of the minimal core complex and the structure of the conserved N-terminal domain of syntaxin have been solved (Fernandez et al., 1998; Sutton et al., 1998; Fig. 4.3). From these and various other studies, a picture of the core complex emerges that suggests its critical role in membrane fusion. First, the complex is formed via a four-helix bundle—one α-helix each from VAMP/synaptobrevin and syntaxin, and two α-helixes from SNAP-25. All of the helixes are parallel to each other, which indicates that the formation of the complex will bring the membranes close together, perhaps even resulting in a merging of the bilayers. This mechanism has been referred to as the "zipper model" of membrane fusion (Geppert and Südhof, 1997; Hanson et al., 1997; Lin and Scheller, 1998; Poirier et al., 1998). Second, the helical bundle conformation buries the hydrophobic residues within the core of the complex. Midway along the α-helix, a glutamine from syntaxin, two glutamines from SNAP-25, and an arginine from VAMP/synaptobrevin form salt bridges that are also buried within the core of the helical bundle. The resulting structure is very stable, with an unfolding transition over 95 °C, which makes it heat stable

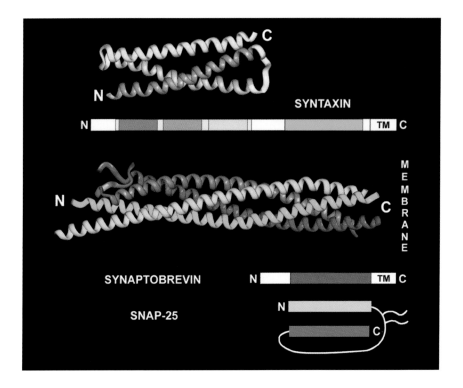

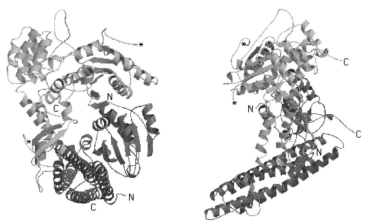

Figure 4.3. Three-dimensional structures of the N-terminal domain of syntaxin, the minimal SNARE core complex, and the munc18a–syntaxin 1A complex.

The three α-helices of the N-terminal domain are shown in blue, and the four α-helices of the core complex in green (syntaxin), yellow (VAMP/synaptobrevin), and orange and red (SNAP-25). The amino (N) and carboxyl (C) termini are indicated; TM, transmembrane region. Modified from Rizo and Südhof (1998).

The bottom two structures are two views of the munc18a–syntaxin 1A complex. Munc18a is divided into three domains, shown in yellow, blue, and green. Syntaxin 1A is depicted as three domains, colored purple, red, and orange. Modified from Misura et al. (2000).

(Hayashi et al., 1994). It has been suggested that there is a large free-energy change upon core complex formation, which could be used to drive membrane fusion.

Assembly of the core complex is certainly a highly regulated event. Studies in vitro suggest that the rate-limiting step may be formation of the dimeric complex between SNAP-25 and syntaxin (Nicholson et al., 1998). The default conformation of syntaxin is a "closed" conformation in which the N-terminal domain is folded back onto the C-terminal SNARE motif, while assembly of the core complex requires the "open" conformation (Dulubova et al., 1998). This closed conformation may arise because the three-helical N-terminal domain of syntaxin weakly binds to its C-terminal SNARE motif (Calakos et al., 1994). The closed conformation of syntaxin is associated with a soluble protein called munc18a or nsec1 that was initially uncovered by affinity chromatography on immobilized syntaxin (Hata et al., 1993; Pevsner et al., 1994). Munc18a is homologous to a large number of proteins involved in membrane traffic, ranging from a *Caenorhabditis elegans* mutant that is paralyzed and gave munc18a its name (Brenner, 1974; Hosono et al., 1992; Gengyo-Ando et al., 1993) to sequences genetically identified in yeast (sec1, sly1, vps45, and vps 33; Fig. 4.4). SNAP-25 binding to syntaxin can be competed off by munc18/nsec1, suggesting that the syntaxin–munc18a complex with the closed conformation and the core complex with the open conformation of syntaxin are mutually exclusive (Pevsner et al., 1994; Dulubova et al., 1998). This suggestion has recently been confirmed by structural analysis of the complex (Misura et al., 2000; Fig. 4.3).

Like SNAREs and rab proteins, proteins homologous to munc18a/nsec1 have been identified in every intracellular membrane fusion process studied (Fig. 4.5; see Jahn and Südhof, 1999, for review). At the plasma membrane of vertebrates, three different isoforms (munc18a/nsec1, munc18b, and munc18c) are expressed in a tissue-specific manner (Hata and Südhof, 1995; Tellam et al., 1995). As sec1 was the first member of this family identified genetically in yeast, while munc18a was the first protein studied biochemically, we refer to this protein family as SM proteins for sec1/munc18 proteins. Because of their homologies, it is likely that SM proteins, SNAREs, and rab proteins perform similar functions in all of the various membrane fusion reactions in a cell (Figs. 4.4 and 4.5). One current hypothesis, based on yeast genetics, is that an SM protein bound to a syntaxin SNARE interacts with Rab GTPases and/or their effectors during the initial stages of vesicle target recognition (Tall et al., 1999). The regulated dissociation of the SM protein from the syntaxin SNARE would allow, or perhaps facilitate, an interaction of the helical domain of syntaxin with SNAP-25 and VAMP/synaptobrevin, resulting in the formation of the core complex. This, in turn, would drive membrane fusion, as diagrammed schematically in Fig. 4.6 (Hanson et al., 1997; Weber et al., 1998; see Geppert and Südhof, 1997, for review). It is important to note, however, that this is a

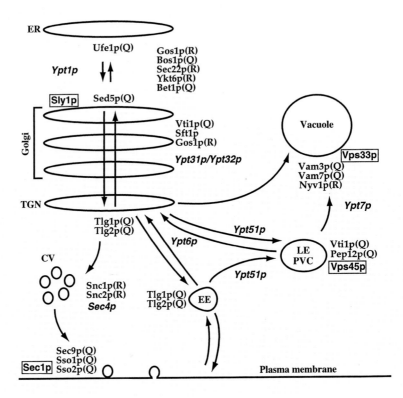

Figure 4.4. Subcellular localization of proteins involved in membrane trafficking in Saccharomyces cerevisiae. There are three families of proteins with members involved at different steps of the yeast pathway: the SNAREs (Q and R), Rab/Ypt, and SM (sec1/munc18) families. The Q-SNARE family is composed of Ufe1p (Lewis and Pelham, 1996), Bos1p and Bet1p (Newman et al., 1990), Vti1p (Fischer von Mollard et al., 1997), Tlg1p and Tlg2p (Holthius et al., 1998), Vam3p (Wada et al., 1997), Vam7p (Sato et al., 1998), Pep12p (Becherer et al., 1996), Sec9p (Brennwald et al., 1994), Sso1p and Sso2p (Aalto et al., 1993), and Sed5p (Hardwick and Pelham, 1992). The R-SNARE family includes Sec22p (Newman et al., 1990), Ykt6p (McNew et al., 1997), Gos1p (McNew et al., 1998), Snc1p and Snc2P (Protopopov et al., 1993), Nyv1p (Nichols et al., 1997), and Sft1p (unclear if R- or Q-SNARE; Banfield et al., 1995). The SM family (boxed) is composed of Sec1p (Aalto et al., 1991), Sly1p (Dascher et al., 1991; Ossig et al., 1991), Vps33p (Banta et al., 1990; Wada et al., 1990), and Vps45p (Cowles et al., 1994; Piper et al., 1994). Finally, members of the Ypt/Rab family (see Lazar et al., 1997, for review) are shown in their putative subcellular localization and/or at the trafficking step in which they are thought to be involved. CV, constitutive secretory vesicles; EE, early endosomes; ER, endoplasmic reticulum; LE, late endosomes; PVC, prevacuolar compartment; TGN, trans-Golgi network.

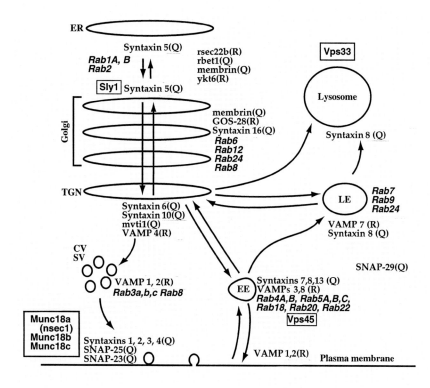

Figure 4.5. Subcellular localization of proteins involved in membrane trafficking in mammalian cells. There are three families of proteins with members involved at different steps of the pathway: the SNARE (Q and R), Rab/Ypt, and SM (sec1/munc18) protein families. Members of the Q-SNARE family include syntaxins 1–5 (Bennett et al., 1992; Bennett and Scheller, 1993), syntaxin 6 (Bock et al., 1996, 1997), syntaxin 7 (H. Wang et al., 1997; Wong et al., 1998), syntaxin 8 and SNAP-29 (Steegmaier et al., 1998); syntaxin 10 (Tang et al., 1998), syntaxin 13 (Advani et al., 1998; Prekeris et al., 1998), syntaxin 16 (Simonsen et al., 1998), mvti1 (Advani et al., 1998), SNAP-25 (Oyler et al., 1989), SNAP-23 (Ravichandran et al., 1996), membrin (Hay et al., 1997), and rbet1 (Hay et al., 1996). The R-SNARE family is composed of VAMP/synaptobrevins 1 and 2 (Elferink et al., 1989), cellubrevin (McMahon et al., 1993), VAMPs 4, 7, and 8 (Advani et al., 1998), GOS-28 (Nagahama et al., 1996), and rsec22b (Hay et al., 1997). The sec1/munc18 family (boxed) includes three isoforms: munc18-1/18a (also referred to as nsec1 or rbsec1), munc18-2/18b, and munc18c (Hata et al., 1993; Pevsner et al., 1994; Hata and Südhof, 1995; Tellam et al., 1995), Sly1 (Peterson et al., 1996), Vps33 and Vps45 (Pevsner et al., 1994). Finally, the Rab family members (see Takai et al., 1996, for review) are shown in their putative subcellular localization and/or at the trafficking step in which they are likely to be involved. CV, constitutive secretory vesicles; EE, early endosomes; ER, endoplasmic reticulum; LE, late endosomes; SV, synaptic vesicles; TGN, trans-Golgi network.

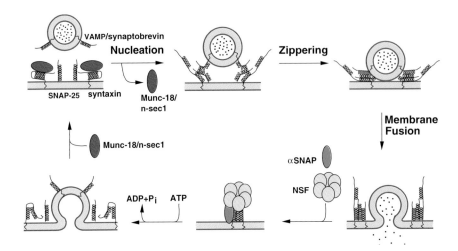

Figure 4.6. A current model of the role of SNAREs in synaptic vesicle exocytosis. Syntaxin is bound to munc18a/nsec1. Once free of munc18a/nsec1, the core complex initiates formation. This core complex consists of two helixes contributed by SNAP-25, one helix contributed by syntaxin, and one helix provided by VAMP/synaptobrevin (only one of the two SNAP-25 helixes is shown in the figure). This complex forces the two lipid bilayers into close apposition, perhaps leading to membrane fusion. After fusion, stable core complexes are part of the plasma membrane. The actions of NSF and α-SNAP dissociate the core complex so that the components can recycle to the appropriate location for another round of vesicle fusion. Although it is possible that the munc18a/syntaxin 1 complex precedes the core complex, this has not actually been demonstrated.

speculative model, and alternative ideas abound, some even going so far as to cast doubt on the significance of the interaction of SM proteins with syntaxins. One of the problems here has been that, apart from the synapse, few direct interactions of an SM protein with a syntaxin have been demonstrated; even yeast SSO1, which is closely related to syntaxin 1, has been proposed not to bind sec1 directly (Carr et al., 1999).

   In view of the stability of the core complex, how is it dissociated after fusion so that the individual components can recycle and take part in another round of membrane fusion? The high stability of the complex dictates that energy will be required for its dissociation. This dissociation is accomplished by NSF, an ATPase that requires SNAP in order to bind to the core complex. NSF acts like a classical chaperone with a broad substrate specificity that dissociates any SNARE core complex (Fig. 4.6). Upon ATP hydrolysis by NSF, the core complex is disassembled, allowing VAMP/synaptobrevin to recycle back to the vesicle, while the majority of SNAP-25 and syntaxin molecules remain associated with the plasma membrane (Söllner et al., 1993b; Pevsner et al., 1994).

   Although a great deal remains to be learned about the regulation of the cycle of SNARE core complex assembly and dissociation, particularly in cells with highly regulated forms of exocytosis such as neurons, the proposed cycle itself is really quite simple (Fig. 4.6). A set of proteins pair across membranes to drive fusion. Three proteins, including an ATPase, regulate aspects of the assembly and disassembly of these proteins. Soon after the characterization of syntaxin 1A as a plasma membrane protein important for secretion in neurons, studies in yeast of trafficking from the endoplasmic reticulum to the Golgi apparatus uncovered a protein called sed5p that exhibits sequence similarity to syntaxin 1A. Interestingly, sed5p is localized in the intermediate Golgi compartment and is critical for transport between the endoplasmic reticulum and the Golgi apparatus (Hardwick and Pelham, 1992). With two syntaxins in hand, one syntaxin critical for vesicular transport between the endoplasmic reticulum and the Golgi apparatus in yeast and another syntaxin important for exocytosis of neurotransmitter in mammalian neurons, two important points became evident. First, the conservation of syntaxins suggested that these proteins function in a similar manner in all eukaryotic organisms, and that neurotransmitter exocytosis should be thought of as an exquisitely regulated form of membrane fusion, comparable to fusion mechanisms that operate in all cells (Bennett and Scheller, 1993). Second, the presence of two syntaxins at different points within the secretory pathway suggested that perhaps a larger set of syntaxins and VAMP/synaptobrevins would specifically localize to particular membrane compartments. If this were indeed the case, perhaps the specific pairing of these proteins could, at least in part, account for the specificity of membrane trafficking. A mechanism for the specificity could be as simple as pairing to form only particular sets of core complexes, or it might involve higher-order interactions.

A clear prediction that can be made from the hypothesis that these proteins are important in determining the organization of membrane compartments in cells is that there will be a number of VAMP/synaptobrevin, syntaxin, and SNAP-25 homologues, and that the proteins will specifically localize to particular organelles or membranes. Several techniques have been used to characterize mammalian SNARE proteins. As the proteins form tight core complexes, they can be readily immunoprecipitated and the components of the complexes characterized (see, e.g., Hay et al., 1997). Experiments using yeast two-hybrid methods have made use of this property, and protein interaction screens with VAMP/synaptobrevin, syntaxin, or SNAP-25 have resulted in the isolation of new homologues of these molecules (Steegmaier et al., 1998). Finally, the large number of sequences in the mammalian data base and the conservation of the coiled-coil domain of SNAREs should make it possible to identify these proteins based on sequence similarity alone (Weimbs et al., 1997; Advani et al., 1998).

In all, more than 15 syntaxins, 10 VAMP/synaptobrevins, and 3 SNAP-25 homologues have been identified. Once the sequences were determined, it was necessary to determine the subcellular localization of the proteins within the cell and to try to understand the particular trafficking step in which the molecules function. A summary of the localization of some of the known vesicle trafficking proteins of the SNARE, Rab, and SM families is presented in Fig. 4.4 for yeast and in Fig. 4.5 for mammals. The proteins localize very specifically to particular membrane compartments, and there are sufficient numbers to mediate most aspects of the specificity of membrane trafficking (Bock et al., 1997; Hay et al., 1998; Prekeris et al., 1998).

The next issue was to determine the characteristics of SNARE protein-protein interactions. Is the specificity of binding interactions consistent with the function and localization of the proteins? A large number of combinations of different VAMP/synaptobrevin, syntaxin, and SNAP-25 complexes were generated and their thermal stabilities determined. All combinations of SNAREs formed stable complexes and the level of stability did not follow any recognizable pattern consistent with known trafficking pathways (Fasshauer et al., 1999; Yang et al., 1999). From these studies it was generally concluded that information for the organization of membrane compartments is not contained in the ability of various combinations of SNAREs to pair. Yet the large number of SNAREs, as well as their precise and restricted localization, does suggest an important role in vesicle targeting. The mechanistic role of SNAREs in the specificity of targeting thus remains unclear.

How Does Calcium Trigger Neurotransmitter Release?

Calcium triggers neurotransmitter release in less than 1 msec, perhaps less than 100 μsec, in a reaction that requires the binding of multiple calcium ions to a low-affinity binding protein. Synaptic vesicles and active zones contain a number of calcium-binding proteins that might be candidates for this role (e.g. calmodulin, rabphilin, synapsin I). The first and still the best candidate, however, is synaptotagmin (Perin et al., 1990). Synaptotagmin is a synaptic vesicle protein that was initially observed by chance in a monoclonal antibody screen for active zone proteins (Matthew et al., 1981). When the structure of synaptotagmin was elucidated, it was found to contain a single N-terminal transmembrane region and two stretches of homology to the C_2 sequence of protein kinase C (Perin et al., 1990; see Fig. 4.7). Subsequent work established that the C_2-like sequences in synaptotagmin represent independently folding domains that bind multiple calcium ions (Davletov and Südhof, 1993) and interact with a number of potential targets. Furthermore, more than 10 synaptotagmins with differential expression patterns have been observed in brain (see Rizo and Südhof, 1998, for review). Not all synaptotagmins are on synaptic vesicles, and some are also expressed outside the brain, albeit at low levels (Li et al., 1995; Butz et al., 1999). In addition, many other proteins contain C_2 domains, particularly proteins that function in signal transduction or in synaptic membrane trafficking (Fig. 4.7). As initially shown for synaptotagmin (Davletov and Südhof, 1993), the

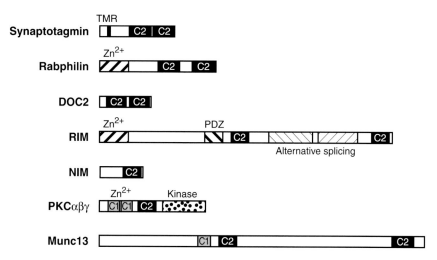

Figure 4.7. Comparative structures of C_2 domain proteins involved in exocytosis. The domain structures of the various proteins are shown schematically. Each protein is part of a gene family of multiple proteins that includes synaptotagmins I–XII, rabphilin, DOC2A and DOC2B, RIM1 and RIM2, NIM2 and NIM3, multiple isoforms of protein kinase C (PKC), and munc13-1, -2, and -3. Selected domains in addition to the C_2 domains are identified: PDZ, PDZ domain; TMR, transmembrane region; Zn^{2+}, zinc-finger domain.

C_2 domains of all of these proteins constitute independently folding domains. In many of these proteins, the C_2 domains are also calcium-binding modules that confer on their resident proteins calcium-dependent properties. In fact, C_2 domains are now known to be probably the second most widely distributed calcium-binding motif after the EF hand (Rizo and Südhof, 1998).

Mutation of synaptotagmins in mice, flies, and worms has demonstrated that synaptotagmins play an important role in regulating neurotransmitter release, while not being essential for release as such. This role was defined most clearly in mutant mice. Electrophysiological tools available for cultured neurons made it possible to define the point of action of synaptotagmin in exocytosis (Geppert et al., 1994b). These studies revealed that synaptotagmin is essential only for fast, calcium-triggered release but not for any other step in the exocytotic reaction; it thus is required for the calcium-sensor step. Although this result does not establish that calcium binding to synaptotagmin triggers release, the following evidence supports this hypothesis: (1) Synaptotagmin binds multiple calcium ions via its two C_2 domains. It does so with a relatively low affinity and with a divalent cation specificity that resembles the properties of the exocytotic calcium sensor. (2) As an abundant synaptic vesicle protein, synaptotagmin is located at the right place to function as a calcium sensor. (3) Synaptotagmin binds to phospholipids, syntaxin, and itself as a function of calcium binding. This suggests the possibility that calcium-induced interaction of synaptotagmin with the SNARE core complex and lipids, in conjunction with its multimerization, could trigger fusion-pore opening. However, until the effects of mutations in both calcium-binding modules of synaptotagmin are tested, it will be impossible to be certain whether synaptotagmin is the calcium sensor itself or a component of the calcium sensor. Of the other known calcium-binding proteins at the synapse, calmodulin has too high an affinity for calcium to fit the role, and mouse knockouts of rabphilin and synapsin have shown that they are not essential for calcium triggering (Rosahl et al., 1995; Schlüter et al., 1999). However, the current absence of a suitable candidate does not exclude the possibility that such a candidate will yet be found; for example, it is possible that an unusual, and as yet undefined, calcium-binding site associated with SNARE core complexes is involved.

Function of Rab3 GTP-Binding Proteins in Synaptic Vesicle Exocytosis

Whenever an action potential invades a nerve terminal, calcium floods the synaptic cytosol; nevertheless, transmitter release is induced unreliably, and usually only one of several "ready" vesicles is stimulated (Korn et al., 1994). It is unclear how release is restricted and how vesicles are selected for release. It is possible that the intrinsic probability of the calcium response by the calcium sensor, be it synaptotagmin or some other protein, is so low that multiple signals are required, on the average, to elicit a response. A second consideration is the removal of calcium; the presence of calcium buffers and pumps may limit the time during which calcium is available for action after its entry into the terminals. It seems certain, however, that there are also active regulatory processes that limit the amount of release. At least one of these processes is mediated by rab3A, a low-molecular-weight GTP-binding protein of synaptic vesicles (Fischer von Mollard et al., 1990). Like SNAREs and SM proteins, Rab GTPases function throughout the secretory pathway in yeast and in mammalian cells (Figs. 4.4 and 4.5).

Rab3A belongs to a subfamily of the overall family of rab proteins. This subfamily includes rab3B, -3C, and -3D, in addition to rab3A. Rab3A and -3C are localized to synaptic vesicles in brain and to endocrine secretory granules, whereas rab3D appears to be primarily expressed in fat cells and exocrine glands (Fischer von Mollard et al., 1990, 1994; Baldini et al., 1992; Ohnishi et al., 1996). Little is known about rab3B. Like other rab proteins, rab3 molecules are attached to the membrane by a posttranslational modification of the C-terminal residues consisting of two geranylgeranyl chains. These hydrophobic additions confer on rab3 proteins the properties of an intrinsic membrane protein of synaptic vesicles (Johnston et al., 1991). Unlike a "real" membrane protein, however, rab3's can dissociate from the synaptic vesicle as a function of activity. Eukaryotic cells express an essential protein called GDI, which is capable of recognizing GDP-complexed rab proteins and removing them from the membrane by enveloping their hydrophobic C-terminal modifications (Ullrich et al., 1993). As a result, rab3's undergo a cycle of membrane binding and dissociation that parallels the synaptic vesicle cycle (Fischer von Mollard et al., 1991). This dynamic association of rab3 with synaptic vesicles is schematically shown in Fig. 4.8 in conjunction with the interactions of rab3 with putative effectors. Normally, rab3's are on the synaptic vesicles in the GTP-bound form. Sometime during or after exocytosis, GTP is hydrolyzed to GDP, resulting in the recognition of the GDP-rab3 by GDI and its subsequent removal from the vesicle. The GDP-rab3 then becomes a substrate for a rab3-GDP/GTP exchange protein that mediates nucleotide exchange with rebinding of rab3's to synaptic vesicles. Although this activity-dependent cycle of association and dissociation for a rab protein has only been shown at the synapse (Fischer von Mollard et al., 1991), it is probable that all rab proteins are subject to the same pattern of cycling. It seems likely that this cycling ensures that rab proteins act only at a particular step in the trafficking pathway of an organelle.

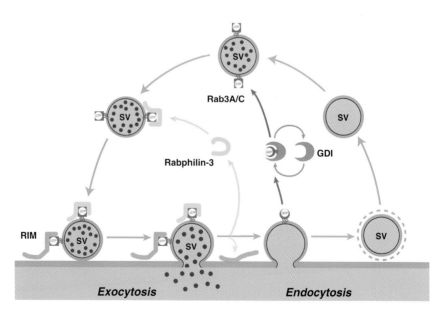

Figure 4.8. The rab3 cycle. The association of rab3A and rab3C in the cycle of synaptic vesicles (SV) is shown schematically. GTP-rab3 is thought to recruit rabphilin to synaptic vesicles and to bind to RIM when the vesicles are attached to the presynaptic plasma membrane. During or after exocytosis, GTP-rab3 is hydrolyzed to GDP-rab3, leading to dissociation of rabphilin and RIM and removal of GDP-rab3 from the vesicles by GDI. The cycle then starts anew when GDP-rab3 is rebound to free vesicles in the nerve terminal under GDP-to-GTP exchange. Adapted from Südhof (1997).

Very little is known about the functions of rab3 proteins except for rab3A. Analysis of mice lacking rab3A has shown that nerve terminals lacking this GTP-binding protein release more neurotransmitter per calcium signal than normal nerve terminals (Geppert et al., 1994a, 1997). This phenomenon leads to an accelerated rundown of the synaptic response during repetitive stimulation. Thus rab3A is to be considered a negative regulator of synaptic vesicle exocytosis, a conclusion that agrees well with transfection experiments in chromaffin cells (Holz et al., 1994; Johannes et al., 1994). Even though the absence of rab3A does not appear to cause morbidity in mice, the regulatory function of rab3A is physiologically important. In the absence of rab3A, short-term synaptic plasticity is relatively unimpaired, but presynaptic LTP or LTD in the mossy fibers cannot be normally induced (Castillo et al., 1997; Tzounopoulos et al., 1998). This finding suggests that this type of long-term synaptic plasticity normally acts according to a pathway that requires rab3A, and it incidentally supports the view that these plastic changes are presynaptic; in addition it confirms the negative regulatory function of rab3A. It seems likely that at least rab3B and -3C are partly or completely redundant with rab3A in the nervous system, so that the rab3A knockout phenotype does not reflect the full range of rab3A functions but only the rate-limiting functions of this abundant GTP-binding protein. This issue remains to be addressed experimentally.

How does rab3A function in regulating release? By analogy to other regulatory GTP-binding proteins, such as ras, the most likely mechanism is the GTP-dependent interaction with a putative effector. There are two known putative effectors for rab3A in brain, rabphilin and RIM (schematically shown in Fig. 4.7; Shirataki et al., 1993; Y. Wang et al., 1997). Interestingly, both effectors bind to rab3A as a function of GTP via a homologous N-terminal zinc finger domain whose three-dimensional structure has been solved in the complex of rabphilin with rab3A (Ostermeier and Brünger, 1998). Apart from the fact that both RIM and rabphilin contain C_2 domains in addition to the zinc finger domain, the two proteins are dissimilar. Rabphilin is a soluble protein that is recruited to synaptic vesicles as a function of rab3's (Li et al., 1994; Stahl et al., 1996). As a result, it cycles on and off synaptic vesicles together with rab3's. RIM, by contrast, is an insoluble protein that is a component of the presynaptic active zone. The distinct stage-dependent interactions of rabphilin and RIM with GTP-rab3 during the synaptic vesicle cycle are diagrammed in Fig. 4.8. In addition to the difference in subcellular localization, rabphilin and RIM have differences in protein architecture (Fig. 4.7) in that RIM has multiple domains that are absent from rabphilin, including a PDZ domain. Finally the C_2 domains of rabphilin bind calcium with a relatively high affinity (Ubach et al., 1999) whereas the sequences of the RIM C_2 domains are different from those of other C_2 domains, and at present it is unclear that RIM binds calcium.

It is also unknown just how the two putative effectors of rab3's relate to each other. Although overexpression and microinjection experiments in chromaffin cells have indicated a major role for rabphilin in regulating release, knockouts of rabphilin in mice have no significant phenotype; in particular, they do not phenocopy the rab3A knockout mice (Schlüter et al., 1999). Some of the major challenges for our understanding of the function of rab3 proteins at the synapse will be to determine if the various rab3 proteins are functionally redundant (beyond what has already been shown for rab3A); to discover which effectors they use; and to investigate how these effectors regulate transmitter release.

Protein Dynamics in the Synaptic Vesicle Cycle: Role of Synapsins

The above description of rab3's should make it apparent that a synaptic vesicle is not a constant entity: as it moves through the synaptic vesicle cycle, it changes its protein composition. In addition to rab3 and rabphilin, another protein family cycles on and off the vesicle during activity: these are the synapsins (Hosaka et al., 1999). The synapsins comprise a family of at least five proteins that form homo- and heterodimers on the vesicle surface, and together they are among the most abundant and enigmatic proteins of synaptic vesicles. Initially discovered as an abundant brain substrate for cAMP-dependent protein kinase (PKA) (Greengard, 1987; Johnson et al., 1972), synapsins are peripheral membrane proteins that coat synaptic vesicles. Synapsins are composed of mosaics of domains, the largest and most conserved of which is the central C domain (Fig. 4.9; Südhof et al., 1989b). The C domain is preceded by two small domains (the A and B domains that are present in all synapsins) and is followed by a combination of variable domains that differ among synapsins. Interestingly, the C domains of synapsins are structurally homologous to ATP synthases, and they are known to bind ATP with high affinity (Esser et al., 1998; Hosaka and Südhof, 1998a,b); however, no enzyme activity has as yet been demonstrated in these domains. Synapsins are probably the most abundant substrates for PKA in the brain; this kinase phosphorylates all synapsins at a single N-terminal site in the A domain (Fig. 4.9). This is the only conserved phosphorylation site in all synapsins. In the nonphosphorylated state, synapsins are bound to synaptic vesicles as dimers, probably partly via a phospholipid-binding activity associated with the A domain (Hosaka and Südhof, 1999; Hosaka et al., 1999). Phosphorylation of the A domain inhibits phospholipid binding and causes dissociation of synapsins. Thus synapsins, like rab3's and rabphilin, cycle on and off synaptic vesicles as a function of activity; however, the type of activity is different: whereas rab3 and rabphilin cycle strictly in parallel with exocytosis, the synapsins cycle on and off depending on the activation of protein kinases.

What is the function of the synapsins? In spite of a large body of work, this is still uncertain. Biochemically, synapsins bind to a large number of proteins in vitro, including actin microfilaments, spectrin, tubulin,

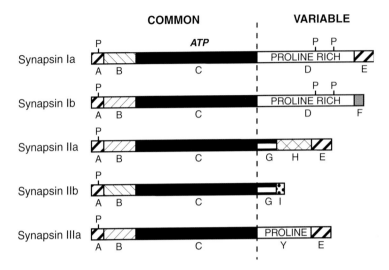

Figure 4.9. Mosaic domain structure of synapsins. Three synapsin genes (synapsin I, II, and III) generate at least five transcripts by alternative splicing. The synapsin I and II transcripts are abundant whereas the synapsin III transcripts are scarce. All synapsins share three common N-terminal domains that account for most of their sequences. These common N-terminal domains are followed by various combinations of C-terminal domains that differ among synapsins. The short conserved N-terminal A domain of synapsins is a phospholipid-binding domain that is regulated by phosphorylation at a single site (indicated by P). This is the only phosphorylation site conserved in all synapsins, and it regulates the association of synapsins with synaptic vesicles. The central C domain of synapsins is an ATP-binding module that is closely related to bacterial ATP synthases (Esser et al., 1998; Hosaka and Südhof, 1998a,b). The C-terminal variable domains contain multiple phosphorylation sites for several kinases that selectively phosphorylate only a subset of synapsins. Adapted from Südhof et al. (1989b).

and neurofilaments. This property has led to the idea that synapsins may link synaptic vesicles to all elements of the cytoskeleton, but this notion does not fit well when one considers the abundance of synapsins on synaptic vesicles and the relative paucity of cytoskeletal elements at the active zone. Another idea is that synapsins directly regulate the release reaction. This idea derives from electrophysiological analyses of mice lacking synapsins I and II. In such mice, short-term synaptic plasticity is selectively impaired; this finding suggests that synapsins may have a role in regulating release probability in conjunction with phosphorylation (Rosahl et al., 1995; Jovanovic et al., 2000). Whatever the precise roles of synapsins (and rab3 proteins), it is now clear that in addition to their "permanent components" synaptic vesicles contain transiently bound proteins that associate with the vesicles in a highly regulated manner. Thus the picture of synaptic vesicles that emerges is that they are organelles whose composition is modulated depending on the regulatory state of the nerve terminal.

General Implications for Membrane Traffic

Neurotransmitter release is brought about by a specialized membrane trafficking pathway, the synaptic vesicle cycle. As the process that initiates synaptic transmission, neurotransmitter release is of central importance for nervous system function. In order to achieve the special requirements of synaptic secretion, the synaptic vesicle cycle differs in some important respects from other intracellular trafficking pathways. The major difference is the high degree of regulation of intracellular trafficking in the nerve terminal: the exclusive targeting of exocytosis to active zones, the high speed with which calcium triggers release, and the restriction of exocytosis at any given time to a single synaptic vesicle. Despite these differences, however, synaptic vesicle trafficking shares the basic characteristics of other intracellular trafficking pathways and operates through the same fundamental mechanisms.

Not only are the special properties of the synaptic vesicle cycle important for nervous system function, they have also provided insights into general intracellular trafficking mechanisms. Indeed many of the trafficking proteins that are now known to function ubiquitously were originally described at the synapse. Because the synaptic pathway is so highly regulated and synaptic transmission can be measured with such high temporal resolution, it has been possible to study different stages of exocytosis at the synapse mechanistically. Thus the study of the synaptic vesicle pathway lies at the interface between cell biology and neurobiology, and it has benefited—and undoubtedly will continue to benefit—both disciplines.

Although several molecules and mechanisms that are now known to be generally involved in membrane traffic were originally discovered at the synapse (e.g., SNAREs), there are some puzzling differences between synaptic and nonsynaptic membrane traffic that must be resolved. For example, several intracellular fusion reactions have been shown to be absolutely dependent on calcium. In the case of vacuole fusion and fusion of endoplasmic reticulum membranes, calcium release from the corresponding organelles is required (Sullivan et al., 1993; Peters and Mayer, 1998). At first glance this may appear similar to synaptic membrane fusion, which is so tightly regulated by calcium. However, synaptic vesicle exocytosis can be stimulated in the complete absence of calcium; hypertonic sucrose effectively triggers release even when the cytosol of the axon terminal is filled with BAPTA (Rosenmund and Stevens, 1996). Thus, paradoxically, the most highly calcium-regulated fusion event appears to be intrinsically calcium-independent—at least when stimulated by nonphysiological means—while constitutive fusion reactions appear to be calcium-dependent.

Another puzzling discrepancy between synaptic and nonsynaptic fusion concerns the role of rab proteins. The major rab protein at the synapse, rab3A, is required for the regulation of a late step in synaptic vesicle exocytosis, a step that is clearly placed after membrane attachment

(Geppert et al., 1997). The evidence for this finding is based not only on the rather sensitive measurements on single synaptic boutons in hippocampal cultures, but also on the lack of mossy fiber LTP and LTD (which represent a change in release probability) in the absence of rab3A. In yeast, however, rab proteins are clearly essential for membrane attachment at an early step of fusion (Cao et al., 1998; Ungermann et al., 1998). It seems rather unlikely that rab proteins could have very different functions in different membrane trafficking reactions. Perhaps the current analysis is simply incomplete in both systems. For example, at the synapse it is possible that rab3-like proteins have multiple, partially redundant functions. In the absence of rab3A but the continued presence of other rab3's, possibly the kinetically most sensitive functions are affected first.

The elucidation of mechanisms that are common to all fusion reactions, and the definitions of specialized features that differentiate particular fusion reactions, is one of the fascinating challenges for the future in this field. Finally, an ultimate challenge will be to understand fully, at the molecular level, the presynaptic modulation of neurotransmitter release that contributes to the mechanisms of learning and memory.

References

Aalto, M. K., Ruohonen, L., Hosono, K., Keränen, S. (1991). Cloning and sequencing of the yeast *Saccharomyces cerevisiae* SEC1 gene localized on chromosome IV. *Yeast* 7:643–650.

Aalto, M. K., Ronne, H., Keränen, S. (1993). Yeast syntaxins Sso1p and Sso2p belong to a family of related membrane proteins that function in vesicular transport. *EMBO J.* 12:4095–4104.

Advani, R. J., Bae, H.-R., Bock, J. B., Chao, D. S., Doung, Y.-C., Prekeris, R., Yoo, J.-S., Scheller, R. H. (1998). Seven novel mammalian SNARE proteins localize to distinct membrane compartments. *J. Biol. Chem.* 273:10,317–10,324.

Augustin, I., Rosenmund, C., Südhof, T. C., Brose, N. (1999). Munc-13 is essential for fusion competence of glutamatergic synaptic vesicles. *Nature* 400: 457–461.

Bajjalieh, S. M., Frantz, G. D., Weimann, J. M., McConnell, S. K., Scheller, R. H. (1994). Differential expression of synaptic vesicle protein 2 (SV2) isoforms. *J. Neurosci.* 14:5223–5235.

Baldini, G., Hohl, T., Lin, H. Y., Lodish, H. F. (1992). Cloning of a Rab3 isotype predominantly expressed in adipocytes. *Proc. Natl. Acad. Sci. USA* 89:5049–5052.

Banfield, D. K., Lewis, M. J., Pelham, H. R. (1995). A SNARE-like protein required for traffic through the Golgi complex. *Nature* 375:806–809.

Banta, L. M., Vida, T. A., Herman, P. K., Emr, S. D. (1990). Characterization of yeast Vps33p, a protein required for vacuolar protein sorting and vacuole biogenesis. *Mol. Cell Biol.* 10:4638–4649.

Barnstable, C. J., Hofstein, R., Akagawa, K. (1985). A marker of early amacrine cell development in rat retina. *Brain Res.* 352:286–290.

Becherer, K. A., Rieder, S. E., Emr, S. D., Jones, E. W. (1996). Novel syntaxin homolog, Pep12p, required for the sorting of lumenal hydrolases to the lysosome-like vacuole in yeast. *Mol. Biol. Cell* 7:579–594.

Bennett, M. K., Scheller, R. H. (1993). The molecular machinery for secretion is conserved from yeast to neurons. *Proc. Natl. Acad. Sci. USA* 90:2559–2563.

Bennett, M. K., Calakos, N., Scheller, R. H. (1992). Syntaxin: a synaptic protein implicated in docking of synaptic vesicles at presynaptic active zones. *Science* 257:255–259.

Betz, W. J., Bewick, G. S. (1992). Optical analysis of synaptic vesicle recycling at the frog neuromuscular junction. *Science* 255: 200–203.

Binz, T., Blasi, J., Yamasaki, S., Baumeister, A., Link, E., Südhof, T. C., Jahn, R., Niemann, H. (1994). Cleavage of members of the synaptobrevin/VAMP family by types D and F botulinal neurotoxins and tetanus toxin. *J. Biol. Chem.* 269:1617–1620.

Blasi, J., Chapman, E. R., Link, E., Binz, T., Yamasaki, S., De Camillip, P., Südhof, T. C., Niemann, H., Jahn, R. (1993a). Botulinum neurotoxin A selectively cleaves the synaptic protein SNAP-25. *Nature* 365:160–163.

Blasi, J., Chapman, E. R., Yamasaki, S., Binz, T., Niemann, H., Jahn, R. (1993b). Botulinum neurotoxin C1 blocks neurotransmitter release by means of cleaving HPC-1/syntaxin. *EMBO J.* 12:4821–4828.

Bock, J. B., Lin, R. C., Scheller, R. H. (1996). A new syntaxin family member implicated in targeting of intracellular transport vesicles. *J. Biol. Chem.* 271: 17,961–17,965.

Bock, J. B., Klumperman, J., Davanger, S., Scheller, R. H. (1997). Syntaxin 6 functions in trans-Golgi network vesicle trafficking. *Mol. Biol. Cell* 8:1261–1271.

Brenner, S. (1974). The genetics of *Caenorhabditis elegans*. *Genetics* 77:71–94.

Brennwald, P., Kearns, B., Champion, K., Keränen, S., Bankaitis, V., Novick, P. (1994). Sec9 is a SNAP-25-like component of a yeast SNARE complex that may be the effector of Sec4 function in exocytosis. *Cell* 79:245–258.

Brose, N., Hofmann, K., Hata, Y., Südhof, T. C. (1995). Mammalian homologues of *C. elegans unc-13* gene define novel family of C_2-domain proteins. *J. Biol. Chem.* 270:25,273–25,280.

Buchner, E., Gundersen, C. B. (1997). The DnaJ-like cysteine string protein and exocytotic neurotransmitter release. *Trends Neurosci.* 20:223–227.

Butz, S., Fernandez-Chacon, R., Schmitz, F., Jahn, R., Südhof, T. C. (1999). The subcellular localizations of atypical synaptotagmins: synaptotagmin III is enriched in synapses and synaptic plasma membranes but not in synaptic vesicles. *J. Biol. Chem.* 274:18,290–18,296.

Calakos, N., Bennett, M. K., Peterson, K. E., Scheller, R. H. (1994). Protein-protein interactions contributing to the specificity of intracellular vesicular trafficking. *Science* 263:1146–1149.

Cao, X., Ballew, N., Barlowe, C. (1998). Initial docking of ER-derived vesicles requires Uso1p and Ypt1p but is independent of SNARE proteins. *EMBO J.* 17:2156–2165.

Carr, C. M., Grote, E., Munson, M., Hughson, F. M., Novick, P. J. (1999). Sec1p binds to SNARE complexes and concentrates at sites of secretion. *J. Cell Biol.* 146:333–344.

Castillo, P. E., Janz, R., Südhof, T. C., Malenka, R. C., Nicoll, R. A. (1997). The synaptic vesicle protein Rab3A is essential for mossy fiber long term potentiation in the hippocampus. *Nature* 388:590–593.

Chamberlain, L. H., Burgoyne, R. D. (1996). Identification of a novel cysteine string protein variant and expression of cysteine string proteins in non-neuronal cells. *J. Biol. Chem.* 271:7320–7323.

Chen, Y. A., Scales, S. J., Patel, S. M., Doung, Y.-C., Scheller, R. H. (1999). SNARE complex formation is triggered by Ca^{2+} and drives membrane fusion. *Cell* 97:165–174.

Cowles, C. R., Emr, S. D., Horazdovsky, B. F. (1994). Mutations in the *VPS45* gene, a *SEC1* homologue, result in vacuolar protein sorting defects and accumulation of membrane vesicles. *J. Cell Sci.* 107:3449–3459.

Dascher, C., Ossig, R., Gallwitz, D., Schmitt, H. D. (1991). Identification and structure of four yeast genes (*SLY*) that are able to suppress the functional loss of *YPT1*, a member of the *RAS* superfamily. *Mol. Cell Biol.* 11:872–885.

Davletov, B., Südhof, T. C. (1993). A single C_2-domain from synaptotagmin I is sufficient for high affinity Ca^{2+}/phospholipid-binding. *J. Biol. Chem.* 268: 26,386–26,390.

Dodge, F. A., Jr., Rahamimoff, R. (1967). Co-operative action of calcium ions in transmitter release at the neuromuscular junction. *J. Physiol.* 193:419–432.

Dulubova, I., Sugita, S., Hill, S., Hosaka, M., Fernandez, I., Südhof, T. C., Rizo, J. (1998). A conformational switch in syntaxin during exocytosis: role of munc18. *EMBO J.* 18:4372–4382.

Elferink, L. A., Trimble, W. S., Scheller, R. H. (1989). Two vesicle-associated membrane protein genes are differentially expressed in the rat central nervous system. *J. Biol. Chem.* 264:11,061–11,064.

Esser, L., Wang, C.-R., Hosaka, M., Smagula, C. S., Südhof, T. C., Deisenhofer, J. (1998). Synapsin I is structurally similar to ATP-utilizing enzymes. *EMBO J.* 17:977–984.

Fasshauer, D., Antonin, W., Margittai, M., Pabst, S., Jahn, R. (1999). Mixed and non-cognate SNARE complexes. Characterization of assembly and biophysical properties. *J. Biol. Chem.* 274:15,440–15,446.

Fatt, P., Katz, B. (1952). Spontaneous subthreshold activity at motor nerve endings. *J. Physiol.* 117:109–128.

Fernandez, I., Ubach, J., Dulubova, I., Zhang, X., Südhof, T. C., Rizo, J. (1998). Three-dimensional structure of an evolutionarily conserved N-terminal domain of syntaxin 1A. *Cell* 94:841–849.

Fernandez-Chacon, R., Südhof, T. C. (1999). Genetics of synaptic vesicle function: towards the complete functional anatomy of an organelle. *Annu. Rev. Physiol.* 61:753–776.

Fischer von Mollard, G., Mignery, G. A., Baumert, M., Perin, M. S., Hanson, T. J., Burger, P. M., Jahn, R., Südhof, T. C. (1990). Rab3 is a small GTP-binding protein exclusively localized to synaptic vesicles. *Proc. Natl. Acad. Sci. USA* 87: 1988–1992.

Fischer von Mollard, G., Südhof, T. C., Jahn, R. (1991). A small GTP-binding protein (rab3A) dissociates from synaptic vesicles during exocytosis. *Nature* 349:79–81.

Fischer von Mollard, G. A., Clementi, E., Raichman, M., Südhof, T. C., Ullrich, A., Meldolesi, J. (1994). Stable expression of truncated inositol 1,4,5-trisphosphate receptor subunits in 3T3 fibroblasts. *J. Biol. Chem.* 269:19,216–19,224.

Fischer von Mollard, G., Nothwehr, S. F., Stevens, T. H. (1997). The yeast v-SNARE Vti1p mediates two vesicle transport pathways through interactions with the t-SNAREs Sed5p and Pep12p. *J. Cell Biol.* 137:1511–1524.

Foletti, D. L., Prekeris, R., Scheller, R. H. (1999). Generation and maintenance of neuronal polarity: mechanisms of transport and targeting. *Neuron* 23: 641–644.

Forgac, M. (1999). Structure and properties of the vacuolar H^+-ATPases. *J. Biol. Chem.* 274:12,951–12,954.

Fykse, E. M., Fonnum, F. (1996). Amino acid neurotransmission: dynamics of vesicular uptake. *Neurochem. Res.* 21:1053–1060.

Fykse, E. M., Takei, K., Walch-Solimena, C., Geppert, M., Jahn, R., De Camilli, P., Südhof, T. C. (1993). Relative properties and localizations of synaptic vesicle protein isoforms: the case of the synaptophysins. *J. Neurosci.* 13:4997–5007.

Gengyo-Ando, K., Kamiya, Y., Yamakawa, A., Kodaira, K., Nishiwaki, K., Miwa, J., Hori, I., Hosono, R. (1993). The *C. elegans* unc-18 gene encodes a protein expressed in motor neurons. *Neuron* 11:703–711.

Geppert, M., Südhof, T. C. (1997). Rab3 and synaptotagmin: the yin and yang of synaptic membrane fusion. *Annu. Rev. Neurosci.* 21:75–95.

Geppert, M., Archer, B. T., III, Südhof, T. C. (1991). Synaptotagmin II: a novel differentially distributed form of synaptotagmin. *J. Biol. Chem.* 266:13,548–13,552.

Geppert, M., Bolshakov, V., Siegelbaum, S. A., Takei, K., De Camilli, P., Hammer, R. E., Südhof, T. C. (1994a). The role of rab3A in neurotransmitter release. *Nature* 369:493–497.

Geppert, M., Goda, Y., Hammer, R. E., Li, C., Rosahl, T. W., Stevens, C. F., Südhof, T. C. (1994b). Synaptotagmin I: A major Ca^{2+} sensor for transmitter release at a central synapse. *Cell* 79:717–727.

Geppert, M., Goda, Y., Stevens, C. F., Südhof, T. C. (1997). Rab3A regulates a late step in synaptic vesicle fusion. *Nature* 387:810–814.

Greengard, P. (1987). Neuronal phosphoproteins. Mediators of signal transduction. *Mol. Neurobiol.* 1:81–119.

Haass, N. K., Kartenbeck, M. A., Leube, R. E. (1996). Pantophysin is a ubiquitously expressed synaptophysin homologue and defines constitutive transport vesicles. *J. Cell Biol.* 134:731–746.

Hanson, P. I., Roth, R., Morisaki, H., Jahn, R., Heuser, J. E. (1997). Structure and conformational changes in NSF and its membrane receptor complexes visualized by quick-freeze/deep-etch electron microscopy. *Cell* 90:523–535.

Hardwick, K. G., Pelham, H. R. (1992). SED5 encodes a 39-kD integral membrane protein required for vesicular transport between the ER and the Golgi complex. *J. Cell Biol.* 119:513–521.

Hata, Y., Südhof, T. C. (1995). A novel ubiquitous form of munc18 interacts with multiple syntaxins. *J. Biol. Chem.* 270:13,022–13,028.

Hata, Y., Slaughter, C. A., Südhof, T. C. (1993). Synaptic vesicle fusion complex contains unc-18 homologue bound to syntaxin. *Nature* 366:347–351.

Hay, J. C., Hirling, H., Scheller, R. H. (1996). Mammalian vesicle trafficking proteins of the endoplasmic reticulum and Golgi apparatus. *J. Biol. Chem.* 271: 5671–5679.

Hay, J. C., Chao, D. S., Kuo, C. S., Scheller, R. H. (1997). Protein interactions regulating vesicle transport between the endoplasmic reticulum and Golgi apparatus in mammalian cells. *Cell* 89:149–158.

Hay, J. C., Klumperman, J., Oorschot, V., Steegmaier, M., Kuo, C. S., Scheller, R. H. (1998). Localization, dynamics, and protein interactions reveal distinct roles for ER and Golgi SNAREs. *J. Cell. Biol.* 141:1489–1502.

Hayashi, T., McMahon, H. T., Yamasaki, S., Binz, T., Hata, Y., Südhof, T. C., Niemann, H. (1994). Synaptic vesicle membrane fusion complex: action of clostridial neurotoxins on assembly. *EMBO J.* 13:5051–5061.

Heinemann, C., Chow, R. H., Neher, E., Zucker, R. S. (1994). Kinetics of the secretory response in bovine chromaffin cells following flash photolysis of caged Ca^{2+}. *Biophys. J.* 267:2546–2557.

Heuser, J. E., Reese, T. S. (1973). Evidence for recycling of synaptic vesicle membrane during transmitter release at the frog neuromuscular junction. *J. Cell Biol.* 57:315–344.

Holthuis, J. C., Nichols, B. J., Dhruvakumar, S., Pelham, H. R. (1998). Two syntaxin homologues in the TGN/endosomal system of yeast. *EMBO J.* 17:113–126.

Holz, R. W., Brondyk, W. H., Senter, R. A., Kuizon, L., Macara, I. G. (1994). Evidence for the involvement of Rab3A in Ca^{2+}-dependent exocytosis from adrenal chromaffin cells. *J. Biol. Chem.* 269:10,229–10,234.

Hosaka, M., Südhof, T. C. (1998a). Synapsins I and II are ATP-binding proteins with differential Ca^{2+} regulation. *J. Biol. Chem.* 273:1425–1429.

Hosaka, M., Südhof, T. C. (1998b). Synapsin III, a novel synapsin with an unusual regulation by Ca^{2+}. *J. Biol. Chem.* 273:13,371–13,374.

Hosaka, M., Südhof, T. C. (1999). Homo- and heterodimerization of synapsins. *J. Biol. Chem.* 274:16,747–16,753.

Hosaka, M., Hammer, R. E., Südhof, T. C. (1999). A phospho-switch controls the dynamic association of synapsins with synaptic vesicles. *Neuron* 24:377–387.

Hosono, R., Hekimi, S., Kamiya, Y., Sassa, T., Murakami, S., Nishiwaki, K., Miwa, J., Taketo, A., Kodaira, K. I. (1992). The unc-18 gene encodes a novel protein affecting the kinetics of acetylcholine metabolism in the nematode *Caenorhabditis elegans*. *J. Neurochem.* 58:1517–1525.

Inoue, A., Obata, K., Akagawa, K. (1992). Cloning and sequence analysis of cDNA for a neuronal cell membrane antigen, HPC-1. *J. Biol. Chem.* 267:10,613–10,619.

Jahn, R., Südhof, T. C. (1993). Synaptic vesicle traffic: rush hour in the nerve terminal. *J. Neurochem.* 61:12–21.

Jahn, R., Südhof, T. C. (1999). Membrane fusion and exocytosis. *Annu. Rev. Biochem.* 68:863–911.

Janz, R., Südhof, T. C. (1998). Cellugyrin, a novel ubiquitous form of synaptogyrin that is phosphorylated by pp60^c-src. *J. Biol. Chem.* 273:2851–2857.

Janz, R., Südhof, T. C. (1999). SV2C is a synaptic vesicle protein with an unusually restricted localization: anatomy of a synaptic vesicle protein family. *Neuroscience* 94:1279–1290.

Janz, R., Goda, Y., Geppert, M., Missler, M., Südhof, T. C. (1999a). SV2A and SV2B function as redundant Ca^{2+} regulators in neurotransmitter release. *Neuron* 24:1003–1016.

Janz, R., Hofmann, K., Südhof, T. C. (1999b). SVOP, an evolutionarily conserved synaptic vesicle protein, suggests novel transport functions of synaptic vesicles. *J. Neurosci.* 18:9269–9281.

Janz, R., Südhof, T. C., Hammer, R. E., Unni, V., Siegelbaum, S. A., Bolshakov, V. Y. (1999c). Essential roles in synaptic plasticity for synaptogyrin I and synaptophysin I. *Neuron* 24:687–700.

Johannes, L., Lledo, P. M., Roa, M., Vincent, J. D., Henry, J. P., Darchen, F. (1994). The GTPase Rab3a negatively controls calcium-dependent exocytosis in neuroendocrine cells. *EMBO J.* 13:2029–2037.

Johnson, E. M., Ueda, T., Maeno, H., Greengard, P. (1972). Adenosin 3′:5′-monophosphate-dependent phosphorylation of a specific protein in synaptic membrane fractions from rat cerebrum. *J. Biol. Chem.* 247:5650–5652.

Johnston, P. A., Cameron, P. L., Stukenbrok, H., Jahn, R., DeCamilli, P., Südhof, T. C. (1989). Synaptophysin is targeted to similar microvesicles in CHO- and PC12-cells. *EMBO J.* 8:2863–2872.

Johnston, P. A., Archer, B. T., III, Robinson, K., Mignery, G. A., Jahn, R., Südhof, T. C. (1991). Rab3A attachment to the synaptic vesicle membrane mediated by a conserved polyisoprenylated carboxy-terminal sequence. *Neuron* 7:101–109.

Jovanovic, J. N., Czernik, A. J., Fienberg, A. A., Greengard, P., Sihra, T. S. (2000). Synapsins as mediators of BDNF-enhanced neurotransmitter release. *Nature Neurosci.* 3:323–329.

Katz, B. (1969). *The release of neural transmitter substances.* Liverpool: Liverpool University Press.

Knaus, P., Marqueze-Pouey, B., Scherer, H., Betz, H. (1990). Synaptoporin, a novel putative channel protein of synaptic vesicles. *Neuron* 5:453–462.

Korn, H., Sur, C., Charpier, S., Legendre, P. I., Faber, D. S. (1994). The one-vesicle hypothesis and multivesicular release. *Adv. Second Mess. Phosphoprotein Res.* 29:301–322.

Kutay, U., Ahnert-Hilger, G., Hartmann, E., Wiedenmann, B., Rapoport, T. A. (1995). Transport route for synaptobrevin via a novel pathway of insertion into the endoplasmic reticulum membrane. *EMBO J.* 14:217–223.

Lazar, T., Götte, M., Gallwitz, D. (1997). Vesicular transport: how many Ypt/Rab GTPases make a eukaryotic cell? *Trends Biochem. Sci.* 22:468–472.

Lewis, M. J., Pelham, H. R. (1996). SNARE-mediated retrograde traffic from the Golgi complex to the endoplasmic reticulum. *Cell* 85:205–215.

Li, C., Takei, K., Geppert, M., Daniell, L., Stenius, K., Chapman, E. R., Jahn, R., De Camilli, P., Südhof, T. C. (1994). Synaptic targeting of rabphilin-3A, a synaptic vesicle Ca^{2+}/phospholipid-binding protein, depends on rab3A/3C. *Neuron* 13:885–898.

Li, C., Ullrich, B., Zhang, J. Z., Anderson, R. G. W., Brose, N., Südhof, T. C. (1995). Ca^{2+}-dependent and Ca^{2+}-independent activities of neural and non-neural synaptotagmins. *Nature* 375:594–599.

Lin, R. C., Scheller, R. H. (1998). Structural organization of the synaptic exocytosis core complex. *Neuron* 19:1087–1094.

Link, E., Edelmann, L., Chou, J. H., Binz, T., Yamasaki, S., Eisel, U., Baumert, M., Südhof, T. C., Niemann, H., Jahn, R. (1992). Tetanus toxin action: inhibition of neurotransmitter release linked to synaptobrevin proteolysis. *Biochem. Biophys. Res. Commun.* 189:1017–1023.

Linstedt, A. D., Vetter, M. L., Bishop, J. M., Kelly, R. B. (1992). Specific association of the proto-oncogene product pp60c-src with an intracellular organelle, the PC12 synaptic vesicle. *J. Cell Biol.* 117:1077–1084.

Liu, Y., Edwards, R. H. (1997). The role of vesicular transport proteins in synaptic transmission and neural degeneration. *Annu. Rev. Neurosci.* 20:125–156.

Llinas, R., Sugimori, M., Silver, R. B. (1992). Microdomains of high calcium concentration in a presynaptic terminal. *Science* 256:677–679.

Lowe, A. W., Madeddu, L., Kelly, R. B. (1988). Endocrine secretory granules and neuronal synaptic vesicles have three integral membrane proteins in common. *J. Cell Biol.* 106:51–59.

McIntire, S. L., Reimer, R. J., Schuske, K., Edwards, R. H., Jorgensen, E. M. (1997). Identification and characterization of the vesicular GABA transporter. *Nature* 389:870–876.

McMahon, H. T., Ushkaryov, Y. A., Edelmann, L., Link, E., Binz, T., Niemann, H., Jahn, R., Südhof, T. C. (1993). Cellubrevin is a ubiquitous tetanus-toxin substrate homologous to a putative synaptic vesicle fusion protein. *Nature* 364:346–349.

McNew, J. A., Sogaard, M., Lampen, N. M., Machida, S., Ye, R. R., Lacomis, L., Tempst, P., Rothman, J. E., Söllner, T. H. (1997). Ykt6p, a prenylated SNARE essential for endoplasmic reticulum–Golgi transport. *J. Biol. Chem.* 272: 17,776–17,783.

McNew, J. A., Coe, J. G., Sogaard, M., Zemelman, B. V., Wimmer, C., Hong, W., Söllner, T. H. (1998). Gos1p, a *Saccharomyces cerevisiae* SNARE protein involved in Golgi complex transport. *FEBS Lett.* 435:89–95.

Matteoli, M., Takei, K., Perin, M. S., Südhof, T. C., DeCamilli, P. (1992). Exo-endocytotic recycling of synaptic vesicles in developing processes of cultured hippocampal neurons. *J. Cell Biol.* 117:849–861.

Matthew, W. D., Tsavaler, L., Reichardt, L. F. (1981). Identification of a synaptic vesicle-specific membrane protein with a wide distribution in neuronal and neurosecretory tissue. *J. Cell Biol.* 91:257–269.

Misura, K. M. S., Scheller, R. H., and Weis, W. I. (2000). Three-dimensional structure of the neuronal sec1–syntaxin 1A complex. *Nature* 404:355–362.

Nagahama, M., Orci, L., Ravazzola, M., Amherdt, M., Lacomis, L., Tempst, P., Rothman, J. E., Söllner, T. H. (1996). A v-SNARE implicated in intra-Golgi transport. *J. Cell Biol.* 133:507–516.

Neher, E. (1998). Vesicle pools and Ca²⁺ microdomains: new tools for understanding their roles in neurotransmitter release. *Neuron* 20:389–399.

Newman, A. P., Shim, J., Ferro-Novick, S. (1990). *BET1, BOS1,* and *SEC22* are members of a group of interacting yeast genes required for the transport from the endoplasmic reticulum to the Golgi complex. *Mol. Cell. Biol.* 10: 3405–3414.

Nichols, B. J., Ungermann, C., Pelham, H. R., Wickner, W. T., Haas, A. (1997). Homotypic vacuolar fusion mediated by t- and v-SNAREs. *Nature* 387:199–202.

Nicholson, K. L., Munson, M., Miller, R. B., Filip, T. J., Fairman, R., Hughson, F. M. (1998). Regulation of SNARE complex assembly by an N-terminal domain of the t-SNARE Sso1p. *Nat. Struct. Biol.* 5:793–802.

Ohnishi, H., Ernst, S. A., Wys, N., McNiven, M., Williams, J. A. (1996). Rab3D localizes to zymogen granules in rat pancreatic acini and other exocrine glands. *Am. J. Physiol.* 271:G531–G538.

Ossig, R., Dascher, C., Trepte, H. H., Schmitt, H. D., Gallwitz, D. (1991). The yeast *SLY* gene products, suppressors of defects in the essential GTP-binding Ypt1 protein, may act in endoplasmic reticulum–to–Golgi transport. *Mol. Cell Biol.* 11:2980–2993.

Ostermeier, C., Brünger, A. T. (1998). Structural basis of Rab effector specificity: crystal structure of the small G protein Rab3A complexed with the effector domain of rabphilin-3A. *Cell* 96:363–374.

Oyler, G. A., Higgins, G. A., Hart, R. A., Battenberg, E., Billingsley, M., Bloom, F. E., Wilson, M. C. (1989). The identification of a novel synaptosomal-associated protein, SNAP-25, differentially expressed by neuronal subpopulations. *J. Cell Biol.* 109:3039–3052.

Pellizzari, R., Rossetto, O., Schiavo, G., Montecucco, C. (1999). Tetanus and botulinum neurotoxins: mechanism of action and therapeutic uses. *Philos. Trans. R. Soc. London Ser. B.* 354:259–268.

Perin, M. S., Fried, V. A., Mignery, G. A., Jahn, R., Südhof, T. C. (1990). Phospholipid binding by a synaptic vesicle protein homologous to the regulatory region of protein kinase C. *Nature* 345:260–263.

Peters, A., Palay, S. L., Webster, H. de F. (1991). *The Fine Structure of the Nervous System.* Oxford: Oxford University Press.

Peters, C., Mayer, A. (1998). Ca²⁺/calmodulin signals the completion of docking and triggers a late step of vacuole fusion. *Nature* 396:575–580.

Peterson, M. R., Hsu, S.-C., Scheller, R. H. (1996). A mammalian homologue of SLY1, a yeast gene required for transport from endoplasmic reticulum to Golgi. *Gene* 169:293–294.

Pevsner, J., Hsu, S.-C., Scheller, R. H. (1994). N-sec1: a neural-specific syntaxin-binding protein. *Proc. Natl. Acad. Sci. USA* 91:1445–1449.

Piper, R. C., Whitters, E. A., Stevens, T. H. (1994). Yeast Vps45p is a Sec1p-like protein required for the consumption of vacuole-targeted, post-Golgi transport vesicles. *Eur. J. Cell Biol.* 65:305–318.

Poirier, M. A., Xiao, W., Macosko, J. C., Chan, C., Shin, Y. K., Bennett, M. K. (1998). The synaptic SNARE complex is a parallel four-stranded helical bundle. *Nat. Struct. Biol.* 5:765–769.

Prekeris, R., Klumperman, J., Chen, Y. A., Scheller, R. H. (1998). Syntaxin13 mediates cycling of plasma membrane proteins via tubulovesicular recycling endosomes. *J. Cell Biol.* 143:957–971.

Protopopov, V., Govindan, B., Novick, P., Gerst, J. E. (1993). Homologs of the synaptobrevin/VAMP family of synaptic vesicle proteins function on the late secretory pathway in *S. cerevisiae. Cell* 74:855–861.

Ranjan, R., Bronk, P., Zinsmaier, K. E. (1998). Cysteine string protein is required for calcium secretion coupling of evoked neurotransmission in *Drosophila* but not for vesicle recycling. *J. Neurosci.* 18:956–964.

Ravichandran, V., Chawla, A., Roche, P. A. (1996). Identification of a novel syntaxin- and synaptobrevin/VAMP-binding protein, SNAP-23, expressed in non-neuronal tissues. *J. Biol. Chem.* 271:13,300–13,303.

Rizo, J., Südhof, T. C. (1998). Progress in membrane fusion: from structure to function. *Nat. Struct. Biol.* 5:839–842.

Rosahl, T. W., Spillane, D., Missler, M., Herz, J., Selig, D. K., Wolff, J. R., Hammer, R. E., Malenka, R. C., Südhof, T. C. (1995). Essential functions of synapsins I and II in synaptic vesicle regulation. *Nature* 375:488–493.

Rosenmund, C., Stevens, C. F. (1996). Definition of the readily releasable pool of vesicles at hippocampal synapses. *Neuron* 16:1197–1207.

Sabatini, B. L., Regehr, W. G. (1999). Timing of synaptic transmission. *Annu. Rev. Physiol.* 61:521–542.

Sato, T. K., Darsow, T., Emr, S. D. (1998). Vam7p, a SNAP-25-like molecule, and Vam3p, a syntaxin homolog, function together in yeast vacuolar protein trafficking. *Mol. Cell Biol.* 18:5308–5319.

Schiavo, G., Benfenati, F., Poulain, B., Rossetto, O., Polverino de Laureto, P., DasGupta, B. R., Montecucco, C. (1992). Tetanus and botulinum-B neurotoxins block neurotransmitter release by proteolytic cleavage of synaptobrevin. *Nature* 359:832–835.

Schiavo, G., Rossetto, O., Catsicas, S., Polverino de Laureto, P., DasGupta, B. R., Benfenati, F., Montecucco, C. (1993). Identification of the nerve terminal targets of botulinum neurotoxin serotypes A, D, and E. *J. Biol. Chem.* 268:23,784–23,787.

Schiavo, G., Malizio, C., Trimble, W. S., Polverino de Laureto, P., Milan, G., Sugiyama, H., Johnson, E. A., Montecucco, C. (1994). Botulinum G neurotoxin cleaves VAMP/synaptobrevin at a single Ala-Ala peptide bond. *J. Biol. Chem.* 269:20,213–20,216.

Schlüter, O. M., Schnell, E., Verhage, M., Tzonopoulos, T., Nicoll, R. A., Janz, R., Malenka, R. C., Geppert, M., Südhof, T. C. (1999). Rabphilin knock-out mice reveal rat rabphilin is not required for rab3 function in regulating neurotransmitter release. *J. Neurosci.* 19:5834–5846.

Shirataki, H., Kaibuchi, K., Sakoda, T., Kishida, S., Yamaguchi, T., Wada, K., Miyazaki, M., Takai, Y. (1993). Rabphilin-3A, a putative target protein for smg p25A/rab3A p25 small GTP-binding protein related to synaptotagmin. *Mol. Cell. Biol.* 13:2061–2068.

Simonsen, A., Bremnes, B., Ronning, E., Aasland, R., Stenmark, H. (1998). Syntaxin-16, a putative Golgi t-SNARE. *Eur. J. Cell Biol.* 75:223–231.

Söllner, T., Bennet, M. K., Whiteheart, S. W., Scheller, R. H., Rothman, J. E. (1993a). A protein assembly-disassembly pathway in vitro that may correspond to sequential steps of synaptic vesicle docking, activation, and fusion. *Cell* 75:409–418.

Söllner, T., Whiteheart, S. W., Brunner, M., Erdjument-Bromage, H., Geromanos, S., Tempst, P., Rothman, J. E. (1993b). SNAP receptors implicated in vesicle targeting and fusion. *Nature* 362:318–324.

Stahl, B., Chou, J. H., Li, C., Südhof, T. C., Jahn, R. (1996). Rab3 reversibly recruits rabphilin to synaptic vesicles by a mechanism analogous to raf recruitment by ras. *EMBO J.* 15:1799–1809.

Steegmaier, M., Yang, B., Yoo, J.-S., Huang, B., Shen, M., Yu, S., Luo, Y., Scheller, R. H. (1998). Three novel proteins of the syntaxin/SNAP-25 family. *J. Biol. Chem.* 273:34,171–34,179.

Südhof, T. C. (1995). The synaptic vesicle cycle: a cascade of protein-protein interactions. *Nature* 375:645–653.

Südhof, T. C. (1997). Function of Rab3A GDP/GTP exchange. *Neuron* 18:519–522.

Südhof, T. C., Baumert, M., Perin, M. S., Jahn, R. (1989a). A synaptic vesicle membrane protein is conserved from mammals to Drosophila. *Neuron* 2:1475–1481.

Südhof, T. C., Czernik, A, J., Kao, H., Takei, K., Johnston, P. A., Horiuchi, A., Wagner, M., Kanazir, S. D., Perin, M. S., DeCamilli, P., Greengard, P. (1989b). Synapsins: mosaics of shared and individual domains in a family of synaptic vesicle phosphoproteins. *Science* 245:1474–1480.

Südhof, T. C., De Camilli, P., Niemann, H., Jahn, R. (1993). Membrane fusion machinery: insights from synaptic proteins. *Cell* 75:1–4.

Sullivan, K. M., Busa, W. B., Wilson, K. L. (1993). Calcium mobilization is required for nuclear vesicle fusion in vitro: implications for membrane traffic and IP3 receptor function. *Cell* 73:1411–1422.

Sutton, B., Fasshauer, D., Jahn, R., Brünger, A. T. (1998). Crystal structure of a SNARE complex involved in synaptic exocytosis at 2.4 Å resolution. *Nature* 395:347–353.

Takai, Y., Sasaki, T., Shiratake, H., Nakanishi, H. (1996). Rab3A small GTP-binding protein in Ca^{2+}-dependent exocytosis. *Genes to Cells* 1:615–632.

Takeshima, H., Shimuta, M., Komazaki, S., Ohmi, K., Nishi, M., Iino, M., Miyata, A., Kangawa, K. (1998). Mitsugumin29, a novel synaptophysin family member from the triad junction in skeletal muscle. *Biochem. J.* 331:317–322.

Tall, G. G., Hama, H., DeWald, D. B., Horazdovsky, B. F. (1999). The phosphatidylinositol 3-phosphate binding protein Vac1p interacts with a Rab GTPase and a Sec1p homologue to facilitate vesicle-mediated vacuolar protein sorting. *Mol. Biol. Cell* 10:1873–1889.

Tang, B. L., Low, D. Y., Tan, A. E., Hong, W. (1998). Syntaxin 10: a member of the syntaxin family localized to the trans-Golgi network. *Biochem. Biophys. Res. Comm.* 242:345–350.

Tellam, J. T., McIntosh, S., James, D. E. (1995). Molecular identification of two novel Munc-18 isoforms expressed in non-neuronal tissues. *J. Biol. Chem.* 270:5857–5863.

Trimble, W. S., Cowan, D. M., Scheller, R. H. (1988). VAMP-1: a synaptic vesicle–associated integral membrane protein. *Proc. Natl. Acad. Sci. USA* 85:4538–4542.

Tzounopoulos, T., Janz, R., Südhof, T. C., Nicoll, R. A., Malenka, R. C. (1998). A role for cAMP in long term depression at hippocampal mossy fiber synapses. *Neuron* 21:837–845.

Ubach, J., Garcia, J., Nittler, M. P., Südhof, T. C., Rizo, J. (1999). The C$_2$B-domain of rabphilin: structural variations in a janus-faced domain. *Nature Cell Biol.* 1:106–112.

Ullrich, O., Stenmark, H., Alexandrov, K., Huber, L. A., Kaibuchi, K., Sasaki, T., Takai, Y., Zerial, M. (1993). Rab GDP dissociation inhibitor as a general regulator for the membrane association of rab proteins. *J. Biol. Chem.* 268: 18,143–18,150.

Ungermann, C., Sato, K., Wickner, W. (1998). Defining the functions of trans-SNARE pairs. *Nature* 396:543–548.

Usdin, T. B., Eiden, L. E., Bonner, T. I., Erickson, J. D. (1995). Molecular biology of the vesicular ACh transporter. *Trends Neurosci.* 18:218–224.

Wada, Y., Kitamoto, K., Kanbe, T., Tanaka, K., Anraku, Y. (1990). The *SLP1* gene of *Saccharomyces cerevisiae* is essential for vacuolar morphogenesis and function. *Mol. Cell Biol.* 10:2214–2223.

Wada, Y., Nakamura, N., Ohsumi, Y., Hirata, A. (1997). Vam3p, a new member of syntaxin related protein, is required for vacuolar assembly in the yeast *Saccharomyces cerevisiae. J. Cell Sci.* 110:1299–1306.

Wang, H., Frelin, L., Pevsner, J. (1997). Human syntaxin 7: a Pep12p/Vps6p homologue implicated in vesicle trafficking to lysosomes. *Gene* 199:39–48.

Wang, Y., Okamoto, M., Schmitz, R., Hofmann, K., Südhof, T. C. (1997). RIM: a putative rab3 effector in regulating synaptic vesicle fusion. *Nature* 388:593–598.

Weber, T., Zemelman, B. V., McNew, J. A., Westermann, B., Gmachl, M., Parlati, F., Sollner, T. H., Rothman, J. E. (1998). SNAREpins: minimal machinery for membrane fusion. *Cell* 92:759–772.

Weimbs, T., Low, S. H., Chapin, S. J., Mostov, K. E., Bucher, P., Hofmann, K. (1997). A conserved domain is present in different families of vesicular fusion proteins: a new superfamily. *Proc. Natl. Acad. Sci. USA* 94:3046–3051.

Whittaker, V. P. (1968). The storage of transmitters in the central nervous system. *Biochem. J.* 109:20P–21P.

Wong, S. H., Xu, Y., Zhang, T., Hong, W. (1998). Syntaxin 7, a novel syntaxin member associated with the early endosomal compartment. *J. Biol. Chem.* 273:375–380.

Yang, B., Gonzalez, L., Jr., Prekeris, R., Steegmaier, M., Advani, R. J., Scheller, R. H. (1999). SNARE interactions are not selective. Implications for membrane fusion specificity. *J. Biol. Chem.* 274:5649–5653.

Yoshida, A., Oho, C., Omori, A., Kuwahara, R., Ito, T., Takahashi, M. (1992). HPC-1 is associated with synaptotagmin and omega-conotoxin receptor. *J. Biol. Chem.* 267:24,925–24,928.

Zhang, J. Z., Davletov, B. A., Südhof, T. C., Anderson, R. G. W. (1994). Synaptotagmin is a high affinity for clathrin AP2: implications for membrane recycling. *Cell* 78:751–760.

Zucker, R. S. (1989). Short-term synaptic plasticity. *Annu. Rev. Neurosci.* 12:13–31.

5 Synaptic Vesicle Endocytosis

Pietro De Camilli, Vladimir I. Slepnev, Oleg Shupliakov, and Lennart Brodin

he convergence of electron microscopy and electrophysiology in the 1950s led to the formulation of the quantal hypothesis of neurotransmitter release (see Cowan and Kandel, and Regehr and Stevens, this volume). This discovery prompted an important question: How could axon endings sustain high rates of secretion for prolonged periods without exhausting their supply of synaptic vesicles? Considering the great distance of nerve terminals from perikarya in the majority of neurons, some form of local membrane recycling appeared to be necessary (Gray and Willis, 1970).

Synaptic Vesicles Undergo Recycling

Conclusive evidence for recycling was provided in the early 1970s, with the introduction of extracellular endocytotic tracers in electron microscopy that could be detected by cytochemical techniques after internalization. A stimulus-dependent uptake of the extracellular marker horseradish peroxidase into synaptic vesicles was demonstrated in terminals of photoreceptors (Holtzman and Peterson, 1969) and then in elegant studies at the neuromuscular junction (Ceccarelli et al., 1973; Heuser and Reese, 1973). Furthermore, correlative electron microscopy and electrophysiological studies demonstrated a reversible shift of synaptic vesicle membranes to the cell surface after massive stimulation, suggesting that the same membrane pool could shuttle between the plasma membrane and vesicles (Ceccarelli et al., 1973; Heuser and Reese, 1973; see Ceccarelli and Hurlbut, 1980b, for review). These findings converged with the emergence of membrane recycling as a general principle in cell biology. Parallel studies on other systems had found that transport of molecules between two sequential compartments of the secretory and endocytotic pathways occurs via vesicular carriers that continuously shuttle between the two compartments (Palade, 1975).

A striking genetic demonstration of the physiological importance of recycling in the maintenance of a functional pool of synaptic vesicles came from the phenotypic characterization of the temperature-sensitive *Drosophila shibire* mutants (*shi*[ts]) (Grigliatti et al., 1973; Poodry et al., 1973). *Shi*[ts] flies are normal at the permissive temperature of 19°C, but become rapidly paralyzed at the restrictive temperature of 29°C. At 29°C exocytosis occurs normally, but endocytosis is selectively impaired, leading to a rapid depletion of synaptic vesicles, a concomitant increase in the surface area of the plasma membrane, and the presence of numerous "frozen," deeply invaginated pits in this membrane (Fig. 5.1). Upon a shift to the permissive temperature, the normal structure and function of synapses, as well as normal motor behavior, are rapidly reestablished, consistent with a regeneration of new synaptic vesicles from the vesicle membrane previously incorporated into the plasma membrane (Koenig and Ikeda, 1989; Ramaswami et al., 1994).

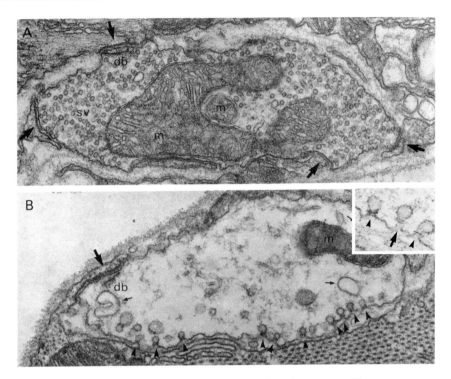

Figure 5.1. Electron micrographs of synapses of shibire mutant flies.
(A) Synapses fixed at the permissive temperature (19°C).
(B) Synapses fixed after 8 min at the restrictive temperature (29°C). Note the depletion of synaptic vesicles in B and the presence of numerous collared, deeply invaginated pits along the plasma membrane. A few larger invaginations, whose neck portions are not within the plane of section, are also seen (small arrows). Release sites, which are characterized by pre- and postsynaptic membrane densities, and a presynaptic dense body are indicated by large arrows. The inset shows examples of collared pits at higher magnification. db, Presynaptic dense body; m, mitochondria; sv, synaptic vesicles.
Scale bar = 250 nm (inset = 120 nm).
Reproduced from Koenig and Ikeda (1989) by permission from the Society for Neuroscience.

Vesicle reformation by recycling may be not only a mechanism to regenerate vesicles after each round of exocytosis, but also the main mechanism through which a mature synaptic vesicle is built. Strong evidence indicates that newly synthesized synaptic vesicle proteins assemble into mature synaptic vesicles only in nerve terminals. For example, focal blocks in transport along the axon result in the accumulation, proximal to the block, of organelles larger and more pleomorphic than synaptic vesicles (Tsukita and Ishikawa, 1980). In transfected cultured neurons, different green fluorescent protein–labeled synaptic vesicle proteins travel down the axon with different kinetics and in different sets of carriers (Hirokawa et al., 1998; Nakata et al., 1998). Furthermore, studies on the fate of newly synthesized synaptophysin in PC12 cells have shown that this major synaptic vesicle protein is not targeted to synaptic-like vesicles directly after synthesis. Rather, it appears in this compartment only after a lag phase during which it shuttles between the surface and intracellular membranes (Regnier-Vigouroux et al., 1991; Hannah et al., 1999).

Thus elucidation of the mechanisms underlying synaptic vesicle recycling will further our understanding of synaptic vesicle biogenesis and, more generally, shed light on the fundamental question of how a vesicle carrier of precise molecular composition and size is generated.

New Probes to Study Recycling

In the 1990s advances in molecular biology and video-enhanced light microscopy provided new evidence for synaptic vesicle recycling, enabling researchers to obtain dynamic information about this process from studies on cultured neurons. Two main approaches have been developed to visualize synaptic vesicle recycling using light microscopy–based techniques. The cloning of synaptic vesicle proteins has facilitated the generation of antibodies (equipped with fluorescent probe tags) directed against lumenal epitopes of synaptic vesicles. When these antibodies are added to the culture medium, they bind to the lumenal face of vesicle membrane, which becomes exposed to the extracellular space as a result of exocytosis (Valtorta et al., 1988; Matteoli et al., 1992). They are then taken up by endocytosis, incorporated into newly formed synaptic vesicles, and subsequently recycled in parallel with the vesicles, as shown by electron microscopy (Fig. 5.2; Kraszewski et al., 1995). The specificity of the antibodies ensures that the fluorescent signal is selective for retrieved synaptic vesicles. The trafficking of recycling synaptic vesicles in living nerve terminals can thus be monitored with high specificity using this method (Matteoli et al., 1992; Kraszewski et al., 1995, 1996).

A second method, which has already greatly advanced our knowledge of the kinetics of synaptic vesicle cycling, is based on fluorescent amphipathic styryl dyes (Betz et al., 1996). Of these molecules, FM1-43 is the most widely used for neurobiological applications (Fig. 5.3). FM1-43 inserts into the outer leaflet of the entire plasma membrane and is taken up into endocytotic vesicles. Upon removal from the medium, the bound dye rapidly dissociates from the cell surface, while internalized dye is discharged from the cell only when dye-containing organelles undergo exocytosis. Given the overwhelming predominance of the endocytosis of synaptic vesicles over any other endocytotic process in nerve terminals, the uptake of the FM1-43 at the synapse closely reflects the endocytosis of synaptic vesicles, as confirmed by electron microscopy (Henkel et al., 1996). Conversely, following a stimulation-dependent dye load of the synapse, the majority of the internalized dye can be released by synaptic vesicle exocytosis.

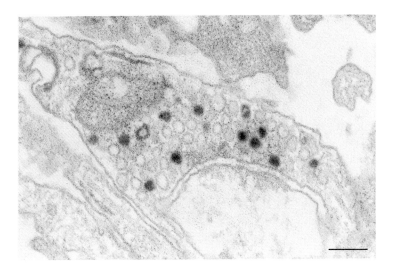

Figure 5.2. Electron micrograph demonstrating receptor-mediated uptake of an extracellular tracer into recycling synaptic vesicles. The culture was exposed to a depolarizing medium (55 mM K⁺) for 10 min (to stimulate exocytosis and compensatory endocytosis) in the presence of antibodies directed against the lumenal domain of the synaptic vesicle protein synaptotagmin 1. The antibodies had been conjugated to horseradish peroxidase. The black label in the lumen of some synaptic vesicles represents peroxidase reaction product. Scale bar = 200 nm. Provided by Laurie Daniell, Olaf Mundigl, and Pietro De Camilli.

The use of styryl dyes has made it possible to define the kinetic parameters of different steps in the cycle and to analyze which fraction of the vesicle pool participates in recycling under various experimental conditions. In addition, these dyes have been employed for monitoring synaptic vesicle exocytosis independently of the release of neurotransmitter, for the analysis of vesicle recycling at distinct synapses of a single neuron, and for the study of synaptic vesicle mobility within the terminal (Cochilla et al., 1999). The sensitivity of FM1-43-based approaches is sufficient to record quantal events, presumably reflecting dye trapped in single synaptic vesicles (Ryan et al., 1997; Murthy and Stevens, 1998).

Kinetic Parameters of Recycling

Studies with styryl dyes have made it possible to define several kinetic parameters of synaptic vesicle recycling both at the neuromuscular junction and at central synapses. Clearly the recycling time varies dramatically, depending upon the intensity and duration of the stimulus. The time required for a vesicle to be retrieved, transported, and prepared for a new round of release can be shorter than 40 sec (Betz and Bewick, 1992; Ryan and Smith, 1995). The half time for the endocytotic step,

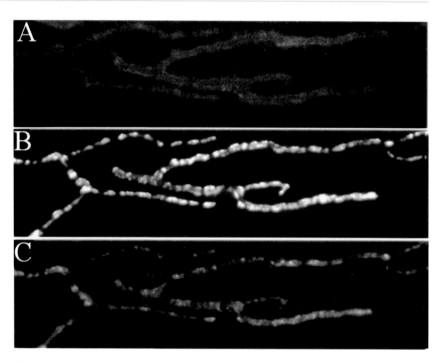

Figure 5.3. Frog motor nerve terminal stained with FM1-43.

(A) Passive staining. The preparation was exposed to 4 mM FM1-43 without stimulation for 5 min and then washed. Little dye was taken up.

(B) Activity-dependent staining. The preparation was again exposed to FM1-43 for 5 min while being stimulated continuously at 10 Hz. Each spot marks a cluster of several hundred stained synaptic vesicles.

(C) Activity-dependent destaining. The preparation was stimulated again for 5 min (no dye added). The destaining reflects release of dye from previously labeled vesicles that have undergone exocytosis again.

Scale bar = 10 μm. From Nicholls et al. (1992) with permission from Sinauer Associates, Inc.

when measured after a brief burst of exocytosis, is approximately 20 sec (Ryan et al., 1996; Wu and Betz, 1996). This time may, however, be even shorter. By monitoring the loss of styryl dyes that differ in their off-rates from membranes, the time constant of endocytosis at hippocampal synapses is estimated to be only a few seconds (Klingauf et al., 1998). The average time of recycling becomes much longer after prolonged strong stimulation, presumably because of a saturation of the endocytotic machinery (Wu and Betz, 1996, 1998). Under such conditions, the endocytotic process cannot cope with the rate of secretion, and nerve terminals may become depleted of synaptic vesicles (Ceccarelli et al., 1973; Heuser and Reese, 1973; Fesce et al., 1986; Gad et al., 1998).

A different method for monitoring the kinetic parameters of endocytosis in living nerve terminals involves measuring membrane capacitance via a patch pipette (Fig. 5.4). As the capacitance is proportional to the membrane surface area, endocytosis is reflected as a decrease in capacitance, which can be tracked in real time. This technique has been extensively applied to the study of compensatory endocytosis, which follows the exocytosis of peptide-containing secretory granules (Henkel and Almers, 1996; Smith and Neher, 1997). However, this technique is not sufficiently sensitive to detect changes produced by the exocytosis of a single synaptic vesicle. Furthermore, its use for monitoring synaptic vesicle exo- and endocytosis has been limited so far to ribbon synapses, which offer a special advantage for this type of analysis but have unique structural and functional properties. Studies performed using this method on the giant terminals of goldfish bipolar neurons (Von Gersdorff and Matthews, 1994a) and saccular hair cells (Parsons et al., 1994) have confirmed the very fast time course of endocytosis. The time constant of the capacitance decrease following exocytosis is as brief as 2 sec.

Following endocytosis, synaptic vesicle membranes must go through several steps before becoming available again for exocytosis as neurotransmitter-filled vesicles. The time required to go through these steps is estimated to be about 30 sec (Betz and Bewick, 1992; Ryan et al., 1993), or even less under some conditions (Klingauf et al., 1998). Some of this time is required for vesicle docking prior to the fusion reaction. Thus the regeneration of a new synaptic vesicle from an endocytotic vesicle must be very fast, a finding that is hardly compatible with its fusion with, and subsequent budding from, an endosome.

Regulation of Synaptic Vesicle Recycling

A strict coupling between exo- and endocytotic events is essential at nerve terminals because of the very high rates of vesicle turnover. Even a slight imbalance between the two processes soon leads to expansion (or shrinkage) of the terminal plasma membrane and to an altered size of the vesicle pool (Fig. 5.5) (Heuser and Miledi, 1971; Ceccarelli and Hurlbut, 1975; Gennaro et al., 1978; Ceccarelli and Hurlbut, 1980a; Haimann et al., 1985; Segal et al., 1985; Torri-Tarelli et al., 1990). More generally, an exocytosis-endocytosis coupling aimed at a homeostatic control of the composition and area of the plasma membrane is a property of all cells. In the nerve terminal, this coupling may in principle be obtained in two ways. Exo- and endocytosis could be induced separately but respond to the same triggering signal, the Ca^{2+} influx, given the requirement of extracellular Ca^{2+} for the endocytotic reaction as demonstrated by several studies (Ceccarelli and Hurlbut, 1980a; Ramaswani et al., 1994; Gad et al., 1998). Alternatively endocytosis could be initiated by a signal linked to the preceding exocytosis, such as the presence of synaptic vesicle–associated molecules in the plasma membrane and the increase in surface area. Experiments performed on the lamprey reticulospinal synapse support the latter possibility, which has the advantage that it would

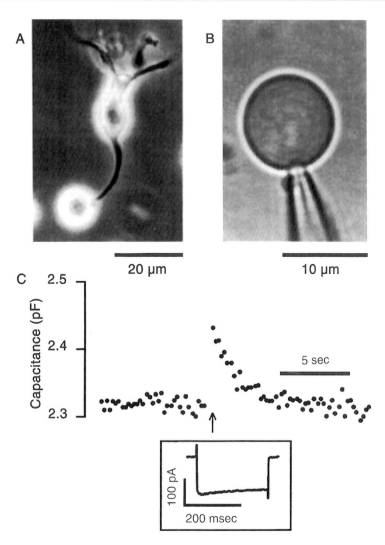

Figure 5.4. *Capacitance measurements from a giant synaptic terminal of a bipolar neuron from goldfish retina.*

(A) Phase-contrast micrograph of an intact, isolated bipolar neuron.

(B) Bright-field micrograph of an isolated synaptic terminal, with attached whole-cell patch pipette. From Von Gersdorff and Matthews (1994a); copyright 1994 Macmillan Magazines Ltd.

(C) Change in capacitance elicited in an isolated bipolar cell synaptic terminal by depolarization. Timing of the 250-msec depolarization from –60 mV to 0 mV is indicated by the arrow. The Ca^{2+} current activated by the depolarization is shown in the inset (expanded time scale). The capacitance increases abruptly in response to depolarization (exocytosis) followed by a slower return to resting capacitance (endocytosis). Reproduced from Matthews (1996), copyright 1996, with permission from Elsevier Science.

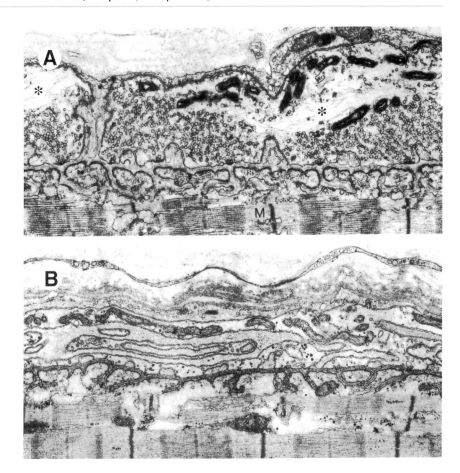

Figure 5.5. Electron micrographs demonstrating a shift of synaptic vesicle membranes to the plasma membrane at the frog neuromuscular junction after stimulation by exposure to La^{3+}.

(A) Control terminal soaked for 1 hr in Ca^{2+}-free solution.

(B) Terminal soaked for 1 hr in Ca^{2+}-free solution with 100 μM La^{3+}. This treatment potently stimulates the asynchronous release of neurotransmitter but blocks endocytosis. Note that the expanded plasma membrane forms deep infoldings. M, muscle cell. The portions of the axon terminal enriched in microtubules and neurofilaments are denoted by asterisks (*).

Reproduced from Segal et al. (1985) with permission of the Biophysical Society.

ensure the selective reuptake from the plasma membrane of components delivered by exocytosis and also preserve the compositional difference between the plasma membrane and the vesicles.

These experiments capitalized on, and provided further strong experimental support for, the critical dependence of the endocytotic reaction on extracellular Ca^{2+} (Ceccarelli and Hurlbut, 1980a; Segal et al., 1985). Following intense (nonphysiological) stimulation, which caused massive incorporation of vesicles into the plasma membrane, the endocytotic process could be blocked at an early step by removing extracellular Ca^{2+}. The block could then be released (Fig. 5.6) by re-adding even low concentrations (in the micromolar range) of extracellular Ca^{2+}. Retrieval of membranes by endocytosis continued until the vesicle pool had been restored, and then it ceased (Gad et al., 1998). The endocytotic process can thus be temporally uncoupled from exocytosis and from the electrical stimulus, implying that it is driven by the presence of vesicle membrane components in the plasma membrane. The essential role of at least low Ca^{2+} in the endocytotic reaction may reflect a need for cytosolic Ca^{2+}, because this requirement was observed only after manipulations that could be expected to produce depletion of intracellular Ca^{2+} stores (Ceccarelli and Hurlbut, 1980a; Ramaswami et al., 1994; Gad et al., 1998; cf. Ryan et al., 1993, 1996).

The next question is whether action potential–evoked Ca^{2+} influx can facilitate the coupling between exo- and endocytosis and regulate the *rate* of the endocytotic reaction. Such a regulation would appear physiologically relevant, as the requirement of quick retrieval is highest under conditions of high-frequency firing and a high rate of secretion. There is evidence for a critical role of Ca^{2+}-dependent dephosphorylation in the assembly of the endocytotic machinery (Robinson et al., 1994; Slepnev et al., 1998). Furthermore, an increase in the rate of vesicle recycling as a function of increased Ca^{2+} influx has been observed in hippocampal synapses (Klingauf et al., 1998). Studies of rat brain synaptosomes have also supported a stimulatory effect of Ca^{2+} influx (Marks and McMahon, 1998).

It should be noted, however, that Ca^{2+} is not the only, and perhaps not always the major, determinant of the time course of endocytosis. Wu and Betz (1996) compared the intracellular Ca^{2+} concentration with the time course of endocytosis after a burst of exocytosis at the frog neuromuscular junction. They found that the endocytosis rate correlates better with the duration of the stimulus train than with the instantaneous Ca^{2+} level. Finally, capacitance measurement studies in goldfish bipolar terminals have revealed an entirely different type of Ca^{2+}-dependent regulation. Elevation of the intracellular Ca^{2+} level to a few hundred nanomolar strongly inhibits endocytosis (Von Gersdorff and Matthews, 1994b), which may reflect a specialized feature of these ribbon-type terminals. Elevation of intracellular Ca^{2+} in terminals of hippocampal neurons does not inhibit endocytosis (Reuter and Porzig, 1995). The molecular basis of the Ca^{2+}-dependent regulation of synaptic vesicle endocytosis will be discussed later in this chapter.

A burst of endocytotic activity that may exceed (and not necessarily be correlated with) the exocytotic response was observed in some nonneuronal cells after depolarization-induced Ca^{2+} entry (rapid endocytosis or excess retrieval) (Thomas et al., 1994; Artalejo et al., 1995; Henkel and Almers, 1996; Smith and Neher, 1997). Even in this case the surface area of the plasma membrane subsequently returns to the baseline level. It is not known whether such Ca^{2+}-triggered rapid endocytosis can occur at synapses (Smith and Neher, 1997; Engisch and Nowycky, 1998).

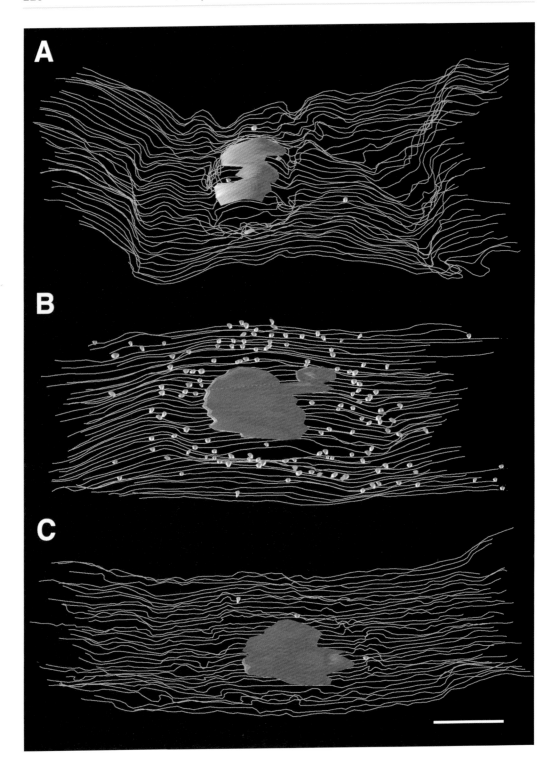

Clathrin-Mediated Endocytosis
versus "Kiss-and-Run"

The importance of clathrin-mediated endocytosis in the recycling of synaptic vesicles is well established (Fig. 5.7). This process is a specialized form of the endocytosis that all cells use for various purposes, including the internalization of receptors to terminate signaling cascades (e.g., growth factor receptors) or the uptake of important nutrients (e.g., iron by the transferrin receptor, lipoproteins via the LDL receptor) (Mellman, 1996). The presence of a dense fuzzy coat (a former description of the clathrin coat) around a subpopulation of small vesicles in nerve endings was reported as early as the 1960s (Gray, 1961; Kanaseki and Kadota, 1969; Gray and Willis, 1970) and the endocytotic nature of these coats was first described in a landmark study on yolk protein endocytosis in developing mosquito oocytes (Roth and Porter, 1964). Some of the observations supporting a role of the clathrin-mediated pathway in synaptic vesicle endocytosis include the following:

1. Clathrin-coated pits are often observed in nerve terminals, generally at the outer margin of the active zone, and their number is increased when fixation is performed during a peak of endocytotic activity (Heuser and Reese, 1973; Heuser 1989; Gad et al., 1998) (Fig. 5.8).

2. Clathrin and the clathrin adaptor complexes, which participate in clathrin-mediated budding from the plasma membrane (AP-2 and AP180), are enriched in the nervous system, where they are concentrated in nerve terminals (Maycox et al., 1992; Ball et al., 1995; David

Figure 5.6. (opposite) A burst of endocytotic activity around synaptic sites is triggered by the restoration of extracellular Ca²⁺ to a synaptic preparation depleted of vesicles. Ca^{2+} Three-dimensional reconstruction of lamprey reticulospinal synapses. Dotted lines indicate the profile of the plasma membrane in different sections; red areas represent active zones (the synaptic area); purple dots indicate coated pits. A spinal cord preparation was stimulated at 20 Hz for 20 min to deplete synaptic vesicles and then incubated for 90 min in Ca^{2+}-free solution to block endocytosis. The spinal cord was then cut into pieces that were subjected to different treatments.

(A) A synapse in a part of an axon that had been fixed immediately after the incubation in Ca^{2+}-free solution (this synapse was depleted of synaptic vesicles).

(B) A synapse incubated for 40 sec in a physiological solution containing 2.6 mM Ca^{2+} before being fixed.

(C) A synapse incubated in Ca^{2+}-containing solution for 15 min before fixation (this synapse had reacquired a normal content of synaptic vesicles). The massive presence of clathrin-coated pits around the active zone in *B* reflects a burst of synaptic vesicle endocytosis induced by the presence of synaptic vesicle membranes in that region of the plasma membrane. Each reconstruction was obtained from 35–45 ultrathin sections.

Scale bar = 1 µm. Reproduced from Gad et al. (1998) with permission of Cell Press.

et al., 1996). This abundance has made the nervous system a tissue of choice for the purification of clathrin-coated vesicles and their components (Kadota and Kadota, 1973; Pearse, 1975).

3. Incubation of lysed synaptosomes in the presence of brain cytosol, ATP, and GTPγS (conditions that enhance coat formation and block uncoating) results in the massive generation of clathrin-coated endocytotic buds (Takei et al., 1995, 1996).

4. Genetically induced disruption of genes encoding clathrin adaptors (α-adaptin subunit of AP-2, or AP180) in *Drosophila* produces drastic impairment of synaptic vesicle recycling (Gonzalez-Gaitan and Jackle, 1997; Zhang et al., 1998).

5. Microinjections into giant axons of antibodies or peptides that perturb the function of the clathrin coat induce a powerful inhibition of synaptic vesicle membrane endocytosis and synaptic transmission (Shupliakov et al., 1997; Ringstad et al., 1999) (Fig. 5.9).

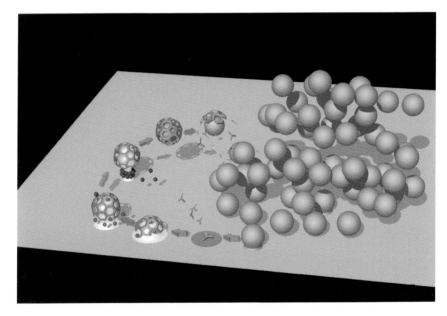

Figure 5.7. Model of clathrin-mediated synaptic vesicle recycling at the synapse. Following fusion of synaptic vesicles with the plasma membrane (gray), retrieval occurs outside the active zone (not indicated). This involves binding of clathrin adaptors to the membrane and recruitment of clathrin triskelia. The clathrin-coated membrane invaginates and forms a coated pit with a narrow neck. The fission machinery (red spheres), which presumably includes dynamin and dynamin-binding proteins, assembles around the neck of the coated pit. The fission reaction liberates the coated vesicle, which rapidly sheds its coat. The newly uncoated vesicle is loaded with neurotransmitter and returns to the release site to join the clustered vesicle pool. Modified from Brodin et al. (1997) with permission of Blackwell Science.

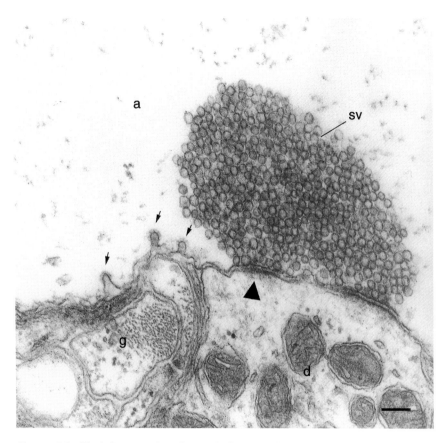

Figure 5.8. Clathrin-coated endocytotic intermediates at a stimulated synapse.
Electron micrograph of a synapse in a reticulospinal axon in the lamprey spinal
cord. The specimen was fixed during ongoing action potential stimulation at 5 Hz.
Clathrin-coated pits (arrows) are present at the plasma membrane in the vicinity
of the active zone (arrowhead). a, Axon; d, dendrite; g, glial element; sv, synaptic
vesicles. Scale bar = 0. 2 μm O. Shupliakov and L. Brodin (unpublished results).

Although a role for clathrin-mediated endocytosis in the recycling of
synaptic vesicles is undisputed, it has been proposed that this process
may coexist with a more direct pathway, which may predominate under
conditions of moderate stimulation. This pathway, referred to as "kiss-
and-run," implies that the retrieval of a recently fused vesicle occurs by
the rapid closure of the fusion pore, before its widening and the subsequent collapse of the vesicle into the plasma membrane (Ceccarelli et al.,
1973; Valtorta et al., 1990; Fesce et al., 1994). This model has attracted
new interest with the demonstration that peptide-containing secretory
granules can indeed secrete part of their content through a transient
fusion pore (Chow et al., 1992; Alvarez de Toledo et al., 1993; Alés et al.,
1999; see Cochilla et al., 1999, and Fesce and Meldolesi, 1999, for reviews). Some observations from dye imaging studies at synapses may also

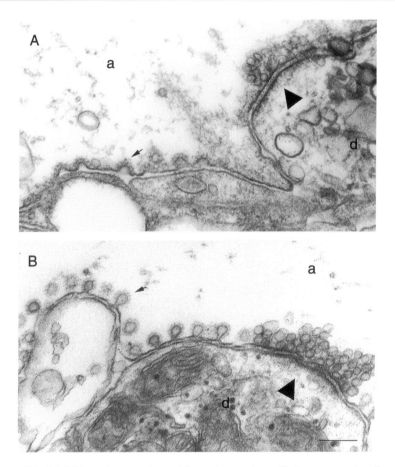

Figure 5.9. Inhibition of synaptic vesicle endocytosis at distinct stages by disruption of the function of "accessory" endocytotic proteins.

(A) A synapse in a lamprey axon microinjected with anti-endophilin antibodies followed by action potential stimulation. The number of synaptic vesicles in the cluster is markedly reduced (compare with Fig. 5.8). Shallow clathrin-coated pits have accumulated at the plasma membrane around the active zone (arrowhead). Actin-like filaments are also present. Reproduced from Ringstad et al. (1999) with permission of Cell Press.

(B) A synapse in an axon microinjected with a GST fusion protein containing the SH3 domain of amphiphysin, followed by stimulation. The number of synaptic vesicles is similarly reduced, but in this case deeply invaginated clathrin-coated pits with narrow necks accumulate at the plasma membrane. A, axon; d, dendrite. Reproduced with permission from Shupliakov et al. (1997); copyright 1997 American Association for the Advancement of Science.

Scale bar = 0. 2 μm.

be compatible with a direct retrieval process. Henkel and Betz (1995) showed that treatment of the neuromuscular junction with the kinase inhibitor staurosporine partially prevents the loss of FM1-43 from motor nerve terminals, without a corresponding reduction of the synaptic response in the target cell. A possible explanation for this finding is that staurosporine may cause a switch to a direct retrieval mechanism (fusion pore–dependent), which may allow the transmitter, but not the membrane probe, to escape. An alternative explanation is that staurosporine decreases the mobility of vesicles in the nerve terminal and favors a preferential recycling of the vesicles that are closer to the presynaptic plasma membrane (Kraszewski et al., 1996). More recently, Klingauf et al. (1998) have reported that dye-labeled vesicles can recycle repeatedly before they completely lose their dye, an observation that could be explained by ultrafast, direct retrieval.

Direct evidence for the kiss-and-run model of synaptic vesicle recycling has not yet been obtained. One question raised by this model is the fate of SNARE complexes after fusion (see Südhof and Scheller, this volume). SNARE complex formation is not a reversible reaction in equilibrium. The efficient segregation of the v-SNARE synaptobrevin (primarily on vesicles) from the t-SNAREs SNAP-25 and syntaxin (primarily on the plasma membrane) implies that the NSF-dependent disassembly of the complex occurs before endocytosis. Whether the kinetics of SNARE complex disassembly are compatible with the kinetics of the opening and closure of a fusion pore remains an open question. An attractive feature of the kiss-and-run model is that rapid closure of a transient fusion pore provides a simple mechanism to preserve the identity of the vesicle membrane at each exo- or endocytotic event. The classical model of clathrin-mediated recycling postulates a collapse of the vesicle into the plasma membrane, followed by the selective reuptake of its components into a vesicular bud. Until recently, these steps had been convincingly documented only after sustained stimulation, when there is massive incorporation of synaptic vesicles into the plasma membrane. However, microinjections of peptides that perturb clathrin-mediated endocytosis have shown that this type of recycling also occurs in nerve terminals that are firing at low frequency (Shupliakov et al., 1997). The one question left open by these microinjection studies is the origin of the few synaptic vesicles that persist long after the majority of the presynaptic vesicle pool has been depleted (Shupliakov et al., 1997; Ringstad et al., 1999). This pool may indeed reflect a kiss-and-run pathway of recycling. However, it may also reflect an incomplete block of the clathrin-dependent pathway or a slow supply of vesicles by axoplasmic transport.

In conclusion, whether kiss-and-run occurs and whether it accounts for a significant fraction of synaptic vesicle recycling are issues that deserve further experimental scrutiny. This chapter accordingly focuses on clathrin-mediated recycling, the only mechanism of synaptic vesicle endocytosis for which there is direct experimental evidence.

One or Two Vesicle Budding Steps?

The average size of clathrin-coated pits and vesicles in nerve endings during the recovery from stimulation is similar to the size of synaptic vesicles (Zhang et al., 1999). Thus, following clathrin-mediated retrieval at the plasma membrane and shedding of the coat, the vesicle could take up neurotransmitter and then return directly to the release site as a mature synaptic vesicle (the one-vesicle-budding-step model) (Fig. 5.10).

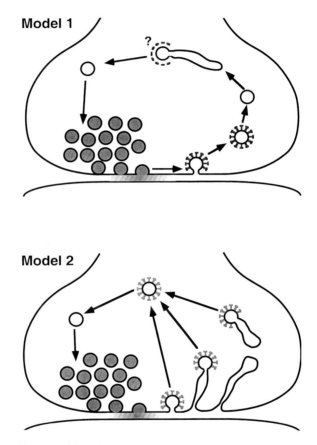

Figure 5.10. Two models of synaptic vesicle recycling. Model 1 (a two-vesicle-budding-steps model) proposes an endosomal sorting station downstream of clathrin-coated vesicles and a second vesicle-budding step from the endosomes. Model 2 (a one-vesicle-budding-step model; see Fig. 5.7) proposes that synaptic vesicles derive directly from clathrin-coated vesicles after shedding of the coat. According to this model, endosome-like structures visible in nerve terminals represent deep plasma membrane invaginations that have lost continuity with the cell surface. These invaginations form during intense exocytotic activity (Heuser and Reese, 1973). Model 2 also implies that clathrin-mediated budding occurs in parallel from the cell surface (outer profile or deep invaginations) or from these internalized plasma membrane fragments. Reproduced from De Camilli and Takei (1996) with permission from Cell Press.

Alternatively, newly internalized vesicles could follow the typical receptor-mediated recycling pathway (Mellman, 1996), which involves an endosomal intermediate and a second vesicle budding step (the two-vesicle-budding-steps model).

Massive stimulation of neurotransmitter release in the presence of extracellular tracers leads to a reduction in the number of synaptic vesicles and to the appearance of numerous endosome-like cisternae labeled by the tracer. Upon interruption of the stimulus these changes are reversed and newly labeled synaptic vesicles appear. These observations suggest a precursor-product–like relationship between cisternae and synaptic vesicles (Heuser and Reese, 1973) and are in principle consistent with the two-vesicle-budding-steps model. More recent studies, however, have indicated that the endosome-like cisternae that accumulate in nerve terminals after massive stimulation do not represent classical endosomes, that is, organelles downstream of vesicular traffic that originates from the cell surface (Takei et al., 1996; see also Miller and Heuser, 1984). Clathrin-coated buds similar in size and in protein composition to those present on the plasma membrane can be detected on these membranes (Heuser and Reese, 1973; Takei et al., 1996; Schmidt et al., 1997). Furthermore, tracing of their profiles in serial ultrathin sections has revealed their continuity with the outer profile of the nerve terminal (Takei et al., 1996; Gad et al., 1998). This finding has led to the following hypotheses: (1) these compartments represent portions of the plasma membrane internalized by bulk endocytosis, and (2) new synaptic vesicles bud in parallel from these structures and from the plasma membrane proper via clathrin-mediated endocytosis (Takei et al., 1996). Thus budding of vesicles from this compartment is compatible with a single-vesicle-budding-step process mediated by the clathrin coat. This model is also consistent with morphological studies of *shibire* nerve terminals during recovery from the temperature-sensitive endocytotic block. Release of the block does not result in the massive pinching-off of small vesicles from the expanded plasma membrane, but rather in the formation of large vacuoles, and only later in the generation of synaptic vesicles from these vacuoles (Koenig and Ikeda, 1989).

The single-vesicle-budding-step model has recently been supported by studies of quantal dye uptake and release. Murthy and Stevens (1998) found that the quantal steps in the release of FM1-43 were equal to those measured for its uptake. Had an endosomal intermediary been involved, a dilution of the dye (with smaller released quanta) would have occurred. An intermediate endosomal sorting station would imply a second set of docking, fusion, sorting, and budding events in addition to those involved in exo- and endocytosis. These additional steps are hardly compatible with the extremely fast kinetics of the synaptic vesicle cycle. Some findings, however, suggest caution in excluding additional recycling pathways. The GTPase Rab5, which is generally implicated in endosomal traffic, is present in synaptic vesicles, and Rab5 mutants do affect the synaptic vesicle cycle (de Hoop et al., 1994). Furthermore, an endosomal intermediate and a budding step requiring the coat protein AP-3, but not clathrin, have been implicated in the biogenesis of synaptic-like microvesicles in neuroendocrine cells (Faundez et al., 1998). AP-3 is present in nerve terminals (De Camilli and Takei, unpublished observations), but its role in synaptic vesicle re-formation at the synapse remains unclear, and mice defective in this coat protein do not exhibit major defects in synaptic vesicle recycling (Kantheti et al., 1998). The differentiation of nerve terminals may correlate with the maturation of an AP-3-independent direct pathway of recycling.

Clearly, bona fide endosomes must exist in nerve terminals as in every other peripheral cell compartment. Together with elements of the smooth endoplasmic reticulum (Takei et al., 1992), they may account for the cisternae and tubules visible in nerve endings at rest. They are needed as a target destination for endocytotic traffic not related to synaptic vesicles, as a sorting compartment for material destined for retrograde transport to the cell body, and possibly as a sorting compartment for material delivered from the cell body by anterograde flow. Accumulation in these organelles of viruses (Lewis and Lentz, 1998), toxins (Verderio et al., 1999), and membrane proteins (Marquez-Sterling et al., 1997) that are not enriched in synaptic vesicles is well documented. The relationship between synaptic vesicle recycling and these organelles deserves further investigation.

The potential existence of two modes of synaptic vesicle membrane reuptake from the plasma membrane has been suggested by electron microscopic studies of retinulae cells of the eye in *Drosophila* (Koenig and Ikeda, 1996). When mutant *shibire* flies are shifted from the restrictive temperature of 29°C to 26°C (rather than to the permissive temperature of 19°C), the endocytotic block is only partially released. Under these conditions, plasma membrane invaginations first occur in the vicinity of the synaptic dense body, which marks the active zone in *Drosophila*. Subsequently a second type of membrane invagination appears some distance from the dense body. These two types of invagination may represent intermediates of two alternative endocytotic pathways (Koenig and Ikeda, 1996). However, as the tubular invaginations have been observed only in mutant retinulae cells, their relevance for synaptic vesicle recycling in all neurons under normal conditions is not clear. Even in mammalian retinal cells, peculiar tubular endocytotic intermediates, which are only rarely seen at conventional synapses, are not uncommon (Lovas, 1971; Takei et al., 1998).

The Cycle of the Clathrin Coat

The clathrin coat, which is represented by a honeycomb-like lattice (Fig. 5.11) (Kanaseki and Kadota, 1969; Heuser, 1980), is only one of several coats implicated in vesicle budding reactions within the cell. These coats are thought to act as scaffolds that provide the driving force to

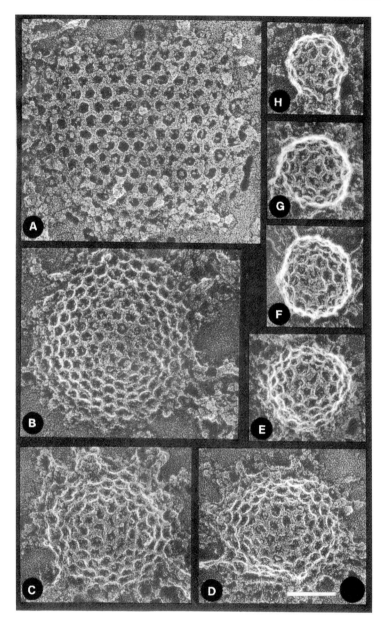

Figure 5.11. A gallery of clathrin-coated pits. Electron micrographs of carbon-platinum replicas of fibroblast inner surfaces prepared by quick-freezing, fracture, and deep etching. The images are arranged according to progressive increases in curvature to suggest a temporal sequence. Scale bar = 0.1 µm. Reproduced from Heuser (1980) by copyright permission of the Rockefeller University Press.

evaginate a flat membrane into a bud of high curvature, as well as sorters that define the molecules to be loaded into the vesicle. The two best-characterized coats besides the clathrin coats are the COPI and COPII coats, which function in traffic connecting the endoplasmic reticulum with the Golgi complex. Although each coat has unique properties and molecular composition, some general principles in their mechanisms of action have emerged (Rothman and Wieland, 1996; Schekman and Orci, 1996; Springer et al., 1999). As a prelude to a more analytic description of the proteins implicated in clathrin coat function, we shall provide an overview of the clathrin-mediated endocytotic reaction in nerve terminals and of some general principles in coat-mediated budding.

Coat Components

The major components of clathrin coats are clathrin and the clathrin adaptors, which link the clathrin lattice to the membrane. The main clathrin adaptor in nerve terminals is the heterotetrameric complex AP-2 (Hirst and Robinson, 1998). Another protein, AP180, is generally referred to as an accessory clathrin adaptor because it co-purifies with synaptic clathrin-coated vesicles and shares with AP-2 all the key properties of adaptors—binding to the membrane, binding to clathrin, and promotion of clathrin assembly (see McMahon, 1999, for review). Clathrin, AP-2, and AP180 interact directly or indirectly with a variety of accessory proteins whose localization at clathrin-coated pits and whose involvement in the coating reaction has been documented. They include dynamin, synaptojanin, amphiphysin, endophilin, Eps15, epsin, intersectin/DAP160, syndapin/pacsin, and auxilin. Coat components and their accessory proteins are recruited to the nerve terminal plasma membrane from a cytosolic pool, but most likely they are never completely free in the cytosol because they are present at higher concentrations in nerve endings than in other regions of the neuron. Furthermore, in nerve endings they are particularly concentrated in the area of cytoplasm surrounding clusters of synaptic vesicles, possibly reflecting their interaction with an organizing cytoskeletal matrix at sites where endocytosis occurs (see the chapter by De Camilli et al. on the structure of synapses, this volume).

Coat Recruitment

Coat assembly nucleation starts with the recruitment and oligomerization of the adaptors and is followed rapidly by clathrin recruitment (Kirchhausen et al., 1997; Schmid, 1997). Both membrane lipids and membrane proteins play an important role in this process. The importance of lipids is demonstrated by the observation that clathrin coats, like COPI and COPII coats (Matsuoka et al., 1998; Spang et al., 1998), can assemble on liposomes in the absence of membrane proteins, and that clathrin coats formed under these conditions are morphologically indistinguishable from clathrin coats observed in living cells (Takei et al., 1998). Lipid bilayers of different composition can function as templates for vesicle budding, but phosphoinositides play an important regulatory role, because specific binding sites in both AP-2 and AP180 recognize their phosphorylated inositol rings. Interactions of intrinsic membrane proteins with coat proteins cooperate with lipid-based interactions and are essential in defining the spatial specificity of recruitment. A major protein-docking site for AP-2 on synaptic vesicle membranes is the C2B domain of synaptotagmin (Zhang et al., 1994; Haucke and De Camilli, 1999). Coat assembly from purified coat proteins can occur on liposomes in the absence of ATP (Takei et al., 1998). However, ATP dependence of coat formation has been observed in various cell-free studies using cellular membranes as a template (Schmid, 1997). This requirement may reflect multiple biochemical reactions, including phosphoinositide synthesis and rearrangements of cytoskeletal membrane scaffolds.

Synergistic interactions and positive cooperativity play an important role in the rapid growth of the coat (Rapoport et al., 1997; Haucke and De Camilli, 1999; Springer et al., 1999; Wieland and Harter, 1999). The assembled coat, in turn, recruits additional membrane proteins in a positive feedback loop. These mechanisms allow coats to function as a molecular sorter for proteins to be accumulated and loaded into the nascent vesicle. Several of the accessory factors that bind clathrin, AP-2, or both—such as amphiphysin, epsin, and Esp15—may cooperate in the growth of the coat by acting as transient additional adaptor molecules. However, these proteins, unlike AP-2, AP180, and clathrin, are not enriched in clathrin-coated vesicles (Chen et al., 1998). If they act as adaptors, they must have a transient, dynamic role. An attractive possibility is that they have a special role in the dynamics of the edges of the clathrin coat, a site where growth and intermolecular rearrangements take place (Cupers et al., 1998; Chen et al., 1999).

Acquisition of Curvature

A major open question is the mechanism by which vesicle coats acquire their curvature. Clathrin lattices can form flat sheets, curved domes, or closed baskets. In curved lattices hexagons are interspersed with pentagons, and the smallest baskets are composed of 12 pentagons and 8 hexagons (Smith et al., 1998). According to a popular model, the clathrin coat assembles first as a flat hexagonal lattice (Heuser, 1980). This model has been supported by electron microscopic observations suggesting a precursor-product relationship between low- and high-curvature states of clathrin. However, the molecular rearrangements required for the transformation of the coat from a planar lattice to a highly curved dome are extremely complex. An alternative view has therefore been proposed, namely that the curvature of a clathrin coat is already defined as the coat starts to oligomerize, such that transition from a low-curvature to a high-curvature state is only possible through a depolymerization-repolymerization process (Musacchio et al., 1999). At present there is little direct evidence to support or exclude either view. Furthermore, it is not clear whether factors intrinsic to the membrane—for example, the clustering of certain lipids or proteins at the coated bud—actively contribute to generate curvature.

Two experimental manipulations have recently been shown to inhibit the acquisition of curvature by clathrin-coated buds: the depletion of cholesterol from cellular membranes (Rodal et al., 1999; Subtil et al., 1999) and the antibody-mediated disruption of endophilin function in living synapses (Ringstad et al., 1999). Endophilin is an enzyme that generates phosphatidic acid from lysophosphatidic acid and arachidonoyl-CoA (Schmidt et al., 1999). It remains to be determined how these manipulations regulate membrane curvature. They may function by directly perturbing the lipid geometry within the bilayer, by affecting protein recruitment to the buds, or indirectly by affecting a cytoskeletal scaffold controlling membrane rigidity.

Regulation of Vesicle Size

A peculiar characteristic of clathrin-coated vesicles implicated in synaptic vesicle recycling is their small and homogeneous size. This size homogeneity is not an intrinsic characteristic of the clathrin coat, because larger clathrin-coated vesicles are observed in many cells, and scattered larger clathrin-coated vesicles are seen even in nerve endings. Most likely, the cocktail of clathrin adaptors is an important determinant of vesicle size, and indeed the accessory clathrin adaptor AP180 seems to play a role in this process (Zhang et al., 1998; Nonet et al., 1999). Synaptic vesicles are among the smallest vesicles observed in cells. One way to guarantee the small, homogeneous size of synaptic vesicles is to generate a membrane scaffold that is capable of evaginating the membrane to the highest possible degree of curvature.

The Fission Reaction

Clathrin coats alone seem to be unable to perform the last step in budding—the fission reaction—once a deeply invaginated bud has formed. A striking accumulation of deeply invaginated pits can be observed at the restrictive temperature in *Drosophila shibire* mutants (Koenig and Ikeda, 1989), which harbor a mutation in the dynamin gene (Chen et al., 1991; van der Bliek and Meyerowitz, 1991) (Fig. 5.1). Furthermore, a variety of experimental manipulations have been shown to arrest clathrin-mediated endocytosis at the stage of invaginated coated pits, clearly indicating the need for additional factors (Takei et al., 1995, 1996; Shupliakov et al., 1997; Ringstad et al., 1999). The most powerful effect was produced by manipulations that impaired dynamin function (the addition of GTPγS in cell-free systems; microinjection into living synapses of peptides that perturb interactions of the targeting domain of dynamin or of GTPγS). These findings provide compelling evidence for a key role of dynamin in the fission reaction. However, the precise mechanism of dynamin's function in fission remains a matter of debate. It is also unclear whether dynamin acts alone or in concert with other proteins, such as amphiphysin (Takei et al., 1999) or endophilin (Ringstad et al., 1999). Other GTPases, for example GTPases of the Rho family that control the actin-based cytoskeleton (Hall, 1998), may contribute to the block in fission produced by GTPγS. It should be noted that "fission factors" (distinct from "budding factors") were not identified in the case of COPII-mediated budding (Schekman and Orci, 1996). In view of the many similarities among these coat-mediated budding steps, this discrepancy remains unexplained.

Uncoating and Recapture by the Synaptic Vesicle Cluster

The final steps in the endocytotic reaction are uncoating, migration of the vesicle to a deeper location in the cytoplasm, and the capture of the vesicle by the synaptic vesicle cluster. Almost immediately after separating from the donor membrane, clathrin-coated vesicles shed their coat. Because this process occurs so rapidly, free-coated vesicles are often difficult to observe. Indeed, the very existence of this intermediate was a subject of heated debate (Willingham and Pastan, 1983; van Deurs et al., 1989).

A key player in the uncoating reaction is the ATPase heat shock protein 70 (Hsc70). Hsc70 is targeted to the assembled clathrin coat by auxilin, an abundant nerve-terminal protein that co-purifies with clathrin-coated vesicles. Auxilin bound to clathrin coats significantly enhances the ATPase activity of Hsc70. Disassembly of the clathrin lattice produces clathrin triskelia as an end product (Ungewickell, 1985; Chappell et al., 1986; Ungewickell et al., 1995). Hsc70 alone does not remove the adaptors, suggesting that additional factors are needed for full uncoating (Hannan et al., 1998). A proper function of the clathrin-mediated endocytotic cycle implies an action of Hsc70 only after the fission reaction. Some events closely coupled to fission have been hypothesized to initiate uncoating (Patzer et al., 1982), but this hypothesis remains to be tested.

Very little is known about the journey from sites of endocytosis to the vesicle cluster. The presence of an actin cytomatrix at sites of endocytosis suggests that actin plays a crucial role in targeting vesicles back to the reserve pool of releasable vesicles (Gustafsson et al., 1998; Brodin, 1999). Moreover, the stages in this journey at which neurotransmitters are pumped into the vesicles and when peripheral proteins required for exocytosis associate with their membranes (such as synapsin, Rab3, and rabphilin; see Südhof and Scheller, this volume) are also poorly defined. Upon reentering the cluster, newly reformed vesicles mix randomly with vesicles of the existing pool at the frog neuromuscular junction and in hippocampal neurons (Betz and Bewick, 1992; Ryan and Smith, 1995;

Kraszewski et al., 1996). In neuromuscular terminals of *Drosophila* larvae, however, studies involving FM1-43 labeling of *shibire* mutants have indicated a preferential targeting to a subregion of the vesicle pool (Kiromi and Kidukori, 1998).

Vectoriality of the Cycle: GTPases and Phosphoinositide Switches

The sequentially ordered nature of the reactions underlying clathrin-mediated endocytosis implies the occurrence of mechanisms controlling its vectoriality. Studies on COPI, COPII, and nonendocytotic clathrin coats have suggested that GTPases function as switches controlling the coat recruitment reaction (Springer et al., 1999). The small GTPase Arf plays an important role in the recruitment of both COPI and the clathrin coats that use AP-1 as the main adaptor complex (i.e., the clathrin coats involved in traffic in the Golgi complex region) (Ostermann et al., 1993; Stamnes et al., 1995; Rothman and Wieland, 1996). Membrane binding of Arf is triggered by GDP-GTP exchange catalyzed by a family of Sec7 domain-containing guanyl nucleotide exchange factors (Goldberg, 1998; Mossessova et al., 1998). Arf-GTP, in turn, regulates coat recruitment via multiple effectors. One effector may be the coat itself, as shown by the direct binding of Arf-GTP to the COPI coat (Goldberg, 1999), but such a direct binding has not yet been demonstrated for the clathrin coat.

Another major effector for Arf is phospholipase D, which through a series of positive feedback loops leads to the generation of $PtdIns(4,5)P_2$ (Liscovitch and Cantley, 1994). Yet other targets are PtdIns 4-kinase and $PtdIns(4)P$ 5-kinase, which synthesizes $PtdIns(4,5)P_2$ (Martin et al., 1996; Godi et al., 1999; Honda et al., 1999). $PtdIns(4,5)P_2$, in turn, potentiates the interaction between coats and membranes (Arneson et al., 1999; Cremona et al., 1999; Gaidarov and Keen, 1999). Whether Arf-family GTPases, or another GTPase switch, participates in clathrin recruitment at synapses remains an open question. GTPγS drastically enhances the formation of clathrin-coated intermediates in nerve terminals (Takei et al., 1996). However, the relative contribution to this effect of enhanced coating and decreased uncoating is not known. If one of the major functions of Arf in clathrin coat recruitment is to stimulate the production of phosphoinositides, such an effect may be achieved primarily by other mechanisms in nerve terminals.

Conversely, GTP hydrolysis is thought to be an important prerequisite in coat shedding (Springer et al., 1999), and recent studies have implicated the polyphosphoinositide phosphatase synaptojanin in clathrin uncoating at the synapse (Cremona et al., 1999; Gad et al., 1999). An attractive hypothesis is that two parallel and interconnected cycles, the GDP-GTP cycle of a GTPase and the phosphorylation-dephosphorylation of phosphoinositides, may act as major regulators of the vectoriality of the clathrin coat cycle.

Molecular Components of the Clathrin Coat Implicated in Synaptic Vesicle Endocytosis

Clathrin

The main structural component of the coat is clathrin (Pearse, 1975), which gave its name to this type of endocytosis. Clathrin is composed of heavy (180-kDa) and light (33- to 35-kDa) chains. The heavy chain is a highly elongated molecule encoded by at least two genes in mammals: CLH-17, which is ubiquitously expressed (Kirchhausen et al., 1987), and CLH-22 (83% identity in amino acid sequence), which is preferentially expressed in skeletal muscle (Kedra et al., 1996; Sirotkin et al., 1996). The light chains are also encoded by a pair of genes, LCa and LCb, that generate ubiquitous and neuron-specific alternative splice variants (Jackson et al., 1987). The heavy chain is responsible for the structural function of the clathrin coat, whereas the light chains are thought to regulate the assembly-disassembly process (Schmid et al., 1984). The dimer of clathrin heavy and light chains further trimerizes to form triskelia (Ungewickell and Branton, 1981), and individual triskelia assemble to form the clathrin lattice.

A first view of the molecular organization of clathrin arrangement within the coat was provided by rotary shadowing (Kirchhausen et al., 1986). More recently, cryoelectron microscopy of assembled cages and crystallography have extended this information to atomic resolution (Smith et al., 1998; ter Haar et al., 1998; Ybe et al., 1999) (Fig. 5.12). The C terminus of the heavy chain (hub) forms the vertices of the polyhedron and is responsible for trimerization. The bulk of the protein, which is composed of seven homologous repeats (clathrin heavy chain repeats), forms a rodlike structure (leg) with a bend in the middle (knee) that separates two domains "proximal" and "distal" to the hub, respectively. The crystal structures of one of the repeats and portions of other repeats show that the rod is formed by an elongated right-handed superhelical coil of short α-helixes (Ybe et al., 1999). A bundle composed of two proximal and two distal domains (both in antiparallel orientation) forms each of the edges connecting the vertices of the polyhedral cage, and each leg of the triskelion participates in two edges. The N-terminal domain projects toward the membrane like a foot, providing a recognition interface for adaptor molecules and other accessory proteins. This domain has a seven-blade β-propeller structure, which represents a polypeptide-binding module also found in other proteins (ter Haar et al., 1998). A lateral groove between two blades of the propeller binds arrestin (ter Haar et al., 1998) and may also interact with other clathrin-binding proteins (including β-adaptin, AP180, amphiphysin, and epsin) that share with arrestin a similar clathrin-binding consensus. The knee is the most flexible region of the triskelion (Musacchio et al., 1999), and it is this flexibility that allows the lattice to adjust to different curvatures and to assemble in different pentagon-hexagon ratios. These

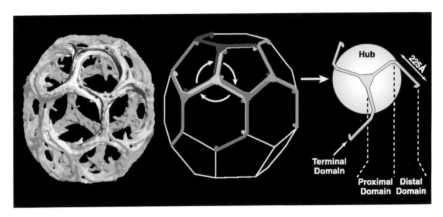

Figure 5.12. Structural organization of the clathrin lattice. The left panel shows a 21-Å-resolution map of a clathrin hexagonal barrel assembled from purified clathrin and AP-2, as revealed by cryoelectron microscopy analysis (Smith et al., 1998). Only the clathrin lattice is shown. The coats of clathrin-coated vesicles are larger. The smallest vesicle-associated coat is estimated to comprise 60 triskelia assembled into 20 hexagons and 12 pentagons. The middle and right panels show schematic views of the clathrin cage and of one individual triskelion. Incorporation and disengagement of triskelia into and from the lattice, respectively, are thought to occur by rotation around the vertexes (as indicated by arrows in the center panel) (Musacchio et al., 1999). Reproduced from Marsh and McMahon (1999); copyright 1999 American Association for the Advancement of Science.

properties of the triskelion make it possible to disengage it smoothly from neighboring triskelia by rotation around its vertex or to permit addition of new triskelia to the growing lattice (Musacchio et al., 1999).

AP-2

While clathrin alone can polymerize into baskets in vitro, its polymerization is significantly enhanced by the clathrin adaptors, whose main function is to link the clathrin lattice to the membrane. The main adaptor on the clathrin-coated vesicles that participate in synaptic vesicle recycling is AP-2, a member of a heterotetrameric family of coat proteins that also includes at least three other members in mammals: AP-1, AP-3, and AP-4 (Hirst and Robinson, 1998). These protein complexes share similar principles of organization (two large and two small subunits tightly bound to each other) and have similar amino acid compositions, but they participate in different vesicular transport reactions.

As mentioned previously, AP-3 has been implicated in a pathway of vesicle biogenesis from endosomes in neuroendocrine cells (Faundez et al., 1998), and it is also involved in a transport pathway to lysosomes (Dell'Angelica et al., 1999b). AP-1 functions in clathrin-dependent export from the Golgi complex. The function of AP-4 is unclear (Dell'Angelica et al., 1999a). The subunits of AP-2 are α- and β2-adaptins (~100 kDa each), μ2 and σ2 (~50 and 20 kDa, respectively) (Fig. 5.13). Based on current information, each subunit is encoded by a single gene in mammals, with the exception of α-adaptin, which is encoded by two distinct genes (Robinson, 1989). Both gene products, αA and αC, are expressed in neurons. Furthermore, both α- and β-adaptins undergo neuron-specific splicing (Ponnambalam et al., 1990; Ball et al., 1995), which results in the insertion of additional amino acids in the hinge regions. The study of purified AP-2 complexes by rotary shadowing reveals a bricklike core (head) with two symmetrical "ears" connected by a flexible "hinge" to the head (Heuser and Keen, 1988). The head contains the N-terminal portions of α- and β2-adaptins plus μ2 and σ2, whereas the ears represent the C-terminal domains of α- and β2-adaptins.

AP-2 functions both as a linker between the plasma membrane and the clathrin lattice and as a sorter for some of the proteins to be internalized. Binding of AP-2 to clathrin is mediated primarily by its β2 subunit. Isolated β2-adaptin binds clathrin and competes with binding of whole AP-2 (Ahle and Ungewickell, 1989). Furthermore, β-adaptin promotes clathrin polymerization into cages (Gallusser and Kirchhausen, 1993; Goodman and Keen, 1995; Shih et al., 1995). These properties were attributed to the binding of the hinge domain of β2 (Shih et al., 1995) to the N-terminal domain of clathrin (Murphy and Keen, 1992). AP-2, however, contains additional binding sites for clathrin (Keen et al., 1991), which may contribute to its clathrin-assembling activity.

Three main types of interaction, which have a synergistic function, have been implicated in the recruitment of AP-2 to the membrane at the synapse. A first interaction is the binding of the N-terminal domain of α-adaptin to membrane phospholipids, primarily phosphoinositides (Beck and Keen, 1991; Gaidarov et al., 1996). α-Adaptin mutants defective in phosphoinositide binding are not recruited to coated pits and interfere with the recruitment of endogenous wild-type AP-2 (Gaidarov and Keen, 1999). Generation of $PtdIns(4,5)P_2$ is a key step in AP-2 recruitment and subsequent assembly of clathrin coats on membranes in vitro (Arneson et al., 1999). Furthermore, reduced catabolism of phosphoinositides at

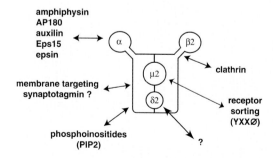

Figure 5.13. Schematic illustration of the clathrin adaptor AP-2 and its interacting partners.

endocytotic sites due to disruption of the polyphosphoinositide phosphatase synaptojanin 1 results in the accumulation of clathrin-coated vesicles (Cremona et al., 1999).

A second interaction, which is thought to have a master role in coat recruitment, is the binding of AP-2 to the C2B domain of the synaptic vesicle protein synaptotagmin. Synaptotagmin isoforms are expressed in all cells, and synaptotagmin may therefore represent the high-affinity, protease-sensitive AP-2 binding site demonstrated in nonneuronal cells (Virshup and Bennett, 1988; Mahaffey et al., 1990; Zhang et al., 1994; Li et al., 1995; Haucke and De Camilli, 1999). In addition, synaptotagmin isoforms have been identified on lysosomes, on which the AP-2/clathrin coat can assemble (Baram et al., 1999). In agreement with a role of synaptotagmin in endocytosis, its mutation in *Caenorhabditis elegans* results in a loss of synaptic vesicles (Fukuda et al., 1995; Jorgensen et al., 1995).

The third interaction is the binding of AP-2 to endocytotic motifs of synaptic vesicle proteins. Studies of receptor-mediated endocytosis have revealed that many proteins internalized via clathrin contain short amino acid sequences in the cytoplasmic domain(s) that are required for their endocytosis. These motifs include tyrosine-containing sequences such as YXX$\emptyset$ (X is any amino acid, $\emptyset$ is a bulky hydrophobic residue) and NPXY, as well as dileucine-containing sequences (Trowbridge et al., 1993; Marks et al., 1996; Kirchhausen et al., 1997; Heilker et al., 1999). The interaction between $\mu2$ and YXX$\emptyset$ peptides has been investigated most extensively, with the characterization at the atomic level in co-crystals of the YXX$\emptyset$-binding region of $\mu2$ with a bound peptide (Owen et al., 1999). Binding of YXX$\emptyset$ motifs to $\mu2$ potentiates the interaction of AP-2 with synaptotagmin, suggesting a mechanism by which a protein cargo of the vesicle may enhance the assembly of endocytotic coat (Haucke and De Camilli, 1999). Two such motifs are present in the cytoplasmic domains of the synaptic vesicle protein SV2.

Finally, AP-2 binds several cytosolic proteins implicated in endocytosis via the ear domain of α-adaptin. These proteins, which are discussed later in this chapter, include AP-180, amphiphysin, Eps15, epsin, and auxilin. They all contain one or multiple D/EPF/W motif(s) in their primary sequence. This motif was shown to contribute to binding, thus suggesting that these proteins share a similar binding site in AP-2. Such a site has been proposed to correspond to the mixed α-/β-platform in the recently described crystal structure of the ear domain (Owen et al., 1999; Traub et al., 1999).

AP180

AP180, also called NP185, F1-20, or AP-3 (note that this protein is different from the heterotetrameric coat protein AP-3) (Ahle and Ungewickell, 1986; Keen, 1987; Kohtz and Puszkin, 1988; Zhou et al., 1992), is a brain-specific protein, highly enriched in nerve terminals. Like AP-2, AP180 co-purifies with brain clathrin-coated vesicles, interacts with clathrin, promotes clathrin assembly into cages (four times more potently than AP-2), and binds membrane phosphoinositides (Lindner and Ungewickell, 1992). AP180 contains two clathrin-binding sites: one in the highly conserved 300-amino-acid N-terminal region (Morris et al., 1993), which presumably binds to the N-terminal domain of clathrin (Murphy et al., 1991), and the other in the C-terminal region (Ye and Lafer, 1995b). The 58-kDa C-terminal fragment is responsible for the clathrin assembly activity of the whole protein (Ye and Lafer, 1995b). AP180 binds AP-2, and the resulting complex promotes clathrin assembly much more efficiently than either protein alone (Hao et al., 1999). The interaction of AP180 with AP-2 maps to the C-terminal portion of AP180, downstream of the DPF motifs, the sequences that play a role in the binding to α-adaptin in other proteins. This region, however, contains several DAF repeats (Hao et al., 1999).

Clathrin cages polymerized in the presence of AP180 appear to be more homogeneous in size than cages formed in the absence of AP180, suggesting a role of AP180 in defining the size of coated vesicles (Ye and Lafer, 1995a). This hypothesis has recently been corroborated by genetic studies in *Drosophila* and *C. elegans*. Genetic disruption of the *Drosophila* homologue of AP180, LAP (Like AP180) results in death at a late stage of development, and the few adults that escape display strong nervous system defects. In the nerve terminals of these mutants, the number of synaptic vesicles is severely reduced, and the remaining vesicles are more variable in size (generally larger) than in normal flies (Zhang et al., 1998). A mutation in the *C. elegans* homologue of AP180, *Unc11*, produces similar defects (Nonet et al., 1999). Although worms lacking UNC-11 are viable, they have motor defects (indicating impaired synaptic transmission), and their synaptic vesicles are heterogeneous in size. Additionally, in these mutants the synaptic vesicle protein synaptobrevin is partially mislocalized to the plasma membrane, suggesting that AP180 may be required to sort this protein into the vesicles (Nonet et al., 1999). As AP180 acts in concert with clathrin, its putative role in defining the size of synaptic vesicles speaks in favor of a direct reformation of synaptic vesicles from the uncoating of clathrin-coated vesicles.

Whereas AP180 is selectively localized to nerve terminals, a homologue of this protein, CALM, is ubiquitously expressed, indicating that this protein family may have a general role in endocytosis (Dreyling et al., 1996; Tebar et al., 1999).

Accessory Factors of Clathrin-Mediated Endocytosis

Clathrin and the clathrin adaptors do not act alone in endocytosis. Although they are sufficient to generate clathrin-coated buds on protein-free liposomes, no fission is observed under such conditions (Takei et al., 1998). Furthermore, the endocytotic reaction in living cells requires

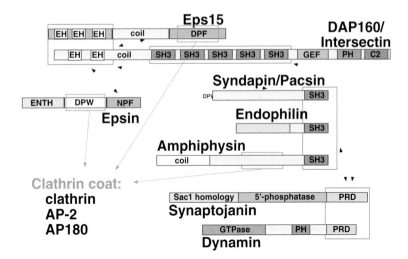

Figure 5.14. *Schematic illustration of the domain structure of accessory proteins of clathrin-mediated endocytosis at the synapse.* Arrows indicate the various interactions.

the function of a variety of other proteins besides clathrin and the adaptors. Considering that this process occurs in a highly regulated fashion and at very precise sites in the complex microenvironment of the cytosol, the existence of a large number of regulatory factors is not unexpected. A summary of the properties of the best-characterized of these factors is given in Fig. 5.14.

Dynamin

The GTPase dynamin was the first protein to be implicated in synaptic vesicle endocytosis, besides the intrinsic components of the clathrin coat. This discovery resulted from its identification as the protein encoded by the *shibire* locus of *Drosophila* (Chen et al., 1991; van der Bliek and Meyerowitz, 1991). Dynamin is the collective name for three mammalian genes that can be alternatively spliced to generate multiple isoforms. The product of one of these genes, dynamin 1, is expressed at high concentration in nervous tissue, where it is selectively concentrated in nerve terminals (Urrutia et al., 1997).

All dynamin isoforms have a similar domain structure. They comprise an N-terminal GTPase domain, a PH (pleckstrin homology) domain, the so-called GED domain (GTPase effector domain), and a proline-rich C-terminal domain (PRD) (Schmid et al., 1998). The GTPase domain is crucial for function—mutations that disrupt GTP binding typically have a dominant-negative effect and produce strong inhibition of clathrin-mediated endocytosis (Herskovits et al., 1993; Damke et al., 1994). Interactions of PH and PRD domains are important for the targeting of dynamin and for the regulation of its GTPase activity (Gout et al., 1993; Lin and Gilman, 1996; Salim et al., 1996; Lin et al., 1997). The PH domain binds phosphoinositides, with a preference for PtdIns(4,5)P_2, and dynamin mutants that harbor mutations in this domain have a dominant-negative effect on clathrin-mediated endocytosis (Achiriloaie et al., 1999; Lee et al., 1999; Vallis et al., 1999). The PRD domain contains several consensus sites for binding to SH3 domains and, accordingly, interacts with a variety of SH3 domain-containing proteins in nerve terminals. These include amphiphysins (David et al., 1996; Grabs et al., 1997; Ramjaun et al., 1997; Wigge et al., 1997a), endophilins (SH3P4, SH3P8, SH3P13) (de Heuvel et al., 1997; Ringstad et al., 1997), Dapl60/intersectin (Roos and Kelly, 1998; Yamabhai et al., 1998; Hussain et al., 1999) and syndapin/pacsin (Qualmann et al., 1999). Several other partners have been documented in nonneuronal cells, such as Grb2 (Gout et al., 1993), cortactin (Grabs et al., 1997), phospholipase C (Scaife et al., 1994; Seedorf et al., 1994), phosphatidylinositol-3 kinase (Gout et al., 1993), and Src (Foster-Barber and Bishop, 1998). At least some of these interactions play a role in the targeting of dynamin to clathrin-coated pits (Grabs et al., 1997; Okamoto et al., 1997; Shupliakov et al., 1997; Wigge et al., 1997b; Roos and Kelly, 1998).

Dynamin has a propensity to oligomerize and monomeric dynamin probably does not exist. The basic structural unit was proposed to be a tetramer (Liu et al., 1996; Muhlberg et al., 1997). Tetramers have a tendency to oligomerize further into open rings and stacks of rings in vitro (Hinshaw and Schmid, 1995; Carr and Hinshaw, 1997). Oligomerization results in a powerful stimulation of GTPase activity (Tuma and Collins, 1994; Warnock et al., 1996). Dynamin rings and spirals can also assemble on lipid membranes, leading to the formation of narrow tubules (Sweitzer and Hinshaw, 1998; Takei et al., 1998). Similar, although thicker, dynamin rings are seen at the neck of the "frozen" endocytotic intermediates in nerve terminals of *shibire* flies (Koenig and Ikeda, 1989). They are also seen around the neck of clathrin-coated pits generated in vitro in the presence of cytosol and GTPγS, a slowly hydrolyzable analogue of GTP (Takei et al., 1995). In these rings, dynamin is associated with SH3 domain-binding partners such as amphiphysin and endophilin (Ringstad et al., 1999; Takei et al., 1999). The property of dynamin to oligomerize relies on its ability to establish multiple intermolecular contacts (Muhlberg et al., 1997; P. M. Okamoto et al., 1999; Smirnova et al., 1999). One such interaction, between the GTPase domain of one molecule and the GED domain of another molecule, stimulates its GTPase activity, suggesting that the GED domain might be the GTPase activating protein for dynamin (Muhlberg et al., 1997; Sever et al., 1999; Smirnova et al., 1999).

The property of dynamin to oligomerize at the neck of invaginated pits had suggested that it may act as a mechano-enzyme harvesting the energy released from GTP hydrolysis for the fission reaction (Hinshaw and Schmid, 1995; Takei et al., 1995). Support for this model came from the demonstration that dynamin-coated lipid tubules undergo fragmentation in the presence of GTP (Sweitzer and Hinshaw, 1998; Takei et al., 1998, 1999). Proteins that bind assembled dynamin in vivo may act as amplifiers of the conformational change in dynamin. Such a model, however, sets aside dynamin from other GTPases, which typically perform their action by interacting with other effectors in their GTP-loaded or "active" conformation (Bourne, 1995). The alternative possibility is that dynamin may act by recruiting and/or activating other proteins at the vesicle neck (De Camilli and Takei, 1996; Roos and Kelly, 1997; Sever et al., 1999). Evidence for such a mechanism has come from the demonstration that some GTPase-defective dynamin mutants enhance receptor-mediated endocytosis rather than inhibiting it (Sever et al., 1999). So far, however, no effectors for the GTPase domain of dynamin (other than dynamin itself) have been identified (Smirnova et al., 1999). Endophilin was proposed to act directly downstream of dynamin and to mediate fission by geometric changes in the membranes produced by its enzymatic activity (Schmidt et al., 1999). However, endophilin binds dynamin through its SH3 domain (Ringstad et al., 1999) and not via the GTP-dependent interaction expected for classical GTPase effectors.

Synaptojanin

Synaptojanin is a polyphosphoinositide phosphatase. The central domain of this protein is homologous to the catalytic region of a variety of inositol 5-phosphatases and acts selectively on the phosphate at the $5'$ position of the inositol ring (McPherson et al., 1996). The N-terminal domain is homologous to the yeast protein Sac1 and, like Sac1, can dephosphorylate other positions of the inositol ring (Guo et al., 1999). $PtdIns(4,5)P_2$ and $PtdIns(3,4,5)P_3$ appear to be the main physiological substrates of this enzyme (Woscholski et al., 1997; Cremona et al., 1999). Of the two known mammalian synaptojanins (McPherson et al., 1996; Nemoto et al., 1997), synaptojanin 1 (more specifically its 145-kDa isoform) is the predominant isoform in mammalian nerve terminals, where it is concentrated on coated endocytotic intermediates (McPherson et al., 1996; Haffner et al., 1997). A splice variant of 170 kDa, which contains a C-terminal extension, is more broadly expressed (Ramjaun and McPherson, 1996). The alternatively spliced C-terminal region of synaptojanin 1 mediates its subcellular targeting. The portion common to both 145- and 170-kDa isoforms binds to several SH3 domain-containing proteins, among which endophilin and amphiphysin are the most important partners at the synapse (McPherson et al., 1996; de Heuvel et al., 1997; Ringstad et al., 1997). The portion unique to the 170-kDa isoform binds directly to clathrin, the clathrin adaptor AP-2, and the clathrin accessory protein Eps15 (Haffner et al., 1997, 2000).

Targeted disruption of the synaptojanin 1 gene in the mouse leads to early death, failure to thrive, and a defect in synaptic transmission.

An increased number of clathrin-coated vesicles was observed in the cytomatrix-rich area surrounding the vesicle cluster at synapses. These results suggest a role of synaptojanin 1 in the uncoating reaction and in the regulation of actin dynamics (Cremona et al., 1999). Consistent with these data, antibody-mediated disruption of synaptojanin 1 function at the giant reticulospinal synapse of the lamprey produces a drastic defect of synaptic vesicle recycling (Gad et al., 1999). In these terminals, the presence of clathrin-coated pits and vesicles is accompanied by a massive increase in the concentration of polymerized actin at endocytotic sites. This effect may be explained by the positive effect of phosphoinositides on actin nucleation (Janmey, 1998). Most likely, synaptojanin 1, via its catabolic effect on phosphoinositides, has multiple sites of action in nerve terminals. In addition to coat proteins and actin, at least two other protein modules implicated in endocytosis bind phosphoinositides—the C2B domain of synaptotagmin (Fukuda et al., 1995) and the PH domain of dynamin.

Amphiphysin

Amphiphysin is a major dynamin-interacting protein in nerve terminals. The amphiphysin family in mammals comprises two very similar proteins, amphiphysin 1 and 2, which form homo- and heterodimers (Wigge et al., 1997a; Slepnev et al., 1998). Although amphiphysin 1 is predominantly expressed in brain (at levels 20- to 30-fold higher than in other tissues; Lichte et al., 1992; David et al., 1996), a number of amphiphysin 2 splice variants (Bin 1, SH3P9) are broadly expressed in other tissues, and one splice variant is expressed at particularly high concentration in skeletal muscle (Sakamuro et al., 1996; Sparks et al., 1996; Butler et al., 1997; Leprince et al., 1997; Ramjaun et al., 1997; Tsutsui et al., 1997; Wigge et al., 1997a). The heterodimer of amphiphysin 1 and 2 is the predominant form in brain (Wigge et al., 1997a).

Amphiphysin has three main domains. The N-terminal domain is responsible for dimerization (Slepnev et al., 1998) and contains a lipid bilayer–binding region (Takei et al., 1999). The central domain binds directly, via two partially overlapping binding sites, to both clathrin (N-terminal region of the heavy chain) and AP-2 (ear domain of α-adaptin) (Wang et al., 1995; McMahon et al., 1997; Ramjaun and McPherson, 1998; Slepnev et al., 1998). Accordingly, this region contains both the clathrin-binding consensus (Dell'Angelica et al., 1998) and a DPF motif (in amphiphysin 1), found in many α-adaptin-binding proteins. The C-terminal region of amphiphysin contains an SH3 domain, which binds with high affinity to both dynamin and synaptojanin (David et al., 1996; Grabs et al., 1997; Ramjaun et al., 1997). These multiple interactions suggest that amphiphysin may act as a multifunctional adaptor. This hypothesis has been supported by the demonstration that amphiphysin can form a tertiary complex with dynamin and α-adaptin or clathrin

(Slepnev et al., 2000). Amphiphysin may also cooperate in the recruitment of coat proteins to the lipid bilayer and in targeting to the coat of the factors needed for fission (dynamin) and uncoating (synaptojanin). Purified amphiphysin evaginates liposomes into narrow tubules and co-assembles with dynamin on the tubules. It also enhances tubule fragmentation by dynamin in these cell-free systems (Takei et al., 1999). It is therefore possible that amphiphysin may assist other endocytotic proteins in the generation of highly curved membrane microdomains and in vesicle separation from the plasma membrane. A critical role of amphiphysin in the fission reaction has been demonstrated by microinjection experiments at the giant reticulospinal synapse of the lamprey. Injection of peptides that compete with the interactions of the amphiphysin SH3 domain produces a potent block of synaptic vesicle recycling, leaving deeply invaginated and constricted coated pits (Shupliakov et al., 1997).

Endophilin

Endophilin is the collective name for a family of three similar proteins referred to as either SH3P4, SH3P8, and SH3P13 or endophilin 1, 2, and 3 (Sparks et al., 1996; de Heuvel et al., 1997; Ringstad et al., 1997). Endophilin 1 (SH3P4) is the most abundant in brain, where it is localized to presynaptic nerve terminals (Ringstad et al., 1997). The primary sequence of the endophilins contains a conserved N-terminal domain connected by a short variable region to a C-terminal SH3 domain. The N-terminal domain has lysophosphatidic acid acyltransferase activity, that is, it catalyzes the incorporation of arachidonic acid into lysophosphatidic acid, leading to the production of phosphatidic acid (Schmidt et al., 1999). The SH3 domain binds both synaptojanin and dynamin, but it clearly prefers synaptojanin (Ringstad et al., 1997). Endophilin 1 is required for synaptic vesicle biogenesis in a cell-free budding assay (Schmidt et al., 1999). Furthermore, microinjection of anti-endophilin antibodies into axons of the lamprey reticulospinal neuron blocks synaptic vesicle recycling (Ringstad et al., 1999). These microinjections produce a striking arrest of the endocytotic reaction at the stage of shallow clathrin-coated pits. The conversion of lysophosphatidic acid to phosphatidic acid could play a direct role in the regulation of bilayer curvature (Schmidt et al., 1999). Alternatively it could act indirectly via a signal transduction mechanism initiated by phosphatidic acid. Disruption of endophilin function in a cell-free system does not affect the formation of deeply invaginated clathrin-coated buds. A principal role of the SH3 domain of endophilin in vivo is the recruitment of synaptojanin 1. The microinjection at the lamprey synapse of peptides that block the binding properties of this SH3 domain mimics the effects produced by synaptojanin disruption. It induces an accumulation of clathrin-coated vesicles and a "hypertrophy" of the actin cytoskeleton at endocytotic sites (Gad et al., 1999).

Other Accessory Factors

A number of other proteins have been implicated as accessory factors in clathrin-mediated endocytosis, including Eps15, epsin, intersectin/ DAP160, and syndapin/pacsin. The involvement of each of these proteins in endocytosis was first suggested by their interaction with either components of the clathrin coat (AP-2 or clathrin) or dynamin. Additional evidence for their endocytotic role includes: (1) their concentration in nerve terminals (Chen et al., 1998; Hussain et al., 1999; Qualmann et al., 1999); (2) their partial localization at clathrin-coated pits in nonneuronal cells (Tebar et al., 1996; Benmerah et al., 1999; Hussain et al., 1999; Qualmann et al., 1999); (3) their co-precipitation with AP-2, clathrin, or dynamin; and (4) their blocking of endocytosis produced by antibody injection or by the injection or overexpression of their fragments (Tebar et al., 1996; Carbone et al., 1997; van Delft et al., 1997; Benmerah et al., 1998; Chen et al., 1998; Benmerah et al., 1999). None of these proteins, however, co-purifies on brain clathrin-coated vesicles with clathrin, AP-2, and AP180 (Chen et al., 1998; Rosenthal et al., 1999), which suggests that their interaction with coat components is both dynamic and transient.

Eps15 is represented by two similar gene products, Eps15 and Eps15R. It was originally identified as a substrate for EGF receptor kinase (Fazioli et al., 1993) and then found to be a major physiological binding partner for AP-2 (Benmerah et al., 1995). One study has demonstrated its concentration at the neck of clathrin-coated pits, suggesting its possible involvement in dynamic rearrangements at the edges of the coat, in fission, or both (Tebar et al., 1996). Its N-terminal region contains three copies of an evolutionarily conserved domain, the EH domain (Eps15 homology domain). The structure of the central region predicts coiled-coil interactions and has been shown to participate in both homo- and heterodimerization (Cupers et al., 1997; Tebar et al., 1997) as well as in the binding to the coiled-coil domain of intersectin (Sengar et al., 1999). The C-terminal domain contains multiple DPF motifs and binds to the ear domain of α-adaptin (Benmerah et al., 1996; Iannolo et al., 1997) and to Crk, a protein adaptor implicated in signaling pathways (Iannolo et al., 1997). The EH domain, whose structure reveals two EF hand motifs and a tightly bound Ca^{2+} (de Beer et al., 1998), represents a protein-protein interaction module. It recognizes sequences with the core consensus NPF (with some exceptions) (Wong et al., 1995; Salcini et al., 1997; Paoluzi et al., 1998; Salcini et al., 1999). This consensus sequence exists in many proteins directly or indirectly implicated in endocytosis and actin function, including synaptojanin 1 (170-kDa isoform; McPherson et al., 1996), the AP180 homologue CALM (Dreyling et al., 1996), and syndapin/pacsin (Plomann et al., 1998; Qualmann et al., 1999). In brain, the main interactor for the EH domains of Eps15 is epsin 1, which contains three NPF motifs in its C-terminal domain (Chen et al., 1998). A less abundant but very similar isoform, epsin 2 (ibp2), has recently been identified (Yamabhai et al., 1998; Rosenthal et al., 1999). The N-terminal region of epsin defines a novel protein domain, referred to as the ENTH domain (Epsin N-terminal homology domain) (Kay et al., 1999; Rosenthal et al., 1999). The central part of epsin binds both the ear domain of α-adaptin AP-2 and the terminal domain of clathrin, as predicted by the presence of multiple copies of the motif D/EPW and of the clathrin-binding consensus (Chen et al., 1998; Hussain et al., 1999; Rosenthal et al., 1999).

Intersectin/DAP160 and syndapin/pacsin have properties that place them at the interface of SH3- and EH-mediated protein networks. Dap160 was first identified as the major dynamin interacting partner in nerve terminals of *Drosophila* (Roos and Kelly, 1998), and its homologues in mammalian cells are referred to as intersectins 1 and 2 (also ESE 1 and 2; SH3P17 and SH3P18) (Sparks et al., 1996; Guipponi et al., 1998; Yamabhai et al., 1998; Hussain et al., 1999; M. Okamoto et al., 1999; Sengar et al., 1999). The intersectins contain two N-terminal EH domains that bind epsin 1 and 2, a coil-coiled region that can bind Eps15 (Sengar et al., 1999), and five SH3 domains, which bind dynamin and synaptojanin with variable preferences (Yamabhai et al., 1998). Alternatively spliced isoforms of intersectins that contain a DH(Dbl) domain (a module with guanine-nucleotide exchange activity for GTPases of the Rho superfamily), a PH domain, and C2 (calcium-binding) domains are abundantly expressed in brain (Guipponi et al., 1998; Hussain et al., 1999). Because of these multiple interactions, intersectin may function as a scaffold linking together multiple components of the endocytotic machinery (Roos and Kelly, 1998; Yamabhai et al., 1998; Sengar et al., 1999).

Syndapin/pacsin (the collective name for the brain-specific syndapin/pacsin 1 and the ubiquitously expressed synapsin/pacsin 2) contains an evolutionarily conserved region at the N terminus, NPF motifs in the central region, and a C-terminal SH3 domain (Merilainen et al., 1997; Plomann et al., 1998; Qualmann et al., 1999; Ritter et al., 1999). This domain binds N-WASP (neuronal Wiskott-Aldrich syndrome protein) in addition to dynamin and synaptojanin (Qualmann et al., 1999). N-WASP is a key regulator of actin nucleation (Rohatgi et al., 1999; Snapper and Rosen, 1999); thus this interaction suggests a critical link of syndapin to actin dynamics. Further evidence for this link comes from the reported localization of syndapin 2 at focal adhesion plaques in chicken cells (Merilainen et al., 1997).

Stoned A and B

A *Drosophila* mutation affecting the behavioral response to stress and first identified 20 years ago (Grigliatti et al., 1973), *stoned*, maps to a bicistronic gene. Both proteins encoded by this gene, stoned A and stoned B, have been implicated in synaptic vesicle recycling. Mutations in the *stoned* locus produce synaptic vesicle depletion, accumulation of endocytotic intermediates, and mislocalization of synaptotagmin (Andrews et al., 1996; Fergestad et al., 1999; Stimson et al., 1998). The *stoned* locus interacts genetically with the *shibire* locus (Petrovich et al., 1993). Stoned B contains a domain homologous to the μ2 subunit of AP-2, suggesting that this protein may act as an additional clathrin adaptor (Andrews et al., 1996). A protein similar to stoned B, UNC-41, has been implicated in synaptic vesicle recycling in *C. elegans* (J. Rand, personal communication). No information is available so far on stoned proteins in mammalian neurons.

General Considerations on Clathrin Accessory Factors

It is clear from the information summarized previously that there is no clear mechanism of action for any of the factors discussed. However, the available data reveal four main themes:

1. Some of these factors function as adaptors, which help both in the recruitment of coat proteins and in the further recruitment to the coat of downstrean factors.
2. Some of them are lipid-metabolizing enzymes, a function that underscores an active role of the lipid bilayer in vesicular transport. Local lipid changes may regulate interactions with coat proteins, affect membrane curvature, or generate signaling molecules.
3. Other factors are also involved in signaling pathways (Pawson, 1995). This finding has interesting implications for clathrin-mediated endocytosis at nonsynaptic sites where clathrin functions primarily in the internalization of receptors. It would appear that the same set of molecules may participate both in mechanistic aspects of endocytosis and in downstream signaling, as originally reported for the arrestins (Lefkowitz, 1998).
4. The function of some factors suggests a link to the actin cytoskeleton. The importance of this link is further strengthened by the existence of yeast homologues for most of these factors, which play a dual role both in endocytosis and in actin function. In yeast a close interdependence between the actin cytoskeleton and endocytosis has been demonstrated (Munn et al., 1995; Wendland et al., 1999; Yang et al., 1999). Actin may be involved in endocytosis in several ways (as discussed subsequently).

A challenging task ahead will be to establish the hierarchy and sequence of action of these various regulatory factors. Initial insight into the regulation of this process has emerged from the analysis of phosphorylation-dephosphorylation reactions. Almost all of the components of the endocytotic machinery are phosphoproteins: amphiphysins (Bauerfeind et al., 1997; Slepnev et al., 1998), AP-2 (Wilde and Brodsky, 1996), AP180 (Murphy et al., 1994), dynamin (Robinson et al., 1993;

Powell and Robinson, 1995), Eps15 and epsin (Fazioli et al., 1993; Chen et al., 1999), and synaptojanin (McPherson et al., 1994). Remarkably, depolarization-dependent Ca^{2+} entry into nerve terminals, which triggers exocytosis and enhances phosphorylation of several proteins implicated in exocytosis, produces a rapid dephosphorylation of proteins of the endocytotic machinery: amphiphysin (Bauerfeind et al., 1997), dynamin (Liu et al., 1994; Powell and Robinson, 1995), Eps15 and epsin (Chen et al., 1999), and synaptojanin (McPherson et al., 1994). Calcineurin (PP2B), a Ca^{2+}-dependent protein phosphatase that associates with dynamin 1 in a Ca^{2+}-dependent manner (Lai et al., 1999), has been implicated in the dephosphorylation reaction (Liu et al., 1994; McPherson et al., 1994; Bauerfeind et al., 1997). Phosphorylation inhibits, while dephosphorylation promotes, the formation of endocytotic complexes. Based on these data, the stimulation-dependent dephosphorylation of endocytotic proteins has been proposed to be a trigger that assembles endocytotic complexes and primes the nerve terminal for a burst of endocytosis (Slepnev et al., 1998). It remains to be seen whether phosphorylation-dephosphorylation acts as a switch at each cycle of exocytosis-endocytosis or simply has a regulatory function.

A variety of approaches—including studies with purified components, semi-intact models, and intact models—will probably be required to address the integrative function of the endocytotic machinery. In vitro approaches will continue to be useful to dissect and reconstitute specific processes. It should be noted, however, that clathrin-mediated endocytosis in vivo depends critically on the specific organization at synaptic release sites. For instance, when endophilin function is perturbed by antibody injection, the plasma membrane area around active zones (where endocytosis occurs normally) contains coated pits that had been trapped at an early stage. In contrast, coated pits that occur on infoldings of the plasma membrane are trapped at a late stage preceding fission (Ringstad et al., 1999). Thus the effect of individual endocytotic components may depend on the specific environment in which they function.

Actin and Endocytosis

One of the most important proteins in endocytosis is actin. In single-cell organisms all forms of endocytosis critically depend on actin (Riezman et al., 1996; Geli and Riezman, 1998). There is also a general consensus regarding the importance of actin for endocytosis in mammalian cells (Lamaze et al., 1996; Mundigl et al., 1998; Witke et al., 1998; Qualman et al., 1999), although the molecular mechanisms underlying its involvement remain elusive.

Some of the mechanisms by which actin may control clathrin-mediated endocytosis include the following. First, the actin cytoskeleton may play an inhibitory role on the assembly of the clathrin coat by generating a nonpermissive cytomatrix, which must be disassembled prior to the formation of a vesicle bud (Huttner, 1998). Second, actin may be needed to allow the nucleation of a deeply invaginated clathrin bud and perhaps to promote its fission. In this context, many connections have emerged between the functions of dynamin and those of actin (Damke et al., 1994; Torre et al., 1994; Mundigl et al., 1998; Witke et al., 1998). One possibility is that dynamin may organize a spiral of actin at the vesicle neck, similar to that proposed for the arrangement of actin around actin patches in yeast (Mulholland et al., 1994). Third, actin may be needed to propel the newly formed vesicles from the periphery of the cell to deeper regions. This effect may be achieved either by myosin motors that carry vesicles along actin tracks (Geli and Riezman, 1996) or by a mechanism of vectorial polymerization-depolarization of actin similar to the one involved in the mobility of internalized *Listeria* (Merrifield et al., 1999). Actin may also play a role in the invagination of the plasma membrane underlying the clathrin-independent formation of intracellular vacuoles and cisternae after massive stimulation. Actin and Rho-family GTPases have been implicated in this form of macropinocytotic activity in other systems (Schmalzing et al., 1995).

Consistent with a critical role of actin in endocytosis is the finding that filamentous actin is most prominent in nerve terminals after stimulation, in the area that surrounds the active zone, that is, the area specialized for endocytosis. Based on studies of the giant lamprey axon, these actin networks are particularly striking when nerve terminals are stimulated following the injection of phalloidin (which stabilizes F-actin) (Brodin, 1999), of antibodies or of peptides that disrupt the function of synaptojanin and therefore increase the phosphoinositide pool (Gad et al., submitted), or of GTPγS (Gustafsson et al., 1998). The requirement of the stimulus indicates a close connection between synaptic vesicle recycling and the onset of actin polymerization. GTPγS may act by affecting small GTPases that control actin function (Hall, 1998), via the stimulation of enzymes inducing polyphosphoinositide synthesis (Liscovitch and Cantley, 1994; Zheng et al., 1994; Martin et al., 1996; Honda et al., 1999), or via both mechanisms. The involvement of small GTPases is supported by the specific localization of Rho-family guanyl nucleotide exchange factors, such as Still-life or the Dbl domain-containing intersectin isoform, at synaptic sites (Sone et al., 1997; Hussain et al., 1999; Sengar et al., 1999). Recently, the ability of the Rho-family GTPase, Cdc42, to stimulate actin polymerization has been linked to N-WASP and to the Arp2/3 complex, with a synergistic effect of $PtdIns(4,5)P_2$. The binding of $PtdIns(4,5)P_2$ and GTP-Cdc42 to the N-terminal region of N-WASP may produce a conformational change of this protein, allowing binding to the Arp2/3 complex (Rohatgi et al., 1999). In addition to regulation by small GTPases and $PtdIns(4,5)P_2$, these processes may also be controlled by protein-protein interaction networks, as indicated by the binding of syndapin to N-WASP (Plomann et al., 1998; Qualmann et al., 1999), and of dynamin to profilin (Witke et al., 1998).

Conclusion

We have discussed the current state of knowledge on synaptic vesicle re-cycling and the underlying mechanisms. We have focused on the clathrin-dependent re-formation of synaptic vesicles because so far this is the only pathway for which there is direct experimental evidence and whose mo-lecular mechanisms are at least partially understood. It will be of con-siderable interest to investigate whether this step can be bypassed—as proposed by the kiss-and-run model of neurotransmitter release. The single-vesicle-budding-step model, which we have favored, also needs additional experimental testing. Studies on the mechanisms underlying the generation of endosome-like structures in stimulated nerve termi-nals and their progressive transformation into synaptic vesicles will help to answer this question. Finally, it will be important to elucidate the relationship, and possible intersection, of the synaptic vesicle recycling pathway with other membrane recycling pathways at the synapse.

We have discussed the field of endocytosis in relation to the function of the presynaptic compartment. Clearly the high concentration in the nervous system of proteins that participate in clathrin-mediated endo-cytosis reflects the important role of synaptic vesicle recycling at the synapse. However, recent studies have brought to center stage a new area of synaptic cell biology—the recycling of postsynaptic receptors (Carroll et al., 1999)—although this process probably represents only a small fraction of the membrane recycling that takes place at the synapse. One should consider the possibility that experimental manipulations affect-ing endocytosis may have both pre- and postsynaptic effects. Thus, even if endocytotic proteins are not highly concentrated in the postsynaptic compartment, they may also be crucially important in regulating synap-tic function at this location.

References

Achiriloaie, M., Barylko, B., Albanesi, J. P. (1999). Essential role of the dynamin pleckstrin homology domain in receptor-mediated endocytosis. *Mol. Cell. Biol.* 19:1410–1415.

Ahle, S., Ungewickell, E. (1986). Purification and properties of a new clathrin assembly protein. *EMBO J.* 5:3143–3149.

Ahle, S., Ungewickell, E. (1989). Identification of a clathrin binding subunit in the HA2 adaptor protein complex. *J. Biol. Chem.* 264:20,089–20,093.

Alés, E., Tabares, L., Poyato, J. M., Valero, V., Lindau, M., Alvarez de Toledo, G. (1999). High calcium concentration shifts the mode of exocytosis to the kiss-and-run mechanism. *Nature Cell Biol.* 1:40–44.

Alvarez de Toledo, G., Fernandez-Chacon, R., Fernandez, J. M. (1993). Release of secretory products during transient vesicle fusion. *Nature* 363:554–558.

Andrews, J., Smith, M., Merakovsky, J., Coulson, M., Hannan, F., Kelly, L. E. (1996). The stoned locus of *Drosophila melanogaster* produces a dicistronic transcript and encodes two distinct polypeptides. *Genetics* 143:1699–1711.

Arneson, L. S., Kunz, J., Anderson, R. A., Traub, L. M. (1999). Coupled inositide phosphorylation and phospholipase D activation initiates clathrin-coat as-sembly on lysosomes. *J. Biol. Chem.* 274:17,794–17,805.

Artalejo, C. R., Henley, J. R., McNiven, M. A., Palfrey, C. H. (1995). Rapid endocytosis coupled to exocytosis in adrenal chromaffin cells involves Ca^{2+}, GTP, and dynamin but not clathrin. *Proc. Natl. Acad. Sci. USA* 92:8328–8332.

Ball, C. L., Hunt, S. P., Robinson, M. S. (1995). Expression and localization of alpha-adaptin isoforms. *J. Cell Sci.* 108:2865–2875.

Baram, D., Adachi, R., Medalia, O., Tuvim, M., Dickey, B. F., Mekori, Y. A., Sagi-Eisenberg, R. (1999). Synaptotagmin II negatively regulates Ca^{2+}-triggered exocytosis of lysosomes in mast cells. *J. Exp. Med.* 189:1649–1658.

Bauerfeind, R., Takei, K., De Camilli, P. (1997). Amphiphysin I is associated with coated endocytic intermediates and undergoes stimulation-dependent dephosphorylation in nerve terminals. *J. Biol. Chem.* 272:30,984–30,992.

Beck, K. A., Keen, J. H. (1991). Interaction of phosphoinositide cycle intermediates with the plasma membrane–associated clathrin assembly protein AP-2. *J. Biol. Chem.* 266:4442–4447.

Benmerah, A., Gagnon, J., Begue, B., Megarbane, B., Dautry-Varsat, A., Cerf-Bensussan, N. (1995). The tyrosine kinase substrate eps15 is constitutively associated with the plasma membrane adaptor AP-2. *J. Cell Biol.* 131:1831–1838.

Benmerah, A., Begue, B., Dautry-Varsat, A., Cerf-Bensussan, N. (1996). The ear of alpha-adaptin interacts with the COOH-terminal domain of the Eps 15 protein. *J. Biol. Chem.* 271:12,111–12,116.

Benmerah, A., Lamaze, C., Begue, B., Schmid, S. L., Dautry-Varsat, A., Cerf-Bensussan, N. (1998). AP-2/Eps15 interaction is required for receptor-mediated endocytosis. *J. Cell Biol.* 140:1055–1062.

Benmerah, A., Bayrou, M., Cerf-Bensussan, N., Dautry-Varsat, A. (1999). Inhibition of clathrin-coated pit assembly by an Eps15 mutant. *J. Cell Sci.* 112:1303–1311.

Betz, W. J., Bewick, G. S. (1992). Optical analysis of synaptic vesicle recycling at the frog neuromuscular junction. *Science* 255:200–203.

Betz, W. J., Mao, F., Smith, C. B. (1996). Imaging exocytosis and endocytosis. *Curr. Opin. Neurobiol.* 6:365–371.

Bourne, H. R. (1995). GTPases:a family of molecular switches and clocks. *Philos. Trans. R. Soc. London Ser. B* 349:283–289.

Brodin, L. (1999). Actin-dependent steps the synaptic vesicle recycling. *Biochimie* 81:S49.

Brodin, L., Low, P., Gad, H., Gustafsson, J., Pieribone, V. A., Shupliakov, O. (1997). Sustained neurotransmitter release: new molecular clues. *Eur. J. Neurosci.* 9:2503–2511.

Butler, M. H., David, C., Ochoa, G. C., Freyberg, Z., Daniell, L., Grabs, D., Cremona, O., De Camilli, P. (1997). Amphiphysin II (SH3P9; BIN1), a member of the amphiphysin/Rvs family, is concentrated in the cortical cytomatrix of axon initial segments and nodes of Ranvier in brain and around T tubules in skeletal muscle. *J. Cell Biol.* 137:1355–1367.

Carbone, R., Fre, S., Iannolo, G., Belleudi, F., Mancini, P., Pelicci, P. G., Torrisi, M. R., Di Fiore, P. P. (1997). Eps15 and eps15R are essential components of the endocytic pathway. *Cancer Res.* 57:5498–5504.

Carr, J. F., Hinshaw, J. E. (1997). Dynamin assembles into spirals under physiological salt conditions upon the addition of GDP and gamma-phosphate analogues. *J. Biol. Chem.* 272:28,030–28,035.

Carroll, R. C., Lissin, D. V., von Zastrow, M., Nicoll, R. A., Malenka, R. C. (1999). Rapid redistribution of glutamate receptors contributes to long-term depression in hippocampal cultures. *Nature Neurosci.* 2:454–460.

Ceccarelli, B., Hurlbut, W. P. (1975). Transmitter release and vesicle hypothesis. In *Golgi Centennial Symposium Proceedings,* M. Santini, ed. New York: Raven Press. pp 529–545.

Ceccarelli, B., Hurlbut, W. P. (1980a). Ca^{2+}-dependent recycling of synaptic vesicles at the frog neuromuscular junction. *J. Cell Biol.* 87:297–303.

Ceccarelli, B., Hurlbut, W. P. (1980b). Vesicle hypothesis of the release of quanta of acetylcholine. *Physiol. Rev.* 60:396–441.

Ceccarelli, B., Hurlbut, W. P., Mauro, A. (1973). Turnover of transmitter and synaptic vesicles at the frog neuromuscular junction. *J. Cell Biol.* 57:499–524.

Chappell, T. G., Welch, W. J., Schlossman, D. M., Palter, K. B., Schlesinger, M. J., Rothman, J. E. (1986). Uncoating ATPase is a member of the 70 kilodalton family of stress proteins. *Cell* 45:3–13.

Chen, H., Fre, S., Slepnev, V. I., Capua, M. R., Takei, K., Butler, M. H., Di Fiore, P. P., De Camilli, P. (1998). Epsin is an EH-domain-binding protein implicated in clathrin-mediated endocytosis. *Nature* 394:793–797.

Chen, H., Slepnev, V. I., Di Fiore, P. P., De Camilli, P. (1999). The interaction of epsin and Eps15 with the clathrin adaptor AP-2 is inhibited by mitotic phosphorylation and enhanced by stimulation-dependent dephosphorylation in nerve terminals. *J. Biol. Chem.* 274:3257–3260.

Chen, M. S., Obar, R. A., Schroeder, C. C., Austin, T. W., Poodry, C. A., Wadsworth, S. C., Vallee, R. B. (1991). Multiple forms of dynamin are encoded by *shibire*, a *Drosophila* gene involved in endocytosis. *Nature* 351:583–586.

Chow, R. H., von Ruden, L., Neher, E. (1992). Delay in vesicle fusion revealed by electrochemical monitoring of single secretory events in adrenal chromaffine cells. *Nature* 356:60–63.

Cochilla, A. J., Angleson, J. K., Betz, W. J. (1999). Monitoring secretory membrane with FM1-43 fluorescence. *Annu. Rev. Neurosci.* 22:1–10.

Cremona, O., Di Paolo, G., Wenk, M. K., Luthi, A., Kim, W., Takei, K., Daniell, L., Nemoto, Y., Shears, S. B., Flavell, R. A., McCormick, D. A., De Camilli, P. (1999). Essential role of phosphoinositide metabolism in synaptic vesicle recycling. *Cell* 99:179–188.

Cupers, P., ter Haar, E., Boll, W., Kirchhausen, T. (1997). Parallel dimers and anti-parallel tetramers formed by epidermal growth factor receptor pathway substrate clone 15. *J. Biol. Chem.* 272:33,430–33,444.

Cupers, P., Jadhav, A. P., Kirchhausen, T. (1998). Assembly of clathrin coats disrupts the association between Eps15 and AP-2 adaptors. *J. Biol. Chem.* 273:1847–1850.

Damke, H., Baba, T., Warnock, D. E., Schmid, S. L. (1994). Induction of mutant dynamin specifically blocks endocytic coated vesicle formation. *J. Cell Biol.* 127:915–934.

David, C., McPherson, P. S., Mundigl, O., de Camilli, P. (1996). A role of amphiphysin in synaptic vesicle endocytosis suggested by its binding to dynamin in nerve terminals. *Proc. Natl. Acad. Sci. USA* 93:331–335.

De Beer, T., Carter, R. E., Lobel-Rice, K. E., Sorkin, A., Overduin, M. (1998). Structure and Asn-Pro-Phe binding pocket of the Eps15 homology domain. *Science* 281:1357–1360.

De Camilli, P., Takei, K. (1996). Molecular mechanisms in synaptic vesicle endocytosis and recycling. *Neuron* 16:481–486.

De Heuvel, E., Bell, A. W., Ramjaun, A. R., Wong, K., Sossin, W. S., McPherson, P. S. (1997). Identification of the major synaptojanin-binding proteins in brain. *J. Biol. Chem.* 272:8710–8716.

De Hoop, M. J., Huber, L. A., Stenmark, H., Williamson, E., Zerial, M., Parton, R. G., Dotti, C. G. (1994). The involvement of the small GTP-binding protein Rab5a in neuronal endocytosis. *Neuron* 13:11–22.

Dell'Angelica, E. C., Klumperman, J., Stoorvogel, W., Bonifacino, J. S. (1998). Association of the AP-3 adaptor complex with clathrin. *Science* 280:431–434.

Dell'Angelica, E. C., Mullins, C., Bonifacino, J. S. (1999a). AP-4, a novel protein complex related to clathrin adaptors. *J. Biol. Chem.* 274:7278–7285.

Dell'Angelica, E. C., Shotelersuk, V., Aguilar, R. C., Gahl, W. A., Bonifacino, J. S. (1999b). Altered trafficking of lysosomal proteins in Hermansky-Pudlak syndrome due to mutations in the beta 3A subunit of the AP-3 adaptor. *Mol. Cell* 3:11–21.

Dreyling, M. H., Martinez-Climent, J. A., Zheng, M., Mao, J., Rowley, J. D., Bohlander, S. K. (1996). The t(10;11)(p13;q14) in the U937 cell line results in the fusion of the AF10 gene and CALM, encoding a new member of the AP-3 clathrin assembly protein family. *Proc. Natl. Acad. Sci. USA* 93:4804–4809.

Engisch, K. L., Nowycky, M. C. (1998). Compensatory and excess retrieval: two types of endocytosis following single step depolarizations in bovine adrenal chromaffin cells. *J. Physiol.* 506:591–608.

Faundez, V., Horng, J. T., Kelly, R. B. (1998). A function for the AP3 coat complex in synaptic vesicle formation from endosomes. *Cell* 93:423–432.

Fazioli, F., Minichiello, L., Matoskova, B., Wong, W. T., Di Fiore, P. P. (1993). Eps15, a novel tyrosine kinase substrate, exhibits transforming activity. *Mol. Cell. Biol.* 13:5814–5828.

Fergestad, T., Davis, W. S., Broadie, K. (1999). The stoned proteins regulate synaptic vesicle recycling in the presynaptic terminal. *J. Neurosci.* 19:5847–5860.

Fesce, F., Meldolesi, J. (1999). Peeping at the vesicle kiss. *Nature Cell Biol.* 1:E3–E4.

Fesce, R., Segal, J. R., Ceccarelli, B., Hurlbut, W. P. (1986). Effects of black widow spider venom and Ca^{2+} on quantal secretion at the frog neuromuscular junction. *J. Gen. Physiol.* 88:59–81.

Fesce, R., Grohovaz, F., Valtorta, F., Meldolesi, J. (1994). Neurotransmitter release: fusion or kiss-and-run? *Trends Cell Biol.* 4:1–6.

Foster-Barber, A., Bishop, J. M. (1998). Src interacts with dynamin and synapsin in neuronal cells. *Proc. Natl. Acad. Sci. USA* 95:4673–4677.

Fukuda, M., Moreira, J. E., Lewis, F. M., Sugimori, M., Niinobe, M., Mikoshiba, K., Llinas, R. (1995). Role of the C2B domain of synaptotagmin in vesicular release and recycling as determined by specific antibody injection into the squid giant synapse preterminal. *Proc. Natl. Acad. Sci. USA* 92:10,708–10,712.

Gad, H., Low, P., Zotova, E., Brodin, L., Shupliakov, O. (1998). Dissociation between Ca^{2+}-triggered synaptic vesicle exocytosis and clathrin-mediated endocytosis at a central synapse. *Neuron* 21:607–616.

Gad, H., Shupliakov, O., Low, P., Kjaerulff, O., Ringstad, N., De Camilli, P., Brodin, L. (1999). Perturbation of synaptojanin function impairs synaptic vesicle endocytosis at a living synapse. *Soc. Neurosci. Abstr.* 25(2):1744.

Gaidarov, I., Keen, J. H. (1999). Phosphoinositide-AP-2 interactions required for targeting to plasma membrane clathrin-coated pits. *J. Cell Biol.* 146:755–764.

Gaidarov, I., Chen, Q., Falck, J. R., Reddy, K. K., Keen, J. H. (1996). A functional phosphatidylinositol 3,4,5-trisphosphate/phosphoinositide binding domain in the clathrin adaptor AP-2 alpha subunit. Implications for the endocytic pathway. *J. Biol. Chem.* 271:20,922–20,929.

Gallusser, A., Kirchhausen, T. (1993). The beta 1 and beta 2 subunits of the AP complexes are the clathrin coat assembly components. *EMBO J.* 12:5237–5244.

Geli, M. I., Riezman, H. (1996). Role of type I myosins in receptor-mediated endocytosis in yeast. *Science* 272:533–535.

Geli, M. I., Riezman, H. (1998). Endocytic internalization in yeast and animal cells: similar and different. *J. Cell Sci.* 111:1031–1037.

Gennaro, J. F., Nastuk, W. L., Rutherford, D. T. (1978). Reversible depletion of synaptic vesicles induced by application of high external potassium to the frog neuromuscular junction. *J. Physiol.* 280:237–247.

Godi, A., Pertile, P., Meyers, R., Marra, P., Di Tillio, G., Iurisci, C., Luini, A., Corda, C., De Matteis, M. A. (1999). Arf mediates recruitment of PtdIns-4-OH kinase-β and stimulates synthesis of PtdIns(4,5)P$_2$ on the Golgi complex. *Nature Cell Biol.* 1:280–288.

Goldberg, J. (1998). Structural basis for activation of ARF GTPase: mechanisms of guanine nucleotide exchange and GTP-myristoyl switching. *Cell* 95:237–248.

Goldberg, J. (1999). Structural and functional analysis of the ARF1-ARFGAP complex reveals a role for coatomer in GTP hydrolysis. *Cell* 96:893–902.

Gonzalez-Gaitan, M., Jackle, H. (1997). Role of *Drosophila* alpha-adaptin in presynaptic vesicle recycling. *Cell* 88:767–776.

Goodman, O. B., Jr., Keen, J. H. (1995). The alpha chain of the AP-2 adaptor is a clathrin binding subunit. *J. Biol. Chem.* 270:23,768–23,773.

Gout, I., Dhand, R., Hiles, I. D., Fry, M. J., Panayotou, G., Das, P., Truong, O., Totty, N. F., Hsuan, J., Booker, G. W., Campbell, I. D., Waterfield, M. D. (1993). The GTPase dynamin binds to and is activated by a subset of SH3 domains. *Cell* 75:25–36.

Grabs, D., Slepnev, V. I., Songyang, Z., David, C., Lynch, M., Cantley, L. C., De Camilli, P. (1997). The SH3 domain of amphiphysin binds the proline-rich domain of dynamin at a single site that defines a new SH3 binding consensus sequence. *J. Biol. Chem.* 272:13,419–13,425.

Gray, E. G. (1961). The granule cells, mossy synapses and Purkinje spine synapses of the cerebellum: light and electron microscopy observations. *J. Anat.* 95: 345–356.

Gray, E. G., Willis, R. A. (1970). On synaptic vesicles, complex vesicles and dense projections. *Brain Res.* 24:149–168.

Grigliatti, T. A., Hall, L., Rosenbluth, R., Suzuki, D. T. (1973). Temperature-sensitive mutations in *Drosophila melanogaster*. XIV. A selection of immobile adults. *Mol. Gen. Genet.* 120:107–114.

Guipponi, M., Scott, H. S., Chen, H., Schebesta, A., Rossier, C., Antonarakis, S. E. (1998). Two isoforms of a human intersectin (ITSN) protein are produced by brain-specific alternative splicing in a stop codon. *Genomics* 53:369–376.

Guo, S., Stolz, L. E., Lemrow, S. M., York, J. D. (1999). SAC1-like domains of yeast SAC1, INP52, and INP53 and of human synaptojanin encode polyphosphoinositide phosphatases. *J. Biol. Chem.* 274:12,990–12,995.

Gustaffson, J., Shupliakov, O., Takei, K., Low, P., De Camilli, P., Brodin, L. (1998). GTPgammaS induces an actin matrix associated with coated intermediates in presynaptic neurons. *Soc. Neurosci. Abstr.* 24(1):823.

Haffner, C., Takei, K., Chen, H., Ringstad, N., Hudson, A., Butler, M. H., Salcini, A. E., Di Fiore, P. P., De Camilli, P. (1997). Synaptojanin 1: localization on coated endocytic intermediates in nerve terminals and interaction of its 170 kDa isoform with Eps15. *FEBS Lett.* 419:175–180.

Haffner, C., Di Paolo, G., Rosenthal, J. A., De Camilli, P. (2000). Direct interaction of the 170 kDa isoform of synaptojanin 1 with clathrin and with the clathrin adaptor AP-2. *Curr. Biol.* 10:471–474.

Haimann, C., Torri-Tarelli, F., Fesce, R., Ceccarelli, B. (1985). Measurement of quantal secretion induced by ouabain and its correlation with depletion of synaptic vesicles. *J. Cell Biol.* 101:1953–1965.

Hall, A. (1998). Rho GTPases and the actin cytoskeleton. *Science* 279:509–514.

Hannah, M. J., Schmidt, A. A., Huttner, W. B. (1999). Synaptic vesicle biogenesis. *Annu. Rev. Cell Dev. Biol.* 15:733–798.

Hannan, L. A., Newmyer, S. L., Schmid, S. L. (1998). ATP- and cytosol-dependent release of adaptor proteins from clathrin-coated vesicles: a dual role for Hsc70. *Mol. Biol. Cell* 9:2217–2229.

Hao, W., Luo, Z., Zheng, L., Prasad, K., Lafer, E. M. (1999). AP180 and AP-2 interact directly in a complex that cooperatively assembles clathrin. *J. Biol. Chem.* 274:22,785–22,794.

Haucke, V., De Camilli, P. (1999). AP-2 recruitment to synaptotagmin stimulated by tyrosine-based endocytic motifs. *Science* 285:1268–1271.

Heilker, R., Spiess, M., Crottet, P. (1999). Recognition of sorting signals by clathrin adaptors. *BioEssays* 21:558–567.

Henkel, A. W., Almers, W. (1996). Fast steps in exocytosis and endocytosis studied by capacitance measurements in endocrine cells. *Curr. Opin. Neurobiol.* 6:350–357.

Henkel, A. W., Betz, W. J. (1995). Staurosporine blocks evoked release of FM 1-43 but not acetylcholine from frog motor nerve terminals. *J. Neurosci.* 15:8246–8258.

Henkel, A. W., Lubke, J., Betz, W. J. (1996). FM1-43 dye ultrastructural localization in and release from frog motor nerve terminals. *Proc. Natl. Acad. Sci. USA* 93:1918–1923.

Herskovits, J. S., Burgess, C. C., Obar, R. A., Vallee, R. B. (1993). Effects of mutant rat dynamin on endocytosis. *J. Cell Biol.* 122:565–578.

Heuser, J. (1980). Three-dimensional visualization of coated vesicle formation in fibroblasts. *J. Cell Biol.* 84:560–583.

Heuser, J. E. (1989). The role of coated vesicles in recycling of synaptic vesicle membrane. *Cell Biol. Int. Rep.* 13:1063–1076.

Heuser, J. E., Keen, J. (1988). Deep-etch visualization of proteins involved in clathrin assembly. *J. Cell Biol.* 107:877–886.

Heuser, J., Miledi, R. (1971). Effects of lanthanum ions on function and structure of frog neuromuscular junctions. *Proc. R. Soc. London Ser. B* 179:247–260.

Heuser, J. E., Reese, T. S. (1973). Evidence for recycling of synaptic vesicle membrane during transmitter release at the frog neuromuscular junction. *J. Cell. Biol.* 57:315–344.

Hinshaw, J. E., Schmid, S. L. (1995). Dynamin self-assembles into rings suggesting a mechanism for coated vesicle budding. *Nature* 374:190–192.

Hirokawa, N., Noda, Y., Okada Y. (1998). Kinesin and dynein superfamily proteins in organelle transport and cell division. *Curr. Opin. Cell Biol.* 10:60–73.

Hirst, J., Robinson, M. S. (1998). Clathrin and adaptors. *Biochim. Biophys. Acta* 1404:173–193.

Holtzman, E., Peterson, E. R. (1969). Uptake of protein by mammalian neurons. *J. Cell Biol.* 40:863–869.

Honda, A., Nogami, M., Yokozeki, T., Yamazaki, M., Nakamura, H., Watanabe, H., Kawamoto, K., Nakayama, K., Morris, A. J., Frohman, M. A., Kanaho, Y. (1999). Phosphatidylinositol 4-phosphate 5-kinase alpha is a downstream effector of the small G protein ARF6 in membrane ruffle formation. *Cell* 24:521–532.

Hussain, N. K., Yamabhai, M., Ramjaun, A. R., Guy, A. M., Baranes, D., O'Bryan, J. P., Der, C. J., Kay, B. K., McPherson, P. S. (1999). Splice variants of intersectin are components of the endocytic machinery in neurons and nonneuronal cells. *J. Biol. Chem.* 274:15,671–15,677.

Huttner, W. B. (1998). Protein and lipid sorting in the secretory and endocytic pathways—receptors and mechanisms. *Semin. Cell Dev. Biol.* 9:491–492.

Iannolo, G., Salcini, A. E., Gaidarov, I., Goodman, O. B., Jr., Baulida, J., Carpenter, G., Pelicci, P. G., Di Fiore, P. P., Keen, J. H. (1997). Mapping of the molecular determinants involved in the interaction between eps15 and AP-2. *Cancer Res.* 57:240–245.

Jackson, A. P., Seow, H. F., Holmes, N., Drickamer, K., Parham, P. (1987). Clathrin light chains contain brain-specific insertion sequences and a region of homology with intermediate filaments. *Nature* 326:154–159.

Janmey, P. A. (1998). The cytoskeleton and cell signaling: component localization and mechanical coupling. *Physiol. Rev.* 78:763–781.

Jorgensen, E. M., Hartwieg, E., Schuske, K., Nonet, M. L., Jin, Y., Horvitz, H. R. (1995). Defective recycling of synaptic vesicles in synaptotagmin mutants of *Caenorhabditis elegans. Nature* 378:196–199.

Kadota, K., Kadota, T. (1973). Isolation of coated vesicles, plain synaptic vesicles, and flocculent material from a crude synaptosome fraction of guinea pig whole brain. *J. Cell Biol.* 58:135–151.

Kanaseki, T., Kadota, K. (1969). The "vesicle in a basket." A morphological study of the coated vesicle isolated from the nerve endings of the guinea pig brain, with special reference to the mechanism of membrane movements. *J. Cell Biol.* 42:202–220.

Kantheti, P., Quiao, X., Diaz, M. E., Peden, A. A., Meyer, G. E., Carskadon, S. L., Kaphamer, D., Sufalko, D., Robinson, M. S., Noebels, J. L., Burmeister, M. (1998). Mutation in AP3 delta in the mocha mouse links endosomal transport to storage deficiency in platelets, melanosomes, and synaptic vesicles. *Neuron* 21:111–122.

Kay, B. K., Yamabhai, M., Wendland, B., Emr, S. D. (1999). Identification of a novel domain shared by putative components of the endocytic and cytoskeletal machinery. *Protein Sci.* 8:435–438.

Kedra, D., Peyrard, M., Fransson, I., Collins, J. E., Dunham, I., Roe, B. A., Dumanski, J. P. (1996). Characterization of a second human clathrin heavy chain polypeptide gene (CLH-22) from chromosome 22q11. *Hum. Mol. Genet.* 5:625–631.

Keen, J. H. (1987). Clathrin assembly proteins: affinity purification and a model for coat assembly. *J. Cell Biol.* 105:1989–1998.

Keen, J. H., Beck, K. A., Kirchhausen, T., Jarrett, T. (1991). Clathrin domains involved in recognition by assembly protein AP-2. *J. Biol. Chem.* 266:7950–7956.

Kirchhausen, T., Harrison, S. C., Heuser, J. (1986). Configuration of clathrin trimers: evidence from electron microscopy. *J. Ultrastruct. Mol. Struct. Res.* 94: 199–208.

Kirchhausen, T., Harrison, S. C., Chow, E. P., Mattaliano, R. J., Ramachandran, K. L., Smart, J., Brosius, J. (1987). Clathrin heavy chain: molecular cloning and complete primary structure. *Proc. Natl. Acad. Sci. USA* 84:8805–8809.

Kirchhausen, T., Bonifacino, J. S., Riezman, H. (1997). Linking cargo to vesicle formation: receptor tail interactions with coat proteins. *Curr. Opin. Cell Biol.* 9:488–495.

Kiromi, H., Kidukori, Y. (1998). Two distinct pools of synaptic vesicles in single presynaptic boutons in a temperature-sensitive *Drosophila* mutant, shibire. *Neuron* 20:917–925.

Klingauf, J., Kavalali, E. T., Tsien, R. W. (1998). Kinetics and regulation of fast endocytosis at hippocampal synapses. *Nature* 394:581–585.

Koenig, J. H., Ikeda, K. (1989). Disappearance and reformation of synaptic vesicle membrane upon transmitter release observed under reversible blockage of membrane retrieval. *J. Neurosci.* 9:3844–3860.

Koenig, J. H., Ikeda, K. (1996). Synaptic vesicles have two distinct recycling pathways. *J. Cell Biol.* 135:797–808.

Kohtz, D. S., Puszkin, S. (1988). A neuronal protein (NP185) associated with clathrin-coated vesicles. Characterization of NP185 with monoclonal antibodies. *J. Biol. Chem.* 263:7418–7425.

Kraszewski, K., Mundigl, O., Daniell, L., Verderio, C., Matteoli, M., De Camilli, P. (1995). Synaptic vesicle dynamics in living cultured hippocampal neurons visualized with CY3-conjugated antibodies directed against the lumenal domain of synaptotagmin. *J. Neurosci.* 15:4328–4342.

Kraszewski, K., Daniell, L., Mundigl, O., De Camilli, P. (1996). Mobility of synaptic vesicles in nerve endings monitored by recovery from photobleaching of synaptic vesicle–associated fluorescence. *J. Neurosci.* 16:5905–5913.

Lai, M. M., Hong, J. J., Ruggiero, A. M., Burnett, P. E., Slepnev, V. I., De Camilli, P., Snyder, S. H. (1999). The calcineurin–dynamin 1 complex as a calcium sensor for synaptic vesicle endocytosis. *J. Biol. Chem.* 274:25,963–25,966.

Lamaze, C., Chuang, T. H., Terlecky, L. J., Bokoch, G. M., Schmid, S. L. (1996). Regulation of receptor-mediated endocytosis by Rho and Rac. *Nature* 382:177–179.

Lee, A., Frank, D. W., Marks, M. S., Lemmon, M. A. (1999). Dominant-negative inhibition of receptor-mediated endocytosis by a dynamin-1 mutant with a defective pleckstrin homology domain. *Curr. Biol.* 9:261–264.

Lefkowitz, R. J. (1998). G protein–coupled receptors. III. New roles for receptor kinases and beta-arrestins in receptor signaling and desensitization. *J. Biol. Chem.* 273:18,677–18,680.

Leprince, C., Romero, F., Cussac, D., Vayssiere, B., Berger, R., Tavitian, A., Camonis, J. H. (1997). A new member of the amphiphysin family connecting endocytosis and signal transduction pathways. *J. Biol. Chem.* 272:15,101–15,105.

Lewis, P., Lentz, T. L. (1998). Rabies virus entry into cultured rat hippocampal neurons. *J. Neurocytol.* 27:559–573.

Li, C., Ullrich, B., Zhang, J. Z., Anderson, R. G., Brose, N., Südhof, T. C. (1995). Ca^{2+}-dependent and -independent activities of neural and non-neural synaptotagmins. *Nature* 375:594–599.

Lichte, B., Veh, R. W., Meyer, H. E., Kilimann, M. W. (1992). Amphiphysin, a novel protein associated with synaptic vesicles. *EMBO J.* 11:2521–2530.

Lin, H. C., Gilman, A. G. (1996). Regulation of dynamin I GTPase activity by G protein betagamma subunits and phosphatidylinositol 4,5-bisphosphate. *J. Biol. Chem.* 271:27,979–27,982.

Lin, H. C., Barylko, B., Achiriloaie, M., Albanesi, J. P. (1997). Phosphatidylinositol (4,5)-bisphosphate-dependent activation of dynamins I and II lacking the proline/arginine-rich domains. *J. Biol. Chem.* 272:25,999–26,004.

Lindner, R., Ungewickell, E. (1992). Clathrin-associated proteins of bovine brain coated vesicles. An analysis of their number and assembly-promoting activity. *J. Biol. Chem.* 267:16,567–16,573.

Liscovitch, M., Cantley, L. C. (1994). Lipid second messengers. *Cell* 77:329–334.

Liu, J. P., Sim, A. T., Robinson, P. J. (1994). Calcineurin inhibition of dynamin I GTPase activity coupled to nerve terminal depolarization. *Science* 265:970–973.

Liu, J. P., Zhang, Q. X., Baldwin, G., Robinson, P. J. (1996). Calcium binds dynamin I and inhibits its GTPase activity. *J. Neurochem.* 66:2074–2081.

Lovas, B. (1971). Tubular networks in the terminal endings of the visual receptor cells in the human, the monkey, the cat and the dog. *Z. Zellforsch.* 121:341–357.

McMahon, H. T. (1999). Endocytosis: an assembly protein for clathrin cages. *Curr. Biol.* 9:R332–R335.

McMahon, H. T., Wigge, P., Smith, C. (1997). Clathrin interacts specifically with amphiphysin and is displaced by dynamin. *FEBS Lett.* 413:319–322.

McPherson, P. S., Takei, K., Schmid, S. L., De Camilli, P. (1994). P145, a major Grb2-binding protein in brain, is co-localized with dynamin in nerve terminals where it undergoes activity-dependent dephosphorylation. *J. Biol. Chem.* 269:30,132–30,139.

McPherson, P. S., Garcia, E. P., Slepnev, V. I., David, C., Zhang, X., Grabs, D., Sossin, W. S., Bauerfeind, R., Nemoto, Y., De Camilli, P. (1996). A presynaptic inositol-5-phosphatase. *Nature* 379:353–357.

Mahaffey, D. T., Peeler, J. S., Brodsky, F. M., Anderson, R. G. (1990). Clathrin-coated pits contain an integral membrane protein that binds the AP-2 subunit with high affinity. *J. Biol. Chem.* 265:16,514–16,520.

Marks, B., McMahon, H. T. (1998). Calcium triggers calcineurin-dependent synaptic vesicle recycling in mammalian nerve terminals. *Curr. Biol.* 8:740–749.

Marks, M. S., Woodruff, L., Ohno, H., Bonifacino, J. S. (1996). Protein targeting by tyrosine- and di-leucine-based signals: evidence for distinct saturable components. *J. Cell Biol.* 135:341–354.

Marquez-Sterling, N. R., Lo, A. C. Y., Sisodia, S. S., Koo, E. H. (1997). Trafficking of cell-surface beta-amyloid precursor protein: evidence that a sorting intermediate participates in synaptic vesicle recycling. *J. Neurosci.* 17:140–151.

Marsh, M., McMahon, H. T. (1999). The structural era of endocytosis. Science 285:215–220.

Martin, A., Brown, F. D., Hodgkin, M. N., Bradwell, A. J., Cook, S. J., Hart, M., Wakelam, M. J. O. (1996). Activation of phospholipase D and phosphatidylinositol 4-phosphate 5-kinase in HL60 membranes is mediated by endogenous Arf but not Rho. *J. Biol. Chem.* 271:17,397–17,403.

Matsuoka, K., Orci, L., Amherdt, M., Bednarek, S. Y., Hamamoto, S., Schekman, R., Yeung, T. (1998). COPII-coated vesicle formation reconstituted with purified coat proteins and chemically defined liposomes. *Cell* 93:263–275.

Matteoli, M., Takei, K., Perin, M. S., Südhof, T. C., De Camilli, P. (1992). Exo-endocytotic recycling of synaptic vesicles in developing processes of cultured hippocampal neurons. *J. Cell Biol.* 117:849–861.

Matthews, G. (1996). Synaptic exocytosis and endocytosis: capacitance measurements. *Curr. Opin. Neurobiol.* 6:358–364.

Maycox, P., Link, E., Reetz, A., Morris, S. A., Jahn, R. (1992). Clathrin-coated vesicles in nervous tissue are involved primarily in synaptic vesicle recycling. *J. Cell Biol.* 118:1379–1388.

Mellman, I. (1996). Endocytosis and molecular sorting. *Annu. Rev. Cell Dev. Biol.* 12:575–625.

Merilainen, J., Lehto, V. P., Wasenius, V. M. (1997). FAP52, a novel, SH3 domain–containing focal adhesion protein. *J. Biol. Chem.* 272:23,278–23,284.

Merrifield, C. J., Moss, S. E., Ballestrem, C., Imhof, B. A., Giese, G., Wunderlich, I., Almers, W. (1999). Endocytic vesicles move at the tips of actin tails in cultured mast cells. *Nature Cell Biol.* 1:72–74.

Miller, T. M., Heuser, J. E. (1984). Endocytosis of synaptic vesicle membrane at the frog neuromuscular junction. *J. Cell Biol.* 98:685–698.

Morris, S. A., Schroder, S., Plessmann, U., Weber, K., Ungewickell, E. (1993). Clathrin assembly protein AP180: primary structure, domain organization and identification of a clathrin binding site. *EMBO J.* 12:667–675.

Mossessova, E., Gulbis, J. M., Goldberg, J. (1998). Structure of the guanine nucleotide exchange factor Sec7 domain of human arno and analysis of the interaction with ARF GTPase. *Cell* 92:415–423.

Muhlberg, A. B., Warnock, D. E., Schmid, S. L. (1997). Domain structure and intramolecular regulation of dynamin GTPase. *EMBO J.* 16:6676–6683.

Mulholland, J., Preuss, D., Moon, A., Wong, A., Drubin, D., Botstein, D. (1994). Ultrastructure of the yeast actin cytoskeleton and its association with the plasma membrane. *J. Cell Biol.* 125:381–391.

Mundigl, O., Ochoa, G. C., David, C., Slepnev, V. I., Kabanov, A., De Camilli, P. (1998). Amphiphysin I antisense oligonucleotides inhibit neurite outgrowth in cultured hippocampal neurons. *J. Neurosci.* 18:93–103.

Munn, A. L., Stevenson, B. J., Geli, M. I., Riezman, H. (1995). End5, end6, and end7: mutations that cause actin delocalization and block the internalization step of endocytosis in *Saccharomyces cerevisiae*. *Mol. Biol. Cell* 6:1721–1742.

Murphy, J. E., Keen, J. H. (1992). Recognition sites for clathrin-associated proteins AP-2 and AP-3 on clathrin triskelia. *J. Biol. Chem.* 267:10,850–10,855.

Murphy, J. E., Pleasure, I. T., Puszkin, S., Prasad, K., Keen, J. H. (1991). Clathrin assembly protein AP-3. The identity of the 155K protein, AP 180, and NP185 and demonstration of a clathrin binding domain. *J. Biol. Chem.* 266:4401–4408.

Murphy, J. E., Hanover, J. A., Froehlich, M., DuBois, G., Keen, J. H. (1994). Clathrin assembly protein AP-3 is phosphorylated and glycosylated on the 50-kDa structural domain. *J. Biol. Chem.* 269:21,346–21,352.

Murthy, V. N., Stevens, C. F. (1998). Synaptic vesicles retain their identity through the endocytic cycle. *Nature* 392:497–501.

Musacchio, A., Smith, C. J., Roseman, A. M., Harrison, S. C., Kirchhausen, T., Pearse, B. M. (1999). Functional organization of clathrin in coats: combining electron cryomicroscopy and X-ray crystallography. *Mol. Cell* 3:761–770.

Nakata, T., Terada, S., Hirokawa, N. (1998). Visualization of the dynamics of synaptic vesicle and plasma membrane proteins in living axons. *J. Cell Biol.* 140:659–674.

Nemoto, Y., Arribas, M., Haffner, C., De Camilli, P. (1997). Synaptojanin 2, a novel synaptojanin isoform with a distinct targeting domain and expression pattern. *J. Biol. Chem.* 272:30,817–30,821.

Nicholls, J., Martin, A. R., Wallace, B. (1992). *From Neuron to Brain*, 3rd ed. Sunderland, Mass.: Sinauer Associates.

Nonet, M. L., Holgado, A. M., Brewer, F., Serpe, C. J., Norbeck, B. A., Holleran, J., Wei, L., Hartwieg, E., Jorgensen, E. M., Alfonso, A. (1999). UNC-11, a *Caenorhabditis elegans* AP180 homologue, regulates the size and protein composition of synaptic vesicles. *Mol. Biol. Cell* 10:2343–2360.

Okamoto, M., Schoch, S., Südhof, T. C. (1999). EHSH1/intersectin, a protein that contains EH and SH3 domains and binds to dynamin and SNAP-25. A protein connection between exocytosis and endocytosis? *J. Biol. Chem.* 274:18,446–18,454.

Okamoto, P. M., Herskovits, J. S., Vallee, R. B. (1997). Role of the basic, proline-rich region of dynamin in Src homology 3 domain binding and endocytosis. *J. Biol. Chem.* 272:11,629–11,635.

Okamoto, P. M., Tripet, B., Litowski, J., Hodges, R. S., Vallee, R. B. (1999). Multiple distinct coiled-coils are involved in dynamin self-assembly. *J. Biol. Chem.* 274:10,277–10,286.

Ostermann, J., Orci, L., Tani, K., Amherdt, M., Ravazzola, M., Elazar, Z., Rothman, J. E. (1993). Stepwise assembly of functionally active transport vesicles. *Cell* 75:1015–1025.

Owen, D. J., Vallis, Y., Noble, M. E., Hunter, J. B., Dafforn, T. R., Evans, P. R., McMahon, H. T. (1999). A structural explanation for the binding of multiple ligands by the alpha-adaptin appendage domain. *Cell* 97:805–815.

Palade, G. E. (1975). Intracellular aspects of the process of protein synthesis. *Science* 189:347–358.

Paoluzi, S., Castagnoli, L., Lauro, I., Salcini, A. E., Coda, L., Fre, S., Confalonieri, S., Pelicci, P. G., Di Fiore, P. P., Cesareni, G. (1998). Recognition specificity of individual EH domains of mammals and yeast. *EMBO J.* 17: 6541–6550.

Parsons, T. D., Lenzi, D., Almers, W., Roberts, W. M. (1994). Calcium-triggered exocytosis and endocytosis in an isolated presynaptic cell: capacitance measurements in saccular hair cells. *Neuron* 15:1085–1096.

Patzer, E. J., Schlossman, D. M., Rothman, J. E. (1982). Release of clathrin from coated vesicles dependent upon a nucleoside triphosphate and a cytosol fraction. *J. Cell Biol.* 93:230–236.

Pawson, T. (1995). Protein modules and signalling networks. *Nature* 373:573–580.

Pearse, B. M. (1975). Coated vesicles from pig brain: purification and biochemical characterization. *J. Mol. Biol.* 97:93–98.

Petrovich, T. Z., Merakovsky, J., Kelly, L. E. (1993). A genetic analysis of the stoned locus and its interaction with dunce, shibire and suppressor of stoned variants of *Drosophila melanogaster. Genetics* 133:955–965.

Plomann, M., Lange, R., Vopper, G., Cremer, H., Heinlein, U. A., Scheff, S., Baldwin, S. A., Leitges, M., Cramer, M., Paulsson, M., Barthels, D. (1998). PACSIN, a brain protein that is upregulated upon differentiation into neuronal cells. *Eur. J. Biochem.* 256:201–211.

Ponnambalam, S., Robinson, M. S., Jackson, A. P., Peiperl, L., Parham, P. (1990). Conservation and diversity in families of coated vesicle adaptins. *J. Biol. Chem.* 265:4814–4820.

Poodry, C. A., Hall, L., Suzuki, D. T. (1973). Developmental properties of Shibire: a pleiotropic mutation affecting larval and adult locomotion and development. *Dev. Biol.* 32:373–386.

Powell, K. A., Robinson, P. J. (1995). Dephosphin/dynamin is a neuronal phosphoprotein concentrated in nerve terminals: evidence from rat cerebellum. *Neuroscience* 64:821–833.

Qualmann, B., Roos, J., DiGregorio, P. J., Kelly, R. B. (1999). Syndapin I, a synaptic dynamin-binding protein that associates with the neural Wiskott-Aldrich syndrome protein. *Mol. Biol. Cell* 10:501–513.

Ramaswami, M., Krishnan, K. S., Kelly, R. B. (1994). Intermediates in synaptic vesicle recycling revealed by optical imaging of *Drosophila* neuromuscular junctions. *Neuron* 13:363–375.

Ramjaun, A. R., McPherson, P. S. (1996). Tissue-specific alternative splicing generates two synaptojanin isoforms with differential membrane binding properties. *J. Biol. Chem.* 271:24,856–24,861.

Ramjaun, A. R., McPherson, P. S. (1998). Multiple amphiphysin II splice variants display differential clathrin binding: identification of two distinct clathrin-binding sites. *J. Neurochem.* 70:2369–2376.

Ramjaun, A. R., Micheva, K. D., Bouchelet, I., McPherson, P. S. (1997). Identification and characterization of a nerve terminal–enriched amphiphysin isoform. *J. Biol. Chem.* 272:16,700–16,706.

Rapoport, I., Miyazaki, M., Boll, W., Duckworth, B., Cantley, L. C., Shoelson, S., Kirchhausen, T. (1997). Regulatory interactions in the recognition of endocytic sorting signals by AP-2 complexes. *EMBO J.* 16:2240–2250.

Regnier-Vigouroux, A., Tooze, S. A., Huttner, W. B. (1991). Newly synthesized synaptophysin is transported to synaptic-like microvesicles via constitutive secretory vesicles and the plasma membrane. *EMBO J.* 10:3589–3601.

Reuter, H., Porzig, H. (1995). Localization and functional significance of the Na$^+$/Ca^{2+} exchanger in presynaptic boutons of hippocampal cells in culture. *Neuron* 15:1077–1084.

Riezman, H., Munn, A., Geli, M. I., Hicke, L. (1996). Actin-, myosin- and ubiquitin-dependent endocytosis. *Experientia* 52:1033–1041.

Ringstad, N., Nemoto, Y., De Camilli, P. (1997). The SH3p4/Sh3p8/SH3p13 protein family: binding partners for synaptojanin and dynamin via a Grb2-like Src homology 3 domain. *Proc. Natl. Acad. Sci. USA* 94:8569–8574.

Ringstad, N., Gad, H., Low, P., Di Paolo, G., Brodin, L., Shupliakov, O., De Camilli, P. (1999). Endophilin/SH3P4 is required for the transition from early to late stages in clathrin-mediated synaptic vesicle endocytosis. *Neuron* 24:143–154.

Ritter, B., Modregger, J., Paulsson, M., Plomann, M. (1999). PACSIN 2, a novel member of the PACSIN family of cytoplasmic adapter proteins. *FEBS Lett.* 454:356–362.

Robinson, M. S. (1989). Cloning of cDNAs encoding two related 100-kD coated vesicle proteins (alpha-adaptins). *J. Cell Biol.* 108:833–842.

Robinson, P. J., Sontag, J. M., Liu, J. P., Fykse, E. M., Slaughter, C., McMahon, H., Südhof, T. C. (1993). Dynamin GTPase regulated by protein kinase C phosphorylation in nerve terminals. *Nature* 365:163–166.

Robinson, P. J., Liu, J. P., Powell, K. A., Fykse, E. M., Südhof, T. C. (1994). Phosphorylation of dynamin I and synaptic-vesicle recycling. *Trends Neurosci.* 17: 348–353.

Rodal, S. K., Skretting, G., Garred, O., Vilhardt, F., van Deurs, B., Sandvig, K. (1999). Extraction of cholesterol with methyl-beta-cyclodextrin perturbs formation of clathrin-coated endocytic vesicles. *Mol. Biol. Cell* 10:961–974.

Rohatgi, R., Ma, L., Miki, H., Lopez, M., Kirchhausen, T., Takenawa, T., Kirschner, M. W. (1999). The interaction between N-WASP and the Arp2/3 complex links Cdc42-dependent signals to actin assembly. *Cell* 97:221–231.

Roos, J., Kelly, R. B. (1997). Is dynamin really a pinchase? *Trends Cell. Biol.* 7:257–259.

Roos, J., Kelly, R. B. (1998). Dap160, a neural-specific Eps15 homology and multiple SH3 domain–containing protein that interacts with *Drosophila* dynamin. *J. Biol. Chem.* 273:19,108–19,119.

Rosenthal, J. A., Chen, H., Slepnev, V. I., Pellegrini, L., Salcini, A. L., Di Fiore, P. P., De Camilli, P. (1999). The epsins define a family of proteins that interact with components of the clathrin coat and contain a new protein module. *J. Biol. Chem.* 274:33,959–33,965.

Roth, T. F., Porter, K. R. (1964). Yolk protein uptake in the oocyte of the moskito *Aedes aegypti. J. Cell Biol.* 20:313–332.

Rothman, J. E., Wieland, F. T. (1996). Protein sorting by transport vesicles. *Science* 272:227–234.

Ryan, T. A., Smith, S. J. (1995). Vesicle pool mobilization during action potential firing at hippocampal synapses. *Neuron* 14:983–989.

Ryan, T. A., Reuter, H., Wendland, B., Schweizer, F. E., Tsien, R. W., Smith, S. J. (1993). The kinetics of synaptic vesicle recycling measured at single presynaptic boutons. *Neuron* 11:713–724.

Ryan, T. A., Smith, S. J., Reuter, H. (1996). The timing of synaptic vesicle endocytosis. *Proc. Natl. Acad. Sci. USA* 93:5567–5571.

Ryan, T. A., Reuter, H., Smith, S. J. (1997). Optical detection of a quantal synaptic turnover. *Nature* 388:478–482.

Sakamuro, D., Elliott, K. J., Wechsler-Reya, R., Prendergast, G. C. (1996). BIN1 is a novel MYC-interacting protein with features of a tumour suppressor. *Nature Genet.* 14:69–77.

Salcini, A. E., Confalonieri, S., Doria, M., Santolini, E., Tassi, E., Minenkova, O., Cesareni, G., Pelicci, P. G., Di Fiore, P. P. (1997). Binding specificity and in vivo targets of the EH domain, a novel protein-protein interaction module. *Genes Dev.* 11:2239–2249.

Salcini, A. E., Chen, H., Iannolo, G., De Camilli, P., Di Fiore, P. P. (1999). Epidermal growth factor pathway substrate 15, Eps15. *Int. J. Biochem. Cell Biol.* 31:805–809.

Salim, K., Bottomley, M. J., Querfurth, E., Zvelebil, M. J., Gout, I., Scaife, R., Margolis, R. L., Gigg, R., Smith, C. I., Driscoll, P. C., Waterfield, M. D., Panayotou, G. (1996). Distinct specificity in the recognition of phosphoinositides by the pleckstrin homology domains of dynamin and Bruton's tyrosine kinase. *EMBO J.* 15:6241–6250.

Scaife, R., Gout, I., Waterfield, M. D., Margolis, R. L. (1994). Growth factor-induced binding of dynamin to signal transduction proteins involves sorting to distinct and separate proline-rich dynamin sequences. *EMBO J.* 13:2574–2582.

Schekman, R., Orci, L. (1996). Coat proteins and vesicle budding. *Science* 271: 1526–1533.

Schmalzing, G., Richter, H. P., Hansen, A., Schwarz, W., Just, I., Aktories, K. (1995). Involvement of the GTP binding protein Rho in constitutive endocytosis in *Xenopus laevis* oocytes. *J. Cell Biol.* 130:1319–1332.

Schmid, S. (1997). Clathrin-coated vesicle formation and protein sorting: an integrated process. *Annu. Rev. Biochem.* 66:511–548.

Schmid, S. L., Braell, W. A., Schlossman, D. M., Rothman, J. E. (1984). A role for clathrin light chains in the recognition of clathrin cages by "uncoating ATPase." *Nature* 311:228–231.

Schmid, S. L., McNiven, M. A., De Camilli, P. (1998). Dynamin and its partners: a progress report. *Curr. Opin. Cell Biol.* 10:504–512.

Schmidt, A., Hannah, M. J., Huttner, W. B. (1997). Synaptic-like microvesicles of neuroendocrine cells originate from a novel compartment that is continuous with the plasma membrane and devoid of transferrin receptor. *J. Cell Biol.* 137:445–458.

Schmidt, A., Wolde, M., Thiele, C., Fest, W., Kratzin, H., Podtelejnikov, A. V., Witke, W., Huttner, W. B., Söling, H. D. (1999). Endophilin I mediates synaptic vesicle formation by transfer of arachidonate to lysophosphatidic acid. *Nature* 401:133–141.

Seedorf, K., Kostka, G., Lammers, R., Bashkin, P., Daly, R., Burgess, W. H., van der Bliek, A. M., Schlessinger, J., Ullrich, A. (1994). Dynamin binds to SH3 domains of phospholipase C gamma and GRB-2. *J. Biol. Chem.* 269:16,009–16,014.

Segal, J. R., Ceccarelli, B., Fesce, R., Hurlbut, W. P. (1985). Miniature endplate potential frequency and amplitude determined by an extension of Campbell's theorem. *Biophys. J.* 47:183–202.

Sengar, A. S., Wang, W., Bishay, J., Cohen, S., Egan, S. E. (1999). The EH and SH3 domain Ese proteins regulate endocytosis by linking to dynamin and Eps15. *EMBO J.* 18:1159–1171.

Sever, S., Muhlberg, A. B., Schmid, S. L. (1999). Impairment of dynamin's GAP domain stimulates receptor-mediated endocytosis. *Nature* 398:481–486.

Shih, W., Gallusser, A., Kirchhausen, T. (1995). A clathrin-binding site in the hinge of the beta 2 chain of mammalian AP-2 complexes. *J. Biol. Chem.* 270: 31,083–31,090.

Shupliakov, O., Low, P., Grabs, D., Gad, H., Chen, H., David, C., Takei, K., De Camilli, P., Brodin, L. (1997). Synaptic vesicle endocytosis impaired by disruption of dynamin-SH3 domain interactions. *Science* 276:259–263.

Sirotkin, H., Morrow, B., DasGupta, R., Goldberg, R., Patanjali, S. R., Shi, G., Cannizzaro, L., Shprintzen, R., Weissman, S. M., Kucherlapati, R. (1996). Isolation of a new clathrin heavy chain gene with muscle-specific expression from the region commonly deleted in velo-cardio-facial syndrome. *Hum. Mol. Genet.* 5:617–624.

Slepnev, V. I., Ochoa, G. C., Butler, M. H., Grabs, D., De Camilli, P. (1998). Role of phosphorylation in regulation of the assembly of endocytic coat complexes. *Science* 281:821–824.

Slepnev, V. I., Ochoa, G. C., Butler, M. H., De Camilli, P. (2000). Tandem arrangement of the clathrin and AP-2 binding domains in amphiphysin 1, and disruption of clathrin coat function mediated by amphiphysin fragments comprising these sites. *J. Biol. Chem.* 275:17,583–17,589.

Smirnova, E., Shurland, D. L., Newman-Smith, E. D., Pishvaee, B., van der Bliek, A. M. (1999). A model for dynamin self-assembly based on binding between three different protein domains. *J. Biol. Chem.* 274:14,942–14,947.

Smith, C., Neher, E. (1997). Multiple forms of endocytosis in bovine adrenal chromaffin cells. *J. Cell Biol.* 139:885–894.

Smith, C. J., Grigorieff, N., Pearse, B. M. (1998). Clathrin coats at 21 Å resolution: a cellular assembly designed to recycle multiple membrane receptors. *EMBO J.* 17:4943–4953.

Snapper, S. B., Rosen, F. S. (1999). The Wiskott-Aldrich syndrome protein (WASP): roles in signaling and cytoskeletal organization. *Annu. Rev. Immunol.* 17:905–929.

Sone, M., Hoshino, M., Suzuki, E., Kuroda, S., Kaibuchi, K., Nakagoshi, H., Saigo, K., Nabeshima, Y., Hama, C. (1997). Still life, a protein in synaptic terminals of *Drosophila* homologous to GDP-GTP exchangers. *Science* 275:543–547.

Spang, A., Matsuoka, K., Hamamoto, S., Schekman, R., Orci, L. (1998). Coatomer, Arf1p, and nucleotide are required to bud coat protein complex I–coated vesicles from large synthetic liposomes. *Proc. Natl. Acad. Sci. USA* 95:11,199–11,204.

Sparks, A. B., Hoffman, N. G., McConnell, S. J., Fowlkes, D. M., Kay, B. K. (1996). Cloning of ligand targets: systematic isolation of SH3 domain–containing proteins. *Nature Biotechnol.* 14:741–744.

Springer, S., Spang, A., Schekman, R. (1999). A primer on vesicle budding. *Cell* 97:145–148.

Stamnes, M. A., Craighead, M. W., Hoe, M. H., Lampen, N., Geromanos, S., Tempst, P., Rothman, J. E. (1995). An integral membrane component of coatomer-coated transport vesicles defines a family of proteins involved in budding. *Proc. Natl. Acad. Sci. USA* 92:8011–8015.

Stimson, D. T., Estes, P. S., Smith, M., Kelly, L. E., Ramaswami, M. (1998). A product of the *Drosophila* stoned locus regulates neurotransmitter release. *J. Neurosci.* 18:9638–9649.

Subtil, A., Gaidarov, I., Kobylarz, K., Lampson, M. A., Keen, J. H., McGraw, T. E. (1999). Acute cholesterol depletion inhibits clathrin-coated pit budding. *Proc. Natl. Acad. Sci. USA* 96:6775–6780.

Sweitzer, S. M., Hinshaw, J. E. (1998). Dynamin undergoes a GTP-dependent conformational change causing vesiculation. *Cell* 93:1021–1029.

Takei, K., Stukenbrok, H., Metcalf, A., Mignery, G. A., Südhof, T. C., Volpe, P., De Camilli, P. (1992). Ca^{2+} stores in Purkinje neurons: endoplasmic reticulum subcompartments demonstrated by the heterogeneous distribution of the InsP3 receptor, Ca^{2+}-ATPase, and calsequestrin. *J. Neurosci.* 12:489–505.

Takei, K., McPherson, P. S., Schmid, S. L., De Camilli, P. (1995). Tubular membrane invaginations coated by dynamin rings are induced by GTP-gamma S in nerve terminals. *Nature* 374:186–190.

Takei, K., Mundigl, O., Daniell, L., De Camilli, P. (1996). The synaptic vesicle cycle: single vesicle budding step involving clathrin and dynamin. *J. Cell Biol.* 133:1237–1250.

Takei, K., Haucke, V., Slepnev, V., Farsad, K., Salazar, M., Chen, H., De Camilli, P. (1998). Generation of coated intermediates of clathrin-mediated endocytosis on protein-free liposomes. *Cell* 94:131–141.

Takei, K., Slepnev, V. I., Haucke, V., De Camilli, P. (1999). Functional partnership between amphiphysin and dynamin in clathrin-mediated endocytosis. *Nature Cell Biol.* 1:33–39.

Tebar, F., Sorkina, T., Sorkin, A., Ericsson, M., Kirchhausen, T. (1996). Eps15 is a component of clathrin-coated pits and vesicles and is located at the rim of coated pits. *J. Biol. Chem.* 271:28,727–28,730.

Tebar, F., Confalonieri, S., Carter, R. E., Di Fiore, P. P., Sorkin, A. (1997). Eps15 is constitutively oligomerized due to homophilic interaction of its coiled-coil region. *J. Biol. Chem.* 272:15,413–15,418.

Tebar, F., Bohlander, S. K., Sorkin, A. (1999). Clathrin assembly lymphoid myeloid leukemia (CALM) protein: localization in endocytic-coated pits, interactions with clathrin, and the impact of overexpression on clathrin-mediated traffic. *Mol. Biol. Cell* 10:2687–2702.

Ter Haar, E., Musacchio, A., Harrison, S. C., Kirchhausen, T. (1998). Atomic structure of clathrin: a beta propeller terminal domain joins an alpha zigzag linker. *Cell* 95:563–573.

Thomas, P., Lee, A. K., Wong, G. J., Almers, W. (1994). A triggered mechanism retrieves membrane in seconds after Ca^{2+}-stimulated exocytosis in single pituitary cell. *J. Cell Biol.* 124:667–675.

Torre, E., McNiven, M. A., Urrutia, R. (1994). Dynamin 1 antisense oligonucleotide treatment prevents neurite formation in cultured hippocampal neurons. *J. Biol. Chem.* 269:32,411–32,417.

Torri-Tarelli, F., Villa, A., Valtorta, F., De Camilli, P., Greengard, P., Ceccarelli, B. (1990). Redistribution of synaptophysin and synapsin I during alpha-latrotoxin-induced release of neurotransmitter at the neuromuscular junction. *J. Cell Biol.* 110:449–459.

Traub, L. M., Downs, M. A., Westrich, J. L., Fremont, D. H. (1999). Crystal structure of the alpha appendage of AP-2 reveals a recruitment platform for clathrin-coat assembly. *Proc. Natl. Acad. Sci. USA* 96:8907–8912.

Trowbridge, I. S., Colawn, J. F., Hopkins, C. R. (1993). Signal-dependent membrane trafficking in the endocytic pathway. *Annu. Rev. Cell Biol.* 9:129–161.

Tsukita, S., Ishikawa, H. (1980). The movement of membranous organelles in axons. Electron microscopic identification of anterogradely and retrogradely transported organelles. *J. Cell Biol.* 84:513–530.

Tsutsui, K., Maeda, Y., Seki, S., Tokunaga, A. (1997). cDNA cloning of a novel amphiphysin isoform and tissue-specific expression of its multiple splice variants. *Biochem. Biophys. Res. Commun.* 236:178–183.

Tuma, P. L., Collins, C. A. (1994). Activation of dynamin GTPase is a result of positive cooperativity. *J. Biol. Chem.* 269:30,842–30,847.

Ungewickell, E. (1985). The 70-kd mammalian heat shock proteins are structurally and functionally related to the uncoating protein that releases clathrin triskelia from coated vesicles. *EMBO J.* 4:3385–3391.

Ungewickell, E., Branton, D. (1981). Assembly units of clathrin coats. *Nature* 289:420–422.

Ungewickell, E., Ungewickell, H., Holstein, S. E., Lindner, R., Prasad, K., Barouch, W., Martin, B., Greene, L. E., Eisenberg, E. (1995). Role of auxilin in uncoating clathrin-coated vesicles. *Nature* 378:632–635.

Urrutia, R., Henley, J. R., Cook, T., McNiven, M. A. (1997). The dynamins: re-dundant or distinct functions for an expanding family of related GTPases? *Proc. Natl. Acad. Sci. USA* 94:377–384.

Vallis, Y., Wigge, P., Marks, B., Evans, P. R., McMahon, H. T. (1999). Importance of the pleckstrin homology domain of dynamin in clathrin-mediated endo-cytosis. *Curr. Biol.* 9:257–260.

Valtorta, F., Jahn, R., Fesce, R., Greengard, P., Ceccarelli, B. (1988). Synapto-physin (p38) at the frog neuromuscular junction: its incorporation into the axolemma and recylcing after intense quantal secretion. *J. Cell Biol.* 107: 2717–2727.

Van Delft, S., Schumacher, C., Hage, W., Verkleij, A. J., van Bergen en Hene-gouwen, P. M. (1997). Association and colocalization of Eps15 with adaptor protein-2 and clathrin. *J. Cell Biol.* 136:811–821.

Van der Bliek, A. M., Meyerowitz, E. M. (1991). Dynamin-like protein encoded by the *Drosophila* shibire gene associated with vesicular traffic. *Nature* 351: 411–414.

Van Deurs, B., Petersen, O. W., Olsnes, S., Sandvig, K. (1989). The ways of endocytosis. *Int. Rev. Cytol.* 117:131–177.

Villa, A., Ceccarelli, B. (1990). Neurotransmitter release and synaptic vesicle recycling. *Neuroscience* 35:477–489.

Von Gersdorff, H., Matthews, G. (1994a). Dynamics of synaptic vesicle fusion and membrane retrieval in synaptic terminals. *Nature* 367:735–739.

Von Gersdorff, H., Matthews, G. (1994b). Inhibition of endocytosis by elevated internal calcium in a synaptic terminal. *Nature* 370:652–655.

Verderio, C., Coco, S., Rossetto, O., Montecucco, C., Matteoli, M. (1999). Inter-nalization and proteolytic action of botulinum toxins in CNS neurons and astrocytes. *J. Neurochem.* 73:372–379.

Virshup, D. M., Bennett, V. (1988). Clathrin-coated vesicle assembly polypep-tides: physical properties and reconstitution studies with brain membranes. *J. Cell Biol.* 106:39–50.

Wang, L. H., Südhof, T. C., Anderson, R. G. (1995). The appendage domain of alpha-adaptin is a high affinity binding site for dynamin. *J. Biol. Chem.* 270: 10,079–10,083.

Warnock, D. E., Hinshaw, J. E., Schmid, S. L. (1996). Dynamin self-assembly stimulates its GTPase activity. *J. Biol. Chem.* 271:22,310–22,314.

Wendland, B., Steece, K. E., Emr, S. D. (1999). Yeast epsins contain an essential N-terminal ENTH domain, bind clathrin and are required for endocytosis. *EMBO J.* 18:4383–4393.

Wieland, F., Harter, C. (1999). Mechanisms of vesicle formation: insights from the COP system. *Curr. Opin. Cell Biol.* 11:440–446.

Wigge, P., Kohler, K., Vallis, Y., Doyle, C. A., Owen, D., Hunt, S. P., McMahon, H. T. (1997a). Amphiphysin heterodimers: potential role in clathrin-mediated endocytosis. *Mol. Biol. Cell* 8:2003–2015.

Wigge, P., Vallis, Y., McMahon, H. T. (1997b). Inhibition of receptor-mediated endocytosis by the amphiphysin SH3 domain. *Curr. Biol.* 7:554–560.

Wilde, A., Brodsky, F. M. (1996). In vivo phosphorylation of adaptors regulates their interaction with clathrin. *J. Cell Biol.* 135:635–645.

Willingham, M. C., Pastan, I. (1983). Formation of receptosomes from plasma membrane coated pits during endocytosis: analysis by serial sections with im-proved membrane labeling and preservation techniques. *Proc. Natl. Acad. Sci. USA* 80:5617–5621.

Witke, W., Podtelejnikov, A. V., Di Nardo, A., Sutherland, J. D., Gurniak, C. B., Dotti, C., Mann, M. (1998). In mouse brain profilin I and profilin II associate with regulators of the endocytic pathway and actin assembly. *EMBO J.* 17: 967–976.

Wong, W. T., Schumacher, C., Salcini, A. E., Romano, A., Castagnino, P., Pelicci, P. G., Di Fiore, P. (1995). A protein-binding domain, EH, identified in the receptor tyrosine kinase substrate Eps15 and conserved in evolution. *Proc. Natl. Acad. Sci. USA* 92:9530–9534.

Woscholski, R., Finan, P. M., Radley, E., Totty, N. F., Sterling, A. E., Hsuan, J. J., Waterfield, M. D., Parker, P. J. (1997). Synaptojanin is the major constitutively active phosphatidylinositol-3,4,5-trisphosphate 5-phosphatase in rodent brain. *J. Biol. Chem.* 272:9625–9628.

Wu, L. G., Betz, W. J. (1996). Nerve activity but not intracellular calcium determines the time course of endocytosis at the frog neuromuscular junction. *Neuron* 17:769–779.

Wu, L. G., Betz, W. J. (1998). Kinetics of synaptic depression and vesicle recycling after tetanic stimulation of frog motor nerve terminals. *Biophys. J.* 74:3003–3009.

Yamabhai, M., Hoffman, N. G., Hardison, N. L., McPherson, P. S., Castagnoli, L., Cesareni, G., Kay, B. K. (1998). Intersectin, a novel adaptor protein with two Eps15 homology and five Src homology 3 domains. *J. Biol. Chem.* 273: 31,401–31,407.

Yang, S., Cope, M. J., Drubin, D. G. (1999). Sla2p is associated with the yeast cortical actin cytoskeleton via redundant localization signals. *Mol. Biol. Cell* 10:2265–2283.

Ybe, J. A., Brodsky, F. M., Hofmann, K., Lin, K., Liu, S. H., Chen, L., Earnest, T. N., Fletterick, R. J., Hwang, P. K. (1999). Clathrin self-assembly is mediated by a tandemly repeated superhelix. *Nature* 399:371–375.

Ye, W., Lafer, E. M. (1995a). Bacterially expressed F1-20/AP-3 assembles clathrin into cages with a narrow size distribution: implications for the regulation of quantal size during neurotransmission. *J. Neurosci. Res.* 41:15–26.

Ye, W., Lafer, E. M. (1995b). Clathrin binding and assembly activities of expressed domains of the synapse-specific clathrin assembly protein AP-3. *J. Biol. Chem.* 270:10,933–10,939.

Zhang, B., Koh, Y. H., Beckstead, R. B., Budnik, V., Ganetzky, B., Bellen, H. J. (1998). Synaptic vesicle size and number are regulated by a clathrin adaptor protein required for endocytosis. *Neuron* 21:1465–1475.

Zhang, B., Ganetzky, B., Bellen, H. J., Murthy, V. N. (1999). Tailoring uniform coats for synaptic vesicles during endocytosis. *Neuron* 23:419–422.

Zhang, J. Z., Davletov, B. A., Südhof, T. C., Anderson, R. G. (1994). Synaptotagmin I is a high affinity receptor for clathrin AP-2: implications for membrane recycling. *Cell* 78:751–760.

Zheng, Y., Bagrodia, S., Cerione, R. A. (1994). Activation of phosphoinositide 3-kinase activity by Cdc42Hs binding to p85. *J. Biol. Chem.* 269:18,727–18,730.

Zhou, S., Sousa, R., Tannery, N. H., Lafer, E. M. (1992). Characterization of a novel synapse-specific protein. II. cDNA cloning and sequence analysis of the F1-20 protein. *J. Neurosci.* 12:2144–2155.

6

The Synaptic Cleft and Synaptic Cell Adhesion

Thomas C. Südhof

ynapses can be viewed as specialized intercellular junctions whose business it is to provide a scaffold for signaling between neurons. As such, they are similar to other intercellular junctions with distinctive intracellular specializations of the plasma membranes that form the junction. Morphologically, synaptic junctions resemble tight junctions with regularly opposed plasma membranes that are coated intracellularly with electron-dense material (see Stevenson and Keon, 1998, for review; see De Camilli et al. on synaptic vesicle endocytosis and Greenberg and Ziff, this volume). However, synapses differ markedly from other intercellular junctions in that they are highly polarized. As described in detail in the chapter by De Camilli et al. on the structure of synapses, central synapses are characterized by three features: presynaptic nerve terminals containing neurotransmitter-filled vesicles; postsynaptic membrane specializations containing clustered neurotransmitter receptors; and synaptic clefts that separate the pre- and postsynaptic membranes. As a central component of synapses, the synaptic cleft performs both mechanical and signaling functions. Mechanically the cleft serves to stabilize the parallel orientation of the pre- and postsynaptic plasma membranes, to connect these membranes to each other at a uniform distance, and to coordinate their relative positions. In terms of signaling, the synaptic cleft is the space in which neurotransmitters act, and it may also mediate nontransmitter signals between the pre- and postsynaptic specializations. With these properties, the synaptic cleft is not only the "glue" that keeps the synapse together, it is also the communication space that mediates synaptic signaling.

The extracellular space between central neurons is usually very small (<10 nm), as is the distance between neurons and glial cells. The intercellular space widens at synapses, with synaptic clefts having an average width of some 20 nm (see Akert et al., 1972, for review). Symmetric inhibitory synapses appear to have a slightly thinner synaptic cleft than asymmetric excitatory synapses. Synaptic clefts of central synapses are quite different from those at neuromuscular junctions, where the synaptic cleft is bisected by a basal lamina (Sanes and Lichtman, 1999). In central synapses the synaptic cleft is much narrower and appears to be relatively "empty," lacking the basal lamina. Electron microscopy, however, has revealed that the synaptic cleft of central synapses is filled with undefined electron-dense material (Gray, 1959). With a width of some 20 nm and a cross-sectional area of $0.075–0.150 \ \mu m^2$, the total volume of the average synaptic cleft is only 20- to 100-fold larger than the interior space of a single synaptic vesicle (Harris and Sultan, 1995). Since synaptic vesicles contain high concentrations of neurotransmitters, exocytosis of a single vesicle is thought to achieve a significant concentration of neurotransmitter in the synaptic cleft as a result.

Examination of osmium-stained synapses suggested that filaments cross the synaptic cleft (van der Loos, 1963; De Robertis, 1964; Gray, 1966), as was also indicated by rapid freeze-etching studies (Landis and Reese, 1983). However, the filaments observed in these studies were not of sufficient abundance to explain the electron-dense cleft material. Other staining methods, particularly those using phosphotungstic acid or bismuth iodide, indicated a more homogeneous material that may be organized into layers parallel to the synaptic plasma membranes (Bloom and Aghajanian, 1966; Pfenninger, 1971a). Studies with proteolytic enzymes showed that the cleft material is largely proteinaceous but may include considerable amounts of carbohydrates (see Pfenninger, 1973, for review). Although the identity of the molecules connecting the pre- and postsynaptic specializations is unclear, these studies together demonstrate that the synaptic cleft contains proteins that may function as the "glue" that connects the two sides.

The idea that the pre- and postsynaptic sides of synapses are tightly bound to each other via the synaptic cleft is supported by observations from synaptosomes, which are pinched-off nerve endings that are capable of regulated neurotransmitter synthesis and release (Gray and Whittaker, 1962). Synaptosomes are formed spontaneously when brain tissue is homogenized in isotonic buffers. Interestingly the homogenization rips a fragment of the postsynaptic plasma membrane with an attached postsynaptic density out of the postsynaptic neuron. As a result the postsynaptic specialization remains bound to the presynaptic active zone in purified synaptosomes, suggesting that the cohesive force linking the pre- and postsynaptic specializations is stronger than the hydrophobic force that acts to retain contiguous plasma membranes. Studies of the cohesive forces between pre- and postsynaptic specializations showed that the pre- and postsynaptic membranes remain stably attached to each other even in the absence of calcium, and that they are not dissociated by high concentrations of salt or low concentrations of urea (Cotman and Taylor, 1972). They can be only partially separated by chaotropic agents (e.g., 1 M $MgCl_2$; Pfenninger, 1971b).

The overall picture that emerges from these findings is that the synaptic cleft contains protein filaments that mechanically connect the pre- and postsynaptic plasma membranes in a calcium-independent manner. These filaments are not abundant, and additional material that may partly consist of carbohydrates probably fills the cleft. It is possible that some of the synaptic cell surface proteins are highly glycosylated, thereby creating the laminated appearance seen in the synaptic cleft with bismuth iodide staining (Pfenninger, 1971a).

The cross-sectional areas of the pre- and postsynaptic specializations and the synaptic cleft vary significantly among central synapses. Nevertheless, for any given synapse these areas are always precisely correlated (Lisman and Harris, 1993). This implies that synapses assemble, grow, and shrink in a concerted way so that the pre- and postsynaptic specializations change in unison. Several studies have observed activity-dependent changes in the size and structure of synapses (e.g., see Vrensen and Cardozo, 1981; Geinisman et al., 1988; Neuhoff et al., 1999). During these changes, the pre- and postsynaptic elements always enlarge or contract together. Most recently, elegant electron microscopy experiments have revealed that during long-term potentiation (LTP) pre- and post-synaptic specializations are subject to precisely coordinated growth (Toni et al., 1999). Initially, within 15–30 min after the induction of LTP, this growth is characterized by the formation of perforated synapses that have bisected pre- and postsynaptic densities and a split synaptic cleft. Later, 30–60 min after LTP induction, a duplication of the postsynaptic dendritic spines and of the presynaptic active zones was observed. These findings imply that the strengthening of synaptic action that occurs during LTP is associated with a duplication of the apparatus for synaptic transmission, a duplication that proceeds in a parallel, concerted fashion for the pre- and postsynaptic specializations and for the synaptic cleft. The only way in which such coordinated growth can be achieved is by a communication of the locations and sizes of the pre- and post-synaptic specializations via transsynaptic cell adhesion molecules that bridge the synaptic cleft. Since in the first 60 min LTP is independent of protein synthesis, this coordinated growth must occur by the regulated assembly of preexisting synaptic components.

Structural and physiological observations make it clear that pre- and postsynaptic plasma membranes are linked by cell adhesion molecules that participate in the following essential synaptic functions:

1. Recognition between the pre- and postsynaptic neurons that is required for the formation and maintenance of appropriate connections.
2. Coordination of the sizes and positions of the pre- and postsynaptic specializations to effect a precise alignment.
3. Transmission of anterograde and retrograde signals between pre- and postsynaptic neurons (e.g., by coupling to tyrosine kinases, as is also often observed with nonsynaptic cell adhesion events) to allow plasticity.
4. Organization of the polarized, asymmetric specializations of synapses that are involved in presynaptic membrane trafficking and post-synaptic signal transduction.
5. Provision of mechanical cohesion between the pre- and postsynaptic specializations, which probably requires coupling of the cell adhesion molecules to the actin cytoskeleton and explains the tight association between these specializations.

In addition to the three large families of classical cell adhesion molecules—the cadherins, integrins, and immunoglobulin-domain proteins—two other families of more recently described neuron-specific cell adhesion molecules—the neuroligins and neurexins—are probably involved. In addition, smaller families of cell adhesion molecules, such as the syndecans, may also play a role in synapses. In evaluating the potential roles of these different cell adhesion molecules, it is important to remember that synapses are inherently asymmetric junctions that are organized in a highly polarized manner. This means that the recognition between pre- and postsynaptic membranes cannot be mediated by the same molecules. The central problem here is that any given neuron is postsynaptic to many different types of presynaptic inputs, but at the same time it acts as a presynaptic neuron of only one class. In most cases the presynaptic inputs that impinge on a neuron are derived from a number of distinct classes of neurons that utilize different transmitters and have characteristic distributions on the postsynaptic neuron. To achieve such a pattern of synaptic connectivity, there must be distinct pre- and postsynaptic cell recognition molecules that signal the specificity of interactions; otherwise it is difficult to see how, for example, a neuron could receive inhibitory and excitatory inputs in a spatially segregated manner, but send out only excitatory signals. These considerations suggest that synaptic cell adhesion is mediated by the cooperation of several distinct cell adhesion proteins rather than a single "master" molecule. And it is likely that mechanical cohesion is mediated by a different interaction than the initial recognition during synaptogenesis, the nucleation of synaptic assembly, or maturation of synapses during the maintenance phase. Finally, the considerable strength and calcium independence of the junctional adhesion should be used as a characteristic to evaluate some of the candidates for synaptic cell adhesion.

Classical Cadherins and Protocadherins

Cadherins constitute a large family of Ca^{2+}-dependent, homotypic cell adhesion molecules. Like most cell adhesion molecules, cadherins are type I transmembrane proteins with short cytoplasmic tails. Cadherins were discovered because they mediate Ca^{2+}-dependent cell adhesion events and are protected from proteolysis by Ca^{2+} (Takeichi, 1977). Classical cadherins consist of an extracellular region comprising five tandem repeats of a typical 110-residue cadherin domain, followed by a single transmembrane region and a highly conserved short cytoplasmic tail (see Takeichi, 1990, for review; Yap et al., 1997; Fig. 6.1A). Classical cadherins belong to the larger family of cadherin-like cell adhesion molecules that also include protocadherins. The defining characteristic of all cadherins is the cadherin domain that is present in multiple copies in the extracellular domains of the proteins. Structurally cadherin domains are composed of a β-sandwich that is remarkably similar to the structure of a subclass of immunoglobulin domains. However, cadherin domains differ from immunoglobulin domains in several ways. For example, cadherin domains require Ca^{2+} for homophilic interactions whereas immunoglobulin domains do not; immunoglobulin domains, by contrast, contain a conserved disulfide bond that is absent from cadherin domains (see Leahy, 1997, for review). Although both classical cadherins and cadherin-related molecules include multiple cadherin domains, classical cadherins are distinguished from other cadherins by their highly conserved cytoplasmic tail, which binds to β-catenin, which in turn binds to α-catenin. Catenins in turn associate with the actin cyto-

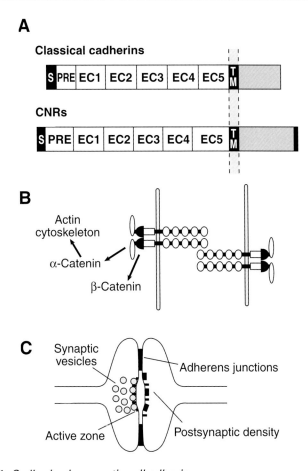

Figure 6.1. Cadherins in synaptic cell adhesion.

(A) Domain structures of classical cadherins and CNR family protocadherins. Classical cadherins and CNRs are composed of a signal peptide (S) followed by a presequence (PRE) and five cadherin domains (EC1–EC5). The cadherins are attached to the membrane by a transmembrane region (TM) and have cytoplasmic domains that differ between various types of cadherins.

(B) Schematic diagram of cell adhesion mediated by homophilic interactions between classical cadherins. Classical cadherins are parallel dimers whose homophilic interactions are mediated by the first cadherin domain. Intracellularly they bind to α- and β-catenins, which in turn are linked to the cytoskeleton.

(C) Schematic drawing of a synapse with a central synaptic cleft that is flanked on both sides by adherens junctions containing classical cadherins.

Modified from Uchida et al. (1996).

skeleton, thereby coupling cadherin-mediated cell adhesion to the cytoskeleton. In addition, β-catenin also performs important signaling functions in the wingless pathway (Wodarz and Nusse, 1998).

Classical cadherins mediate homophilic cell adhesion in most if not all tissues during development. In mature animals they probably form intercellular adherens junctions, primarily in epithelial tissues and brain. The extracellular cadherin domains associate into parallel homo-dimers that bind to similar homodimers on an adjacent cell in a calcium-dependent manner (Fig. 6.1B; Brieher et al., 1996). The specificity of the intercellular interaction is defined by the first cadherin domain, but the whole cadherin-catenin complex is required for actual cell adhesion (Nose et al., 1988, 1990). Although the ectodomains are able to bind to each other, stable cell adhesion occurs only when the cytoplasmic tail is bound to α- and β-catenins and to the actin cytoskeleton (Ozawa et al., 1990). Vertebrates express multiple classical cadherins (>10 genes) whose genes are organized in clusters in the genome. Since classical cadherins appear to be involved only in homotypic interactions, one of their major functions is probably in tissue morphogenesis in early embryonic devel-opment. Several additional functions in the elaboration and maintenance of specialized intercellular junctions in mature organs are also likely to be mediated by classical cadherins.

Classical cadherins are localized to synapses with a distinct, charac-teristic distribution in different brain areas for each isoform (Yamagata et al., 1995; Fannon and Colman, 1996; Wohrn et al., 1999; see Takeichi et al., 1997, Rubenstein et al., 1999, and Sanes and Yamagata, 1999, for re-views). The overlapping, highly reproducible expression patterns of cadherins are independent of axonal inputs and intrinsic to developing brain regions. This situation results in a cadherin-based map of the de-veloping brain. Based on these findings, a model was proposed that cad-herins contribute to establishing synaptic specificity, that they may serve as "postal codes" for outgrowing axons (see references cited earlier). The precise localization of classical cadherins and their interacting proteins, α- and β-catenin, by immunoelectron microscopy has revealed that these proteins are not actually present in the synaptic junction itself, but participate in adherens junctions immediately adjacent to synapses (Fig. 6.1C; Uchida et al., 1996). This observation demonstrates that clas-sical cadherins are in fact parasynaptic and supports the model that classical cadherins participate in the initial contact between neurons that later leads to synaptic junctions. In further agreement with this model, the presence of cadherin-containing adherens junctions adjacent to synapses appears to be under developmental control (Uchida et al., 1996). Most immature central synapses contain adjacent adherens junc-tions, whereas many mature synapses lack them, suggesting that ad-herens junctions may be important in establishing synapses but are lost after a synaptic junction has been successfully formed.

Together these results raise the possibility that homophilic cadherin interactions at adherens junctions prime synapse formation. There is an intriguing similarity here between synaptic junctions and tight junctions. Tight junctions also contain adjacent adherens junctions, which appear to form first and to guide the establishment of tight junctions. Without initial cadherin-based junctions, tight junctions do not develop (see Yap et al., 1997, for review). Based on this result, one would expect that at the synapse a synaptic junction can only form after an adjacent adherens junction has primed it—a prediction that has not yet been experimentally tested. Another prediction would be that ablation of the expression of specific classical cadherins should lead to a reorganization of synaptic connectivity, but this again is an untested hypothesis. In the few cadherin knockouts that have been performed in mice, major developmental effects on tissue morphogenesis are observed (Radice et al., 1997). This is probably a reflection of the fact that most (and maybe all) classical cadherins have other functions in addition to their potential role in synaptogenesis, making it impossible to test the role of cadherins in synapse formation without the use of conditional knockouts.

In addition to the classical cadherins, the cadherin family includes protocadherins, a large heterogeneous group of cadherin-related type I transmembrane proteins. Like classical cadherins, protocadherins contain multiple (usually five or six) extracellular cadherin domains; however, their transmembrane regions and cytoplasmic tails exhibit no similarity. There are several subclasses of protocadherins, some of which comprise large numbers of polymorphic proteins. The most diverse protocadherins may be the cadherin-related neuronal receptors (CNRs; Kohmura et al., 1998), a class of protocadherins that is composed of more than 50 members organized into several subfamilies (Wu and Maniatis, 1999). Another family of interesting protocadherins is made up of the **fat** proteins, which contain 34 tandem cadherin repeats (Mahoney et al., 1991; Dunne et al., 1995). Other cadherin-related proteins perform several important cellular functions in animals. The most prominent of these proteins probably are the desmosomal cadherins called desmogleins and desmocollins, which create desmosomes that differ from intercellular adherens junctions formed by classical cadherins. The desmosomal cadherins are linked intracellularly to intermediate filaments and not to the actin cytoskeleton (Kowalczyk et al., 1999).

In terms of synapse function, CNRs are probably the most interesting protocadherins. In contrast to classical cadherins that are parasynaptic, at least some of the CNRs appear to be synaptic cell adhesion molecules (Kohmura et al., 1998; Wu and Maniatis, 1999). CNRs are only expressed in brain. They are composed of six extracellular cadherin domains, a single transmembrane region, and a cytoplasmic tail that differs between subclasses of CNRs, is unrelated to the cytoplasmic tail of classical cadherins, and may bind the Fyn tyrosine kinase (Kohmura et al., 1998). The diversity of CNRs has an unusual genomic basis: the N-terminal variable six extracellular cadherin domains and single transmembrane region of CNRs are encoded by single large exons in the genome. These exons are arranged in tandem arrays with more than 10 members for each subfamily. All members of each subfamily have the same intracellular C-terminal sequences, which are encoded by a set of constant exons that are located downstream of the array of the N-terminal large exons and are used by all members (Wu and Maniatis, 1999). Three such subfamilies, organized in distinct gene clusters with a total of at least 52 products, were identified; more families are likely to be discovered. In the CNR gene clusters, it is unknown if CNR expression involves the use of multiple promoters or somatic DNA rearrangements. The most likely mechanism for expression of the various CNRs is that individual promoters upstream of each large N-terminal exon drive transcription of these exons, which then splice into the invariable C-terminal exons. Alternatively it is conceivable (but less probable) that genomic DNA rearrangements directly couple the N-terminal exons to the C-terminal exons. It is currently not known if the expression of the different CNRs is spatially and temporally regulated, and if a neuron can express more than a single subfamily and more than a single member of each subfamily.

CNRs have been found by immunoelectron microscopy to be localized to synaptic junctions (Kohmura et al., 1998). Since no quantitative studies were reported, it is unclear if CNRs are exclusively or only partly synaptic. The large number of different CNRs uncovered in cDNA cloning experiments and genomic analyses suggests the possibility that CNRs are differentially expressed in various types of synapses to create a synaptic code. This possibility is based on the idea that CNRs, like classical cadherins, only interact in a homophilic manner. The idea of a synaptic code specified by CNRs is thus based on the same premise as discussed earlier for classical cadherins, with the differences that CNRs may actually be in the synapse, instead of adjacent to synapses like classical cadherins, and there may be more variants of CNRs than of classical cadherins. For such a code to operate, CNRs would have to be expressed in tightly regulated patterns that are distinct for each member, as has been demonstrated for classical cadherins (Sanes and Yamagata, 1999; Serafini, 1999). If all the CNRs were indeed revealed to be synaptic and were found to mediate exclusively homophilic interactions, a recognition code effected by CNRs (or classical cadherins) would be attractive for synaptogenesis, because it would mean that only plasma membranes expressing the same cadherin isotype could interact.

However, a synaptic code specified by homophilic interactions between classical cadherins or CNRs is subject to conceptual limitations. Since most neurons receive synaptic connections from a large number of different inputs, they would have to express a large number of distinct classical cadherins or CNRs to "catch" each of these inputs separately. At the same time, however, the presynaptic terminals formed by the neuron should not contain the same large number of CNRs, which would otherwise lose their information content. It is thus difficult to understand how the neuron could direct only one of the many CNRs it expresses to presynaptic junctions. One solution to this problem would be if pre- and postsynaptic CNRs had identical extracellular domains but differed in their cytoplasmic tails, with each neuron expressing multiple CNRs with different postsynaptic cytoplasmic tails and only a single CNR with a presynaptic cytoplasmic tail. However, no such pattern has yet been found. Until the expression patterns and localizations of different cadherins are precisely mapped for individual neurons, such speculations must remain hypothetical, and the synaptic roles of CNRs and other protocadherins remain to be established. If the various protocadherins are revealed to interact with each other in a mutually exclusive, specific manner, their large numbers would serve to explain the specificity of synapse formation. However, their interactions are homophilic and they cannot explain the polarized nature of the synaptic junction. Furthermore, as calcium-dependent cell adhesion molecules they cannot account for the tight calcium-independent connections between pre- and postsynaptic plasma membranes at the synapse. Finally, CNRs are not conserved in *Drosophila*, whose nervous system is composed of complex neural networks requiring a high degree of synaptic specificity (Rubin et al., 2000). The absence of CNRs in *Drosophila* also suggests that CNRs do not perform an essential role in synaptic specificity.

In addition to synaptogenesis, CNRs may have completely different functions during development. Recently it was discovered that several CNRs represent high-affinity receptors for reelin, an extracellular protein that is essential for normal neuronal migration during development (Senzaki et al., 1999). As discussed previously for the classical cadherins, the possibility emerges that CNRs may perform consecutive functions in development, first as guideposts for neuronal migration and then as synaptic cell adhesion molecules. This appears to be a general rule for many cell adhesion molecules implicated in brain development, as also exemplified by the role of fasciclin II in *Drosophila*, discussed subsequently.

Immunoglobulin Superfamily Adhesion Molecules

The immunoglobulin domain is an autonomously folding module of circa 100–115 residues composed of a β-sandwich consisting of two antiparallel β-sheets with a conserved topology and a characteristic intradomain disulfide bond (see Leahy, 1997, for review). Immunoglobulin domains are found primarily in extracellular proteins, but intracellular immunoglobulin domain proteins (e.g., titin) are also known. The primary function of the immunoglobulin domain is to bind ligands. Many immunoglobulin superfamily proteins that function in cell adhesion have been described; hundreds of such proteins are probably expressed in vertebrate brains. Immunoglobulin cell adhesion molecules are typically attached to the cell surface by a single transmembrane region or a glycolipid anchor, and they often contain several immunoglobulin domains. In addition, most immunoglobulin cell adhesion molecules include multiple copies of another cell adhesion domain, the fibronectin type III repeat, which is also an autonomously folding domain of some 100 residues that is composed of an antiparallel β-sheet sandwich (see Leahy, 1997, for review).

Most of the immunoglobulin cell adhesion molecules that have been studied so far mediate calcium-independent cell adhesion and signaling processes. They function as either homophilic or heterophilic cell

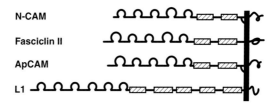

Figure 6.2. Schematic diagram of immunoglobulin cell adhesion molecules. The drawings depict the design of vertebrate N-CAM, *Drosophila* fasciclin II, *Aplysia* ApCAM, and vertebrate L1. Open circles indicate immunoglobulin domains and boxes indicate fibronectin type III repeat domains. The plasma membrane is indicated by a vertical line on the right; note that N-CAM and ApCAM occur in alternatively spliced variants containing either transmembrane regions and cytoplasmic tails or glycolipid anchors. Modified from Mayford et al. (1992).

adhesion molecules, and they often bind to multiple ligands. This characteristic is probably best exemplified by the L1 protein, which is essential for several neuronal migration and axonal guidance processes during brain development (see Brümmendorf et al., 1998, for review). L1 is composed of six immunoglobulin domains followed by five fibronectin type III repeats, a transmembrane region, and a cytoplasmic tail (Fig. 6.2). L1 mediates homophilic cell adhesion by binding to itself, but in addition it binds to the immunoglobulin cell adhesion molecules N-CAM, axonin 1, and contactin, and also to integrins and neurocan, which do not have immunoglobulin domains. Interestingly, the multiple ligands of L1 do not appear to bind to distinct immunoglobulin domains. Instead recent studies suggest that almost all of the six immunoglobulin domains and the five fibronectin type III repeats are required for the homo- and heterophilic interactions of L1 (De Angelis et al., 1999; Oleszewski et al., 1999). It seems likely, therefore, that in L1 and possibly other CAMs, the arrays of autonomously folding immunoglobulin domains and fibronectin III repeats assemble into an interacting set of modules. This creates composite binding sites that involve most of these domains and are directed toward several ligands. As described previously, this is in striking contrast to the classical cadherins, in which only the first of the five cadherin domains appears to be required for homophilic interactions.

Most of the immunoglobulin domain proteins in brain whose functions have been identified are involved in axon fasciculation and axonal pathfinding. Essential interactions between many different immunoglobulin domain proteins have been characterized in complementary genetic experiments and in vitro assays, resulting in the description of several exquisitely specific cell adhesion events during neuronal development (see Goodman, 1996, and Chisholm and Tessier-Lavigne, 1999, for reviews). In view of the large number of known functions of immunoglobulin cell adhesion molecules in axonal pathfinding, and the variety, conservation, and abundance of immunoglobulin cell adhesion molecules in brain, it is perhaps surprising that few of them have been linked to synaptic cell adhesion. One reason for this may be that these proteins have multiple sequential functions in brain development and may be involved in both axonal pathfinding and synaptic cell adhesion. Since axonal pathfinding comes first during development and is much easier to measure in vitro than synapse function, it is possible that synaptic roles for immunoglobulin cell adhesion molecules may be more widespread than currently envisioned.

Arguably the best-studied immunoglobulin cell adhesion molecule is N-CAM, one of the first such proteins discovered by Edelman and his colleagues in their initial search for tissue-specific cell adhesion molecules (Fig. 6.2; Thiery et al., 1977). N-CAM deservedly received a large amount of attention, not only because it efficiently mediates homophilic cell adhesion between neurons but also because it carries a unique carbohydrate modification consisting of polysialic acid. The polysialylation of N-CAM is developmentally regulated and decreases dramatically after birth. Polysialic acid inhibits cell adhesion by N-CAM, suggesting a regulatory role to prevent inappropriate adhesion during development (see Rutishauser and Landmesser, 1996, for review). Considerable evidence has suggested a role for N-CAM in cell migration, axonal outgrowth and fasciculation, and synaptogenesis. However, N-CAM is not enriched in synapses, and mice lacking N-CAM exhibit a phenotype that is surprisingly discrete in view of the abundance of the protein and its stature as the only polysialylated protein in brain (Tomasiewicz et al., 1993; Cremer et al., 1994). In these mice, major developmental changes in several brain areas are observed, especially in the olfactory system. However, these changes are not lethal, and most synapses appear to function normally. The only well-defined synaptic abnormality is a disorganization of the mossy fiber projection to the hippocampal CA3 region. This is accompanied by a loss of mossy fiber LTP, but other parameters of synaptic transmission in the mossy fiber terminals are normal (Cremer et al., 1998). Furthermore, N-methyl-D-aspartate (NMDA)–dependent LTP in other synapses also appears to be unaffected. It is unclear if the impairment in mossy fiber LTP reflects a requirement for N-CAM in the synaptic reorganization associated with LTP or a developmental change that inactivates the capacity of the synapse to undergo LTP.

Overall the existing data point to an important role of N-CAM in neuronal migration and axonal elongation but suggest that its function is not directly related to synapses. It is possible that N-CAM function at synapses is redundant with that of other immunoglobulin cell adhesion molecules expressed in brain, although the fact that N-CAM is the only cell adhesion molecule that is polysialylated argues against this possibility. In addition to N-CAM, several other immunoglobulin cell adhesion molecules have been related to synaptic function, but few have actually been localized there. The most extensive data were probably obtained for a transmembrane protein called p84/SHPS-1, which comprises three extracellular immunoglobulin domains and appears to be highly enriched in synapses (Comu et al., 1997). However, expression of p84/ SHPS-1 does not appear to be restricted to brain, so a specifically synaptic role for it is difficult to envision (Fujioka et al., 1996).

The evidence linking immunoglobulin superfamily cell adhesion molecules to synapses is better in invertebrates in the case of ApCAM in *Aplysia* and of fasciclin II in *Drosophila* (Bailey et al., 1992; Mayford et al., 1992; Schuster et al., 1996a,b). In *Aplysia* induction of long-term facilitation in the gill-withdrawal reflex results in the growth of new synaptic connections. A search for proteins whose levels are changed as a function of this learning paradigm identified an immunoglobulin cell adhesion molecule called ApCAM that is closely related to N-CAM (Fig. 6.2; Bailey et al., 1992; Mayford et al., 1992). ApCAM probably mediates homophilic interactions and functions in axon fasciculation and axonal pathfinding in addition to its role at the synapse. Long-term facilitation in *Aplysia* requires ApCAM downregulation, first by endocytosis and then by transcriptional mechanisms. Endocytosis of ApCAM is triggered via its phosphorylation by MAP kinase, suggesting a signal transduction pathway for inducing synaptic sprouting. Together these data suggest that ApCAM at the synapse serves to demarcate and restrict the area of synaptic contact, and that removal of ApCAM allows spontaneous growth of the synapses by the assembly of additional areas of synaptic contact.

This hypothesis has been supported and expanded by independent results obtained in *Drosophila* with fasciclin II, a homophilic cell adhesion molecule that, like ApCAM, also closely resembles N-CAM (see Goodman and Shatz, 1993, and Goodman, 1996, for reviews). Fasciclin II performs a major function in axon fasciculation during axonal pathfinding (Lin and Goodman, 1994; Lin et al., 1994). However, in addition to axonal pathfinding, fasciclin II is important in regulating the size of neuromuscular synapses in *Drosophila*. In these synapses, activity-dependent increases of synaptic effectiveness have been observed that correlate with synaptic sprouting. Fasciclin II is present on both pre- and postsynaptic membranes and binds with its intracellular C-terminal tail to a PDZ domain–containing multidomain protein called dlg (discs large) (Thomas et al., 1997; Zito et al., 1997). Dlg is a member of the membrane-associated guanylate kinase (MAGUK) protein family, members of which contain an SH3 domain and an enzymatically inactive guanylate kinase sequence in addition to PDZ domains. Fasciclin II and dlg are not required for initial synapse formation but are essential for the activity-dependent remodeling and sprouting of neuromuscular synapses (Schuster et al., 1996a,b). Fasciclin II has to be localized on both sides of the synapse, probably because it functionally interacts with itself. Interestingly, the precise levels of fasciclin II at the synapse seem to be critical for regulating synapse size: in null mutants synapses are eliminated, whereas in hypomorphs expressing 50% of wild-type levels, synapses sprout inappropriately. Thus fasciclin II levels appear to titrate synapse growth, being essential for maintenance of synapses and at the same time inhibiting inappropriate growth. Indeed fasciclin II is downregulated during activity-dependent synaptic sprouting, and this down-regulation is essential for sprouting (Schuster et al., 1996a,b).

These data support the hypothesis derived from the ApCAM work that strong homophilic cell adhesion functions to "freeze" the synapse at a certain size (Martin and Kandel, 1996). As a result, activity-dependent downregulation of the levels of synaptic fasciclin II and dlg activates synapse sprouting. However, it should be noted that according to this model fasciclin II or ApCAM does not serve as the synaptic "glue" that connects pre- and postsynaptic specializations, nor does it function as the synaptic recognition molecule that primes synapse assembly. Instead this model assigns to the immunoglobulin cell adhesion molecules another important role—that of a "fence" encircling the active zone and postsynaptic density so that it does not grow out of bounds. Although most of the data obtained in *Drosophila* and *Aplysia* are in agreement with this model, it does not explain why fasciclin II null mutants face synapse elimination—a question that points to a more complex role of the fasciclin II–dlg system in synapses than that of just a mechanical hurdle to synapse growth.

Interestingly, independent support for this general model has been obtained in vertebrate synapses, suggesting that the model may be widely applicable. This supporting evidence is of two types. First, more and more results demonstrate that synaptic facilitation in vertebrates is associated with increases in the size of the synaptic contact. In vertebrate central neurons, however, the synapses do not simply become larger. Instead they split into two in order to increase the extent of synaptic contact. This process leads to the formation of perforated synapses, which later divide into truly separate synapses. As mentioned previously, the most definitive evidence for an activity-dependent formation of perforated synapses was contained in recent studies that described how a short pharmacological treatment of cultured hippocampal neurons that enhances synaptic activity led to a major increase in the number of perforated synapses (Neuhoff et al., 1999), and how induction of LTP in hippocampal slices results initially in the formation of perforated and then of split synapses (Toni et al., 1999). Second, several studies have demonstrated that induction of LTP is associated with a secretion of proteases, especially tPA, and that inhibition of tPA can impair LTP (e.g., Qian et al., 1993; Tsirka et al., 1995; Gualandris et al., 1996; Baranes et al., 1998). This raises the possibility that proteases serve to remove cell adhesion barriers to synaptic expansion—an idea that is strongly supported by the fact that tPA inhibitors also block the activity-dependent formation of perforated synapses observed in cultured hippocampal neurons (Neuhoff et al., 1999).

However, although this is intriguing evidence, it is far from conclusive. The precise proteases and cell adhesion molecules involved have not been identified, nor has it been possible to define mutations in mice that have a phenotype similar to the fasciclin II and dlg mutations. N-CAM, fasciclin II, and ApCAM share extensive sequence homology, with all three cell adhesion molecules displaying the same organization of extracellular domains (Fig. 6.2). However, there are also major differences between these proteins that make it difficult to combine their properties into a single model. Fasciclin II, the protein whose functions are best defined, binds intracellularly to dlg in an essential interaction via a type I PDZ domain–binding consensus sequence at the C terminus (Thomas et al., 1997; Zito et al., 1997). However, neither ApCAM nor N-CAM displays this consensus sequence, or is known to interact with a MAGUK. Although many neuronal PDZ domain proteins with important functions at the synapse are known, none binds to N-CAM. Furthermore, the N-CAM mutation in mice does not have a prominent synaptic phenotype similar to that observed for dlg.

Neurexins as Candidate Synaptic Cell Adhesion Molecules

Neurexins are neuron-specific proteins that were discovered as cell surface receptors for α-latrotoxin, a presynaptic neurotoxin that causes massive synaptic vesicle exocytosis (Ushkaryov et al., 1992). The structures and properties of neurexins, however, suggest that they have a wider synaptic function in neuronal cell adhesion and cell recognition (see Missler et al., 1998b, for review). This suggestion is based on the large numbers of neurexin isoforms (>1000) generated by alternative splicing, their extracellular interactions (which include the binding of a subset of neurexins to a family of neuronal cell adhesion molecules called neuroligins), and the formation of intracellular PDZ domain protein complexes that assemble on the neurexins.

Vertebrates contain at least three neurexin genes (neurexins 1, 2, and 3). Each gene has two independent promoters that direct transcription of the larger α-neurexins and the shorter β-neurexins (Ushkaryov et al., 1992, 1993, 1994). Structurally α- and β-neurexins are transmembrane proteins that resemble cell surface receptors and cell adhesion molecules, with a large extracellular region and a short intracellular sequence (Fig. 6.3). α- and β-Neurexins differ from each other in their N-terminal extracellular domains but contain identical transmembrane regions and intracellular C-terminal sequences.

In α-neurexins the N-terminal extracellular sequences are composed of three overall repeats, each of which consists of a central EGF-like domain flanked by distantly related LNS domains (LNS domains are defined later in this chapter). Following the three overall LNS-EGF-LNS domain repeats, α-neurexins contain a serine-threonine–rich sequence that is O-glycosylated, a single transmembrane region, and a short cytoplasmic tail. In β-neurexins, the N terminus is composed of a short β-specific sequence, after which the β-neurexin mRNAs splice into the α-neurexin sequence at the beginning of the last LNS domain and share with the α-neurexins all remaining C-terminal sequences (Fig. 6.3). Both

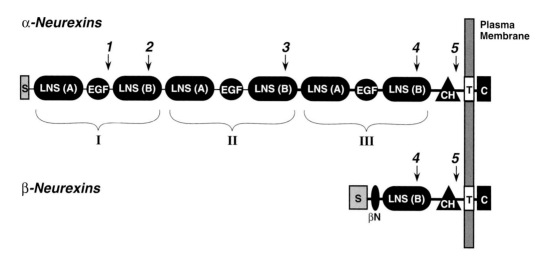

Figure 6.3. Domain structures of α- and β-neurexins. Positions of LNS and EGF-like domains are depicted in the context of the three overall LNS-EGF-LNS domain repeats in α-neurexins. Arrows indicate sites of alternative splicing, numbered 1–5. Domains are identified as follows: C, cytoplasmic tail; CH, carbohydrate attachment domain; EGF, EGF-like domains; LNS, LNS domains; βN, β-neurexin-specific sequence; S, signal peptide; T, transmembrane region. Modified from Missler et al. (1998b).

α- and β-neurexins contain cleaved N-terminal signal peptides, with a classical signal sequence for α-neurexins and an unusual nonhydrophobic sequence for β-neurexins (Ushkaryov et al., 1994). Since β-neurexins lack the N-terminal five LNS domains and the EGF-like sequences of α-neurexins, β-neurexins can be viewed as N-terminally truncated α-neurexins that contain a short β-specific N-terminal sequence.

Neurexin transcripts are subject to extensive alternative splicing, resulting in potentially thousands of isoforms (Ullrich et al., 1995). To be physiologically meaningful, such alternative splicing should not be random (i.e., different neurons should express distinct combinations of splice variants), should be evolutionarily conserved, and should be similar for different neurexin genes. All three conditions appear to be met. The transcripts of all three neurexin genes are alternatively spliced at evolutionarily conserved positions (numbered arrows in Fig. 6.3). The alternatively spliced inserts are also evolutionarily conserved, with more than 10 distinct variants for some splice sites. In situ hybridizations have demonstrated that neurexins are expressed in overlapping but distinct patterns, with a pronounced regional regulation of at least some of the alternative splice forms (Ullrich et al., 1995). This is particularly striking for splice site 4, which is expressed in two variants with a prominent regional regulation of expression (Ichtchenko et al., 1995). The different sites of alternative splicing are used independently of each other. Overall these findings indicate that neurexins are expressed in hundreds, and possibly thousands, of isoforms that are present in different brain regions in a characteristic combination of multiple variants, which could provide a combinatorial code.

Neurexins are evolutionarily conserved not only in mammals but also in *Caenorhabditis elegans* and presumably other invertebrates. However, classical neurexins should not be confused with neurexinlike proteins, such as *Drosophila* neurexin IV, whose name misleadingly suggests that it is a member of the neurexin family. This *Drosophila* protein is required for septate junctions and exhibits a distant resemblance to neurexins (Baumgartner et al., 1996). The similarity between neurexin IV and genuine neurexins is restricted to the fact that all these proteins are cell surface receptors containing, among others, LNS- and EGF-like domains and a single transmembrane region (see Missler and Südhof, 1998a, for review). Unlike classical neurexins, neurexin IV is also expressed in non-neuronal cells, is not alternatively spliced or polymorphic, has a different domain structure, and does not bind to the same ligands as neurexins. The closest vertebrate homologue of neurexin IV is CASPR, a fascinating protein involved in glia-axon interactions (Peles et al., 1997a,b).

The three overall LNS-EGF-LNS domain repeats in the extracellular region are the defining feature of α-neurexins, whereas the single N-terminal LNS domain is characteristic of β-neurexins (Fig. 6.3). The LNS domain was named for laminin A, neurexins, and sex hormone–binding globulin, because it was first identified as a repeat in the G-domain of laminin; was shown to constitute an independently folding, functional domain in neurexins; and was independently identified as a sequence motif in sex hormone–binding globulin (see Joseph and Baker, 1992, and Missler and Südhof, 1998a, for reviews). LNS domains are modules of some 160–200 residues found in many other extracellular sequences, including those of agrin, slit, protein S, and perlecan. Thus LNS domains are present equally in serum transport proteins, extracellular matrix components, and neuronal signaling molecules (Ushkaryov et al., 1992). It is thought that in most of these proteins LNS domains constitute a ligand-binding domain, as demonstrated for example for the binding of dystroglycan to the LNS domains of laminin A and of neuroligin to the LNS domain of β-neurexins. Interestingly, like the neurexin LNS domains, the agrin LNS domains are subject to extensive alternative splicing that regulates acetylcholine receptor clustering mediated by agrin (Ferns et al., 1992; Ichtchenko et al., 1995; Ullrich et al., 1995; Burgess et al., 1999).

The crystal structures of two different LNS domains have been determined: those of the sixth LNS domain of neurexin 1α and of the fifth LNS domain of laminin α2 (Hohenester et al., 1999; Rudenko et al., 1999). These structures have shown that LNS domains are composed of a β-sandwich with 14 β-strands that are folded into a stable convex-concave structure (Fig. 6.4). Mapping of the three sites of alternative splicing in neurexin LNS domains onto the three-dimensional structure of the domain reveals that all sites are localized to the same surface of the domain, despite the wide separation of the relative position of these sites in the primary sequences. Furthermore, the sites of alterna-

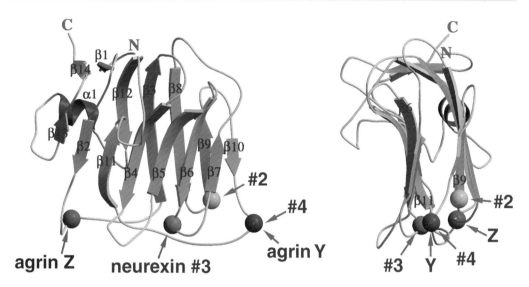

Figure 6.4. Three-dimensional structure of the sixth LNS domain of neurexin Iα, corresponding to the single LNS domain in neurexin Iβ (Fig. 6.3). The domain consists of a β-sandwich composed of 14 β-strands, with an overall convex-concave shape. The locations of alternatively spliced sequences of neurexin LNS domains in the three-dimensional structure are indicated. Splice sites #2, #3, and #4 are located in the second, fourth, and sixth LNS domains, respectively (see Fig. 6.3), in different positions of the consensus sequence for LNS domains, but are found on the same surface of the LNS domain structure as shown. In addition, the two sites of alternative splicing in LNS domains in agrin that regulate its activity are also indicated (designated agrin Y and agrin Z) and are also present on the same surface of the domain. Modified from Rudenko et al. (1999).

tive splicing in agrin also localize to the same area of the LNS domain as the alternatively spliced sequences in neurexins (Rudenko et al., 1999), suggesting that there is a general mechanism by which LNS domains are regulated by alternative splicing.

Neurexins were originally purified as receptors for α-latrotoxin, a component of black widow spider venom that has been a valuable tool in studying neurotransmitter release for more than 30 years (see Rosenthal and Meldolesi, 1989, for review). Despite considerable study, there is still uncertainty as to how α-latrotoxin triggers neurotransmitter release. Classical studies by Bruno Ceccarelli and colleagues demonstrated that α-latrotoxin binds to specific receptors that are concentrated on the presynaptic membrane. The toxin then triggers massive exocytosis in a reaction that requires magnesium but not calcium (Ceccarelli and Hurlbut, 1980; Valtorta et al., 1984). Since release induced by α-latrotoxin is inhibited by tetanus toxin, the toxin acts through the normal SNARE-dependent fusion machinery (Dreyer et al., 1987; Capogna et al., 1996). The initial rationale for the isolation of α-latrotoxin-binding proteins was to learn how this enigmatic toxin triggers neurotransmitter release even in the absence of Ca^{2+}. The expectation was that high-affinity

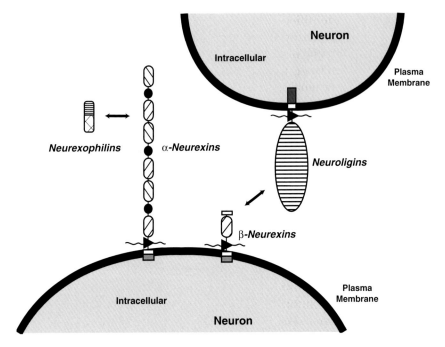

Figure 6.5. Extracellular ligands for neurexins: neurexophilins for α-neurexins and neuroligins for β-neurexins. The second LNS domain of α-neurexins binds to neurexophilins, which form a family of secreted proteins with a structure similar to neuropeptides. By contrast β-neurexins bind to neuroligins in an interaction that causes cell adhesion and depends on the alternative splicing of β-neurexins.

binding of α-latrotoxin to a single synaptic receptor would activate the receptor, thereby triggering release. Thus it was thought that the identification of the receptor would provide insight into the mechanism of synaptic vesicle exocytosis. Recent results on the function of neurexins as α-latrotoxin receptors and the mechanism of α-latrotoxin action have led to a revision of this view: α-latrotoxin is now thought to use the cell surface receptor only to localize it to synapses, rather than as an effector (Ichtchenko et al., 1998; Sugita et al., 1998, 1999). Furthermore, not only one but two classes of receptors were identified (neurexins and CLs) that bind α-latrotoxin either Ca^{2+}-dependently (neurexins) or -independently (CLs).

The neurexins clearly evolved in order to provide receptors for some endogenous ligands. Gene families encoding two such ligands have been defined: neurexophilins, which bind only to α-neurexins, and neuroligins, which bind only to β-neurexins (Fig. 6.5). The neuroligins will be discussed subsequently since they constitute synaptic cell adhesion molecules in their own right. Neurexophilins, on the other hand, are not cell surface proteins but display a structure that is very similar to the structures of neuropeptides. Neurexophilins were discovered when neurexins were initially purified on immobilized α-latrotoxin in a tight complex with a 29-kDa protein. Upon cloning this protein was identified as a novel neuropeptide-like protein that was then named neurexophilin (Petrenko et al., 1993, 1996). Neurexophilins make up a family of at least four closely related genes that are expressed primarily in brain (Missler and Südhof, 1998b). All neurexophilins contain an N-terminal signal peptide, a proteolytically processed N-terminal region that is not conserved among different isoforms, and two C-terminal highly conserved domains. In situ hybridizations have shown that neurexophilins are synthesized in subsets of neurons; in the case of neurexophilin 1, these correspond to inhibitory interneurons (Petrenko et al., 1996). However, little else is known about the localization of neurexophilins in brain. For example, it is not known at present if they are secreted constitutively or only in a regulated manner. Expression studies have shown that neurexophilins are proteolytically processed only in neuronal cells, and that all neurexophilins (with the possible exception of neurexophilin 4) bind to α-neurexins in a tight and specific interaction (Missler et al., 1998a). Neurexophilins bind only to the second of the six LNS domains of α-neurexins, providing further evidence that these domains are autonomously functioning binding modules (Missler et al., 1998a). Since α-neurexins are present in all neurons, neurexophilins could be specialized ligands for α-neurexins that may initiate an as yet uncharacterized signaling cascade, or they could function as regulators of α-neurexin binding to another ligand, for example, an as yet unidentified CAM.

Based on their structure as cell surface proteins, their extensive alternative splicing (which results in hundreds of isoforms), their differential distributions, and their interactions with multiple synaptic proteins (e.g., neuroligins, CASK, PSD-95), neurexins are prime candidates for a role as synaptic cell adhesion molecules. Furthermore, deletion of α-neurexins in mice is lethal (Missler and Südhof, unpublished observations). Knocking out two of the three α-neurexin genes results in a significant decrease in survival, and triple knockout animals die immediately after birth. This finding suggests that the α-neurexins serve overlapping, partially redundant functions that are essential for postnatal survival, which would be consistent with a role in synaptic function. At this point, the best evidence for a function of neurexins at the synapse is derived from their role as α-latrotoxin receptors; in this they are an essential component of the cellular machinery that enables the toxin to trigger synaptic exocytosis (Geppert et al., 1998; Sugita et al., 1999). This hypothesis is supported by subcellular fractionation experiments that demonstrate that neurexins are highly enriched in purified synaptic plasma membranes that contain synaptic junctions (Butz et al., 1998), and by light microscopy immunofluorescence studies that reveal their enrichment at synapses (Ushkaryov et al., 1992). The interactions of neurexins with other synaptic proteins, neuroligins and CASK, provide additional support for this hypothesis. However, it has not yet been established unequivocally that neurexins have a direct role at synapses, as opposed to the vicinity of the synapse, nor is it clear what their natural role is.

Neuroligins: Connectors of Cell Adhesion to Neurotransmitter Receptors and Signaling Molecules

Neuroligins were discovered as endogenous ligands for β-neurexins (Ichtchenko et al., 1995, 1996). Like neurexins, neuroligins are neuron-specific cell surface proteins that are encoded by three homologous genes. Neuroligins also have relatively large extracellular sequences with an O-glycosylation domain, a single transmembrane region, and a comparatively short C-terminal cytoplasmic tail (Fig. 6.5). The domain structure of the extracellular sequences of neuroligins, however, differs from that of neurexins: they are almost entirely composed of a single domain that is highly homologous to acetylcholinesterase and other esterases, but they lack the active-site serine of these enzymes (Ichtchenko et al., 1995, 1996). This domain probably adopts a fold that is very similar to the acetylcholinesterase fold (Tsigelny et al., 2000). All three β-neurexins bind to all three neuroligins. Interestingly, the binding of β-neurexins to neuroligins is tightly controlled by alternative splicing. β-Neurexins bind only to neuroligin if they lack an insert in splice site 4, which is located in the single LNS domain of β-neurexins; neurexins that have an insert at this position are inactive (Ichtchenko et al., 1995, 1996). The binding of neuroligins to β-neurexins is Ca^{2+}-dependent, like the binding of α-latrotoxin to neurexins. The interaction of β-neurexins with neuroligins on separate cells results in cell adhesion and is the basis for identifying these proteins as cell-adhesion molecules (Nguyen and Südhof, 1997). The use of an esterase domain in cell adhesion molecules has also been observed in *Drosophila*, in which two cell-adhesion molecules called neurotactin and gliotactin have been described (Hortsch et al., 1990; Auld et al., 1995). Although the function of neurotactin is unknown, gliotactin is known to be essential for the formation of normal intercellular junctions in the blood-nerve barrier in *Drosophila* (Auld et al., 1995).

The developmental time course of neuroligin expression tightly parallels synaptogenesis, with a peak, in mice, during the second postnatal week. Precise ultrastructural localization of neuroligin 1 using monoclonal antibodies has revealed that neuroligin 1 is highly enriched at synaptic junctions, with a distribution of immunogold particles suggestive of a postsynaptic localization (Song et al., 1999). Double immunofluorescence experiments indicate that neuroligin 1 may be excluded from inhibitory GABAergic synapses, which suggests that at least neuroligin 1 may be specifically limited to a subset of synapses. Knockouts of neuroligins in mice have revealed that the neuroligins are essential for survival (Brose and Südhof, unpublished observations). Deletion of all three neuroligins is lethal immediately after birth, in a manner similar to that observed with other synaptic knockouts (Geppert et al., 1994; Verhage et al., 2000). Together these data demonstrate that neuroligins are synaptic cell adhesion molecules, that they are probably localized to a functional subset of synapses, and that the β-neurexin–neuroligin junction has a role in synaptic function. By virtue of the preferential localization of neuroligin 1 and the regulation of the binding of neuroligins to β-neurexins by alternative splicing, it seems likely that the neuroligin–β-neurexin interaction mediates a recognition event between neurons at the synapse, and possibly in establishing what type of synapse is being formed. It is also clear, however, that this interaction is not critical for holding synapses together, because the neuroligin–β-neurexin complex is Ca^{2+}-dependent, whereas the adhesion between pre- and postsynaptic elements at synaptic junctions is known to be Ca^{2+}-independent.

Coating of the Neurexin-Neuroligin Junction

The interaction of neuroligins with β-neurexins forms a stable intercellular junction between neurons. In other intercellular junctions—most notably in adherens junctions established by homophilic binding of cadherins or in tight junctions assembled from occludins and claudins—the cytoplasmic tails of the cell adhesion molecules bind to intracellular coat proteins. These in turn couple the cell adhesion event mediated by the cell adhesion molecules to the intracellular cytoskeleton and to signal transduction proteins (Yap et al., 1997; Itoh et al., 1999). In the case of tight junctions, the intracellular coat is composed of PDZ domain proteins called MAGUKs in which the PDZ domains are instrumental in recruiting various proteins to the junctions. For cadherins, by contrast, a single adaptor protein complex composed of α- and β-catenins couples the cell adhesion molecules to intracellular targets. In both cases the intracellular coupling targets the cytoskeleton in addition to signal transduction proteins, and junction formation requires the association of intracellular binding proteins. Based on results described in detail later in this chapter, a model for the intercellular junction created by the interaction of β-neurexins with neuroligins was proposed (Fig. 6.6). This model suggests that, like occludin and claudins in tight junctions, neurexins and neuroligins are coated intracellularly by PDZ domain proteins. However, unlike tight junctions, the neurexin-neuroligin junction is asymmetric and exhibits two specialized sides, as expected for a synapse. It is noteworthy here that the C termini of the neurexins and neuroligin protein families are highly conserved between isoforms. Furthermore, the neurexin C termini are related to the C termini of erythrocyte membrane protein glycophorin C as well as a number of unidentified sequences in the data bases, whereas the neuroligin C termini resemble those of NMDA receptors clustered at glutamatergic synapses.

The first insight into which proteins might coat the β-neurexin–neuroligin junction was obtained when yeast two-hybrid screens with the cytoplasmic tail of neurexins isolated a novel MAGUK protein called CASK (Hata et al., 1996). CASK is composed of a large N-terminal domain

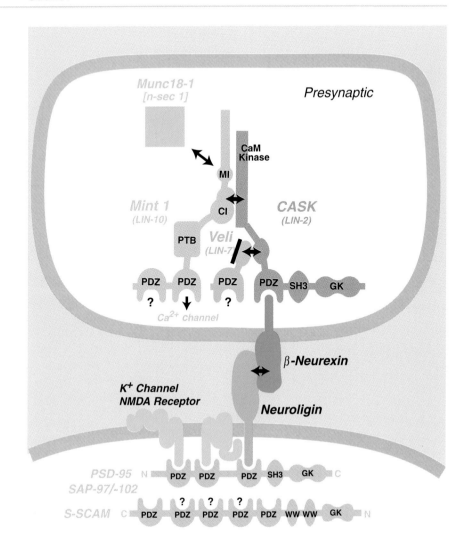

Figure 6.6. Model of the intercellular junctions formed by the β-neurexin–neuroligin interaction and coated by PSD-95 and S-SCAM on the postsynaptic side and by CASK, mint 1, and velis on the presynaptic side. The localization of this junction to synapses is a model based on the presence of CASK, neuroligin 1, and PSD-95 at this structure. Modified from Butz et al. (1998).

that is highly homologous to the catalytic domain of Ca^{2+}/calmodulin-dependent protein kinase II (CaMKII) but catalytically inactive; central PDZ and SH3 domains; and a C-terminal guanylate kinase sequence (Fig. 6.6). Neurexin binding to CASK is mediated by the very C terminus of neurexins, suggesting that a PDZ domain interaction is involved. In the brain CASK is primarily expressed in neurons, but it is also present at low levels in nonneuronal cells that do not express neurexins (Hata et al., 1996). The CASK PDZ domain probably binds to several other cell surface proteins in addition to neurexins, including the syndecans, a family of proteoglycans that are also expressed outside the brain (Cohen et al., 1998; Hsueh et al., 1998), to NG2, and to glycophorin C (Biederer and Südhof, unpublished observations). Immunocytochemistry has shown that CASK is a plasma membrane protein that is enriched in synapses (Hsueh et al., 1998). A gene homologous to CASK was discovered as the *lin-2* gene in *C. elegans* (Hoskins et al., 1996). *Lin-2* is a member of a group of three genes called *lin-2, lin-7,* and *lin-10,* mutations which result in a mislocation of the EGF receptor in developing epithelial cells that give rise to the vulva (Simske et al., 1996). As a consequence all of these mutations lead to a vulvaless phenotype.

When binding partners for CASK were investigated by immunoprecipitation, a protein called mint 1 was unexpectedly found to form a stoichiometric, tight complex with CASK in brain (Butz et al., 1998). Mint 1 is a member of a protein family composed of two neuron-specific proteins, mint 1 and mint 2, and one ubiquitously expressed protein, mint 3 (Okamoto and Südhof, 1998). Prior to the finding that mint 1 is a binding partner for CASK, mints were identified in a number of diverse contexts. Partial sequences of murine mint 2 and human mint 1 were originally characterized by positional cloning as candidate genes for Friedreich's ataxia, and they were named X11 because they were thought to represent orthologues (Duclos et al., 1993; Duclos and Koenig, 1995). Although the involvement of mints in Friedreich's ataxia was later ruled out, mints/X11s were subsequently shown to bind in vitro to APP, the Alzheimer's amyloid precursor protein (Zhang et al., 1997). They were rediscovered independently and first fully cloned in a screen for munc18-1–interacting proteins (Okamoto and Südhof, 1997). Since human and murine X11 are products of distinct genes and not orthologues, they were renamed mints 1 and 2 in order to prevent confusion. Since munc18-1 is essential for exocytosis (Verhage et al., 2000), the binding of mints to munc18-1 indicates a possible function in synaptic vesicle exocytosis.

Mints have an intriguing domain structure composed of variable N-terminal and highly conserved C-terminal domains (Fig. 6.6). The N-terminal region contains a short sequence that binds munc18-1 and is present in mints 1 and 2 but not mint 3 (Okamoto and Südhof, 1997, 1998). Since munc18-1 (also known as nsec1/rbsec1; see Südhof and Scheller, this volume) is essential for synaptic vesicle exocytosis, the binding of munc18-1 to mints implicates the latter in synaptic vesicle exocytosis. Although there are no recognizable domains in the N-terminal mint sequences, they are highly conserved among vertebrates, suggesting that each mint may have isoform-specific activities. The C-terminal domains of mints, on the other hand, are almost identical among the three mints and consist of one PTB and two PDZ domains. Studies on the interaction of mints with CASK showed that only mint 1 binds to CASK via a sequence that is located between the munc18-interacting domain and the PTB domain and is not present in mints 2 and 3. At the same time the *C. elegans* homologue of mint 1 was identified as the product of the *lin-10* gene (Kaech et al., 1998). Furthermore, vertebrate homologues of *lin-7*—the third member of the gene triad, which in *C. elegans* gives rise to the same vulvaless phenotype—were identified as velis (for vertebrate *lin-7* homologues; Butz et al., 1998; Jo et al., 1999). Interestingly, in brain velis are as tightly bound to CASK as mint 1, identifying a tripartite complex that is composed of CASK, mint 1, and velis (Butz et al., 1998). The central molecule in this complex is CASK, which binds to both mint 1 (via the catalytically inactive CaMK domain of CASK) and velis (via a novel domain between the CaMK domain and the PDZ domain). The entire trimeric complex binds to neurexins via the PDZ domain of CASK, suggesting that on the neurexin side the β-neurexin–neuroligin junction is coated by a PDZ domain protein complex (Fig. 6.6). Since the CASK PDZ domain also binds to other cell surface proteins, it seems likely that the same complex is coupled to other cell surface molecules.

Although the current data about mint 1, CASK, and velis are intriguing, they raise the question of what the tripartite complex actually does. In *C. elegans* the complex is required for the polarized localization of an EGF receptor homologue to a particular part of the cell, indicating that the complex could function either in protein sorting or in establishing plasma membrane specializations (Simske et al., 1996). In vertebrates biochemical data have suggested two possible functions for the tripartite complex. First, both mint 1 and CASK bind to the C-terminal cytoplasmic tails of synaptic Ca^{2+} channels in vitro (Maximov et al., 1999). This interaction is mediated by the SH3 domain of CASK and one of the two PDZ domains of mint 1. As a result of this interaction, the tripartite complex could serve to recruit Ca^{2+} channels to sites of synaptic cell adhesion, mediated by the binding of β-neurexins to neuroligins. Second, affinity chromatography on recombinant cytoplasmic tails of neurexins has shown that CASK and mint 1 bind in a quantitative reaction, dependent on the C terminus of neurexins. Coomassie blue staining, however, demonstrated that spectrin and actin were assembled on the immobilized neurexins (Biederer and Südhof, unpublished observations). Interestingly, assembly of actin and spectrin was inhibited by ATP, suggesting that it is physiologically regulated. These data link the neurexins and the tripartite complex to the actin cytoskeleton. What might be the mechanism that links this complex to actin microfilaments? The C-terminal domains of CASK are highly homologous to p55, an erythrocyte MAGUK that binds to the cytoplasmic tail of the membrane protein glycophorin C. The p55/glycophorin C complex, in turn, is coupled to the actin cytoskeleton via protein 4.1 (see Chishti, 1998, for review). Since the C termini of neurexins and glycophorin C are related and both bind CASK, it seems likely that the neurexin/CASK complex is also linked to the actin cytoskeleton via protein 4.1. Such an interaction could anchor the cytoskeleton to the plasma membrane at the site of an intercellular junction created by extracellular interactions of neurexins.

A PDZ domain protein coat was also discovered for neuroligins. Yeast two-hybrid screens with the C-terminal tails of neuroligins have revealed that they interact cytoplasmically with the postsynaptic density protein PSD-95 and other related proteins (Irie et al., 1997; Hirao et al., 1998). Like CASK and dlg, PSD-95 is a MAGUK (Cho et al., 1992; Kistner et al., 1993). PSD-95 also binds to NMDA receptors, K^+ channels (Kim et al., 1995; Kornau et al., 1995), and at least 10 other proteins, including citron, a rho target, and ras synGAP p135 (Chen et al., 1998; Kim et al., 1998; Zhang et al., 1999). The co-localization of neuroligins and PSD-95 and their co-purification on immobilized β-neurexins suggest that their interaction is physiological. These results indicate that neuroligins recruit PSD-95 to locations of intercellular junctions between neurons, with PSD-95 and other PDZ domain proteins in turn recruiting NMDA receptors, p135 synGAP, citron, and others. In striking similarity to CASK, PSD-95 has also been linked to the cytoskeleton, reinforcing the symmetry between the neurexin and neuroligin sides of the β-neurexin–neuroligin junction (Naisbitt et al., 1999).

A model for the junction created by cell adhesion between β-neurexins and neuroligins can be postulated based on these combined results, as depicted in Fig. 6.6. This model envisions the junction as coated with distinct complexes of PDZ domain proteins on both sides, resulting in an intrinsic polarization of the junction. As a result of protein-protein interactions, multiple PDZ domains are clustered on both sides of the junction together with other protein interaction domains (SH3 domains, guanylate kinase domains, CaMKII domains). Interestingly, the model suggests that the PDZ domain protein complexes on both sides have similar functions, namely to couple the β-neurexin–neuroligin junction to the actin cytoskeleton and to recruit channels and receptors to the junction. This model should serve as a working hypothesis to guide future experiments. However, its heuristic value is that it ties together various properties of synaptic function that hitherto have been difficult to connect biochemically—the polarized nature of synapses; the co-localization of channels and receptors, and cell adhesion; the coupling to the cytoskeleton; and, last but not least, the specific regulation of the β-neurexin–neuroligin junction by alternative splicing of β-neurexins.

Other Candidate Cell Adhesion Molecules

CIRL/Latrophilins

Although the initial search for α-latrotoxin receptors identified only neurexins, subsequent studies revealed that neurons express a second, distinct cell surface protein that functions as a toxin receptor, a protein originally called CIRL or latrophilin (Krasnoperov et al., 1997; Lelianova et al., 1997) and now referred to as CL1. CL1 is also a member of a protein family encoded by at least three genes (Sugita et al., 1998). All CLs are G-protein-linked receptors with seven transmembrane regions and unusually large extra- and intracellular sequences with multiple domains. As α-latrotoxin receptors, CLs bind the toxin independent of Ca^{2+}, whereas the neurexins bind only in the presence of Ca^{2+} (Sugita et al., 1999). The large extracellular sequences of CLs contain multiple domains that in other proteins are involved in cell adhesion (e.g., a lectin domain). Since CLs function as α-latrotoxin receptors, it seems likely that they are localized close to synaptic sites, as has already been argued for neurexins. However, unlike neurexins, CLs are also expressed outside the brain—in fact, CL2 is primarily expressed in nonneuronal tissues (Sugita et al., 1998). This suggests that CLs have a more general signaling role. As in neurexins, the C-terminal intracellular sequences of CLs are not required for α-latrotoxin action; this suggests that these proteins serve to recruit α-latrotoxin to neurons for a subsequent action of the toxin that is receptor-independent (Sugita et al., 1998, 1999). Overall very little is currently known about CLs except that their structure is compatible with a cell adhesion function, and their role as α-latrotoxin receptors (which they share with neurexins) implies their localization at synapses.

Syndecans

Syndecans are a family of four related cell surface proteoglycans that are highly modified by heparan sulfates. They are type I transmembrane proteins whose extracellular sequences exhibit little homology or evolutionary conservation besides the fact that they are glycosylated. By contrast the transmembrane regions and cytoplasmic tails of all syndecans are closely related (see Carey, 1997, for review). This structure suggests that their major function is to display heparan sulfate chains on the cell surface and to couple them to an intracellular function mediated by the transmembrane and cytoplasmic sequences.

Syndecans are expressed in all cells in varying combinations. In brain syndecan 1 is expressed in glial cells, and syndecans 2 and 3 are expressed in neurons. Syndecan 2 appears to be enriched in synapses and syndecan 3 on axons (Hsueh and Sheng, 1999); syndecan 2, however, is also expressed in nonneuronal cells and so has no specific synaptic function (see Carey, 1997, for review). A large number of potential extracellular ligands have been reported for the syndecans, including growth factors and extracellular matrix proteins, suggesting that syndecans are co-receptors for these proteins. Intracellularly syndecans bind to syntenin (Grootjahns et al., 1997) and to CASK (Cohen et al., 1998; Hsueh et al., 1998) and may be coupled to the cytoskeleton (Burridge et al., 1988) in an interaction similar to that described for the neurexins. However, the intracellular interactions of syndecans must be differentially regulated, despite their similar binding activities, as different syndecans exhibit different localizations in the same cell even though their intracellular sequences are very similar (Carey, 1997). Although it is unlikely that syndecan 2 is a major component of the synaptic cleft (because of its low abundance in the brain), it could contribute to the carbohydrates that have been observed in the synaptic cleft by cytochemistry (Pfenninger, 1973). However, since no extracellular matrix components or growth factors are known to be present in the synaptic cleft, it is difficult to envision a function for syndecan 2 here, and it is possible that syndecans have as yet undiscovered roles.

Integrins

Integrins constitute a large family of dimeric cell surface proteins that perform critical functions in cell-cell communication, focal adhesions, and cell-matrix interactions. Integrins mediate heterophilic cell adhesion, making them structurally well suited for a role in synaptic cell adhesion, but little is known about their role at synapses. The most persuasive evidence for such a role is derived from *Drosophila* mutants; a mutation called *volado*, which exhibits impaired learning behavior, has been identified in a gene for a novel integrin (Grotewiel et al., 1998). Integrin mutants in *Drosophila* display general changes in synaptic properties at neuromuscular junctions (Beumer et al., 1999), suggesting that integrins may actually function at the synapse. However, the precise localization and binding partners of *volado* gene product or other integrins at synapses are unknown, so it is difficult to know if these effects are direct.

Conclusion: Too Many Unanswered Questions

It is clear that we are only beginning to identify the molecular components that keep synapses together and are far from understanding their mechanism of action. In view of the abundance of synapses and the large number of potential synaptic cell adhesion molecules, it is surprising that so few such molecules have been definitively linked to synapses. One reason for this slow progress could be that many of the currently known molecules may have functions as synaptic cell adhesion proteins that have not yet been recognized. Paradigmatic of this possibility is fasciclin II in *Drosophila*, which performs an essential role in both axon fasciculation and synapse structure formation, although its role in both axonal fasciculation dominates the phenotype of constitutive mutants. The function of fasciclin II at synapses was only discovered with conditional mutants, and it could not have been understood without such mutants (Schuster et al., 1996a,b). Furthermore, one of the most elegant and parsimonious mechanisms by which neurotransmitter identities at a synapse could be established would be if neurotransmitter receptors also served as cell recognition molecules. Is it possible that the extracellular sequences of the NMDA or GABA receptors also function to interact with a presynaptic cell adhesion molecule? If the interacting presynaptic molecule was specific for glutamate or GABA, respectively, such a mechanism would allow the selective clustering of the correct receptor opposite the presynaptic release sites for that particular transmitter. Such possibilities have not yet been tested.

The view of synapses that emerges from the currently available data is that they share some properties with tight junctions. In both cases the junction is probably formed by cell adhesion molecules that interact intracellularly with PDZ domain proteins; these in turn are coupled to the actin cytoskeleton. Another parallel is that the establishment of both types of junctions appears to involve initial contacts mediated by cadherins at adherens junctions—contacts that are later dispensable. For both tight junctions and synapses, the cadherin junctions move to the periphery of the final junction and may even disappear (Uchida et al., 1996; Yap et al., 1997). A further similarity is that both tight junctions and synapses form a focal target not only for the intracellular cytoskeleton but also for membrane traffic (Grindstaff et al., 1998). However, it is important to remember that there are also major differences between the two types of junctions; of these, the most important is that synapses are structurally and functionally asymmetric, whereas tight junction are symmetric.

Among the many questions that remain to be addressed are the following:

1. How do neurons initially recognize each other prior to establishing synapses? Clearly neurotransmitter signaling plays only a limited role, if any, in this process because synapses appear to form normally even in the absence of triggered neurotransmitter release, as recently shown in mice with induced mutations in munc18-1, which have no detectable neurotransmitter secretion (Verhage et al., 2000). During synaptic assembly the presynaptic neuron must not only specifically recognize the appropriate postsynaptic neuron, it must also recognize distinct parts of the postsynaptic neuron (e.g., its distal dendrites, dendritic shaft, or cell body) in order to form synapses at the correct site. That is, neurons do not simply recognize each other, they recognize each other spatially.

2. What is the exact protein composition of the synaptic cleft? The morphologically defined material must correspond to specific molecules whose nature is still unknown.

3. What mediates the precise correlation of pre- and postsynaptic sides? These are always exactly positioned opposite each other, and they always change size in unison. The most plausible mechanism is cell adhesion, in which the cell adhesion molecules nucleate the intracellular assembly of pre- and postsynaptic specializations. Although it is tempting to speculate that neurexins and neuroligins contribute to

this function, based on their intracellular assembly reactions, at present there is no direct evidence for this speculation.

4. What are the signals that trigger dynamic changes in the correlated assembly of the pre- and postsynaptic specializations? Such signals are important not only for the initial synapse assembly but also for the large changes observed during synaptic plasticity. An especially fascinating phenomenon is the generation of new synapses by division of existing synapses during LTP (Toni et al., 1999); these changes must occur from presynthesized components since they occur before the protein synthesis–dependent phase of LTP.

These questions are embedded in larger overall issues related to brain function. For example, why do some neurons readily form dendritic spines whereas others do not? How is the neurotransmitter type of a presynaptic neuron determined? What determines the shape of a neuron—a feature that is so characteristic for each neuronal type? In the end, all of these properties cannot simply be genetically encoded but must result from intercellular interactions that are presumably mediated by cell adhesion molecules. It follows from this conclusion that cell adhesion molecules are required not only for the formation, maintenance, and dynamics of synapses but also for overall neuronal development.

References

Akert, K., Pfenninger, K. H., Sandri, C., Moor, H. (1972). Freeze etching and cytochemistry of vesicles and membrane complexes in synapses of the central nervous system. In *Structure and Function of Synapses*, Pappas, G. D., and Purpura, D. P., eds. New York: Raven Press. pp. 67–86.

Auld, V. J., Fetter, R. D., Broadie, K., Goodman, C. S. (1995). Gliotactin, a novel transmembrane protein on peripheral glia, is required to form the blood-nerve barrier in Drosophila. *Cell* 81:757–767.

Bailey, C. H., Chen, M., Keller, F., Kandel, E. R. (1992). Serotonin-mediated endocytosis of apCAM: an early step of learning-related synaptic growth in Aplysia. *Science* 256:645–649.

Baranes, D., Lederfein, D., Huang, Y. Y., Chen, M., Bailey, C. H., Kandel, E. R. (1998). Tissue plasminogen activator contributes to the late phase of LTP and to synaptic growth in the hippocampal mossy fiber pathway. *Neuron* 21: 813–825.

Baumgartner, S., Littleton, J. T., Broadie, K., Bhat, M. A., Harbecke, R., Lengyel, J. A., Chiquet-Ehrismann, R., Prokop, A., Bellen, H. J. (1996). A Drosophila neurexin is required for septate junction and blood-nerve barrier formation and function. *Cell* 87:1059–1068.

Beumer, K. J., Rohrbough, J., Prokop, A., Broadie, K. (1999). A role for PS integrins in morphological growth and synaptic function at the postembryonic neuromuscular junction of Drosophila. *Development* 126:5833–5846.

Bloom, F. E., Aghajanian, G. K. (1966). Cytochemistry of synapses: selective staining for electron microscopy. *Science* 154:1575–1577.

Brieher, W., Yap, A. S., Gumbiner, B. M. (1996) Lateral dimerization is required for the homophilic binding activity of C-cadherin. *J. Cell Biol.* 135:487–496.

Brümmendorf, T., Kenwrick, S., Rathjen, F. G. (1998) Neural cell recognition molecule L1: from cell biology to human hereditary brain malformations. *Curr. Opin. Neurobiol.* 8:87–97.

Burgess, R. W., Nguyen, Q. T., Son, Y. J., Lichtman, J. W., Sanes, J. R. (1999). Alternatively spliced isoforms of nerve- and muscle-derived agrin: their roles at the neuromuscular junction. *Neuron* 23:33–44.

Burridge, K., Fath, K., Kelly, T., Nuckolls, G., Turner, C. (1988). Focal adhesions: transmembrane junctions between the extracellular matrix and the cytoskeleton. *Annu. Rev. Cell Biol.* 4:487–525.

Butz, S., Okamoto, M., Südhof, T. C. (1998). A tripartite protein complex with the potential to couple synaptic vesicle exocytosis to cell adhesion in brain. *Cell* 94:773–782.

Capogna, M., Gahwiler, B. H., Thompson, S. M. (1996). Calcium-independent actions of alpha-latrotoxin on spontaneous and evoked synaptic transmission in the hippocampus. *J. Neurophysiol.* 76:149–158.

Carey, D. J. (1997). Syndecans: multifunctional cell-surface co-receptors. *Biochem. J.* 327:1–16.

Ceccarelli, B., Hurlbut, W. P. (1980). Ca^{2+}-dependent recycling of synaptic vesicles at the frog neuromuscular junction. *J. Cell Biol.* 87:297–303.

Chen, H. J., Rojas-Soto, M., Oguni, A., Kennedy, M. B. (1998). A synaptic Ras-GTPase activating protein (p135 SynGAP) inhibited by CaM kinase II. *Neuron* 120:895–904.

Chisholm, A., Tessier-Lavigne, M. (1999). Conservation and divergence of axon guidance mechanisms. *Curr. Neurobiol.* 9:603–615.

Chishti, A. H. (1998). Function of p55 and its nonerythroid homologues. *Curr. Opin. Hematol.* 5:116–121.

Cho, K.-O., Hunt, C. A., Kennedy, M. B. (1992). The rat brain postsynaptic density fraction contains a homolog of the Drosophila discs-large tumor suppressor protein. *Neuron* 9:929–942.

Cohen, A. R., Woods, D. F., Marfatia, S. M., Walther, Z., Chishti, A. H., Anderson, J. M. (1998). Human CASK/Lin-2 binds syndecan-2 and protein 4.1 and localizes to the basolateral membrane of epithelial cells. *J. Cell Biol.* 142: 129–138.

Comu, S., Weng, W., Olinsky, S., Ishwad, P., Mi, Z., Hempel, J., Watkins, S., Lagenaur, C. F., Narayanan, V. (1997). The murine P84 neural adhesion molecule is SHPS-1, a member of the phosphatase-binding protein family. *J. Neurosci.* 17:8702–8710.

Cotman, C. W., Taylor, D. J. (1972). Isolation and structural studies on synaptic complexes from rat brain. *J. Cell Biol.* 55:696–711.

Cremer, H., Lange, R., Christoph, A., Plomann, M., Vopper, G., Roes, J., Brown, R., Baldwin, S., Kraemer, P., Scheff, S. (1994). Inactivation of the N-CAM gene in mice results in size reduction of the olfactory bulb and deficits in spatial learning. *Nature* 367:455–459.

Cremer, H., Chazal, G., Carleton, A., Goridis, C., Vincent, J. D., Lledo, P. M. (1998). Long-term but not short-term plasticity at mossy fiber synapses is impaired in neural cell adhesion molecule deficient mice. *Proc. Natl. Acad. Sci. USA* 95:13,242–13,247.

De Angelis, E., MacFarlane, J., Du, J. S., Yeo, G., Hicks, R., Rathjen, F. G., Kenwrick, S., Brümmendorf, T. (1999). Pathological missense mutations of neural cell adhesion molecule L1 affect homophilic and heterophilic binding activities. *EMBO J.* 18:4744–4753.

De Robertis, E. (1964). Histophysiology of synapses and neurosecretion. New York: Pergamon Press.

Dreyer, F., Rosenberg, F., Becker, C., Bigalke, H., Penner, R. (1987). Differential effects of various secretagogues on quantal transmitter release from mouse motor nerve terminals treated with botulinum A and tetanus toxin. *Naunyn-Schmiedebergs Arch. Pharmacol.* 335:1–7.

Duclos, F., Koenig, M. (1995). Comparison of primary structure of a neuron-specific protein, X11, between human and mouse. *Mamm. Genome* 6:57–58.

Duclos, F., Boschert, U., Sirugo, G., Mandel, J.-L., Hen, R., Koenig, M. (1993). Gene in the region of the Friedreich ataxia locus encodes a putative trans-membrane protein expressed in the nervous system. *Proc. Natl. Acad. Sci. USA* 90:109–113.

Dunne, J., Hanby, A. M., Poulsom, R., Jones, T. A., Sheer, D., Chin, W. G., Da, S. M., Zhao, Q., Beverley, P. C., Owen, M. J. (1995). Molecular cloning and tissue expression of FAT, the human homologue of the Drosophila fat gene that is located on chromosome 4q34-q35 and encodes a putative adhesion molecule. *Genomics* 30:207–223.

Fannon, A. M., Colman, D. R. (1996). A model for central synaptic junctional complex formation based on the differential adhesive specificities of the cadherins. *Neuron* 17:423–434.

Ferns, M., Hoch, W., Campanelli, J. T., Rupp, F., Hall, Z. W., Scheller, R. H. (1992). RNA splicing regulates agrin-mediated acetylcholine receptor clustering activity on cultured myotubes. *Neuron* 8:1079–1086.

Fujioka, Y., Matozaki, T., Noguchi, T., Iwamatsu, A., Yamao, T., Takahashi, N., Tsuda, M., Takada, T., Kasuga, M. (1996). A novel membrane glycoprotein, SHPS-1, that binds the SH2-domain-containing protein tyrosine phosphatase SHP-2 in response to mitogens and cell adhesion. *Mol. Cell Biol.* 16:6887–6899.

Geinisman, Y., Morrell, F., deToledo-Morrell, L. (1988). Remodeling of synaptic architecture during hippocampal "kindling." *Proc. Natl. Acad. Sci. USA* 85:3260–3264.

Geppert, M., Goda, Y., Hammer, R. E., Li, C., Rosahl, T. W., Stevens, C. F., Süd-hof, T. C. (1994). Synaptotagmin I: a major Ca^{2+} sensor for transmitter re-lease at a central synapse. *Cell* 79:717–727.

Geppert, M., Khvotchev, M., Krasnoperov, V., Goda, Y., Missler, M., Hammer, R. E., Ichtchenko, K., Petrenko, A. G., Südhof, T. C. (1998). Neurexin Iα is a major α-latrotoxin receptor that cooperates in α-latrotoxin action. *J. Biol. Chem.* 273:1705–1710.

Goodman, C. S. (1996). Mechanisms and molecules that control growth cone guidance. *Annu. Rev. Neurosci.* 19:341–377.

Goodman, C. S., Shatz, C. J. (1993). Developmental mechanisms that generate precise patterns of neuronal connectivity. *Cell* 72:77–98.

Gray, E. G. (1959). Axo-somatic and axo-dendritic synapses of the cerebral cor-tex: an electron microscope study. *J. Anat.* 95:101–106.

Gray, E. G. (1966). Problems of interpreting the fine structure of vertebrate and invertebrate synapses. *Int. Rev. Gen. Exp. Zool.* 2:139–170.

Gray, E. G., Whittaker, V. P. (1962). The isolation of nerve endings from brain. *J. Anat.* 96:79–88.

Grindstaff, K. K., Yeaman, C., Anandasabapathy, N., Hsu, S. C., Rodriguez-Boulan, E., Scheller, R. H., Nelson, W. J. (1998). Sec6/8 complex is re-cruited to cell-cell contacts and specifies transport vesicle delivery to the basal-lateral membrane in epithelial cells. *Cell* 93:731–740.

Grootjans, J. J., Zimmermann, P., Reekmans, G., Smets, A., Degeest, G., Durr, J., David, G. (1997). Syntenin, a PDZ protein that binds syndecan cytoplasmic domains. *Proc. Natl. Acad. Sci. USA* 94:13,683–13,688.

Grotewiel, M. S., Beck, C. D., Wu, K. H., Zhu, X. R., Davis, R. L. (1998). Integrin-mediated short-term memory in *Drosophila. Nature* 391:455–460.

Gualandris, A., Jones, T. E., Strickland, S., Tsirka, S. E. (1996). Membrane depolarization induces calcium-dependent secretion of tissue plasminogen activator. *J. Neurosci.* 16:2220–2225.

Harris, K. M., Sultan, P. (1995). Variation in number, location, and size of synaptic vesicles provides an anatomical basis for the nonuniform probability of release at hippocampal CA1 synapses. *Neuropharmacology* 34:1387–1395.

Hata, Y., Butz, S., Südhof, T. C. (1996). CASK: A novel dlg/PSD95 homologue with an N-terminal CaM kinase domain identified by interaction with neurexins. *J. Neurosci.* 16:2488–2494.

Hirao, K., Hata, Y., Ide, N., Takeuchi, M., Irie, M., Yao, I., Deguchi, M., Toyoda, A., Südhof, T. C., Takai, Y. (1998). A novel multiple PDZ domain–containing molecule interacting with NMDA receptors and neuronal cell adhesion proteins. *J. Biol. Chem.* 273:21,105–21,110.

Hohenester, E., Tisi, D., Talts, J. F., Timpl, R. (1999). The crystal structure of a laminin G–like module reveals the molecular basis of alpha-dystroglycan binding to laminins, perlecan, and agrin. *Mol. Cell* 4:783–792.

Hortsch, M., Patel, N. H., Bieber, A. J., Traquina, Z. R., Goodman, C. S. (1990). *Drosophila* neurotactin, a surface glycoprotein with homology to serine esterases, is dynamically expressed during embryogenesis. *Development* 110: 1327–1340.

Hoskins, R., Hajnal, A. F., Harp, S. A., Kim, S. K. (1996). The *C. elegans* vulval induction gene *lin-2* encodes a member of the MAGUK family of cell junction proteins. *Development* 122:97–111.

Hsueh, Y.-P., Sheng, M. (1999). Regulated expression and subcellular localization of syndecan heparan sulfate proteoglycans and the syndecan-binding protein CASK/LIN-2 during rat brain development. *J. Neurosci.* 19:7415–7425.

Hsueh, Y.-P., Yang, F.-C., Kharazia, V., Naisbitt, S., Cohen, A. R., Weinberg, R. J., Sheng, M. (1998). Direct interaction of CASK/Lin-2 and syndecan heparan sulfate proteoglycan and their overlapping distribution in neuronal synapses. *J. Cell Biol.* 142:139–151.

Ichtchenko, K., Hata, Y., Nguyen, T., Ullrich, B., Missler, M., Moomaw, C., Südhof, T. C. (1995). Neuroligin 1: A splice-site specific ligand for β-neurexins. *Cell* 81:435–443.

Ichtchenko, K., Nguyen, T., Südhof, T. C. (1996). Structures, alternative splicing, and neurexin binding of multiple neuroligins. *J. Biol. Chem.* 271:2676–2682.

Ichtchenko, K., Khvotchev, M., Kiyatkin, N., Simpson, L., Sugita, S., Südhof, T. C. (1998). α-Latrotoxin action probed with recombinant toxin: receptors recruit α-latrotoxin but do not transduce an exocytotic signal. *EMBO J.* 17:6188–6199.

Irie, M., Hata, Y., Takeuchi, M., Ichtchenko, K., Toyoda, A., Hirao, K., Takai, Y., Rosahl, T. W., Südhof, T. C. (1997). Binding of neuroligins to PSD-95. *Science* 277:1511–1515.

Itoh, M., Furuse, M., Morita, K., Kubota, K., Saitou, M., Tsukita, S. (1999). Direct binding of three tight junction–associated MAGUKs, ZO-1, ZO-2, and ZO-3, with the COOH termini of claudins. *J. Cell Biol.* 147:1351–1363.

Jo, K., Derin, R., Li, M., Bredt, D. S. (1999). Characterization of MALS/Velis-1, -2, and -3: a family of mammalian LIN-7 homologs enriched at brain synapses in association with the postsynaptic density-95/NMDA receptor postsynaptic complex. *J. Neurosci.* 19:4189–4199.

Joseph, D. R., Baker, M. E. (1992). Sex hormone–binding globulin, androgen-binding protein, and vitamin K–dependent protein S are homologous to laminin A, merosin, and *Drosophila* crumbs protein. *FASEB J.* 6:2477–2481.

Kaech, S. M., Whitfield, C. W., Kim, S. K. (1998). The LIN-2/LIN-7/LIN-10 complex mediates basolateral membrane localization of the *C. elegans* EGF receptor LET-23 in vulval epithelial cells. *Cell* 94:761–771.

Kim, E., Niethammer, M., Rothschild, A., Jan, Y. N., Sheng, S. (1995). Clustering of the Shaker-type K$^+$ channels by direct interaction with the PSD-95/SAP90 family of membrane-associated guanylate kinases. *Nature* 378:85–88.

Kim, J. H., Liao, D., Lau, L. F., Huganir, R. L. (1998). SynGAP: a synaptic RasGAP that associates with the PSD-95/SAP90 protein family. *Neuron* 20:683–691.

Kistner, U., Wenzel, B. M., Veh, R. W., Cases-Langhoff, C., Garner, A. (1993). SAP90, a rat presynaptic protein related to the product of the *Drosophila* tumor suppressor gene dlg-A. *J. Biol. Chem.* 268:4580–4583.

Kohmura, N., Senzaki, K., Hamada, S., Kai, N., Yasuda, R., Watanabe, M., Ishii, H., Yasuda, M., Mishina, M., Yagi, T. (1998). Diversity revealed by a novel family of cadherins expressed in neurons at a synaptic complex. *Neuron* 20:1137–1151.

Kornau, H. C., Schenker, L. L., Kennedy, M. B., Seeburg, P. H. (1995). Domain interaction between NMDA receptor subunits and the postsynaptic density protein PSD-95. *Science* 269:1737–1740.

Kowalczyk, A. P., Bornslaeger, E. A., Norvell, S. M., Palka, H. L., Green, K. J. (1999). Desmosomes: intercellular adhesive junctions specialized for attachment of intermediate filaments. *Int. Rev. Cytol.* 185:237–302.

Krasnoperov, V. G., Bittner, M. A., Beavis, R., Kuang, Y., Salnikow, K. V., Chepurny, O. G., Little, A. R., Plotnikov, A. N., Wu, D., Holz, R. W., Petrenko, A. G. (1997). α-Latrotoxin stimulates exocytosis by the interaction with a neuronal G protein–coupled receptor. *Neuron* 18:925–937.

Landis, D. M., Reese, T. S. (1983). Cytoplasmic organization in cerebellar dendritic spines. *J. Cell Biol.* 97:1169–1178.

Leahy, D. J. (1997). Implications of atomic-resolution structures for cell adhesion. *Annu. Rev. Cell Dev. Biol.* 13:363–393.

Lelianova, V. G., Davletov, B. A., Sterling, A., Rahman, M. A., Grishin, E. V., Totty, N. F., and Ushkaryov, Y. A. (1997). α-Latrotoxin receptor, latrophilin, is a novel member of the secretin family of G protein–coupled receptors. *J. Biol. Chem.* 272:21,504–21,508.

Lin, D. M., Goodman, C. S. (1994). Ectopic and increased expression of Fasciclin II alters motoneuron growth cone guidance. *Neuron* 13:507–523.

Lin, D. M., Fetter, R. D., Kopczynski, C., Grenningloh, G., Goodman, C. S. (1994). Genetic analysis of Fasciclin II in *Drosophila*: defasciculation, refasciculation, and altered fasciculation. *Neuron* 13:1055–1069.

Lisman, J., Harris, K. M. (1993). Quantal analysis and synaptic anatomy—integrating two views of hippocampal plasticity. *Trends Neurosci.* 16:141–147.

Mahoney, P. A., Weber, U., Onofrechuk, P., Biessmann, H., Bryant, P. J., Goodman, C. S. (1991). The fat tumor suppressor gene in *Drosophila* encodes a novel member of the cadherin gene superfamily. *Cell* 67:853–868.

Martin, K. C., Kandel, E. R. (1996). Cell adhesion molecules, CREB, and the formation of new synaptic connections. *Neuron* 17:567–570.

Maximov, A., Südhof, T. C., Bezprozvanny, I. (1999). Association of neuronal voltage-gated calcium channels with modular adaptor proteins. *J. Biol. Chem.* 274: 24,453–24,456.

Mayford, M., Barzilai, A., Keller, F., Schacher, S., Kandel, E. R. (1992). Modulation of an NCAM-related adhesion molecule with long-term synaptic plasticity in *Aplysia*. *Science* 256:638–644.

Missler, M., Südhof, T. C. (1998a). Neurexins: three genes and 1001 products. *Trends Genet.* 14:20–26.

Missler, M., Südhof, T. C. (1998b). Neurexophilins form a conserved family of neuropeptide-like glycoproteins. *J. Neurosci.* 18:3630–3638.

Missler, M., Hammer, R. E., Südhof, T. C. (1998a). Neurexophilin binding to α-neurexins: a single LNS-domain functions as independently folding ligand-binding unit. *J. Biol. Chem.* 273:34,716–34,723.

Missler, M., Fernandez-Chacon, R., Südhof, T. C. (1998b). The making of neurexins. *J. Neurochem.* 71:1339–1347.

Naisbitt, S., Kim, E., Tu, J. C., Xiao, B., Sala, C., Valtschanoff, J., Weinberg, R. J., Worley, P. F., Sheng, M. (1999). Shank, a novel family of postsynaptic density proteins that binds to the NMDA receptor/PSD-95/GKAP complex and cortactin. *Neuron* 23:569–582.

Neuhoff, H., Roeper, J., Schweizer, M. (1999). Activity-dependent formation of perforated synapses in cultured hippocampal neurons. *Eur. J. Neurosci.* 11: 4241–4250.

Nose, A., Nagafuchi, A., Takeichi, M. (1988). Expressed recombinant cadherins mediate cell sorting in model systems. *Cell* 54:993–1002.

Nose, A., Tsuji, K., Takeichi, M. (1990). Localization of specificity determining sites in cadherin cell adhesion molecules. *Cell* 61:147–155.

Nguyen, T., Südhof, T. C. (1997). Binding properties of neuroligin 1 and neurexin 1β reveal function as heterophilic CAMs. *J. Biol. Chem.* 272:26,032–26,039.

Okamoto, M., Südhof, T. C. (1997). Mints: munc18-interacting proteins in synaptic vesicle exocytosis. *J. Biol. Chem.* 272:31,459–31,464.

Okamoto, M., Südhof, T. C. (1998). Mint 3: a ubiquitous mint isoform that does not bind to munc18-1 or 2. *Eur. J. Cell Biol.* 77:161–165.

Oleszewski, M., Beer, S., Katich, S., Geiger, C., Zeller, Y., Rauch, U., Altevogt, P. (1999). Integrin and neurocan binding to L1 involves distinct Ig domains. *J. Biol. Chem.* 274:24,602–24,610.

Ozawa, M., Ringwald, M., Kemler, R. (1990). Uvomorulin-catenin complex formation is regulated by a specific domain in the cytoplasmic region of the cell adhesion molecule. *Proc. Natl. Acad. Sci. USA* 87:4246–4250.

Peles, E., Nativ, M., Lustig, M., Grumet, M., Schilling, J., Martinez, R., Plowman, G. D., Schlessinger, J. (1997a). Identification of a novel contactin-associated transmembrane receptor with multiple domains implicated in protein-protein interactions. *EMBO J.* 16:978–988.

Peles, E., Joho, K., Plowman, G. D., Schlessinger, J. (1997b). Close similarity between *Drosophila* neurexin IV and mammalian Caspr protein suggests a conserved mechanism for cellular interactions. *Cell* 88:745–746.

Petrenko, A. G., Lazaryeva, V. D., Geppert, M., Tarasyuk, T. A., Moomaw, C., Khokhlatchev, A. V., Ushkaryov, Y. A., Slaughter, C., Nasimov, I. V., Südhof, T. C. (1993). Polypeptide composition of the α-latrotoxin receptor. *J. Biol. Chem.* 268:1860–1867.

Petrenko, A. G., Ullrich, B., Missler, M., Krasnoperov, V., Rosahl, T. W., Südhof, T. C. (1996). Structure and evolution of neurexophilin. *J. Neurosci.* 16:4360–4369.

Pfenninger, K. H. (1971a). The cytochemistry of synaptic densities. I. An analysis of the bismuth iodide impregnation method. *J. Ultrastruct. Res.* 34:103–122.

Pfenninger, K. H. (1971b). The cytochemistry of synaptic densities. II. Proteinaceous components and mechanisms of synaptic connectivity. *J. Ultrastruct. Res.* 35:451–475.

Pfenninger, K. H. (1973). *Synaptic Morphology and Cytochemistry.* Stuttgart: Gustav Fischer Verlag.

Qian, Z., Gilbert, M. E., Colicos, M. A., Kandel, E. R., Kuhl D. (1993). Tissue-plasminogen activator is induced as an immediate-early gene during seizure, kindling and long-term potentiation. *Nature* 361:453–457.

Radice, G. L., Rayburn, H., Matsunami, H., Knudsen, K. A., Takeichi, M., Hynes, R. O. (1997). Developmental defects in mouse embryos lacking N-cadherin. *Dev. Biol.* 181:64–78.

Rosenthal, L., Meldolesi, J. (1989). α-Latrotoxin and related toxins. *Pharmacol. Ther.* 42:115–134.

Rubenstein, J. L., Anderson, S., Shi, L., Miyashita-Lin, E., Bulfone, A., Hevner, R. (1999). Genetic control of cortical regionalization and connectivity. *Cerebral Cortex* 9:524–532.

Rubin, G. M., Yandell, M. D., Wortman, J. R., Gabor Miklos, G. L., Nelson, C. R., Hariharan, I. K., Fortini, M. E., Li, P. W., Apweiler, R., Fleischmann, W., Cherry, J. M., Henikoff, S., Skupski, M. P., Misra, S., Ashburner, M., Birney, E., Boguski, M. S., Brody, T., Brokstein, P., Celniker, S. E., Chervitz, S. A., Coates, D., Cravchik, A., Gabrielian, A., Galle, R. F., Gelbart, W. M., George, R. A., Goldstein, L. S., Gong, F., Guan, P., Harris, N. L., Hay, B. A., Hoskins, R. A., Li, J., Li, Z., Hynes, R. O., Jones, S. J., Kuehl, P. M., Lemaitre, B., Littleton, J. T., Morrison, D. K., Mungall, C., O'Farrell, P. H., Pickeral, O. K., Shue, C., Vosshall, L. B., Zhang, J., Zhao, Q., Zheng, X. H., Zhong, F., Zhong, W., Gibbs, R., Venter, J. C., Adams, M. D., Lewis, S. (2000). Comparative genomics of the eukaryotes. *Science* 287:2204–2215.

Rudenko, G., Nguyen, T., Chelliah, Y., Südhof, T. C., Deisenhofer, J. (1999). The structure of the ligand-binding domain of neurexin 1β: regulation of LNS domain function by alternative splicing. *Cell* 99:93–101.

Rutishauser, U., Landmesser L. (1996). Polysialic acid in the vertebrate nervous system: a promoter of plasticity in cell-cell interactions. *Trends Neurosci.* 19: 422–427.

Sanes, J., Lichtman, J. W. (1999). Development of the vertebrate neuromuscular junction. *Annu. Rev. Neurosci.* 22:389–442.

Sanes, J. R., Yamagata, M. (1999). Formation of lamina-specific synaptic connections. *Curr. Opin. Neurobiol.* 9:79–87.

Schuster, C. M., Davis, G. W., Fetter, R. D., Goodman, C. S. (1996a). Genetic dissection of structural and functional components of synaptic plasticity. I. Fasciclin II controls synaptic stabilization and growth. *Neuron* 17:641–654.

Schuster, C. M., Davis, G. W., Fetter, R. D., Goodman, C. S. (1996b). Genetic dissection of structural and functional components of synaptic plasticity. II. Fasciclin II controls presynaptic structural plasticity. *Neuron* 17:655–667.

Senzaki, K., Ogawa, M., Yagi, T. (1999). Proteins of the CNR family are multiple receptors for Reelin. *Cell* 99:635–647.

Serafini, T. (1999). Finding a partner in a crowd: neuronal diversity and synaptogenesis. *Cell* 98:133–136.

Simske, J. S., Kaech, S. M., Harp, S. A., Kim, S. K. (1996). LET-23 receptor localization by the cell junction protein LIN-7 during *C. elegans* vulval induction. *Cell* 85:195–204.

Song, J.-Y., Ichtchenko, K., Südhof, T. C., Brose, N. (1999). Neuroligin 1 is a postsynaptic cell-adhesion molecule of excitatory synapses. *Proc. Natl. Acad. Sci. USA* 96:1100–1125.

Stevenson, B. R., Keon, B. H. (1998). The tight junction: morphology to molecules. *Annu. Rev. Cell Dev. Biol.* 14:89–109.

Sugita, S., Ichtchenko, K., Khvotchev, M., Südhof, T. C. (1998). α-Latrotoxin receptor CIRL/Latrophilin 1 (CL1) defines an unusual family of ubiquitous G protein–linked receptors. G-protein coupling not required for triggering exocytosis. *J. Biol. Chem.* 273:32,715–32,724.

Sugita, S., Khvochtev, M., Südhof, T. C. (1999). Neurexins are functional α-latrotoxin receptors. *Neuron* 22:489–496.

Takeichi, M. (1977). Functional correlation between cell adhesive properties and some cell surface proteins. *J. Cell Biol.* 75:464–474.

Takeichi, M. (1990). Cadherins: a molecular family important in selective cell-cell adhesion. *Annu. Rev. Biochem.* 59:237–252.

Takeichi, M., Uemura, T., Iwai, Y., Uchida, N., Inoue, T., Tanaka, T., Suzuki, S. C. (1997). Cadherins in brain patterning and neural network formation. *Cold Spring Harbor. Symp. Quant. Biol.* 62:505–510.

Thiery, J. P., Brackenbury, R., Rutishauser, U., Edelman, G. M. J. (1977). Adhesion among neural cells of the chick embryo. II. Purification and characterization of a cell adhesion molecule from neural retina. *J. Biol. Chem.* 252: 6841–6851.

Thomas, U., Kim, E., Kuhlendahl, S., Koh, Y. H., Gundelfinger, E. D., Sheng, M., Garner, C. C., Budnik, V. (1997). Synaptic clustering of the cell adhesion molecule fasciclin II by discs-large and its role in the regulation of presynaptic structure. *Neuron* 19:787–799.

Tomasiewicz, H., Ono, K., Yee, D., Thompson, C., Goridis, C., Rutishauser, U., Magnuson, T. (1993). Genetic deletion of a neural CAM variant (N-CAM-180) produces distinct defects in the central nervous system. *Neuron* 11:1163–1174.

Toni, N., Buchs, P.-A., Nikonenko, I., Bron, C. R., Muller, D. (1999). LTP promotes formation of multiple spine synapses between a single axon terminal and a dendrite. *Nature* 402:421–425.

Tsigelny, I., Shindyalov, I. N., Bourne, P. E., Südhof, T. C., Taylor, P. (2000). Common EF-hand motifs in cholinesterase and neuroligins suggest a role for calcium binding in cell surface associations. *Protein Sci.* 9:180–185.

Tsirka, S. E., Gualandris, A., Amaral, D. G., Strickland, S. (1995). Excitotoxin-induced neuronal degeneration and seizure are mediated by tissue plasminogen activator. *Nature* 377:340–344.

Uchida, N., Honjo, Y., Johnson, K. R., Wheelock, M. J., Takeichi, M. J. (1996). The catenin/cadherin adhesion system is localized in synaptic junctions bordering transmitter release zones. *J. Cell Biol.* 135:767–779.

Ullrich, B., Ushkaryov, Y. A., Südhof, T. C. (1995). Cartography of neurexins: more than 1000 isoforms generated by alternative splicing and expressed in distinct subsets of neurons. *Neuron* 14:497–507.

Ushkaryov, Y. A., Südhof, T. C. (1993). Neurexin IIIα: extensive alternative splicing generates membrane-bound and soluble forms in a novel neurexin. *Proc. Natl. Acad. Sci. USA* 90:6410–6414.

Ushkaryov, Y. A., Petrenko, A. G., Geppert, M., Südhof, T. C. (1992). Neurexins: synaptic cell surface proteins related to the α-latrotoxin receptor and laminin. *Science* 257:50–56.

Ushkaryov, Y. A., Hata, Y., Ichtchenko, K., Moomaw, C., Afendis, S. Slaughter, C. A., Südhof, T. C. (1994). Conserved domain structure of β-neurexins. *J. Biol. Chem.* 269:11,987–11,992.

Valtorta, F., Maddedu, L., Meldolesi, J., Ceccarelli, B. (1984). Specific localization of the α-latrotoxin receptor in the nerve terminal plasma membrane. *J. Cell Biol.* 99:124–144.

Van der Loos, H. (1963). Fine structure of synapses in the cerebral cortex. *Z. Zellforsch.* 60: 815–825.

Verhage, V., Maia, A. S., Plomp, J. J., Brussaard, A. B., Heeroma, J. H., Vermeer, H., Toonen, R. F., Hammer, R. E., van den Berg, T. K., Missler, M., Geuze, H., Südhof, T. C. (2000). Synaptic assembly of the brain in the absence of neurotransmitter secretion. *Science* 287:864–869.

Vrensen, G., Cardozo, J. N. (1981). Changes in size and shape of synaptic connections after visual training: an ultrastructural approach of synaptic plasticity. *Brain Res.* 218:79–97.

Wodarz, A., Nusse, R. (1998). Mechanisms of Wnt signaling in development. *Annu. Rev. Cell Dev. Biol.* 14:59–88.

Wohrn, J. C., Nakagawa, S., Ast, M., Takeichi, M., Redies, C. (1999). Combinatorial expression of cadherins in the tectum and the sorting of neurites in the tectofugal pathways of the chicken embryo. *Neuroscience* 90:985–1000.

Wu, Q., Maniatis, T. (1999). A striking organization of a large family of human neural cadherin-like cell adhesion genes. *Cell* 97:779–790.

Yamagata, M., Herman, J. P., Sanes, J. R. (1995). Lamina-specific expression of adhesion molecules in developing chick optic tectum. *J. Neurosci.* 15:4556–4571.

Yap, A. S., Brieher, W. M., Gumbiner, B. M. (1997). Molecular and functional analysis of cadherin-based adherens junctions. *Annu. Rev. Cell Dev. Biol.* 13: 119–146.

Zhang, W., Vazquez, L., Apperson, M., Kennedy, M. B. (1999). Citron binds to PSD-95 at glutamatergic synapses on inhibitory neurons in the hippocampus. *J. Neurosci.* 19:96–108.

Zhang, Z., Lee, C. H., Mandiyan, V., Borg, J. P., Margolis, B., Schlessinger, J., Kuriyan, E. Y. (1997). Sequence-specific recognition of the internalization motif of the Alzheimer's amyloid precursor protein by the X11 PTB domain. *EMBO J.* 16:6141–6150.

Zito, K., Fetter, R. D., Goodman, C. S., Isacoff, E. Y. (1997). Synaptic clustering of Fasciclin II and Shaker: essential targeting sequences and role of Dlg. *Neuron* 19:1007–1016.

7

The Postsynaptic Specialization

Morgan H.-T. Sheng

The presynaptic terminal is specialized for ultrarapid release of neurotransmitter in response to the arrival of an action potential (see Südhof and Scheller, this volume). The released neurotransmitter diffuses across the narrow synaptic cleft (circa 20–30 nm wide) to impinge on the postsynaptic cell. The postsynaptic membrane is specialized for the reception of the neurotransmitter signal and its transmission to the rest of the postsynaptic neuron. To accomplish this function, the postsynaptic specialization contains specific receptors for the neurotransmitter (e.g., glutamate receptors or γ-aminobutyric acid [GABA] receptors); signal transduction proteins that couple receptor activation to intracellular second messenger pathways; cytoskeletal elements that anchor the neurotransmitter receptors at the postsynaptic site; and adhesion molecules that mediate the proper alignment of the postsynaptic membrane with the presynaptic terminal. In addition, the postsynaptic specialization contains enzymes that modulate the activity of receptors (e.g., protein kinases and phosphatases), as well as ion channels that are not directly regulated by neurotransmitters but that control postsynaptic excitability (e.g., voltage-gated ion channels).

The best-understood postsynaptic specialization is that of the vertebrate neuromuscular junction (NMJ), which utilizes acetylcholine for its transmitter (Sanes and Lichtman, 1999). The NMJ will not be covered in this chapter, which instead focuses on postsynaptic specializations of central neuron-to-neuron synapses. Neuronal synapses are smaller than NMJs and show much greater heterogeneity in the brain and on any given postsynaptic cell; hence, they have been technically more difficult to study. However, the development of ever more sophisticated imaging and molecular techniques has fueled rapid progress in the elucidation of the architecture of neuronal postsynaptic specializations. As with the NMJ, knowledge of neuronal synapses has progressed through morphological, pharmacological, and electrophysiological levels to an increasingly molecular level of understanding, and it is at the molecular level that this chapter is pitched.

Classification of Postsynaptic Specializations

Just as presynaptic terminals are not uniform in structure or chemical makeup, postsynaptic specializations are heterogeneous in terms of morphology, molecular composition, and function. This heterogeneity occurs within a single postsynaptic neuron, the most obvious example being the coexistence of excitatory and inhibitory synapses on the same cell. The soma and dendrites are generally considered the postsynaptic compartment of the neuron, though axons can, and often do, receive presynaptic innervation. Inhibitory synapses are typically formed on more proximal dendrites and on the cell body, including the initial axon segment. Excitatory synapses, on the other hand, are usually formed on more distal dendritic branches, particularly on small protrusions of these branches called dendritic spines.

Excitatory synapses predominantly utilize glutamate as the neurotransmitter, and postsynaptic specializations of excitatory synapses show a specific concentration of glutamate receptors. When viewed by electron microscopy (EM), excitatory postsynaptic specializations are characterized by a fuzzy, dense thickening of the postsynaptic membrane, termed the postsynaptic density (PSD), which gives the synapse an "asymmetric" (type 1) appearance. Inhibitory synapses use GABA or glycine as the neurotransmitter, and the appropriate receptors for these neurotransmitters (GABA and glycine receptors) are specifically concentrated in the postsynaptic membrane of these synapses. Inhibitory synapses lack a highly developed PSD (though they do show some postsynaptic membrane thickening) and are hence classified as "symmetric" (type 2) in morphology. Thus neuronal postsynaptic specializations can be classified by their ultrastructural appearance and by their specific array of neurotransmitter receptors, both of which correlate with their function (excitatory versus inhibitory). Beyond this broad classification, much heterogeneity exists within postsynaptic specializations of a particular class. In general, much more is known about excitatory synapses at the molecular level than about inhibitory synapses, and this is reflected in the balance of this chapter.

Dendritic Spines as a Specialized Postsynaptic Compartment

The Structure of Dendritic Spines

In many neurons, the surface area of the dendritic tree is greatly increased by the presence of dendritic spines, which protrude from the main shaft of the dendrite at high density (up to some 200 spines per 100 μm dendritic length in mature neurons; see Harris and Kater, 1994, and Shepherd, 1996, for review) (see Fig. 7.1). Most principal neurons (e.g., pyramidal neurons) bear dendritic spines, but many neurons (for instance, most GABAergic interneurons) do not. The aspiny neurons form excitatory synapses on their dendritic shafts; such shaft synapses also occur as a minority on spiny neurons. The vast majority of excitatory synapses in the mature brain occur on spines, and a typical mature spine accommodates a single synapse located on its head. Thus dendritic spines represent the major unitary postsynaptic compartment for excitatory input.

Dendritic spines come in a large variety of shapes (thin, stubby, mushroom, and branched), and in a wide range of sizes (ranging in volume from 0.01 mm^3 to 0.8 mm^3) (see Harris and Kater, 1994, and Harris, 1999, for review). The great heterogeneity of dendritic spines is apparent on a single dendrite (Fig. 7.1), but the functional significance of the variation in spine size and shape is still unclear.

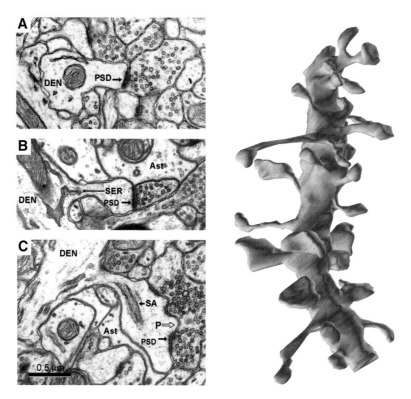

Figure 7.1. The morphology of dendritic spines. Left: Electron micrographs of representative spines: *(A)* A stubby spine (green), *(B)* a thin spine (purple), and *(C)* a mushroom-shaped spine (blue). The presynaptic terminal is shaded red, the astrocyte (Ast) yellow. DEN, dendrite; P, perforation in the PSD; PSD, postsynaptic density; SA, spine apparatus; SER, smooth endoplasmic reticulum. Scale bar = 0.5 m.

Right: Three-dimensional reconstruction from serial electron micrographs of a similar dendrite. Representative spines are color-matched to the electron micrographs. Courtesy of Kristen Harris and Karen Szumowski.

Within spines there are found a variety of organelles, including smooth endoplasmic reticulum (SER), which is thought to extend into the spine from SER in the dendritic shaft (Spacek and Harris, 1997). The SER in spines functions at least in part as an intracellular calcium store from which calcium can be released in response to synaptic stimulation. Particularly common in the larger spines is a structure known as the spine apparatus, an organelle characterized by stacks of SER membranes surrounded by densely staining amorphous material of unknown molecular identity. The role of the spine apparatus is unknown, though some speculate that it acts as a repository or a relay for membrane proteins trafficking to or from the synapse. Vesicles of "coated" or smooth appearance are observed occasionally in spines, as are multivesicular bodies, all consistent with local membrane trafficking processes, which presumably must occur in dendritic spines. As polyribosomes have been detected in spines (Steward et al., 1996), it is possible that protein translation occurs within this postsynaptic compartment as well as in dendritic shafts.

Within dendritic spines, the PSD is found at the site of the synaptic junction. The PSD occupies roughly 10% of the surface area of the spine and is exactly aligned with the presynaptic active zone. It contains the glutamate receptors and their associated signaling and cytoskeletal proteins. Cell-cell adherens junctions like those found in nonneuronal cells can be seen morphologically at the edge of about half of the PSDs. These synaptic adherens junctions are probably based on cadherins, homophilic calcium-dependent adhesion molecules that likely play a major role in maintaining adhesion of pre- and postsynaptic membranes (see Serafini, 1999, and Shapiro and Colman, 1999, for review).

The predominant cytoskeleton within dendritic spines appears to be filamentous actin (F-actin). Tubulin and microtubule-associated proteins are also found in the PSD and in spines, but ultrastructurally defined microtubules are generally sparse or lacking (Harris and Kater, 1994). The enrichment of F-actin in spines allows these structures to be brightly labeled by fluorescently tagged phalloidin, a drug that binds to F-actin. Presumably the shape of spines is determined largely by the actin cytoskeleton, which in turn is regulated by small GTPases such as Rac, Rho, or CDC42. Indeed, transgenic mice expressing constitutively active Rac1 develop much smaller dendritic spines (Luo et al., 1996).

The Function of Dendritic Spines

What is the physiological significance of dendritic spines in neuronal function? The fact that each dendritic spine usually accommodates a single synapse suggests that the significance of spines relates to the creation of a local synapse-specific compartment, rather than the mere expansion of postsynaptic surface area (Shepherd, 1996). The consensus is that spines function as microcompartments that can segregate as well as integrate synaptic signals; they increase the compartmentalization and hence the information-processing power of dendrites (Shepherd, 1996). Spines can act as semiautonomous chemical compartments because they are separated from the dendritic shaft by up to a few microns and their connection to the dendritic cytoplasm is often through a thin neck. Calcium imaging has conclusively demonstrated that spines can function as integrative compartments for control of intracellular calcium and that they are at least in part isolated from the parent dendrite (Yuste and Denk, 1995; Finch and Augustine, 1998; Schiller et al., 1998; Takechi et al., 1998). The resistance of the neck, and impedance mismatch between spine and dendritic shaft, may also contribute to electrical compartmentalization of spines, but this is a controversial issue that has been difficult to address directly.

Development of Spines

Filopodia (long, thin processes with a pointed tip that are generally devoid of organelles) rapidly protrude and retract from dendrites, especially during early stages of synaptogenesis (Dailey and Smith, 1996; Ziv and Smith, 1996; Fiala et al., 1998). Dendritic filopodia often bear synapses and are believed to be precursors of synaptic spines. Filopodia are most abundant in the brain during the first postnatal week in vivo but are gradually replaced by shaft synapses and stubby spines. With further development, the number of shaft synapses and stubby spines decreases, and synapses on thin and mushroom-shaped spines predominate in the adult brain (Fiala et al., 1998). Quantitative analysis of the developmental appearance of the various spine types in brain has led to the proposal that dendritic filopodia participate in the formation of shaft synapses by recruitment of the presynaptic contact to the dendrite. Subsequently, mature spines are believed to develop from the early shaft synapses (Harris, 1999).

Spine formation is of clinical relevance, because developmental abnormalities of dendritic spines are sometimes found in conditions that lead to mental retardation. For instance, in a mouse genetic model of fragile X syndrome, cortical dendrites show excessive filopodia-like protrusions, as though normal spine maturation is defective (Comery et al., 1997).

Plasticity of Spines

Advances in video microscopy and green fluorescent protein (GFP) technology have revealed the dynamic plasticity of dendritic spines, even in mature neurons. Modifications of spine number and shape have been documented under various physiological and pathological conditions. Over a time course of seconds to minutes, the majority of spines (visualized by GFP-tagged actin) change their shape as a result of remodeling of the actin cytoskeleton (Fischer et al., 1998). Over a time course of many hours, a substantial fraction (10–20%) of spines appear or disappear in pyramidal neurons of cortical slices, and their stability is affected by the neurotrophin brain-derived neurotrophic factor (BDNF) (Horch et al., 1999). Cyclical emergence and disappearance of spines has been found to occur during the estrous cycle of female rats (Woolley et al., 1996). Local synaptic activity can actually cause the appearance of new spines or filopodia in an N-methyl-D-aspartate (NMDA)–receptor-dependent fashion (Engert and Bonhoeffer, 1999; Maletic-Savatic et al., 1999). On the other hand, exogenous application of NMDA can cause the loss of dendritic spines in a concentration-dependent manner (Halpain et al., 1998), and low levels of α-amino-3-hydroxy-5-methyl-4-isoxazolepropionate (AMPA) receptor activation are required to maintain spines (McKinney et al., 1999). Thus the regulation of spine structure is likely to be complex and involve both activity-dependent and -independent mechanisms.

The dynamic behavior of spines has attracted particular attention because they are the only neuronal structures that convincingly show experience-dependent morphological changes in the mammalian brain. Regulated changes in spine morphology and number may reflect mechanisms for converting short-term changes in synaptic activity into lasting alterations in structure, connectivity, and function of synapses.

Molecular Organization of the Postsynaptic Density

The PSD was originally defined ultrastructurally as an electron-dense thickening (circa 40–50 nm thick and up to a few hundred nanometers wide) associated with the cytoplasmic side of the postsynaptic membrane of excitatory synapses. Biochemically the PSD is relatively insoluble in nonionic detergents and can be purified to a considerable degree, yielding disclike structures (Carlin et al., 1980). Ultrastructurally PSDs contain weblike filaments that appear to hold together particulate components; the molecular basis of these substructures is not established. Functionally the PSD can be regarded as an organelle specialized for glutamatergic postsynaptic signal transduction, consisting of a matrix of proteins having receptor, enzymatic, regulatory, scaffolding, and cytoskeletal functions. As it is the keystone of the postsynaptic specialization of excitatory synapses and contains the critical molecules involved in synaptic plasticity, the PSD has been studied intensively in recent years (see Kennedy, 1997, and Ziff, 1997, for review).

Numerous molecules have been identified as components of the PSD based on its biochemical purification. In addition to glutamate receptors and associated proteins, these components include cytoskeletal elements, such as actin, tubulin, and brain spectrin; heat shock and vesicle transport proteins; regulatory molecules, such as calmodulin, the subunit of calmodulin-dependent protein kinase II (CaMKII), protein kinase C (PKC), protein kinase A (PKA), protein phosphatase-1 (PP1), and Fyn tyrosine kinase; and receptor tyrosine kinases, such as TrkB, the receptor for BDNF (see Ziff, 1997, for review). It is important to distinguish between those proteins that are merely present in the PSD and those that are specifically enriched in the PSD. Some components of the PSD (e.g., N-ethylmaleimide-sensitive fusion protein [NSF], CaMKII) show activity-dependent or ischemia-induced increases in their association with the PSD (Hu et al., 1998; Shen and Meyer, 1999). The molecular mechanisms of protein targeting to the PSD, and their regulation by synaptic activity, are important avenues of research in the postsynaptic specialization.

Glutamate Receptors and Their Associated Proteins

Our understanding of the molecular structure of the NMJ derives largely from the study of acetylcholine receptor clustering at the muscle endplate. In neuronal excitatory synapses, glutamate receptors are the cardinal components of the postsynaptic specialization, and recent progress in deciphering the molecular organization of the PSD is based largely on studies of glutamate receptors and their associated proteins. Because of their central importance (both functionally and historically), it is useful to discuss the molecular organization of the PSD from the point of view of glutamate receptors.

Postsynaptic glutamate receptors can be broadly classified as ionotropic receptors or metabotropic receptors (mGluRs). Ionotropic glutamate receptors can be further divided into NMDA receptors, AMPA receptors, kainate receptors, and δ receptors. Among the mGluRs, the group I mGluRs mGluR1α and mGluR5, which are linked to phospholipase C (PLC) and phosphoinositide turnover, are predominant at postsynaptic sites. In recent years, it has become clear that glutamate receptors are targeted to postsynaptic domains in neurons, indeed even to specific subdomains within the postsynaptic specialization (Takumi et al., 1999a). Understanding the molecular mechanisms that underlie this postsynaptic targeting was a major motivation behind the search for cytoplasmic glutamate receptor-binding proteins.

Binding to specific intracellular proteins is likely to be important for immobilization and clustering of glutamate receptors, for their localization at postsynaptic sites, for their ability to transmit signals to appropriate cytoplasmic pathways, and for functional modulation of the receptors by kinases, phosphatases, and other regulatory proteins. In the past few years, the successful identification of many glutamate receptor-binding proteins (particularly by the use of the yeast two-hybrid system) has revealed an unexpectedly complex molecular architecture within the PSD.

The NMDA Receptor–PSD-95 Complex

Association of NMDA Receptors with the PSD

Among the glutamate receptors, NMDA receptors are biochemically the most tightly associated with the PSD. Viewed by EM, NMDA receptors are a highly consistent presence in excitatory synapses of the forebrain; in contrast, a significant fraction of excitatory synapses lack AMPA receptors (Nusser et al., 1998a; Petralia et al., 1999; Takumi et al., 1999b). Thus the NMDA receptor can be considered a cardinal component of the postsynaptic specialization and of key functional importance.

NMDA receptors are anchored in the PSD through specific protein-protein interactions that link them to the subsynaptic cytoskeleton and to intracellular signaling pathways. Cytoskeletal interactions are functionally relevant because NMDA receptor activity is influenced by the actin cytoskeleton (Rosenmund and Westbrook, 1993; Paoletti and Ascher, 1994). For transmembrane receptors, the specificity of signal transduction is often determined by the nature of the proteins that interact with the cytoplasmic domains of the receptor. This concept is likely to apply to NMDA receptor signaling because calcium influx through NMDA receptors stimulates a variety of intracellular events (e.g., synaptic plasticity, neurotoxicity, and transcriptional responses in the nucleus) that are not seen with other modes of calcium entry into the cell (Dingledine et al., 1999). Recent studies have uncovered many specific protein interactions mediated by the cytoplasmic tails of NMDA receptor subunits and have provided the first insights into the molecular organization of the PSD.

NMDA Receptors Interact with PSD-95

NMDA receptors are heteromeric (probably tetrameric) complexes composed of NR1 and NR2 subunits (Dingledine et al., 1999). The four different NR2 subunits (NR2A to -2D) contain long cytoplasmic tails (up to 644 amino acid residues), the C termini of which end in the conserved sequence -ESDV or -ESEV. This short C-terminal peptide motif mediates binding to PSD-95/SAP90, a protein that was first isolated as an abundant constituent of the PSD (Fig. 7.2) (Cho et al., 1992; Kistner et al., 1993; Kornau et al., 1995, 1997; Niethammer et al., 1996; Sheng, 1996; O'Brien et al., 1998).

PSD-95/SAP90 belongs to the MAGUK superfamily of proteins, which are characterized by the presence of PDZ domains, an SH3 domain, and a guanylate kinase–like (GK) domain (Fig. 7.2). PDZ domains are modular protein domains of some 90 amino acids that are specialized for binding to C-terminal peptides in a sequence-specific fashion (Doyle et al., 1996; Ponting et al., 1997; Songyang et al., 1997). However, other modes of interaction are also possible with PDZ domains, including binding to internal sequences that fold into a "beta-finger" (Hillier et al., 1999). PSD-95 has three PDZ domains in its N-terminal region, and recognition of the -ESDV C-terminal sequence of NR2 subunits is mediated by the first two PDZ domains (PDZ1 and PDZ2). Three other members of the PSD-95 family in mammals have been cloned, including PSD-93/chapsyn-110 (Brenman et al., 1996b; Kim et al., 1996), SAP97/hDlg (Lue et al., 1994; Müller et al., 1995), and SAP102 (Müller et al., 1996). All the family members except SAP97 (Müller et al., 1995) appear to be components of the PSD and to be associated with NMDA receptors in synapses.

The in vivo functional significance of PSD-95 binding to NMDA receptors remains incompletely understood. PSD-95 may be involved in

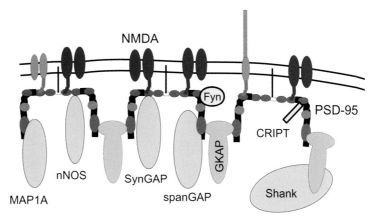

Figure 7.2. The NMDA receptor–PSD-95 complex. The cytoplasmic tail of NMDA
receptor NR2 subunits (blue) binds to either of the first two PDZ domains of PSD-
95. Other membrane proteins, such as receptors and ion channels (gray), also
interact with the PDZ domains of PSD-95. A subset of identified cytoplasmic
components of the PSD-95 complex and their sites of binding are shown (see text
for details). PSD-95 is shown multimerized via its N-terminal region (Hsueh et al.,
1997). Palmitoylation (black line) of the N terminus may contribute to the membrane and synaptic targeting of PSD-95 (Craven et al., 1999). The PDZ domains of
PSD-95 are indicated in red, SH3 domains in green, and guanylate kinase–like
domains in purple.

the synaptic targeting of NMDA receptors, in the coupling of NMDA receptors to signaling proteins, or in the anchoring of NMDA receptors to
the postsynaptic cytoskeleton, or in a combination of these functions.

Synaptic Targeting by PSD-95

PSD-95 and NMDA receptors are both specifically localized to the PSD
of excitatory synapses. An attractive idea based on the interaction of
PSD-95 and NMDA receptors is that PSD-95 is important for the postsynaptic localization of NMDA receptors. This hypothesis was supported
by genetic experiments in *Drosophila* on Discs large (Dlg), the fly homologue of PSD-95 (Woods and Bryant, 1991). In *Drosophila*, Dlg is concentrated in the NMJ, a glutamatergic synapse, where it co-localizes with
the Shaker K$^+$ channel and the Fasciclin II (FasII) cell adhesion molecule, two integral membrane proteins whose C termini bind directly to
the PDZ domains of Dlg. Synaptic localization of Shaker and FasII is
disrupted in *Dlg* mutants (Tejedor et al., 1997; Thomas et al., 1997;
Zito et al., 1997). Moreover, the C-terminal PDZ-binding motif of FasII
and Shaker is sufficient to confer synaptic targeting on a heterologous
protein in wild-type but not in *Dlg* mutant flies (Zito et al., 1997). Morphology of the *Drosophila* NMJ is also abnormal in *Dlg* mutants, particularly with respect to development of the subsynaptic reticulum, a complex infolding of the postsynaptic membrane (Lahey et al., 1994). These

genetic studies show that Dlg is important in vivo for synapse development and for synaptic localization of its membrane protein–binding
partners. Several *Drosophila* glutamate receptors have been cloned, but
they appear not to bind to Dlg, so it remains unclear how they are targeted to postsynaptic sites in the NMJ.

By extrapolation from genetic studies in *Drosophila*, it was natural to
speculate that the PSD-95 family of proteins in mammals is involved in
the targeting of NMDA receptors to the postsynaptic specialization.
However, a knockout of the *PSD95* gene in mice did not cause any detectable defect in synaptic localization of NMDA receptors, although
downstream signaling functions of NMDA receptors were drastically altered (Migaud et al., 1998). (However, as there are multiple members in
the PSD-95 family, it could be argued that close relatives could compensate for loss of PSD-95.) Dominant interfering approaches with peptides
that compete for PDZ binding also argued that PSD-95 and its relatives
are not essential for normal targeting of NMDA receptors (Passafaro et
al., 1999). Similarly, mice with targeted deletions of the cytoplasmic tails
of NR2A, NR2B, and NR2C (rendering them unable to bind to PSD-95)
also had apparently normal synaptic localization of NMDA receptors
(Sprengel et al., 1998). On the other hand, another study did find an
impaired synaptic localization of NR2B in mice expressing a "tailless"
NR2B (Mori et al., 1998). However, these results are complicated by the
deleterious effects of this mutation on brain development and organismal survival. Thus the balance of the evidence in mammals suggests that
the NR2 interaction with PSD-95 proteins is not required for synaptic
targeting of NMDA receptors. Therefore PSD-95 may not be involved in
the postsynaptic anchoring of NR2 subunits. However, an alternative
explanation is that additional redundant mechanisms exist for the proper
localization of NMDA receptors, such as via interactions with NR1. It
may be informative to study the targeting of other PSD-95-interacting
proteins to see whether their synaptic distribution is disrupted by the
loss of PSD-95 function. PSD-95 itself has multiple determinants within
its primary structure that are required to target it to postsynaptic sites
(Arnold and Clapham, 1999; Craven et al., 1999).

Assembly of a Postsynaptic Signaling Complex by PSD-95

Much stronger evidence exists that PSD-95 is critical for intracellular
signaling by the NMDA receptor. The *PSD95* knockout mice showed dramatic changes in NMDA receptor–dependent synaptic plasticity, shifting
the threshold between long-term potentiation (LTP) and long-term depression (LTD), greatly enhancing LTP magnitude, and disrupting spatial learning (Migaud et al., 1998). Unexpectedly, the mutant phenotype
suggests that PSD-95 normally has a restraining influence on LTP; the
mechanism is unknown, but perhaps PSD-95 links NMDA receptors to
negative regulators of synaptic transmission such as protein phosphatases.
Further evidence that PSD-95 is important for NMDA receptor signaling

comes from mouse mutants that have targeted deletions of the cytoplasmic tails of NR2A, NR2B, and NR2C (Sprengel et al., 1998). These mutations have essentially the same phenotype as deletions of the entire genes (in the case of NR2A and NR2C, without obviously affecting NMDA receptor expression or channel activity). However, it should be borne in mind that these targeted mutations delete the entire cytoplasmic tails (400–600 amino acids) of the NR2 subunits, therefore removing much more than just the PSD-95-binding site at the C terminus. Other protein interactions are undoubtedly lost in these tailless NR2 subunits; for instance, CaMKII can bind to the proximal region of the NR2B cytoplasmic tail as well as to the NR1 subunit (Strack and Colbran, 1998; Leonard et al., 1999).

Probably the most important role of PSD-95 in the function of NMDA receptors is its ability to serve as a "scaffold" for a signaling complex linked to the receptor (Fig. 7.2). The multidomain PSD-95 protein binds to a variety of cytoplasmic molecules that are involved in downstream signaling of NMDA receptors. For instance, neuronal nitric oxide synthase (nNOS) has been shown to bind to PDZ2 of PSD-95 via a PDZ-PDZ interaction (Brenman et al., 1996a). nNOS is a calcium/calmodulin-regulated enzyme that is selectively activated by calcium influx through NMDA receptors (as opposed to calcium entry through voltage-gated calcium channels). This specific coupling can be explained by the mutual binding of the NMDA receptor and nNOS to PSD-95, thereby bringing nNOS close to the mouth of the NMDA receptor channel (Fig. 7.2). This idea is supported by antisense suppression of PSD-95, which inhibits nNOS activation in response to NMDA receptor stimulation (Sattler et al., 1999). Thus by assembling a specific protein complex PSD-95 brings nNOS into a calcium signaling microdomain regulated by NMDA receptor activation. PSD-95 probably functions in an analogous manner with respect to its other interacting proteins, but the physiological significance of these other interactions is less well understood.

A growing list of signaling proteins is being identified that interact directly with PSD-95 and that presumably associate indirectly with NMDA receptors (Fig. 7.2). Many of these PSD-95 binding proteins are specifically enriched in the postsynaptic specialization. Examples include regulators or effectors of Ras and Rho GTPases. SynGAP, a GTPase activating protein for Ras, has a C terminus that interacts with all three PDZ domains of PSD-95 (Chen et al., 1998; Kim et al., 1998b). SynGAP is an abundant PSD protein whose role may be to inactivate Ras that is activated locally by NMDA receptor stimulation. It is also possible that SynGAP is involved in Ras modulation following Ras activation by tyrosine kinases, which are also present in the postsynaptic specialization. Other roles of SynGAP in postsynaptic function and plasticity remain to be clarified—it is a large protein that probably has additional functions unrelated to its RasGAP activity. Citron, an effector for Rho, can also bind PSD-95, specifically via PDZ3 (Furuyashiki et al., 1999; Zhang et al.,

1999). As Rho-type GTPases are involved in regulation of the cytoskeleton, Citron may be involved in NMDA receptor-dependent modulation of postsynaptic actin. Interestingly, Citron is concentrated selectively in the glutamatergic synapses of inhibitory interneurons (Zhang et al., 1999), illustrating the molecular heterogeneity of postsynaptic specializations that exists between different neuronal cell types.

Protein kinases and phosphatases are enriched postsynaptically where they can act on postsynaptic receptors and ion channels. An emerging theme in cell biology is that these enzymes are often targeted to their substrates by association with specific anchoring proteins. PSD-95 (and other proteins of the PSD, as discussed subsequently) may play such a role at the postsynaptic specialization. Nonreceptor tyrosine kinases of the Src family, such as Src and Fyn, are implicated in NMDA receptor modulation (Salter, 1998) and appear to be components of the NMDA receptor-associated protein complex (Yu et al., 1997; Tezuka et al., 1999). Association of Fyn with NMDA receptors appears to be mediated by PSD-95, which binds to the SH2 domain of Fyn via its PDZ3 domain (Tezuka et al., 1999). This ternary complex formation supported by PSD-95 enhances the phosphorylation of NR2A by Fyn in heterologous cells. Thus PSD-95 family proteins likely play a role in bringing Src-like tyrosine kinases to the PSD, specifically into the NMDA receptor complex (Fig. 7.2).

The GK domain of PSD-95 family proteins is catalytically inactive; instead, it acts as another site for protein-protein interaction that can recruit cytoplasmic signaling proteins to the NMDA receptor complex (Fig. 7.2). The GK domain binds to an abundant family of proteins in the PSD, termed GKAP/SAPAP/DAP, whose function is unclear (Kim et al., 1997; Naisbitt et al., 1997; Satoh et al., 1997; Takeuchi et al., 1997). In addition the GK domain binds to BEGAIN, a protein of unknown function (Deguchi et al., 1998); SpanGAP, a putative GTPase-activating protein for the small GTPase Rap (D. Pak and M. Sheng, unpublished observations); and MAP1A, a microtubule-binding protein (Brenman et al., 1998). It is unclear at present what these GK-binding proteins are doing for postsynaptic function in general, or for NMDA receptors in particular. Some of these may link NMDA receptors to other proteins in the PSD. For instance, GKAP binds to the Shank family of proteins, which in turn binds to Homer and cortactin (Naisbitt et al., 1999; Tu et al., 1999). This chain of protein-protein interactions could couple NMDA receptors to intracellular calcium stores, because Homer interacts with inositol 1,4,5-trisphosphate receptors (IP3R) and appears to be involved in excitation-calcium coupling (see Fig. 7.5 and the subsequent section on mGluRs). There is physiological evidence that NMDA receptors can communicate with intracellular calcium stores (Emptage et al., 1999), which might be explained by the Homer-Shank connection to the NMDA receptor complex.

Thus PSD-95 scaffolds a specific protein complex in the immediate vicinity of the NMDA receptor (Fig. 7.2), but this complex is also connected into the overall network of protein-protein interactions that underpins the PSD (Fig. 7.5). Further studies are needed to characterize the functional significance of the PSD-95-based complex in the post-synaptic specialization. It should be emphasized that in addition to NMDA receptors, PSD-95 probably organizes other membrane proteins (e.g., adhesion molecules, receptor tyrosine kinases, and ion channels) in the postsynaptic specialization; thus PSD-95-associated proteins may serve signaling functions that are not exclusively related to NMDA receptors. For instance, PSD-95 has been reported to bind kainate receptors, a less well-characterized class of ionotropic glutamate receptor that also exists at postsynaptic sites (Garcia et al., 1998). The number of proteins assigned to the PSD-95-based protein complex in the PSD will probably continue to grow in the coming years, whereas our understanding of the functional significance of these proteins in the postsynaptic specialization may continue to lag for some time.

Anchoring to the Postsynaptic Cytoskeleton via PSD-95

Anchoring to the subsynaptic cytoskeleton can be considered a final step in the process of postsynaptic targeting of glutamate receptors. By binding to cytoskeletal elements, a scaffold protein such as PSD-95 can indirectly connect NMDA receptors to the cytoskeleton. Members of the PSD-95 family of proteins have been shown to bind in vitro to band 4.1, an actin/spectrin-binding protein (Lue et al., 1994), and 4.1-like proteins are present in synapses (Walensky et al., 1999). Such an interaction has the potential to link NMDA receptors indirectly to F-actin, which is the predominant cytoskeleton in dendritic spines.

PSD-95 also interacts with microtubule-associated proteins. This is somewhat surprising, as microtubules are generally thought to be sparse or absent from dendritic spines, where PSD-95 is largely situated. Nevertheless, tubulin is present in PSD preparations, and microtubule-associated proteins such as MAP2 have been localized using immuno-cytochemistry at synapses (Kelly and Cotman, 1978; Caceres et al., 1984; Walsh and Kuruc, 1992). As mentioned earlier, the PSD-95 GK domain binds to MAP1A (Brenman et al., 1998). In addition, the third PDZ domain of PSD-95 interacts with CRIPT, a small polypeptide that binds directly to microtubules (Niethammer et al., 1998; Passafaro et al., 1999). The interaction of PSD-95 family proteins with microtubule-binding proteins such as CRIPT or MAP1A may link NMDA receptors indirectly to a postsynaptic tubulin-based cytoskeleton. It is controversial whether tubulin contributes to the cytoskeletal organization of the PSD in dendritic spines, though it may do so in an atypical way (Harris and Kater, 1994; Lai et al., 1998). However, microtubule anchoring could certainly be relevant for the minor fraction of excitatory synapses that occur on microtubule-rich dendritic shafts (such as the aspiny excitatory synapses of inhibitory interneurons).

PSD-95 Interactions with Other Membrane Proteins

Although the prior discussion focused on NMDA receptors (which are major components of the PSD), it should be emphasized that PSD-95 also interacts with other classes of membrane proteins through the PDZ–C-terminus mode of binding (Fig. 7.2). These include voltage-gated K^+ channels of the Shaker family (Kim et al., 1995), inward-rectifying K^+ channels (Cohen et al., 1996), the plasma membrane calcium pump PMCA4b (Kim et al., 1998a), and the putative cell adhesion molecule neuroligin (Irie et al., 1997). At least some of these interacting proteins have been localized in the PSD in mammalian brain, suggesting that they also interact with PSD-95 at postsynaptic sites. As noted earlier, Shaker K^+ channels and FasII have been shown to interact in vivo with Dlg in the *Drosophila* NMJ, an invertebrate glutamatergic synapse. Probably additional classes of membrane proteins will be identified that bind to PSD-95 in vertebrate excitatory synapses. That multiple membrane proteins interact with PSD-95 should not be surprising, since the post-synaptic membrane presumably contains ion channels, receptors, transporters, and adhesion molecules in addition to glutamate receptors. Thus for these presumably less abundant components of the postsynaptic specialization, PSD-95 may play a targeting and scaffolding role analogous to that discussed for NMDA receptors. This hypothesis is consistent with the high abundance of PSD-95 family proteins in the PSD. Nevertheless, despite being major components of the PSD, PSD-95 and its close relatives are not the only important scaffold proteins at the postsynaptic membrane. As discussed subsequently, interactions of other subunits of glutamate receptors have uncovered additional postsynaptic organizing proteins.

Other Interactions of the NMDA Receptor

NMDA receptors contain the essential NR1 subunit in addition to the NR2 subunits that bind to PSD-95 family proteins. Although it does not bind to PSD-95, the C-terminal tail of NR1 does interact with several other cytoplasmic proteins (Fig. 7.3). Thus from the NR1 subunit stems another branch of the network of protein interactions that integrates the NMDA receptor into the postsynaptic specialization. Like NR2-PSD-95 interactions, these NR1-mediated interactions may play a role in synaptic targeting of NMDA receptors and NMDA receptor signaling and modulation.

Alternative splicing of the cytoplasmic tail of NR1 (Hollmann et al., 1993) determines the set of NR1-interacting proteins. α-Actinin, an actin-binding protein of the spectrin superfamily, binds to the membrane proximal segment (termed C0) of NR1's cytoplasmic tail that is common to all splice variants (Wyszynski et al., 1997, 1998a; Fig. 7.3). As α-actinin and F-actin are enriched in the PSD, α-actinin interaction with

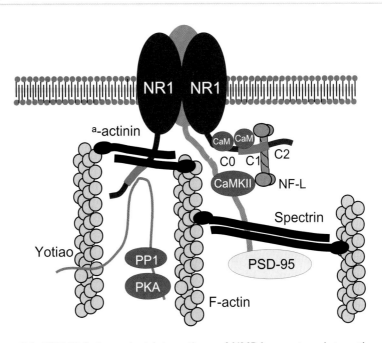

Figure 7.3. PSD-95-independent interactions of NMDA receptors. Interactions of the NR1 subunit (dark blue) are highlighted; the NR2 subunit (cyan) is shown in the background. C0 is the membrane-proximal region, and C1 and C2 are alternatively spliced segments of the NR1 cytoplasmic tail. Black filled ovals represent actin-binding domains of α-actinin and spectrin, which are depicted as antiparallel dimers. CaM, Ca²⁺/calmodulin; CaMKII, calmodulin-dependent kinase type II; NF-L, neurofilament-L; PKA, protein kinase A; PP1, protein phosphatase 1.

NR1 likely contributes to NMDA receptor-cytoskeletal anchoring at postsynaptic sites. Ca^{2+}/calmodulin binds to two sites in the NR1 tail, the C0 segment and the alternatively spliced C1 segment (Ehlers et al., 1996). The binding of Ca^{2+}/calmodulin to NR1 inhibits NMDA receptor opening and reduces mean channel open time (Ehlers et al., 1996). The calmodulin and α-actinin binding sites overlap in C0, and these proteins compete in vitro for binding to NR1 (Wyszynski et al., 1997). Competitive binding of calmodulin and α-actinin to the C0 segment appears to mediate calcium-dependent inactivation of NMDA receptors (Zhang et al., 1998; Krupp et al., 1999). Thus the actinin-NR1 interaction is intimately tied to the gating of NMDA receptors and could explain the observed dependence of NMDA receptor activity on the F-actin cytoskeleton (Rosenmund and Westbrook, 1993).

The C1 exon segment of the NR1 tail is not required for calcium-dependent inactivation of NMDA receptors despite binding calmodulin, but it does contain several PKC phosphorylation sites that play a role in the clustering of NR1 in heterologous cells (Ehlers et al., 1995). Two proteins, Yotiao (Lin et al., 1998) and neurofilament subunit NF-L (Ehlers et al., 1998), interact specifically with splice variants of NR1 containing the C1 exon (Fig. 7.3). Yotiao is an A-kinase anchoring protein that binds to both PKA and PP1 (Westphal et al., 1999) and is thus another example of a postsynaptic scaffold protein. It may function in synapses to organize a serine/threonine kinase-phosphatase complex anchored to NMDA receptors, thus facilitating NMDA receptor regulation by these enzymes.

Spectrin, a well-known actin-binding protein, binds to the cytoplasmic domains of NR1 and also has affinity for NR2A and NR2B (Wechsler and Teichberg, 1998). The spectrin-binding site in NR2B is distinct from the α-actinin and PSD-95 binding regions (Fig. 7.3). Brain spectrin is abundant in the PSD, and it could be a major cytoskeletal component of the cortical cytoskeleton under the postsynaptic membrane. Spectrin binding may offer another mode for attaching NMDA receptors to the postsynaptic actin cytoskeleton. Spectrin interaction with NR2B is sensitive to tyrosine phosphorylation and calcium, whereas the binding of spectrin to NR1 is inhibited by PKC/PKA phosphorylation and calmodulin (Wechsler and Teichberg, 1998). These findings suggest possible mechanisms for activity-dependent regulation of NMDA receptor anchoring to the cytoskeleton.

CaMKII is reported to bind to proximal regions of the NR2B cytoplasmic tail as well as to the NR1 subunit (Strack and Colbran, 1998; Leonard et al., 1999; Fig. 7.3). αCaMKII is one of the most abundant proteins of the PSD, but the functional significance of its direct interaction with NMDA receptor subunits is unclear, though NMDA receptors are substrates for CaMKII (Omkumar et al., 1996). αCaMKII can also localize to postsynaptic sites by forming heteromers with βCaMKII, which contains an F-actin targeting determinant (Shen et al., 1998).

In conclusion, the NMDA receptor can be considered a defining feature of glutamatergic synapses and an integral component of the PSD. NMDA receptor subunits interact with a multitude of intracellular proteins, either directly or indirectly via scaffold proteins such as PSD-95. The immediate network of protein interactions that anchors and envelopes NMDA receptors in the PSD (the NMDA receptor–PSD-95 complex; Fig. 7.2) can be regarded as a modular subarchitecture of the postsynaptic specialization. The molecular composition of this protein assembly can be altered to suit the different needs of various neuronal cell types.

AMPA Receptors

Synaptic Targeting of AMPA Receptors

Like NMDA receptors, AMPA receptors are also typically concentrated at postsynaptic sites of excitatory synapses. Recent evidence suggests, however, that synaptic AMPA receptor content is much more heterogeneous than that of NMDA receptors. Many excitatory synapses contain

NMDA receptors but not AMPA receptors, especially early in development, and the content of AMPA receptors in AMPA receptor–positive synapses is quite variable (Nusser et al., 1998a; Petralia et al., 1999; Takumi et al., 1999b). It is also becoming appreciated that a substantial fraction of AMPA receptors lies within intracellular compartments. Synapses that contain NMDA receptors but lack AMPA receptors may be morphological correlates of functionally "silent" synapses that have NMDA receptor– but not AMPA receptor–mediated responses (Malenka and Nicoll, 1997). Such "morphologically silent" synapses are also found in cultured neurons, where the synaptic distribution of AMPA receptors can be altered by activity (Carroll et al., 1999; Liao et al., 1999). Recent imaging studies provide direct evidence for a rapid activity-regulated delivery of AMPA receptors to dendritic spines (Shi et al., 1999). Thus the synaptic targeting of AMPA receptors appears to be regulated on a much shorter time scale than that of NMDA receptors, and its underlying mechanisms should reveal a more dynamic side of the postsynaptic specialization.

AMPA receptors are typically composed of heteromeric combinations of GluR1–4 subunits (Hollmann and Heinemann, 1994; Dingledine et al., 1999), whose membrane topology is similar to that of NMDA receptor subunits. In analogous fashion, the C-terminal cytoplasmic tails of AMPA receptor subunits also interact with intracellular proteins (Fig. 7.4). However, AMPA receptors bind to a distinctly different set of cytoplasmic

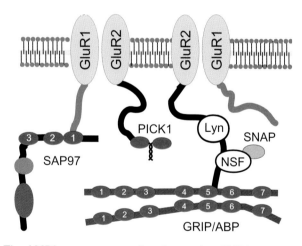

Figure 7.4. The AMPA receptor–associated complex. AMPA receptors are shown as heteromers of GluR1 and GluR2. The GluR2 C terminus binds to PDZ5 of GRIP/ABP and to the single PDZ domain of PICK1. PICK1 is depicted as a dimer via association of coiled-coil domains. GRIP/ABP is shown dimerized via PDZ4–6. Other binding partners for the PDZ domains of GRIP are discussed in the text. NSF and associated SNAP proteins bind to the GluR2 cytoplasmic tail in an ATP-dependent manner. Lyn tyrosine kinase may also interact with GluR2 (Hayashi et al., 1999). The GluR1 C terminus interacts with PDZ1 of SAP97, a member of the PSD-95 family of proteins.

proteins than do NMDA receptors. These differential protein interactions presumably reflect the differential regulation of NMDA and AMPA receptor channels. As with NMDA receptors, many AMPA receptor interactions are mediated by the binding of subunit C termini to specific PDZ-containing scaffold proteins.

Interactions with PDZ Proteins

Most AMPA receptor-binding proteins have been identified through their interactions with the GluR2/3 subunits. GluR2 and GluR3 subunits share a C-terminal sequence (-SVKI) that interacts with the fifth PDZ domain of GRIP (now termed GRIP1), a protein containing seven PDZ domains (Dong et al., 1997; Wyszynski et al., 1998b; Fig. 7.4). A protein with six PDZ domains (AMPA receptor binding protein or ABP) was also isolated by its binding to GluR2/3 (Srivastava et al., 1998). ABP appears to be a splice variant of a GRIP-related protein (also called GRIP2) that contains seven PDZs; ABP lacks the N terminus and PDZ7 of GRIP2 (Bruckner et al., 1999; Dong et al., 1999; Wyszynski et al., 1999). Although a large fraction of GluR2/3 appears to be biochemically associated with GRIP in vivo (Wyszynski et al., 1999), the function of the GluR2/3-GRIP interaction is still unclear. GRIP is enriched in synapses in the brain, but to only a modest degree compared with PSD-95. GRIP also differs from PSD-95 in being relatively abundant in intracellular compartments in dendrites and cell bodies of neurons, suggesting that GRIP may be involved in trafficking of AMPA receptors, rather than, or in addition to, synaptic anchoring (Dong et al., 1999; Wyszynski et al., 1999). The fact that overexpression of the C-terminal tail of GluR2 in neurons inhibits synaptic clustering of AMPA receptors (Dong et al., 1997) is consistent with either an anchoring or a trafficking role for GRIP. Blocking GluR2-GRIP interactions also prevents activation of silent synapses, suggesting that binding to GRIP is involved in recruitment of functional AMPA receptors to the synapse (Li et al., 1999).

GRIP and ABP, containing seven and six PDZ domains, respectively, have the capacity to assemble a large protein complex around AMPA receptors. GRIP binds to EphB2 and EphA7, members of the large family of Eph receptor tyrosine kinases, and to the EphrinB ligands for Eph receptors (Torres et al., 1998; Bruckner et al., 1999). Eph receptor–ephrin interactions are involved in axon guidance, cell migration, and establishment of tissue boundaries (Flanagan and Vanderhaeghen, 1998). Liprins, proteins that bind to the LAR family of receptor tyrosine phosphatases (Serra-Pagès et al., 1998), also bind to GRIP, utilizing PDZ6 (M. Wyszynski and M. Sheng, unpublished observations). LAR tyrosine phosphatases are involved in axon guidance during neural development (Van Vactor, 1998). Interestingly, LAR tyrosine phosphatases, Eph receptors, and ephrins appear to concentrate in synapses of mature neurons (Torres et al., 1998; M. Wyszynski and M. Sheng, unpublished observations). The physiological significance of these receptor tyrosine kinases and phosphatases in the postsynaptic specialization is unclear, as is the way in which GRIP-mediated interactions with these proteins are relevant to AMPA receptors. It should be emphasized that GRIP expression predates AMPA receptors during development, and, surprisingly, it also seems to localize in GABAergic axon terminals. These observations suggest that GRIP has additional functions unrelated to AMPA receptors and postsynaptic specializations.

In addition to GRIP/ABP, the C-terminal sequence of GluR2/3 mediates binding to PICK1 (Xia et al., 1999), another PDZ-containing protein previously shown to bind PKC (Staudinger et al., 1995; Fig. 7.4). Thus, like NMDA receptors, AMPA receptor subunits interact specifically with more than one PDZ-containing protein. PICK1 co-localizes with GluR2 in synapses and is capable of clustering GluR2 in heterologous cells (Xia et al., 1999), perhaps via coiled-coil dimerization of PICK1. As PKCα is enriched in synapses, the possibility exists that PICK1 may recruit PKC to AMPA receptors (in the same way that PSD-95 recruits Src family kinases to NMDA receptors), but this remains to be demonstrated. The relative importance of PICK1 and GRIP/ABP in AMPA receptor anchoring and trafficking in vivo is also unknown.

Interaction with NSF

As with NMDA receptors, C terminus–PDZ interactions are not the only means of associating AMPA receptors with intracellular proteins. Surprisingly, GluR2 binds to NSF, an ATPase involved in membrane fusion and vesicle trafficking (Nishimune et al., 1998; Osten et al., 1998; Song et al., 1998). NSF binding is mediated by a membrane proximal segment of GluR2's cytoplasmic tail, distinct from the C terminus that binds to GRIP or PICK1 (Fig. 7.4). Surface expression of AMPA receptors is inhibited by peptides that block the GluR-NSF interaction, suggesting that NSF is involved in the insertion of AMPA receptors into the postsynaptic membrane (Noel et al., 1999). Inhibition of NSF activity prevents LTP (Lledo et al., 1998), and the abundance of NSF in the PSD appears to be dynamically regulated (Hu et al., 1998). It is possible that the NSF-GluR2 interaction is relevant to synaptic plasticity by regulating the vesicle trafficking or protein unfolding of AMPA receptors (see Lin and Sheng, 1998, for review).

Kainate Receptors and δ Receptors

Kainate receptors, which can be pre- or postsynaptic, represent a third class of glutamate-gated ion channel (Dingledine et al., 1999). They are made up of subunits (GluR5–7, KA1, and KA2) that are homologous to AMPA receptor subunits. The cytoplasmic domains of GluR6 and KA2 bind to PDZ1 and to the SH3 and GK domains of PSD-95, respectively (Garcia et al., 1998). Co-expression with PSD-95 alters the desensitization properties of kainate receptors in heterologous expression systems. Obviously, binding to PSD-95 could be a mechanism for localizing and integrating kainate receptors in the PSD of glutamatergic synapses, but this remains to be confirmed.

Another member of the ionotropic glutamate receptor superfamily is GluRδ, distantly related (circa 25% identity) to NMDA and AMPA/kainate receptors. GluRδ2, the best-studied member of this family, is expressed specifically in cerebellar Purkinje cells and localized in the PSD. GluRδ2 binds to PSD-93/chapsyn-110 in vitro and co-localizes with PSD-93 in parallel fiber–Purkinje cell synapses in vivo, suggesting a possible role for this close relative of PSD-95 in anchoring of GluRδ2-containing glutamate receptors (Roche et al., 1999). If one includes the kainate and δ receptors, then it could be argued that the PSD-95 family of MAGUK proteins plays a role in the postsynaptic organization of three classes of ionotropic glutamate receptors (NMDA, kainate, and δ).

Metabotropic Glutamate Receptors

Glutamate acts on postsynaptic G-protein-coupled metabotropic receptors in addition to the ionotropic receptors discussed previously. mGluRs are divided into three classes based on G-protein coupling and pharmacology. Members of group 1 (mGluR1 and -5) are predominantly postsynaptic and activate PLC and intracellular calcium release, whereas group 2 (mGluR2 and -3) and group 3 (mGluR-4, -6, -7, and -8) receptors function at both pre- and postsynaptic sites and negatively couple to adenylyl cyclase. Recent evidence suggests that differential subcellular targeting among the mGluR family is probably determined by sequences in the cytoplasmic C-terminal tails (Stowell and Craig, 1999). The following discussion focuses on the postsynaptic group 1 mGluRs (mGluR1 and -5).

Unlike NMDA receptors and AMPA receptors, which are distributed across the width of the PSD, group 1 mGluRs are concentrated in a ring around the periphery of the PSD (Nusser et al., 1994; Luján et al., 1997; Fig. 7.5). The mechanism of segregation of ionotropic and metabotropic receptors at the subsynaptic level is unknown, but it probably depends on differential interactions of these membrane proteins with cytoplasmic proteins. Worley and co-workers discovered an interaction between mGluR1α (a splice variant of mGluR1), mGluR5, and the cytoplasmic protein Homer (Brakeman et al., 1997; Fig. 7.5). This binding occurs between the EVH domain of Homer and an internal sequence motif (PPXXF) in the cytoplasmic tail of mGluR1/5 (Tu et al., 1998; Xiao et al., 1998).

The originally identified Homer gene (now termed Homer1a) is an immediate early gene that is induced by synaptic activity (Brakeman et al., 1997). Subsequently, a family of Homer proteins was described that contain a coiled-coil domain that mediates self-association (Kato et al., 1998; Xiao et al., 1998). These "CC-Homers" multimerize to form multi-

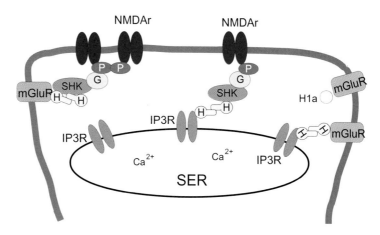

Figure 7.5. Postsynaptic interactions that link together NMDA receptors, mGluRs, and IP3 receptors. NMDA receptors and mGluRs are shown respectively at the center and around the periphery of the postsynaptic specialization. IP_3 receptors (IP3R) are present in the smooth endoplasmic reticulum (SER), an intracellular store for Ca^{2+}. Homer1a (H1a) is an immediate early gene product that cannot multimerize; it therefore interferes with the crosslinking of mGluR and IP3R by multimeric Homer (H). G, GKAP; mGluR, group 1 metabotropic glutamate receptor; P, PSD-95; SHK, Shank.

valent complexes that can crosslink multiple mGluR1α/5 molecules or link mGluR1α/5 to other proteins containing the PPXXF Homer-binding motif (Xiao et al., 1998). Homer1a cannot multimerize because it lacks the coiled-coil domain; instead, it behaves as an activity-inducible dominant negative that disrupts CC-Homer-mediated protein complexes (Xiao et al., 1998; Fig. 7.5). Several other proteins have been noted to contain the PPXXF Homer-binding consensus, including the IP3R, a downstream effector in the mGluR1/5-PLC signaling pathway. mGluR1α can be co-immunoprecipitated as a complex with Homer and IP3R from rat cerebellum (Tu et al., 1999), consistent with a biochemical linkage between group 1 mGluRs and the IP3R. Moreover, overexpression of the interfering Homer1a in Purkinje neurons impaired mGluR-evoked intracellular calcium release (Tu et al., 1998). These findings argue that the physical linkage of mGluR1α/5 to IP3R via CC-Homer complexes is physiologically important for postsynaptic calcium responses to mGluR stimulation. These findings have led to a model in which Homer brings IP3R into close proximity with the group 1 mGluRs, thereby allowing for more efficient coupling between these receptors (Fig. 7.5). The idea is that stimulation of mGluR1α/5 leads to highly localized production of IP3 at postsynaptic sites, such that the IP3R has to be in close proximity. The mGluR-Homer-IP3R complex is therefore another example of a postsynaptic signaling microdomain based on a series of protein-protein interactions (cf. the NMDA receptor–PSD-95–nNOS complex, discussed earlier; Figs 7.2 and 7.5).

IP3Rs are concentrated in the SER, an intracellular calcium store that extends into dendritic spines and that often approaches the postsynaptic specialization (Spacek and Harris, 1997). So the morphological basis exists in dendritic spines for a close interaction between postsynaptic mGluRs and intracellular calcium compartments (Fig. 7.5). In this context, it is of interest that ryanodine receptors, another class of intracellular calcium release channel, also contain the PPXXF consensus for Homer binding.

Homer also binds to the Shank family of proteins, which contains a PPXXF motif in its proline-rich domain (Tu et al., 1999). As Shank is a component of the NMDA receptor complex via binding to GKAP (Naisbitt et al., 1999), the Homer-Shank interaction potentially links the group 1 mGluRs to the NMDA receptor and its associated proteins. In addition, the group 1 mGluRs may interact directly with Shank. The cytoplasmic tail of mGluR5 ends with a sequence (-SSSL) suggestive of a PDZ-binding motif. This C-terminal sequence can bind to the PDZ domain of Shank, which recognizes the -T/SXL C-terminal sequence (Naisbitt et al., 1999; Tu et al., 1999). These direct and indirect interactions with Shank may contribute to the anchoring of group I mGluRs at postsynaptic sites and to their crosstalk with NMDA receptors (Fig. 7.5).

Is the binding to Homer and/or Shank important for determining the specific perisynaptic location of group 1 mGluRs? This question has not been addressed directly. One argument against such a targeting function is that although Homer and Shank are enriched in synapses, they are found throughout the PSD. This is in contrast to mGluR1/5, which are arranged around the periphery of the PSD. Thus the specific subsynaptic segregation of group 1 mGluRs cannot be explained simply by binding to Homer and Shank.

Overview of Glutamate Receptor Complexes in the Postsynaptic Specialization

As detailed previously, a dauntingly complicated picture has emerged of the interactions of glutamate receptors with cytoplasmic proteins. This seems particularly true of the NMDA receptors, which play diverse roles in postsynaptic signaling as a result of their calcium permeability. NMDA receptors utilize both NR1 and NR2 subunits to participate in multiple specific sets of interactions with cytoplasmic proteins (Figs. 7.2 and 7.3). These NMDA receptor-interacting proteins may have direct effects on receptor channel activity involving such proteins as α-actinin and calmodulin, or they may function as adaptor or scaffold proteins (such as PSD-95) that connect the receptor to a much more complex network of postsynaptic molecules. NMDA receptors do not associate with microtubules or actin directly but use several intermediary proteins. Their main mode of anchoring appears to be to the actin cytoskeleton; this can be through interactions with actin-binding proteins, such as α-actinin and spectrin, or more indirectly through scaffold proteins (e.g., via PSD-95-mediated interactions). The involvement of the actin cytoskeleton in NMDA receptor localization is attested to by the fact that depolymerization of F-actin by latrunculin A causes a 40% reduction in the number of synaptic NMDA receptor clusters (Allison et al., 1998). The incomplete effect of actin depolymerization is consistent with complex interactions between NMDA receptors and the cytoskeleton. NMDA receptors may also associate with microtubules, albeit indirectly via PSD-95 and CRIPT and MAP1A, and they may even interact with neurofilaments via less well-defined mechanisms.

Biochemically, AMPA receptors are easier to solubilize than NMDA receptors, and a large fraction is associated with intracellular compartments (Molnár et al., 1993; Baude et al., 1995; Nusser et al., 1998a; Petralia et al., 1999). Correlated with this biochemical behavior is the fact that relatively few interactions with cytoskeletal elements have been uncovered for AMPA receptors or AMPA receptor-binding proteins. Nevertheless, the different subunits of AMPA receptors do mediate interactions with various cytoplasmic proteins, including the multi-PDZ scaffolds GRIP and ABP. The binding of NSF to AMPA receptor GluR2 subunits in particular seems to suggest the dynamic nature of the trafficking and regulation of AMPA receptors.

Surprisingly, metabotropic and NMDA-type glutamate receptors may be physically linked via the Homer-Shank-GKAP assembly, which bridges group 1 mGluRs to the NMDA receptor–PSD-95 complex (Naisbitt et al., 1999; Tu et al., 1999). Such a network of synaptic protein interactions may functionally couple NMDA receptors to the IP3R or ryanodine receptor and thus contribute to the activity-dependent release of Ca^{2+} from intracellular stores (Emptage et al., 1999; see Svoboda and Mainen, 1999, for review). The C termini of mGluR1α/5 bind to the PDZ domain of Shank; thus all postsynaptic glutamate receptors participate in PDZ-based interactions. By binding to Homer via the PPXXF motif, mGluRs are also similar to AMPA and NMDA receptor subunits in utilizing internal segments of their cytoplasmic tail to associate with non-PDZ proteins. Thus much of the available "space" on the intracellular domains of glutamate receptors is utilized for interactions with cytoplasmic proteins.

Each class of glutamate receptor interacts at close range with a different set of cytoplasmic proteins, thus defining specialized protein microdomains within the PSD (Fig. 7.6). An important question is: how are the NMDA, AMPA, and metabotropic receptor complexes integrated into the overall architecture of the PSD? It is becoming clear that the networks of protein interactions that emanate from different glutamate receptors ultimately overlap and converge, for instance on the Shank family of proteins. Scaffold proteins like Shank that exist in the deeper parts of the PSD may play a higher-order organizational role to link together the different glutamate receptor complexes in the postsynaptic specialization (Naisbitt et al., 1999). It remains unclear at the ultrastructural level whether AMPA and NMDA receptors segregate into distinct

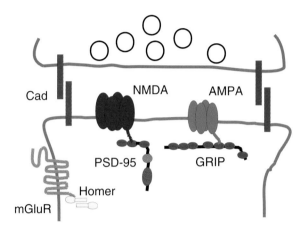

Figure 7.6. General organization of the excitatory postsynaptic specialization. NMDA and AMPA receptors are distributed centrally across the PSD, whereas mGluRs are distributed around the periphery. Each class of glutamate receptor interacts directly with a distinct scaffold protein. Homophilic interactions are shown between presynaptic and postsynaptic cadherins localized at the edge of the synapse.

islands within the PSD (as mGluRs are segregated to the periphery), or whether they are intimately mingled in the postsynaptic membrane (Fig. 7.6). At the functional level, it will be important to determine whether the biochemical interactions that link the various glutamate receptor complexes also mediate crosstalk between different glutamate receptors.

We are presently in a qualitative descriptive phase in the analysis of the molecular organization of the PSD. It will be critical to determine the stoichiometry and geometry of interactions involving glutamate receptors and other proteins if we are to appreciate the functional architecture of this postsynaptic specialization. It should also be clear that at present we have a rather static view of postsynaptic structure. An important future challenge is to uncover the dynamic developmental and activity-dependent regulation of the protein interactions underlying the PSD.

Molecular Heterogeneity of Excitatory Postsynaptic Specializations

As mentioned earlier, excitatory synapses are predominantly made onto dendritic spines; these occur abundantly on principal neurons such as pyramidal neurons of the forebrain, Purkinje cells of the cerebellum, and medium spiny cells of the striatum. On the other hand, aspiny neurons receive excitatory input on their dendritic shafts. Thus excitatory postsynaptic specializations can be classified as spiny or nonspiny. Within each class, the dimensions of excitatory synapses as measured by EM show great variation, as do the shapes of their PSDs. In addition to this morphological diversity, molecular and biochemical differences exist among excitatory synapses. For instance, α-actinin is found exclusively in spiny synapses (Wyszynski et al., 1998a), whereas Citron is selectively enriched in aspiny synapses (Zhang et al., 1999). Curiously, AMPA receptors are extractable by the nonionic detergent Triton X-100 from pyramidal neurons but not from GABAergic neurons (Allison et al., 1998), a finding that suggests that there are different cytoskeletal anchoring mechanisms in spiny versus nonspiny synapses.

Even glutamate receptors, a defining feature of excitatory synapses, show marked intersynapse differences. For instance, there is marked heterogeneity in AMPA receptor content between synapses of different size and at different stages of brain development (Nusser et al., 1998a; Petralia et al., 1999; Takumi et al., 1999b). On the other hand, most, if not all, excitatory synapses in principal neurons of the forebrain contain NMDA receptors at a relatively constant level, and synaptic accumulation of NMDA receptors occurs early in development. Curiously, this picture is reversed in cerebellar Purkinje neurons, which express AMPA receptors and metabotropic receptors in their spiny postsynaptic specializations, but no NMDA receptors.

Even within the same cell, there can be heterogeneity in synaptic expression of glutamate receptor subunits. In the fusiform cells of the dorsal cochlear nucleus, the AMPA receptor subunit, GluR4, and the metabotropic receptor, mGluR1α, are found only at auditory nerve synapses on basal dendrites but not at parallel fiber synapses on apical dendrites (Rubio and Wenthold, 1997). In Purkinje cells, postsynaptic GluRδ2 occurs at both parallel fiber synapses and climbing fiber synapses early in development but is restricted to parallel fiber synapses in adult animals (Roche et al., 1999). Thus cellular sorting mechanisms must exist for differential targeting of glutamate receptor subtypes and other postsynaptic proteins to specific subsets of synapses in the same neuron. In addition to these mechanisms, more subtle means must exist for regulating the levels of glutamate receptors and associated proteins in a synapse-specific fashion, perhaps dependent on synaptic activity. There is now accumulating evidence that the postsynaptic distribution of glutamate receptors can be modulated by neural activity, on a time scale of minutes to weeks (Craig, 1998). The molecular basis for such regulation is unknown, but as methods for analyzing individual synapses become more quantitative, it is to be expected that the nature and significance of the molecular heterogeneity of postsynaptic specializations will become apparent.

Adhesion Molecules of Excitatory Synapses

Adhesion molecules of the postsynaptic membrane are of interest because they presumably mediate the apposition of pre- and postsynaptic membranes, which is critical for synaptic development, maintenance, and function. In addition, selective expression of pairs of interacting adhesion molecules may confer specificity on presynaptic-postsynaptic contact, in this way contributing to the guidance of axons to their correct postsynaptic targets.

Cadherins are an expanding family of homophilic cell adhesion molecules that have emerged as prime candidates for mediating synaptic adhesion and specifying synaptic connectivity (see Serafini, 1999, and Shapiro and Colman, 1999, for review). N-cadherin and its associated cytoplasmic proteins α- and β-catenins are present in neuronal synapses, adjacent to the active zone presynaptically, and flanking the PSD postsynaptically (Fannon and Colman, 1996; Uchida et al., 1996; Fig. 7.6). They are thus well placed to "lock in" nascent synaptic contacts (Fannon and Colman, 1996). Because of their homophilic interactions, selective expression of a subset of the cadherin superfamily in neurons of a particular circuit could mediate the specific connectivity of that neuronal pathway.

Synaptic adhesion is very likely mediated by multiple sets of cell adhesion molecules. In addition to cadherins, some integral components of the PSD have also been proposed to play an adhesive role. In particular, such proteins could help to align the PSD with the presynaptic active zone. In the best-developed model, the postsynaptic membrane protein neuroligin binds across the synaptic cleft to β-neurexin, a presynaptic membrane protein implicated in transmitter release (Irie et al., 1997; Song et al., 1999; see Südhof and Scheller, this volume). This notion is made more attractive by the fact that the cytoplasmic tail of neuroligin binds to PSD-95 (Irie et al., 1997), whereas that of neurexin binds to CASK, a MAGUK protein that is at least partly presynaptic in location (Hata et al., 1996; Hsueh et al., 1998). Other candidate postsynaptic adhesion molecules include densin-180, an O-sialoglycoprotein of the PSD (Apperson et al., 1996), and syndecan-2, a cell surface heparan sulfate proteoglycan that also binds to CASK (Hsueh et al., 1998).

Postsynaptic Specializations of Inhibitory Synapses

Reflecting their distinct morphology and function, the postsynaptic specializations of inhibitory synapses differ fundamentally at the molecular level from excitatory synapses (Fig. 7.7). However, as with excitatory synapses and glutamate receptors, understanding of the molecular structure of inhibitory postsynaptic specializations has evolved largely from studies of GABA and glycine receptors and their associated proteins.

GABA_A Receptors and Associated Proteins

Fast inhibitory transmission in the brain is mediated mainly by ionotropic GABA (principally $GABA_A$) receptors, which are chloride channels with a diverse subunit composition drawn from at least six different gene

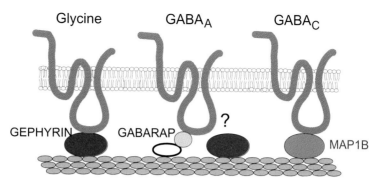

Figure 7.7. Anchoring of neurotransmitter receptors at inhibitory synapses. Ionotropic glycine, GABA$_A$, and GABA$_C$ receptors (only single subunits of these pentameric receptors are shown) interact with distinct microtubule-binding proteins via the major intracellular loop between TM3 and TM4. GABARAP is depicted as binding to a hypothetical microtubule-associated protein (white). Gephyrin (purple) is also present at GABAergic synapses, but the mechanism of GABA$_A$ receptor interaction with gephyrin is unknown.

families. In general, GABA$_A$ receptors are segregated from NMDA and AMPA receptors and concentrated specifically in the postsynaptic membrane of GABAergic synapses (Craig et al., 1994). However, detailed ultrastructural studies show a complicated differential subcellular distribution among GABA$_A$ receptor subtypes, with some subunits being extrasynaptic in location (Nusser ct al., 1998b).

Recent findings indicate that GABA$_A$ receptors interact with cytoplasmic proteins via their intracellular domains, even though GABA$_A$ receptors have a membrane topology that is quite different from that of glutamate receptors (Fig. 7.7). Like the nicotinic acetylcholine receptor, GABA$_A$ receptor subunits expose their C terminus on the extracellular side of the membrane. Their major cytoplasmic domain is the loop between the third and fourth transmembrane segments (TM3 and TM4). An 18-amino-acid segment of the intracellular loop of $\gamma2$ (the most abundant GABA$_A$ subunit in the CNS) binds to the GABA$_A$ receptor-associated protein (GABARAP) (Wang et al., 1999). GABARAP and GABA$_A$ receptors co-immunoprecipitate from brain extracts and co-localize in cultured cortical neurons. GABARAP is a small, 117-amino-acid polypeptide with circa 30% identity to light chain-3 of microtubule-associated proteins MAP1A and MAP1B. This sequence similarity suggests that GABARAP may also be a component of some MAP complexes, and indeed GABARAP fractionates with microtubules (Wang et al., 1999). Remarkably, MAP1B itself was isolated as a specific binding protein for the $\rho1$ subunit of ionotropic GABA$_C$ receptors (Hanley et al., 1999), a GABA receptor subtype expressed almost exclusively in the retina. This interaction is mediated by the intracellular loop of $\rho1$, which binds to a region adjacent to the microtubule-binding domain of MAP1B. As MAP1B binds in vitro to actin as well as microtubules, the interaction of ρ1 and MAP1B could potentially link GABA$_C$ receptors to both actin and tubulin cytoskeletons. MAP1B and GABARAP may play a role in the synaptic localization of ionotropic GABA receptors, or in their downstream signaling mechanisms, as has been proposed for PSD-95 and NMDA receptors. Further studies are required to test the physiological significance of MAP1B and GABARAP in the organization of GABAergic postsynaptic specializations.

Glycine Receptors and Associated Proteins

The molecular structure of the glycinergic postsynaptic region is presently better understood than that of GABAergic synapses. It has been known for several years that ionotropic glycine receptors (pentamers of a β subunit and various α subunits) bind to gephyrin, a microtubule-binding protein (see Kuhse et al., 1995, for review; Fig. 7.7). The gephyrin interaction is mediated by the intracellular loop between TM3 and TM4 of the glycine receptor β subunit (Meyer et al., 1995). The importance of gephyrin in synaptic localization of glycine receptors has been confirmed in knockout mice (Feng et al., 1998). An unexpected finding to emerge from this study was that gephyrin is also involved in molybdenum metabolism in vivo, though how this activity relates to postsynaptic function is unclear. It has also been known for some years that gephyrin is present in GABAergic as well as glycinergic synapses, but no direct biochemical interaction has been shown between gephyrin and GABA$_A$ receptor subunits (Fig. 7.7). Recent genetic evidence in mice, however, shows that gephyrin and GABA$_A$ receptors are dependent on each other for their proper postsynaptic localization, suggesting some indirect interaction (Essrich et al., 1998). Thus gephyrin may yet be revealed to be a master organizing molecule for inhibitory postsynaptic specializations.

Based on known interactions with GABARAP, MAP1B, and gephyrin, it appears that the primary mode of cytoskeletal attachment for inhibitory ionotropic receptors is via binding to specific microtubule-associated proteins (Fig. 7.7). Involvement of microtubule-binding proteins in the anchoring of GABA and glycine receptors correlates with the fact that inhibitory synapses form primarily on the shafts of dendrites (in which microtubules are abundant). The specific interaction of GABA$_A$, GABA$_C$, and glycine receptors with different microtubule-binding proteins is consistent with the idea that GABARAP, MAP1B, and gephyrin mediate the segregation of these various receptors among different subsets of inhibitory synapses. However, this idea is probably too simplistic, especially given the complex subcellular distribution of GABA$_A$ receptor subtypes (Nusser et al., 1998b). It remains to be seen whether these microtubule-associated proteins also function as adaptors for binding to other signal transduction proteins, in addition to their likely roles as cytoskeletal anchors. (Presumably, inhibitory ionotropic receptors are also associated directly or indirectly with a specific set of membrane proteins and modulatory enzymes at postsynaptic sites.) Obviously, much remains to be discovered about the organization of the postsynaptic specialization in GABAergic and glycinergic synapses.

Conclusion

From the many studies discussed in this chapter, a daunting picture is emerging of the molecular complexity of postsynaptic specializations. Neurotransmitter receptors (which are fundamental components of postsynaptic processes) utilize their cytoplasmic domains to interact with a variety of intracellular proteins. The receptor is thus anchored by, and integrated into, a sophisticated protein network that supports the receptor's postsynaptic actions and that modulates the receptor's activity. Each class of synapse appears to be constructed around a specific framework of proteins whose composition can be varied in time and space to give rise to heterogeneity of structure and function.

Within individual synapses, different subclasses of neurotransmitter receptors (e.g., AMPA, NMDA, and mGluRs) are segregated by different protein interactions into distinct molecular environments that correspond to highly localized signaling microdomains (Fig. 7.6). Examples include the PSD-95-based protein complex, which brings (among other things) calcium-regulated molecules into the sphere of influence of the NMDA receptor calcium channel. Thus at the subsynaptic level there is molecular and functional specialization.

This chapter has focused at the molecular level on the neurotransmitter receptors, because these are the sine qua non of postsynaptic specializations and because their study has revealed so much about the organization of postsynaptic complexes. However, it should be emphasized that other membrane proteins are likely to play critical roles in postsynaptic function. These include not only cell adhesion molecules but also voltage-gated ion channels, ion transporters, neurotransmitter transporters, and receptor tyrosine kinases, many of which remain to be identified. Receptor tyrosine kinases may be of special significance for synapse development, by analogy to the NMJ, where the postsynaptic receptor tyrosine kinase MuSK is of primary importance in setting up the postsynaptic specialization (Sanes and Lichtman, 1999). Identification and characterization of postsynaptic receptor tyrosine kinases in neurons is likely to provide clues to the mechanisms of development of neuron-neuron synapses.

In recent years, we have begun to glimpse something of the molecular architecture of the postsynaptic specialization, although our view at present is rather sketchy and static. The major challenges ahead are to quantify the stoichiometry of neurotransmitter receptors and associated proteins at each synapse; to visualize the geometry of postsynaptic protein complexes by microscopic and structural biological approaches; to uncover the dynamic regulation of the postsynaptic site during development and in response to activity; and to dissect out the functions of each component of the postsynaptic apparatus. Only with these advances will we truly be able to appreciate the functional architecture of the postsynaptic specialization.

Acknowledgments

I am grateful to Kristen Harris and Karen Szumowski for providing Fig. 7.1. Research reported in this chapter is supported by the National Institutes of Health (grant NS35050). M.S. is an assistant investigator of the Howard Hughes Medical Institute.

References

Allison, D. W., Gelfand, V. I., Spector, I., Craig, A. M. (1998). Role of actin in anchoring postsynaptic receptors in cultured hippocampal neurons: differential attachment of NMDA versus AMPA receptors. *J. Neurosci.* 18:2423–2436.

Apperson, M. L., Moon, I. S., Kennedy, M. B. (1996). Characterization of densin-180, a new brain-specific synaptic protein of the *O*-sialoglycoprotein family. *J. Neurosci.* 16:6839–6852.

Arnold, D. B., Clapham, D. E. (1999). Molecular determinants for subcellular localization of PSD-95 with an interacting K+ channel. *Neuron* 23:149–157.

Baude, A., Nusser, Z., Molnár, E., McIlhinney, A. J., Somogyi, P. (1995). High-resolution immunogold localization of AMPA type glutamate receptor subunits at synaptic and non-synaptic sites in rat hippocampus. *Neuroscience* 69:1031–1055.

Brakeman, P. R., Lanahan, A. A., O'Brien, R., Roche, K., Barnes, C. A., Huganir, R. L., Worley, P. F. (1997). Homer, a protein that selectively binds metabotropic glutamate receptors. *Nature* 386:284–288.

Brenman, J. E., Chao, D. S., Gee, S. H., McGee, A. W., Craven, S. E., Santillano, D. R., Wu, Z., Huang, F., Xia, H., Peters, M. F., Froehner, S. C., Bredt, D. S. (1996a). Interaction of nitric oxide synthase with the postsynaptic density protein PSD-95 and a1-syntrophin mediated by PDZ domains. *Cell* 84:757–767.

Brenman, J. E., Christopherson, K. S., Craven, S. E., McGee, A. W., Bredt, D. S. (1996b). Cloning and characterization of postsynaptic density 93, a nitric oxide synthase interacting protein. *J. Neurosci.* 16:7407–7415.

Brenman, J. E., Topinka, R. J., Cooper, E. C., McGee, A. W., Rosen, J., Milroy, T., Ralston, H. J., Bredt, D. S. (1998). Localization of postsynaptic density-93 to dendritic microtubules and interaction with microtubule-associated protein 1A. *J. Neurosci.* 18:8805–8813.

Bruckner, K., Pablo Labrador, J., Scheiffele, P., Herb, A., Seeburg, P. H., Klein, R. (1999). EphrinB ligands recruit GRIP family PDZ adaptor proteins into raft membrane microdomains. *Neuron* 22:511–524.

Caceres, A., Binder, L. I., Payne, M. R., Bender, P., Rebhun, L., Steward, O. (1984). Differential subcellular localization of tubulin and the microtubule-associated protein MAP2 in brain tissue as revealed by immunocytochemistry with monoclonal hybridoma antibodies. *J. Neurosci.* 4:394–410.

Carlin, R., Grab, D., Cohen, R., Siekevitz, P. (1980). Isolation and characterization of postsynaptic densities from various brain regions: enrichment of different types of postsynaptic densities. *J. Cell Biol.* 86:831–845.

Carroll, R. C., Lissin, D. V., von Zastrow, M., Nicoll, R. A., Malenka, R. C. (1999). Rapid redistribution of glutamate receptors contributes to long-term depression in hippocampal cultures. *Nature Neurosci.* 2:454–460.

Chen, H. J., Rojas-Soto, M., Oguni, A., Kennedy, M. B. (1998). A synaptic Ras-GTPase activating protein (p135 SynGAP) inhibited by CaM kinase II. *Neuron* 20:895–904.

Cho, K.-O., Hunt, C. A., Kennedy, M. B. (1992). The rat brain postsynaptic density fraction contains a homolog of the Drosophila discs-large tumor suppressor protein. *Neuron* 9:929–942.

Cohen, N. A., Brenman, J. E., Snyder, S., Bredt, D. S. (1996). Binding of the inward rectifier K+ channel Kir 2.3 to PSD-95 is regulated by protein kinase A phosphorylation. *Neuron* 17:759–767.

Comery, T. A., Harris, J. B., Willems, P. J., Oostra, B. A., Irwin, S. A., Weiler, I. J., Greenough, W. T. (1997). Abnormal dendritic spines in fragile X knockout mice: maturation and pruning deficits. *Proc. Natl. Acad. Sci. USA* 94:5401–5405.

Craig, A. M. (1998). Activity and synaptic receptor targeting: the long view. *Neuron* 21:459–462.

Craig, A. M., Blackstone, C. D., Huganir, R. L., Banker, G. (1994). Selective clustering of glutamate and γ-aminobutyric acid receptors opposite terminals releasing the corresponding neurotransmitters. *Proc. Natl. Acad. Sci. USA* 91:12,373–12,377.

Craven, S. E., El-Husseini, A. E., Bredt, D. S. (1999). Synaptic targeting of the postsynaptic density protein PSD-95 mediated by lipid and protein motifs. *Neuron* 22:497–509.

Dailey, M. E., Smith, S. J. (1996). The dynamics of dendritic structure in developing hippocampal slices. *J. Neurosci.* 16:2983–2994.

Deguchi, M., Hata, Y., Takeuchi, M., Ide, N., Hirao, K., Yao, I., Irie, M., Toyoda, A., Takai, Y. (1998). BEGAIN (brain-enriched guanylate kinase–associated protein), a novel neuronal PSD-95/SAP90-binding protein. *J. Biol. Chem.* 273:26,269–26,272.

Dingledine, R., Borges, K., Bowie, D., Traynelis, S. F. (1999). The glutamate receptor ion channels. *Pharmacol. Rev.* 51:7–61.

Dong, H., O'Brien, R. J., Fung, E. T., Lanahan, A. A., Worley, P. F., Huganir, R. L. (1997). GRIP: a synaptic PDZ domain–containing protein that interacts with AMPA receptors. *Nature* 386:279–284.

Dong, H., Zhang, P., Song, I., Petralia, R. S., Liao, D., Huganir, R. L. (1999). Characterization of the glutamate receptor–interacting proteins GRIP1 and GRIP2. *J. Neurosci.* 19:6930–6941.

Doyle, D. A., Lee, A., Lewis, J., Kim, E., Sheng, M., MacKinnon, R. (1996). Crystal structures of a complexed and peptide-free membrane protein–binding domain: molecular basis of peptide recognition by PDZ. *Cell* 85:1067–1076.

Ehlers, M. D., Tingley, W. G., Huganir, R. L. (1995). Regulated subcellular distribution of the NR1 subunit of the NMDA receptor. *Science* 269:1734–1737.

Ehlers, M. D., Zhang, S., Bernhardt, J. P., Huganir, R. L. (1996). Inactivation of NMDA receptors by direct interaction of calmodulin with the NR1 subunit. *Cell* 84:745–755.

Ehlers, M. D., Fung, E. T., O'Brien, R. J., Huganir, R. L. (1998). Splice variant—specific interaction of the NMDA receptor subunit NR1 with neuronal intermediate filaments. *J. Neurosci.* 18:720–730.

Emptage, N., Bliss, T. V., Fine, A. (1999). Single synaptic events evoke NMDA receptor-mediated release of calcium from internal stores in hippocampal dendritic spines. *Neuron* 22:115–124.

Engert, F., Bonhoeffer, T. (1999). Dendritic spine changes associated with hippocampal long-term synaptic plasticity. *Nature* 399:66–70.

Essrich, C., Lopez, M., Benson, J. A., Fritschy, J.-M., Luscher, B. (1998). Post-synaptic clustering of major GABAa receptor subtypes requires the g_2 subunit and gephyrin. *Nature Neurosci.* 1:563–571.

Fannon, A. M., Colman, D. R. (1996). A model for central synaptic junctional complex formation based on the differential adhesive specificities of the cadherins. *Neuron* 17:423–434.

Feng, G., Tintrup, H., Kirsch, J., Nichol, M., Kuhse, J., Betz, H., Sanes, J. R. (1998). Dual requirement for gephyrin in glycine receptor clustering and molybdoenzyme activity. *Science* 282:1321–1324.

Fiala, J. C., Feinberg, M., Popov, V., Harris, K. M. (1998). Synaptogenesis via dendritic filopodia in developing hippocampal area CA1. *J. Neurosci.* 18:8900–8911.

Finch, E. A., Augustine, G. J. (1998). Local calcium signalling by inositol-1,4,5-trisphosphate in Purkinje cell dendrites. *Nature* 396:753–756.

Fischer, M., Kaech, S., Knutti, D., Matus, A. (1998). Rapid actin-based plasticity in dendritic spines. *Neuron* 20:847–854.

Flanagan, J. G., Vanderhaeghen, P. (1998). The ephrins and Eph receptors in neural development. *Annu. Rev. Neurosci.* 21:309–345.

Furuyashiki, T., Fujisawa, K., Fujita, A., Madaule, P., Uchino, S., Mishina, M., Bito, H., Narumiya, S. (1999). Citron, a Rho-target, interacts with PSD-95/SAP-90 at glutamatergic synapses in the thalamus. *J. Neurosci.* 19:109–118.

Garcia, E. P., Mehta, S., Blair, L. A., Wells, D. G., Shang, J., Fukushima, T., Fallon, J. R., Garner, C. C., Marshall, J. (1998). SAP90 binds and clusters kainate receptors causing incomplete desensitization. *Neuron* 21:727–739.

Halpain, S., Hipolito, A., Saffer, L. (1998). Regulation of F-actin stability in dendritic spines by glutamate receptors and calcineurin. *J. Neurosci.* 18:9835–9844.

Hanley, J. G., Koulen, P., Bedford, F., Gordon-Weeks, P. R., Moss, S. J. (1999). The protein MAP-1B links GABAc receptors to the cytoskeleton at retinal synapses. *Nature* 397:66–69.

Harris, K. (1999). Structure, development, and plasticity of dendritic spines. *Curr. Opin. Neurobiol.* 9:343–348.

Harris, K., Kater, S. (1994). Dendritic spines: cellular specializations imparting both stability and flexibility to synaptic function. *Annu. Rev. Neurosci.* 17:341–371.

Hata, Y., Butz, S., Südhof, T. C. (1996). CASK: a novel *Dlg*/PSD95 homolog with an N-terminal calmodulin-dependent protein kinase domain identified by interaction with neurexins. *J. Neurosci.* 16:2488–2494.

Hayashi, T., Umemori, H., Mishina, M., Yamamoto, T. (1999). The AMPA receptor interacts with and signals through the protein tyrosine kinase Lyn. *Nature* 397:72–76.

Hillier, B. J., Christopherson, K. S., Prehoda, K. E., Bredt, D. S., Lim, W. A. (1999). Unexpected modes of PDZ domain scaffolding revealed by structure of nNOS-syntrophin complex. *Science* 284:812–815.

Hollmann, M., Heinemann, S. (1994). Cloned glutamate receptors. *Annu. Rev. Neurosci.* 17:31–108.

Hollmann, M., Boulter, J., Maron, C., Beasley, L., Sullivan, J., Pecht, G., Heinemann, S. (1993). Zinc potentiates agonist-induced currents at certain splice variants of the NMDA receptor. *Neuron* 10:943–954.

Horch, H. W., Kruttgen, A., Portbury, S. D., Katz, L. C. (1999). Destabilization of cortical dendrites and spines by BDNF. *Neuron* 23:353–364.

Hsueh, Y.-P., Kim, E., Sheng, M. (1997). Disulfide-linked head-to-head multimer-ization in the mechanism of ion channel clustering by PSD-95. *Neuron* 18: 803–814.

Hsueh, Y.-P., Yang, F.-C., Kharazia, V., Naisbitt, S., Cohen, A. R., Weinberg, R. J., Sheng, M. (1998). Direct interaction of CASK/LIN-2 and syndecan heparan sulfate proteoglycan and their overlapping distribution in neuronal synapses. *J. Cell Biol.* 142:139–151.

Hu, B.-R., Park, M., Martone, M. E., Fischer, W. H., Ellisman, M. H., Zivin, J. A. (1998). Assembly of proteins to postsynaptic densities after transient cere-bral ischemia. *J. Neurosci.* 18:625–633.

Irie, M., Hata, Y., Takeuchi, M., Ichtchenko, K., Toyoda, A., Hirao, K., Takai, Y., Rosahl, T. W., Südhof, T. C. (1997). Binding of neuroligins to PSD-95. *Science* 277:1511–1515.

Kato, A., Ozawa, F., Saitoh, Y., Fukazawa, Y., Sugiyama, H., Inokuchi, K. (1998). Novel members of the Vesl/Homer family of PDZ proteins that bind meta-botropic glutamate receptors. *J. Biol. Chem.* 273:23,969–23,975.

Kelly, P. T., Cotman, C. W. (1978). Synaptic proteins. Characterization of tubu-lin and actin and identification of a distinct postsynaptic density polypep-tide. *J. Cell Biol.* 79:173–183.

Kennedy, M. B. (1997). The postsynaptic density at glutamatergic synapses. *Trends Neurosci.* 20:264–268.

Kim, E., Niethammer, M., Rothschild, A., Jan, Y. N., Sheng, M. (1995). Cluster-ing of shaker-type K^+ channels by interaction with a family of membrane-associated guanylate kinases. *Nature* 378:85–88.

Kim, E., Cho, K.-O., Rothschild, A., Sheng, M. (1996). Heteromultimerization and NMDA receptor–clustering activity of chapsyn-110, a member of the PSD-95 family of proteins. *Neuron* 17:103–113.

Kim, E., Naisbitt, S., Hsueh, Y.-P., Rao, A., Rothschild, A., Craig, A. M., Sheng, M. (1997). GKAP, a novel synaptic protein that interacts with the guanylate kinase–like domain of the PSD-95/SAP90 family of channel clustering mole-cules. *J. Cell Biol.* 136:669–678.

Kim, E., DeMarco, S. J., Marfatia, S. M., Chishti, A. H., Sheng, M., Strehler, E. E. (1998a). Plasma membrane Ca^{2+} ATPase isoform 4b binds to membrane-associated guanylate kinase (MAGUK) proteins via their PDZ (PSD-95/Dlg/ZO-1) domains. *J. Biol. Chem.* 273:1591–1595.

Kim, J. H., Liao, D., Lau, L. F., Huganir, R. L. (1998b). SynGAP: a synaptic Ras-GAP that associates with the PSD-95/SAP90 protein family. *Neuron* 20:683–691.

Kistner, U., Wenzel, B. M., Veh, R. W., Cases-Langhoff, C., Garner, A. M., Ap-peltauer, U., Voss, B., Gundelfinger, E. D., Garner, C. C. (1993). SAP90, a rat presynaptic protein related to the product of the *Drosophila* tumor suppres-sor gene *Dlg-A*. *J. Biol. Chem.* 268:4580–4583.

Kornau, H.-C., Schenker, L. T., Kennedy, M. B., Seeburg, P. H. (1995). Domain interaction between NMDA receptor subunits and the postsynaptic density protein PSD-95. *Science* 269:1737–1740.

Kornau, H.-C., Seeburg, P. H., Kennedy, M. B. (1997). Interaction of ion chan-nels and receptors with PDZ domain proteins. *Curr. Opin. Neurobiol.* 7:368–373.

Krupp, J. J., Vissel, B., Thomas, C. G., Heinemann, S. F., Westbrook, G. L. (1999). Interactions of calmodulin and alpha-actinin with the NR1 subunit modulate Ca^{2+}-dependent inactivation of NMDA receptors. *J. Neurosci.* 19: 1165–1178.

Kuhse, J., Betz, H., Kirsch, J. (1995). The inhibitory glycine receptor: architecture, synaptic localization and molecular pathology of a postsynaptic ion-channel complex. *Curr. Opin. Neurobiol.* 5:318–323.

Lahey, T., Gorczyca, M., Jia, X.-X., Budnik, V. (1994). The *Drosophila* tumor suppressor gene *Dlg* is required for normal synaptic bouton structure. *Neuron* 13:823–835.

Lai, S. L., Ling, S. C., Kuo, L. H., Shu, Y. C., Chow, W. Y., Chang, Y. C. (1998). Characterization of granular particles isolated from postsynaptic densities. *J. Neurochem.* 71:1694–1701.

Leonard, A. S., Lim, I. A., Hemsworth, D. E., Horne, M. C., Hell, J. W. (1999). Calcium/calmodulin-dependent protein kinase II is associated with the N-methyl-D-aspartate receptor. *Proc. Natl. Acad. Sci. USA* 96:3239–3244.

Li, P., Kerchner, G. A., Sala, C., Wei, F., Huettner, J. E., Sheng, M., Zhuo, M. (1999). AMPA receptor—PDZ interactions in facilitation of spinal sensory synapses. *Nature Neurosci.* 2:972–977.

Liao, D., Zhang, X., O'Brien, R., Ehlers, M. D., Huganir, R. L. (1999). Regulation of morphological postsynaptic silent synapses in developing hippocampal neurons. *Nature Neurosci.* 2:37–43.

Lin, J. W., Sheng, M. (1998). NSF and AMPA receptors get physical. *Neuron* 21:267–270.

Lin, J. W., Wyszynski, M., Madhavan, R., Sealock, R., Kim, J. U., Sheng, M. (1998). Yotiao, a novel protein of neuromuscular junction and brain that interacts with specific splice variants of NMDA receptor subunit NR1. *J. Neurosci.* 18:2017–2027.

Lledo, P.-M., Zhang, X., Südhof, T. C., Malenka, R. C., Nicoll, R. A. (1998). Postsynaptic membrane fusion and long-term potentiation. *Science* 279:399–403.

Lue, R. A., Marfatia, S. M., Branton, D., Chishti, A. H. (1994). Cloning and characterization of hDlg: the human homologue of the Drosophila Discs large tumor suppressor binds to protein 4.1. *Proc. Natl. Acad. Sci. USA* 91:9818–9822.

Luján, R., Roberts, J. D. B., Shigemoto, R., Ohishi, H., Somogyi, P. (1997). Differential plasma membrane distribution of metabotropic glutamate receptors mGluR1a, mGluR2 and mGluR5 relative to neurotransmitter release sites. *J. Chem. Neuroanat.* 13:219–241.

Luo, L., Hensch, T. K., Ackerman, L., Barbel, S., Jan, L. Y., Jan, Y. N. (1996). Differential effects of the Rac GTPase on Purkinje cell axons and dendritic trunks and spines. *Nature* 379:837–840.

McKinney, R. A., Capogna, M., Durr, R., Gahwiler, B. H., Thompson, S. M. (1999). Miniature synaptic events maintain dendritic spines via AMPA receptor activation. *Nature Neurosci.* 2:44–49.

Malenka, R. C., Nicoll, R. A. (1997). Silent synapses speak up. *Neuron* 19:473–476.

Maletic-Savatic, M., Malinow, R., Svoboda, K. (1999). Rapid dendritic morphogenesis in CA1 hippocampal dendrites induced by synaptic activity. *Science* 283:1923–1927.

Meyer, G., Kirsch, J., Betz, H., Langosch, D. (1995). Identification of a gephyrin binding motif on the glycine receptor β subunit. *Neuron* 15:563–572.

Migaud, M., Charlesworth, P., Dempster, M., Webster, L. C., Watabe, A. M., Mahkinson, M., He, Y., Ramsay, M. F., Morris, R. G., Morrison, J. H., O'Dell, T. J., Grant, S. G. (1998). Enhanced long-term potentiation and impaired learning in mice with mutant postsynaptic density-95 protein. *Nature* 396:433–439.

Molnár, E., Baude, A., Richmond, S. A., Patel, P. B., Somogyi, P., McIlhinney, R. A. J. (1993). Biochemical and immunocytochemical characterization of antipeptide antibodies to a cloned GluR1 glutamate receptor subunit: cellular and subcellular distribution in the rat forebrain. *Neuroscience* 53:307–326.

Mori, H., Manabe, T., Watanabe, M., Satoh, Y., Suzuki, N., Toki, S., Nakamura, K., Yagi, T., Kushiya, E., Takahashi, T., Inoue, Y., Sakimura, K., Mishina, M. (1998). Role of the carboxy-terminal region of the GluR epsilon2 subunit in synaptic localization of the NMDA receptor channel. *Neuron* 21:571–580.

Müller, B. M., Kistner, U., Veh, R. W., Cases-Langhoff, C., Becker, B., Gundelfinger, E. D., Garner, C. C. (1995). Molecular characterization and spatial distribution of SAP97, a novel presynaptic protein homologous to SAP90 and the *Drosophila* discs-large tumor suppressor protein. *J. Neurosci.* 15:2354–2366.

Müller, B. M., Kistner, U., Kindler, S., Chung, W. J., Kuhlendahl, S., Fenster, S. D., Lau, L. F., Veh, R. W., Huganir, R. L., Gundelfinger, E. D., Garner, C. C. (1996). SAP102, a novel postsynaptic protein that interacts with the cytoplasmic tail of the NMDA receptor subunit NR2B. *Neuron* 17:255–265.

Naisbitt, S., Kim, E., Weinberg, R. J., Rao, A., Yang, F.-C., Craig, A. M., Sheng, M. (1997). Characterization of guanylate kinase–associated protein, a postsynaptic density protein at excitatory synapses that interacts directly with postysynaptic density-95/synapse-associated protein 90. *J. Neurosci.* 17:5687–5696.

Naisbitt, S., Kim, E., Tu, J. C., Xiao, B., Sala, C., Valtschanoff, J., Weinberg, R. J., Worley, P. F., Sheng, M. (1999). Shank, a novel family of postsynaptic density proteins that binds to the NMDA receptor/PSD-95/GKAP complex and cortactin. *Neuron* 23:569–582.

Niethammer, M., Kim, E., Sheng, M. (1996). Interaction between the C terminus of NMDA receptor subunits and multiple members of the PSD-95 family of membrane-associated guanylate kinases. *J. Neurosci.* 16:2157–2163.

Niethammer, M., Valtschanoff, J. G., Kapoor, T. M., Allison, D. W., Weinberg, R. J., Craig, A. M., Sheng, M. (1998). CRIPT, a novel postsynaptic protein that binds to the third PDZ domain of PSD-95/SAP90. *Neuron* 20:693–707.

Nishimune, A., Isaac, J. T., Molnár, E., Noel, J., Nash, S. R., Tagaya, M., Collingridge, G. L., Nakanishi, S., Henley, J. M. (1998). NSF binding to GluR2 regulates synaptic transmission. *Neuron* 21:87–97.

Noel, J., Ralph, G. S., Pickard, L., Williams, J., Molnár, E., Uney, J. B., Collingridge, G. L., Henley, J. M. (1999). Surface expression of AMPA receptors in hippocampal neurons is regulated by an NSF-dependent mechanism. *Neuron* 23:365–376.

Nusser, Z., Mulvihill, E., Streit, P., Somogyi, P. (1994). Subsynaptic segregation of metabotropic and ionotropic glutamate receptors as revealed by immunogold localization. *Neuroscience* 61:421–427.

Nusser, Z., Lujan, R., Laube, G., Roberts, J. D., Molnar, E., Somogyi, P. (1998a). Cell type and pathway dependence of synaptic AMPA receptor number and variability in the hippocampus. *Neuron* 21:545–559.

Nusser, Z., Sieghart, W., Somogyi, P. (1998b). Segregation of different GABA$_A$ receptors to synaptic and extrasynaptic membranes of cerebellar granule cells. *J. Neurosci.* 18:1693–1703.

O'Brien, R., Lau, L., Huganir, R. (1998). Molecular mechanisms of glutamate receptor clustering at excitatory synapses. *Curr. Opin. Neurobiol.* 8:364–369.

Omkumar, R. V., Kiely, M. J., Rosenstein, A. J., Min, K. T., Kennedy, M. B. (1996). Identification of a phosphorylation site for calcium/calmodulin dependent protein kinase II in the NR2B subunit of the *N*-methyl-D-aspartate receptor. *J. Biol. Chem.* 271:31,670-31,678.

Osten, P., Srivastava, S., Inman, G. J., Vilim, F. S., Khatri, L., Lee, L. M., States, B. A., Einheber, S., Milner, T. A., Hanson, P. I., Ziff, E. B. (1998). The AMPA receptor GluR2 C terminus can mediate a reversible, ATP-dependent interaction with NSF and alpha- and beta-SNAPs. *Neuron* 21:99–110.

Paoletti, P., Ascher, P. (1994). Mechanosensitivity of NMDA receptors in cultured mouse central neurons. *Neuron* 13:645–655.

Passafaro, M., Sala, C., Niethammer, M., Sheng, M. (1999). Microtubule binding by CRIPT and its potential role in the synaptic clustering of PSD-95. *Nature Neurosci.* 2:1063–1069.

Petralia, R. S., Esteban, J. A., Wang, Y.-X., Partridge, J. G., Zhao, H.-M., Wenthold, R. J., Malinow, R. (1999). Selective acquisition of AMPA receptors over postnatal development suggests a molecular basis for silent synapses. *Nature Neurosci.* 2:31–36.

Ponting, C. P., Phillips, C., Davies, K. E., Blake, D. J. (1997). PDZ domains: targeting signalling molecules to sub-membranous sites. *BioEssays* 19:469–479.

Roche, K. W., Ly, C. D., Petralia, R. S., Wang, Y.-X., McGee, A. W., Bredt, D. S., Wenthold, R. J. (1999). Postsynaptic density-93 interacts with the delta2 glutamate receptor subunit at parallel fiber synapses. *J. Neurosci.* 19:3926–3934.

Rosenmund, C., Westbrook, G. L. (1993). Calcium-induced actin depolymerization reduces NMDA channel activity. *Neuron* 10:805–814.

Rubio, M. E., Wenthold, R. J. (1997). Glutamate receptors are selectively targeted to postsynaptic sites in neurons. *Neuron* 18:939–950.

Salter, M. W. (1998). Src, N-methyl-D-aspartate (NMDA) receptors, and synaptic plasticity. *Biochem. Pharmacol.* 56:789–798.

Sanes, J. R., Lichtman, J. W. (1999). Development of the vertebrate neuromuscular junction. *Annu. Rev. Neurosci.* 22:389–442.

Satoh, K., Yanai, H., Senda, T., Kohu, K., Nakamura, T., Okumura, N., Matsumine, A., Kobayashi, S., Toyoshima, K., Akiyama, T. (1997). DAP-1, a novel protein that interacts with the guanylate kinase–like domains of hDLG and PSD-95. *Genes Cells* 2:415–424.

Sattler, R., Xiong, Z., Lu, W. Y., Hafner, M., MacDonald, J. F., Tymianski, M. (1999). Specific coupling of NMDA receptor activation to nitric oxide neurotoxicity by PSD-95 protein. *Science* 284:1845–1848.

Schiller, J., Schiller, Y., Clapham, D. E. (1998). NMDA receptors amplify calcium influx into dendritic spines during associative pre- and postsynaptic activation. *Nature Neurosci.* 1:114–118.

Serafini, T. (1999). Finding a partner in a crowd: neuronal diversity and synaptogenesis. *Cell* 98:133–136.

Serra-Pagès, C., Medley, Q. G., Tang, M., Hart, A., Streuli, M. (1998). Liprins, a family of LAR transmembrane protein-tyrosine phosphatase–interacting proteins. *J. Biol. Chem.* 273:15,611–15,620.

Shapiro, L., Colman, D. R. (1999). The diversity of cadherins and implications for a synaptic adhesive code in the CNS. *Neuron* 23:427–430.

Shen, K., Meyer, T. (1999). Dynamic control of CaMKII translocation and localization in hippocampal neurons by NMDA receptor stimulation. *Science* 284: 162–166.

Shen, K., Teruel, M. N., Subramanian, K., Meyer, T. (1998). CaMKIIbeta functions as an F-actin targeting module that localizes CaMKIIalpha/beta heterooligomers to dendritic spines. *Neuron* 21:593–606.

Sheng, M. (1996). PDZs and receptor/channel clustering: rounding up the latest suspects. *Neuron* 17:575–578.

Shepherd, G. M. (1996). The dendritic spine: a multifunctional integrative unit. *J. Neurophysiol.* 75:2197–2210.

Shi, S. H., Hayashi, Y., Petralia, R. S., Zaman, S. H., Wenthold, R. J., Svoboda, K., Malinow, R. (1999). Rapid spine delivery and redistribution of AMPA receptors after synaptic NMDA receptor activation. *Science* 284:1811–1816.

Song, I., Kamboj, S., Xia, J., Dong, H., Liao, D., Huganir, R. L. (1998). Interaction of the N-ethylmaleimide-sensitive factor with AMPA receptors. *Neuron* 21:393–400.

Song, J. Y., Ichtchenko, K., Südhof, T. C., Brose, N. (1999). Neuroligin 1 is a postsynaptic cell-adhesion molecule of excitatory synapse. *Proc. Natl. Acad. Sci. USA* 96:1100–1105.

Songyang, Z., Fanning, A. S., Fu, C., Xu, J., Marfatia, S. M., Chishti, A. H., Crompton, A., Chan, A. C., Anderson, J. M., Cantley, L. C. (1997). Recognition of unique carboxyl-terminal motifs by distinct PDZ domains. *Science* 275: 73–77.

Spacek, J., Harris, K. (1997). Three-dimensional organization of smooth endoplasmic reticulum in hippocampal CA1 dendrites and dendritic spines of the immature and mature rat. *J. Neurosci.* 17:190–203.

Sprengel, R., Suchanek, B., Amico, C., Brusa, R., Burnasheve, N., Rozov, A., Hvalby, O., Jensen, V., Paulsen, O., Andersen, P., Kim, J. J., Thompson, R. F., Sun, W., Webster, L. C., Grant, S. G., Eilers, J., Konnerth, A., Li, J., McNamara, J. O., Seeburg, P. H. (1998). Importance of the intracellular domain of NR2 subunits for NMDA receptor function in vivo. *Cell* 92:279–289.

Srivastava, S., Osten, P., Vilim, F., Khatri, L., Inman, G., States, B., Daly, C., DeSouza, S., Abagyan, R., Valtschanoff, J. G., Weinberg, R. J., Ziff, E. B. (1998). Novel anchorage of GluR2/3 to the postsynaptic density by the AMPA receptor-binding protein ABP. *Neuron* 21:581–591.

Staudinger, J., Zhou, J., Burgess, R., Elledge, S. J., Olson, E. N. (1995). PICK1: a perinuclear binding protein and substrate for protein kinase C isolated by the yeast two-hybrid system. *J. Cell Biol.* 128:263–271.

Steward, O., Falk, P. M., Torre, E. R. (1996). Ultrastructural basis for gene expression at the synapse: synapse-associated polyribosome complexes. *J. Neurocytol.* 25:717–734.

Stowell, J. N., Craig, A. M. (1999). Axon/dendrite targeting of metabotropic glutamate receptors by their cytoplasmic carboxy-terminal domains. *Neuron* 22:525–536.

Strack, S., Colbran, R. J. (1998). Autophosphorylation-dependent targeting of calcium/calmodulin-dependent protein kinase II by the NR2B subunit of the N-methyl-D-aspartate receptor. *J. Biol. Chem.* 273:20,689–20,692.

Svoboda, K., Mainen, Z. F. (1999). Synaptic [Ca^{2+}]: intracellular stores spill their guts. *Neuron* 22:427–430.

Takechi, H., Eilers, J., Konnerth, A. (1998). A new class of synaptic response involving calcium release in dendritic spines. *Nature* 396:757–760.

Takeuchi, M., Hata, Y., Hirao, K., Toyoda, A., Irie, M., Takai, Y. (1997). SAPAPs, a family of PSD-95/SAP90-associated proteins localized at postsynaptic density. *J. Biol. Chem.* 272:11,943–11,951.

Takumi, Y., Matsubara, A., Rinvik, E., Ottersen, O. P. (1999a). The arrangement of glutamate receptors in excitatory synapses. *Ann. N.Y. Acad. Sci.* 868:474–482.

Takumi, Y., Ramirez-Leon, V., Laake, P., Rinvik, E., Ottersen, O. P. (1999b). Different modes of expression of AMPA and NMDA receptors in hippocampal synapses. *Nature Neurosci.* 2:618–624.

Tejedor, F. J., Bokhari, A., Rogero, O., Gorczyca, M., Zhang, J., Kim, E., Sheng, M., Budnik, V. (1997). Essential role for *Dlg* in synaptic clustering of Shaker K[+] channels *in vivo*. *J. Neurosci.* 17:152–159.

Tezuka, T., Umemori, H., Akiyama, T., Nakanishi, S., Yamamoto, T. (1999). PSD-95 promotes fyn-mediated tyrosine phosphorylation of the N-methyl-D-aspartate receptor subunit NR2A. *Proc. Natl. Acad. Sci. USA* 96:435–440.

Thomas, U., Kim, E., Kuhlendahl, S., Ho Koh, Y., Gundelfinger, E. D., Sheng, M., Garner, C. C., Budnik, V. (1997). Synaptic clustering of the cell adhesion molecule Fasciclin II by discs-large and its role in the regulation of pre-synaptic structure. *Neuron* 19:787–799.

Torres, R., Firestein, B. L., Dong, H., Staudinger, J., Olson, E. N., Huganir, R. L., Bredt, D. S., Gale, N. W., Yancopoulos, G. D. (1998). PDZ proteins bind, cluster, and synaptically colocalize with Eph receptors and their ephrin ligands. *Neuron* 21:1453–1463.

Tu, J. C., Xiao, B., Yuan, J. P., Lanahan, A. A., Leoffert, K., Li, M., Linden, D. J., Worley, P. F. (1998). Homer binds a novel proline-rich motif and links group 1 metabotropic glutamate receptors with IP3 receptors. *Neuron* 21:717–726.

Tu, J. C., Xiao, B., Naisbitt, S., Yuan, J. P., Petralia, R. S., Brakeman, P., Doan, A., Aakalu, V. K., Lanahan, A. A., Sheng, M., Worley, P. F. (1999). Coupling of mGluR/Homer and PSD-95 complexes by the Shank family of postsynaptic density proteins. *Neuron* 23:583–592.

Uchida, N., Honjo, Y., Johnson, K. R., Wheelock, M. J., Takeichi, M. (1996). The catenin/cadherin adhesion system is localized in synaptic junctions bordering transmitter release zones. *J. Cell Biol.* 135:767–779.

Van Vactor, D. (1998). Protein tyrosine phosphatases in the developing nervous system. *Curr. Opin. Cell Biol.* 10:174–181.

Walensky, L. D., Blackshaw, S., Liao, D., Watkins, C. C., Weier, H. U., Parra, M., Huganir, R. L., Conboy, J. G., Mohandas, N., Snyder, S. H. (1999). A novel neuron-enriched homolog of the erythrocyte membrane cytoskeletal protein 4.1. *J. Neurosci.* 19:6457–6467.

Walsh, M. J., Kuruc, N. (1992). The postsynaptic density: constituent and associated proteins characterized by electrophoresis, immunoblotting, and peptide sequencing. *J. Neurochem.* 59:667–678.

Wang, H., Bedford, F. K., Brandon, N. J., Moss, S. J., Olsen, R. W. (1999). GABAa-receptor-associated protein links GABAa receptors and the cytoskeleton. *Nature* 397:69–72.

Wechsler, A., Teichberg, V. (1998). Brain spectrin binding to the NMDA receptor is regulated by phosphorylation, calcium and calmodulin. *EMBO J.* 17:3931–3939.

Westphal, R. S., Tavalin, S. J., Lin, J. W., Alto, N. M., Fraser, I. D., Langeberg, L. K., Sheng, M., Scott, J. D. (1999). Regulation of NMDA receptors by an associated phosphatase-kinase signaling complex. *Science* 285:93–96.

Woods, D. F., Bryant, P. J. (1991). The discs-large tumor suppressor gene of Drosophila encodes a guanylate kinase homolog localized at septate junctions. *Cell* 66:451–464.

Woolley, C. S., Wenzel, H. J., Schwartzkroin, P. A. (1996). Estradiol increases the frequency of multiple synapse boutons in the hippocampal CA1 region of the adult female rat. *J. Comp. Neurol.* 373:108–117.

Wyszynski, M., Lin, J., Rao, A., Nigh, E., Beggs, A. H., Craig, A. M., Sheng, M. (1997). Competitive binding of alpha-actinin and calmodulin to the NMDA receptor. *Nature* 385:439–442.

Wyszynski, M., Kharazia, V., Shanghvi, R., Rao, A., Beggs, A. H., Craig, A. M., Weinberg, R., Sheng, M. (1998a). Differential regional expression and ultra-structural localization of alpha-actinin-2, a putative NMDA receptor-anchoring protein, in rat brain. *J. Neurosci.* 18:1383–1392.

Wyszynski, M., Kim, E., Yang, F.-C., Sheng, M. (1998b). Biochemical and immunocytochemical characterization of GRIP, a putative AMPA receptor anchoring protein, in rat brain. *Neuropharmacology* 37:1335–1344.

Wyszynski, M., Valtschanoff, J. G., Naisbitt, S., Dunah, A. W., Kim, E., Standaert, D. G., Weinberg, R., Sheng, M. (1999). Association of AMPA receptors with a subset of glutamate receptor–interacting protein in vivo. *J. Neurosci.* 19: 6528–6537.

Xia, J., Zhang, X., Staudinger, J., Huganir, R. L. (1999). Clustering of AMPA receptors by the synaptic PDZ domain–containing protein PICK1. *Neuron* 22:179–187.

Xiao, B., Tu, J. C., Petralia, R. S., Yuan, J. P., Doan, A., Breder, C. D., Ruggiero, A., Lanahan, A. A., Wenthold, R. J., Worley, P. F. (1998). Homer regulates the association of group 1 metabotropic glutamate receptors with multivalent complexes of homer-related, synaptic proteins. *Neuron* 21:707–716.

Yu, X., Askalan, R., Keil, G., Salter, M. (1997). NMDA channel regulation by channel-associated protein tyrosine kinase Src. *Science* 275:674–678.

Yuste, R., Denk, W. (1995). Dendritic spines as basic functional units of neuronal integration. *Nature* 375:682–684.

Zhang, S., Ehlers, M. D., Bernhardt, J. P., Su, C. T., Huganir, R. L. (1998). Calmodulin mediates calcium-dependent inactivation of *N*-methyl-D-aspartate receptors. *Neuron* 21:443–453.

Zhang, W., Vazquez, L., Apperson, M., Kennedy, M. B. (1999). Citron binds to PSD-95 at glutamatergic synapses on inhibitory neurons in the hippocampus. *J. Neurosci.* 19:96–108.

Ziff, E. B. (1997). Enlightening the postsynaptic density. *Neuron* 19:1163–1174.

Zito, K., Fetter, R. D., Goodman, C. S., Isacoff, E. Y. (1997). Synaptic clustering of Fasciclin II and shaker: essential targeting sequences and role of Dlg. *Neuron* 19:1007–1016.

Ziv, N. E., Smith, S. J. (1996). Evidence for a role of dendritic filopodia in synaptogenesis and spine formation. *Neuron* 17:91–102.

8

Signal Transduction in the Postsynaptic Neuron
Activity-Dependent Regulation of Gene Expression

Michael E. Greenberg and Edward B. Ziff

The rapid propagation of electrical signals between neurons is mediated by the binding of neurotransmitter molecules to their receptors, which then function as ligand-gated channels to regulate the influx and efflux of ions into and from the postsynaptic cell (Mayer and Miller, 1990). Although the importance of ion fluxes in fast synaptic transmission and in modulating the biophysical properties of neuronal membranes has been known for some time, it has become clear over the past decade that the transsynaptic release of neurotransmitter can have profound and long-lasting consequences for the postsynaptic neuron. These include effects on neuronal differentiation, survival, axon outgrowth, and changes in synaptic strength (see Ghosh and Greenberg, 1995, for review). In the central nervous system, presynaptic release of the excitatory neurotransmitter glutamate induces Ca^{2+} influx into the postsynaptic neuron through neurotransmitter- and voltage-gated Ca^{2+} channels. This influx of Ca^{2+} plays a role in promoting the survival of developing neurons, and in the mature nervous system it is critical for neuronal adaptive responses that are believed to be a major mechanism of information storage in the brain (Bailey et al., 1992; Franklin and Johnson, 1992).

There are at least four means of elevating postsynaptic Ca^{2+} levels. The major type of Ca^{2+}-conducting ligand-gated ion channel expressed by the postsynaptic neuron is a class of glutamate receptors termed NMDA receptors because they are specifically activated by the agonist N-methyl-D-aspartate (NMDA) (see Sucher et al., 1996, for review). NMDA receptor gating is dependent on the convergence of two signals: the binding of glutamate as the agonist and membrane depolarization to release Mg^{2+}, which blocks the channel at resting potentials (Mori and Mishina, 1995; Ozawa et al., 1998). NMDA receptors have a high permeability to Ca^{2+} relative to other subtypes of glutamate receptor. The blockade of Ca^{2+} influx through the NMDA receptor leads to a loss of neuronal activity–dependent synaptic enhancement, indicating that NMDA receptors are critical mediators of synaptic plasticity (Perkel et al., 1993).

Another route of Ca^{2+} entry into neurons is through voltage-sensitive Ca^{2+} channels (VSCCs) (Catterall, 1995). Although a variety of types of VSCCs are expressed in neurons, neurotransmitter-initiated Ca^{2+} influx through L-type VSCCs elicits a particularly diverse array of cellular responses (Rosen et al., 1995). L-VSCCs have a relatively high activation range, a slow inactivation rate, and a high single-channel conductance, making them ideal channels for passing the large amounts of Ca^{2+} that are necessary to increase Ca^{2+} concentrations in the cytoplasm and nucleus sufficiently to initiate cellular responses (Catterall, 1995; Clapham, 1995).

AMPA receptors are glutamate-regulated cation channels that are selectively activated by α-amino-3-hydroxy-5-methyl-4-isoxazolepropionate (AMPA) and that provide the majority of fast excitatory synaptic transmission (see Kaczmarek et al., 1997, and O'Brien et al., 1998, for review). These glutamate receptors are composed of four types of subunits,

GluR1–4 (GluRA–D). Ca^{2+} can enter the cell through AMPA receptors unless the receptor contains the prevalent form of the GluR2 subunit, which is encoded by GluR2 mRNA that has been modified by RNA editing (Seeburg et al., 1998). Editing alters the coding capacity of the message, leading to substitution of an Arg for Gln in the membrane reentrant hairpin loop that forms the channel pore. Co-assembly of an edited GluR2 subunit into an AMPA receptor has a dominant effect in blocking Ca^{2+} passage through the channel.

Finally, glutamate can promote an increase in the level of cytoplasmic Ca^{2+} through stimulation of a class of G protein–coupled receptors called metabotropic glutamate receptors (mGluRs). Of particular significance are the type 1 metabotropic receptors, which are phosphoinositide linked and whose activation induces the release of Ca^{2+} from intracellular stores.

Calcium-Dependent Neuronal Responses

Transcription-Independent Responses

Transsynaptic stimulation that leads to the activation of NMDA receptors and/or L-VSCCs is followed by a series of dramatic Ca^{2+}-dependent cellular responses. These rapid changes include alterations in ion channel properties that appear to be triggered in part by protein kinase–catalyzed phosphorylation of channel subunits (Jonas and Kaczmarek, 1996; Smart, 1997). Stimulation of the NMDA receptor and the activation of L-VSCCs lead to the activation of a variety of distinct protein kinases, including calmodulin-dependent protein kinases (CaMK) I–IV, components of the Ras/Erk/Rsk signaling pathway, protein kinase C (PKC), and tyrosine kinases (see Ghosh and Greenberg, 1995; and Finkbeiner and Greenberg, 1996, 1998, for review). The importance of these various kinases for neuronal adaptive responses has been extensively studied in hippocampal slices, where, depending on the frequency of NMDA receptor stimulation, one sees an enhancement in synaptic efficacy termed long-term potentiation (LTP) (Matthies et al., 1990; Malenka and Nicoll, 1999) or a decrease in synaptic efficacy referred to as long-term depression (LTD) (Bear, 1999). There is increasing evidence that LTP and LTD are at least partially mediated by the protein kinase–catalyzed phosphorylation of specific ion channel subunits, leading to ion channel modulation or changes in the cell surface expression of ion channel subunits (Soderling, 1993; Swope et al., 1999).

Transcription-Dependent Responses

In addition to inducing rapid cellular responses, such as ion channel modulation, the transsynaptic release of glutamate has long-lasting effects that are dependent on the induction of new gene expression (see Sheng and Greenberg, 1990; Ghosh and Greenberg, 1995; and Fink-

Transcription factors	Receptors
c-fos	GABA-R delta
c-jun	NMDA-R(NR2A)
zif/268	p75NGF-R
CREM	TrkA
fosB	GluR1
junB	GluR3
nur77	MHCl
	trkC

Soluble factors that regulate synaptic function	Neuromodulators and enzymes
BDNF	CRH
PC3	NP-Y
Narp	tyrosine hydroxylase
NGF	VIP
	secretogranin II
	dynorphin
	somatostatin

Signaling molecules and enzymes	Other regulators of synaptic function
KID1	SNAP-25
GAP-43	synaptotagmin IV
rheb	homer
nNOS	Arc
COX2	cpg15/neuritin
MKP1	Kv3.1 K channel
calbindin-D28	clathrin
	Na channel alpha

Figure 8.1. Summary of Ca^{2+}-regulated genes. Shown is a subset of the large number of Ca^{2+}-regulated genes that have been identified.

beiner and Greenberg, 1998, for review) (Fig. 8.1). Glutamate-induced Ca^{2+} influx triggers changes in gene expression that appear to be critical during development. For example, during the development of the peripheral and central nervous systems, neuronal activity–dependent changes occur in the expression of certain neuropeptides, neuro-trophins, neurotransmitter synthesizing enzymes, and components of the synaptic machinery that appear to be important for proper development. In addition, long-lasting forms of LTP and LTD have been identified that require new gene expression and protein synthesis (Huang and Kandel, 1994). Given the widely held view that LTP and LTD may represent models for information storage in the brain, there has been considerable interest in identifying and characterizing the regulation and function of the specific genes that are induced during the long-lasting forms of LTP and LTD.

Synaptic Tagging

As a single neuron can form over a thousand synapses, it is perplexing how changes in gene expression that are typically thought to be regulated in the cell nucleus can selectively alter the efficacy of only a few of the many synapses that the neuron forms. One possibility is that the synaptic stimulation locally tags the synapse and simultaneously sends a signal to the nucleus to induce new gene transcription (Frey and Morris, 1997). The products of the newly transcribed genes might then be selectively targeted to the potentiated synapses, where they might contribute to further synaptic modulation. Alternatively, the products of the newly transcribed genes might be targeted to all of the synapses that the neuron forms but might be capable of functioning only at synapses that have already been tagged by the initial release of neurotransmitter. Another possibility that has recently gained support is that neurotransmitters regulate gene expression posttranscriptionally by regulating the stability or translation of specific mRNAs that are localized and translated within the dendritic region of the postsynaptic neuron (Huang, 1999). In the following sections we review current evidence suggesting that both transcriptional and posttranscriptional mechanisms operate to control neurotransmitter-dependent changes in gene expression that contribute to the modulation of synaptic function.

The c-fos Proto-Oncogene and Immediate Early Genes

The first indication that that the transsynaptic release of neurotransmitter could lead to the activation of new gene transcription came from studies of the c-*fos* proto-oncogene (Greenberg et al., 1986; Morgan and Curran, 1986). Transcription is activated rapidly and transiently in response to a wide variety of extracellular stimuli, including neurotransmitters and other agents that lead to membrane depolarization and the activation of L-type VSCCs (Sheng and Greenberg, 1990). The c-*fos* gene encodes a transcription factor, Fos, which forms a heterodimer with members of the Jun family of transcription factors (see Curran and Franza, 1988; Sheng and Greenberg, 1990; and Curran and Morgan, 1995, for review). Fos/Jun heterodimers bind to the consensus DNA sequence element (5′-ATGAc/gTCAT-3′), termed the AP-1 site. By binding to AP-1 sites within the regulatory region of their target genes, Fos/Jun complexes regulate subsequent programs of gene expression that are often cell type and stimulus specific. The activation of Fos/Jun complexes in mature neurons in response to neuronal activity may lead to the subsequent activation of later programs of gene transcription that are important for the adaptive response of the neuron to the initial stimulus. It has been difficult to establish how genes such as c-*fos* and c-*jun* that are activated in such a diverse set of circumstances might mediate specific responses in particular cell types. The emerging view is that Fos/Jun complexes do not function alone, but rather that they activate gene transcription by cooperating with other transcriptional regulators whose pattern of expression is restricted to specific cell types.

The importance of Fos/Jun complexes as mediators of the adaptive responses of mature neurons to synaptic stimulation is underscored by numerous examples in which neuronal activity–dependent cellular and organismal responses have been found to be correlated with the induction of c-*fos* transcription, specifically in the subset of neurons that are the critical mediators of the activity-dependent response. For example, specific external stimuli that elicit responses—such as long-lasting LTP and LTD, circadian entrainment, epileptic kindling, addiction to drugs of abuse, stress responses, and maternal nurturing—induce c-*fos* transcription specifically in the region of the brain that mediates the organism's response to the external stimulus (Kornhauser et al., 1990; Hope et al., 1992; Brown et al., 1996; Watanabe et al., 1996; Hiroi et al., 1997). Several of these studies used mice in which genes encoding specific members of the Fos family have been deleted, revealing the critical importance of c-fos and its relative FosB for specific neuronal adaptive responses.

The c-*fos* gene is now known to be just one member of a family of 50–100 genes termed immediate early genes (IEGs) that are activated rapidly by many extracellular stimuli in diverse cell types, including neurotransmitters that stimulate Ca^{2+} influx into the postsynaptic neuron (see Lanahan and Worley, 1998, for review). Numerous studies have established that environmental stimulation of the intact nervous system leads to the induction of many different IEGs in the regions of the brain that mediate the organism's response to the stimulus. Thus there is considerable interest in defining the functions of the various IEG-encoded proteins and in elucidating the mechanisms by which Ca^{2+} influx triggers IEG transcription. The best-characterized IEGs are transcription factors that, like c-*fos*, regulate subsequent programs of late-response gene transcription that may be critical for neuronal adaptive responses. Among the IEG-encoded transcription factors are members of the Fos family (c-fos, FosB, Fra-1, and Fra-2) and Jun family (c-jun and junB), as well as a family of zinc finger–containing transcription factors (e.g., Egr1/Zif268) and Nur77, an orphan member of the family of steroid receptors (Sheng and Greenberg, 1990). A feature of the IEGs that encode transcription factors is that these genes are typically induced in many different cell types by a wide array of stimuli. A current challenge in the field is to determine how these transcriptional regulators mediate neuron-specific responses.

Examples of Neuron-Specific IEGs

In contrast to the IEGs that encode transcriptional regulators, a number of IEGs have been identified that are selectively activated in neurons. These genes encode either soluble factors that affect neuronal function, receptors involved in neuronal function, or enzymes that mediate the synthesis of neurotransmitters (Fig. 8.1) (Sheng and Greenberg, 1990; Lanahan and Worley, 1998). Of the neuron-specific IEGs, several have drawn attention because they encode proteins that are believed to modulate synaptic function. Several neuron-specific IEGs encode proteins

that play an important role in promoting the development and organization of central nervous system synapses. These include the IEG products
BDNF, Homer, Arc, and Narp.

BDNF

One of the best characterized of the neuron-specific, Ca^{2+}-regulated
IEGs is brain-derived neurotrophic factor (BDNF) (Zafra et al., 1990;
Zafra et al., 1992; Ghosh et al., 1994). BDNF is a member of the neurotrophin family that plays a critical role in regulating the survival and
differentiation of selective populations of neurons during development.
Recently BDNF has also been shown to be an important mediator of
neuronal adaptive responses (Kang and Schuman, 1995a,b, 1996). BDNF
expression is induced by neuronal activity (Zafra et al., 1990, 1992;
Ghosh et al., 1994). Activation of L-VSCCs or NMDA receptors leads to
a significant increase in the level of BDNF mRNA and also stimulates the
release of the BDNF protein. Stimuli that induce LTP in the hippocampus as well as physiological stimuli such as visual experience lead to a
large increase in the level of BDNF mRNA, specifically in the regions of
the nervous system that have been stimulated (Castren et al., 1992; Patterson et al., 1992; Dragunow et al., 1993). Several recent studies have
shown that BDNF modulates synaptic function in mature neurons. Importantly, synaptic transmission and LTP are impaired in the hippocampus of BDNF knockout mice (Patterson et al., 1996; Korte et al., 1998).
The application of BDNF to the *Xenopus* neuromuscular junction leads
to an enhancement of the spontaneous release of acetylcholine (Lohof et
al., 1993). In addition, BDNF induces long-lasting LTP in young hippocampal slices that normally show only short-term potentiation (Figurov
et al., 1996). Thus there is considerable interest in defining how neuronal activity regulates BDNF transcription and how BDNF acts to regulate synaptic function.

Homer

Homer (vesl) has a single PDZ domain that binds to the carboxy terminus
of type 1, phosphoinositide-linked, mGluRs (Brakeman et al., 1997). PDZ
domains are globular protein domains of 80–90 amino acids that bind
the carboxy termini of integral membrane proteins as well as other
proteins (for review see Sheng and Pak, 1999). Homer is encoded in
two forms, one of which has a coiled-coil domain that allows Homer to
homodimerize. In its dimeric, coiled-coil-containing form, Homer can
cluster type 1 metabotropic glutamate receptors. The short form, which
lacks the coiled-coil domain, is induced by synaptic activity and acts as a
dominant-negative peptide to displace the coiled-coil form and modify
the properties of metabotropic receptors at the synapse (Kato et al., 1997;
Kato et al., 1998; Xiao et al., 1998). Homer also binds to IP3 receptors,
suggesting that it may link the spatial localization of type 1 mGluRs to their

intracellular target for Ca^{2+} release (Tu et al., 1999). A protein called Shank may also link mGluR/Homer complexes to NMDA receptors through binding to the NMDA receptor–associated postsynaptic density protein, PSD-95 (Tu et al., 1999). The modulation of receptor interactions by the induction of the short form of Homer may provide an example of how activity-dependent changes in gene transcription might lead to the reorganization of the structure of a synapse.

Arc

Another Ca^{2+}-regulated gene encodes Arc, a protein that is associated with the neuronal cytoskeleton (Lyford et al., 1995). Arc mRNA displays the intriguing property of targeting specifically to dendrites at postsynaptic positions that have been activated by synaptic stimulation (Steward et al., 1998). This activity-dependent localization does not require new protein synthesis and appears to depend upon signals in the mRNA structure itself. The targeting of Arc mRNA to activated synaptic sites may provide a basis for selective expression of new proteins at activated synapses.

Narp

An additional IEG induced by synaptic activity encodes Narp, a secreted peptide of the pentraxin family first described as a stimulator of neurite outgrowth in cultures of cortical neurons (Tsui et al., 1996). Narp directs the clustering of AMPA receptors when co-expressed in heterologous cells (O'Brien et al., 1999), and can also induce the synaptic clustering of AMPA receptors in cultured neurons. Thus the regulation of *Narp* gene expression by synaptic activity may exert trans-acting influences on glutamate receptor localization at neighboring cell synapses.

Calcium Regulation of Transcription

The c-fos Proto-Oncogene

Given the critical functions of Ca^{2+}-regulated genes, considerable attention has been focused on elucidating the mechanisms by which membrane depolarization and Ca^{2+} influx through L-type VSCCs stimulate transcription. Elucidation of these Ca^{2+}-regulated signal transduction pathways began with the analysis of c-*fos* transcription in the pheochromocytoma cell line PC12 (Sheng et al., 1988, 1990; Sheng and Greenberg, 1990). Exposure of PC12 cells to elevated levels of KCl that lead to membrane depolarization and the activation of L-type VSCCs induces c-*fos* transcription within minutes. This activation of c-*fos* transcription is rapid and transient, with the level of c-*fos* mRNA returning to its prestimulation level within 30 min. Membrane depolarization induction of c-*fos* transcription requires Ca^{2+} influx through L-type VSCCs, as the activation of c-*fos* transcription is blocked by antagonists of the L-type VSCCs or by the addition of ethylene glycol-bis(β-aminoethyl ether)-*N,N,N',N'*-tetraacetic acid (EGTA) to the culture media. In nerve growth factor–differentiated PC12 cells, agonists of nicotinic acetylcholine receptors, which cause a passive flow of ions through the ligand-gated acetylcholine receptor and an opening of L-type VSCCs, stimulate c-*fos* transcription (Greenberg et al., 1986). Likewise, exposure of cortical, hippocampal, striatal, or cerebellar neurons to NMDA induces c-*fos* transcription in a Ca^{2+}-dependent manner (Szekely et al., 1989, 1990; Zafra et al., 1990; Bading et al., 1993; Lerea and McNamara, 1993). These and other findings established that a variety of neurotransmitters that trigger Ca^{2+} influx from the extracellular medium effectively activate c-*fos* transcription.

Promoter deletion analysis in PC12 cells and in primary neuronal cultures revealed the DNA regulatory elements within the c-*fos* promoter that mediate Ca^{2+} induction of c-*fos* transcription (Sheng et al., 1988, 1990; Sheng and Greenberg, 1990; Miranti et al., 1995; Xia et al., 1996). One key regulatory sequence (5'-TGACGTTT-3'), designated the Ca^{2+} response element (CaRE), is located approximately 60 nucleotides 5' of the initiation site of c-*fos* mRNA synthesis. Subtle mutations within the c-*fos* CaRE lead to a loss of Ca^{2+} induction of a reporter gene that contains the c-*fos* CaRE within its regulatory region. The c-*fos* CaRE is similar in sequence to the cAMP regulatory element (CRE), a DNA regulatory element that mediates the transcriptional response to elevated levels of cAMP. Mutational analysis of the CRE (5'-TGACGTCA-3') and the c-*fos* CaRE revealed that they are functionally indistinguishable and suggested that the same DNA regulatory element mediates both Ca^{2+} and cAMP responses (Sheng et al., 1990).

DNA mobility shift analyses using nuclear extracts from untreated and membrane-depolarized PC12 cells revealed the presence of a protein that specifically binds to the c-*fos* CaRE whether or not the PC12 cells were membrane depolarized. Purification of this nuclear protein revealed that it is a member of the cAMP response element binding protein (CREB) family of transcription factors. Notably, CREB binds to the wild-type c-*fos* CaRE, but not to subtle mutants of the c-*fos* CaRE that inhibit the ability of the CaRE to confer a Ca^{2+} response on a reporter gene (Sheng et al., 1990). This finding suggests that CREB or a closely related family member is likely to be the transcriptional regulator that mediates Ca^{2+} induction of c-*fos* transcription in vivo.

CREB

CREB is the most thoroughly studied of the Ca^{2+}-regulated transcription factors and has been implicated in neuronal adaptive responses in species ranging from mollusks to man (see Frank and Greenberg, 1994, and Silva et al., 1998, for review). As CREB is a regulator of IEGs that encode transcription factors (e.g., c-*fos*, c-*jun*, *egr-1*) and neuron-specific IEGs (e.g., BDNF), there has been considerable interest in the biological function of CREB in mature neurons. CREB has been implicated in long-term facilitation in *Aplysia* neurons (Kaang et al., 1993; Bartsch et al., 1995; Martin and Kandel, 1996), in long-term memory in flies (Yin et al., 1994; Yin and Tully, 1996), and in long-lasting LTP in the mouse (Bourtchuladze et al., 1994; Kogan et al., 1997; Silva et al., 1998). CREB may also play an important role in neuronal differentiation and survival during development (Ginty et al., 1994; Bonni et al., 1995, 1999; Bonni and Greenberg, 1997).

CREB is a member of the basic leucine zipper (bZIP) family of transcription factors (Struhl, 1989; Vinson et al., 1989; Vinson et al., 1993) (Fig. 8.2). The bZIP proteins, such as CREB and CREB's close relatives ATF1 and CREM, dimerize via their leucine zipper domain, which is a heptad repeat of leucine residues within an amphipathic α-helical region (Vinson et al., 1989, 1993). Binding of bZIP proteins to their DNA target sequence is mediated by a basic stretch of amino acids that lies just amino-terminal to the leucine zipper (Vinson et al., 1989, 1993). Of the various bZIP proteins—which include CREB, ATF1, CREM, c-fos, c-Jun, c-Myc, and C/EBP—the members of the CREB family (e.g., CREB, ATF1, and CREM) form a subgroup of proteins that are highly related structurally and form homo- and heterodimers with one another (Hai et al., 1988, 1989; Hai and Curran, 1991; Rehfuss et al., 1991). Although the various CREB family dimers bind the consensus DNA sequence

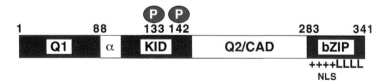

Figure 8.2. The principal domains of the CREB protein. The kinase-inducible domain (KID) is the region targeted by intracellular kinase cascades activated by extracellular signals; phosphorylation of residues within the KID, including Ser-133 and Ser-142, regulates the ability of CREB to associate with the co-activator protein CBP. The KID is flanked by two glutamine-rich regions, Q1 and Q2/CAD, which contact elements of the basal transcription machinery and contribute to constitutive CREB activity (in contrast to the KID, which is critical for stimulus-induced CREB activity). The ability of CREB to bind DNA sequences in the promoter of target genes is conferred by the bZIP domain located within its C terminus. The leucine zipper region is composed of a patch of leucine residues (L), which permit CREB to dimerize with other bZIP transcription factors. The nuclear localization signal (NLS) is composed of positively charged residues (+) within the basic region of the bZIP domain. The α domain is composed of a 14-residue stretch of amino acids, which is present in an alternatively spliced form of CREB, CREBα. Reproduced from Shaywitz and Greenberg (1999) with permission from the *Annual Review of Biochemistry,* © 1999 by Annual Reviews; http://www.Annual-Reviews.org.

5'-TGACGTCA-3' within the regulatory region of their target genes, it is likely that subtle changes within the consensus DNA binding sites, or the sequence of the DNA that flanks the core site, may determine which of the CREB family dimers actually bind to particular CaREs that lie within the regulatory regions of Ca^{2+}-responsive genes (Shaywitz and Greenberg, 1999).

Although there has been some controversy regarding the mechanism by which extracellular stimuli activate CREB, the prevailing view is that extracellular stimuli induce the phosphorylation of prebound CREB and that phosphorylation leads to the induction of CREB's ability to activate transcription (Shaywitz and Greenberg, 1999). This was demonstrated by fusing CREB to a heterologous DNA binding domain—that of the yeast transcription factor Gal4—and then testing the ability of the CREB-Gal4 fusion to confer Ca^{2+} responsiveness on a reporter gene containing the Gal4 DNA binding site within its regulatory region (Sheng et al., 1991). After transfection of the CREB-Gal4 fusion protein and Gal4 reporter gene into neurons, Ca^{2+} influx was found to stimulate CREB-Gal4-dependent reporter gene transcription (Sheng et al., 1991). Ca^{2+} influx through L-type VSCCs activates CREB by inducing a cascade of events leading to the phosphorylation of CaRE-bound CREB dimers at a specific amino acid residue, serine 133 (Dash et al., 1991; Sheng et al., 1991; Deisseroth et al., 1996, 1998; Finkbeiner and Greenberg, 1996; Impey et al., 1998, 1999).

An antibody has been developed that specifically recognizes the Ser-133 phosphorylated form of CREB (Ginty et al., 1993; Hagiwara et al., 1993). Studies with this antibody have shown that many environmental stimuli that induce a neuronal response also induce CREB Ser-133 phosphorylation and that this phosphorylation generally correlates with IEG activation (Ginty et al., 1993; Jeon et al., 1997; Ji and Rupp, 1997; Lee et al., 1999; Sakaguchi et al., 1999). Moreover, the induction of CREB Ser-133 phosphorylation occurs specifically in the regions of the brain that mediate the organism's response to the particular environmental stimulus being tested. The importance of CREB Ser-133 phosphorylation for CREB-dependent transcription and biological responses has also been tested directly by mutating CREB Ser-133 to a nonphosphorylatable residue, such as alanine, and then examining the effect of this mutation on various CREB-dependent events (Sheng et al., 1991). The Ser133Ala mutation abolishes Ca^{2+} induction of CREB-dependent transcription, indicating that the phosphorylation of CREB at Ser-133 is required for Ca^{2+} induction of CREB-dependent transcription in neurons (Sheng et al., 1991).

When CREB Ser133Ala mutants (CREB M1) are introduced into cells, they function as dominant-negative mutants. When overexpressed in cells, CREB M1 competes with endogenous CREB for binding to CARE/CREs within the regulatory regions of CREB target genes (Tao et al., 1998). However, because CREB M1 cannot be activated by phosphorylation at Ser-133, the presence of CREB M1 in cells inhibits transcription of CREB-regulated genes. CREB M1 is therefore a useful tool for testing the importance of CREB for various biological responses. Together with other dominant-negative CREB mutants, CREB M1 has been used to implicate CREB as a mediator of synaptic facilitation in *Aplysia*, long-term memory in *Drosophila*, and long-lasting LTD in rodents (for review see Frank and Greenberg, 1994; and Silva et al., 1998).

Ser-133 is located within a 61-amino-acid region of CREB known as the kinase-inducible domain (KID) (Gonzalez et al., 1989, 1991). The KID (residues 100–160) contains multiple serine and threonine residues in addition to Ser-133 that are potential sites for phosphorylation. When fused to the Gal4 DNA binding domain, the KID alone is capable of conferring Ca^{2+} responsiveness on a Gal4-driven reporter gene. Ser-133 phosphorylation is also critical in this context (Quinn, 1993; Brindle et al., 1995). This finding suggested that the KID of CREB plays a crucial role in the activation of gene transcription and fueled the effort to identify proteins that interact with the KID and thereby mediate the CREB-dependent transcription.

CBP

A large co-adaptor protein, the CREB-binding protein (CBP), specifically interacts with the Ser-133-phosphorylated CREB (Chrivia et al., 1993). CBP appears to function as a molecular bridge that allows transcription factors that bind within the promoter region of genes to recruit and stabilize the RNA polymerase II transcription complex near the site of initiation of mRNA synthesis. CBP and its close relative p300 function as adaptor proteins for a wide range of transcription factors in addition to CREB (see Giordano and Avantaggiati, 1999, for review). Although many of the interactions that CBP or p300 undergo with transcriptional regulators appear to be constitutive, the interaction of CREB with CBP appears to require phosphorylation of CREB at Ser-133 (see Shaywitz and Greenberg, 1999, for review). This phosphorylation serves to bring CBP to the promoters of CREB target genes, where CBP facilitates the assembly of the Pol II transcription complex (Shaywitz and Greenberg, 1999) (Fig. 8.3).

In addition to recruiting Pol II, CBP has a clear effect on chromatin structure. CBP is itself a histone acetylase and also associates with another histone acetylase, termed p/CAF (Bannister and Kouzarides, 1996; Ogryzko et al., 1996; Yang et al., 1996). When recruited to the promoters of inactive genes, histone acetylases such as CBP and p/CAF catalyze the acetylation of lysine residues in the N terminus of histones, leading to alterations in chromatin structure that render the DNA template more accessible to the transcriptional machinery (Struhl, 1998). Thus, by triggering CREB phosphorylation at Ser-133, Ca^{2+} influx induces the assembly of an active transcription complex on the promoters of CREB-regulated IEGs. These transcription complexes include histone acetylases that promote a dramatic change in chromatin structure,

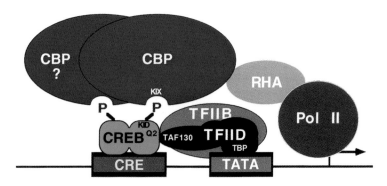

Figure 8.3. Multiple domains of CREB contribute to transcriptional activation. Different domains of CREB bind distinct co-activators and basal transcription factors to activate transcription. The figure depicts a CREB dimer bound to its cognate CaRE/CRE element on the promoter of a CREB target gene. Downstream of the CaRE/CRE is the TATA box, which binds the multiprotein TFIID basal transcription factor (via the TBP protein). Another factor within TFIID, TAF130, binds to the Q2 domain of CREB. The Q2 domain of CREB has also been shown to interact with TFIIB, which is a part of the basal transcription machinery as well. A distinct domain of CREB, the KID, contributes to signal-induced transcriptional activation. When phosphorylated at Ser-133, the KID of CREB can bind to the KIX domain of the CBP. It is presently unclear whether CBP associates with Ser-133-phosphorylated CREB as a dimer. CBP associates indirectly with Pol II via the RNA helicase A (RHA) protein. Therefore, recruitment of CBP to Ser-133-phosphorylated CREB results in recruitment and stabilization of Pol II on the promoter of CREB target genes, while the Q2 domain interacts with other elements of the basal transcription machinery that are required for transcription, such as TFIID and TFIIB. Reproduced from Shaywitz and Greenberg (1999) with permission from the *Annual Review of Biochemistry,* © 1999 by Annual Reviews; http://www.AnnualReviews.org.

resulting in the robust transcription of genes that minutes before were silent.

Ca^{2+}-Regulated CREB Kinases

Given the clear importance of Ser-133 phosphorylation for CREB activation and function in mature neurons, there has been considerable interest in defining the kinase cascades that trigger this critical event. The prevailing view is that there are multiple Ca^{2+}-regulated kinase cascades that catalyze CREB Ser-133 phosphorylation in membrane-depolarized cells (see Finkbeiner and Greenberg, 1996, for review) (Fig. 8.4). Many kinases whose activity is directly or indirectly enhanced by Ca^{2+} and calmodulin are capable of phosphorylating CREB at Ser-133 in vitro. They include CaMKI, II, and IV; the ribosomal S6 kinases (pp90Rsk1-3); PKC; the cAMP-dependent protein kinase (PKA); the serine/threonine kinase Akt; and the MAP kinase–activated protein kinase. Of these kinases, two (CamKIV and pp90[rsks]) have been shown to mediate Ca^{2+}-dependent CREB phosphorylation in neurons (Matthews et al., 1994;

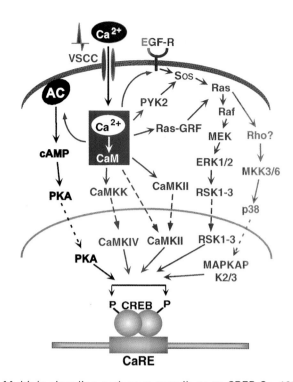

Figure 8.4. Multiple signaling pathways contribute to CREB Ser-133 phosphorylation in response to Ca^{2+} influx. In neuronal cells, electrical activity leads to
membrane depolarization, opening voltage-sensitive Ca^{2+} channels (VSCCs) in the
plasma membrane and resulting in influx of extracellular Ca^{2+}. Inside the cell, calcium activates many kinases, some of which directly phosphorylate CREB at Ser-
133. Upon entry, Ca^{2+} binds to calmodulin (CaM). The Ca^{2+}/CaM complex (shaded
box) can activate the PKA pathway (blue) by directly stimulating calcium-sensitive
adenylyl cyclases, leading to generation of cAMP and the activation of PKA. PKA
can then translocate to the nucleus, where it phosphorylates CREB at Ser-133. Ca^{2+}/
CaM also activates members of the Ca^{2+}/calmodulin-dependent kinase (CaMK)
family (black), all of which can phosphorylate CREB at Ser-133. Ca^{2+}/CaM directly
activates CaMKI (not shown), CaMKII, and CaMKIV. Ca^{2+}/CaM can also activate
CaMKK, which can then directly activate both CaMKIV and CaMKI (not shown).
Nuclear translocation of Ca^{2+}/CaM may account for the activation of CaMKIV and
CaMKII. CaMKIV is localized predominantly to the nucleus while isoforms of CaMKII
are found both in the nucleus and in the cytoplasm. In addition, certain CaMKII
isoforms may translocate from the cytoplasm to the nucleus. Ca^{2+}/CaM also activates the Ras/MAPK pathway (red). Ca^{2+} activation of Ras may occur through multiple mechanisms. Ca^{2+} influx can lead to the ligand-independent activation of the
EGF-receptor (EGF-R), which then leads to activation of the guanine-nucleotide exchange factor Sos and Ras activation. Activation of Ras stimulates the Raf, MEK,
and ERK1/2 kinase cascade. The MAP kinases ERK1/2 directly activate members of
the pp90RSK family of protein kinases (RSK1-3). Activated RSKs then translocate
to the nucleus, where they phosphorylate CREB at Ser-133. Ca^{2+}/CaM can also activate Ras by activating Ras-GRF, a Ca^{2+}-activated guanine-nucleotide exchange
factor. The calcium-activated tyrosine kinase PYK2 can also activate GEFs and lead
to stimulation of the Ras pathway. Dashed lines indicate translocation from the
cytoplasm to the nucleus. Reproduced from Shaywitz and Greenberg (1999) with
permission from the *Annual Review of Biochemistry*, © 1999 by Annual Reviews;
http://www.AnnualReviews.org.

Sun et al., 1994; Braun and Schulman, 1995; Enslen et al., 1995; Finkbeiner and Greenberg, 1998; Impey et al., 1999). More thorough analysis is required to determine which of these kinases mediates Ca^{2+}-dependent CREB Ser-133 phosphorylation under particular circumstances. It is possible that in a given cell type Ca^{2+} may activate several distinct signaling pathways, each of which leads to CREB Ser-133 phosphorylation. This process may facilitate the stoichiometric phosphorylation of CREB, but it could also allow for some variation in the duration of CREB phosphorylation if different CREB kinases are activated with distinct kinetics (Finkbeiner et al., 1997; Impey et al., 1998).

The mechanisms by which Ca^{2+} influx leads to the activation of CaMKIV and pp90[rsks] and the role of these kinases as mediators of CREB Ser-133 phosphorylation within neurons have been the subject of considerable investigation. Ca^{2+} influx through L-type VSCCs leads to the activation of a CaM kinase kinase, which phosphorylates and activates CaMKIV (Braun and Schulman, 1995; Tokumitsu et al., 1995). Activated CamKIV is localized within the nucleus, where it is believed to catalyze the phosphorylation of CREB at Ser-133 (Nakamura et al., 1995). In hippocampal neurons, the expression of CaMKIV appears to be critical for CREB Ser-133 phosphorylation, as the inhibition of CaMKIV expression using an RNA antisense strategy led to a reduction in CREB Ser-133 phosphorylation (Bito et al., 1996). Although this finding is consistent with the possibility that CaMKIV directly phosphorylates CREB at Ser-133, an alternative possibility is that CaMKIV indirectly mediates CREB phosphorylation by inducing the activation of an upstream activator of an alternative kinase cascade that culminates in CREB Ser-133 phosphorylation.

One such cascade may be the Ras/MAPK signaling pathway (Rosen et al., 1994; Rosen and Greenberg, 1996). Ca^{2+} influx into neurons activates Ras by several distinct signaling pathways involving the Ca^{2+}-regulated kinase PYK2, the ligand-independent activation of the EGF receptor, and activation of a neuron-specific Ca^{2+}-sensitive GTP/GDP exchange factor (GRF) that binds and activates Ras in a Ca^{2+}-dependent manner (see Finkbeiner and Greenberg, 1996, for review). Once activated, Ras interacts with and induces the kinase activity of the Ser/Thr kinase c-Raf, which then triggers the sequential activation of the kinases MEK, Erks, and Rsks. Activated Rsks translocate to the nucleus, where they phosphorylate transcription factors, including CREB at Ser-133 (Campbell et al., 1995; Xing et al., 1996; Finkbeiner et al., 1997; Impey et al., 1999).

Several observations suggest that the Ras/Erk signaling pathway, at least under some circumstances, mediates Ca^{2+}-dependent CREB Ser-133 phosphorylation. First, inhibition of MEK by pharmacological or genetic means leads to a reduction in CREB Ser-133 phosphorylation in Ca^{2+}-stimulated neurons (Impey et al., 1998). Second, there is evidence from experiments using Ca^{2+} chelators that bind Ca^{2+} with different affinity constants that Ca^{2+} regulates CREB Ser-133 phosphorylation by acting at the inner surface of the plasma membrane very close to the site of Ca^{2+} entry (Deisseroth et al., 1996). This observation is consistent with the idea that Ca^{2+} activates Ras, as Ras is localized at the inner surface of the plasma membrane in close proximity to the site of Ca^{2+} entry.

The duration of CREB Ser-133 phosphorylation is controlled not only by the kinetics of CREB kinase activation and shut-off but also through the action of several phosphatases. There is pharmacological evidence that protein phosphatase 1 (PP1) dephosphorylates CREB at Ser-133 in hippocampal neurons (Bito et al., 1996). However, the calmodulin-regulated Ser/Thr phosphatase calcineurin (CaN) may also be involved, as CaN indirectly stimulates PP1 activity (Cohen, 1989). Consistent with this possibility, the inhibition of CaN extends the duration of CREB Ser-133 phosphorylation in hippocampal and striatal neurons (Bito et al., 1996; Liu and Graybiel, 1996).

Alternative Mechanisms of CREB Activation

Despite ample evidence that Ca^{2+} induction of CREB Ser-133 phosphorylation is critical for the induction of CREB-dependent IEG transcription, there is evidence suggesting that CREB Ser-133 phosphorylation may not be sufficient to activate transcription under all circumstances (Bonni et al., 1995; Brindle et al., 1995). Although Ca^{2+} induction of CREB Ser-133 phosphorylation is necessary for BDNF transcription, CREB is phosphorylated at Ser-133 at times when BDNF transcription is shut off (Tao et al., 1998). This suggests that there may be other sites on CREB or CBP whose phosphorylation is regulated by Ca^{2+} influx and may be important for CREB-dependent transcription. Consistent with this idea, Ca^{2+} influx induces the phosphorylation of CREB at sites in addition to Ser-133 that may play an important role in regulating CREB-dependent gene transcription (Enslen et al., 1994; Sun et al., 1994).

Like CaMKIV, overexpression of constitutively active CaMKII in cells triggers CREB phosphorylation at Ser-133 (Enslen et al., 1994; Matthews et al., 1994; Sun et al., 1994). However, in contrast to CaMKIV, constitutively active CaMKII fails to induce CREB-dependent transcription. The failure of CaMKII to activate CREB-mediated transcription has been attributed to CaMKII's ability to induce CREB phosphorylation at Ser-142 as well as Ser-133. When CREB Ser-142 is mutated to an alanine, constitutively active CaMKII induces CREB-dependent transcription, suggesting that phosphorylation at Ser-142 inhibits CREB-mediated transcription (Enslen et al., 1994; Matthews et al., 1994; Sun et al., 1994).

However, membrane depolarization of neurons induces CREB phosphorylation at Ser-142 at times when CREB-dependent gene transcription is activated rather than inhibited (Kornhauser and Greenberg, unpublished observations). This observation raises the seemingly contradictory possibility that phosphorylation at Ser-142 might actually activate rather than inhibit CREB activity. There is recent evidence that once CREB is phosphorylated at Ser-142, it becomes a target for an additional kinase that catalyzes CREB phosphorylation at Ser-143 (Kornhauser and Green-berg, unpublished observations). The phosphorylation of CREB at this additional site renders CREB active. The inhibitory effect of Ser-142 phosphorylation reflects the effect of phosphorylation of CREB at Ser-142 under conditions in which Ser-143 is not phosphorylated. A striking finding is that when CREB becomes phosphorylated at serines 133, 142, and 143, this triply phosphorylated form of CREB may no longer form a complex with CBP, although it is transcriptionally active (Kornhauser and Greenberg, unpublished observations). This finding suggests that by stimulating CREB phosphorylation at all three sites, Ca^{2+} influx alters the mechanism by which CREB activates transcription.

Neurotransmitters that stimulate adenylyl cyclase and lead to PKA activation induce CREB phosphorylation at Ser-133, but not at Ser-142 and Ser-143 (Enslen et al., 1994; Matthews et al., 1994; Sun et al., 1994). Under these circumstances CREB activates transcription by a CBP-dependent mechanism. Depending on the nature of the stimulus (e.g., the route of Ca^{2+} entry into the neuron and the duration of the increase in intracellular Ca^{2+} levels), Ca^{2+} influx may lead to CREB phosphory-lation exclusively at Ser-133, or it might induce CREB phosphorylation at all three sites. Depending on its phosphorylation status CREB would then induce transcription in either a CBP-independent or -dependent fashion. When CREB is phosphorylated at Ser-133, it interacts with CBP and activates a subset of target genes that may be different from the subset of genes that are activated by the triply phosphorylated form of CREB that no longer interacts with CBP. By controlling the phosphory-lation status of CREB, Ca^{2+} entry into neurons may elicit different long-term neuronal adaptive responses. The particular program of gene ex-pression elicited by CREB bound to CBP may be determined in part by CBP's interaction with other transcriptional regulators that bind to CREB-regulated genes. When CREB is phosphorylated at serines 133, 142, and 143 and cannot complex with CBP, CREB's mechanism of action is likely to be distinct and may be reflected in CREB's ability to interact with different transcriptional activators, co-adaptor proteins, or both. Characterization of the mechanism by which CREB activates tran-scription independently of CBP, and identification of the targets of CREB whose transcription is regulated under these circumstances, will undoubtedly be the object of future studies.

SRE and p62^{TCF}

The analysis of the promoters of Ca^{2+}-regulated genes has revealed sev-eral additional transcriptional regulators that contribute to Ca^{2+} induc-tion of IEG transcription. In addition to the CaRE, a DNA regulatory element within the c-*fos* promoter termed the serum response element (SRE) was found to mediate Ca^{2+}-dependent transcriptional responses in PC12 cells and primary neurons (Misra et al., 1994; Miranti et al., 1995; Xia et al., 1996; Johnson et al., 1997). The SRE was originally identified as the sequence within the c-*fos* promoter that is critical for the induction of c-*fos* transcription in serum-stimulated fibroblasts (Treisman, 1985; Gilman et al., 1986; Greenberg et al., 1987). Subsequent studies revealed that the c-*fos* SRE mediates c-fos induction in response to a wide array of growth factors (Rivera and Greenberg, 1990; Treisman, 1992). The SRE contains a region of dyad symmetry (5′-GATGTCCA-TATTAGGACATC-3′) that is critical to its function. Subtle mutations within the dyad symmetry element block the ability of the SRE to drive growth factor or Ca^{2+} induction of reporter gene expression. The 20-bp SRE can be divided into two regions: an inner core that contains the sequence 5′-CCATATTAGG-3′ and outer arms that contain a region of extended dyad symmetry (see Rivera and Greenberg, 1990, and Treisman, 1992, for review). By means of DNA mobility shift analyses, a protein complex composed of two proteins, serum response factor (SRF) and a 62-kDa ternary complex factor ($p62^{TCF}$), was identified that specifically binds to the SRE and mediates the c-*fos* response to extracellular stimuli (Rivera and Greenberg, 1990; Treisman, 1992). SRF is a member of the MADS (MCM1/Agamous/Deficiens/SRF) domain family of transcription factors, a group of transcriptional regulators whose DNA-binding domain is conserved throughout evolution (Nurrish and Treisman, 1995). The MADS domain refers to the conserved sequence motif, the MADS box, which corresponds to the N-terminal two-thirds of the DNA binding domain of these transcriptional regulators. SRF binds as a dimer to the inner core of the SRE via its MADS domain and serves as a docking site for $p62^{TCF}$ (Schroter et al., 1990; Hipskind et al., 1991). $p62^{TCF}$, a member of the Ets family of transcriptional regulators, forms contacts with both SRF and the outer arms of the SRE. $p62^{TCF}$ does not appear to bind to the SRE in the absence of SRF.

In addition to SRF and $p62^{TCF}$, a host of other nuclear factors have been identified that can interact with the SRE. It may be that there is a dynamic exchange of factors that bind to the SRE within a cell and that particular extracellular stimuli stabilize the binding of particular complexes to the SRE, thus allowing c-*fos* to respond to a diverse array of signal transduction pathways. A striking finding is that sequences that are highly similar to the inner core of the c-*fos* SRE are found within the regulatory regions of a large number of other IEGs. This suggests that there may be a common mechanism by which extracellular stimuli activate c-*fos* and other IEGs (see Herdegen and Leah, 1998, for review). Likewise, sequences that are highly similar to the 60-bp c-*fos* CaRE are also found within the regulatory regions of a variety of different IEGs (see Herdegen and Leah, 1998, for review). This finding suggests that the proteins that bind to the CaRE and the SRE (e.g., CREB, SRF, and $p62^{TCF}$) are likely to regulate a diverse array of IEGs.

The activity of both SRF and $p62^{TCF}$ is regulated by Ca^{2+} influx into neurons (Xia et al., 1996; Johnson et al., 1997). SRE mutants that bind SRF, but not $p62^{TCF}$, are still capable of conferring a Ca^{2+} response, suggesting that SRF itself might be a direct target of Ca^{2+}-activated kinases.

Ca^{2+} influx induces the phosphorylation of SRF at Ser-103 (Rivera et al., 1993; Misra et al., 1994; Miranti et al., 1995). Although phosphorylation at Ser-103 enhances the binding of SRF to the SRE in vitro, there is no evidence that phosphorylation of SRF at this site affects SRF function in neurons. Whether SRF's ability to mediate a Ca^{2+} response reflects a Ca^{2+}-dependent change in SRF or an SRF-associated protein remains to be determined.

In contrast to SRF, it is clear that p62^{TCF}'s transcriptional activation function is activated by Ca^{2+} influx into neurons, and that this is a consequence of p62^{TCF} phosphorylation by MAP kinases at multiple sites within the p62^{TCF} transactivation domain (Xia et al., 1996; Johnson et al., 1997). It is not yet clear how phosphorylation of p62^{TCF} leads to transcriptional activation, although it is known that p62^{TCF} functions via an interaction with CBP (Janknecht and Nordheim, 1996). One possibility is that CBP facilitates cooperative interactions between p62^{TCF}, SRF, CREB, and the pol II transcription machinery, thereby leading to IEG induction.

MEF2

The MEF2 transcription factors (also termed "related to serum response factors" [RSRFs]) are members of a family of MADS domain transcription factors that are related to SRF and are selectively expressed in muscle and postmitotic neurons (Brand, 1997). Although MEF2 A, B, C, and D form homo- and heterodimers with one another, they fail to interact with SRF, and they recognize a DNA regulatory element, designated the MRE, that is distinct from the SRE (Pollock and Treisman, 1991). The MRE is present within the regulatory region of a number of IEGs that are responsive to Ca^{2+} influx, and Ca^{2+} influx through L-type VSCCs stimulates MEF2-dependent transcription (Mao et al., 1999). Ca^{2+} influx induces MEF2 activity by inducing dephosphorylation of MEF2 at several sites whose phosphorylation inhibits the binding of MEF2 to the MRE (Mao and Widemann, 1999). In addition, Ca^{2+} influx activates the p38 MAP kinase, which then phosphorylates MEF2 at several sites within the MEF2 transactivation domain, thereby stimulating MEF2-dependent transcription. These findings suggest that MEF2s function in a manner analogous to CREB in mediating Ca^{2+} induction of IEGs in neurons, and they raise the possibility that MEF2s, like CREB, might be mediators of adaptive neuronal responses.

DREAM

In addition to its ability to stimulate the activity of transcriptional activators, Ca^{2+} influx into neurons relieves the inhibitory effect of a transcriptional repressor termed DREAM (Carrion et al., 1999). A downstream regulatory element (DRE) has been identified within the first intron of the c-*fos* gene and within the prodynorphin gene, whose function is to suppress the transcription of these genes under conditions in which the intracellular level of Ca^{2+} is low. It has been proposed that when Ca^{2+} levels are low, DREAM binds to the DRE and represses the transcription of Ca^{2+}-responsive genes. Ca^{2+} influx results in the direct binding of Ca^{2+} to DREAM, which induces a conformational change in DREAM, a loss of DREAM's interaction with the DRE, and the derepression of DREAM-regulated genes. At the same time that Ca^{2+} influx triggers the inactivation of DREAM, it also induces the activity of several protein kinase cascades that phosphorylate and activate a host of Ca^{2+}-dependent transcriptional activators, including CREB, p62[TCF], and MEF2. Thus, by inhibiting negative regulatory factors and inducing positive regulators of transcription, Ca^{2+} influx into neurons effectively stimulates the transcription of a variety of genes that mediate adaptive neuronal responses.

BDNF *Transcriptional Regulators*

A feature of c-*fos* regulation is that c-*fos* transcription can be induced in a wide range of cell types by many different stimuli. However, a significant number of IEGs are selectively activated by Ca^{2+} influx into postmitotic neurons (Zafra et al., 1990, 1992; Ghosh et al., 1994; Yamagata et al., 1994; Lanahan and Worley, 1998; Steward et al., 1998; O'Brien et al., 1999; Yamagata et al., 1999). This finding indicates that there are likely to be Ca^{2+}-regulated transcription factors that are selectively expressed and/or activated in neurons. Of the well-characterized transcriptional regulators, members of the CREB and p62[TCF] families are widely expressed and are therefore unlikely to function on their own in conferring a neuron-specific Ca^{2+} response. However, there may be specific mechanisms, such as the regulation of CREB Ser-142 phosphorylation, that may allow CREB and other ubiquitously expressed transcription factors to activate the expression of specific genes in Ca^{2+}-stimulated neurons.

Analysis of the regulatory elements within the *BDNF* gene, a Ca^{2+}-regulated gene whose transcription is largely restricted to neurons, has provided some insight into a mechanism by which Ca^{2+} influx can selectively induce transcription in neurons (Timmusk et al., 1993; Shieh et al., 1998; Tao et al., 1998). *BDNF* is a complex gene that extends over 40 kilobases of genomic DNA. There are four distinct *BDNF* promoters, each of which gives rise to two different mRNA transcripts yielding a total of eight distinct *BDNF* mRNAs (Fig. 8.5). Each of the *BDNF* mRNAs encodes an identical BDNF protein, raising the question of why there are so many *BDNF* transcripts if they each encode the same protein. As the different *BDNF* mRNAs have distinct 5′ and 3′ untranslated regions, it is possible that the *BDNF* mRNAs are differentially localized within the neurons, for example to dendrites or axons. Another possibility is that there is differential regulation of the stability or translation of the various *BDNF* mRNAs within the cell body or dendrites. There is also

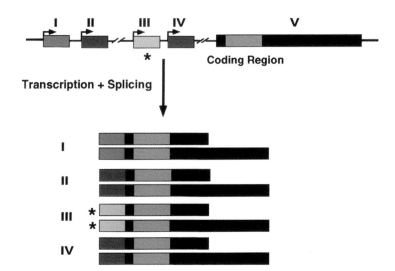

Figure 8.5. Schematic representation of the BDNF gene and mature BDNF transcripts. The BDNF coding region is indicated by the orange boxes. The 3' untranslated regions encoded by BDNF exon V are indicated by black boxes. The 5' untranslated regions encoded by 5' exons are indicated by the green, yellow, red, and purple boxes (Timmusk et al., 1993). Reproduced from Tao et al. (1998) with permission of Cell Press.

evidence that the four *BDNF* promoters are differentially regulated, with promoter III being the most responsive to Ca^{2+} influx in embryonic cortical cultures (Tao et al., 1998).

Deletion analysis of *BDNF* promoter III has revealed the presence of three DNA regulatory elements that are critical for Ca^{2+} induction of *BDNF* transcription (Shieh et al., 1998; Tao et al., 1998; Tao and Greenberg, unpublished observations). The most promoter-proximal element binds CREB, which has been widely implicated in the Ca^{2+} regulation of *BDNF* transcription. Given the evidence that *BDNF* can potentiate synaptic function, this raises the possibility that *BDNF* is one of the key gene targets of CREB, which mediates CREB's effects on synaptic potentiation (Fig. 8.6). In addition to the CREB binding site, there are at least two distal elements within *BDNF* promoter III that are critical for Ca^{2+} induction of *BDNF* transcription. These *BDNF* regulatory elements may bind novel transcription factors that may account for the neural specificity of *BDNF* transcription (Tao et al., unpublished observations).

The Mode of Ca^{2+} Entry Matters

Although Ca^{2+} entry through L-type VSCCs activates *BDNF* transcription robustly and for a prolonged period, Ca^{2+} influx through NMDA receptors only induces *BDNF* transcription weakly and transiently (Ghosh et al., 1994). This may reflect the fact that Ca^{2+} influx through L-type VSCCs is much more effective at inducing CREB-dependent transcription than

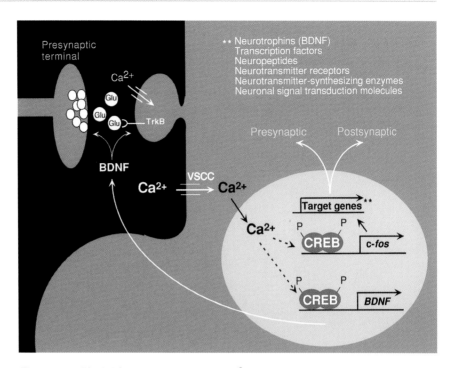

Figure 8.6. Model for how products of Ca^{2+}-regulated genes might affect synaptic function. Glutamate is released from the presynaptic neuron and binds to its receptor on the postsynaptic neuron, leading to synaptic potentiation. If glutamate binding to its receptor induces sufficient membrane depolarization to stimulate the opening of L-type VSCCs, the ensuing Ca^{2+} influx sends a signal to the nucleus, which leads to CREB phosphorylation and activation. Activated CREB induces the transcription of target genes such as *BDNF* and c-*fos*. Together with c-*jun*, c-*fos* activates a variety of target genes that encode proteins that could function directly at synapses. CREB also activates *BDNF* transcription directly. BDNF may then act through its receptor TrkB at the potentiated synapse.

is Ca^{2+} influx through the NMDA receptor channel (Hardingham et al., 1999). It is not yet clear how the route of Ca^{2+} entry into a neuron differentially affects CREB activation. The simplest explanation is that the amount of Ca^{2+}, or the duration of the increase in intracellular Ca^{2+} concentration, determines whether or not CREB is activated. However, available evidence suggests that differences in Ca^{2+} dynamics do not explain the observed difference. The possibility that Ca^{2+} activates specific signal transduction molecules that are localized in close proximity to its site of entry has gained support from recent studies showing that NMDA receptors form complexes with PDZ domain–containing proteins. These proteins function as scaffolds but may also have a signal transduction role (Sheng and Pak, 1999). The hypothesis that the cytoplasmic domains of the L-type VSCCs and NMDA receptors associate with distinct signaling molecules that sense the Ca^{2+} signal and convey it to the nucleus is currently under investigation (Fig. 8.7).

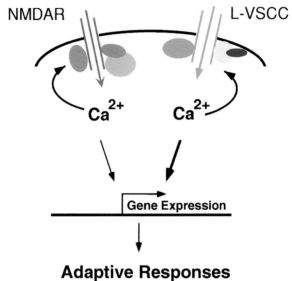

Figure 8.7. *The mode of Ca²⁺ entry into a neuron may dictate the signaling pathways that regulate gene expression.* The pathway activated by Ca²⁺ may be determined by signaling proteins that associate with the cytoplasmic domains of the NMDA receptor (NMDAR) and L-type VSCC channel subunits.

Conveying Transcriptional Events to Potentiated Synapses

The capacity of neurons to regulate gene expression in response to synaptic activity raises several fundamental questions. How does activity-induced gene expression serve neuronal function? How does the unusual architecture of the neuron affect the mechanisms for activity-dependent gene expression? And how are new gene products specifically targeted to potentiated synapses?

New protein synthesis is necessary for long-term memory storage (Goelet et al., 1986), and a paradigm for activity-dependent synaptic plasticity that may underlie memory storage is LTP (Bliss and Collingridge, 1993). Kandel and co-workers have shown that induction of LTP in the CA1 region of the rat hippocampus by high-frequency stimulation or a cAMP analogue is blocked by inhibitors of protein synthesis (Nguyen et al., 1994). Gene induction during LTP has features similar to other examples of activity-regulated gene control in that LTP induces CRE-regulated gene transcription by a phosphorylation-dependent mechanism (Impey et al., 1996). Following LTP, a complex pattern of changes in the levels of individual proteins has been observed (Fazeli et al., 1993). Because the mechanisms that regulate synaptic plasticity are not well understood, it is difficult to predict which proteins are needed for activity-dependent changes in synaptic strength. Candidates include structural components of the synaptic junction, proteins that modify the architecture or transmission properties of synapses, and proteins that alter the specificity of signaling molecules. A number of genes that are activated during the induction of LTP have been identified (Nedivi et al., 1993). In addition to IEGs that encode transcription factors (discussed previously), genes induced during LTP include those for tissue plasminogen activator (Qian et al., 1993), pim kinase (Konietzko et al., 1999), prostaglandin H synthase (Yamagata et al., 1993), and the aforementioned Arc protein (Lyford et al., 1995).

In neurons that make a large number of synaptic connections, LTP is not uniform among the synapses but is confined to a subpopulation of activated synapses (Schuman, 1997a,b). The fact that protein synthesis generally takes place in the cell body, remote from the location of the potentiated synapses, complicates the problem of delivering the new protein products, specifically to the synapses destined for synaptic modification. In one scheme, protein trafficking could route new peptides to the appropriate locations. However, recent studies suggest an alternative mechanism for the delivery of new protein products to potentiated synapses. Frey and Morris (1997) have suggested that synaptic activity leads to the formation of a synaptic tag that marks the activated synapses. Although the protein products of some of the Ca^{2+}-regulated genes might be transported to each of the neuron synapses, these newly synthesized proteins might be capable of functioning only at the tagged synapses.

An alternative mechanism for targeted gene expression is dendritic mRNA localization (Kuhl and Skehel, 1998; Steward et al., 1998). Although the majority of neuronal mRNAs are confined to the cell body, a select population of mRNAs is specifically localized in dendrites. This targeting is best illustrated for the IEG mRNA that encodes Arc. Most mRNAs of IEGs remain in the cell body; however, the Arc mRNA is transported at a rate of some 300 μm/hr to dendrites (Wallace et al., 1998). Because transport takes place in the presence of protein synthesis inhibitors, it is the Arc mRNA itself, not its translation product, that is responsible for the targeting. The Arc mRNA appears to incorporate aspects of the synaptic tagging hypothesis, because not only is it dendritic, it is also transported preferentially to the vicinity of synapses that have been activated (Steward et al., 1998).

mRNA translation may also be synapse-specific and activity-dependent. The α-CamKII mRNA, another mRNA that is targeted to dendrites (Mayford et al., 1996), contains two cytoplasmic polyadenylation elements that contain the sequence UUUUUAU and bind the CPE binding protein (CPEB) (Wu et al., 1998). CPEB regulates the activity-dependent lengthening of the poly(A) tail at the 3' end of the CaMKII mRNA, leading to the activation of translation of the CaMKII mRNA. The selective translation of mRNAs such as the CaMKII mRNA provides neurons with yet another means to target specific proteins to activated synapses.

Although the modulation of synaptic transmission can occur very rapidly when compared to the time scale of gene regulation, it appears that long-term changes in synaptic transmission, such as those that underlie certain forms of learning and memory, cannot take place without new gene expression. The elaborate asymmetry of the neuron—in particular the separation between the compartments of synaptic transmission, which must be the ultimate targets of new gene products that provide synaptic plasticity, and the locus of gene regulation, the nucleus—has led to novel mechanisms of signaling and control. As our understanding of the synapse and its control increases, new features of this novel regulation are sure to emerge. Not only will these disclose fundamental features of neuron control, they may also shed light on mechanisms of brain dysfunction and disease and provide opportunities for pharmacological intervention.

References

Bading, H., Ginty, D. D., Greenberg, M. E. (1993). Regulation of gene expression in hippocampal neurons by distinct calcium signaling pathways. *Science* 260:181–186.

Bailey, C. H., Chen, M., Keller, F., Kandel, E. R. (1992). Serotonin-mediated endocytosis of apCAM: an early step of learning-related synaptic growth in Aplysia. *Science* 256:645–649.

Bannister, A. J., Kouzarides, T. (1996). The CBP co-activator is a histone acetyltransferase. *Nature* 384:641–643.

Bartsch, D., Ghirardi, M., Skehel, P. A., Karl, K. A., Herder, S. P., Chen, M., Bailey, C. H., Kandel, E. R. (1995). Aplysia CREB2 represses long-term facilitation: relief of repression converts transient facilitation into long-term functional and structural change. *Cell* 83:979–992.

Bear, M. F. (1999). Homosynaptic long-term depression: a mechanism for memory? *Proc. Natl. Acad. Sci. USA* 96:9457–9458.

Bito, H., Deisseroth, K., Tsien, R. W. (1996). CREB phosphorylation and dephosphorylation: a Ca^{2+}- and stimulus duration–dependent switch for hippocampal gene expression. *Cell* 87:1203–1214.

Bliss, T. V. P., Collingridge, G. L. (1993). A synaptic model of memory: long-term potentiation in the hippocampus. *Nature* 361:31–39.

Bonni, A., Greenberg, M. E. (1997). Neurotrophin regulation of gene expression. *Can. J. Neurosci.* 24:272–283.

Bonni, A., Ginty, D. D., Dudek, H., Greenberg, M. E. (1995). Serine 133–phosphorylated CREB induces transcription via a cooperative mechanism that may confer specificity to neurotrophin signals. *Mol. Cell. Neurosci.* 6:168–183.

Bonni, A., Brunet, A., West, A. E., Datta, S. R., Takasu, M., Greenberg, M. E. (1999). Cell survival promoted by the Ras-MAPK signaling pathway by transcription-dependent and -independent mechanisms. *Science* 286:1358–1362.

Bourtchuladze, R., Frenguelli, B., Blendy, J., Cioffi, D., Schutz, G., Silva, A. J. (1994). Deficient long-term memory in mice with a targeted mutation of the cAMP-responsive element-binding protein. *Cell* 79:59–68.

Brakeman, P., Lanahan, A., O'Brian, R., Roche, K., Barnes, C., Huganir, R. L., Worley, P. F. (1997). Homer: a protein that selectively binds metabotropic glutamate receptors. *Nature* 386:284–288.

Brand, N. J. (1997). Myocyte enhancer factor 2 (MEF2). *Int. J. Biochem. Cell Biol.* 29:1467–1470.

Braun, A. P., Schulman, H. (1995). The multifunctional calcium/calmodulin-dependent protein kinase. *Annu. Rev. Neurosci.* 57:417–445.

Brindle, P., Nakajima, T., Montminy, M. (1995). Multiple protein kinase A–regulated events are required for transcriptional induction by cAMP. *Proc. Natl. Acad. Sci. USA* 92:10,521–10,525.

Brown, J. R., Ye, H., Bronson, R. T., Dikkes, P., Greenberg, M. E. (1996). A defect in nurturing in mice lacking the immediate early gene fosB. *Cell* 86: 297–309.

Campbell, J. S., Seger, R., Graves, J. D., Graves, L. M., Jensen, A. M., Krebs, E. G. (1995). The MAP kinase cascade. *Recent Prog. Horm. Res.* 50:131–159.

Carrion, A., Link, W., Ledo, F., Mellström, B., Naranjo, J. (1999). DREAM is a Ca^{2+}-regulated transcriptional repressor. *Nature* 398:80–84.

Castren, E., Zafra, F., Thoenen, H., Lindholm, D. (1992). Light regulates expression of brain-derived neurotrophic factor mRNA in rat visual cortex. *Proc. Natl. Acad. Sci. USA* 89:9444–9448.

Catterall, W. A. (1995). Structure and function of voltage-gated ion channels. *Annu. Rev. Biochem.* 64:493–531.

Chrivia, J. C., Kwok, R. P., Lamb, N., Hagiwara, M., Montminy, M. R., Goodman, R. H. (1993). Phosphorylated CREB binds specifically to the nuclear protein CBP. *Nature* 365:855–859.

Clapham, D. E. (1995). Calcium signaling. *Cell* 80:259–268.

Cohen, P. (1989). The structure and regulation of protein phosphatases. *Annu. Rev. Biochem.* 58:453–508.

Curran, T., Franza, B. R., Jr. (1988). Fos and Jun: the AP-1 connection. *Cell* 55: 395–397.

Curran, T., Morgan, J. I. (1995). Fos: an immediate-early transcription factor in neurons. *J. Neurobiol.* 26:403–412.

Dash, P. K., Karl, K. A., Colicos, M. A., Prywes, R., Kandel, E. R. (1991). cAMP response element–binding protein is activated by Ca^{2+}/calmodulin- as well as cAMP-dependent protein kinase. *Proc. Natl. Acad. Sci. USA* 88:5061–5065.

Deisseroth, K., Bito, H., Tsien, R. W. (1996). Signaling from synapse to nucleus: postsynaptic CREB phosphorylation during multiple forms of hippocampal synaptic plasticity. *Neuron* 16:89–101.

Deisseroth, K., Heist, E. K., Tsien, R. W. (1998). Translocation of calmodulin to the nucleus supports CREB phosphorylation in hippocampal neurons. *Nature* 392:198–202.

Dragunow, M., Beilharz, E., Mason, B., Lawlor, P., Abraham, W., Gluckman, P. (1993). Brain-derived neurotrophic factor expression after long-term potentiation. *Neurosci. Lett.* 160:232–236.

Enslen, H., Sun, P., Brickey, D., Soderling, S. H., Klamo, E., Soderling, T. R. (1994). Characterization of Ca^{2+}/calmodulin-dependent protein kinase. IV. Role in transcriptional regulation. *J. Biol. Chem.* 269:15,520–15,527.

Enslen, H., Tokumitsu, H., Soderling, T. R. (1995). Phosphorylation of CREB by CaM-kinase IV activated by CaM-kinase IV kinase. *Biochem. Biophys. Res. Commun.* 207:1038–1043.

Fazeli, M. S., Corbet, J., Dunn, M. J., Dolphin, A. C., Bliss, T. V. (1993). Changes in protein synthesis accompanying long-term potentiation in the dentate gyrus in vivo. *J. Neurosci.* 13:1346–1353.

Figurov, A., Pozzo-Miller, L. D., Olafsson, P., Wang, T., Lu, B. (1996). Regulation of synaptic responses to high-frequency stimulation and LTP by neurotrophins in the hippocampus. *Nature* 381:706–709.

Finkbeiner, S., Greenberg, M. E. (1996). Ca^{2+}-dependent routes to Ras: Mechanisms for neuronal survival, differentiation, and plasticity? *Neuron* 16:233–236.

Finkbeiner, S., Greenberg, M. E. (1998). Ca^{2+} channel–regulated neuronal gene expression. *J. Neurobiol.* 37:171–189.

Finkbeiner, S., Tavazoie, S. F., Maloratsky, A., Jacobs, K. M., Harris, K. M., Greenberg, M. E. (1997). CREB: a major mediator of neuronal neurotrophin responses. *Neuron* 19:1031–1047.

Frank, D. A., Greenberg, M. E. (1994). CREB: a mediator of long-term memory from mollusks to mammals. *Cell* 79:5–8.

Franklin, J. L., Johnson, E. M. (1992). Suppression of programmed neuronal death by sustained elevation of cytoplasmic calcium. *Trends Neurosci.* 15:501–508.

Frey, U., Morris, R. G. (1997). Synaptic tagging and long-term potentiation. *Nature* 385:533–536.

Ghosh, A., Greenberg, M. E. (1995). Calcium signaling in neurons: molecular mechanisms and cellular consequences. *Science* 268:239–247.

Ghosh, A., Carnahan, J., Greenberg, M. E. (1994). Requirement for BDNF in activity-dependent survival of cortical neurons. *Science* 263:1618–1623.

Gilman, M. Z., Wilson, R. N., Weinberg, R. A. (1986). Multiple protein-binding sites in the 5′-flanking region regulate c-fos expression. *Mol. Cell. Biol.* 6:4305–4316.

Ginty, D. D., Bonni, A., Greenberg, M. E. (1994). Nerve growth factor activates a Ras-dependent protein kinase that stimulates c-*fos* transcription via phosphorylation of CREB. *Cell* 77:713–725.

Ginty, D. D., Kornhauser, J. M., Thompson, M. A., Bading, H., Mayo, K. E., Takahashi, J. S., Greenberg, M. E. (1993). Regulation of CREB phosphorylation in the suprachiasmatic nucleus by light and a circadian clock. *Science* 260:238–241.

Giordano, A., Avantaggiati, M. L. (1999). p300 and CBP: partners for life and death. *J. Cell. Physiol.* 181:218–230.

Goelet, P., Castellucci, V. F., Schacher, S., Kandel, E. R. (1986). The long and short of long-term memory—a molecular framework. *Nature* 322:419–422.

Gonzalez, G. A., Yamamoto, K. K., Fischer, W. H., Karr, D., Menzel, P., Biggs, W., 3d, Vale, W. W., Montminy, M. R. (1989). A cluster of phosphorylation sites on the cyclic AMP–regulated nuclear factor CREB predicted by its sequence. *Nature* 337:749–752.

Gonzalez, G. A., Menzel, P., Leonard, J., Fischer, W. H., Montminy, M. R. (1991). Characterization of motifs which are critical for activity of the cyclic AMP–responsive transcription factor CREB. *Mol. Cell. Biol.* 11:1306–1312.

Greenberg, M., Ziff, E. B., Greene, L. A. (1986). Stimulation of neuronal acetylcholine receptors induces rapid gene transcription. *Science* 234:80–83.

Greenberg, M. E., Siegfried, Z., Ziff, E. B. (1987). Mutation of the c-*fos* gene dyad symmetry element inhibits serum inducibility of transcription in vivo and the nuclear regulatory factor binding in vitro. *Mol. Cell. Biol.* 7:1217–1225.

Hagiwara, M., Brindle, P., Harootunian, A., Armstrong, R., Rivier, J., Vale, W., Tsien, R., Montminy, M. R. (1993). Coupling of hormonal stimulation and transcription via the cyclic AMP–responsive factor CREB is rate limited by nuclear entry of protein kinase A. *Mol. Cell. Biol.* 13:4852–4859.

Hai, T., Curran, T. (1991). Cross-family dimerization of transcription factors Fos/Jun and ATF/CREB alters DNA binding specificity. *Proc. Natl. Acad. Sci. USA* 88:3720–3724.

Hai, T. W., Liu, F., Allegretto, E. A., Karin, M., Green, M. R. (1988). A family of immunologically related transcription factors that includes multiple forms of ATF and AP-1. *Genes Dev* 2:1216–1226.

Hai, T. W., Liu, F., Coukos, W. J., Green, M. R. (1989). Transcription factor ATF cDNA clones: an extensive family of leucine zipper proteins able to selectively form DNA-binding heterodimers. *Genes Dev.* 3:2083–2090.

Hardingham, G. E., Chawla, S., Cruzalegui, F. H., Bading, H. (1999). Control of recruitment and transcription-activating function of CBP determines gene regulation by NMDA receptors and L-type calcium channels. *Neuron* 22:789–798.

Herdegen, T., Leah, J. D. (1998). Inducible and constitutive transcription factors in the mammalian nervous system: control of gene expression by Jun, Fos and Krox, and CREB/ATF proteins. *Brain Res. Brain Res. Rev.* 28:370–490.

Hipskind, R. A., Rao, V. N., Mueller, C. G. F., Reddy, E. S. P., Nordheim, A. (1991). Ets-related protein Elk-1 is homologous to the c-*fos* regulatory factor p62TCF. *Nature* 354:531–534.

Hiroi, N., Brown, J. R., Haile, C. N., Ye, H., Greenberg, M. E., Nestler, E. J. (1997). FosB mutant mice: loss of chronic cocaine induction of Fos-related proteins and heightened sensitivity to cocaine's psychomotor and rewarding effects. *Proc. Natl. Acad. Sci. USA* 94:10,397–10,402.

Hope, B., Kosofsky, B., Hyman, S. E., Nestler, E. J. (1992). Regulation of immediate early gene expression and AP-1 binding in the rat nucleus accumbens by chronic cocaine. *Proc. Natl. Acad. Sci. USA* 89:5764–5768.

Huang, E. P. (1999). Synaptic plasticity: regulated translation in dendrites. *Curr. Biol.* 9:R168–R170.

Huang, Y. Y., Kandel, E. R. (1994). Recruitment of long-lasting and protein kinase A–dependent long-term potentiation in the CA1 region of hippocampus requires repeated tetanization. *Learn. Mem.* 1:74–82.

Impey, S., Mark, M., Villacres, E. C., Poser, S., Chavkin, C., Storm, D. R. (1996). Induction of CRE-mediated gene expression by stimuli that generate long-lasting LTP in area CA1 of the hippocampus. *Neuron* 16:973–982.

Impey, S., Obrietan, K., Wong, S. T., Poser, S., Yano, S., Wayman, G., Deloume, J. C., Chan, G., Storm, D. R. (1998). Cross talk between ERK and PKA is required for Ca^{2+} stimulation of CREB-dependent transcription and ERK nuclear translocation. *Neuron* 21:869–883.

Impey, S., Obrietan, K., Storm, D. R. (1999). Making new connections: role of ERK/MAP kinase signaling in neuronal plasticity. *Neuron* 23:11–14.

Janknecht, R., Nordheim, A. (1996). MAP kinase–dependent transcriptional coactivation by Elk-1 and its cofactor CBP. *Biochem. Biophys. Res. Commun.* 228:831–837.

Jeon, S. H., Seong, Y. S., Juhnn, Y. S., Kang, U. G., Ha, K. S., Kim, Y. S., Park, J. B. (1997). Electroconvulsive shock increases the phosphorylation of cyclic AMP response element binding protein at Ser-133 in rat hippocampus but not in cerebellum. *Neuropharmacology* 36:411–414.

Ji, R. R., Rupp, F. (1997). Phosphorylation of transcription factor CREB in rat spinal cord after formalin-induced hyperalgesia: relationship to c-fos induction. *J. Neurosci.* 17:1776–1785.

Johnson, C. M., Hill, C. S., Chawla, S., Treisman, R., Bading, H. (1997). Calcium controls gene expression via three distinct pathways that can function independently of the Ras/mitogen-activated protein kinases (ERKs) signaling cascade. *J. Neurosci.* 17:6189–6202.

Jonas, E. A., Kaczmarek, L. K. (1996). Regulation of potassium channels by protein kinases. *Curr. Opin. Neurobiol.* 6:318–323.

Kaang, B. K., Kandel, E. R., Grant, S. G. (1993). Activation of cAMP-responsive genes by stimuli that produce long-term facilitation in Aplysia sensory neurons. *Neuron* 10:427–435.

Kaczmarek, L., Kossut, M., Skangiel-Kramska, J. (1997). Glutamate receptors in cortical plasticity: molecular and cellular biology. *Physiol. Rev.* 77:217–255.

Kang, H., Schuman, E. M. (1995a). Long-lasting neurotrophin-induced enhancement of synaptic transmission in the adult hippocampus. *Science* 267:1658–1662.

Kang, H. J., Schuman, E. M. (1995b). Neurotrophin-induced modulation of synaptic transmission in the adult hippocampus. *J. Physiol.* 89:11–22.

Kang, H., Schuman, E. M. (1996). A requirement for local protein synthesis in neurotrophin-induced hippocampal synaptic plasticity. *Science* 273:1402–1406.

Kato, A., Ozawa, F., Saitoh, Y., Hirai, K., Inokuchi, K. (1997). vesl, a gene encoding VASP/Ena family related protein, is upregulated during seizure, long-term potentiation and synaptogenesis. *FEBS Lett.* 412:183–189.

Kato, Y., Tapping, R. I., Huang, S., Watson, M. H., Ulevitch, R. J., Lee, J.-D. (1998). Bmk1/Erk5 is required for cell proliferation induced by epidermal growth factor. *Nature* 395:713–716.

Kogan, J. H., Frankland, P. W., Blendy, J. A., Coblentz, J., Marowitz, Z., Schutz, G., Silva, A. J. (1997). Spaced training induces normal long-term memory in CREB mutant mice. *Curr. Biol.* 7:1–11.

Konietzko, U., Kauselmann, G., Scafidi, J., Staubli, U., Mikkers, H., Berns, A., Schweizer, M., Waltereit, R., Kuhl, D. (1999). Pim kinase expression is induced by LTP stimulation and required for the consolidation of enduring LTP. *EMBO J.* 18:3359–3369.

Kornhauser, J. M., Nelson, D. E., Mayo, K. E., Takahashi, J. S. (1990). Photic and circadian regulation of c-*fos* gene expression in the hamster suprachiasmatic nucleus. *Neuron* 5:127–134.

Korte, M., Kang, H., Bonhoeffer, T., Schuman, E. (1998). A role for BDNF in the late-phase of hippocampal long-term potentiation. *Neuropharmacology* 37:553–559.

Kuhl, D., Skehel, P. (1998). Dendritic localization of mRNAs. *Curr. Opin. Neurobiol.* 8:600–606. [Erratum appears in *Curr. Opin. Neurobiol.* (1999) 9:142.]

Lanahan, A., Worley, P. (1998). Immediate-early genes and synaptic function. *Neurobiol. Learn. Mem.* 70:37–43.

Lee, M. M., Badache, A., DeVries, G. H. (1999). Phosphorylation of CREB in axon-induced Schwann cell proliferation. *J. Neurosci. Res.* 55:702–712.

Lerea, L. S., McNamara, J. O. (1993). Ionotropic glutamate receptor subtypes activate c-fos transcription by distinct calcium-requiring intracellular signaling pathways. *Neuron* 10:31–41.

Liu, F. C., Graybiel, A. M. (1996). Spatiotemporal dynamics of CREB phosphorylation: transient versus sustained phosphorylation in the developing striatum. *Neuron* 17:1133–1144.

Lohof, A. M., Ip, N. Y., Poo, M. M. (1993). Potentiation of developing neuromuscular synapses by the neurotrophins NT-3 and BDNF. *Nature* 363:350–353.

Lyford, G. L., Yamagata, K., Kaufmann, W. E., Barnes, C. A., Sanders, L. K., Copeland, N. G., Gilbert, D. J., Jenkins, N. A., Lanahan, A. A., Worley, P. F. (1995). Arc, a growth factor and activity-regulated gene, encodes a novel cytoskeleton-associated protein that is enriched in neuronal dendrites. *Neuron* 14:433–445.

Malenka, R. C., Nicoll, R. A. (1999). Long-term potentiation—a decade of progress? *Science* 285:1870–1874.

Mao, A., Widemann, M. (1999). Calcineurin enhances MEF2 DNA binding activity in calcium-dependent survival of cerebellar granule neurons. *J. Biol. Chem.* 274:31,102–31,107.

Mao, Z., Bonni, A., Xia, F., Nadal-Vicens, M., Greenberg, M. E. (1999). Neuronal activity–dependent cell survival mediated by transcription factor MEF2. *Science* 286:785–790.

Martin, K. C., Kandel, E. R. (1996). Cell adhesion molecules, CREB, and the formation of new synaptic connections. *Neuron* 17:567–570.

Matthews, R. P., Guthrie, C. R., Wailes, L. M., Zhao, X., Means, A. R., McKnight, G. S. (1994). Calcium/calmodulin-dependent protein kinase types II and IV differentially regulate CREB-dependent gene expression. *Mol. Cell. Biol.* 14: 6107–6116.

Matthies, H., Frey, U., Reymann, K., Krug, M., Jork, R., Schroeder, H. (1990). Different mechanisms and multiple stages of LTP. *Adv. Exp. Med. Biol.* 268: 359–368.

Mayer, M. L., Miller, J. J. (1990). Excitatory amino acid receptors, second messengers and regulation of intracellular Ca^{2+} in mammalian neurons. *Trends Pharmacol. Sci.* 11:36–42.

Mayford, M., Baranes, D., Podsypanina, K., Kandel, E. R. (1996). The 3′-untranslated region of CaMKII alpha is a cis-acting signal for the localization and translation of mRNA in dendrites. *Proc. Natl. Acad. Sci. USA* 93:13,250–13,255.

Miranti, C. K., Ginty, D. D., Huang, G., Chatila, T., Greenberg, M. E. (1995). Calcium activates serum response factor–dependent transcription by a Ras- and Elk-1-independent mechanism that involves a Ca^{2+}/calmodulin-dependent kinase. *Mol. Cell. Biol.* 15:3672–3684.

Misra, R. P., Bonni, A., Miranti, C. K., Rivera, V. M., Sheng, M., Greenberg, M. E. (1994). L-type voltage-sensitive calcium channel activation stimulates gene expression by a serum response factor–dependent pathway. *J. Biol. Chem.* 269:25,483–25,493.

Morgan, J. I., Curran, T. (1986). Role of ion flux in the control of c-fos expression. *Nature* 322:552–555.

Mori, H., Mishina, M. (1995). Structure and function of the NMDA receptor channel. *Neuropharmacology* 34:1219–1237.

Nakamura, Y., Okuno, S., Sato, F., Fujisawa, H. (1995). An immunohistochemical study of Ca^{2+}/calmodulin-dependent protein kinase IV in the rat central nervous system: light and electron microscopic observations. *Neuroscience* 68: 181–194.

Nedivi, E., Hevroni, D., Naot, D., Israeli, D., Citri, Y. (1993). Numerous candidate plasticity-related genes revealed by differential cDNA cloning. *Nature* 363:718–722.

Nguyen, P. V., Abel, T., Kandel, E. R. (1994). Requirement of a critical period of transcription for induction of a late phase of LTP. *Science* 265:1104–1107.

Nurrish, S. J., Treisman, R. (1995). DNA binding specificity determinants in MADS-box transcription factors. *Mol. Cell. Biol.* 15:4076–4185.

O'Brien, R. J., Lau, L. F., Huganir, R. L. (1998). Molecular mechanisms of glutamate receptor clustering at excitatory synapses. *Curr. Opin. Neurobiol.* 8: 364–369.

O'Brien, R. J., Xu, D., Petralia, R. S., Steward, O., Huganir, R. L., Worley, P. (1999). Synaptic clustering of AMPA receptors by the extracellular immediate-early gene product Narp. *Neuron* 23:309–323.

Ogryzko, V. V., Schiltz, R. L., Russanova, V., Howard, B. H., Nakatani, Y. (1996). The transcriptional coactivators p300 and CBP are histone acetyltransferases. *Cell* 87:953–959.

Ozawa, S., Kamiya, H., Tsuzuki, K. (1998). Glutamate receptors in the mammalian central nervous system. *Prog. Neurobiol.* 54:581–618.

Patterson, S. L., Grover, L. M., Schwartzkroin, P. A., Bothwell, M. (1992). Neurotrophin expression in rat hippocampal slices: a stimulus paradigm inducing LTP in CA1 evokes increases in BDNF and NT-3 mRNAs. *Neuron* 9:1081–1088.

Patterson, S. L., Abel, T., Deuel, T. A., Martin, K. C., Rose, J. C., Kandel, E. R. (1996). Recombinant BDNF rescues deficits in basal synaptic transmission and hippocampal LTP in BDNF knockout mice. *Neuron* 16:1137–1145.

Perkel, D., Petrozzino, J., Nicoll, R., Connor, J. (1993). The role of Ca^{2+} entry via synaptically activated NMDA receptors in the induction of long-term potentiation. *Neuron* 11:817–823.

Pollock, R., Treisman, R. (1991). Human SRF-related proteins: DNA-binding properties and potential regulatory targets. *Genes Dev.* 5:2327–2341.

Qian, Z., Gilbert, M. E., Colicos, M. A., Kandel, E. R., Kuhl, D. (1993). Tissue-plasminogen activator is induced as an immediate-early gene during seizure, kindling and long-term potentiation. *Nature* 361:453–457.

Quinn, P. G. (1993). Distinct activation domains within cAMP response element-binding protein (CREB) mediate basal and cAMP-stimulated transcription. *J. Biol. Chem.* 268:16,999–17,009.

Rehfuss, R. P., Walton, K. M., Loriaux, M. M., Goodman, R. H. (1991). The cAMP-regulated enhancer-binding protein ATF-1 activates transcription in response to cAMP-dependent protein kinase A. *J. Biol. Chem.* 266:18,431–18,434.

Rivera, V. M., Greenberg, M. E. (1990). Growth factor–induced gene expression: the ups and downs of c-fos regulation. *New Biologist* 2:751–758.

Rivera, V., Miranti, C., Misra, R., Ginty, D., Chen, R.-H., Blenis, J., Greenberg, M. E. (1993). A growth factor-induced kinase phosphorylates the serum response factor at a site that regulates its DNA-binding activity. *Mol. Cell. Biol.* 13:6260–6273.

Rosen, L. B., Greenberg, M. E. (1996). Stimulation of growth factor receptor signal transduction by activation of voltage-sensitive calcium channels. *Proc. Natl. Acad. Sci. USA* 93:1113–1118.

Rosen, L. B., Ginty, D. D., Weber, M. J., Greenberg, M. E. (1994). Membrane depolarization and calcium influx stimulate MEK and MAP kinase via activation of Ras. *Neuron* 12:1207–1221.

Rosen, L. B., Ginty, D. D., Greenberg, M. E. (1995). Calcium regulation of gene expression. *Adv. Second Mess. Phosph. Res.* 30:225–253.

Sakaguchi, H., Wada, K., Maekawa, M., Watsuji, T., Hagiwara, M. (1999). Song-induced phosphorylation of cAMP response element-binding protein in the songbird brain. *J. Neurosci.* 19:3973–3981.

Schroter, H., Mueller, C. G. F., Meese, K., Nordheim, A. (1990). Synergism in ternary complex formation between the dimeric glycoprotein p67/SRF, polypeptide p62/TCF and the c-fos serum response element. *EMBO J.* 9: 1123–1130.

Schuman, E. (1997a). Growth factors sculpt the synapse. *Science* 275:1277–1278.

Schuman, E. M. (1997b). Synapse specificity and long-term information storage. *Neuron* 18:339–342.

Seeburg, P. H., Higuchi, M., Sprengel, R. (1998). RNA editing of brain glutamate receptor channels: mechanism and physiology. *Brain Res. Brain Res. Rev.* 26: 217–229.

Shaywitz, A. J., Greenberg, M. E. (1999). CREB: a stimulus-induced transcription factor activated by a diverse array of extracellular signals. *Annu. Rev. Biochem.* 68:821–861.

Sheng, M., Greenberg, M. E. (1990). The regulation and function of c-fos and other immediate early genes in the nervous system. *Neuron* 4:477–485.

Sheng, M., Pak, D. T. (1999). Glutamate receptor anchoring proteins and the molecular organization of excitatory synapses. *Ann. N.Y. Acad. Sci.* 868:483–493.

Sheng, M., Dougan, S. T., McFadden, G., Greenberg, M. E. (1988). Calcium and growth factor pathways of c-*fos* transcriptional activation require distinct upstream regulatory sequences. *Mol. Cell. Biol.* 8:2787–2796.

Sheng, M., McFadden, G., Greenberg, M. E. (1990). Membrane depolarization and calcium induce c-*fos* transcription via phosphorylation of transcription factor CREB. *Neuron* 4:571–582.

Sheng, M., Thompson, M. A., Greenberg, M. E. (1991). CREB: a Ca^{2+}-regulated transcription factor phosphorylated by calmodulin-dependent kinases. *Science* 252:1427–1430.

Shieh, P. B., Hu, S.-C., Bobb, K., Timmusk, T., Ghosh, A. (1998). Identification of a signaling pathway involved in calcium regulation of *BDNF* expression. *Neuron* 20:727–740.

Silva, A. J., Kogan, J. H., Frankland, P. W., Kida, S. (1998). CREB and memory. *Annu. Rev. Neurosci.* 21:127–148.

Smart, T. G. (1997). Regulation of excitatory and inhibitory neurotransmitter-gated ion channels by protein phosphorylation. *Curr. Opin. Neurobiol.* 7:358–367.

Soderling, T. R. (1993). Calcium/calmodulin-dependent protein kinase II: role in learning and memory. *Mol. Cell. Biochem.* 127–128:93–101.

Steward, O., Wallace, C. S., Lyford, G. L., Worley, P. F. (1998). Synaptic activation causes the mRNA for the IEG Arc to localize selectively near activated postsynaptic sites on dendrites. *Neuron* 21:741–751.

Struhl, K. (1989). Helix-turn-helix, zinc-finger, and leucine-zipper motifs for eukaryotic transcriptional regulatory proteins. *Trends Biochem. Sci.* 14:137–140.

Struhl, K. (1998). Histone acetylation and transcriptional regulatory mechanisms. *Genes Dev.* 12:599–606.

Sucher, N. J., Awobuluyi, M., Choi, Y. B., Lipton, S. A. (1996). NMDA receptors: from genes to channels. *Trends Pharmacol. Sci.* 17:348–355.

Sun, P., Enslen, H., Myung, P. S., Maurer, R. A. (1994). Differential activation of CREB by Ca^{2+}/calmodulin-dependent protein kinases type II and type IV involves phosphorylation of a site that negatively regulates activity. *Genes Dev.* 8:2527–2539.

Swope, S. L., Moss, S. I., Raymond, L. A., Huganir, R. L. (1999). Regulation of ligand-gated ion channels by protein phosphorylation. *Adv. Second Mess. Phosph. Res.* 33:49–78.

Szekely, A. M., Barbaccia, M. L., Alho, H., Costa, E. (1989). In primary cultures of cerebellar granule cells the activation of N-methyl-D-aspartate-sensitive glutamate receptors induces c-fos mRNA expression. *Mol. Pharmacol.* 35:401–408.

Szekely, A. M., Costa, E., Grayson, D. R. (1990). Transcriptional program coordination by N-methyl-D-aspartate-sensitive glutamate receptor stimulation in primary cultures of cerebellar neurons. *Mol. Pharmacol.* 38:624–633.

Tao, X., Finkbeiner, S., Arnold, D. B., Shaywitz, A. J., Greenberg, M. E. (1998). Ca^{2+} influx regulates *BDNF* transcription by a CREB family transcription factor–dependent mechanism. *Neuron* 20:709–726.

Timmusk, T., Palm, K., Metsis, M., Reintam, T., Paalme, V., Saarma, M., Persson, H. (1993). Multiple promoters direct tissue-specific expression of the rat BDNF gene. *Neuron* 10:475–489.

Tokumitsu, H., Enslen, H., Soderling, T. R. (1995). Characterization of a Ca^{2+}/calmodulin-dependent protein kinase cascade. *J. Biol. Chem.* 270:19,320–19,324.

Treisman, R. (1985). Transient accumulation of c-*fos* RNA following serum stimulation requires a conserved 5′ element and c-*fos* 3′ sequences. *Cell* 42:889–902.

Treisman, R. (1992). The serum response element. *Trends Biochem. Sci.* 17:423–426.

Tsui, C., Copeland, N., Gilbert, D., Jenkins, N., Barnes, C., Worley, P. (1996). Narp, a novel member of the pentraxin family, promotes neurite outgrowth and is regulated by neuronal activity. *J. Neurosci.* 16:2463–2478.

Tu, J. C., Xiao, B., Naisbitt, S., Yuan, J. P., Petralia, R. S., Brakman, P., Doan, A., Aakalu, U. K., Lanahan, A. A. (1999). Coupling of mGluR/Homer and PSD-95 complexes by the Shank family of postsynaptic density proteins. *Neuron* 23:583–592.

Vinson, C. R., Sigler, P. B., McKnight, S. L. (1989). Scissors-grip model for DNA recognition by a family of leucine zipper proteins. *Science* 246:911–916.

Vinson, C. R., Hai, T., Boyd, S. M. (1993). Dimerization specificity of the leucine zipper–containing bZIP motif on DNA binding: prediction and rational design. *Genes Dev.* 7:1047–1058.

Wallace, C. S., Lyford, G. L., Worley, P. F., Steward, O. (1998). Differential intracellular sorting of immediate early gene mRNAs depends on signals in the mRNA sequence. *J. Neurosci.* 18:26–35.

Watanabe, Y., Johnson, R. S., Butler, L. S., Binder, D. K., Spiegelman, B. M., Papaioannou, V. E., McNamara, J. O. (1996). Null mutation of c-fos impairs structural and functional plasticities in the kindling model of epilepsy. *J. Neurosci.* 16:3827–3836.

Wu, L., Wells, D., Tay, J., Mendis, D., Abbott, M. A., Barnitt, A., Quinlan, E., Heynen, A., Fallon, J. R., Richter, J. D. (1998). CPEB-mediated cytoplasmic polyadenylation and the regulation of experience-dependent translation of alpha-CaMKII mRNA at synapses. *Neuron* 21:1129–1139.

Xia, Z., Dudek, H., Miranti, C. K., Greenberg, M. E. (1996). Calcium influx via the NMDA receptor induces immediate early gene transcription by a MAP kinase/ERK-dependent mechanism. *J. Neurosci.* 16:5425–5436.

Xiao, B., Tu, J. C., Petralia, R. S., Yuan, J. P., Doan, A., Breder, C. D., Ruggiero, A., Lanahan, A. A., Wenthold, R. J. (1998). Homer regulates the association of group 1 metabotropic glutamate receptors with multivalent complexes of homer-related, synaptic proteins. *Neuron* 21:707–716.

Xing, J., Ginty, D. D., Greenberg, M. E. (1996). Coupling of the RAS-MAPK pathway to gene activation by RSK2, a growth factor–regulated CREB kinase. *Science* 273:959–963.

Yamagata, K., Andreasson, K., Kaufmann, W., Barnes, C., Worley, P. (1993). Expression of a mitogen-inducible cyclooxygenase in brain neurons: regulation by synaptic activity and glucocorticoids. *Neuron* 11:371–386.

Yamagata, K., Sanders, L. K., Kaufmann, W. E., Yee, W., Barnes, C. A., Nathans, D. (1994). rheb, a growth factor– and synaptic activity–regulated gene, encodes a novel Ras-related protein. *J. Biol. Chem.* 269:16,333–16,339.

Yamagata, K., Andreasson, K. I., Sugiura, H., Maru, E., Dominique, M., Irie, Y., Miki, N., Hayashi, Y., Yoshioka, M., Kaneko, K., Kato, H., Worley, P. F. (1999). Arcadlin is a neural activity–regulated cadherin involved in long term potentiation. *J. Biol. Chem.* 274:19,473–19,479.

Yang, X. J., Ogryzko, V. V., Nishikawa, J., Howard, B. H., Nakatani, Y. (1996). A p300/CBP-associated factor that competes with the adenoviral oncoprotein E1A. *Nature* 382:319–324.

Yin, J. C., Tully, T. (1996). CREB and the formation of long-term memory. *Curr. Opin. Neurobiol.* 6:264–268.

Yin, J. C. P., Wallach, J. S., Del Vecchio, M., Wilder, E. L., Zhou, H., Quinn, W. G., Tully, T. (1994). Induction of a dominant negative CREB transgene specifically blocks long-term memory in Drosophila. *Cell* 79:49–58.

Zafra, F., Hengerer, B., Leibrock, J., Thoenen, H., Lindholm, D. (1990). Activity dependent regulation of BDNF and NGF mRNAs in the rat hippocampus is mediated by non-NMDA glutamate receptors. *EMBO J.* 9:3545–3550.

Zafra, F., Lindholm, D., Castren, E., Hartikka, J., Thoenen, H. (1992). Regulation of brain-derived neurotrophic factor and nerve growth factor mRNA in primary cultures of hippocampal neurons and astrocytes. *J. Neurosci.* 12: 4793–4799.

9

Synaptic Plasticity
Diverse Targets and Mechanisms for Regulating Synaptic Efficacy

Robert C. Malenka and Steven A. Siegelbaum

393

T he strength of the connection between a presynaptic and a postsynaptic neuron often exhibits a remarkable degree of plasticity. Synaptic transmission can be either enhanced or depressed, and these alterations span a wide range of time scales, from a transient few milliseconds to enduring modifications that persist for days, weeks, or perhaps longer. Such changes in the efficacy of synaptic transmission are likely to be important for a number of aspects of neural function. Transient modifications have been associated with short-term adaptation to sensory inputs, changes in behavioral states associated with arousal, and short-term memory. More lasting changes have been associated with neuronal development in the immature nervous system and with long-term memory in the mature nervous system. Not surprisingly, given these diverse functions and time scales, a large number of mechanisms for synaptic plasticity have been described. Here we provide a brief overview of some of the various forms of plasticity that have been characterized at both invertebrate and vertebrate synapses, focusing on plasticity in the central nervous system. We consider in depth two forms of synaptic plasticity that have been implicated in learning and memory and that span the time scale from seconds to weeks: presynaptic heterosynaptic facilitation in *Aplysia* and long-term potentiation in the mammalian hippocampus.

A Variety of Forms of Synaptic Plasticity

As illustrated in Fig. 9.1, the various forms of synaptic plasticity can be classified according to three criteria: source of induction, site of expression, and molecular basis of induction.

Source of Induction

Alterations in synaptic strength can occur through either intrinsic or extrinsic mechanisms. In homosynaptic plasticity, it is the intrinsic activity of the synapse itself that subsequently alters its own functional state. These functional changes can be triggered by biochemical processes localized within either the presynaptic terminal or the postsynaptic cell. In heterosynaptic plasticity, in contrast, a synapse between two neurons is modified by the extrinsic action of a third neuron. This modulatory neuron can alter synaptic transmission either by a directed synaptic action or through the diffuse release of a transmitter or hormone, to alter the function of the first synapse.

Site of Expression

For both homosynaptic and heterosynaptic forms of plasticity, changes in synaptic function can occur either at the level of the presynaptic terminal or at the postsynaptic membrane. Presynaptic forms of plasticity involve either an increase (e.g., presynaptic facilitation) or a decrease

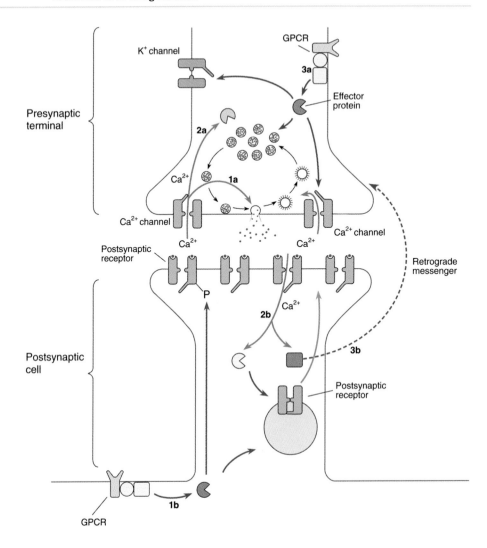

Figure 9.1. A variety of mechanisms and sites for synaptic plasticity. Calcium entry into the presynaptic terminal through voltage-gated Ca^{2+} channels directly triggers release (1a) and acts as a second messenger for short-term homosynaptic plasticity (2a). Activation of G protein–coupled receptors (GPCRs) also alters release (3a) by modulating Ca^{2+} and K^+ channels, altering Ca^{2+} influx, and acting directly on the release machinery and vesicle recycling. Postsynaptic plasticity can be produced in response to stimulation of postsynaptic GPCRs (1b) or by postsynaptic Ca^{2+} influx (2b). These postsynaptic signaling cascades modify fast postsynaptic responses mediated by ionotropic receptors through phosphorylation of receptors in the plasma membrane or by altering the cycling of receptors to and from a cytoplasmic pool. Postsynaptic cells can also generate retrograde messengers that alter release from the presynaptic cell (3b).

(e.g., presynaptic inhibition) in transmitter release. Postsynaptic forms of plasticity have also been associated with an increase or decrease in the response of the postsynaptic cell to a fixed amount of transmitter.

Molecular Basis for Induction

A wide range of mechanisms has been implicated in the induction of the different forms of plasticity. But all involve some form of second messenger that carries the information from the surface of the cell to its interior. The briefest forms of plasticity, short-term homosynaptic processes, may involve a direct action of a residual elevation of calcium in the presynaptic terminal as a result of prior presynaptic activity. Here Ca^{2+} acts not only as a carrier of positive charge and the direct trigger of transmitter release but also as a second messenger. Longer-lasting forms of hetero- and homosynaptic plasticity can occur through activation of G protein–coupled receptors or protein kinases that may target pre- or postsynaptic proteins. These forms of plasticity can last from seconds to many minutes. Finally, more permanent changes in synaptic transmission depend on the recruitment of gene transcription and new protein synthesis. Such changes can last for days, weeks, or perhaps even a lifetime.

We focus first on those forms of plasticity that are generally thought to involve alterations in the amount of transmitter release from presynaptic terminals. We then consider postsynaptic mechanisms of plasticity.

Mechanisms of Presynaptic Plasticity

Presynaptic changes in release occur through two broad types of mechanisms. The first involves a change in the amplitude of the transient rise in $[Ca^{2+}]_i$ in the presynaptic terminal elicited by a presynaptic action potential. Most commonly this results from an alteration of Ca^{2+} influx, due either to direct Ca^{2+} channel modulation or indirect effects on presynaptic excitability. The second type of mechanism occurs at some site that is downstream of presynaptic Ca^{2+} elevation and is due to the modulation of some stage of the synaptic vesicle cycle. Such changes may involve an alteration of some early phase in the release process, for example a modification in the size of the pool of synaptic vesicles available for release. Alternatively there could be a direct modification of some late stage in release, perhaps involving the fusion apparatus itself.

Short-Term Enhancement and Depression of Release

It has been known for over 50 years that activity of the presynaptic nerve can lead to short-term changes in synaptic transmission over a time scale ranging from tens of milliseconds to several minutes. These forms of plasticity can involve either increases in synaptic transmission, termed short-term enhancement, or decreases in synaptic transmission, termed

short-term depression. These forms of plasticity are due to altered transmitter release resulting from changes intrinsic to the presynaptic terminal
that are induced either by presynaptic Ca^{2+} influx or by the transmitter
release process itself. As short-term plastic changes have already been
discussed to some extent in the context of basic mechanisms of release
(Regehr and Stevens, this volume) and have been reviewed elsewhere
(Zucker, 1989, 1999; Fisher et al., 1997), we will consider them here only
briefly, to provide a context for other, longer-lasting forms of plasticity.

Paired Pulse Facilitation and Depression

When two presynaptic stimuli are delivered within a short interval, the
response to the second stimulus can be either enhanced or depressed
relative to the response to the first. The sign of the change depends on
both the identity of the synapse and the duration of the interstimulus
interval. Paired pulse depression is almost invariably observed at the very
shortest interstimulus intervals (<20 msec). It may result from inactivation of voltage-gated presynaptic sodium or calcium channels (Forsythe
et al., 1998), a transient depletion of docked vesicles (Liley and North,
1953), or a transient decrease in release probability of the releasable pool
of vesicles (Wu and Borst, 1999).

At longer interstimulus intervals (20–500 msec), many synapses show
paired pulse facilitation (PPF). This facilitation is likely to be a result of
Ca^{2+} influx during the first action potential. It was initially proposed
that PPF results from the summation of a low level of residual Ca^{2+} following the first action potential (around 1 μM) with the Ca^{2+} influx during the second action potential, leading to an enhanced release (Katz and
Miledi, 1968). In this simple view, the residual Ca^{2+} binds to the Ca^{2+}
sensor that directly triggers exocytosis. As the relation between calcium
and release is highly nonlinear, a small rise in resting calcium could in
principle cause a substantial facilitation. However, we now know that this
sensor has a very low affinity for Ca^{2+}, requiring 50–100 μM Ca^{2+} for
efficient exocytosis: even a highly nonlinear release process could not
account for the ability of 1 μM residual calcium to cause the twofold
facilitation often observed. Rather, the residual calcium may act at a
separate, higher-affinity modulatory site. At some synapses postsynaptic
Ca^{2+} influx may also contribute to PPF (Wang and Kelly, 1997), perhaps
through the activation of a retrograde signal.

Whether a particular class of synapse displays an enhancement or
depression of transmitter release is likely to depend on the initial state
of that synapse. Thus, as short-term plasticity is largely due to changes
in release probability, the tendency of a synapse to show facilitation or
depression depends on the initial probability of release at that synapse
(Bolshakov and Siegelbaum, 1995; Dobrunz and Stevens, 1997). At
synapses that start with an initial high probability of transmitter release,
synaptic depression predominates. In part this is because the initial high
probability of release acts as a ceiling, limiting the potential for any

further enhancement (as release probability cannot be greater than 1). In addition, the high probability of release results in a greater rate of depletion of synaptic vesicles upon repetitive stimulation. In contrast, synapses that start with a low probability of release tend to display a larger degree of facilitation. The heterogeneity in release properties, even among synapses between similar types of neurons, has been correlated with the heterogeneity in active zone morphology. Synapses that show a high degree of facilitation (and thus a lower initial probability of release) display a lower packing density of synaptic vesicles (Bower and Haberly, 1986) and a smaller active zone area (Pierce and Lewin, 1994; Schikorski and Stevens, 1997).

Enhancement and Depression Elicited by Tetanic Stimulation

Longer-lasting forms of plasticity are observed during and following tetanic stimuli of prolonged trains of action potentials. Four distinct kinetic components of synaptic enhancement have been observed that develop and decay with distinct time courses (Fig. 9.2; Magleby and Zengel, 1982; Zengel and Magleby, 1982). The two most rapid phases of enhancement are termed facilitation. These components, F1 and F2, develop rapidly during a train of stimuli and then decay equally quickly once the stimulation ends, with time constants of tens of milliseconds and hundreds of milliseconds, respectively. Augmentation is a slightly slower phase of enhancement. It requires several seconds to turn on during a tetanus and then decays following the tetanus over a period of a few seconds. Finally, posttetanic potentiation (PTP), the slowest component of enhancement, requires tens of seconds of stimulation to be recruited and persists for several minutes following a tetanus.

These forms of plasticity are all thought to be induced by a residual rise in presynaptic Ca^{2+} concentration that persists following the period of presynaptic stimulation (Kamiya and Zucker, 1994; see Zucker, 1999, for review). The various kinetic components of enhancement are likely to reflect, at least in part, the different kinetic components in the time course of the buildup and decay of presynaptic Ca^{2+} (Swandulla et al., 1991; Kamiya and Zucker, 1994). The two components of facilitation are likely to reflect relatively rapid equilibration of calcium near the site of exocytosis. The slower processes of augmentation and PTP may depend on the slower buildup of bulk Ca^{2+} concentration in the core of the terminal. For example, the slow kinetics of PTP is thought to depend on the slow time course of loading of the mitochondria with Ca^{2+} during a tetanus, followed by the slow time course at which Ca^{2+} subsequently leaks out of the mitochondria into the cytoplasm (Tang and Zucker, 1997). Although it is generally agreed that the expression of short-term enhancement is presynaptic, there is some evidence, especially at *Aplysia* synapses, that its induction may depend on a rise in postsynaptic calcium, which then signals back to the presynaptic terminal by some unknown mechanism (Bao et al., 1997).

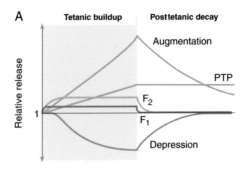

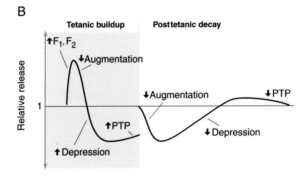

Figure 9.2. Short-term presynaptic plasticity.

(A) Several distinct kinetic phases of short-term homosynaptic plasticity of transmitter release develop during a period of tetanic stimulation (shaded area) and then decay following the tetanus. (The tetanus lasts for 10 sec in this illustration.) Facilitation is the most rapid process. Two distinct kinetic components of facilitation, F1 and F2, turn on rapidly during a tetanus and then decay rapidly after stimulation. Augmentation turns on and decays with somewhat slower kinetics, and posttetanic potentiation (PTP) is the slowest component. There is also a short-term depression of release that is superimposed on these phases of enhancement. The amplitudes of the components are arbitrary. Adapted from Zengel and Magleby (1982).

(B) The net effect on the relative magnitude of release due to the overlapping processes of facilitation, augmentation, PTP, and short-term depression. The approximate time course of buildup and decay of the various components is indicated. The beginning of the trace has been shifted arbitrarily from the start of the tetanus for display.

Adapted from Zucker (1989).

Very often a short-term depression of transmission can be seen superimposed on the enhancement of synaptic transmission during and following tetanic stimulation. This short-term depression is believed to be produced by the transient depletion of the readily releasable pool of docked vesicles (Zucker, 1989; Stevens and Tsujimoto, 1995; Dobrunz et al., 1997) or by the persistent inactivation of voltage-gated Ca^{2+} channels (Forsythe et al., 1998). The rate of recovery from short-term depression is enhanced by presynaptic Ca^{2+} entry (Dittman and Regehr, 1998; Stevens and Wesseling, 1998; Wang and Kaczmarek, 1998). At *Aplysia* synapses, short-term depression does not appear to involve a depletion of releasable vesicles or a decrease in presynaptic Ca^{2+} influx, but it may involve an inactivation of the release process itself (Eliot et al., 1994; Armitage and Siegelbaum, 1998).

Molecular Mechanisms for Short-Term Plasticity

One very attractive mechanism for certain forms of activity-dependent presynaptic plasticity involves the activation of the calcium/calmodulin-dependent protein kinases, CaMKI and CaMKII, as a result of presynaptic Ca^{2+} influx. These kinases are localized in both the presynaptic terminals and postsynaptic membrane. One prominent presynaptic substrate is the synapsins, a family of proteins encoded by three separate genes: synapsins I, II, and III. These proteins in their dephosphorylated form associate with synaptic vesicles and may serve to anchor the vesicles to the cytoskeleton (Greengard et al., 1993; see Südhof and Scheller, this volume). In vitro studies demonstrated that phosphorylation of synapsin I can enhance vesicle mobility and that injection of either CaMKII or phosphorylated synapsin I into giant presynaptic terminals of the squid enhances transmitter release (Llinas et al., 1985; Lin et al., 1990). This finding led to the proposal that phosphorylation of synapsin I by CaMKII underlies Ca^{2+}-dependent short-term forms of presynaptic plasticity, such as PTP (Greengard et al., 1993). However, this mechanism cannot be a general one for the synapsin family members because only synapsin I is phosphorylated by CaMKII. In addition, CaMKII inhibitors do not necessarily inhibit PTP (see Zucker, 1999).

More recent studies point to the importance of a conserved N-terminal serine residue, present in all synapsin isoforms, that is phosphorylated by cAMP-dependent protein kinase (PKA) and CaMKI (Hosaka et al., 1999). Phosphorylation at this site leads to the dissociation of synaptic vesicles from the membrane and controls the association of all synapsins with synaptic vesicles. Analysis of synapsin knockout mice provides support for important, overlapping roles of synapsin I and synapsin II in short-term plasticity (Rosahl et al., 1995). Thus, whereas deletion of either isoform alone has minimal effects on synaptic transmission, removal of both synapsin I and synapsin II leads to a profound decrease in PTP and enhancement of the rate of synaptic depression.

Insights into the molecular mechanisms of short-term depression have been provided by studying mutant mice lacking specific synaptic vesicle proteins through homologous recombination. Deletion of the small GTP-binding protein rab3a leads to an increase in the rate of depression of synaptic transmission during repetitive stimulation (Geppert et al., 1994). This effect is thought to be due to an increase in the number of synaptic vesicles released from an active zone per action potential, leading to an increased rate of vesicle depletion (Geppert et al., 1997). Deletion of synapsin II alone or of synapsins I and II in combination also leads to a marked increase in the rate of development and extent of short-term depression (Rosahl et al., 1995). Here the effect is likely to be due to a reduction in the size of the total pool of synaptic vesicles, which may be decreased by as much as 50%.

Control of Transmitter Release by Activation of Presynaptic Receptors

In addition to the short-term forms of homosynaptic plasticity elicited by mechanisms intrinsic to the presynaptic terminal discussed previously, transmitter release is often regulated by extrinsic signals owing to the action of transmitters and hormones on presynaptic receptors. The presynaptic receptors can be ionotropic, such as nicotinic ACh receptors or $GABA_A$ receptors (MacDermott et al., 1999), or metabotropic G protein–coupled receptors (see Miller, 1998, for review). These presynaptic receptors can be autoreceptors, activated homosynaptically by transmitter released from the terminal itself, or heteroreceptors, activated heterosynaptically by transmitter released from a modulatory neuron at axo-axonic synapses.

The first example of this form of plasticity to be characterized was presynaptic inhibition at the crayfish opener muscle neuromuscular junction (Dudel and Kuffler, 1961). Release of γ-aminobutyric acid (GABA) from an inhibitory neuron produces an inhibitory postsynaptic potential in the muscle cell because of the activation of postsynaptic $GABA_A$ receptor chloride channels. In addition, the size of the excitatory postsynaptic potential (EPSP) elicited by glutamate released from the excitatory input is reduced. Using quantal analysis, Dudel and Kuffler demonstrated that the reduction in EPSP amplitude was due, in part, to a decrease in transmitter release, resulting from a decrease in the mean number of quanta released per action potential. This presynaptic inhibition is caused by the opening of presynaptic $GABA_A$ receptor–activated Cl^- channels, which depress presynaptic excitability and act to shunt the presynaptic action potential (Dudel and Kuffler, 1961; Baxter and Bittner, 1991).

Activation of ionotropic receptors can also cause presynaptic facilitation. In particular, stimulation of presynaptic nicotinic receptors can enhance release of acetylcholine (ACh) (McGehee et al., 1995) and glutamate (Gray et al., 1996), possibly because of the ability of these receptors to conduct Ca^{2+} and directly elevate presynaptic Ca^{2+} levels.

Presynaptic Inhibition

One of the most common forms of short-term heterosynaptic plasticity is presynaptic inhibition produced through the activation of presynaptic metabotropic G protein–coupled receptors (Wu and Saggau, 1997; Miller, 1998). Metabotropic receptors linked to presynaptic inhibition are coupled to the G_i, G_o, or $G_{q/11}$ class of heterotrimeric G proteins. G protein activation has been shown to depress release through one of two mechanisms: the rise in presynaptic calcium triggered by a presynaptic action potential can be inhibited, or the presynaptic release machinery can be directly modified.

Early on it was noted that transmitters that cause presynaptic inhibition also inhibit current flow through voltage-gated calcium channels (Dunlap and Fischbach, 1978; Mudge et al., 1979; Shapiro et al., 1980). This finding led to the hypothesis that a depression of Ca^{2+} influx is responsible for the depressed release. These initial studies, however, were limited to recordings of calcium channel currents from cell bodies because the presynaptic terminals were too small to permit direct electrophysiological measurements. Thus it was not clear whether calcium currents at the site of transmitter release in the presynaptic terminals were also inhibited. More recent studies have provided convincing support for the hypothesis that a decrease in Ca^{2+} influx in synaptic terminals can contribute to at least certain forms of presynaptic inhibition (Wu and Saggau, 1997).

Norepinephrine (NE) inhibits its own release by acting on presynaptic adrenergic ($\alpha 2$) receptors (Langer, 1997). In voltage clamp recordings from sympathetic neuron cell bodies, Lipscombe et al. (1989) found that activation of $\alpha 2$ adrenergic receptors inhibits the opening of the N-type class of calcium channels—the same class of channel that specifically mediates calcium-dependent transmitter release from the synaptic terminals of these neurons (Hirning et al., 1988). Lipscombe et al. further showed that the action of NE to inhibit Ca^{2+} influx is necessary for presynaptic inhibition. Thus NE has no effect on release that is elicited when presynaptic Ca^{2+} levels are elevated directly by a Ca^{2+} ionophore, bypassing the voltage-gated Ca^{2+} channels.

Several studies have used Ca^{2+} imaging techniques to show that Ca^{2+} influx into presynaptic terminals is diminished during many (but not all) forms of presynaptic inhibition (Wu and Saggau, 1994, 1997). Patch clamp recordings from the giant presynaptic terminals of the chick calyx have directly shown inhibition of presynaptic calcium channels associated with G protein activation (Stanley and Mirotznik, 1997). Interestingly, this effect is blocked upon cleavage of the presynaptic t-SNARE protein syntaxin with Botulinum toxin, suggesting a close link between Ca^{2+} channel modulation and the release machinery.

Inhibition of presynaptic calcium channels through G protein–coupled receptors is also frequently accompanied by an activation of K^+ channels through the same G protein–coupled receptor. In principle K^+ channel activation in the presynaptic terminal could also contribute to presynaptic inhibition by shunting the presynaptic action potential, further inhibiting Ca^{2+} influx. However, there is little convincing evidence for a causal role for this K^+ channel activation in presynaptic inhibition (Miller, 1998). For example, in the hippocampus activation of a number of G protein–coupled receptors by various transmitters causes presynaptic inhibition at CA3-CA1 excitatory synapses and also causes an increase in an inward-rectifying K^+ current mediated by channels containing the GIRK2 subunit of inwardly rectifying K$^+$ channels. However, GIRK2 knockout mice exhibit normal presynaptic inhibition, although the increase in K$^+$ current is blocked (Lüscher et al., 1997).

Modulation of calcium influx is not the only mechanism by which neurotransmitters inhibit transmitter release from presynaptic terminals. Presynaptic inhibition can also occur through mechanisms that act downstream of calcium entry by the modulation of some step in synaptic vesicle cycling. Inhibition at a site downstream of Ca^{2+} influx has been most convincingly demonstrated through experiments in which the inhibitory transmitter is found to depress release in response to direct elevation of presynaptic Ca^{2+} levels by procedures that bypass Ca^{2+} influx through voltage-gated channels. Thus in the land snail *Helisoma* the tetrapeptide FMRFamide causes presynaptic inhibition of both action potential–evoked release and of release evoked by direct elevation of intraterminal Ca^{2+} using ultraviolet photolysis of caged Ca^{2+} chelators (Man-Son-Hing et al., 1989). At synapses between hippocampal neurons, agonist activation of adenosine receptors, GABA$_B$ receptors, or mu-opioid receptors causes presynaptic inhibition of evoked release. At least part of this action must be independent of changes in Ca^{2+} influx, because these transmitters also decrease the frequency of miniature synaptic events elicited by a rise in presynaptic Ca^{2+} induced by application of a Ca^{2+} ionophore (Capogna et al., 1996). This downstream action, at least for GABA$_B$ receptor stimulation, does not appear to involve a change in the rate of extent of vesicle endocytosis, as measured by the rate of uptake of the marker of synaptic vesicle turnover, FM1-43 (Isaacson and Hille, 1997).

Often a transmitter can depress release by both inhibiting presynaptic Ca^{2+} currents and acting at a site downstream of Ca^{2+} influx. At *Aplysia* sensory-motor neuron synapses, FMRFamide causes presynaptic inhibition of evoked transmitter release (Abrams et al., 1984) through such a dual effect. Voltage clamp recordings show that FMRFamide decreases a component of Ca^{2+} current that triggers evoked release (Edmonds et al., 1990), and Ca^{2+} imaging studies show a decrease in action potential–evoked Ca^{2+} levels in the presynaptic cell body, neurites, and growth cones (Blumenfeld et al., 1990). However, FMRFamide also reduces the frequency of spontaneous miniature excitatory postsynaptic potentials (mEPSPs) (Dale and Kandel, 1990). This latter effect is not mediated by a decrease in presynaptic Ca^{2+} levels as FMRFamide reduces spontaneous release even in the absence of external Ca^{2+} or after presynaptic injection of Ca^{2+} chelators. At hippocampal synapses adenosine both inhibits Ca^{2+} influx into presynaptic terminals and exerts a downstream action to inhibit release, as discussed earlier (Scholz and Miller, 1992; Capogna et al., 1996).

Most studies of presynaptic inhibition have relied on the exogenous application of neurotransmitters and pharmacological agonists to activate presynaptic receptors and reduce release. This approach leads to questions about the physiological relevance of these inhibitory effects.

However, in several instances presynaptic inhibition has been demonstrated in response to physiological stimuli (as opposed to the application of exogenous agonists). In the hippocampus the firing of a Schaffer collateral input to a CA1 pyramidal neuron activates presynaptic adenosine (A1) receptors to inhibit the EPSP elicited by the subsequent firing of a second independent Schaffer collateral input (Mitchell et al., 1993). Adenosine also appears to be released tonically in hippocampal slices (Dunwiddie, 1985). Thus application of A1 adenosine receptor antagonists increases basal synaptic transmission, presumably reflecting the antagonism of the effects of basal adenosine release (Wu and Saggau, 1994). Glutamate, the major excitatory transmitter, can limit its own release by activating presynaptic metabotropic receptors (mGluRs) that cause presynaptic inhibition. Interestingly, this effect of glutamate may only be observed following large amounts of glutamate release using relatively high-frequency trains of presynaptic impulses (Scanziani et al., 1997). In *Aplysia*, transmitter release from the L10 neuron onto its follower cells has been found to be inhibited by stimulation of a set of interneurons that release histamine, which decreases the magnitude of the presynaptic voltage-dependent Ca^{2+} current and increases the magnitude of a K^+ current (Kretz et al., 1986a,b).

Presynaptic Facilitation of Release

Heterosynaptic activation of presynaptic metabotropic receptors can also facilitate transmitter release, although the number of well-characterized examples is much less than for presynaptic inhibition. Some of the most thoroughly studied facilitative actions involve the effects of serotonin (5-hydroxytryptamine or 5-HT), which enhances release at both invertebrate and vertebrate synapses. Some of these actions will be considered in detail when we discuss presynaptic facilitation at *Aplysia* sensory-motor neuron synapses.

Similar to presynaptic inhibition, at least two broad classes of mechanisms have been associated with presynaptic facilitation: enhanced presynaptic calcium levels and direct modulation of the release process. In some instances both mechanisms can contribute to presynaptic facilitation.

There are three potential mechanisms for enhancing presynaptic Ca^{2+} levels: (1) direct modulation of presynaptic voltage-gated Ca^{2+} channels; (2) indirect increases in Ca^{2+} influx resulting from an increase in action potential duration owing to modulation of presynaptic potassium channels; and (3) receptor-stimulated release of Ca^{2+} from intracellular stores. Although a number of studies have shown that transmitters can increase the magnitude of Ca^{2+} currents, most cases involve increases in current carried by L-type Ca^{2+} channels, which do not directly participate in rapid release (e.g., Edmonds et al., 1990). Similarly, whereas many transmitters have been shown to release Ca^{2+} from internal stores (Berridge, 1998), there have been only a few examples in which this mechanism has been shown to enhance release at presynaptic terminals, and in these the facilitation occurs only under special circumstances (e.g., Cochilla and Alford, 1998). To date the most thoroughly studied mechanisms involve enhanced Ca^{2+} influx caused by action potential broadening.

Artificially prolonging the presynaptic action potential, using either pharmacological agents to block K^+ channels or presynaptic current injections, can significantly enhance both presynaptic Ca^{2+} entry and transmitter release (Hochner et al., 1986a; Augustine, 1990; Jackson et al., 1991; Qian and Saggau, 1999). The enhancement of Ca^{2+} influx reflects both an increase in the probability that a voltage-gated calcium channel will activate during an action potential and the increased duration of opening of those Ca^{2+} channels that do activate. Because transmitter release is a steep function of presynaptic calcium, small (circa 20%) changes in action potential duration can lead to large (100%) increases in release (Hochner et al., 1986a; Augustine, 1990).

Behavioral Sensitization Caused by Presynaptic Facilitation at Aplysia Sensory-Motor Neuron Synapses

In the marine snail *Aplysia,* behavioral sensitization, a simple form of learning that gives rise to short- and long-term memory, is mediated by presynaptic facilitation at synapses between mechanoreceptor sensory neurons and their follower motor neurons (Byrne and Kandel, 1996). Strong, noxious stimuli to the animal's head or tail lead to sensitization by releasing 5-HT from modulatory interneurons, resulting in an increase in synaptic transmission at sensory-motor neuron synapses and an enhancement of tail and siphon withdrawal. The increase in synaptic transmission is due to an enhancement of transmitter release that results from at least two separate actions of 5-HT: enhancement of Ca^{2+} influx into the presynaptic terminal and modulation of the release machinery downstream from Ca^{2+} influx (see Fig. 9.3).

Initial studies found that 5-HT increases the duration of the action potential in the cell body of the presynaptic sensory neuron (Klein and Kandel, 1978). Voltage clamp experiments showed that the primary effect of 5-HT was to decrease a slowly activating outward K^+ current (Klein

Figure 9.3. (opposite) Short-term facilitation of transmitter release in Aplysia. 5-HT acts at two types of GPCRs to activate parallel kinase cascades. PKA phosphorylates two types of K^+ channels, $I_{K,S}$ and $I_{K,V}$, decreasing current through both conductance pathways. The decrease in $I_{K,S}$ mediates an enhanced excitability. The decrease in $I_{K,V}$ mediates an increase in spike duration, which enhances calcium influx and release. PKA also directly facilitates some aspect of the release process (dashed lines). Activation of PKC leads to a decrease in $I_{K,V}$ and a direct facilitation of release. At a nondepressed synapse (*A*), the PKA-dependent events predominate. At a depressed synapse (*B*), the PKC-dependent process predominates. Adapted from Byrne and Kandel (1996).

A. Nondepressed synapse

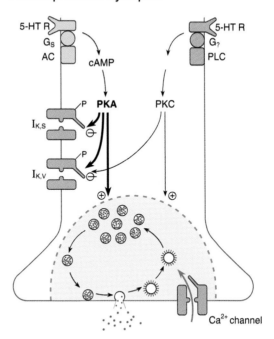

B. Depressed synapse

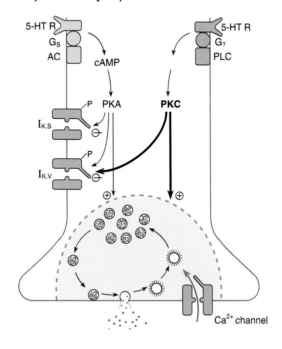

and Kandel, 1980; Klein et al., 1982). This finding led to a simple hypothesis in which presynaptic facilitation was thought to result from an increase in Ca^{2+} influx into the presynaptic terminal as a result of increased action potential duration. (Although a subsequent voltage clamp study did find that 5-HT also directly enhances a voltage-gated Ca^{2+} current, the L-type Ca^{2+} current component that was modulated does not directly participate in triggering fast release from these terminals [Edmonds et al., 1990].) Patch clamp experiments then identified a weakly voltage-dependent background K^+ channel, the S-type K^+ channel, that was open at the resting potential and was closed in an all-or-none manner by 5-HT. This effect of 5-HT was mediated by intracellular cAMP, which activates PKA (Kandel and Schwartz, 1982). The activated PKA closes the S channel by directly phosphorylating either the channel itself or a protein tightly coupled to the channel (Shuster et al., 1985).

Although S channel closure was initially thought to underlie the increase in action potential duration, its more important role is to decrease resting K^+ conductance and increase the excitability of the sensory neurons. The increase in action potential duration, which contributes to presynaptic facilitation, is largely due to the decrease in a second K^+ current, the delayed rectifier, $I_{K,V}$ (Baxter and Byrne, 1989; Goldsmith and Abrams, 1992; Hochner and Kandel, 1992). The decrease in $I_{K,V}$ by 5-HT is mediated by the activation of both PKA and protein kinase C (PKC) (Goldsmith and Abrams, 1992; Hochner and Kandel, 1992; Sugita et al., 1992). The PKC pathway appears to play a specific role in a late phase of spike broadening during prolonged (>3 min) applications of 5-HT, whereas PKA phosphorylation contributes more to an early phase of spike broadening. The relative contributions of the PKA and PKC pathways to presynaptic facilitation will be discussed later in this chapter.

Experimental support for the hypothesis that spike broadening does indeed contribute to presynaptic facilitation is provided by experiments in which the depolarization waveform is directly controlled in voltage clamp experiments (Hochner et al., 1986a). When release is evoked by voltage clamp depolarizations of the sensory neuron cell body (which is electrotonically close to the terminals), the synaptic response varies with the duration of the presynaptic depolarization raised to the second to third power, a steep power law relation. Thus a small 10–20% change in action potential duration in response to 5-HT is *sufficient* to cause a significant increase in release, similar to the 1.5- to 2-fold enhancement typically observed with 5-HT. Hochner et al. (1986a) also showed that spike broadening is *necessary* for facilitation, because 5-HT cannot enhance release that is evoked by a voltage clamp depolarization of fixed duration.

According to the spike duration hypothesis, 5-HT application acts to enhance Ca^{2+} influx into the presynaptic terminals, and this enhanced influx should quantitatively account for the amount of presynaptic facilitation. By imaging intracellular Ca^{2+} levels in presynaptic terminals at cultured *Aplysia* sensory-motor neuron synapses, Eliot et al. (1993) confirmed that 5-HT does indeed increase Ca^{2+} influx into presynaptic terminals. The extent of presynaptic facilitation was found to be correlated with the degree of enhancement of the presynaptic Ca^{2+} transient raised to the second to third power, consistent with the known nonlinear dependence of release on presynaptic Ca^{2+} levels.

Enhanced Ca^{2+} influx into presynaptic terminals, however, is not sufficient to explain all the facilitative effects of 5-HT (Byrne and Kandel, 1996). Thus the ability of spike broadening and enhanced Ca^{2+} influx to facilitate release depends on the prior history of synaptic activity. The experiments that demonstrated the importance of spike broadening were performed on synapses that had been stimulated infrequently so as not to induce homosynaptic depression. Although 5-HT facilitates release at depressed synapses, this effect cannot be accounted for by an increase in spike duration. Thus direct lengthening of the presynaptic depolarization at depressed synapses, using either voltage clamp depolarizations of increasing length or K^+ channel blockers to broaden the action potential, causes little or no enhancement of release (Hochner et al., 1986b). This finding suggests that 5-HT must have a second action that is independent of changes in spike duration and Ca^{2+} influx.

Further evidence for a Ca^{2+}-independent component to presynaptic facilitation comes from a study of the effects of 5-HT on the frequency of spontaneous mEPSPs (Dale and Kandel, 1990). The enhanced mEPSP frequency does not require changes in presynaptic Ca^{2+} levels, as shown by the lack of effect of removal of external Ca^{2+} or injection of Ca^{2+} chelators into the sensory neuron (Dale and Kandel, 1990). This spike duration–independent action of 5-HT can also predominate at nondepressed synapses under certain conditions. In cultured neurons (Klein, 1994) and during early stages of development (Stark and Carew, 1999), 5-HT can enhance release without any noticeable increase in action potential duration. Thus 5-HT must act by both a Ca^{2+}-dependent mechanism (due to spike broadening) and a mechanism that is downstream of Ca^{2+} influx, involving a modulation of the release machinery.

What are the second messenger cascades responsible for the spike duration-dependent and -independent components of presynaptic facilitation? Biochemical studies show that 5-HT activates both the PKA (Kandel and Schwartz, 1982) and PKC (Braha et al., 1990; Sacktor and Schwartz, 1990) pathways and that both pathways appear to contribute to presynaptic facilitation (Braha et al., 1990; Sugita et al., 1992). However, the relative roles of PKA and PKC in enhancing release depend on the state of the synapse (Ghirardi et al., 1992). At nondepressed synapses, facilitation is largely blocked by specific inhibitors of PKA. Inhibitors of the PKC pathway have little effect. In contrast, at depressed synapses, PKA inhibitors are much less effective at blocking presynaptic facilitation. Thus the PKA pathway is primarily involved in enhancing release from nondepressed synapses whereas the PKC pathway is predominant in enhancing release from depressed synapses (Fig. 9.3).

The facilitation of synaptic transmission (at depressed or non-depressed synapses) in response to a single modulatory tail shock in vivo, or a single 5-min application of 5-HT in vitro, lasts for many minutes but eventually decays back to baseline. Repeated training protocols or repeated exposures to 5-HT, however, can produce long-term facilitation (LTF) of synaptic transmission owing to enhanced transmitter release lasting 24 h or more (Frost et al., 1985; Dale et al., 1988). One component of LTF, most prominent at 1–12 h, depends on the proteolysis of the regulatory subunit of PKA, mediated by the ubiquitin-proteasome pathway, which produces a prolonged, cAMP-independent activation of PKA (Hegde et al., 1997; Chain et al., 1999). Unlike short-term facilitation, LTF also involves a persistent growth of new synaptic boutons and an increase in the total area of synaptic contact (Bailey and Chen, 1988a,b). Moreover, unlike short-term facilitation, LTF requires new protein synthesis and the CREB-mediated transcriptional activation of several immediate early genes (Dash et al., 1990; Alberini et al., 1994; Bartsch et al., 1995, 1998). One of the immediate early gene products is ubiquitin C-terminal hydrolase, which may also contribute to prolonged activation of PKA (Hegde et al., 1997). A protein synthesis–dependent increase in the number of synaptic contacts also occurs during long-term facilitation induced by high-frequency stimulation at the crayfish neuromuscular junction (Wojtowicz et al., 1994).

Presynaptic Facilitation in Invertebrates and Mammalian Brain

Direct activation of PKA, PKC, or both enhances release at many other synapses in both invertebrates and vertebrates (see Capogna, 1998). What are the presynaptic targets that are responsible for this phosphorylation-dependent presynaptic facilitation? Are they the same as in *Aplysia* or are different mechanisms recruited? So far, most other examples of presynaptic facilitation do not appear to rely on K^+ channel modulation to alter calcium influx. Although a direct enhancement of the presynaptic calcium current by PKC does contribute to presynaptic facilitation in at least one instance (Parfitt and Madison, 1993), presynaptic facilitation more commonly occurs through a downstream mechanism that is likely to involve a modification of the release machinery (Delaney et al., 1991; Capogna et al., 1995; Trudeau et al., 1996; Chen and Regehr, 1997).

Such a downstream mechanism has been convincingly shown to contribute to the presynaptic facilitation produced by 5-HT at the crayfish neuromuscular junction. The effect of 5-HT depends on the combined actions of cAMP and the hydrolysis products of phosphatidylinositol (Dixon and Atwood, 1989). Ca^{2+} imaging studies demonstrate that 5-HT potentiates release without causing any detectable change in presynaptic Ca^{2+} levels, either at rest or in response to action potentials (Delaney et al., 1991). To determine the site of this downstream mechanism, Wang and Zucker (1998) used two methods to estimate the size of the pool of releasable synaptic vesicles. First, they measured the rate at which repetitive presynaptic stimulation leads to a depression of release, which is thought to reflect the rate of vesicle depletion. Second, they monitored the rate at which stimulation released FM1-43, a fluorescent dye that is specifically taken up into synaptic vesicles from the extracellular medium by endocytosis and then released back into the medium following subsequent exocytosis (see the chapter on synaptic vesicle endocytosis by De Camilli et al.). Both sets of measurements indicated that 5-HT increases the size of the pool of vesicles that are available for release. This effect is thought to be due to either an increase in the number of releasable vesicles per active zone, an increase in the total number of functional release sites, or both. Recently Beaumont and Zucker (2000) have provided evidence that at least part of the facilitation with 5-HT is due to the cAMP-dependent modulation of a hyperpolarization-activated cation channel in the presynaptic terminal. Surprisingly, their results suggest that the activation of this channel may somehow directly couple to vesicle mobilization.

Evidence both against and in support of changes in the number of functional release sites has been obtained with cAMP at different synapses. At inhibitory synapses in hippocampal neuronal cultures, cAMP elevation enhances release with no detectable change in the number of active presynaptic terminals, as assessed by the number of boutons that take up FM1-43 in an activity-dependent manner (Trudeau et al., 1996). Not only does cAMP enhance release evoked by action potentials, which is Ca^{2+} dependent, it also enhances release evoked by ruthenium red, which acts at a late step in the release process that is downstream of Ca^{2+} entry and independent of Ca^{2+} binding to its receptor. This implies that cAMP acts to enhance the probability of vesicle fusion independent of changes in presynaptic calcium (Trudeau et al., 1996).

Evidence from three independent systems, however, indicates that cAMP *can* enhance the number of functional release sites under certain conditions. Using cultured cerebellar granule cells, Chavis et al. (1998) assayed presynaptic function based on the ability of terminals to take up an antibody that reacts with a luminal epitope of synaptotagmin. This epitope becomes transiently exposed to the extracellular environment upon exocytosis, when it is free to bind externally applied antibodies. cAMP elevation with forskolin enhances the number of active presynaptic boutons that become stained with the antibody (Chavis et al., 1998). In these experiments the presynaptic terminals were not stimulated, so antibody uptake presumably occurs as a result of spontaneous release. Therefore it is not clear if these newly labeled boutons can support evoked release.

Independent studies at two types of hippocampal synapses provide evidence that the number of presynaptic terminals that show evoked release can be enhanced by cAMP. At synapses between cultured hippocampal CA3 and CA1 pyramidal cells, a brief exposure to a membrane-permeant analogue of cAMP increases the number of presynaptic boutons that undergo depolarization-dependent labeling with FM1-43 (Ma et al., 1999). This increase in the number of active boutons requires 1–2 h to develop fully and requires new protein synthesis. The increase in number of presumed release sites is thought to result from the activation of preexisting, presynaptically silent boutons because the total number of axonal varicosities (i.e., neurite swellings, which are thought to represent presynaptic boutons) is not altered by cAMP. A turning on of preexisting but presynaptically silent synapses by cAMP has also been reported, using electrophysiological probes of synapse function, at mossy fiber synapses formed by cultured dentate granule cells (Tong et al., 1996).

Molecular Targets of Presynaptic Facilitation

What are the presynaptic molecular targets involved in the Ca^{2+}-independent facilitative actions of cAMP? Several synaptic vesicle proteins have been identified that are phosphorylated by PKA, including the synapsins, α-SNAP, and rabphilin, the latter a protein that interacts with the small GTP-binding protein rab3a (Südhof and Scheller, this volume). Rab3a has attracted much attention as its deletion in mice leads to a decrease in cAMP-dependent mossy fiber long-term potentiation. To study the relation between rab3a and the cAMP-dependent facilitation of release more directly, the effect of deletion of rab3a was investigated on the synaptic enhancement seen with forskolin. In synaptosomes prepared from the CA3 region of wild-type mice, forskolin enhances release evoked by any one of three separate procedures: (1) KCl depolarization, (2) application of the Ca^{2+} ionophore ionomycin, and (3) application of hypertonic sucrose-containing solutions. Deletion of rab3a blocks the ability of forskolin to potentiate release evoked by direct elevation of Ca^{2+} with ionomycin. However, it does not affect the ability of forskolin to potentiate release evoked by either depolarization or hypertonic solution (Castillo et al., 1997; Lonart et al., 1998). These results point to multiple actions of cAMP to facilitate release—and only some of these actions require rab3a.

Since rab3a is not phosphorylated by PKA, some other effector that is the target of cAMP must be involved in facilitation. Rabphilin is one potential candidate protein that both interacts with rab3a and is phosphorylated by PKA. However, rabphilin knockout mice show normal mossy fiber LTP, so rabphilin cannot be the only effector for cAMP-dependent facilitation (Schlüter et al., 1999). A large number of other presynaptic proteins can be phosphorylated by different protein kinases (Turner et al., 1999), including the synapsins (Greengard et al., 1993), the soluble NSF attachment protein, α-SNAP (Hirling and Scheller, 1996), and the t-SNARE SNAP-25. The latter is a weak substrate for PKA (Hirling and Scheller, 1996) but a strong substrate for PKC phosphorylation (Shimazaki et al., 1996).

Studies of presynaptic facilitation at inhibitory hippocampal synapses provide evidence implicating SNAP-25 based on its selective cleavage with Botulinum toxin A (Trudeau et al., 1998). As discussed previously for this system (Trudeau et al., 1996), cAMP potentiates both Ca^{2+}-dependent release evoked by action potentials and Ca^{2+}-independent release evoked by ruthenium red. Cleavage of SNAP-25 reduces the frequency of spontaneous release and shifts the normal relation between external Ca^{2+} and evoked release toward higher Ca^{2+} concentrations, suggesting that SNAP-25 regulates the efficacy of the fusion process. Following cleavage of SNAP-25, cAMP no longer stimulates Ca^{2+}-independent release, although it does enhance action potential–evoked release. Similar to the effects of rab3a deletion, these results point to multiple actions of cAMP: one may be to enhance a Ca^{2+}-dependent step in release and another may be to enhance a late step in fusion that occurs after Ca^{2+} binding.

One other interesting candidate protein in presynaptic facilitation is the v-SNARE synaptobrevin. At the *Drosophila* neuromuscular junction, cAMP elevation leads to an increase in frequency of miniature excitatory synaptic currents (Zhang et al., 1999). Flies deficient in synaptobrevin show a defect in fast release in response to presynaptic action potentials and a reduction in frequency of spontaneous miniature excitatory postsynaptic currents (mEPSCs). In addition, the mutant flies no longer exhibit an increase in mEPSC frequency in response to cAMP, although elevation of intracellular Ca^{2+} can still enhance mEPSC frequency (Yoshihara et al., 1999).

The foregoing results, based on genetic deletion of rab3a and synaptobrevin and toxin cleavage of SNAP-25, are difficult to reconcile with a simple linear scheme in which an upstream calcium-dependent binding step is coupled to a single downstream, calcium-independent step that mediates the final fusion reaction. Rather the evidence points to parallel mechanisms for calcium-independent and calcium-dependent fusion reactions that are both regulated by cAMP but through distinct molecular mechanisms. The results from *Drosophila* suggest that facilitation requires the presence of intact synaptobrevin, which enhances a Ca^{2+}-independent component of release. The action of cAMP at mammalian synapses appears more complex. First, cAMP may act in a rab3a/SNAP-25–independent manner to potentiate release at an early stage of the release process, perhaps by enhancing Ca^{2+} influx through voltage-gated calcium channels. Second, cAMP might act in a rab3a/SNAP-25-dependent manner at a late stage of the release process, downstream of voltage-gated calcium channels, to enhance release independently of the earlier Ca^{2+}-entry steps.

Postsynaptic Mechanisms

So far we have focused on the forms of synaptic plasticity that result from changes in transmitter release from presynaptic terminals. However, there are also forms of synaptic plasticity that result from changes in the postsynaptic membrane that modify the postsynaptic response to a fixed amount of transmitter. Such postsynaptic changes can be produced by a change in the number of functional receptors in the postsynaptic membrane or a change in the efficacy with which ligands activate a relatively fixed number of postsynaptic receptors. Perhaps the most common mechanism of postsynaptic plasticity results from the direct phosphorylation of an ionotropic receptor by serine/threonine or tyrosine protein kinases (see Swope et al., 1999, for review).

Phosphorylation and Modulation of Nicotinic Acetylcholine Receptors

The nicotinic acetylcholine receptor (nAChR) was the first receptor channel shown to be regulated by protein phosphorylation (see Swope et al., 1999, for review). The pentameric receptor is composed of four types of homologous subunits: two α subunits and one β, γ, and δ subunit each. Each subunit contains an extracellular amino terminus, four transmembrane segments (M1–M4), and an extracellular carboxy terminus. The receptors contain two ACh-binding sites that are formed by the α subunits in conjunction with a neighboring γ or δ subunit (see Karlin and Akabas, 1995, for review).

Protein phosphorylation has been most thoroughly studied for the nAChR from the electric organ of the electric fish, *Torpedo californica,* as this tissue provides a ready source for biochemical purification of large quantities of receptor protein. Several different types of protein kinases have been shown to phosphorylate this receptor on distinct subunits (Swope et al., 1999). All the phosphorylation sites have been localized to the major intracellular loop of the receptor that connects the third and fourth transmembrane segments (M3 and M4). PKA phosphorylates single serine residues in both the γ (Ser-353) and δ (Ser-361) subunits. PKC phosphorylates the δ subunit at two distinct serine residues that differ from the PKA site (Ser-362 and Ser-377/379). Protein tyrosine kinases phosphorylate single tyrosine residues in the β (Tyr-355), γ (Tyr-364), and δ (Tyr-372) subunits. The major functional effect of phosphorylation is to increase the rate of receptor desensitization (Middleton et al., 1986; Hopfield et al., 1988), leading to a decrease in receptor activity. However, for recombinant embryonic-type muscle nAChRs, PKA enhances the rate of *recovery* from a slow component of desensitization, which would enhance receptor function (Paradiso and Brehm, 1998).

Although most studies have used direct activators of kinase cascades, such as forskolin or phorbol esters, to study phosphorylation-dependent modulation of nAChR function, in some cases the physiological pathways underlying receptor phosphorylation have been well characterized. At the neuromuscular junction, the neuropeptide CGRP is co-released with ACh from the motor neuron terminals. CGRP acts on a G protein–coupled receptor (GPCR) in the muscle membrane to stimulate cAMP synthesis, leading to a rise in cAMP levels in the muscle and the PKA-dependent phosphorylation of the nAChR on its γ and δ subunits (Miles et al., 1989). The phosphorylation produces an increase in the rate of nAChR receptor desensitization (Mulle et al., 1988). ACh itself may serve as a modulator, stimulating the PKC pathway, possibly as a result of Ca^{2+} influx through the nAChR (Ross et al., 1988; Miles et al., 1994). PKC activation, like PKA and tyrosine kinase activation, leads to an enhanced rate of desensitization. However, the physiological importance of the enhanced rate of receptor desensitization remains an open question because the synaptic action of ACh is normally so brief that desensitization, even after its acceleration by receptor phosphorylation, should be minimal. Finally, a pathway leading to AChR tyrosine phosphorylation has been identified that involves agrin, an extracellular matrix protein that is important for the clustering of AChR at the neuromuscular junction (Wallace et al., 1991). Agrin may enhance phosphorylation by stimulating a muscle receptor tyrosine kinase (Glass et al., 1996).

Phosphorylation and Modulation of GABA_A Receptors

Ionotropic $GABA_A$ receptors are also targets for several different protein kinases, which modulate receptor function (see Smart, 1997, and Swope et al., 1999, for reviews). The $GABA_A$ receptors are pentameric proteins composed of homologous subunits with a transmembrane topology similar to that of the nAChRs. Native $GABA_A$ receptors are likely to contain α, β, and γ types of subunits. Like the subunits of nAChRs, the $GABA_A$ receptor subunits are phosphorylated at sites in the M3–M4 intracellular loop. A single serine residue, conserved among β subunit isoforms (Ser-409 in the mouse β1 subunit), is promiscuously phosphorylated by PKA, PKC, CaMKI, and cGMP-dependent protein kinase. The γ subunit is also phosphorylated on serine residues by PKC and CaMKII as well as on tyrosine residues by the nonreceptor tyrosine kinase v-Src.

The reported effects of protein phosphorylation on $GABA_A$ receptor channel function are somewhat variable. Recombinant receptors studied in heterologous expression systems often show a decrease in GABA-activated current in response to phosphorylation. PKA-mediated phosphorylation of a serine residue in the β subunit causes a downregulation of receptor function as well as an enhancement of desensitization (Moss et al., 1992). Similar inhibitory effects of PKA-mediated phosphorylation on native $GABA_A$ receptor function in many different neurons have been reported (Swope et al., 1999). However, in the retina and in cerebellar Purkinje neurons, PKA activation through GPCR stimulation increases the magnitude of GABA-activated currents (Parfitt et al., 1990; Veruki and Yeh, 1992). Whether this is due to a direct effect of phosphorylation of the $GABA_A$ receptor or an indirect effect due to phosphorylation of some regulatory protein remains to be determined.

Activation of PKC, similar to the most commonly observed effects of PKA, inhibits $GABA_A$ receptor function, in some instances by phosphorylating the same serine residue in the β subunit that is targeted by PKA. The major inhibitory effect of PKC phosphorylation on GABA$_A$ receptor function, however, is due to phosphorylation of a serine residue in the γ subunit (Krishek et al., 1994). In contrast to the inhibitory effects of serine/threonine kinases, tyrosine phosphorylation of residues in the β and γ subunits enhances GABA-activated current amplitude (Moss et al., 1995; Wan et al., 1997). Despite the large number of examples of effects on GABA$_A$ receptor function of purified kinases and exogenous kinase activators (such as forskolin and phorbol esters), here again, as with the nAChR, relatively little is known about whether (or how) these kinases are recruited under physiological conditions.

Phosphorylation and Modulation of Ionotropic Glutamate Receptors

Ionotropic glutamate receptors represent the third major class of synaptic receptor channels that are targets of protein kinases. There are three major types of ionotropic glutamate receptors: α-amino-3-hydroxy-5-methyl-4-isoxazolepropionic acid (AMPA) receptors, kainate receptors, and N-methyl-D-aspartate (NMDA) receptors (Hollmann and Heinemann, 1994; Dingledine et al., 1999). All three types of receptor can be modulated through direct phosphorylation by serine/threonine and tyrosine kinases. The AMPA receptors are encoded by four homologous subunits, GluR1–4 (or GluRA–D). The NMDA receptors are composed of NR1 and one or more of the NR2A–D type of subunits. Kainate receptors are encoded by GluR5–7 and KA1,2 subunits.

The glutamate receptor family is structurally distinct from the pentameric nAChR/GABA receptor family (Dingledine et al., 1999). The receptors are likely to be tetramers composed of subunits with a very different transmembrane topology compared to the four transmembrane segment nAChRs and GABA receptors (Hollmann et al., 1994). All the subunits are believed to have a large extracellular amino terminus and three membrane-spanning segments, M1, M3, and M4. The M2 segment forms a pore-lining loop that dips into and out of the intracellular side of the membrane, similar to the pore-forming P region of K$^+$ channels. According to this topology, the carboxy terminus is on the cytoplasmic side of the membrane, in contrast to the ionotropic receptors for ACh and GABA, which have extracellular carboxy termini. The glutamate-binding site is formed as a bilobed structure by sequences from both the extracellular amino terminus and the extracellular M3–M4 loop (Armstrong et al., 1998).

Biochemical studies have demonstrated direct phosphorylation of AMPA, NMDA, and kainate receptor subunits (Lee and Huganir, 1999). Although there was some initial confusion as to the localization of the sites owing to the novel transmembrane topology of these receptors, a number of well-characterized sites have now been confirmed in the intracellular carboxy terminus of the receptor. Thus PKA and PKC each phosphorylate a distinct serine residue in the C terminus of the GluR1 AMPA receptor subunit.

Non-NMDA Receptors

Some of the first evidence for the functional modulation of current flow through ionotropic glutamate receptors was provided by studies demonstrating that dopamine enhances non-NMDA receptor currents in retinal horizontal cells through PKA-dependent phosphorylation (Knapp and Dowling, 1987; Liman et al., 1989). Subsequent studies on non-NMDA receptors showed that direct activation of PKA also enhanced agonist-induced currents in central neurons (Greengard et al., 1991; Wang et al., 1991). Interestingly, the effects of PKA appear to involve the action of a protein kinase A attachment protein that anchors PKA to the membrane near the non-NMDA receptor (Rosenmund et al., 1994).

Basal phosphorylation of the AMPA receptors by PKA may be important for maintaining normal levels of receptor function. Thus during whole-cell recording there is often a rundown of non-NMDA receptor currents that can be prevented by dialysis with purified PKA (Greengard et al., 1991; Wang et al., 1991; Rosenmund et al., 1994), suggesting that receptor dephosphorylation results in a decline in current. The major site of PKA-dependent phosphorylation has been localized to Ser-845 in the C-terminal tail of the GluR1 subunit of the AMPA receptors (Roche et al., 1996). Recombinant kainate-type receptors have been shown to be phosphorylated by PKA on the GluR6-type subunit at Ser-684 of the intracellular carboxy terminus (Raymond et al., 1993; Wang et al., 1993). This phosphorylation also enhances agonist-induced currents through the receptors.

The AMPA receptor is also regulated by CaMKII, which enhances the magnitude of glutamate-activated currents (McGlade-McCulloh et al., 1993). Biochemical studies demonstrate that the enhancement of receptor function by CaMKII results from the direct phosphorylation of a single serine residue (Ser-831) in the carboxy terminus of the GluR1 subunit (Barria et al., 1997a; Mammen et al., 1997). This phosphorylation enhances AMPA-activated current by increasing the frequency of large conductance channel openings, as observed in single-channel recordings of recombinant GluR1-containing AMPA receptors (Derkach et al., 1999). The phosphorylation of AMPA receptors by PKA and CaMKII has been implicated in long-term hippocampal synaptic plasticity.

NMDA Receptors

The NMDA receptors are also regulated by a variety of protein kinases (see MacDonald et al., 1998, for review). In hippocampal neurons, basal levels of PKA-dependent phosphorylation appear important for normal

NMDA receptor function, similar to that reported for the AMPA receptors. This basal phosphorylation is regulated by phosphoprotein phosphatases that are closely associated with the NMDA receptor in the membrane. Thus treatment of cell-free patches of membrane with phosphatase inhibitors enhances the magnitude of the NMDA receptor currents (Wang et al., 1994). The activity of the phosphatases can, moreover, be regulated by neuronal activity. Thus calcium influx during synaptic transmission downregulates the activity of the NMDA receptor by activating the calcium-dependent phosphatase calcineurin (Raman et al., 1996). This inhibitory effect can be prevented by activation of β-adrenergic receptors, which stimulates the activity of PKA and thus maintains the NMDA receptor in its phosphorylated state (Raman et al., 1996). The endogenous kinases and phosphatases that regulate NMDA receptor function appear to be physically associated with the receptor through the action of an intermediary protein, yotiao, which binds to PKA, phosphatase I, and the C terminus of the NR2 subunit (Westphal et al., 1999). Finally, NMDA receptor function can also be downregulated independently of changes in its state of phosphorylation by the direct binding of calcium/calmodulin to an inhibitory site on the carboxy terminus of the NR1 subunit (Ehlers et al., 1996; Zhang et al., 1998).

Functional and biochemical data also implicate tyrosine phosphorylation in the control and regulation of NMDA receptor function. In hippocampal neurons, tyrosine kinase inhibitors reduce the amplitude of NMDA receptor currents, whereas inhibitors of tyrosine phosphatases enhance current amplitude, suggesting that endogenous tyrosine phosphorylation and dephosphorylation events regulate receptor function (Wang and Salter, 1994). This regulation is apparently due to the action of the Src and Fyn nonreceptor tyrosine kinases that are closely associated with the channel in the membrane (Yu et al., 1997; Tezuka et al., 1999). Fyn appears to interact with the NMDA receptor by binding to the postsynaptic protein PSD-95, which associates with the C-terminal tails of the NR2 subunits (Tezuka et al., 1999). Src appears to increase current through the NMDA receptor by phosphorylating three residues contained in the NR2a receptor C terminus. This phosphorylation reduces a tonic inhibition of the NMDA receptors caused by the binding of extracellular Zn^{2+} to a site on the receptor (Zheng et al., 1998).

Long-Term Potentiation: An Intensively Studied Model of Synaptic Plasticity

NMDA Receptor–Dependent LTP

Perhaps no form of synaptic plasticity has generated more interest or more vigorous discussion than NMDA receptor–dependent LTP in the CA1 region of the hippocampus. The excitement surrounding this long-lasting increase in synaptic strength derives mainly from four sources:

1. There is compelling evidence from lesion studies in rodents and higher primates, including humans, that the hippocampus is a critical component of a neural system involved in the initial storage of certain forms of long-term memory (Zola-Morgan and Squire, 1993).

2. Several properties of LTP make it, theoretically, an attractive cellular mechanism for information storage (Bliss and Lomo, 1973; Nicoll et al., 1988; Bliss and Collingridge, 1993). Like memories, LTP can be generated rapidly and is prolonged and strengthened with repetition. It exhibits input specificity such that LTP is elicited primarily at the synapses stimulated by afferent activity but not at adjacent synapses on the same postsynaptic cell. This presumably dramatically increases the storage capacity within neural circuits. LTP is also associative, meaning that temporally pairing activity in a "weak" input (incapable of generating LTP by itself) with activation of a strong input (capable of generating LTP at adjacent synapses on the same postsynaptic cell) results in LTP of the weak input. This associative property is reminiscent of classical conditioning and has been incorporated into many neural network models of information storage.

3. Perhaps most important, LTP is readily triggered in in vitro slice preparations of the hippocampus, making it accessible to rigorous experimental analysis. Indeed much of what we know about the detailed cellular mechanisms of LTP, the main topic of this section, derives from studies on two types of excitatory synapses in hippocampal slices: synapses between the Schaffer collateral-commissural axons and the apical dendrites of CA1 pyramidal cell and synapses between the mossy fiber terminals of dentate granule cells and the dendrites of CA3 pyramidal cells.

4. LTP has been observed at virtually every excitatory synapse in the mammalian brain that has been studied. This includes excitatory synapses in different regions of the hippocampus; all areas and layers of the cortex, including visual, somatosensory, motor, and prefrontal; the amygdala; the thalamus; the neostriatum and nucleus accumbens; the ventral tegmental area; and the cerebellum. Thus LTP, which (as will be discussed in later sections) is not a unitary phenomenon, appears to be a ubiquitous property of mammalian excitatory synapses and may subserve a variety of functions in addition to its postulated role in learning and memory.

Induction of LTP: Role of NMDA Receptors and Ca^{2+}

LTP was first described at excitatory synapses in the dentate gyrus in vivo by Bliss and his colleagues in the early 1970s (Bliss and Gardner-Medwin, 1973; Bliss and Lomo, 1973). Over the next decade some of its basic properties were elucidated, with the major breakthroughs in our understanding of its molecular mechanisms being the demonstration of the key role of NMDA receptors in triggering LTP and the elucidation of the biophysical properties of these receptors (Collingridge et al., 1983;

Mayer et al., 1984; Nowak et al., 1984; MacDermott et al., 1986). It is now well established that during low-frequency synaptic transmission the neurotransmitter glutamate binds to different subtypes of ionotropic receptor that are often, but not always, co-localized on individual dendritic spines.

As mentioned previously, the major subtypes of these receptors are AMPA and NMDA receptors. (Kainate receptors are also found at some excitatory synapses but will not be discussed further here.) AMPA receptors have a channel permeable to monovalent cations (Na^+, K^+) and provide the majority of the inward current responsible for generating synaptic responses when the cell is close to its resting membrane potential. In contrast, NMDA receptors exhibit a strong voltage dependence owing to the blocking of their channels at negative voltages by extracellular Mg^{2+}. As a result, the receptors contribute little to the postsynaptic response during low-frequency synaptic activity. However, when the cell is depolarized Mg^{2+} dissociates from its binding site within the NMDA receptor channel, allowing Ca^{2+}, as well as Na^+, to enter the dendritic spine (Fig. 9.4). The resultant rise in intracellular Ca^{2+} is a necessary, and perhaps sufficient, trigger for LTP. Thus the only requirement for triggering LTP appears to be the activation of postsynaptic NMDA receptors by synaptically released glutamate during postsynaptic depolarization. Experimentally this is normally accomplished either by stimulating afferents at high frequency (25–100 Hz) or by directly depolarizing the CA1 cell with current injection while maintaining low-frequency afferent stimulation (a so-called pairing protocol).

The evidence in support of this model for the initial triggering of LTP is compelling. Specific antagonists of the NMDA receptor have minimal effects on basal synaptic transmission (Collingridge et al., 1983; Bliss and Collingridge, 1993). Preventing the rise in Ca^{2+} by loading postsynaptic cells with Ca^{2+} chelators blocks LTP (Lynch et al., 1983; Malenka et al., 1988), whereas direct increases in the level of postsynaptic Ca^{2+} can mimic LTP (Malenka et al., 1988; Yang et al., 1999). And imaging studies have demonstrated that NMDA receptor activation causes a large increase in the Ca^{2+} level within dendritic spines (Regehr and Tank, 1990; Alford et al., 1993; Perkel et al., 1993; Yuste and Denk, 1995). The exact properties of the Ca^{2+} transient that is required to trigger LTP

Figure 9.4. (opposite) Model for the induction of NMDA receptor–dependent LTP. During normal synaptic transmission, glutamate (GLU) is released from the presynaptic bouton and acts on both AMPA receptors (AMPAR) and NMDA receptors (NMDAR). However, Na^+ flows through the AMPA receptor, but not the NMDA receptor, because Mg^{2+} blocks the channel of the NMDA receptor. Depolarization of the postsynaptic cell relieves the Mg^{2+} block of the NMDA receptor channel, allowing Na^+ and Ca^{2+} to flow into the dendritic spine by means of the NMDA receptor. The resultant rise in Ca^{2+} within the dendritic spine is the critical trigger for LTP. Adapted from Malenka and Nicoll (1999).

Normal synaptic transmission

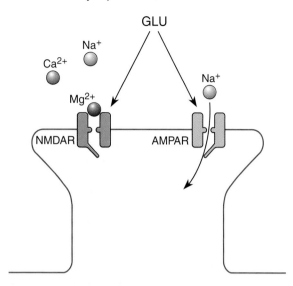

During depolarization

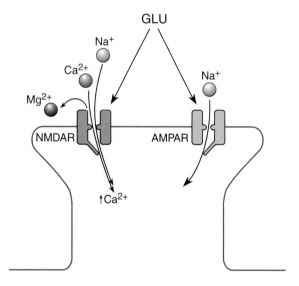

are presently unknown, but it appears that a transient lasting only 1–3 sec is sufficient (Malenka et al., 1992).

Another important issue that remains unresolved is whether an NMDA receptor–dependent rise in Ca^{2+} alone is sufficient to trigger LTP or whether additional factors provided by synaptic activity are required. It is clear that various neurotransmitters found in the hippocampus, such as ACh, can modulate the ability to trigger LTP, but thus far there is no compelling evidence that any neurotransmitter, other than glutamate, is absolutely required. Synaptically released glutamate can, of course, activate mGluRs, which are found at most excitatory synapses and might play an important role in LTP (see Anwyl, 1999, for review). Their activation, however, does not appear to be absolutely required for the generation of LTP, at least in CA1 pyramidal cells.

NMDA receptor activation and increases in postsynaptic Ca^{2+} that do not reach the threshold for eliciting LTP can generate either short-term potentiation (STP) (Malenka and Nicoll, 1993) that decays to baseline over the course of 5–20 min or long-term depression (LTD) (Bear and Linden, this volume). Thus any experimental manipulation that influences the magnitude or dynamics of the postsynaptic Ca^{2+} signal within dendritic spines may have an effect on the form of the synaptic plasticity caused by a given pattern of synaptic activation. A rise in postsynaptic Ca^{2+} can also occur as a consequence of activation of voltage-dependent Ca^{2+} channels, and this can trigger either LTP, STP, or LTD (Johnston et al., 1992). However, perhaps because of the different subcellular localization of Ca^{2+} channels, the LTP due to activation of these channels may utilize mechanisms distinct from those involved in NMDA receptor–dependent LTP (Cavus and Teyler, 1996).

Signal Transduction Mechanisms in LTP

A large number of signaling molecules have been suggested to play a role in translating the Ca^{2+} signal that is required to trigger LTP into an increase in synaptic strength. For only a few of these, however, has compelling evidence for a mandatory role in LTP been presented. A critical distinction should be made between those molecules that play a direct active role in the induction of LTP and those that play a modulatory or permissive role. For example, any treatment that up- or downregulates the activity of NMDA receptors will necessarily enhance or depress the ability to induce LTP, although such processes need not necessarily be recruited to induce LTP. Similarly any process that affects postsynaptic depolarization, for example the opening or closing of K^+ channels, will also alter the ability to generate LTP. We can therefore distinguish several key criteria that must be met for a signaling molecule to be implicated in LTP. First, it should be produced or activated by LTP-inducing stimuli but not by stimuli that fail to induce LTP. Second, blockade of the pathway in which the molecules participate should block the induction of LTP. And third, activation of the pathway should lead to activation of LTP.

One limitation of much of the research on this topic is that experiments—which might critically distinguish molecules that are key components of the signal transduction machinery for LTP from others that modulate the ability to generate LTP—have not been performed. For example, simple tests as to whether or not a manipulation affects NMDA receptor function should always be performed. Similarly studies of inhibitors should be performed not only on tetanus-induced LTP but also on pairing-induced LTP, in which the postsynaptic depolarization is directly provided by a current or voltage clamp. The following brief discussion will therefore be limited to those molecules for which the evidence for involvement in the triggering of LTP is reasonably strong.

There is general agreement that CaMKII is a key component of the molecular machinery responsible for LTP. It is found in high concentrations in the postsynaptic density (PSD) and has the interesting biochemical property that, when autophosphorylated on threonine 286, its activity is no longer dependent on calcium-calmodulin, thus allowing its activity to outlast the Ca^{2+} signal that originally activated the enzyme (Kennedy et al., 1990; Braun and Schulman, 1995). Among the first evidence that CaMKII plays a critical role in LTP were reports that inhibiting its activity by directly loading CA1 pyramidal cells with inhibitors of CaMKII or genetically knocking out a critical CaMKII subunit blocks the ability to generate LTP (Malenka et al., 1989; Malinow et al., 1989; Silva et al., 1992). Subsequently it was found that acutely increasing the postsynaptic concentration of constitutively active CaMKII increases synaptic strength and occludes LTP (Pettit et al., 1994; Lledo et al., 1995). These latter findings are particularly important because, as mentioned earlier, experimental inhibition of LTP can occur via any number of mechanisms that may or may not involve disruption of the LTP signal transduction machinery.

However, certain puzzling results suggest that, even with the CaMKII pathway, the mechanisms may not be straightforward. For example, prolonged overexpression of a constitutively active CaMKII mutant in transgenic mice did not occlude LTP but instead inhibited the LTP that normally is elicited by 5- to 10-Hz afferent stimulation (Mayford et al., 1995, 1996). Moreover, analysis of an α CaMKII knockout mouse, on a different genetic background from the initial knockout mouse, shows that a substantial fraction of LTP (circa 50% of wild-type levels) can be preserved in the absence of this enzyme (Hinds et al., 1998). These results suggest that other kinases besides CaMKII may be important and that CaMKII may have actions quite apart from its role in the induction of LTP. Since it took several weeks for the full expression of the transgene for CaMKII or for the development of the CaMKII knockout mice, it is possible that compensatory biochemical changes had occurred at the relevant synapses (Lisman et al., 1997).

Biochemical measurements also support a role for CaMKII in LTP. Autophosphorylation of CaMKII occurs after the triggering of LTP (Fukunaga et al., 1995; Barria et al., 1997b), as does an increase in the phosphorylation level of one of its major substrates in the PSD, the AMPA receptor GluR1 subunit (Barria et al., 1997b). A final piece of evidence supporting a key role for CaMKII involved the elegant demonstration that LTP was blocked in mice in which endogenous CaMKII had been replaced with a form of CaMKII containing a point mutation at threonine 286, the normal autophosphorylation site (Giese et al., 1998).

Several other protein kinases have also been suggested to play important roles in the triggering of LTP, but the experimental evidence for these kinases is considerably weaker than that for CaMKII. The initial activation of PKA, perhaps via activation of a calmodulin-dependent adenylyl cyclase, has been suggested to boost the activity of CaMKII indirectly by decreasing competing protein phosphatase activity (Blitzer et al., 1998; Makhinson et al., 1999). This is presumably accomplished via PKA-dependent phosphorylation of inhibitor-1, an endogenous protein phosphatase I inhibitor (Shenolikar and Nairn, 1991). (The potential role of PKA in the late protein synthesis phase of LTP is discussed by Greenberg and Ziff, this volume.)

PKC may play a role analogous to that of CaMKII, as PKC inhibitors have been reported to block LTP and loading PKC into CA1 pyramidal cells can enhance synaptic transmission (Hu et al., 1987; Linden and Routtenberg, 1989; Klann et al., 1991; Sacktor et al., 1993). But whether the PKC-mediated enhancement of synaptic transmission utilizes the same mechanisms as LTP has not been determined. Additional protein kinases implicated in LTP include two tyrosine kinases, Fyn and Src (Grant et al., 1992; Lu et al., 1998), and the serine/threonine mitogen-activated protein kinase (MAPK) (English and Sweatt, 1996, 1997). Src may participate in LTP by enhancing NMDA receptor function during the synaptic triggering of LTP (Lu et al., 1998), whereas MAPK may act by down-regulating K^+ channel activity (Adams et al., 1999).

Another potentially important class of signaling molecules that may play important roles in LTP are retrograde messengers. These are substances that are produced in the postsynaptic cell and diffuse across the synaptic cleft to modify presynaptic function. Because the importance of such messengers is dependent on the conclusion that LTP involves presynaptic increases in transmitter release, we will discuss these biochemical pathways after the experimental evidence for the pre- versus postsynaptic site of LTP expression has been presented.

Expression Mechanisms

Over the past decade, no question concerning LTP has generated more confusion or controversy than the seemingly simple issue of whether LTP is due primarily to a pre- or a postsynaptic modification. The great challenge in addressing this issue derives largely from the technical difficulties inherent in attempting to examine changes at individual synapses that are embedded in a complex network in which each cell may receive up to 10,000 or more synapses. The simplest postsynaptic modification that could cause LTP would be a change in AMPA receptor function or number, whereas the simplest presynaptic change would be an increase in the probability of neurotransmitter release. (Of course, the growth of new synaptic contacts is a third possibility, but this will not be discussed further here.)

The first attempts to address whether LTP involved presynaptic modifications employed measurements of extracellular glutamate levels, which were reported to increase after the induction of LTP (Dolphin et al., 1982; Bliss et al., 1986). Thus early on in the study of LTP a requirement for some form of retrograde messenger was postulated. However, subsequent attempts to replicate this finding have failed (Aniksztejn et al., 1989; Jay et al., 1999), and it remains uncertain whether the approach used accurately measures synaptically released glutamate.

Most studies that have examined whether LTP is expressed pre- or postsynaptically have used electrophysiological assays. Again the results from different laboratories have been inconsistent, although a degree of consensus has recently been reached by some investigators in the field. One experimental approach has taken advantage of the fact that AMPA and NMDA receptors are often co-localized at individual synapses. Therefore, manipulations that increase transmitter release might be expected to cause an equal increase in the synaptic currents mediated by these two receptor subtypes. Most investigators have found that LTP causes a greater increase in the AMPA receptor–mediated EPSC than the NMDA receptor–mediated EPSC (Kauer et al., 1988; Muller et al., 1988; Perkel and Nicoll, 1993; Liao et al., 1995; Durand et al., 1996; but see Clark and Collingridge, 1995). This result does not preclude an increase in transmitter release occurring during LTP, but it does suggest that some postsynaptic modification has occurred.

Changes in transmitter release probability during the LTP should influence the various forms of short-term plasticity that were discussed earlier. Some investigators find that these phenomena are minimally affected by LTP (Muller and Lynch, 1989; Manabe et al., 1993; Asztely et al., 1996; Pananceau et al., 1998; Selig et al., 1999), whereas other investigators report an alteration (Kuhnt and Voronin, 1994; Schulz et al., 1994). Another approach that has been used is based on the assumption that if LTP is primarily due to an increase in the probability of release, then synapses at which the probability of release is extremely high (i.e., approaching 1) should exhibit minimal LTP. It has been suggested that this is in fact the case in very young hippocampal synapses, at which LTP could only be detected when release probability (which was already close to 1 under basal conditions) was reduced by lowering external Ca^{2+} (Bolshakov and Siegelbaum, 1995; but see Malinow and Mainen, 1996). In older hippocampal slices, pharmacologically increasing the probability of release has been found to have no effect on LTP (Hjelmstad et al., 1997).

Two further approaches have been taken in an attempt to measure the release of glutamate more directly. One took advantage of the fact that glial cells tightly ensheath synapses and, as a result of activation of their electrogenic glutamate transporters, they can generate a measurable current that is directly proportional to the amount of synaptically released glutamate (Diamond et al., 1998; Lüscher et al., 1998). The other approach took advantage of use-dependent antagonists of the NMDA receptor or of a mutant AMPA receptor that lacks the GluR2 subunit. These antagonists decrease the EPSC at a rate that is directly proportional to transmitter release probability (Hessler et al., 1993; Rosenmund et al., 1993). All these measures were affected in the predicted fashion by manipulations known to increase transmitter release globally but were not altered after induction of LTP (Manabe and Nicoll, 1994; Diamond et al., 1998; Lüscher et al., 1998; Mainen et al., 1998). However, such approaches necessarily assume that the released glutamate being assayed is representative of the glutamate released from synapses that undergo LTP.

Many of the findings reviewed thus far argue against a large increase in the probability of transmitter release causing LTP (but see the subsequent discussion). In addition, a number of electrophysiological and biochemical measures of postsynaptic function have been found to increase during LTP. If we assume that the amount of glutamate in each presynaptic vesicle is relatively fixed, then an increase in the amplitude of mEPSCs would indicate an increase in the function of AMPA receptors, their number, or both. Such an increase has been reported during LTP (Oliet et al., 1996) as well as following brief applications of NMDA or repetitive activation of Ca^{2+} channels, all of which should load dendritic spines with Ca^{2+} (Manabe et al., 1992; Wyllie et al., 1994). An even more direct way of monitoring changes in AMPA receptors is to measure the responses generated by direct application of agonists. During LTP, such responses have been reported to increase, albeit gradually over the course of tens of minutes (Davies et al., 1989). LTD (see Bear and Linden, this volume) has also been examined using these approaches, and, like LTP, it was found to be accompanied by a decrease in the amplitude of mEPSCs (Oliet et al., 1996; Carroll et al., 1999a) and a decrease in the response to glutamate (Kandler et al., 1998; Dodt et al., 1999).

Taken together, the evidence presented thus far makes a strong case for the view that an increase in AMPA receptor responsiveness is at least one of the major mechanisms contributing to LTP. How does this occur? One possibility is the phosphorylation of the AMPA receptor GluR1 subunit. As discussed previously, AMPA receptors are heteromers that, in CA1 pyramidal cells, are composed primarily of GluR1 and GluR2 subunits. Both in expression systems and in native hippocampal cells, GluR1 is phosphorylated on serine 831 by CaMKII whereas PKA phosphorylates serine 845 (Roche et al., 1996; Barria et al., 1997a; Mammen et al., 1997). LTP is accompanied by an increase in phosphorylation of serine 831, an effect that is blocked by a CaMKII inhibitor (Barria et al., 1997b). This phosphorylation increases the single-channel conductance of homomeric GluR1 AMPA receptors (Derkach et al., 1999). As an increase in the single-channel conductance of AMPA receptors also occurs during LTP (Benke et al., 1998), a likely mechanism contributing to LTP is

CaMKII-dependent phosphorylation of the AMPA receptor subunit GluR1. Support for this hypothesis comes from the finding that genetic knockout of GluR1 prevents the generation of LTP in mouse CA1 pyramidal cells (Zamanillo et al., 1999).

Analysis of Quantal Events and Silent Synapses

Although the evidence reviewed thus far makes a strong case for postsynaptic changes contributing to LTP, other data argue that LTP is also due to an increase in the probability of transmitter release. The principal evidence in support of this notion derives from experiments that have taken advantage of the probabilistic nature of transmitter release. As discussed by Regehr and Stevens (this volume), the action potential–dependent release of a quantum of neurotransmitter from an individual synapse occurs only about 10–40% of the time. Therefore, if a very small number of synapses are activated, one records a mixture of quantal EPSCs, or "successes," and "failures." An extensively replicated finding is that LTP causes a decrease in the proportion of synaptic failures (see Kullmann and Siegelbaum, 1995, for review). Because these failures were assumed to be due to failures of neurotransmitter release, it was concluded that LTP involves an increase in the probability of transmitter release. Consistent with this conclusion was the finding that the coefficient of variation (CV) of EPSCs (which is thought to be inversely proportional to the quantal content, i.e., the average number of synapses that are activated and release neurotransmitter with each stimulation), decreased during LTP (Kullmann and Siegelbaum, 1995).

Even stronger evidence in support of the view that LTP is due to an increase in transmitter release probability comes from experiments in which responses that were presumably generated from only a single synapse were studied. In these experiments, following the induction of LTP, the proportion of failures decreased, but more importantly the amplitude of the EPSC "successes" was unchanged (Stevens and Wang, 1994; Bolshakov and Siegelbaum, 1995). If any postsynaptic modifications had occurred in these experiments, the average amplitude of the EPSC "successes" should have increased. Thus these observations were taken as evidence that LTP is due to a presynaptic increase in transmitter release.

However, like many of the results on LTP reviewed thus far, some of these results have not been replicated by other workers (Isaac et al., 1996), and certain aspects of the experiments have been questioned on technical grounds (Malinow and Mainen, 1996). There also appears to be a significant contradiction in the findings, one that is perhaps best encapsulated in the following question: how can the LTP-induced change in the failure rate and decrease in the CV be reconciled with all the findings that argue for a postsynaptic change during LTP? The first clue to a possible resolution of this dilemma came from the observation that the CV of AMPA receptor–mediated EPSCs was greater than the CV of NMDA receptor–mediated EPSCs (Kullmann, 1994). This finding suggested that synaptically released glutamate activated more synapses containing NMDA receptors than synapses containing AMPA receptors. The simplest explanation for this observation is that some synapses express only NMDA receptors whereas others express both AMPA and NMDA receptors. Synapses with only NMDA receptors would be functionally silent at hyperpolarized membrane potentials; thus, when transmitter is released from the corresponding presynaptic bouton, they would not yield a response. However, LTP at such "silent" synapses could occur if there was activity-induced expression of AMPA receptors. Theoretically such a change could explain the decrease in failure rate and CV during LTP. (But it is important to repeat that this LTP-induced conversion of "silent" to "functional" synapses cannot explain the experiments on single synapses in which no change in quantal EPSC amplitude was observed.)

There is now reasonably strong evidence to support several aspects of the silent synapse hypothesis:

1. It is possible to record EPSCs that are mediated solely by NMDA receptors (Fig. 9.5), and by applying an LTP induction protocol at such synapses one can cause the rapid appearance of AMPA receptor–mediated EPSCs (Isaac et al., 1995; Liao et al., 1995; Durand et al., 1996).

2. Electron microscopic analysis using immunogold labeling fails to detect AMPA receptors at a significant proportion of synapses in developing hippocampus and hippocampal cultures, whereas all synapses contain NMDA receptors (Rao and Craig, 1997; Gomperts et al., 1998; Nusser et al., 1998; Liao et al., 1999; Petralia et al., 1999; Takumi et al., 1999). In addition, synapses that do not express NMDA receptor–dependent LTP always appear to contain a substantial number of AMPA receptors (Nusser et al., 1998; Takumi et al., 1999).

3. Activation of NMDA receptors results in an increase in fluorescence in dendritic spines of cells expressing a green fluorescent protein–GluR1 fusion protein (Shi et al., 1999). Conversely NMDA receptor–dependent LTD in hippocampal cultures causes a loss of AMPA receptors from synapses (Carroll et al., 1999a). These results suggest that changes in the pattern of activity at individual synapses can cause a redistribution of AMPA receptors into or out of synapses.

4. AMPA and NMDA receptors interact with different clustering proteins at the synapse (see Sheng, this volume), suggesting that they are probably regulated independently.

5. Interference with membrane fusion in the postsynaptic cell impairs LTP (Lledo et al., 1998), and proteins involved in membrane fusion can interact with AMPA receptors (Nishimune et al., 1998; Osten et al., 1998; Song et al., 1998). These findings are consistent with (but certainly do not prove) the idea that membrane fusion may be an important mechanism for the delivery of AMPA receptors to the cell surface.

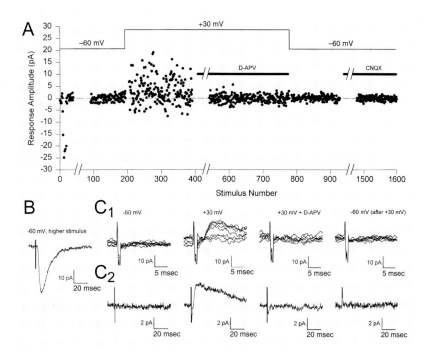

Figure 9.5. Demonstration of the occurrence of synaptic responses mediated only by NMDA receptors.

(A) Time course of the experiment. The CA1 pyramidal cell was held at –60 mV, and, after obtaining a small EPSC, stimulus intensity was reduced so that no EPSCs were detected for 100 consecutive stimuli. The cell was then depolarized to +30 mV and stimulation now evoked responses that were completely blocked by application of D-APV (25 μM), indicating that they were mediated by NMDA receptors. The cell was then returned to –60 mV, where again no EPSCs could be detected, as evidenced by the lack of effect of the AMPA receptor antagonist CNQX (10 μM), which was applied at the end of the experiment.

(B) Sample of the EPSC (average of 10 consecutive responses) recorded at the beginning of the experiment.

(C) Examples of 8 consecutive traces (*C₁*) or the average of 100 consecutive traces (*C₂*) taken at the indicated times during the course of the experiment.

Reproduced from Isaac et al. (1995) with permission of Cell Press.

Many of the data discussed thus far are consistent with the simple model (Fig. 9.6) that LTP, at least initially, is caused by both phosphorylation of AMPA receptors and the delivery or clustering of AMPA receptors within the synaptic plasma membrane (Malenka and Nicoll, 1999). These events presumably would occur both at synapses that contain functional AMPA receptors and at ones that are functionally silent because they do not express AMPA receptors on their synaptic surfaces. The mechanisms that control the activity-dependent delivery of AMPA receptors are unknown. It has been demonstrated, however, that AMPA receptors can undergo a dynamin-dependent endocytosis in response to

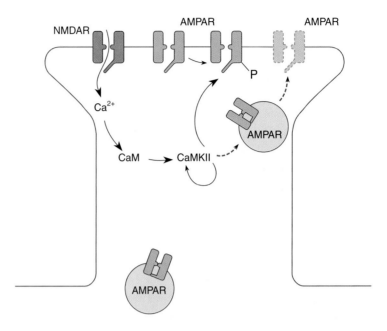

Figure 9.6. Model for the postsynaptic mechanisms that may contribute to the expression of NMDA receptor–dependent LTP. An increase in Ca^{2+} within the dendritic spine binds to calmodulin (CaM) to activate CaMKII, which undergoes autophosphorylation, thus maintaining its activity after Ca^{2+} returns to basal levels. CaMKII phosphorylates AMPA receptors (AMPAR) already present in the synaptic plasma membrane, thus increasing their single-channel conductance. CaMKII may also influence the subsynaptic localization of AMPA receptors such that more AMPA receptors are delivered to the synaptic plasma membrane. The localization of the "reserve" AMPA receptors is unclear. For purposes of illustration, they are indicated as a pool of receptor–containing vesicles. Before the triggering of LTP, some synapses may be functionally silent in that they contain no AMPA receptors in the synaptic plasma membrane. Adapted from Malenka and Nicoll (1999).

direct ligand activation or NMDA receptor activation (Carroll et al., 1999b) and that there is a pool of AMPA receptors that may cycle rapidly into and out of the PSD (Lüscher et al., 1999). Furthermore, an elegant series of experiments has demonstrated that recombinant homomeric AMPA receptors composed of GluR1 subunits are inserted into the synaptic plasma membrane following the overexpression of CaMKII or the induction of LTP in cultured CA1 pyramidal cells (Hayashi et al., 2000). Thus the molecular substrates for the delivery and removal of AMPA receptors appear to be present at excitatory synapses. An additional major issue that remains to be resolved is the precise localization of the reserve AMPA receptors that, according to this hypothesis, are recruited to the

PSD during LTP. They could be recruited from the cytoplasm of the dendritic spines, from membrane adjacent to the PSD, or from dendritic shafts at the base of spines. Current anatomical studies are most consistent with the latter possibility.

Finally, even this simple silent synapse model is subject to alternate interpretations, based on the fact that NMDA receptors are much more sensitive to glutamate than are AMPA receptors. According to the spill-over hypothesis, glutamate released at one synapse may diffuse to a neigh-boring synapse (spill over), where, owing to its low concentration, it selectively activates NMDA receptors (Kullmann and Siegelbaum, 1995; Kullmann et al., 1996). In a variant of this idea, Choi et al. (2000) pro-vide suggestive data that the selective activation of NMDA receptors at silent synapses is due to the incomplete fusion of a synaptic vesicle with the presynaptic plasma membrane, causing glutamate to trickle out of the vesicle at a relatively low rate, thus limiting its concentration in the synap-tic cleft. According to this model, LTP results from a switch from incom-plete fusion to complete fusion, providing a higher cleft concentration of glutamate sufficient to activate AMPA receptors.

Retrograde Messengers

Although the postsynaptic model just presented can account for much of the experimental literature, it does not preclude the occurrence of significant presynaptic modifications contributing to LTP. Indeed several experimental results strongly suggest that presynaptic changes do occur. As mentioned earlier, the finding that LTP caused a change in failure rate without a change in the amplitude of quantal EPSCs can only be explained by an increase in transmitter release probability (Stevens and Wang, 1994; Bolshakov and Siegelbaum, 1995). In addition, direct measures of vesicle cycling in hippocampal cultures exhibit an activity- and NMDA receptor–dependent increase (Malgaroli et al., 1995; Ryan et al., 1996).

Because of the possibility that NMDA receptor–dependent LTP in-volves presynaptic changes, much effort has been directed toward identi-fying the putative retrograde messenger that must exist as this form of LTP is triggered in the postsynaptic cell. Before discussing the most popular candidates that could play this role, it is important to note that several criteria must be met before any given substance can be consid-ered a candidate retrograde messenger during LTP. First, the substance must be produced in dendrites in response to NMDA receptor activation. Second, inhibition of its production should block LTP. Third, applica-tion of the substance during synaptic activation but in the presence of an NMDA receptor antagonist should elicit a synaptic enhancement that occludes LTP.

The substance that has received most attention in this regard is nitric oxide (NO) (Snyder and Ferris, this volume). It has the attractive prop-erty that the enzyme responsible for its synthesis, nitric oxide synthase

(NOS), can be stimulated by increases in calcium-calmodulin concentration due to NMDA receptor activation (Schuman and Madison, 1994). Consistent with its potential importance in LTP, initial studies reported that inhibitors of NOS block the generation of LTP (Böhme et al., 1991; O'Dell et al., 1991; Schuman and Madison, 1991). Subsequent work, however, has shown that this blockade of LTP is highly dependent on the experimental conditions and that robust LTP can be elicited in the absence of NOS activity. Given the caveat about the interpretation of studies that report impairment of LTP, experiments that test whether direct application of NO enhances synaptic transmission and occludes LTP are of paramount importance. Such experiments have been performed, but, unfortunately, like many of the findings in the LTP field, the published reports are contradictory. One group of investigators has found that NO can cause a synaptic enhancement in both hippocampal slices and cultures, an effect that appeared to be due to an NO-mediated stimulation of presynaptic guanylyl cyclase and cGMP-dependent protein kinase (Zhuo et al., 1994; Arancio et al., 1995, 1996). A second group has found that knockout of the neuronal and endothelial isoforms of NO synthase (both of which are normally expressed in hippocampus) largely inhibits LTP at Schaffer collateral-commissural synapses on the apical dendrites of CA1 neurons (Son et al., 1996). On the other hand, several groups have not replicated some of the key observations in these studies (Murphy et al., 1994; Selig et al., 1996). Furthermore, it has been reported that normal LTP can be elicited in the CA1 region in mice lacking both isoforms of cGMP-dependent protein kinase (Kleppisch et al., 1999). Thus at the moment it is difficult to conclude that the production of NO is absolutely required for the triggering of LTP (unless, of course, there are multiple forms of LTP that utilize distinct mechanisms).

Some evidence in support of heterogeneity in LTP-generating mechanisms comes from comparisons of LTP in the stratum radiatum of CA1 (where Schaffer collaterals synapse onto the apical dendrites of CA1 neurons) and LTP in stratum oriens (where the Schaffer collaterals synapse onto basal dendrites of the same neurons). Interestingly, the groups of investigators who found that blockade of LTP in the stratum radiatum is blocked by NOS inhibitors and in NOS knockout mice have reported that LTP in the stratum oriens is not sensitive to NOS manipulations (Haley et al., 1996; Son et al., 1996). If this is correct, it would appear that there can be distinct differences within single CA1 neurons in terms of the importance of this retrograde signaling pathway.

Other candidate retrograde messengers in LTP include arachidonic acid, platelet-activating factor, and carbon monoxide (Williams et al., 1989, 1993; Stevens and Wang, 1993; Zhuo et al., 1993; Kato and Zorumski, 1996; Snyder and Ferris, this volume). The roles of these substances in LTP, however, have been explored much less thoroughly than that of NO, and the current experimental evidence in support of any one of these substances as a critical retrograde messenger is not compelling.

Conclusion

The initial triggering mechanisms for NMDA receptor–dependent LTP are now well established, and significant progress has been made in elucidating some of the postsynaptic modifications that contribute to the expression of LTP. There remains, however, considerable debate and confusion about what sorts of presynaptic and postsynaptic changes occur immediately after the triggering of LTP. Much of the confusion derives from the fact that independent laboratories, apparently performing essentially the same experiment, obtain different results. The reasons for these discrepancies are not clear, but two general classes of explanation are possible. The first is that, because of either poor experimental design or technical inadequacies, some of the published data are wrong. This must be the case if there is only one form of NMDA receptor–dependent LTP that is due to one specific subset of molecular mechanisms. Alternatively it is conceivable that strong activation of NMDA receptors can lead to an LTP that can be generated by any one of several independent mechanisms. The existence of different forms of LTP is most clearly demonstrated at different synapses. For example, LTP at both hippocampal mossy fiber synapses and cerebellar granule cell–Purkinje cell synapses clearly differs from LTP in the CA3–CA1 pathway in that it does not require NMDA receptor activation. A corollary of this explanation is that the specific form of LTP being studied in any individual laboratory depends on a number of poorly controlled and unknown variables (such as the temperature when the experiment is done, the age of the tissue used, and the method of preparation).

No matter what the explanation, it is clear that if presynaptic changes occur immediately after the induction of LTP, the identity of the retrograde messenger(s) is critical, and its identification would enormously facilitate progress in understanding the presynaptic molecular changes that contribute to LTP. Furthermore, as many chapters in this book have pointed out, the synapse is a structural unit with important protein-protein interactions occurring between and within pre- and postsynaptic elements. Thus it seems likely that long-lasting synaptic modifications will involve structural alterations both in dendritic spine morphology and in the presynaptic boutons. If the synaptic change is triggered by postsynaptic activity, as is the case for NMDA receptor–dependent LTP, then some retrograde communication must occur. The challenge for future research in this area remains to understand in greater molecular detail the post- and presynaptic changes that occur in response to NMDA receptor activation and exactly how retrograde communication is accomplished. As discussed in the chapter by Greenberg and Ziff, our increased understanding of how synaptic activity affects gene transcription and the targeting of newly synthesized proteins should greatly facilitate this effort.

NMDA Receptor–Independent LTP

Although most of the effort toward understanding long-lasting forms of synaptic plasticity has focused on NMDA receptor–dependent LTP, it is clear that at certain synapses a robust form of LTP can be generated that does not require NMDA receptor activation. The NMDA receptor-independent form of LTP that has been most extensively examined occurs at mossy fiber synapses in the hippocampus, which are formed between the axons of dentate gyrus granule cells (the mossy fibers) and the proximal dendrites of CA3 pyramidal cells (Nicoll and Malenka, 1995). A similar form of LTP also appears to occur in the cerebellum at synapses between parallel fibers and Purkinje cells (Salin et al., 1996) as well as at corticothalamic synapses (Castro-Alamancos and Calcagnotto, 1999).

In contrast to NMDA receptor–dependent LTP, there is little controversy about the site of expression for mossy fiber LTP, which virtually all investigators agree is presynaptic. There is, however, some disagreement about whether mossy fiber LTP is triggered pre- or postsynaptically. The first indication that LTP at mossy fiber synapses may have different properties than the much more extensively studied NMDA receptor-dependent LTP was that it could be triggered in the presence of NMDA receptor antagonists (Harris and Cotman, 1986; Zalutsky and Nicoll, 1990). This finding suggested either that mossy fiber LTP was triggered presynaptically or that a postsynaptic rise in Ca^{2+} was required but that this was achieved via some source other than NMDA receptors, such as voltage-dependent Ca^{2+} channels or metabotropic glutamate receptors. In principle the answer to this issue should be relatively straightforward to determine. For example, does loading CA3 pyramidal cells with a Ca^{2+} chelator block mossy fiber LTP? However, there remains a vigorous disagreement about the results of this sort of manipulation. Most investigators find that mossy fiber LTP is independent of postsynaptic Ca^{2+} and membrane potential (Zalutsky and Nicoll, 1990; Ito and Sugiyama, 1991; Katsuki et al., 1991; Castillo et al., 1994; Langdon et al., 1995). However, one group disagrees with this conclusion (Williams and Johnston, 1989; Johnston et al., 1992), arguing instead that mossy fiber LTP can be triggered by postsynaptic Ca^{2+} increases due to activation of either voltage-dependent Ca^{2+} channels or metabotropic glutamate receptors that release Ca^{2+} from intracellular stores (Yeckel et al., 1999).

Elegant experiments that support the hypothesis that mossy fiber LTP is triggered presynaptically by a tetanus-induced rise in presynaptic Ca^{2+} took advantage of the fact that the synapses formed by associational-commissural fibers on the distal dendrites of CA3 pyramidal cells express a standard form of NMDA receptor–dependent LTP. Recording from a single CA3 cell and monitoring the synaptic responses to both mossy fiber and associational-commissural fiber stimulation, Zalutsky and Nicoll (1990) showed that filling the cell with a Ca^{2+} chelator or clamping the cell at a hyperpolarized membrane potential blocked the LTP at the distal synapses but had no effect on the LTP at mossy fiber synapses. As both manipulations should have a much stronger effect on the proximal mossy fiber synapses, these results provide compelling data in support of the presynaptic triggering of LTP. That a rise in presynaptic Ca^{2+} levels is important for triggering mossy fiber LTP is supported by the finding that blockade of postsynaptic glutamate receptors does not affect the triggering of mossy fiber LTP but that it is blocked by removal of extracellular Ca^{2+} (Castillo et al., 1994).

There is little disagreement that mossy fiber LTP is due to a presynaptic enhancement of transmitter release. The two main lines of evidence in support of this conclusion are that (1) the magnitude of paired pulse facilitation decreases during mossy fiber LTP (Staubli et al., 1990; Zalutsky and Nicoll, 1990; Huang et al., 1994; Xiang et al., 1994) and (2) a large increase in the probability of release, as measured by the use-dependent block of NMDA receptors by MK-801, occurs (Weisskopf and Nicoll, 1995).

How does the transient rise of Ca^{2+} in the presynaptic terminal initiate mossy fiber LTP? Most current evidence suggests that this rise of Ca^{2+} activates a calcium/calmodulin-dependent adenylyl cyclase and that the subsequent rise in cAMP and activation of PKA are required for triggering mossy fiber LTP (see Nicoll and Malenka, 1995, for review). Thus inhibitors of PKA block mossy fiber LTP (Huang et al., 1994; Weisskopf et al., 1994), pharmacological manipulations that increase cAMP levels cause a presynaptic enhancement of mossy fiber synaptic transmission that occludes mossy fiber LTP (Hopkins and Johnston, 1988; Weisskopf et al., 1994), and mice lacking certain isoforms of PKA or the type I adenylyl cyclase (which is calcium-calmodulin dependent) do not express mossy fiber LTP (Huang et al., 1995; Villacres et al., 1998).

The critical substrates of PKA that may mediate mossy fiber LTP are beginning to be identified, and it appears that one important action of PKA is to modify some aspect of the synaptic vesicle cycle or the vesicle release machinery itself. Although modulation of Ca^{2+} channel function during mossy fiber LTP cannot be ruled out, the relative contribution of N- and P-type channels to synaptic transmission does not change during LTP (Castillo et al., 1994), suggesting either that LTP upregulates Ca^{2+} entry through both types of channel to the same extent or that Ca^{2+} channel modulation is not involved. On the other hand, analysis of knockout mice has revealed at least one promising presynaptic candidate and ruled out several others. One prominent set of presynaptic substrates for PKA are the synapsins, but mice lacking both synapsins I and II exhibit normal mossy fiber LTP (Spillane et al., 1995). In contrast, mice lacking the synaptic vesicle protein rab3a do not express this form of LTP (Castillo et al., 1997). Surprisingly, these mice showed apparently normal synaptic enhancement in response to pharmacological activation of adenylyl cyclase. A possible explanation for this apparent discrepancy is that PKA can enhance neurotransmitter release via two independent mechanisms, one involving the modulation of voltage-dependent Ca^{2+} channels and one involving a direct effect on the release machinery, as previously discussed (Lonart et al., 1998). Rab3a interacts with at least two other presynaptic proteins, rabphilin and Rim, both of which are substrates for PKA (see Südhof and Scheller, this volume). The finding that knockout mice lacking rabphilin exhibit normal mossy fiber LTP (Schlüter et al., 1999) suggests that Rim may be one of the key PKA substrates mediating mossy fiber LTP.

In parallel with the progress in understanding NMDA receptor–dependent LTP, we are beginning to understand the molecular details by which presynaptic forms of LTP are generated. However, it remains unclear what role this form of LTP plays and, more generally, why different synapses utilize different forms of LTP. Further elucidation of the molecular mechanisms underlying these different forms of synaptic plasticity should provide important information that will aid in finding the answers to these questions.

A Perspective

Synaptic plasticity is clearly a diverse set of mechanisms that provide for short- to long-term regulation of synaptic transmission. Although much progress has been made in delineating the various forms of pre- and postsynaptic plasticity, a number of significant challenges remain. We still do not have a clear molecular understanding of even the simplest forms of short-term plasticity. In most cases, the physiological and behavioral functions of synaptic plasticity are unknown. And the potential contribution of disorders in plasticity to human genetic and acquired neurological diseases has yet to be explored. When it comes to long-term plasticity in the mammalian brain, our understanding is also far from complete. Controversies surrounding both the molecular and cellular mechanisms of long-term plasticity in the CNS may stem from complexities associated with the cell biological properties of central synapses that do not apply to simpler synapses, such as the neuromuscular junction. Alternatively, the more complex cytoarchitecture of central synapses may introduce certain systematic experimental errors into the analysis of synaptic function. As pointed out in the chapter by Cowan and Kandel, the history of the study of synaptic transmission, starting with Cajal and Golgi, has been beset by scientific controversies that could only be settled after the introduction of more powerful methodologies with sufficient resolution to provide precise, quantitative measurements. How long it will be before such advances resolve the outstanding problems of synaptic plasticity is an open question.

References

Abrams, T. W., Castellucci, V. F., Camardo, J. S., Kandel, E. R., Lloyd, P. E. (1984). Two endogenous neuropeptides modulate the gill and siphon withdrawal reflex in Aplysia by means of presynaptic facilitation involving cyclic AMP–dependent closure of a serotonin-sensitive potassium channel. *Proc. Natl. Acad. Sci. USA* 81:7956–7960.

Adams, J. P., Anderson, A. E., Dinely, K. T., Qian, Y., Pfaffinger, P. J., Sweatt, J. D. (1999). Regulation of K channel phosphorylation in the hippocampus by PKA, MAPK, PKC and CaMKII. *Soc. Neurosci. Abstr.* 25:1312.

Alberini, C. M., Ghirardi, M., Metz, R., Kandel, E. R. (1994). C/EBP is an immediate-early gene required for the consolidation of long-term facilitation in Aplysia. *Cell* 76:1099–1114.

Alford, S., Frenguelli, B. G., Schofield, J. G., Collingridge, G. L. (1993). Characterization of Ca^{2+} signals induced in hippocampal CA1 neurones by the synaptic activation of NMDA receptors. *J. Physiol.* 469:693–716.

Aniksztejn, L., Roisin, M. P., Amsellem, R., Ben-Ari, Y. (1989). Long-term potentiation in the hippocampus of the anaesthetized rat is not associated with a sustained enhanced release of endogenous excitatory amino acids. *Neuroscience* 28:387–392.

Anwyl, R. (1999). Metabotropic glutamate receptors: electrophysiological properties and role in plasticity. *Brain Res. Brain Res. Rev.* 29:83–120.

Arancio, O., Kandel, E. R., Hawkins, R. D. (1995). Activity-dependent long-term enhancement of transmitter release by presynaptic 3′,5′-cyclic GMP in cultured hippocampal neurons. *Nature* 376:74–80.

Arancio, O., Kiebler, M., Lee, C. J., Lev-Ram, V., Tsien, R. Y., Kandel, E. R., Hawkins, R. D. (1996). Nitric oxide acts directly in the presynaptic neuron to produce long-term potentiation in cultured hippocampal neurons. *Cell* 87:1025–1035.

Armitage, B. A., Siegelbaum, S. A. (1998). Presynaptic induction and expression of homosynaptic depression at Aplysia sensorimotor neuron synapses. *J. Neurosci.* 18:8770–8779.

Armstrong, N., Sun, Y., Chen, G. Q., Gouaux, E. (1998). Structure of a glutamate-receptor ligand-binding core in complex with kainate. *Nature* 395:913–917.

Asztely, F., Xiao, M. Y., Gustafsson, B. (1996). Long-term potentiation and paired-pulse facilitation in the hippocampal CA1 region. *NeuroReport* 7:1609–1612.

Augustine, G. J. (1990). Regulation of transmitter release at the squid giant synapse by presynaptic delayed rectifier potassium current. *J. Physiol.* 431: 343–364.

Bailey, C. H., Chen, M. (1988a). Long-term memory in Aplysia modulates the total number of varicosities of single identified sensory neurons. *Proc. Natl. Acad. Sci. USA* 85:2372–2377.

Bailey, C. H., Chen, M. (1988b). Long-term sensitization in Aplysia increases the number of presynaptic contacts onto the identified gill motor neuron L7. *Proc. Natl. Acad. Sci. USA* 85:9356–9359.

Bao, J. X., Kandel, E. R., Hawkins, R. D. (1997). Involvement of pre- and postsynaptic mechanisms in posttetanic potentiation at Aplysia synapses. *Science* 275:969–973.

Barria, A., Derkach, V., Soderling, T. (1997a). Identification of the $Ca^{2+}/$ calmodulin-dependent protein kinase II regulatory phosphorylation site in the alpha-amino-3-hydroxyl-5-methyl-4-isoxazole-propionate-type glutamate receptor. *J. Biol. Chem.* 272:32,727–32,730.

Barria, A., Muller, D., Derkach, V., Griffith, L. C., Soderling, T. R. (1997b). Regulatory phosphorylation of AMPA-type glutamate receptors by CaM-KII during long-term potentiation. *Science* 276:2042–2045.

Bartsch, D., Ghirardi, M., Skehel, P. A., Karl, K. A., Herder, S. P., Chen, M., Bailey, C. H., Kandel, E. R. (1995). Aplysia CREB2 represses long-term facilitation: relief of repression converts transient facilitation into long-term functional and structural change. *Cell* 83:979–992.

Bartsch, D., Casadio, A., Karl, K. A., Serodio, P., Kandel, E. R. (1998). CREB1 encodes a nuclear activator, a repressor, and a cytoplasmic modulator that form a regulatory unit critical for long-term facilitation. *Cell* 95:211–223.

Baxter, D. A., Bittner, G. D. (1991). Synaptic plasticity at crayfish neuromuscular junctions: presynaptic inhibition. *Synapse* 7:244–251.

Baxter, D. A., Byrne, J. H. (1989). Serotonergic modulation of two potassium currents in the pleural sensory neurons of Aplysia. *J. Neurophysiol.* 62:665–679.

Beaumont, V., Zucker, R. S. (2000). Enhancement of synaptic transmission by cyclic AMP modulation of presynaptic Ih channels. *Nature Neurosci.* 3:133–141.

Benke, T. A., Luthi, A., Isaac, J. T., Collingridge, G. L. (1998). Modulation of AMPA receptor unitary conductance by synaptic activity. *Nature* 393:793–797.

Berridge, M. J. (1998). Neuronal calcium signaling. *Neuron* 21:13–26.

Bliss, T. V., Collingridge, G. L. (1993). A synaptic model of memory: long-term potentiation in the hippocampus. *Nature* 361:31–39.

Bliss, T. V., Gardner-Medwin, A. R. (1973). Long-lasting potentiation of synaptic transmission in the dentate area of the unanaesthetized rabbit following stimulation of the perforant path. *J. Physiol.* 232:357–374.

Bliss, T. V., Lomo, T. (1973). Long-lasting potentiation of synaptic transmission in the dentate area of the anaesthetized rabbit following stimulation of the perforant path. *J. Physiol.* 232:331–356.

Bliss, T. V., Douglas, R. M., Errington, M. L., Lynch, M. A. (1986). Correlation between long-term potentiation and release of endogenous amino acids from dentate gyrus of anaesthetized rats. *J. Physiol.* 377:391–408.

Blitzer, R. D., Connor, J. H., Brown, G. P., Wong, T., Shenolikar, S., Iyengar, R., Landau, E. M. (1998). Gating of CaMKII by cAMP-regulated protein phosphatase activity during LTP. *Science* 280:1940–1942.

Blumenfeld, H., Spira, M. E., Kandel, E. R., Siegelbaum, S. A. (1990). Facilitatory and inhibitory transmitters modulate calcium influx during action potentials in Aplysia sensory neurons. *Neuron* 5:487–499.

Böhme, G. A., Bon, C., Stutzmann, J. M., Doble, A., Blanchard, J. C. (1991). Possible involvement of nitric oxide in long-term potentiation. *Eur. J. Pharmacol.* 199:379–381.

Bolshakov, V. Y., Siegelbaum, S. A. (1995). Regulation of hippocampal transmitter release during development and long-term potentiation. *Science* 269:1730–1734.

Bower, J. M., Haberly, L. B. (1986). Facilitating and nonfacilitating synapses on pyramidal cells: a correlation between physiology and morphology. *Proc. Natl. Acad. Sci. USA* 83:1115–1119.

Braha, O., Dale, N., Hochner, B., Klein, M., Abrams, T. W., Kandel, E. R. (1990). Second messengers involved in the two processes of presynaptic facilitation that contribute to sensitization and dishabituation in Aplysia sensory neurons. *Proc. Natl. Acad. Sci. USA* 87:2040–2044.

Braun, A. P., Schulman, H. (1995). The multifunctional calcium/calmodulin-dependent protein kinase: from form to function. *Annu. Rev. Physiol.* 57:417–445.

Byrne, J. H., Kandel, E. R. (1996). Presynaptic facilitation revisited: state and time dependence. *J. Neurosci.* 16:425–435.

Capogna, M. (1998). Presynaptic facilitation of synaptic transmission in the hippocampus. *Pharmacol. Ther.* 77:203–223.

Capogna, M., Gahwiler, B. H., Thompson, S. M. (1995). Presynaptic enhancement of inhibitory synaptic transmission by protein kinases A and C in the rat hippocampus in vitro. *J. Neurosci.* 15:1249–1260.

Capogna, M., Gahwiler, B. H., Thompson, S. M. (1996). Presynaptic inhibition of calcium-dependent and -independent release elicited with ionomycin, gadolinium, and alpha-latrotoxin in the hippocampus. *J. Neurophysiol.* 75: 2017–2028.

Carroll, R. C., Lissin, D. V., von Zastrow, M., Nicoll, R. A., Malenka, R. C. (1999a). Rapid redistribution of glutamate receptors contributes to long-term depression in hippocampal cultures. *Nature Neurosci.* 2:454–460.

Carroll, R. C., Beattie, E. C., Xia, H., Lüscher, C., Altschuler, Y., Nicoll, R. A., Malenka, R. C., Von Zastrow, M. (1999b). Dynamin-dependent endocytosis of ionotropic glutamate receptors. *Proc. Natl. Acad. Sci. USA* 96:14,112–14,117.

Castillo, P. E., Weisskopf, M. G., Nicoll, R. A. (1994). The role of Ca^{2+} channels in hippocampal mossy fiber synaptic transmission and long-term potentiation. *Neuron* 12:261–269.

Castillo, P. E., Janz, R., Südhof, T. C., Tzounopoulos, T., Malenka, R. C., Nicoll, R. A. (1997). Rab3A is essential for mossy fibre long-term potentiation in the hippocampus. *Nature* 388:590–593.

Castro-Alamancos, M. A., Calcagnotto, M. E. (1999). Presynaptic long-term potentiation in corticothalamic synapses. *J. Neurosci.* 19:9090–9097.

Cavus, I., Teyler, T. (1996). Two forms of long-term potentiation in area CA1 activate different signal transduction cascades. *J. Neurophysiol.* 76:3038–3047.

Chain, D. G., Casadio, A., Schacher, S., Hegde, A. N., Valbrun, M., Yamamoto, N., Goldberg, A. L., Bartsch, D., Kandel, E. R., Schwartz, J. H. (1999). Mechanisms for generating the autonomous cAMP-dependent protein kinase required for long-term facilitation in Aplysia. *Neuron* 22:147–156.

Chavis, P., Mollard, P., Bockaert, J., Manzoni, O. (1998). Visualization of cyclic AMP–regulated presynaptic activity at cerebellar granule cells. *Neuron* 20: 773–781.

Chen, C., Regehr, W. G. (1997). The mechanism of cAMP-mediated enhancement at a cerebellar synapse. *J. Neurosci.* 17:8687–8694.

Choi, S., Klingauf, J., Tsien, R. W. (2000). Postfusional regulation of cleft glutamate concentration during LTP at "silent synapses." *Nature Neurosci.* 3:330–336.

Clark, K. A., Collingridge, G. L. (1995). Synaptic potentiation of dual-component excitatory postsynaptic currents in the rat hippocampus. *J. Physiol.* 482:39–52.

Cochilla, A. J., Alford, S. (1998). Metabotropic glutamate receptor–mediated control of neurotransmitter release. *Neuron* 20:1007–1016.

Collingridge, G. L., Kehl, S. J., McLennan, H. (1983). The antagonism of amino acid–induced excitations of rat hippocampal CA1 neurones in vitro. *J. Physiol.* 334:19–31.

Dale, N., Kandel, E. R. (1990). Facilitatory and inhibitory transmitters modulate spontaneous transmitter release at cultured Aplysia sensorimotor synapses. *J. Physiol.* 421:203–222.

Dale, N., Schacher, S., Kandel, E. R. (1988). Long-term facilitation in Aplysia involves increase in transmitter release. *Science* 239:282–285.

Dash, P. K., Hochner, B., Kandel, E. R. (1990). Injection of cAMP-responsive element into the nucleus of Aplysia sensory neurons blocks long-term facilitation. *Nature* 345:718–721.

Davies, S. N., Lester, R. A., Reymann, K. G., Collingridge, G. L. (1989). Temporally distinct pre- and post-synaptic mechanisms maintain long-term potentiation. *Nature* 338:500–503.

Delaney, K., Tank, D. W., Zucker, R. S. (1991). Presynaptic calcium and serotonin-mediated enhancement of transmitter release at crayfish neuromuscular junction. *J. Neurosci.* 11:2631–2643.

Derkach, V., Barria, A., Soderling, T. R. (1999). Ca^{2+}/calmodulin-kinase II enhances channel conductance of alpha-amino-3-hydroxy-5-methyl-4-isoxazole-propionate type glutamate receptors. *Proc. Natl. Acad. Sci. USA* 96:3269–3274.

Diamond, J. S., Bergles, D. E., Jahr, C. E. (1998). Glutamate release monitored with astrocyte transporter currents during LTP. *Neuron* 21:425–433.

Dingledine, R., Borges, K., Bowie, D., Traynelis, S. F. (1999). The glutamate receptor ion channels. *Pharmacol. Rev.* 51:7–61.

Dittman, J. S., Regehr, W. G. (1998). Calcium dependence and recovery kinetics of presynaptic depression at the climbing fiber to Purkinje cell synapse. *J. Neurosci.* 18:6147–6162.

Dixon, D., Atwood, H. L. (1989). Adenylate cyclase system is essential for long-term facilitation at the crayfish neuromuscular junction. *J. Neurosci.* 9:4246–4252.

Dobrunz, L. E., Stevens, C. F. (1997). Heterogeneity of release probability, facilitation, and depletion at central synapses. *Neuron* 18:995–1008.

Dobrunz, L. E., Huang, E. P., Stevens, C. F. (1997). Very short-term plasticity in hippocampal synapses. *Proc. Natl. Acad. Sci. USA* 94:14,843–14,847.

Dodt, H., Eder, M., Frick, A., Zieglgänsberger, W. (1999). Precisely localized LTD in the neocortex revealed by infrared-guided laser stimulation. *Science* 286:110–113.

Dolphin, A. C., Errington, M. L., Bliss, T. V. (1982). Long-term potentiation of the perforant path in vivo is associated with increased glutamate release. *Nature* 297:496–498.

Dudel, J., Kuffler, S. W. (1961). Presynaptic inhibition at the crayfish neuromuscular junction. *J. Physiol.* 155:543–562.

Dunlap, K., Fischbach, G. D. (1978). Neurotransmitters decrease the calcium component of sensory neurone action potentials. *Nature* 276:837–839.

Dunwiddie, T. V. (1985). The physiological role of adenosine in the central nervous system. *Int. Rev. Neurobiol.* 27:63–139.

Durand, G. M., Kovalchuk, Y., Konnerth, A. (1996). Long-term potentiation and functional synapse induction in developing hippocampus. *Nature* 381:71–75.

Edmonds, B., Klein, M., Dale, N., Kandel, E. R. (1990). Contributions of two types of calcium channels to synaptic transmission and plasticity. *Science* 250:1142–1147.

Ehlers, M. D., Zhang, S., Bernhadt, J. P., Huganir, R. L. (1996). Inactivation of NMDA receptors by direct interaction of calmodulin with the NR1 subunit. *Cell* 84:745–755.

Eliot, L. S., Kandel, E. R., Siegelbaum, S. A., Blumenfeld, H. (1993). Imaging terminals of Aplysia sensory neurons demonstrates role of enhanced Ca^{2+} influx in presynaptic facilitation. *Nature* 361:634–637.

Eliot, L. S., Kandel, E. R., Hawkins, R. D. (1994). Modulation of spontaneous transmitter release during depression and posttetanic potentiation of Aplysia sensory-motor neuron synapses isolated in culture. *J. Neurosci.* 14:3280–3292.

English, J. D., Sweatt, J. D. (1996). Activation of p42 mitogen-activated protein kinase in hippocampal long term potentiation. *J. Biol. Chem.* 271:24,329–24,332.

English, J. D., Sweatt, J. D. (1997). A requirement for the mitogen-activated protein kinase cascade in hippocampal long term potentiation. *J. Biol. Chem.* 272:19,103–19,106.

Fisher, S. A., Fischer, T. M., Carew, T. J. (1997). Multiple overlapping processes underlying short-term synaptic enhancement. *Trends Neurosci.* 20:170–177.

Forsythe, I. D., Tsujimoto, T., Barnes-Davies, M., Cuttle, M. F., Takahashi, T. (1998). Inactivation of presynaptic calcium current contributes to synaptic depression at a fast central synapse. *Neuron* 20:797–807.

Frost, W. N., Castellucci, V. F., Hawkins, R. D., Kandel, E. R. (1985). Monosynaptic connections made by the sensory neurons of the gill- and siphon-withdrawal reflex in Aplysia participate in the storage of long-term memory for sensitization. *Proc. Natl. Acad. Sci. USA* 82:8266–8269.

Fukunaga, K., Muller, D., Miyamoto, E. (1995). Increased phosphorylation of Ca^{2+}/calmodulin-dependent protein kinase II and its endogenous substrates in the induction of long-term potentiation. *J. Biol. Chem.* 270:6119–6124.

Geppert, M., Bolshakov, V. Y., Siegelbaum, S. A., Takei, K., De Camilli, P., Hammer, R. E., Südhof, T. C. (1994). The role of Rab3A in neurotransmitter release. *Nature* 369:493–497.

Geppert, M., Goda, Y., Stevens, C. F., Südhof, T. C. (1997). The small GTP-binding protein Rab3A regulates a late step in synaptic vesicle fusion. *Nature* 387:810–814.

Ghirardi, M., Braha, O., Hochner, B., Montarolo, P. G., Kandel, E. R., Dale, N. (1992). Roles of PKA and PKC in facilitation of evoked and spontaneous transmitter release at depressed and nondepressed synapses in Aplysia sensory neurons. *Neuron* 9:479–489.

Giese, K. P., Fedorov, N. B., Filipkowski, R. K., Silva, A. J. (1998). Autophosphorylation at Thr286 of the alpha calcium-calmodulin kinase II in LTP and learning. *Science* 279:870–873.

Glass, D. J., Bowen, D. C., Stitt, T. N., Radziejewski, C., Bruno, J., Ryan, T. E., Gies, D. R., Shah, S., Mattsson, K., Burden, S. J., DiStefano, P. S., Valenzuela, D. M., DeChiara, T. M., Yancopoulos, G. D. (1996). Agrin acts via a MuSK receptor complex. *Cell* 85:513–523.

Goldsmith, B. A., Abrams, T. W. (1992). cAMP modulates multiple K^+ currents, increasing spike duration and excitability in Aplysia sensory neurons. *Proc. Natl. Acad. Sci. USA* 89:11,481–11,485.

Gomperts, S. N., Rao, A., Craig, A. M., Malenka, R. C., Nicoll, R. A. (1998). Postsynaptically silent synapses in single neuron cultures. *Neuron* 21:1443–1451.

Grant, S. G., O'Dell, T. J., Karl, K. A., Stein, P. L., Soriano, P., Kandel, E. R. (1992). Impaired long-term potentiation, spatial learning, and hippocampal development in fyn mutant mice. *Science* 258:1903–1910.

Gray, R., Rajan, A. S., Radcliffe, K. A., Yakehiro, M., Dani, J. A. (1996). Hippocampal synaptic transmission enhanced by low concentrations of nicotine. *Nature* 383:713–716.

Greengard, P., Jen, J., Nairn, A. C., Stevens, C. F. (1991). Enhancement of the glutamate response by cAMP-dependent protein kinase in hippocampal neurons. *Science* 253:1135–1138.

Greengard, P., Valtorta, F., Czernik, A. J., Benfenati, F. (1993). Synaptic vesicle phosphoproteins and regulation of synaptic function. *Science* 259:780–785.

Haley, J. E., Schaible, E., Pavlidis, P., Murdock, A., Madison, D. V. (1996). Basal and apical synapses of CA1 pyramidal cells employ different LTP induction mechanisms. *Learn. Mem.* 3:289–295.

Harris, E. W., Cotman, C. W. (1986). Long-term potentiation of guinea pig mossy fiber responses is not blocked by N-methyl-D-aspartate antagonists. *Neurosci. Lett.* 70:132–137.

Hayashi, Y., Shi, S.-H., Esteban, J. A., Piccini, A., Poncer, J.-C., and Malinow, R. (2000). Driving AMPA receptors into synapses by LTP and CaMKII: requirements for GluR1 and PDZ domain interactions. *Science* 287:2262–2267.

Hegde, A. N., Inokuchi, K., Pei, W., Casadio, A., Ghirardi, M., Chain, D. G., Martin, K. C., Kandel, E. R., Schwartz, J. H. (1997). Ubiquitin C-terminal hydrolase is an immediate-early gene essential for long-term facilitation in Aplysia. *Cell* 89:115–126.

Hessler, N. A., Shirke, A. M., Malinow, R. (1993). The probability of transmitter release at a mammalian central synapse. *Nature* 366:569–572.

Hinds, H. L., Tonegawa, S., Malinow, R. (1998). CA1 long-term potentiation is diminished but present in hippocampal slices from alpha-CaMKII mutant mice. *Learn. Mem.* 5:344–354.

Hirling, H., Scheller, R. H. (1996). Phosphorylation of synaptic vesicle proteins: modulation of the alpha SNAP interaction with the core complex. *Proc. Natl. Acad. Sci. USA* 93:11,945–11,949.

Hirning, L. D., Fox, A. P., McCleskey, E. W., Olivera, B. M., Thayer, S. A., Miller, R. J., Tsien, R. W. (1988). Dominant role of N-type Ca^{2+} channels in evoked release of norepinephrine from sympathetic neurons. *Science* 239:57–61.

Hjelmstad, G. O., Nicoll, R. A., Malenka, R. C. (1997). Synaptic refractory period provides a measure of probability of release in the hippocampus. *Neuron* 19:1309–1318.

Hochner, B., Kandel, E. R. (1992). Modulation of a transient K^+ current in the pleural sensory neurons of Aplysia by serotonin and cAMP: implications for spike broadening. *Proc. Natl. Acad. Sci. USA* 89:11,476–11,480.

Hochner, B., Klein, M., Schacher, S., Kandel, E. R. (1986a). Action potential duration and the modulation of transmitter release from the sensory neurons of Aplysia in presynaptic facilitation and behavioral sensitization. *Proc. Natl. Acad. Sci. USA* 83:8410–8414.

Hochner, B., Klein, M., Schacher, S., Kandel, E. R. (1986b). Additional component in the cellular mechanism of presynaptic facilitation contributes to behavioral dishabituation in Aplysia. *Proc. Natl. Acad. Sci. USA* 83:8794–8798.

Hollmann, M., Heinemann, S. (1994). Cloned glutamate receptors. *Annu. Rev. Neurosci.* 17:31–108.

Hollmann, M., Maron, C., Heinemann, S. (1994). N-glycosylation site tagging suggests a three transmembrane domain topology for the glutamate receptor GluR1. *Neuron* 13:1331–1343.

Hopfield, J. F., Tank, D. W., Greengard, P., Huganir, R. L. (1988). Functional modulation of the nicotinic acetylcholine receptor by tyrosine phosphorylation. *Nature* 336:677–680.

Hopkins, W. F., Johnston, D. (1988). Noradrenergic enhancement of long-term potentiation at mossy fiber synapses in the hippocampus. *J. Neurophysiol.* 59: 667–687.

Hosaka, M., Hammer, R. E., Südhof, T. C. (1999). A phospho-switch controls the dynamic association of synapsins with synaptic vesicles. *Neuron* 24:377–387.

Hu, G. Y., Hvalby, O., Walaas, S. I., Albert, K. A., Skjeflo, P., Andersen, P., Greengard, P. (1987). Protein kinase C injection into hippocampal pyramidal cells elicits features of long term potentiation. *Nature* 328:426–429.

Huang, Y. Y., Li, X. C., Kandel, E. R. (1994). cAMP contributes to mossy fiber LTP by initiating both a covalently mediated early phase and macromolecular synthesis–dependent late phase. *Cell* 79:69–79.

Huang, Y. Y., Kandel, E. R., Varshavsky, L., Brandon, E. P., Qi, M., Idzerda, R. L., McKnight, G. S., Bourtchouladze, R. (1995). A genetic test of the effects of mutations in PKA on mossy fiber LTP and its relation to spatial and contextual learning. *Cell* 83:1211–1222.

Isaac, J. T., Nicoll, R. A., Malenka, R. C. (1995). Evidence for silent synapses: implications for the expression of LTP. *Neuron* 15:427–434.

Isaac, J. T., Hjelmstad, G. O., Nicoll, R. A., Malenka, R. C. (1996). Long-term potentiation at single fiber inputs to hippocampal CA1 pyramidal cells. *Proc. Natl. Acad. Sci. USA* 93:8710–8715.

Isaacson, J. S., Hille, B. (1997). GABA(B)-mediated presynaptic inhibition of excitatory transmission and synaptic vesicle dynamics in cultured hippocampal neurons. *Neuron* 18:143–152.

Ito, I., Sugiyama, H. (1991). Roles of glutamate receptors in long-term potentiation at hippocampal mossy fiber synapses. *NeuroReport* 2:333–336.

Jackson, M. B., Konnerth, A., Augustine, G. J. (1991). Action potential broadening and frequency-dependent facilitation of calcium signals in pituitary nerve terminals. *Proc. Natl. Acad. Sci. USA* 88:380–384.

Jay, T. M., Zilkha, E., Obrenovitch, T. P. (1999). Long-term potentiation in the dentate gyrus is not linked to increased extracellular glutamate concentration. *J. Neurophysiol.* 81:1741–1748.

Johnston, D., Williams, S., Jaffe, D., Gray, R. (1992). NMDA-receptor-independent long-term potentiation. *Annu. Rev. Physiol.* 54:489–505.

Kamiya, H., Zucker, R. S. (1994). Residual Ca^{2+} and short-term synaptic plasticity. *Nature* 371:603–606.

Kandel, E. R., Schwartz, J. H. (1982). Molecular biology of learning: modulation of transmitter release. *Science* 218:433–443.

Kandler, K., Katz, L. C., Kauer, J. A. (1998). Focal photolysis of caged glutamate produces long-term depression of hippocampal glutamate receptors. *Nature Neurosci.* 1:119–123.

Karlin, A., Akabas, M. H. (1995). Toward a structural basis for the function of nicotinic acetylcholine receptors and their cousins. *Neuron* 15:1231–1244.

Kato, K., Zorumski, C. F. (1996). Platelet-activating factor as a potential retrograde messenger. *J. Lipid Med. Cell Signal.* 14:341–348.

Katsuki, H., Kaneko, S., Tajima, A., Satoh, M. (1991). Separate mechanisms of long-term potentiation in two input systems to CA3 pyramidal neurons of rat hippocampal slices as revealed by the whole-cell patch-clamp technique. *Neurosci. Res.* 12:393–402.

Katz, B., Miledi, R. (1968). The role of calcium in neuromuscular facilitation. *J. Physiol.* 195:481–492.

Kauer, J. A., Malenka, R. C., Nicoll, R. A. (1988). A persistent postsynaptic modification mediates long-term potentiation in the hippocampus. *Neuron* 1:911–917.

Kennedy, M. B., Bennett, M. K., Bulleit, R. F., Erondu, N. E., Jennings, V. R., Miller, S. G., Molloy, S. S., Patton, B. L., Schenker, L. J. (1990). Structure and regulation of type II calcium/calmodulin-dependent protein kinase in central nervous system neurons. *Cold Spring Harbor Symp. Quant. Biol.* 55: 101–110.

Klann, E., Chen, S. J., Sweatt, J. D. (1991). Persistent protein kinase activation in the maintenance phase of long-term potentiation. *J. Biol. Chem.* 266:24,253–24,256.

Klein, M. (1994). Synaptic augmentation by 5-HT at rested Aplysia sensorimotor synapses: independence of action potential prolongation. *Neuron* 13: 159–166.

Klein, M., Kandel, E. R. (1978). Presynaptic modulation of voltage-dependent Ca^{2+} current: mechanism for behavioral sensitization in *Aplysia californica*. *Proc. Natl. Acad. Sci. USA* 75:3512–3516.

Klein, M., Kandel, E. R. (1980). Mechanism of calcium current modulation underlying presynaptic facilitation and behavioral sensitization in Aplysia. *Proc. Natl. Acad. Sci. USA* 77:6912–6916.

Klein, M., Camardo, J. S., Kandel, E. R. (1982). Serotonin modulates a specific potassium current in the sensory neurons that show presynaptic facilitation in Aplysia. *Proc. Natl. Acad. Sci. USA* 79:5713–5717.

Kleppisch, T., Pfeifer, A., Klatt, P., Ruth, P., Montkowski, A., Fässler, R., Hofmann, F. (1999). Long-term potentiation in the hippocampal CA1 region of mice lacking cGMP-dependent kinases is normal and susceptible to inhibition of nitric oxide synthase. *J. Neurosci.* 19:48–55.

Knapp, A. G., Dowling, J. E. (1987). Dopamine enhances excitatory amino acid–gated conductances in cultured retinal horizontal cells. *Nature* 325: 437–439.

Kretz, R., Shapiro, E., Bailey, C. H., Chen, M., Kandel, E. R. (1986a). Presynaptic inhibition produced by an identified presynaptic inhibitory neuron. II. Presynaptic conductance changes caused by histamine. *J. Neurophysiol.* 55:131–146.

Kretz, R., Shapiro, E., Kandel, E. R. (1986b). Presynaptic inhibition produced by an identified presynaptic inhibitory neuron. I. Physiological mechanisms. *J. Neurophysiol.* 55:113–130.

Krishek, B. J., Xie, X., Blackstone, C., Huganir, R. L., Moss, S. J., Smart, T. G. (1994). Regulation of $GABA_A$ receptor function by protein kinase C phosphorylation. *Neuron* 12:1081–1095.

Kuhnt, U., Voronin, L. L. (1994). Interaction between paired-pulse facilitation and long-term potentiation in area CA1 of guinea-pig hippocampal slices: application of quantal analysis. *Neuroscience* 62:391–397.

Kullmann, D. M. (1994). Amplitude fluctuations of dual-component EPSCs in hippocampal pyramidal cells: implications for long-term potentiation. *Neuron* 12:1111–1120.

Kullmann, D. M., Siegelbaum, S. A. (1995). The site of expression of NMDA receptor-dependent LTP: new fuel for an old fire. *Neuron* 15:997–1002.

Kullmann, D. M., Erdemli, G., Asztely, F. (1996). LTP of AMPA and NMDA receptor–mediated signals: evidence for presynaptic expression and extrasynaptic glutamate spill-over. *Neuron* 17:461–474.

Langdon, R. B., Johnson, J. W., Barrionuevo, G. (1995). Posttetanic potentiation and presynaptically induced long-term potentiation at the mossy fiber synapse in rat hippocampus. *J. Neurobiol.* 26:370–385.

Langer, S. Z. (1997). 25 years since the discovery of presynaptic receptors: present knowledge and future perspectives. *Trends Pharmacol. Sci.* 18:95–99.

Lee, H.-K., Huganir, R. L. (1999). Phosphorylation of glutamate receptors. In *Handbook of Experimental Pharmacology*, Vol. 141, Monyer, H., Jonas, P., eds. Berlin: Springer-Verlag. pp. 99–119.

Liao, D., Hessler, N. A., Malinow, R. (1995). Activation of postsynaptically silent synapses during pairing-induced LTP in CA1 region of hippocampal slice. *Nature* 375:400–404.

Liao, D., Zhang, X., O'Brien, R., Ehlers, M. D., Huganir, R. L. (1999). Regulation of morphological postsynaptic silent synapses in developing hippocampal neurons. *Nature Neurosci.* 2:37–43.

Liley, A. W., North, K. A. K. (1953). An electrical investigation of effects of repetitive stimulation on mammalian neuromuscular junction. *J. Neurophysiol.* 16: 509–527.

Liman, E. R., Knapp, A. G., Dowling, J. E. (1989). Enhancement of kainate-gated currents in retinal horizontal cells by cyclic AMP–dependent protein kinase. *Brain Res.* 481:399–402.

Lin, J. W., Sugimori, M., Llinas, R. R., McGuinness, T. L., Greengard, P. (1990). Effects of synapsin I and calcium/calmodulin-dependent protein kinase II on spontaneous neurotransmitter release in the squid giant synapse. *Proc. Natl. Acad. Sci. USA* 87:8257–8261.

Linden, D. J., Routtenberg, A. (1989). The role of protein kinase C in long-term potentiation: a testable model. *Brain. Res. Brain. Res. Rev.* 14:279–296.

Lipscombe, D., Kongsamut, S., Tsien, R. W. (1989). Alpha-adrenergic inhibition of sympathetic neurotransmitter release mediated by modulation of N-type calcium-channel gating. *Nature* 340:639–642.

Lisman, J., Malenka, R. C., Nicoll, R. A., Malinow, R. (1997). Learning mechanisms: the case for CaM-KII. *Science* 276:2001–2002.

Lledo, P. M., Hjelmstad, G. O., Mukherji, S., Soderling, T. R., Malenka, R. C., Nicoll, R. A. (1995). Calcium/calmodulin-dependent kinase II and long-term potentiation enhance synaptic transmission by the same mechanism. *Proc. Natl. Acad. Sci. USA* 92:11,175–11,179.

Lledo, P. M., Zhang, X., Südhof, T. C., Malenka, R. C., Nicoll, R. A. (1998). Postsynaptic membrane fusion and long-term potentiation. *Science* 279:399–403.

Llinas, R., McGuinness, T. L., Leonard, C. S., Sugimori, M., Greengard, P. (1985). Intraterminal injection of synapsin I or calcium/calmodulin-dependent protein kinase II alters neurotransmitter release at the squid giant synapse. *Proc. Natl. Acad. Sci. USA* 82:3035–3039.

Lonart, G., Janz, R., Johnson, K. M., Südhof, T. C. (1998). Mechanism of action of rab3A in mossy fiber LTP. *Neuron* 21:1141–1150.

Lu, Y. M., Roder, J. C., Davidow, J., Salter, M. W. (1998). Src activation in the induction of long-term potentiation in CA1 hippocampal neurons. *Science* 279:1363–1367.

Lüscher, C., Jan, L. Y., Stoffel, M., Malenka, R. C., Nicoll, R. A. (1997). G-protein-coupled inwardly rectifying K$^+$ channels (GIRKS) mediate postsynaptic but not presynaptic transmitter actions in hippocampal neurons. *Neuron* 19:687–695.

Lüscher, C., Malenka, R. C., Nicoll, R. A. (1998). Monitoring glutamate release during LTP with glial transporter currents. *Neuron* 21:435–441.

Lüscher, C., Xia, H., Beattie, E., Carroll, R., Von Zastrow, M., Malenka, R. C., Nicoll, R. A. (1999). Role of AMPA receptor cycling in synaptic transmission and plasticity. *Neuron* 24:649–658.

Lynch, G., Larson, J., Kelso, S., Barrionuevo, G., Schottler, F. (1983). Intracellular injections of EGTA block induction of hippocampal long-term potentiation. *Nature* 305:719–721.

Ma, L., Zablow, L., Kandel, E. R., Siegelbaum, S. A. (1999). Cyclic AMP induces functional presynaptic boutons in hippocampal CA3–CA1 neuronal cultures. *Nature Neurosci.* 2:24–30.

MacDermott, A. B., Mayer, M. L., Westbrook, G. L., Smith, S. J., Barker, J. L. (1986). NMDA-receptor activation increases cytoplasmic calcium concentration in cultured spinal cord neurones. *Nature* 321:519–522.

MacDermott, A. B., Role, L. W., Siegelbaum, S. A. (1999). Presynaptic ionotropic receptors and the control of transmitter release. *Annu. Rev. Neurosci.* 22:443–485.

MacDonald, J. F., Xiong, X. G., Lu, W. Y., Raouf, R., Orser, B. A. (1998). Modulation of NMDA receptors. *Prog. Brain. Res.* 116:191–208.

Magleby, K. L., Zengel, J. E. (1982). A quantitative description of stimulation-induced changes in transmitter release at the frog neuromuscular junction. *J. Gen. Physiol.* 30:613–638.

Mainen, Z. F., Jia, Z., Roder, J., Malinow, R. (1998). Use-dependent AMPA receptor block in mice lacking GluR2 suggests postsynaptic site for LTP expression. *Nature Neurosci.* 1:579–586.

Makhinson, M., Chotiner, J. K., Watson, J. B., O'Dell, T. J. (1999). Adenylyl cyclase activation modulates activity-dependent changes in synaptic strength and Ca^{2+}/calmodulin-dependent kinase II autophosphorylation. *J. Neurosci.* 19:2500–2510.

Malenka, R. C., Nicoll, R. A. (1993). NMDA-receptor-dependent synaptic plasticity: multiple forms and mechanisms. *Trends Neurosci.* 16:521–527.

Malenka, R. C., Nicoll, R. A. (1999). Long-term potentiation—a decade of progress? *Science* 285:1870–1874.

Malenka, R. C., Kauer, J. A., Zucker, R. S., Nicoll, R. A. (1988). Postsynaptic calcium is sufficient for potentiation of hippocampal synaptic transmission. *Science* 242:81–84.

Malenka, R. C., Kauer, J. A., Perkel, D. J., Mauk, M. D., Kelly, P. T., Nicoll, R. A., Waxham, M. N. (1989). An essential role for postsynaptic calmodulin and protein kinase activity in long-term potentiation. *Nature* 340:554–557.

Malenka, R. C., Lancaster, B., Zucker, R. S. (1992). Temporal limits on the rise in postsynaptic calcium required for the induction of long-term potentiation. *Neuron* 9:121–128.

Malgaroli, A., Ting, A. E., Wendland, B., Bergamaschi, A., Villa, A., Tsien, R. W., Scheller, R. H. (1995). Presynaptic component of long-term potentiation visualized at individual hippocampal synapses. *Science* 268:1624–1628.

Malinow, R., Mainen, Z. F. (1996). Long-term potentiation in the CA1 hippocampus. *Science* 271:1604–1606.

Malinow, R., Schulman, H., Tsien, R. W. (1989). Inhibition of postsynaptic PKC or CaMKII blocks induction but not expression of LTP. *Science* 245:862–866.

Mammen, A. L., Kameyama, K., Roche, K. W., Huganir, R. L. (1997). Phosphorylation of the alpha-amino-3-hydroxy-5-methylisoxazole-4-propionic acid receptor GluR1 subunit by calcium/calmodulin-dependent kinase II. *J. Biol. Chem.* 272:32,528–32,533.

Manabe, T., Nicoll, R. A. (1994). Long-term potentiation: evidence against an increase in transmitter release probability in the CA1 region of the hippocampus. *Nature* 265:1888–1892.

Manabe, T., Renner, P., Nicoll, R. A. (1992). Postsynaptic contribution to long-term potentiation revealed by the analysis of miniature synaptic currents. *Nature* 355:50–55.

Manabe, T., Wyllie, D. J., Perkel, D. J., Nicoll, R. A. (1993). Modulation of synaptic transmission and long-term potentiation: effects on paired pulse facilitation and EPSC variance in the CA1 region of the hippocampus. *J. Neurophysiol.* 70:1451–1459.

Man-Son-Hing, H., Zoran, M. J., Lukowiak, K., Haydon, P. G. (1989). A neuromodulator of synaptic transmission acts on the secretory apparatus as well as on ion channels. *Nature* 341:237–239.

Mayer, M. L., Westbrook, G. L., Guthrie, P. B. (1984). Voltage-dependent block by Mg^{2+} of NMDA responses in spinal cord neurones. *Nature* 309:261–263.

Mayford, M., Wang, J., Kandel, E. R., O'Dell, T. J. (1995). CaMKII regulates the frequency-response function of hippocampal synapses for the production of both LTD and LTP. *Cell* 81:891–904.

Mayford, M., Bach, M. E., Huang, Y. Y., Wang, L., Hawkins, R. D., Kandel, E. R. (1996). Control of memory formation through regulated expression of a CaMKII transgene. *Science* 274:1678–1683.

McGehee, D. S., Heath, M. J. S., Gelber, S., Devay, P., Role, L. W. (1995). Nicotine enhancement of fast excitatory synaptic transmission in CNS by presynaptic receptors. *Science* 269:1692–1696.

McGlade-McCulloh, E., Yamamoto, H., Tan, S.-E., Brickey, D. A., Soderling, T. R. (1993). Phosphorylation and regulation of glutamate receptors by calcium/calmodulin-dependent protein kinase II. *Nature* 362:640–642.

Middleton, P., Jaramillo, F., Schuetze, S. (1986). Forskolin increases the rate of acetylcholine receptor desensitization at rat soleus endplates. *Proc. Natl. Acad. Sci. USA* 83:4967–4971.

Miles, K., Greengard, P., Huganir, R. L. (1989). Calcitonin gene–related peptide regulates phosphorylation of the nicotinic acetylcholine receptor in rat myotubes. *Neuron* 2:1517–1524.

Miles, K., Audigier, S. S., Greengard, P., Huganir, R. L. (1994). Autoregulation of phosphorylation of the nicotinic acetylcholine receptor. *J. Neurosci.* 14:3271–3279.

Miller, R. J. (1998). Presynaptic receptors. *Annu. Rev. Pharmacol. Toxicol.* 38:201–227.

Mitchell, J. B., Lupica, C. R., Dunwiddie, T. V. (1993). Activity-dependent release of endogenous adenosine modulates synaptic responses in the rat hippocampus. *J. Neurosci.* 13:3439–3447.

Moss, S. J., Smart, T. G., Blackstone, C. D., Huganir, R. L. (1992). Functional modulation of $GABA_A$ receptors by cAMP-dependent protein phosphorylation. *Science* 257:661–665.

Moss, S. J., Gorrie, G. H., Amato, A., Smart, T. G. (1995). Modulation of $GABA_A$ receptors by tyrosine phosphorylation. *Nature* 377:344–348.

Mudge, A. W., Leeman, S. E, Fischbach, G. D. (1979). Enkephalin inhibits release of substance P from sensory neurons in culture and decreases action potential duration. *Proc. Natl. Acad. Sci. USA* 76:526–530.

Mulle, C., Benoit, P., Pinset, C., Roa, M., Changeux, J.-P. (1988). Calcitonin gene–related peptide enhances the rate of desensitization of the nicotinic acetylcholine receptor in cultured mouse muscle cells. *Proc. Natl. Acad. Sci. USA* 85:5728–5732.

Muller, D., Lynch, G. (1989). Evidence that changes in presynaptic calcium currents are not responsible for long-term potentiation in hippocampus. *Brain Res.* 479:290–299.

Muller, D., Joly, M., Lynch, G. (1988). Contributions of quisqualate and NMDA receptors to the induction and expression of LTP. *Science* 242:1694–1697.

Murphy, K. P., Williams, J. H., Bettache, N., Bliss, T. V. (1994). Photolytic release of nitric oxide modulates NMDA receptor–mediated transmission but does not induce long-term potentiation at hippocampal synapses. *Neuropharmacology* 33:1375–1385.

Nicoll, R. A., Malenka, R. C. (1995). Contrasting properties of two forms of long-term potentiation in the hippocampus. *Nature* 377:115–118.

Nicoll, R. A., Kauer, J. A., Malenka, R. C. (1988). The current excitement in long-term potentiation. *Neuron* 1:97–103.

Nishimune, A., Isaac, J. T., Molnar, E., Noel, J., Nash, S. R., Tagaya, M., Collingridge, G. L., Nakanishi, S., Henley, J. M. (1998). NSF binding to GluR2 regulates synaptic transmission. *Neuron* 21:87–97.

Nowak, L., Bregestovski, P., Ascher, P., Herbet, A., Prochiantz, A. (1984). Magnesium gates glutamate-activated channels in mouse central neurones. *Nature* 307:462–465.

Nusser, Z., Lujan, R., Laube, G., Roberts, J. D., Molnar, E., Somogyi, P. (1998). Cell type and pathway dependence of synaptic AMPA receptor number and variability in the hippocampus. *Neuron* 21:545–559.

O'Dell, T. J., Hawkins, R. D., Kandel, E. R., Arancio, O. (1991). Tests of the roles of two diffusible substances in long-term potentiation: evidence for nitric oxide as a possible early retrograde messenger. *Proc. Natl. Acad. Sci. USA* 88: 11,285–11,289.

Oliet, S. H., Malenka, R. C., Nicoll, R. A. (1996). Bidirectional control of quantal size by synaptic activity in the hippocampus. *Science* 271:1294–1297.

Osten, P., Srivastava, S., Inman, G. J., Vilim, F. S., Khatri, L., Lee, L. M., States, B. A., Einheber, S., Milner, T. A., Hanson, P. I., Ziff, E. B. (1998). The AMPA receptor GluR2 C terminus can mediate a reversible, ATP-dependent interaction with NSF and alpha- and beta-SNAPs. *Neuron* 21:99–110.

Pananceau, M., Chen, H., Gustafsson, B. (1998). Short-term facilitation evoked during brief afferent tetani is not altered by long-term potentiation in the guinea-pig hippocampal CA1 region. *J. Physiol.* 508:503–514.

Paradiso, K., Brehm, P. (1998). Long-term desensitization of nicotinic acetylcholine receptors is regulated via protein kinase A–mediated phosphorylation. *J. Neurosci.* 18:9227–9237.

Parfitt, K. D., Madison, D. V. (1993). Phorbol esters enhance synaptic transmission by a presynaptic, calcium-dependent mechanism in rat hippocampus. *J. Physiol.* 471:245–268.

Parfitt, K. D., Hoffer, B. J., Bickford-Wimer, P. C. (1990). Potentiation of gamma-aminobutyric acid–mediated inhibition by isoproterenol in the cerebellar cortex: receptor specificity. *Neuropharmacology* 29:909–916.

Perkel, D. J., Nicoll, R. A. (1993). Evidence for all-or-none regulation of neurotransmitter release: implications for long-term potentiation. *J. Physiol.* 471: 481–500.

Perkel, D. J., Petrozzino, J. J., Nicoll, R. A., Connor, J. A. (1993). The role of Ca^{2+} entry via synaptically activated NMDA receptors in the induction of long-term potentiation. *Neuron* 11:817–823.

Petralia, R. S., Esteban, J. A., Wang, Y. X., Partridge, J. G., Zhao, H. M., Wenthold, R. J., Malinow, R. (1999). Selective acquisition of AMPA receptors over postnatal development suggests a molecular basis for silent synapses. *Nature Neurosci.* 2:31–36.

Pettit, D. L., Perlman, S., Malinow, R. (1994). Potentiated transmission and prevention of further LTP by increased CaMKII activity in postsynaptic hippocampal slice neurons. *Science* 266:1881–1885.

Pierce, J. P., Lewin, G. R. (1994). An ultrastructural size principle. *Neuroscience* 58:441–446.

Qian, J., Saggau, P. (1999). Modulation of transmitter release by action potential duration at the hippocampal CA3–CA1 synapse. *J. Neurophysiol.* 81:288–298.

Raman, I. M., Tong, G., Jahr, C. E. (1996). Beta-adrenergic regulation of synaptic NMDA receptors by cAMP-dependent protein kinase. *Neuron* 16:415–421.

Rao, A., Craig, A. M. (1997). Activity regulates the synaptic localization of the NMDA receptor in hippocampal neurons. *Neuron* 19:801–812.

Raymond, L. A., Blackstone, C. D., Huganir, R. L. (1993). Phosphorylation and modulation of recombinant GluR6 glutamate receptors by cAMP-dependent protein kinase. *Nature* 361:637–641.

Regehr, W. G., Tank, D. W. (1990). Postsynaptic NMDA receptor–mediated calcium accumulation in hippocampal CA1 pyramidal cell dendrites. *Nature* 345: 807–810.

Roche, K. W., O'Brien, R. J., Mammen, A. L., Bernhardt, J., Huganir, R. L. (1996). Characterization of multiple phosphorylation sites on the AMPA receptor GluR1 subunit. *Neuron* 16:1179–1188.

Rosahl, T. W., Spillane, D., Missler, M., Herz, J., Selig, D. K., Wolff, J. R., Hammer, R. E., Malenka, R. C., Südhof, T. C. (1995). Essential functions of synapsins I and II in synaptic vesicle regulation. *Nature* 375:488–493.

Rosenmund, C., Clements, J. D., Westbrook, G. L. (1993). Nonuniform probability of glutamate release at a hippocampal synapse. *Science* 262:754–757.

Rosenmund, C., Carr, D. W., Bergeson, S. E., Nilaver, G., Scott, J. D., Westbrook, G. L. (1994). Anchoring of protein kinase A is required for modulation of AMPA/kainate receptors on hippocampal neurons. *Nature* 368:853–856.

Ross, A., Rapuano, M., Prives, J. (1988). Induction of phosphorylation and cell surface redistribution of acetylcholine receptors by phorbol ester and carbamylcholine in cultured chick muscle cells. *J. Cell. Biol.* 107:1139–1145.

Ryan, T. A., Ziv, N. E., Smith, S. J. (1996). Potentiation of evoked vesicle turnover at individually resolved synaptic boutons. *Neuron* 17:125–134.

Sacktor, T. C., Schwartz, J. H. (1990). Sensitizing stimuli cause translocation of protein kinase C in Aplysia sensory neurons. *Proc. Natl. Acad. Sci. USA* 87: 2036–2039.

Sacktor, T. C., Osten, P., Valsamis, H., Jiang, X., Naik, M. U., Sublette, E. (1993). Persistent activation of the zeta isoform of protein kinase C in the maintenance of long-term potentiation. *Proc. Natl. Acad. Sci. USA* 90:8342–8346.

Salin, P. A., Malenka, R. C., Nicoll, R. A. (1996). Cyclic AMP mediates a presynaptic form of LTP at cerebellar parallel fiber synapses. *Neuron* 16:797–803.

Scanziani, M., Salin, P. A., Vogt, K. E., Malenka, R. C., Nicoll, R. A. (1997). Use-dependent increases in glutamate concentration activate presynaptic metabotropic glutamate receptors. *Nature* 385:630–634.

Schikorski, T., Stevens, C. F. (1997). Quantitative ultrastructural analysis of hippocampal excitatory synapses. *J. Neurosci.* 17:5858–5867.

Schlüter, O. M., Schnell, E., Verhage, M., Tzonopoulos, T., Nicoll, R. A., Janz, R., Malenka, R. C., Geppert, M., Südhof, T. C. (1999). Rabphilin knock-out mice reveal that rabphilin is not required for rab3 function in regulating neurotransmitter release. *J. Neurosci.* 19:5834–5846.

Scholz, K. P., Miller, R. J. (1992). Inhibition of quantal transmitter release in the absence of calcium influx by a G protein–linked adenosine receptor at hippocampal synapses. *Neuron* 8:1139–1150.

Schulz, P. E., Cook, E. P., Johnston, D. (1994). Changes in paired-pulse facilitation suggest presynaptic involvement in long-term potentiation. *J. Neurosci.* 14:5325–5337.

Schuman, E. M., Madison, D. V. (1991). A requirement for the intercellular messenger nitric oxide in long-term potentiation. *Science* 254:1503–1506.

Schuman, E. M., Madison, D. V. (1994). Nitric oxide and synaptic function. *Annu. Rev. Neurosci.* 17:153–183.

Selig, D. K., Segal, M. R., Liao, D., Malenka, R. C., Malinow, R., Nicoll, R. A., Lisman, J. E. (1996). Examination of the role of cGMP in long-term potentiation in the CA1 region of the hippocampus. *Learn. Mem.* 3:42–48.

Selig, D. K., Nicoll, R. A., Malenka, R. C. (1999). Hippocampal long-term potentiation preserves the fidelity of postsynaptic responses to presynaptic bursts. *J. Neurosci.* 19:1236–1246.

Shapiro, E., Castellucci, V. F., Kandel, E. R. (1980). Presynaptic inhibition in Aplysia involves a decrease in the Ca^{2+} current of the presynaptic neuron. *Proc. Natl. Acad. Sci. USA* 77:1185–1189.

Shenolikar, S., Nairn, A. C. (1991). Protein phosphatases: recent progress. *Adv. Second Mess. Phosphoprotein Res.* 23:1–121.

Shi, S. H., Hayashi, Y., Petralia, R. S., Zaman, S. H., Wenthold, R. J., Svoboda, K., Malinow, R. (1999). Rapid spine delivery and redistribution of AMPA receptors after synaptic NMDA receptor activation. *Science* 284:1811–1816.

Shimazaki, Y., Nishiki, T., Omori, A., Sekiguchi, M., Kamata, Y., Kozaki, S., Takahashi, M. (1996). Phosphorylation of 25-kDa synaptosome-associated protein. Possible involvement in protein kinase C–mediated regulation of neurotransmitter release. *J. Biol. Chem.* 271:14,548–14,553.

Shuster, M. J., Camardo, J. S., Siegelbaum, S. A., Kandel, E. R. (1985). Cyclic AMP–dependent protein kinase closes the serotonin-sensitive K+ channels of Aplysia sensory neurones in cell-free membrane patches. *Nature* 313:392–395.

Siegelbaum, S., Camardo, J. S., Kandel, E. R. (1982). Serotonin and cAMP close single K+ channels in Aplysia sensory neurones. *Nature* 299:413–417.

Silva, A. J., Wang, Y., Paylor, R., Wehner, J. M., Stevens, C. F., Tonegawa, S. (1992). Alpha calcium/calmodulin kinase II mutant mice: deficient long-term potentiation and impaired spatial learning. *Cold Spring Harbor Symp. Quant. Biol.* 57:527–539.

Smart, T. G. (1997). Regulation of excitatory and inhibitory neurotransmitter-gated ion channels by protein phosphorylation. *Curr. Opin. Neurobiol.* 7:358–367.

Son, H., Hawkins, R. D., Martin, K., Kiebler, M., Huang, P. L., Fishman, M. C., Kandel, E. R. (1996). Long-term potentiation is reduced in mice that are doubly mutant in endothelial and neuronal nitric oxide synthase. *Cell* 87:1015–1023.

Song, I., Kamboj, S., Xia, J., Dong, H., Liao, D., Huganir, R. L. (1998). Interaction of the N-ethylmaleimide-sensitive factor with AMPA receptors. *Neuron* 21:393–400.

Spillane, D. M., Rosahl, T. W., Südhof, T. C., Malenka, R. C. (1995). Long-term potentiation in mice lacking synapsins. *Neuropharmacology* 34:1573–1579.

Stanley, E. F., Mirotznik, R. R. (1997). Cleavage of syntaxin prevents G-protein regulation of presynaptic calcium channels. *Nature* 385:340–343.

Stark, L. L., Carew, T. J. (1999). Developmental dissociation of serotonin-induced spike broadening and synaptic facilitation in Aplysia sensory neurons. *J. Neurosci.* 19:334–346.

Staubli, U., Larson, J., Lynch, G. (1990). Mossy fiber potentiation and long-term potentiation involve different expression mechanisms. *Synapse* 5:333–335.

Stevens, C. F., Tsujimoto, T. (1995). Estimates for the pool size of releasable quanta at a single central synapse and for the time required to refill the pool. *Proc. Natl. Acad. Sci. USA* 92:846–849.

Stevens, C. F., Wang, Y. (1993). Reversal of long-term potentiation by inhibitors of haem oxygenase. *Nature* 364:147–149.

Stevens, C. F., Wang, Y. (1994). Changes in reliability of synaptic function as a mechanism for plasticity. *Nature* 371:704–707.

Stevens, C. F., Wesseling, J. F. (1998). Activity-dependent modulation of the rate at which synaptic vesicles become available to undergo exocytosis. *Neuron* 21:415–424.

Sugita, S., Goldsmith, J. R., Baxter, D. A., Byrne, J. H. (1992). Involvement of protein kinase C in serotonin-induced spike broadening and synaptic facilitation of sensorimotor connections in Aplysia. *J. Neurophysiol.* 68:643–651.

Swandulla, D., Hans, M., Zipser, K., Augustine, G. J. (1991). Role of residual calcium in synaptic depression and posttetanic potentiation: fast and slow calcium signaling in nerve terminals. *Neuron* 7:915–926.

Swope, S. L., Moss, S. I., Raymond, L. A., Huganir, R. L. (1999). Regulation of ligand-gated ion channels by protein phosphorylation. *Adv. Second Mess. Phosphoprotein Res.* 33:49–78.

Takumi, Y., Ramírez-León, V., Laake, P., Rinvik, E., Ottersen, O. P. (1999). Different modes of expression of AMPA and NMDA receptors in hippocampal synapses. *Nature Neurosci.* 2:618–624.

Tang, Y., Zucker, R. S. (1997). Mitochondrial involvement in post-tetanic potentiation of synaptic transmission. *Neuron* 18:483–491.

Tezuka, T., Umemori, H., Akiyama, T., Nakanishi, S., Yamamoto, T. (1999). PSD-95 promotes Fyn-mediated tyrosine phosphorylation of the *N*-methyl-D-aspartate receptor subunit NR2A. *Proc. Natl. Acad. Sci. USA* 96:435–440.

Tong, G., Malenka, R. C., Nicoll, R. A. (1996). Long-term potentiation in cultures of single hippocampal granule cells: a presynaptic form of plasticity. *Neuron* 16:1147–1157.

Trudeau, L. E., Doyle, R. T., Emery, D. G., Haydon, P. G. (1996). Calcium independent activation of the secretory apparatus by ruthenium red in hippocampal neurons: a new tool to assess modulation of presynaptic function. *J. Neurosci.* 16:46–54.

Trudeau, L. E., Fang, Y., Haydon, P. G. (1998). Modulation of an early step in the secretory machinery in hippocampal nerve terminals. *Proc. Natl. Acad. Sci. USA* 95:7163–7168.

Turner, K. M., Burgoyne, R. D., Morgan, A. (1999). Protein phosphorylation and the regulation of synaptic membrane traffic. *Trends Neurosci.* 22:459–464.

Veruki, M. L., Yeh, H. H. (1992). Vasoactive intestinal polypeptide modulates GABA$_A$ receptor function in bipolar cells and ganglion cells of the rat retina. *J. Neurophysiol.* 67:791–797.

Villacres, E. C., Wong, S. T., Chavkin, C., Storm, D. R. (1998). Type I adenylyl cyclase mutant mice have impaired mossy fiber long-term potentiation. *J. Neurosci.* 18:3186–3194.

Wallace, B. G., Qu, Z., Huganir, R. L. (1991). Agrin induces phosphorylation of the nicotinic acetylcholine receptor. *Neuron* 6:869–878.

Wan, Q., Man, H. Y., Braunton, J., Wang, W., Salter, M. W., Becker, L., Wang, Y. T. (1997). Modulation of GABA$_A$ receptor function by tyrosine phosphorylation of beta subunits. *J. Neurosci.* 17:5062–5069.

Wang, C., Zucker, R. S. (1998). Regulation of synaptic vesicle recycling by calcium and serotonin. *Neuron* 21:155–167.

Wang, J. H., Kelly, P. T. (1997). Attenuation of paired-pulse facilitation associated with synaptic potentiation mediated by postsynaptic mechanisms. *J. Neurophysiol.* 78:2707–2716.

Wang, L. Y., Kaczmarek, L. K. (1998). High-frequency firing helps replenish the readily releasable pool of synaptic vesicles. *Nature* 394:384–388.

Wang, L. Y., Salter, M. W., MacDonald, J. F. (1991). Regulation of kainate receptors by cAMP-dependent protein kinase and phosphatases. *Science* 253:1132–1135.

Wang, L. Y., Taverna, F. A., Huang, X. P., MacDonald, J. F., Hampson, D. R. (1993). Phosphorylation and modulation of a kainate receptor (GluR6) by cAMP-dependent protein kinase. *Science* 259:1173–1175.

Wang, L. Y., Orser, B. A., Brautigan, D. L., MacDonald, J. F. (1994). Regulation of NMDA receptors in cultured hippocampal neurons by protein phosphatases 1 and 2A. *Nature* 369:230–232.

Wang, Y. T., Salter, M. W. (1994). Regulation of NMDA receptors by tyrosine kinases and phosphatases. *Nature* 369:233–235.

Weisskopf, M. G., Nicoll, R. A. (1995). Presynaptic changes during mossy fibre LTP revealed by NMDA receptor–mediated synaptic responses. *Nature* 376: 256–259.

Weisskopf, M. G., Castillo, P. E., Zalutsky, R. A., Nicoll, R. A. (1994). Mediation of hippocampal mossy fiber long-term potentiation by cyclic AMP. *Science* 265:1878–1882.

Westphal, R. S., Tavalin, S. J., Lin, J. W., Alto, N. M., Fraser, I. D., Langeberg, L. K., Sheng, M., Scott, J. D. (1999). Regulation of NMDA receptors by an associated phosphatase-kinase signaling complex. *Science* 285:93–96.

Williams, J. H., Errington, M. L., Lynch, M. A., Bliss, T. V. (1989). Arachidonic acid induces a long-term activity-dependent enhancement of synaptic transmission in the hippocampus. *Nature* 341:739–742.

Williams, J. H., Errington, M. L., Li, Y.-G., Lynch, M. A., Bliss, T. V. P. (1993). The search for retrograde messengers in long-term potentiation. *Sem. Neurosci.* 5:149–158.

Williams, S., Johnston, D. (1989). Long-term potentiation of hippocampal mossy fiber synapses is blocked by postsynaptic injection of calcium chelators. *Neuron* 3:583–588.

Wojtowicz, J. M., Marin, L., Atwood, H. L. (1994). Activity-induced changes in synaptic release sites at the crayfish neuromuscular junction. *J. Neurosci.* 14: 3688–3703.

Wu, L. G., Borst, J. G. (1999). The reduced release probability of releasable vesicles during recovery from short-term synaptic depression. *Neuron* 23: 821–832.

Wu, L. G., Saggau, P. (1994). Adenosine inhibits evoked synaptic transmission primarily by reducing presynaptic calcium influx in area CA1 of hippocampus. *Neuron* 12:1139–1148.

Wu, L. G., Saggau, P. (1997). Presynaptic inhibition of elicited neurotransmitter release. *Trends Neurosci.* 20:204–212.

Wyllie, D. J., Manabe, T., Nicoll, R. A. (1994). A rise in postsynaptic Ca^{2+} potentiates miniature excitatory postsynaptic currents and AMPA responses in hippocampal neurons. *Neuron* 12:127–138.

Xiang, Z., Greenwood, A. C., Kairiss, E. W., Brown, T. H. (1994). Quantal mechanism of long-term potentiation in hippocampal mossy-fiber synapses. *J. Neurophysiol.* 71:2552–2556.

Yang, S. N., Tang, Y. G., Zucker, R. S. (1999). Selective induction of LTP and LTD by postsynaptic $[Ca^{2+}]_i$ elevation. *J. Neurophysiol.* 81:781–787.

Yeckel, M. F., Kapur, A., Johnston, D. (1999). Multiple forms of LTP in hippocampal CA3 neurons use a common postsynaptic mechanism. *Nature Neurosci.* 2:625–633.

Yoshihara, M., Ueda, A., Zhang, D., Deitcher, D. L., Schwarz, T. L., Kidokoro, Y. (1999). Selective effects of neuronal-synaptobrevin mutations on transmitter release evoked by sustained versus transient Ca^{2+} increases and by cAMP. *J. Neurosci.* 19:2432–2441.

Yu, X. M., Askalan, R., Keil, G. J., 2nd, Salter, M. W. (1997). NMDA channel regulation by channel-associated protein tyrosine kinase Src. *Science* 275:674–678.

Yuste, R., Denk, W. (1995). Dendritic spines as basic functional units of neuronal integration. *Nature* 375:682–684.

Zalutsky, R. A., Nicoll, R. A. (1990). Comparison of two forms of long-term potentiation in single hippocampal neurons. *Science* 248:1619–1624.

Zamanillo, D., Sprengel, R., Hvalby, O., Jensen, V., Burnashev, N., Rozov, A., Kaiser, K. M., Köster, H. J., Borchardt, T., Worley, P., Lübke, J., Frotscher, M., Kelly, P. H., Sommer, B., Andersen, P., Seeburg, P. H., Sakmann, B. (1999). Importance of AMPA receptors for hippocampal synaptic plasticity but not for spatial learning. *Science* 284:1805–1811.

Zengel, J. E., Magleby, K. L. (1982). Augmentation and facilitation of transmitter release: a quantitative description at the frog neuromuscular junction. *J. Gen. Physiol.* 80:583–611.

Zhang, D., Kuromi, H., Kidokoro, Y. (1999). Activation of metabotropic glutamate receptors enhances synaptic transmission at the Drosophila neuromuscular junction. *Neuropharmacology* 38:645–657.

Zhang, S., Ehlers, M. D., Bernhardt, J. P., Su, C. T., Huganir, R. L. (1998). Calmodulin mediates calcium-dependent inactivation of N-methyl-D-aspartate receptors. *Neuron* 21:443–453.

Zheng, F., Gingrich, M. B., Traynelis, S. F., Conn, P. J. (1998). Tyrosine kinase potentiates NMDA receptor currents by reducing tonic zinc inhibition. *Nature Neurosci.* 1:185–191.

Zhuo, M., Small, S. A., Kandel, E. R., Hawkins, R. D. (1993). Nitric oxide and carbon monoxide produce activity-dependent long-term synaptic enhancement in hippocampus. *Science* 260:1946–1950.

Zhuo, M., Hu, Y., Schultz, C., Kandel, E. R., Hawkins, R. D. (1994). Role of guanylyl cyclase and cGMP-dependent protein kinase in long-term potentiation. *Nature* 368:635–639.

Zola-Morgan, S., Squire, L. R. (1993). Neuroanatomy of memory. *Annu. Rev. Neurosci.* 16:547–563.

Zucker, R. S. (1989). Short-term synaptic plasticity. *Annu. Rev. Neurosci.* 12:13–31.

Zucker, R. S. (1999). Calcium- and activity-dependent synaptic plasticity. *Curr. Opin. Neurobiol.* 9:305–313.

10 The Mechanisms and Meaning of Long-Term Synaptic Depression in the Mammalian Brain

Mark F. Bear and David J. Linden

A widely held assumption among neuroscientists is that experience is capable of persistently modifying the properties of synapses and that this use-dependent modification is central to both neuronal memory storage and the refinement of connections in brain development. This general idea was initially voiced by Sechenov and Cajal and was later formalized by Hebb (1949) in his famous synaptic modification postulate. However, it was not until many years later that an electrophysiological model system emerged that appeared to embody this concept in the mammalian brain. Bliss and Lømo (1973) showed that brief, high-frequency stimulation of a population of axons, the perforant path projection to the dentate gyrus, produced an increase in the strength of these synapses that could last for hours. This phenomenon, called long-term potentiation (LTP), has since been seen to last for days to weeks in chronic preparations. The duration of LTP, together with its initial discovery in the hippocampus (a brain region known from behavioral studies to be important for the storage of declarative memory), produced a surge of interest in LTP as a putative cellular model system for memory. This interest only increased when it became clear that under certain conditions LTP could display some of the formal properties of learning, such as specificity (LTP is confined to activated synapses) and associativity (weak stimulation of an input to a postsynaptic cell will induce LTP only when paired with a neighboring strong input to that same cell).

Although the first studies of LTP relied upon field potential recording in the intact hippocampus, this phenomenon has subsequently been observed in almost every type of glutamatergic synapse in the brain, and it has been extensively studied in reduced preparations, such as brain slices and cultures of embryonic neurons (see Malenka and Siegelbaum, this volume). At the same time that LTP was gradually "escaping" from the hippocampus, it was becoming clear that it was not the only form of use-dependent synaptic modification. The converse phenomenon, long-term depression (LTD), was also initially observed in the hippocampus before being found in other brain regions. At present, it appears likely that there are no synapses that express only LTP or LTD. In most synapses, LTP and LTD are typically evoked by brief, strong stimulation and sustained, weak stimulation, respectively. In their most frequently studied forms, the direction of change in synaptic strength (LTP versus LTD) is believed to be determined by the amount of postsynaptic activity (as indexed by Ca^{2+} influx) that occurs during induction: a small amount of postsynaptic Ca^{2+} influx results in LTD, whereas a larger amount results in LTP (see Linden, 1999, and Zucker, 1999, for review).

If LTP in the mature brain truly underlies memory storage, then what is the function of LTD? One proposal has been that LTD is a "neuronal substrate of forgetting" (Tsumoto, 1993). Although there is no definitive evidence to dispute this view, there is no particular support for it either. A potentially more useful construct is to consider that information is likely to be stored in the brain, at least in part, as an array of synaptic weights. If these synapses are driven to their maximal or minimal strengths, then those elements of the array become limited in their ability to contribute to subsequent plasticity. Thus neural circuits containing synapses that can actively both increase and decrease their strength are at a distinct computational advantage.

Experience-dependent refinement of connections during brain development can also potentially benefit from having both LTP and LTD mechanisms. Synapses that undergo strong, correlated activity can be strengthened and thereby retained, while synapses that have weak, uncorrelated activity can be weakened and ultimately removed. Like memory storage, one could imagine that developmental refinement of connections could proceed using either LTD or LTP alone, but the presence of both allows for faster and more flexible change.

In this chapter we do not attempt to provide a comprehensive overview of LTD at the many synapses in the brain where it has been studied (for a nearly complete review of the early days of the field, see Linden and Connor, 1995). Rather we focus on the two best-understood forms (LTD of the cerebellar parallel fiber–Purkinje cell synapse and LTD at the hippocampal Schaffer collateral/commissural–CA1 pyramidal cell synapse) as case studies to examine both the cellular processes that underlie LTD and its larger role in behavior and development.

LTD of the Cerebellar Parallel Fiber– Purkinje Cell Synapse

Why bother to study synaptic plasticity in an obscure and atypical part of the brain like the cerebellum? The answer is that it is one of only two locations where learning and memory can be understood at the level of circuits (the other being the amygdala). In contrast, the hippocampus, for all its experimental utility, receives information that is so highly processed that its content cannot be easily characterized (what is the nature of the information conveyed by the perforant path?).

Cerebellar Anatomy and Some Useful Models

The cerebellum functions largely to integrate various forms of sensory information to smooth and fine-tune complex voluntary movements and reflexes (see Ito, 1984, for review). Therefore, cerebellar damage in humans is associated not with outright paralysis but rather with dysmetric and ataxic syndromes, as well as impairments in motor learning. In addition, some recent work on human cerebellar lesions complemented by functional imaging studies has implicated the cerebellum in certain forms of nonmotor procedural learning (see Schmahmann, 1997, for review).

The cerebellum accounts for some 10% of the total weight of the human brain, but it may contain more than 50% of the total number of neurons, packed into the most infolded and convoluted structure in the brain. This degree of specialization suggests that throughout evolution fast, accurate coordinated movements have been highly adaptive. To coordinate many joints and muscles, it is necessary that sensory and proprioceptive signals from any location in the body or sensory world be able to influence motor commands to any muscle in the body. This requires a giant switchboard, implemented in the following way. The cerebellar circuitry is essentially composed of a relay station in the deep cerebellar nuclei (DCN) and a cortical "side-loop" (Fig. 10.1A). The neurons of the DCN receive their main excitatory drive from glutamatergic mossy fibers, which are the axons of a large number of precerebellar nuclei. The main outflow of information from this structure is carried by excitatory axons, which originate from the large neurons of the DCN and project to premotor areas including the red nucleus and thalamus. In addition, there are small projection neurons in the DCN that contain γ-aminobutyric acid (GABA) (Kumoi et al., 1988; Batini et al., 1992) and send axons to the inferior olive (Fredette and Mugnaini, 1991).

The sole output of the cortical side-loop is the inhibitory, GABAergic projection from Purkinje cells to the neurons of the DCN (both large and small projection neurons are innervated; see De Zeeuw and Berrebi, 1995; Teune et al., 1998). Purkinje cells receive two major excitatory inputs, which are organized in very different ways. Each Purkinje cell is innervated by a single climbing fiber. This climbing fiber, which originates in the neurons of the inferior olive, will innervate about 10 Purkinje cells. This is potentially the most powerful synaptic contact in the brain, as each Purkinje cell receives about 1400 synapses from a single climbing fiber axon (Strata and Rossi, 1998). Climbing fibers also provide a very weak innervation of the DCN consisting of a few synapses in the most distal dendrites, the function of which is poorly understood. In contrast, each Purkinje cell receives about 200,000 synapses from parallel fibers that are the axons of granule cells. Because of the large number of granule cells (circa 50,000,000,000) and the divergent output of their parallel fibers (each contacts about 1000 Purkinje cells), this synapse is the most abundant of any in the brain. Closing the loop, granule cells receive excitatory synapses from branches of the same mossy fibers, which innervate the DCN directly. Because there are approximately 10,000-fold more granule cells than DCN cells, the innervation of granule cells by mossy fibers is highly convergent.

Putting this circuit together, it appears as if cerebellar output is driven by direct excitatory input from the mossy fibers and is modulated by the inhibitory input from the Purkinje cell axons, which will reflect computations and interactions in the Purkinje cell. These computations will be performed upon very subtle and informationally rich excitatory parallel fiber input and massive, synchronous excitation produced by the climbing fiber. This striking anatomical organization has inspired

A

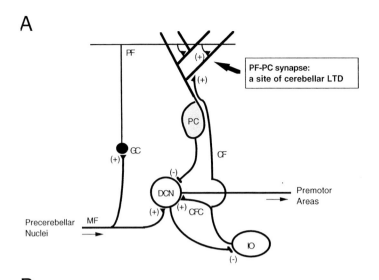

PF-PC synapse:
a site of cerebellar LTD

B

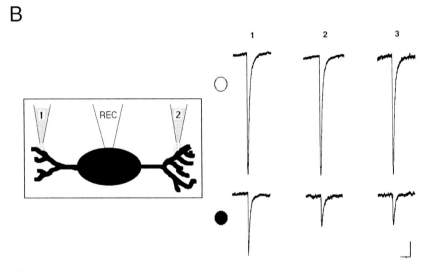

C

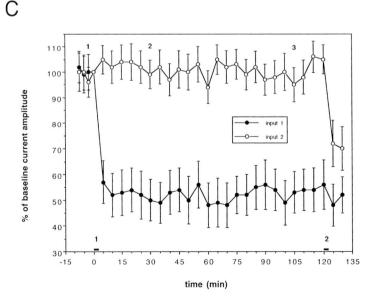

some notable models of motor learning. In particular, Marr (1969) proposed that the parallel fiber–Purkinje cell synapses could provide contextual information, that climbing fiber–Purkinje cell synapses could signal an "error" in motor performance that required alteration of subsequent behavior, and that the conjunction of these two signals could strengthen the parallel fiber–Purkinje cell synapse to create a memory trace for motor learning. This model was modified by Albus (1971), who noted that a decrease in synaptic strength would be more appropriate given the sign-reversing function of the Purkinje cell inhibitory output. Importantly, Albus also noted that this model is analogous to classical conditioning, with the parallel fibers conveying a conditioned stimulus (CS), the climbing fiber conveying an unconditioned stimulus (US), and a depression of the parallel fiber–Purkinje cell synapse giving rise to a conditioned response (CR) via disinhibition of the DCN. To place this model in a behavioral context, let us consider a well-characterized form of classical conditioning, associative eyeblink conditioning in the rabbit. Before training, an airpuff to the eye (US) gives rise to an immediate reflexive blink (the unconditioned response, UR). During training, a neutral stimulus such as a tone (CS) is paired with the airpuff stimulation so that the onset of the tone precedes the airpuff and the two stimuli co-terminate. As the rabbit acquires the association, it performs a blink carefully timed to immediately precede the airpuff (CR). This associative learning can also be actively reversed. In well-trained animals that reliably perform CRs, this response can undergo rapid extinction if tone stimuli are repeatedly presented without airpuffs.

Figure 10.1. (opposite) *Basic cerebellar functional anatomy and LTD of the parallel fiber–Purkinje cell synapse.*

(A) Simplified diagram of cerebellar circuitry. Information flow through the main relay pathway—consisting of precerebellar nuclei, their axons, the mossy fibers (MF), their targets in the deep cerebellar nuclei (DCN), and DCN excitatory axons projecting to premotor centers—is indicated with arrows. Excitatory synapses are denoted with a (+) and inhibitory synapses with a (–). CF, climbing fiber; CFC, climbing fiber collateral; GC, granule cell; IO, inferior olive; PC, Purkinje cell; PF, parallel fibers.

(B) Diagram of the recording configuration showing glutamate application to two nonoverlapping sites on the arbor of a Purkinje cell in culture. REC, perforated patch clamp recording electrode. Current traces from single representative Purkinje cells correspond to the times indicated in C. Scale bars = 2 sec, 50 pA.

(C) Both early and late phases of cerebellar LTD are input specific. Six glutamate-depolarization conjunctive stimuli were applied to site 1 at $t = 0$ min (indicated by a heavy horizontal bar) and to site 2 at $t = 120$ min (also indicated by a heavy horizontal bar). Each point represents the mean $\pm$ SEM of five different Purkinje cells normalized to baseline at $t = 0$ min.

Panels *B* and *C* originally appeared in Linden (1996) and are reproduced with permission of Cell Press.

The Role of the Cerebellum in Associative
Eyeblink Conditioning

There is extensive evidence to support the involvement of cerebellar circuits in associative eyeblink conditioning (see Kim and Thompson, 1997, for review). Similar evidence implicates the cerebellum in other forms of motor learning, such as limb-load adjustment and adaptation of the vestibulo-ocular reflex (VOR; du Lac et al., 1995; De Zeeuw et al., 1998). Extracellular recording showed that populations of cells in the nucleus interpositus (a particular portion of the DCN) discharge during the UR before training and, in well-trained animals, begin to fire during the CS-US interval. This firing is predictive of and correlated with the performance of the CR, suggesting that the CR behavior is expressed in the firing rate and pattern of DCN neurons (McCormick and Thompson, 1984a,b; Berthier and Moore, 1986, 1990). This notion is further supported by the finding that microstimulation in the appropriate region of the nucleus interpositus elicited a strong eyelid response in either trained or untrained animals (McCormick and Thompson, 1984a). Moreover, during training stimulation of mossy and climbing fibers can substitute for the CS and US, respectively (Mauk et al., 1986; Steinmetz et al., 1986, 1989).

The data obtained using lesions and reversible inactivation has been somewhat more complex (see Mauk, 1997, for review). A Marr/Albus model would predict that lesions of the cerebellar cortex would both delete the memory trace in previously trained animals and prevent further learning. Initially it was observed that lesioning either the whole cerebellum (ipsilateral to the trained eye) or the anterior interpositus nucleus completely abolished the CR but *not* the UR (McCormick et al., 1982; McCormick and Thompson, 1984a,b; Yeo et al., 1985a; Steinmetz et al., 1992; but see Welsh and Harvey, 1989). These experiments suggested that cerebellar lesions abolished the memory trace for eyeblink conditioning, but their irreversibility made it difficult to dissociate this interpretation from a performance deficit. A more convincing case was made when experiments showed that reversible inactivation of the DCN with muscimol (a $GABA_A$ receptor agonist) prevented the acquisition of the eyeblink CR but not the performance of the UR (Krupa et al., 1993; Hardiman et al., 1996; Krupa and Thompson, 1997; but see Bracha et al., 1994). In contrast, inactivation of the superior cerebellar peduncle or red nucleus (sites through which excitatory DCN output is conveyed) prevented the expression of the CR during training but not its acquisition, as evidenced by the fact that the CR was present after inactivation (Krupa et al., 1993; Krupa and Thompson, 1995). These studies suggest that the cerebellum and its associated projections are essential for acquisition and expression of the eyeblink CR. More specifically, the memory trace seems to be localized "upstream" of the red nucleus, in the cerebellar cortex, the DCN, or both.

Although there is general agreement that lesions or inactivation of the DCN blocks the acquisition of the eyeblink CR, there has been considerable debate over the specific role of the cerebellar cortex in eyeblink conditioning. Reports using lesions and inactivation of the cerebellar cortex have ranged from those that have found a complete blockade of CR acquisition (Yeo et al., 1985b) to those that have slowed, but not prevented, acquisition (Lavond and Steinmetz, 1989; Yeo and Hardiman, 1992) to those that have found no effect at all (McCormick and Thompson, 1984a,b). Some recent reports point to a potential resolution of this problem. Lesions that included the anterior cerebellar cortex (a region previously thought not to be important) or infusion of picrotoxin (a $GABA_A$ receptor antagonist, the opposite of muscimol) into the DCN to block Purkinje cell input did not abolish the CR entirely, but affected its timing (Perrett et al., 1993; Perrett and Mauk, 1995; Garcia and Mauk, 1998) . Recently a model has been proposed to explain these findings. In this model the memory trace of the eyeblink CR is sequentially stored, initially as a depression of the parallel fiber–Purkinje cell synapse in the cerebellar cortex. This would result in an attenuation of Purkinje cell firing and hence Purkinje cell–DCN synaptic drive, thereby disinhibiting the DCN targets. This disinhibition, when coupled with activation of the mossy fiber–DCN synapse, could then potentiate the latter, resulting in storage of the CR at the mossy fiber–DCN synapse while the timing of the conditioned response is retained in the cerebellar cortex (Raymond et al., 1996; Mauk, 1997; Mauk and Donegan, 1997; Medina and Mauk, 1999).

Potential Cellular Substrates of Associative Eyeblink Conditioning

LTD of the parallel fiber–Purkinje cell synapse has been proposed as a cellular mechanism that could, at least in part, underlie the acquisition of associative eyeblink conditioning. This phenomenon, which was first described by Ito and colleagues (Ito et al., 1982), results when the climbing fiber (corresponding to the US) and parallel fiber (corresponding to the CS) inputs are activated together at low frequencies (1–4 Hz). In addition, repetitive stimulation of parallel fibers at an intermediate frequency can produce LTP of the parallel fiber–Purkinje cell synapse, thus providing a form of bidirectional control (Sakurai, 1987; Hirano, 1990a; Crepel and Jaillard, 1991; Shibuki and Okada, 1992; Salin et al., 1996; Linden, 1997, 1998; Storm et al., 1998; Linden and Ahn, 1999). LTD in the parallel fiber–Purkinje cell synapse requires association of parallel fiber (CS) and climbing fiber (US) activation and would result in decreased firing of the Purkinje cell, causing increased firing of DCN neurons and enhanced expression of the CR. Essentially this is a cellular restatement of the Marr/Albus model. Conversely, repeated activation of the parallel fiber (CS) could, through enhanced inhibition, decrease firing of the DCN, thereby reducing expression of the CR during extinction.

As indicated by the lesion and inactivation studies described previously, it is likely that the parallel fiber–Purkinje cell synapse is not the only site of information storage during cerebellar motor learning. At a cellular level, extinction of the CR, as results from repeated application of a tone CS, could be mediated not only by LTP of the parallel fiber–Purkinje cell synapse but also by the LTD of the mossy fiber–DCN synapse. This idea is consistent with reports that both cortical lesions that include the anterior region (Perrett and Mauk, 1995) and reversible inactivation of the DCN with muscimol (Hardiman et al., 1996; Ramnani and Yeo, 1996) block CR extinction.

Finally, LTP of the mossy fiber–DCN synapse, if it could be conclusively demonstrated, could contribute to expression of the CR under the appropriate circumstances. At present, it has not been straightforward to induce LTP of the mossy fiber–DCN synapse by stimulation of mossy fibers alone (Aizenman and Linden, unpublished observations). In the context of eyeblink conditioning, this actually makes sense as CRs are not acquired from presentations of a CS alone. Mauk and coworkers have proposed a model (discussed earlier) in which disinhibition of the DCN from reduced Purkinje cell input, when coupled with activation of the mossy fiber–DCN synapses, results in LTP of mossy fiber–DCN synapses, constituting a portion of the memory trace of the CR (Raymond et al., 1996; Mauk, 1997; Mauk and Donegan, 1997). A recent computational analysis has suggested that an LTP induction rule for the mossy fiber–DCN synapse that depends upon specific patterns of Purkinje cell input (plus ongoing mossy fiber activity) could constitute a memory trace that is unusually resistant to degradation by ongoing "background" activity in the cerebellar circuit (Medina and Mauk, 1999).

Parallel Fiber LTD Induction

Parametric Requirements

Cerebellar LTD was first described in the intact cerebellum (Ito et al., 1982), and since that time it has been analyzed in acute slice preparations, primary cultures, acutely dissociated Purkinje cells, and macropatches of Purkinje cell dendrite. In slice or in situ, the standard induction protocol consists of stimulating the parallel and climbing fiber inputs together at low frequency (1–4 Hz) for a period of 2–6 min. This results in a selective attenuation of the parallel fiber–Purkinje cell synapse (typically a 20–50% reduction of baseline synaptic strength) that reaches its full extent in about 10 min and persists for the duration of the experiment, typically 1–2 h.

LTD is said to result from co-activation of parallel fibers and climbing fibers, but what are the precise timing constraints on this co-activation? This is an important point, because if parallel fiber LTD underlies associative eyeblink conditioning, then the temporal constraints on CS-US association should be reflected in the temporal constraints on LTD induction. One study, using intracellular recording in rabbit cerebellar slices, has indicated that LTD is optimally induced when climbing fiber stimulation precedes parallel fiber stimulation by 125–250 msec (Ekerot and Kano, 1989). Another study, using a similar preparation, has shown that LTD may be induced by climbing fiber–parallel fiber stimulation with an interval of 50 msec but claims that LTD induced by climbing fiber–parallel fiber pairing will not occur unless disynaptic inhibition is blocked by addition of a $GABA_A$ antagonist (Schreurs and Alkon, 1993). Neither of these intervals (in which the US precedes the CS) will support robust eyeblink conditioning. However, with slightly different stimulation protocols (small trains of parallel fiber stimulation instead of single pulses in one case) pairing of parallel fiber before climbing fiber, at intervals that support eyeblink conditioning, may also be effective in inducing LTD in the absence of $GABA_A$ receptor blockade (Chen and Thompson, 1995; Schreurs et al., 1996). At present it is unclear what pattern of artificial stimulation would best mimic the CS, as the spatio-temporal patterns of parallel fiber activation that correspond to various neutral CSs (e.g., tone, light) have not been well defined.

The traditional description of cerebellar LTD is that it is strictly input-specific: it occurs only in those parallel fiber synapses stimulated during climbing fiber activation (Ito et al., 1982; Ekerot and Kano, 1985). An example of this is shown in Fig. 10.1B,C. Using cultured mouse Purkinje cells, a simplified postsynaptic preparation has been developed for the study of cerebellar LTD in which iontophoretic glutamate pulses and direct Purkinje cell depolarization are substituted for parallel fiber and climbing fiber stimulation, respectively (Linden et al., 1991). LTD induced in this manner may be seen as a reduction of the glutamate-evoked current as measured with a perforated patch electrode attached to the Purkinje cell soma. Following acquisition of baseline responses to glutamate test pulses applied to two nonoverlapping sites in the Purkinje cell dendritic arbor, glutamate-depolarization conjunctive stimulation was applied to site 1 but not site 2 (Linden, 1996). This produced an input-specific depression of the site 1 response that persisted for the duration of the recording ($52 \pm 8.5\%$ of baseline at $t = 10$ min, $52 \pm 7.0\%$ at $t = 130$ min, mean $\pm$ SEM, $n = 5$). Conjunctive stimulation applied to site 2, 120 min after site 1, produced an input-specific depression at this location as well, although this was somewhat smaller in initial amplitude ($70 \pm 8.5\%$ at $t = 130$ min), probably because of a general deterioration in the health of the preparation after 130 min of recording. These studies were designed to activate dendritic sites that were as distant from each other as possible. As such, although they show that LTD does not spread over the entire dendritic arbor of Purkinje cells, they do not address the more subtle point of whether LTD can spread to nearby unstimulated synapses. In fact, this has been shown using synaptic activation in a slice preparation (Hartell, 1996). It will be interesting to see if models of motor learning that incorporate parallel fiber LTD remain functional with various degrees of LTD spreading.

Climbing Fiber Signals

The climbing fiber contributes to LTD induction by causing sufficient postsynaptic depolarization (through activation of α-amino-3-hydroxyl-5-methyl-4-isoxazolepropionic acid receptors [AMPARs]) to strongly activate voltage-sensitive Ca^{2+} channels in the dendrites, thereby causing a complex spike and a large Ca^{2+} influx (Fig. 10.2). In fact, climbing fiber activation may be replaced in the LTD induction protocol by direct depolarization of the Purkinje cell (Crepel and Krupa, 1988; Hirano, 1990b; Linden et al., 1991). Furthermore, LTD induction is blocked by postsynaptic application of a Ca^{2+} chelator (Sakurai, 1990; Linden and Connor, 1991; Konnerth et al., 1992), electrical inhibition of Purkinje cells during parallel fiber–climbing fiber conjunctive stimulation (Ekerot and Kano, 1985; Hirano, 1990b; Crepel and Jaillard, 1991), or removal of external Ca^{2+} (Linden and Connor, 1991). In addition, studies using optical indicators have shown large Ca^{2+} accumulations in Purkinje cell dendrites following climbing fiber stimulation (Ross and Werman, 1987; Knopfel et al., 1990; Konnerth et al., 1992). Although these studies have suggested that Ca^{2+} influx is the sole mediator of climbing fiber action, another view has come from studies that have examined a peptide released from climbing fiber terminals, corticotrophin-releasing factor (CRF). Miyata et al. (1999) have found that LTD induced by either parallel fiber–climbing fiber conjunction or parallel fiber–depolarization conjunction can be blocked by antagonists of the CRF receptor in a slice preparation. Furthermore, parallel fiber–depolarization conjunction fails to induce LTD in slices prepared from rats in which climbing fibers were chemically prelesioned, but this may be restored with exogenous CRF. These observations have led to the suggestion that CRF plays a permissive role in LTD of parallel fiber synapses. At present, it is unclear how this requirement would be manifest in preparations that display cerebellar LTD but lack climbing fibers, such as cerebellar cultures, acutely dissociated Purkinje cells, or dendritic macropatches.

Parallel Fiber Signals

Parallel fiber activation results in glutamate release, which activates glutamate receptors in the Purkinje cell dendrite. Although mature Purkinje cells do not express functional N-methyl-D-aspartate receptors (NMDARs), they are found on both cultured embryonic Purkinje cells and acutely dissociated Purkinje cells in early postnatal life (Linden and Connor, 1991; Rosenmund et al., 1992). Purkinje cells also express AMPARs of the GluR2-containing, Ca^{2+}-impermeable variety (Linden et al., 1993; Tempia et al., 1996) as well as a particular G protein–coupled metabotropic glutamate receptor, mGluR1, at high levels in the dendritic spines where parallel fiber synapses are received (Martin et al., 1992).

parallel fiber terminal

Purkinje cell dendrite

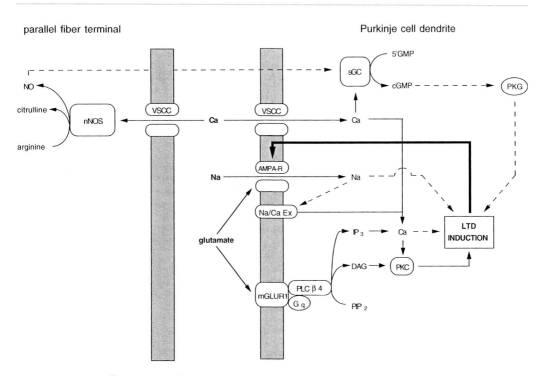

Figure 10.2. Model of LTD induction at the cerebellar parallel fiber–Purkinje cell synapse. LTD induction typically requires co-activation of climbing fiber and parallel fiber synapses at low frequency (1–4 Hz). The climbing fiber (not shown) exerts its effect by activating AMPA receptors and producing a large depolarization, resulting in a widespread postsynaptic Ca^{2+} influx through voltage-sensitive Ca^{2+} channels (VSCCs). Glutamate released from the parallel fiber exerts its effect by activating both a metabotropic glutamate receptor (mGluR1) that is linked via a G protein (G q) to the enzyme phospholipase C β4 (PLC β 4) and an AMPA receptor (AMPA-R) that produces an Na^+ flux. The former signal metabolizes phosphatidylinositol-4,5-bisphosphate (PIP_2) to yield inositol-1,4,5-trisphosphate (IP_3) and 1,2-diacylglycerol (DAG). The function of the AMPAR-mediated Na^+ flux is unclear, but it may involve slowing or reversing the operation of a plasma membrane Na_o/Ca_i exchanger (Na/Ca Ex). IP_3 mobilizes Ca^{2+} from internal stores through action on an IP_3 receptor (not shown). Ca^{2+}, together with DAG, serves to activate protein kinase C (PKC). Although PKC activation is likely to be sufficient to trigger LTD induction in some situations (including cultured Purkinje cells), a second pathway appears to be required in most studies that have utilized cerebellar slices. This involves stimulation of the Ca^{2+}/calmodulin-sensitive enzyme, neuronal nitric oxide synthase (nNOS), which is localized to parallel fiber terminals and interneurons, and the consequent production of the diffusable messenger NO. NO then activates a postsynaptic cascade consisting of soluble guanylyl cyclase (sGC), cGMP, and cGMP-dependent protein kinase (PKG). In this model, mechanisms we regard as well established are represented as solid lines and those that remain controversial are dotted. The thick line illustrates the expression mechanism.

The first evidence indicating that activation of metabotropic receptors was required for parallel fiber LTD induction came from experiments showing that agonists that activated both AMPA and metabotropic receptors (such as glutamate and quisqualate) could substitute for parallel fiber activation during LTD induction but that agonists that failed to activate metabotropic receptors (such as AMPA or aspartate) could not (Kano and Kato, 1987; Linden et al., 1991). Complementary evidence was found that metabotropic receptor antagonists blocked LTD induction (Linden et al., 1991; Hartell, 1994; Narasimhan and Linden, 1996; Lev-Ram et al., 1997a). These results, although they indicated that metabotropic receptor activation was required, did not specify which metabotropic receptor(s) were important for LTD induction. The first findings to address this issue were those of Shigemoto et al. (1994), who demonstrated that specific inactivating antibodies directed against mGluR1 could block LTD induction in cell culture. This result was confirmed and extended by two different groups using mGluR1-knockout mice (Aiba et al., 1994; Conquet et al., 1994).

Activation of mGluR1 results in the activation of phospholipase C (PLC) and the consequent production of two initial products, inositol-1,4,5-trisphosphate (IP_3) and 1,2-diacylglycerol. The former binds to specific intracellular IP3 receptors, resulting in the liberation of Ca^{2+} from internal stores, whereas the latter results in activation of protein kinase C (PKC). Are both of these products required for parallel fiber LTD induction? Purkinje cells express IP_3 receptors, particularly the type I isoform, at unusually high levels (Nakanishi et al., 1991), and it has been shown through photolysis of caged IP_3 that these receptors are functionally coupled to intracellular Ca^{2+} release in situ (Khodakhah and Ogden, 1993; Wang and Augustine, 1995). Several lines of evidence have supported a role for IP_3 receptor activation in the induction of parallel fiber LTD. First, compounds that interfere with IP_3 receptor function block LTD. Application of heparin, a nonspecific inhibitor of the IP_3 receptor (Herbert and Maffrand, 1991; Bezprozvanny et al., 1993), blocks LTD induced by glutamate-depolarization conjunction in cultured Purkinje cells (Kasono and Hirano, 1995) or by parallel fiber–depolarization conjunction in Purkinje cells in cerebellar slices (Khodakhah and Armstrong, 1997). A specific inactivating antibody directed against the IP_3 receptor was similarly effective (Inoue et al., 1998). Application of thapsigargin, which depletes internal Ca^{2+} stores through inhibition of the endoplasmic reticulum Ca^{2+}-ATPase, also blocks LTD induction (Kohda et al., 1995). Thapsigargin would be expected to deplete Ca^{2+} stores gated by both the IP_3 receptor and the ryanodine receptor; the latter mediates Ca^{2+}-induced Ca^{2+} release. Second, photolysis of IP_3 in cultured Purkinje cells can induce LTD when combined with depolarization plus AMPAR activation (Kasono and Hirano, 1995). Similarly, IP_3 photolysis combined with depolarization can induce LTD in slices derived from either wild-type (Khodakhah and Armstrong, 1997) or mGluR1-knockout mice (Daniel et al., 1999). Finally, parallel fiber LTD is blocked in slices derived from a mutant mouse lacking the type I IP_3 receptor (Inoue et al., 1998).

Although these experiments would appear to provide a strong case for the involvement of IP_3 receptors in cerebellar LTD induction, it is worth noting that not all evidence has been consistent with this view. For example, thapsigargin application in slices was found to block LTD induced by bath application of the mGluR agonist 1-aminocyclopentane-1,3-dicarboxylate (ACPD) together with depolarization, but not parallel fiber–depolarization conjunction (Hemart et al., 1995). Furthermore, when Narasimhan et al. (1998) performed ratiometric imaging of free cytosolic Ca^{2+} on both acutely dissociated and cultured Purkinje cells, they determined that the threshold for glutamate pulses to contribute to LTD induction was below the threshold for producing a Ca^{2+} transient. Furthermore, the Ca^{2+} transients produced by depolarization alone and glutamate plus depolarization were not significantly different. In addition, the potent and selective IP_3 receptor channel blocker xestospongin C (an improvement over heparin) was not found to affect the induction of LTD in either acutely dissociated or cultured Purkinje cells at a concentration that was sufficient to block mGluR1-evoked Ca^{2+} mobilization. Finally, replacement of mGluR1 activation by exogenous synthetic diacylglycerol in an LTD induction protocol was successful. At present it is not clear why an IP_3 signaling cascade is not required for induction of cerebellar LTD in these experiments using reduced preparations, while other experiments using both slice and culture preparations have suggested otherwise.

Although voltage-gated Ca^{2+} influx coupled with mGluR1 activation is necessary for LTD induction, present evidence suggests that it is not sufficient. Application of CNQX (an antagonist of AMPARs but not of metabotropic receptors) blocks LTD induction in cultured Purkinje cells (Linden et al., 1991, 1993) and cerebellar slices (Hemart et al., 1995). These findings are consistent with an earlier observation in the intact cerebellum that application of kynurenate (also an antagonist of AMPARs but not of metabotropic receptors) blocks LTD produced by parallel fiber–climbing fiber stimulation (Kano and Kato, 1988). Iontophoretic application of a metabotropic receptor agonist, (1S,3R)-ACPD, coupled with depolarization sufficient to cause Ca^{2+} influx is not sufficient to induce LTD in cultured Purkinje cells when voltage-gated Na^+ channels are blocked with tetrodotoxin (Linden et al., 1993). The AMPAR appears to exert its effect on LTD induction through a specific chemical consequence of Na^+ influx, as replacement of external Na^+ with other permeant cations such as Li^+ or Cs^+ during quisqualate-depolarization conjunction blocked the induction of LTD (Linden et al., 1993). Furthermore, with no blockade of voltage-gated Na^+ channels, LTD may be induced infrequently by (1S,3R)-ACPD–depolarization conjunction. Promoters of voltage-gated Na^+ influx, such as veratridine, increased the probability of induction.

Taken together, these results suggest that Na^+ influx is necessary for LTD induction, and that although influx via voltage-gated Na^+ channels may provide a sufficient signal in some cases, the flux through AMPARs is much more effective. Although it is unclear what intracellular processes are engaged by an increase in postsynaptic Na^+ to contribute to LTD induction, one possibility is that the high levels of intracellular Na^+ slow the Na^+-Ca^{2+} exchanger, thereby reducing the ability of the cell to extrude Ca^{2+} and leading to increased internal Ca^{2+} for a given amount of influx. Alternatively, Na^+ may produce a direct stimulatory effect on the mGluR1–PLC cascade (Gusovsky et al., 1986).

Second Messengers

Two major postsynaptic signals resulting from cerebellar LTD induction are 1,2-diacylglycerol and Ca^{2+}. These signals are known to activate PKC synergistically. The involvement of PKC in LTD induction was suggested by experiments in which PKC inhibitors blocked induction when applied during glutamate-depolarization conjunction in cultured Purkinje cells (Linden and Connor, 1991). Application of these compounds after LTD had been induced had no effect, suggesting that continued PKC activation is not required for LTD to persist. Blockade of LTD induction by PKC inhibitors has since been confirmed using several preparations, including cerebellar slices (Hartell, 1994; Freeman et al., 1998), acutely dissociated Purkinje cells and Purkinje cell dendritic macropatches (Narasimhan and Linden, 1996), and cultured Purkinje cells derived from a transgenic mouse that expresses a PKC inhibitor peptide (De Zeeuw et al., 1998). These observations are complemented by the finding that bath application of PKC-activating phorbol esters induces an LTD-like attenuation of Purkinje cell responses to exogenous glutamate or AMPA (Crepel and Krupa, 1988; Linden and Connor, 1991) that occludes pairing-induced LTD.

In addition to PKC activation, a number of studies have indicated that release of the gaseous second messenger nitric oxide (NO) by the action of the Ca^{2+}/calmodulin-sensitive enzyme NO synthase (NOS) is necessary for parallel fiber LTD induction. They have shown that an LTD-like phenomenon could be induced when climbing fiber stimulation was replaced by bath application of NO via donor molecules such as sodium nitroprusside (Crepel and Jaillard, 1990; Shibuki and Okada, 1991; Daniel et al., 1993; but see Glaum et al., 1992). Likewise, induction of LTD by more conventional means could be blocked by inhibitors of NOS (such as N^G-nitro-L-arginine), agents that bind NO in the extracellular fluid (such as hemoglobin), or genetic deletion of the neuronal isoform of NOS (Lev-Ram et al., 1997b). Application of NO donors, cGMP analogues, or cGMP phosphodiesterase inhibitors directly to the Purkinje cell (via a patch pipette) also resulted in depression of parallel fiber responses (Daniel et al., 1993; Hartell, 1994, 1996) whereas postsynaptic application of a NOS inhibitor did not block LTD induction

(Daniel et al., 1993). In contrast, postsynaptic application of a specific guanylyl cyclase inhibitor was effective in blocking LTD induction (Boxall and Garthwaite, 1996; Lev-Ram et al., 1997a). These findings suggested a model in which climbing fiber activation results in NO production, which then diffuses to the Purkinje cell to activate soluble guanylyl cyclase. However, this model was complicated by the fact that both climbing fibers (Bredt et al., 1990; Vincent and Kimura; 1992; Ikeda et al., 1993) and Purkinje cells (Bredt et al., 1990; Vincent and Kimura, 1992; Crepel et al., 1994) lack NOS.

A proposal that addresses this complication is that climbing fiber–evoked Ca^{2+} influx into Purkinje cell dendrites causes K^+ efflux, which depolarizes adjacent parallel fiber terminals or basket cells, resulting in Ca^{2+} influx and the consequent activation of NOS in these compartments (Daniel et al., 1998). Another approach has been taken by Lev-Ram et al. (1995), who found that photolysis of caged NO loaded into Purkinje cells could substitute for parallel fiber activation in LTD induction. When NO photolysis was followed by direct Purkinje cell depolarization within a 50-msec window, LTD of parallel fiber excitatory postsynaptic currents (EPSCs) ensued. LTD induced in this manner could be blocked by a postsynaptic application of a Ca^{2+} chelator or NO scavenger, but not external application of an NOS inhibitor or an NO scavenger. In contrast, LTD produced by parallel fiber–depolarization conjunction could be blocked by either an internally or externally applied NO scavenger, or an externally applied NOS inhibitor, but not an internally applied NOS inhibitor. These results suggest a model in which activation of parallel fibers causes an anterograde NO signal, which acts inside the Purkinje cell. A subsequent investigation by this group showed that when photolysis of caged Ca^{2+} was used in place of Purkinje cell depolarization, the coincidence requirement for NO and Ca^{2+} pairing was less than 10 msec and the resultant LTD could be blocked by an inhibitor of soluble guanylyl cyclase (Lev-Ram et al., 1997a). However, when caged Ca^{2+} and cGMP were used, the inhibition of guanylyl cyclase could be overcome and the coincidence requirement was lengthened to about 200 msec.

The production of cGMP by this cascade is likely to be exerting its effect through activation of cGMP-dependent protein kinase (PKG). LTD induced by parallel fiber–depolarization conjunctive stimulation (Hartell, 1994) or photolytic NO–depolarization stimulation (Lev-Ram et al., 1997a) may be blocked with PKG inhibitors. The mechanisms by which PKG might contribute to LTD induction are not known; one suggestion has been that phosphorylation of G-substrate by PKG could result in inhibition of protein phosphatases (Ito, 1990).

In contrast to the extensive evidence indicating a requirement for an NO-cGMP-PKG cascade in slice preparations, LTD of glutamate currents produced without synaptic stimulation in cultured Purkinje cells is unaffected by reagents that stimulate (sodium nitroprusside) or inhibit (hemoglobin, N^G-nitro-L-arginine) NO signaling (Linden and Connor, 1992). Furthermore, in cerebellar cultures made from neuronal NOS knockout mice, LTD was indistinguishable from that in cultures from wild-type mice (Linden et al., 1995). In wild-type cultures, neither an activator of soluble guanylate cyclase, nor an inhibitor of type V cGMP-phosphodiesterase, nor inclusion of cGMP analogues in the patch pipette produced an LTD-like effect. Induction of LTD was not blocked by inclusion in the patch pipette of three different PKG inhibitors. These results suggest that an NO-cGMP-PKG cascade is not required for cerebellar LTD induction in culture.

The contradictory nature of experiments with NO-cGMP reagents has led several investigators to examine the possibility that multiple induction mechanisms exist for cerebellar LTD, some NO-dependent and some NO-independent. Hemart et al. (1995) found that although induction of LTD by parallel fiber–depolarization conjunction was blocked by an NOS inhibitor, LTD induced by pairing depolarization with a bath-applied mGluR agonist was not. In contrast, LTD induced by both protocols was blocked by a PKC inhibitor, suggesting that whereas LTD induction in the former case required activation of both NO-cGMP- and PKC-mediated pathways, the latter required PKC, but not NO-cGMP, similar to results seen in culture (Linden end Connor, 1992; Linden et al., 1995). Thus it is possible that the mGluR1-PKC cascade and the NO-cGMP cascade ultimately converge on the same molecular alteration (such as phosphorylation of a particular site on a synaptic protein) through PKC activation and protein phosphatase inhibition, respectively (Daniel et al., 1998). In this scheme, the requirements for mGluR1-PKC and NO-cGMP activation could depend upon other aspects of the biochemical status of the Purkinje cell. For example, if basal phosphatase activity were low, then PKC activation might be sufficient for LTD induction. Conversely, if basal phosphatase activity were high, then PKG activation–phosphatase inhibition might be necessary. If basal PKC activity were also high then PKG activation might even be sufficient for LTD induction.

Parallel Fiber LTD Expression

Although considerable attention has been paid to the molecular mechanisms of cerebellar LTD induction, much less effort has been focused upon its expression. A widely accepted notion has been that LTD is expressed, at least in part, as a downregulation of the postsynaptic sensitivity to AMPA, as LTD may be detected using AMPA or glutamate test pulses in intact (Ito et al., 1982), slice (Crepel and Krupa, 1988), and culture (Linden et al., 1991) preparations. These experiments have been extended with ultrareduced preparations that completely lack functional presynaptic terminals (Linden, 1994), including outside-out dendritic macropatches and acutely dissociated Purkinje cells (Narasimhan and Linden, 1996; Narasimhan et al., 1998), which provide definitive evidence for a postsynaptic locus of expression. Further support for this idea comes from a study showing that the coefficient of variation of parallel fiber EPSCs is altered by manipulations known to act presynaptically (such as transient synaptic attenuation produced by addition of adenosine) but not by induction of LTD (Blond et al., 1997).

One postsynaptic model for LTD expression is that phosphorylation of AMPARs by an enzyme such as PKC is a trigger for some form of AMPAR downregulation. One observation has been put forward in support of this model. Using an antibody that recognizes the AMPAR subunit GluR2 phosphorylated at serine 696 (and possibly some corresponding sites on other subunits), it was shown that bath application of AMPA prior to 8Br-cGMP (a manipulation that is claimed to cause an LTD-like effect in grease gap recordings; Ito and Karachot, 1992) produced a persistent (>30 min) increase in immunoreactivity in Purkinje cell dendrites (Nakazawa et al., 1995). Unfortunately the best current evidence indicates that the phosphorylated residues recognized by this antibody are located on an extracellular loop of AMPARs, significantly complicating their potential role in LTD expression.

The attenuation of postsynaptic AMPAR function that underlies cerebellar LTD could result from several factors acting individually or in combination. These include changes in receptor number or distribution, unitary conductance, kinetics, or glutamate affinity. Until recently there has been little evidence that would allow one to distinguish among these possibilities. Some investigators have suggested that cerebellar LTD is expressed as an increase in the kinetics of AMPAR desensitization based on interactions with the drug aniracetam (Hemart et al., 1994), but this seems unlikely as parallel fiber EPSCs do not change their shape after LTD induction (Khodakhah and Armstrong, 1997). The recent observation that GluR2-containing synaptic AMPARs could be internalized by regulated clathrin-mediated endocytosis (Lüscher et al., 1999; Man et al., 2000) has suggested that changes in receptor number or distribution that utilize this process could underlie the expression of LTD. To test this hypothesis, Wang and Linden (2000) postsynaptically applied dynamin and amphiphysin peptides that interfere with the clathrin endocytotic complex and found that they blocked expression of LTD in cultured Purkinje cells. In addition, prior induction of LTD occluded subsequent attenuation of AMPA responses by stimulation of clathrin-mediated endocytosis. These findings suggest that the expression of cerebellar LTD requires clathrin-mediated internalization of postsynaptic AMPARs. It will be useful to extend this analysis to measure directly AMPAR kinetics, unitary conductance, and glutamate affinity following LTD induction in ultrareduced preparations of Purkinje cells (Narasimhan and Linden, 1996).

A Late Phase of Parallel Fiber LTD

Memory storage in neural circuits appears to involve the consolidation of labile short-term memory into a more permanent, long-term form. Using learning tasks in both vertebrate and invertebrate model systems, it has been shown that this consolidation is blocked by treatments that interfere with protein synthesis. Hippocampal LTP has also been shown to have a late phase, which requires new protein synthesis (see Abraham and Otani, 1991, for review). Furthermore, LTP induction is correlated with the expression of a set of immediate-early genes, several of which are transcription factors. However, until recently it had not been determined if any form of LTD also had a protein synthesis–dependent late phase.

LTD induced in the cultured mouse Purkinje cell when glutamate pulses and Purkinje cell depolarization are co-applied showed an attenuated late phase, with a complete return to baseline values within about 75 min in the presence of the translation inhibitor anisomycin (Linden, 1996). A similar blockade of a late phase of LTD was seen using transcription inhibitors such as actinomycin D and 5,6-dichloro-1-β-D-ribofuranosyl benzimidazole. Finally, LTD was produced using glutamate-depolarization conjunction in a perforated outside-out macropatch of Purkinje cell dendrite, which lacks nuclear material. LTD in this preparation returned to baseline values with a time course similar to that produced by protein synthesis inhibitors in intact cultured Purkinje cells. The effects of transcription- and translation-inhibiting drugs on the late phase of LTD are likely to be specific for the following reasons: (1) application of these drugs *after* conjunctive stimulation minimizes the risk that they interfere with LTD induction processes; (2) two specific signaling mechanisms required for LTD induction (mGluR1 activation and voltage-gated Ca^{2+} channel function) are unaltered by these drugs; (3) application of anisomycin has no effect on a control input; (4) prolonged (118 min) exposure to anisomycin has no effect on the ability of a control input to show the initial stages of LTD; and (5) the effect of anisomycin is restricted to a specific period after conjunctive stimulation. These findings suggest that there is a distinct late phase of cerebellar LTD that is dependent upon protein synthesis in the postsynaptic compartment.

This general result was confirmed by another study, which used a creative approach to allow for the prolonged monitoring of the LTD time course. Murashima and Hirano (1999) induced LTD by using chemical stimulations in cerebellar cultures (consisting of a 10-min application of 50 mM K^+ and 100 μM glutamate in the bath) and then monitored miniature EPSCs from populations of Purkinje cells at various times thereafter. Although this design has the disadvantage of being rather remote from LTD in the intact cerebellum, it has provided the only estimate to date of LTD duration in Purkinje cells, about 48 h. Application of protein synthesis inhibitors shortly after chemical LTD induction reduced the duration of LTD to 1–2 h.

What are the molecular mechanisms by which induction of LTD might regulate gene expression? One candidate mechanism involves the cAMP response element binding protein (CREB), a nuclear protein that regulates the transcription of genes with a cAMP response element site in their promoter. Several lines of evidence suggest that CREB activation is necessary for the consolidation of certain forms of memory and the establishment of a late phase of use-dependent increases in synaptic strength (see Silva et al., 1998, for review). CREB belongs to a large family of transcription factors that contain basic leucine zipper domains and is capable of forming homodimers or heterodimers with its closely related family members ATF-1 or CREM. CREB is an attractive candidate for transducing neuronal activation into gene expression because, like LTP, it is activated by synaptic stimuli that strongly increase internal calcium concentration (Ginty, 1997). This activation is mediated by phosphorylation of the transcriptional regulatory residue Ser-133. Several kinases phosphorylate this residue, including cAMP-dependent protein kinase (PKA), RSK2, MAPKAP kinase-2, and Ca^{2+}/calmodulin-dependent protein kinase IV (CaMKIV). CaMKIV is a particularly interesting activator because it has been shown using hippocampal neurons that either antisense oligonucleotides directed against this protein or expression of a nuclear calmodulin trap (which prevents activation of CaMKIV) can block Ca^{2+}-mediated phosphorylation of CREB on Ser-133 (Bito et al., 1996; Deisseroth et al., 1998). Alternatively the Ras-MAPK-RSK2 pathway may mediate glutamate induction of CREB phosphorylation and CREB-mediated transcription in hippocampal and striatal neurons (Impey et al., 1998; Vanhoutte et al., 1999).

Particle-mediated transfection of cultured Purkinje cells with an expression vector encoding a dominant inhibitory form of CREB resulted in a nearly complete blockade of the late phase of LTD (Ahn et al., 1999). Fura-2 microfluorimetry revealed that these transfections did not significantly alter depolarization-induced Ca^{2+} influx or mGluR-induced Ca^{2+} mobilization, suggesting that their effects were not mediated by inhibition of processes previously shown to be necessary for LTD induction. As CREB is activated by phosphorylation at Ser-133, kinases that phosphorylate this site were inhibited and LTD was assessed. Whereas inhibitors of PKA or the MAPK-RSK cascade were without effect on the late phase, transfection with expression vectors encoding a CaMK inhibitor peptide, dominant-negative forms of CaMKIV, or a calmodulin trap localized to the nucleus produced attenuation of the late phase of LTD and did not significantly alter Ca^{2+} signaling.

These results indicate that activation of CaMKIV and CREB are necessary to establish a late phase of cerebellar LTD. However, caveats should be sounded in interpreting these results. First, although it is tempting to speculate that phosphorylation of CREB on Ser-133 by CaMKIV is required for the late phase of cerebellar LTD, it should be emphasized that there is no direct evidence that this is the case. The requirement for CREB activation could represent the action of yet another pathway that also results in Ser-133 phosphorylation. Likewise, the requirement for CaMKIV could indicate the action of CaMKIV upon some other portion of the transcriptional regulatory apparatus, as has been demonstrated for the transcriptional co-activator protein CBP (Chawla et al., 1998; Hu et al., 1999). Second, although CaMKIV is strongly expressed in the nucleus of neonatal Purkinje cells and in cultures derived from embryonic cerebellum, adult Purkinje cells express CaMKIV only weakly (Sakagami et al., 1992; Sakagami and Kondo, 1993), suggesting that this pathway may not necessarily underlie a late phase of cerebellar LTD throughout the life span.

Interactions between LTP and LTD at Parallel Fiber Synapses

Cerebellar LTP induction appears to require presynaptic (but not postsynaptic) Ca^{2+} influx (Sakurai, 1990; Shibuki and Okada, 1992; Salin et al., 1996; Linden, 1997) and activation of Ca^{2+}-sensitive adenylyl cyclase type I (Storm et al., 1998), an enzyme that is concentrated in granule cell presynaptic terminals. This presynaptic cAMP elevation then activates PKA (Salin et al., 1996) in this same compartment (Linden and Ahn, 1999), leading to LTP induction. Several lines of evidence suggest that the expression of cerebellar LTP is presynaptic as well. First, induction of cerebellar LTP is associated with a decrease in the rate of synaptic failures (Hirano, 1991; Linden, 1997, 1998; Storm et al., 1998) and the extent of paired pulse facilitation (Salin et al., 1996; Linden, 1998). Unfortunately, neither of these forms of evidence is definitive, as postsynaptic scenarios have been proposed in which these parameters could be altered. Second, induction of an LTP-like effect by application of an exogenous cAMP analogue was associated with an increase in presynaptic vesicular cycling as measured using an immunocytochemical technique (Chavis et al., 1998). This LTP-like effect is independent of alterations in axonal excitability or Ca^{2+} influx into presynaptic terminals, suggesting a direct effect on the secretory apparatus (Chen and Regehr, 1997). Third, cerebellar LTP in culture can be detected using either AMPA/ kainate receptor-mediated currents recorded in postsynaptic Purkinje cells, AMPA/kainate receptor-mediated currents recorded in postsynaptic glial cells, or electrogenic glutamate transport currents recorded in postsynaptic glial cells, suggesting a common presynaptic locus of expression (Linden, 1997, 1998). Thus the most parsimonious model for cerebellar LTP induction is that presynaptic Ca^{2+} influx activates Ca^{2+}-sensitive adenylyl cyclase and the resultant cAMP transient activates presynaptic PKA, resulting in a phosphorylation event that ultimately potentiates glutamate release.

Our present understanding of the parallel fiber–Purkinje cell synapse is that LTD is expressed postsynaptically while LTP is expressed presynaptically. In such a scheme, cerebellar LTP and LTD would not truly reverse each other (as has been demonstrated for LTP and LTD at the hippocampal Schaffer collateral–CA1 synapse, as discussed subsequently) but rather would be additive, independent phenomena. To test this notion, a saturation-reversal experiment was performed using granule cell–Purkinje cell pairs in culture and perforated-patch recording (Fig. 10.3). Induction of LTP by granule cell stimulation at 4 Hz for 100 pulses resulted in

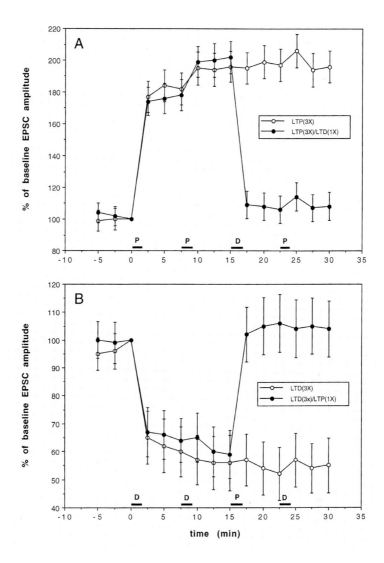

Figure 10.3. LTP and LTD of granule cell–Purkinje cell synapses in a cell culture system do not reverse each other. Recordings were made from granule cell–Purkinje cell pairs in cultures derived from embryonic mouse cerebellum (see Linden, 1997, for methods). Purkinje cell membrane currents were recorded in perforated patch voltage clamp mode at a holding potential of –80 mV. Granule cells were stimulated to fire single action potentials using loose-patch electrodes attached to the soma. Cultures were bathed in a normal HEPES-buffered external saline supplemented with picrotoxin to block $GABA_A$ receptors. EPSC amplitudes were calculated as the mean of 10 consecutive responses, including failures.

(A) Following a period of baseline recording, LTP-inducing stimulation (consisting of 100 pulses at 4 Hz) was applied at $t = 0$ min, and again at $t = 7.5$ and 22.5 min [LTP(3X) group; $n = 6$]. A second group received the same treatment with the addition of LTD-inducing stimulation (consisting of 100 pulses at 1 Hz, each paired with a 50-msec depolarization of the Purkinje cell to 0 mV) at $t = 15$ min. [LTP(3X)/LTD(1X) group; $n = 7$].

(B) The converse experiment to the one illustrated in *A*. LTD(3X), $n = 6$. LTD(3X)/ LTP(1X), $n = 6$.

an increase in synaptic strength. When, in the same cell pairs, this stimulation was repeated 7.5 min and then again 22.5 min later, a small additional degree of potentiation resulted from the second conditioning stimulus and no further potentiation resulted from the third, indicating that LTP was saturated after two conditioning stimuli. If LTD reverses LTP at this synapse, then LTD (induced by granule cell stimulation paired with depolarization) interposed between the second and third LTP stimuli should allow for some LTP at the third stimulus. This is not the case—intervening LTD did not rescue LTP saturation (Fig. 10.3A). Similarly, when a reverse saturation experiment was performed, intervening LTP also failed to rescue saturated LTD (Fig. 10.3B).

These findings indicate that parallel fiber LTP and LTD do not reverse each other. Obviously this does not allow for a simple model of cerebellar information storage, for which reversible, bidirectional control of synaptic strength would be preferable. One possible resolution of this problem is that optimal and rapid flexibility is not required at this synapse because the relevant information is rapidly transferred to a more permanent storage site in the DCN. Another possibility is that there are additional forms of plasticity (presynaptic LTD, postsynaptic LTP) that would allow for reversible, bidirectional control but that have yet to be characterized.

Potential Role in Motor Learning

Is LTD of the parallel fiber–Purkinje cell synapse involved in motor learning? The experimental evidence is suggestive but far from conclusive. Application in the cerebellar molecular layer of drugs that interfere with LTD induction during motor learning tasks would be instructive in this regard. At present only a small number of studies have made use of this strategy, and these have focused upon NO-cGMP signaling—a somewhat complicated aspect of the LTD induction process. Nagao and Ito (1991) have shown that VOR adaptation may be blocked by application of hemoglobin (which functions as an extracellular NO trap) to the subdural space overlying the ipsilateral flocculus in monkey and rabbit. Similarly Li et al. (1995) found that injection of an NOS inhibitor into the vestibulocerebellum of the goldfish also blocked VOR adaptation and that the effect of this NOS inhibitor could be overcome by simultaneous administration of L-arginine (the substrate of NOS). They found that adaptive gain increases (but not decreases) were blocked by this process and that administration of the NOS inhibitor after vestibulo-ocular reflex adaptation had no effect on retention of this form of motor learning. This finding is consistent with a model in which NO is required for the induction of LTD, but not LTP, and in which LTD induction, but not maintenance, requires the continued presence of NO.

A related approach to testing the hypothesized LTD–motor learning connection has been provided by the generation of mutant mice that lack proteins thought to be required for cerebellar LTD (see Chen and Tonegawa, 1997, for review). At present three forms of knockout mice have been reported that have shown impairments in both parallel fiber LTD and motor learning. Mutant mice that lack mGluR1 have severely impaired cerebellar LTD (Aiba et al., 1994; Conquet et al., 1994), consistent with previous studies using mGluR1 antagonists or inactivating antibodies. These mice have normal postsynaptic voltage-gated Ca^{2+} currents and normal paired pulse facilitation and depression of parallel and climbing fiber synapses, respectively. mGluR1 mutant mice are severely ataxic and show impairments in several motor coordination tasks. When associative eyeblink conditioning was performed, these animals showed a partial deficit in CR acquisition (Aiba et al., 1994), which may not reflect an inability to produce a CR, but rather an inability to produce the optimal timing of the eyelid closure.

Two additional knockout mice also have LTD and motor learning deficits. Unfortunately in both cases it is not understood how the missing protein functions in LTD induction. GluRδ2 is a protein expressed almost exclusively in Purkinje cell dendritic spines. While a point mutation in this receptor gives rise to a constitutive cation conductance and the *lurcher* phenotype (Zuo et al., 1997), the function of wild-type GluRδ2 is not clear. A GluRδ2 null mouse has impaired cerebellar LTD (Kashiwabuchi et al., 1995), consistent with reports that have used application of antisense oligonucleotides to suppress expression of this protein in culture preparations (Hirano et al., 1994a,b; Jeromin et al., 1996). Although the cerebellar cortex appears normal by light microscopy, electron microscopic analysis reveals a 50% reduction in the number of parallel fiber–Purkinje cell synapses. This mouse is severely ataxic and is retarded in its ability to achieve a form of vestibular compensation: the righting response under a rotation load following unilateral middle ear destruction (Funabiki et al., 1995).

An even more puzzling LTD defect is found in a mutant mouse that lacks glial fibrillary acidic protein (GFAP), which is of course not expressed in neurons. Since at least the early phase of cerebellar LTD may be expressed in reduced preparations that lack glia (Narasimhan and Linden, 1996; Narasimhan et al., 1998), it is likely that this knockout is exerting its effect indirectly. These mice show normal motor coordination but are severely impaired in parallel fiber LTD and partially impaired in acquisition of associative eyeblink conditioning (Shibuki et al., 1996). Interestingly, these mice show clear improvement with training in a motor coordination task (the rotating rod), indicating that there are forms of motor learning that do not require cerebellar LTD.

Several major problems have complicated the analysis of knockout mice. First, knockout mice have the gene of interest deleted from the earliest stages of development. As a result, these mice often have a complex developmental phenotype. For example, PKCγ, mGluR1, and GluRδ2 (but not GFAP) knockout mice all have cerebellar Purkinje cells that fail to undergo the normal developmental conversion from multiple to mono climbing fiber innervation in early postnatal life (Chen et al., 1995; Kano et al., 1995, 1997; Kurihara et al., 1997). Second, knockout of one gene sometimes produces compensatory upregulation in the expression of other related genes during development. In the CREBα–δ knockout mouse, there is compensatory upregulation of the related transcription factor CREM (Hummler et al., 1994; Blendy et al., 1996). Similarly, it is possible that PKCγ knockout mice do not show impaired LTD or cerebellar motor learning because of compensation by other PKC isoforms in Purkinje cells (Chen et al., 1995). A third complicating factor is that knockout mice have the gene of interest deleted in every cell of the body, not just the cells of interest, hindering efforts to ascribe the knockout's behavioral effects to dysfunction in any one particular structure or cell type. This could be a problem for the analysis of behaviors such as associative eyeblink conditioning and VOR adaptation, which are likely to require use-dependent plasticity at multiple sites, synapses received by both cerebellar Purkinje cells and their targets in the DCN or vestibular nuclei.

To address the latter two complications, an alternative approach has been used (De Zeeuw et al., 1998). Using the promoter of the Purkinje cell–specific gene *pcp2* (also known as L7; Oberdick et al., 1990), transgenic mice have been created in which a selective inhibitor to a broad range of PKC isoforms (House and Kemp, 1987; Linden and Connor, 1991) is chronically overexpressed. This strategy ensures that in L7-PKCI mice PKC inhibition will be restricted to Purkinje cells, and that compensation via upregulation of different PKC isoforms will not succeed in blunting the biochemical effect of the transgene. This transgenic strategy resulted in nearly complete suppression of both cerebellar LTD (as assessed in culture) and adaptation of the VOR in the intact, behaving animal. The phenotype of these animals was remarkably delimited, exhibiting normal motor coordination as measured by their ability to display normal eye movement reflexes (optokinetic reflex and VOR) as well as by several tests of gross motor coordination (rotorod, thin rod). Basal electrophysiological and morphological features of Purkinje cells were unaltered (with the exception that the Purkinje cells remained multiply innervated by climbing fibers).

Although L7-PKCI mice represent a refinement in testing the relationship between parallel fiber LTD and motor learning, several caveats still remain. First, the one basal physiological abnormality found in L7-PKCI mice is that about 50% of the Purkinje cells show a persistent multiple climbing fiber innervation, raising the possibility that this could be a cause of their failure to demonstrate VOR adaptation. However, this is unlikely because the PKCγ knockout mouse shows multiple climbing fiber innervation with normal cerebellar LTD and no motor learning deficit (Chen et al., 1995; Kano et al., 1995). A strategy to alleviate this problem in the future will be to make the L7-PKCI transgene inducible in adult mice. Second, it has recently been reported that the climbing fiber–Purkinje cell synapse undergoes LTD, which, like parallel fiber LTD, requires activation of PKC (Hansel and Linden, 2000). Thus even inhibition of PKC that is restricted to Purkinje cells cannot have its behavioral effects solely ascribed to parallel fiber LTD. Third, it is clear that PKC does not function in Purkinje cells only to produce LTD of various synapses. For example, it is known that voltage-gated K^+ channels are modulated by PKC in Purkinje cells. Could this or some other function of PKC that is unrelated to LTD underlie the VOR adaptation deficit? Although this is not easy to address, future studies will benefit from in vivo recording during the appropriate behavioral tasks to distinguish among these possibilities.

LTD of the Hippocampal Schaffer Collateral–CA1 Synapse

Because it is located many synapses from the sensory periphery, the hippocampus is not the ideal place to connect the mechanisms of synaptic plasticity with their behavioral consequences. However, as in the cerebellar cortex, the simple organization of this structure renders it extraordinarily useful for studies of basic synaptic physiology. Indeed much of what we know about the elementary mechanisms of synaptic transmission in the mammalian brain has derived from work using the hippocampus, particularly the Schaffer collateral–CA1 synapse. Of course the hippocampus does store certain types of information, and it is reasonable to expect that synaptic plasticity plays a role in this process (Riedel et al., 1999). Moreover, available evidence suggests that what has been learned in the hippocampus is also widely applicable to synapses elsewhere in the brain, where changes in synaptic function can be more easily associated with changes in behavior.

The Schaffer collateral–CA1 synapse is perhaps most famous as a model for LTP. Although the original discovery of LTP (in the dentate gyrus, by Terje Lømo) was accidental (the unexpected outcome of experiments designed to study synaptic responses during repetitive stimulation), it was quickly embraced as a potential synaptic mechanism for memory (Bliss and Lømo, 1973). Excitement grew in the mid-1980s, when the properties of LTP in CA1 were shown to satisfy the requirements of Hebb's famous postulate that active synapses strengthen when their activity correlates specifically with a strong postsynaptic response (Wigstrom and Gustafsson, 1985; Kelso et al., 1986; Malinow and Miller, 1986; Sastry et al., 1986). However, theoreticians had concluded years before that "Hebbian" synaptic modifications alone were not likely to be sufficient to account for memory storage; the efficient storage of information by synapses requires bidirectional synaptic modifications—that is, LTD as well as LTP. Thus the search for homosynaptic LTD in the hippocampus, like Ito's search for LTD in the cerebellum, was theoretically motivated; it was not an accident. (The reasons that it was not stumbled upon accidentally will become clear in the following discussion.) The theoretical suggestion was that synapses should depress when their activity *fails* to correlate with a strong postsynaptic response. To realize this situation experimentally, induction of LTD was attempted in CA1 using prolonged trains of presynaptic stimulation, delivered at frequencies (0.5–10 Hz) that fail to evoke a strong postsynaptic response (Dudek and Bear, 1992). Trains of low-frequency stimulation (LFS) are now the standard protocol for induction of homosynaptic LTD in CA1 and at synapses throughout the forebrain.

Remarkable progress has been made since the last major review of hippocampal LTD (Bear and Abraham, 1996). At that time, there was a general consensus that LFS induces LTD, but there were puzzling contradictions regarding the induction and expression mechanisms of LTD. It is now understood that LFS induces at least two, and possibly three (Berretta and Cherubini, 1998), mechanistically distinct forms of LTD in CA1. One form depends on activation of NMDARs; the other depends on activation of mGluR5, a postsynaptic glutamate receptor coupled to phosphoinositide metabolism. Another controversy that raged in 1996 concerned the relevance of LTD to hippocampal function in the adult brain, particularly in vivo. Although perhaps not resolved fully, it has become clear that LTD (like learning) is extraordinarily sensitive to changes in behavioral state. Finally, there has been progress in deducing the relationship of LTD to the "depotentiation" that occurs when LFS is given after LTP has been induced.

Theoretical Framework

The bidirectional modification of excitatory synaptic transmission is not just an abstract theoretical construct. We need to understand mechanisms of bidirectional synaptic plasticity because direct experimental observations have repeatedly shown that synapses in the cerebral cortex are in fact bidirectionally modifiable. The value added by a theoretical structure is that it helps to make sense of what bidirectional synaptic plasticity accomplishes with respect to information storage and provides insight into how it might be implemented.

Neurons throughout the cerebral cortex, including area CA1 of the hippocampus, have stimulus-selective receptive fields. Chronic recording from cortical neurons has shown that as something new is learned, stimulus selectivity changes—some synaptic inputs potentiate whereas others depress. In CA1, for example, neurons show selectivity for positions in space, and this selectivity shifts rapidly as animals learn a new spatial environment (Breese et al., 1989; Wilson and McNaughton, 1993). What does a stable shift in selectivity tell us about memory? Neural network theory suggests that the selectivity shift reflects the creation of new neural representations. The memory is encoded by changing the pattern of synaptic weights across the network of neurons (Bear, 1996).

Now consider what happens when more new information is learned: stimulus selectivity (i.e., the pattern of synaptic weights) shifts further. An implication of this finding is that previously encoded memories can remain stable, even as the pattern of synaptic strengths is again modified to create new representations. According to this way of thinking, memory requires the episodic (if not continual) bidirectional modification of synaptic transmission to fine-tune the patterns of synaptic weights in the neural network. It is important to emphasize, of course, that in the absence of new learned information synaptic weights must remain stable. Passive decay of synaptic weight (i.e., back to an initial value that might be larger or smaller) leads to a loss of the stored representations.

The bidirectional modification of synaptic transmission obviously requires that individual synapses on neurons be capable of some form of LTP and some form of LTD. However, every theory of memory storage that assumes bidirectional synaptic modification places an important constraint on the mechanisms of LTP and LTD: reversibility. Consider the problem that arises if the LTP and LTD mechanisms are distinct and irreversible. Although it is true that synaptic weights could be fine-tuned initially by simple summation of the two independent processes, eventually saturation would occur as the synapses underwent rounds of bidirectional modification (see Fig. 10.3). This problem does not occur if LTP and LTD are inverse processes mechanistically.

Now we come to the question of what distinguishes stimulation conditions that yield synaptic potentiation from those that yield synaptic depression. To encode memory specifically, synaptic modifications must depend on the presynaptic activation of the synapses bringing information into the network. In other words, the modifications must be "homosynaptic." The variables that determine the polarity or sign of the modification, in principle, could be the absolute amount of presynaptic activity, the concurrent level and timing of postsynaptic activity, or some combination of these variables. There are many theoretical "learning rules" based on these variables, but the most useful are those that attempt to account for what has actually been observed experimentally. One very influential proposal was made by Bienenstock, Cooper, and Munro (1982) in what is known as the BCM theory. In order to account for the development and plasticity of neuronal stimulus selectivity, they proposed that active synapses are potentiated when the total postsynaptic response exceeds a critical value—the "modification threshold" (θ_m)—and that active synapses are depressed when the total postsynaptic response is greater than zero but less than θ_m. In addition, it was proposed that the value of θ_m varies as a function of the average integrated postsynaptic activity.

Once the requirements for LTP induction in CA1 had been elucidated, a specific physiological basis for the BCM theory became apparent. A number of proposals were made: (1) the term θ_m corresponds to the critical level of postsynaptic depolarization at which the Ca^{2+} flux through the NMDAR exceeds the threshold for inducing LTP; (2) LTD should be a consequence of presynaptic activity that consistently fails to evoke a postsynaptic Ca^{2+} response large enough to induce LTP; and (3) the postsynaptic threshold for LTP should vary depending on the stimulation history of the postsynaptic neuron (Bear et al., 1987). These hypotheses have now all been validated experimentally (Fig. 10.4).

The BCM theory motivated the search for LTD using LFS (Dudek and Bear, 1992). The rationale was to provide a high level of presynaptic activity that did not evoke a large postsynaptic response. Critical variables for LTD induction in rat hippocampal slices proved to be the stimulation strength (it could not be so strong as to elicit orthodromic action potentials), healthy inhibition, stimulation frequency (<10 Hz), the number of stimuli (typically hundreds), and the age of the animal (greater magnitude before 35 days of age). LTD was found to be a very reliable phenomenon when the appropriate conditions were met. Even so, there was concern that LTD might be an artifact. First, there was heightened skepticism because several previous reports of LTD induction using different protocols had proven difficult to replicate in the hippocampus. Second, the same LFS protocol that induces LTD had previously been reported by others to be ineffective in altering baseline synaptic transmission. And third, synaptic depression could easily be dismissed as a pathological change rather than a form of synaptic plasticity. Most of these concerns faded when it was found that at least one form of LTD depended specifically upon activation of NMDARs and a rise in postsynaptic Ca^{2+}, that the same synapses that showed LTD could subsequently be potentiated, and that LTD could be elicited in vivo (Dudek and Bear, 1992, 1993; Mulkey and Malenka, 1992; Thiels et al., 1994; Heynen et al., 1996; Debanne et al., 1997; Manahan-Vaughan, 1997).

It is now apparent that multiple forms of LTD exist in CA1, possibly at the same synapses. Remarkably, however, all forms of LTD can be elicited using variations of the LFS protocol, specifically under conditions that fail to evoke a large postsynaptic response. Thus the hypothesis of homosynaptic LTD, inspired by the BCM theory, has been amply confirmed. It is also noteworthy that the NMDAR-dependent form of LTD appears to be the functional inverse of LTP, thus satisfying the theoretical requirement that synaptic modifications be both bidirectional and reversible. Thus, at least in theory, the mechanisms of LTD as well as LTP can contribute to the receptive field plasticity underlying memory storage in the hippocampus and elsewhere.

NMDAR-Dependent LTD

Induction by Ca^{2+}

Induction of homosynaptic LTD in CA1 using the standard 1-Hz LFS protocol in vivo, and under most experimental conditions in vitro, is blocked by NMDAR antagonists (Dudek and Bear, 1992; Mulkey and Malenka, 1992; Heynen et al., 1996; Manahan-Vaughan, 1997). Although

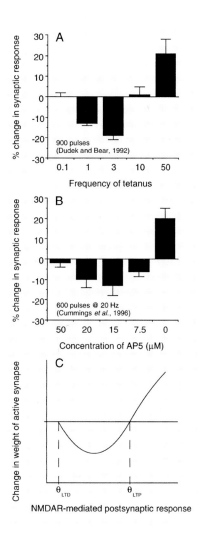

Figure 10.4. Induction of NMDAR-dependent bidirectional plasticity of the Schaffer collateral synapse in CA1.

(A) Summary of the effects of a 900-pulse tetanus delivered at different frequencies. Replotted from Dudek and Bear (1992).

(B) Summary of the effects of a 600-pulse, 20-Hz tetanus in different concentrations of the NMDAR antagonist AP5. Replotted from Cummings et al. (1996).

(C) A synaptic "learning rule" based on data such as those given in A and B. This learning rule is formally similar to that proposed in the BCM theory.

this form of LTD shares with LTP a dependence upon NMDA receptor activation and a rise in postsynaptic calcium ion concentration, there is a systematic difference in the type of stimulation that yields the two types of synaptic modification. This difference is easily demonstrated simply by varying the frequency of tetanic stimulation. In rat CA1, for example, 900 pulses at 0.5–3 Hz typically yield LTD, whereas the same amount of stimulation at frequencies greater than 10 Hz yields LTP (Fig. 10.4A; Dudek and Bear, 1992). The different consequences of stimulation at different frequencies have been attributed to systematic differences in the postsynaptic Ca^{2+} currents through the postsynaptic NMDA receptors. Indeed it is now well established that the critical variables are postsynaptic depolarization and Ca^{2+} entry, not stimulation frequency per se. For example, while 1-Hz stimulation normally produces LTD, postsynaptic hyperpolarization during conditioning prevents any change, and depolarization leads to induction of LTP (Mulkey and Malenka, 1992). Likewise, while high-frequency stimulation normally produces LTP, it produces LTD instead if delivered in the presence of subsaturating concentrations of an NMDA receptor antagonist (Cummings et al., 1996; Fig. 10.4B).

The appropriate activation of postsynaptic NMDARs appears to be sufficient to induce LTD; presynaptic activity is not necessary. This conclusion is supported by the observation that photolysis of caged extracellular glutamate (Kandler et al., 1998; Dodt et al., 1999) and brief bath application of NMDA (Lee et al., 1998; Kamal et al., 1999) induce LTD without concurrent presynaptic stimulation. In agreement with the idea that Ca^{2+} passing through NMDARs is the trigger for LTD, photolysis of caged calcium in the postsynaptic neuron can also induce synaptic depression (Neveu and Zucker, 1996). This finding is important, as it indicates that LTD can be induced by Ca^{2+} entry through the NMDAR without the need to invoke any other calcium-independent signaling or triggering process. Curiously, however, a modest, brief elevation in $[Ca^{2+}]$ was found to induce LTP and LTD with equal probability. Subsequent analysis suggests that LTD is most reliably induced (and LTP is never induced) by a modest (circa 0.7 μM) but prolonged (circa 60 sec) rise in $[Ca^{2+}]$. LTP, in contrast, is most reliably induced by a large (circa 10 μM) and brief (circa 3 sec) increase in $[Ca^{2+}]$ (Yang et al., 1999). Although these findings are all consistent with the proposal that LTD and LTP are triggered by distinct calcium responses during NMDAR activation (Bear et al., 1987; Lisman, 1989; Artola and Singer, 1993; Bear and Malenka, 1994), they argue against a simple relationship between synaptic modification and calcium level. The dynamics of the calcium response are also important determinants of the polarity of synaptic modification.

Perhaps not surprisingly, under certain circumstances voltage-gated Ca^{2+} channels may also contribute to the Ca^{2+} signal that triggers LTD (Christie et al., 1996). However, this role appears to be supplementary, since NMDAR activation is still required to observe homosynaptic LTD under these conditions, and calcium channel activation is not always necessary (Selig et al., 1995b). Nonetheless, the modulation of synaptic plasticity by active dendritic calcium conductances can be striking. Markram et al. (1997) made the remarkable observation that a back-propagating dendritic action potential, precisely timed to occur a few milliseconds before a synaptically evoked excitatory postsynaptic potential (EPSP), could promote induction of LTD in neocortical pyramidal neurons by stimuli that otherwise are ineffective. Similar findings have been reported in hippocampal cultures (Bi and Poo, 1998). Although the mechanism remains to be determined, one appealing hypothesis is that, owing to Ca^{2+}-dependent inactivation of the NMDARs, postsynaptic Ca^{2+} entry via voltage-gated ion channels sums sublinearly with the Ca^{2+} entering through NMDAR to satisfy the conditions required for LTD induction (Linden, 1999; Zucker, 1999). Although this associative mechanism for LTD induction suggests particular neural computations, it is important to keep in mind that LTD can also be elicited at the same synapses with subthreshold stimuli. The more computationally significant finding is that if postsynaptic spikes occur immediately *after* the EPSP, LTD never occurs (owing to enhanced Ca^{2+} entry through NMDARs relieved of their Mg^{2+} block). Instead, LTP usually results. Thus the maxim that LTD results when presynaptic activity consistently fails to correlate with a strong postsynaptic response still holds.

To summarize what has been learned to date, LTD is induced by an elevation of Ca^{2+} that is apparently constrained by three variables: (1) proximity to the postsynaptic membrane, (2) peak concentration, and (3) duration. Synaptic stimulation causes LTD when it yields the appropriate response within the limits defined by these parameters. Although LFS at 1 Hz is usually effective, under different experimental conditions different protocols may be required (e.g., Debanne et al., 1994; Thiels et al., 1994). Ca^{2+} entry through the NMDA receptor is sufficient to induce LTD, but this can also be supplemented by other Ca^{2+} sources (Christie et al., 1996; Reyes and Stanton, 1996).

The Role of Ca^{2+}-Dependent Enzymatic Reactions

Induction of LTP requires activation of Ca^{2+}-dependent serine-threonine protein kinases in the postsynaptic neuron, in particular PKC and Ca^{2+}/calmodulin-dependent protein kinase II (CaMKII) (Malinow et al., 1989; Malenka et al., 1989). Lisman (1989) proposed that LTD might result from activation of a protein phosphatase cascade, leading to dephosphorylation of the same synaptic proteins that are involved in LTP. Subsequent experiments confirmed an essential role for postsynaptic protein phosphatase 1 (PP1) and calcineurin (PP2B) in the induction of LTD with LFS (Mulkey et al., 1993, 1994). So far this model has withstood the test of time.

The basic working hypothesis continues to be that LTD results from dephosphorylation of postsynaptic PP1 substrates. As will be discussed subsequently, there is now direct evidence for dephosphorylation of synaptic proteins following LTD induction protocols (Lee et al., 1998, 2000; Ramakers et al., 1999). Moreover there is evidence that PP1 activity is persistently increased by LTD-inducing stimulation (Thiels et al., 1998). PP1 is regulated by the protein inhibitor-1 (I-1). When I-1 is phosphorylated by PKA, PP1 is inactive. Dephosphorylation of I-1 by PP2B releases PP1 from inhibition (Cohen, 1989; Nairn and Shenolikar, 1992).

PP2B is activated by Ca^{2+}/calmodulin and therefore is believed to be key for translating an increase in $[Ca^{2+}]$ into LTD (Fig. 10.5A).

Before moving on to consider expression mechanisms, it should be noted that there are also data suggesting that Ca^{2+}/calmodulin triggers LTD by activating NOS (Izumi and Zorumski, 1993; Gage et al., 1997). The proposed mechanism of NO action is the retrograde activation of a second messenger cascade in the presynaptic terminal involving soluble guanylyl cyclase and cyclic GMP-dependent protein kinase (Reyes et al., 1999). This scenario is not universally agreed upon, however, since others report no effect of NO inhibitors on LTD (Cummings et al., 1994). At the present time, it appears that this mechanism lies in parallel with that involving postsynaptic phosphatase activation. It is plausible that NO signaling is involved in the mGluR-dependent form of LTD, but this hypothesis remains to be examined explicitly.

Expression Mechanisms

It now seems very clear that a modification of postsynaptic glutamate sensitivity is a major expression mechanism for the NMDAR-dependent form of LTD. As previously mentioned, liberating caged glutamate in CA1, under conditions in which synaptic transmission is blocked, results in LTD of the glutatmate-evoked currents. This LTD is restricted spatially to the site of glutamate release and depends upon NMDAR activation and postsynaptic protein phosphatase activity (Kandler et al., 1998). The possibility remains that this form of LTD is actually expressed at extrasynaptic glutamate receptors and therefore could be mechanistically distinct from LFS-induced LTD. However, very similar findings have been obtained in rat neocortex, where it was also shown that synaptically induced LTD results in a decrease in sensitivity to laser-stimulated photolysis of caged glutamate. Moreover, LFS-induced LTD occluded further synaptic depression by glutamate pulses (Dodt et al., 1999). The close similarities between CA1 and neocortical LTD (Kirkwood et al., 1993; Kirkwood and Bear, 1994) suggest that these findings apply generally to NMDAR-dependent LTD (NMDAR-LTD) in the cerebral cortex.

An interesting picture has emerged recently to account for decreased glutamate sensitivity following LTD. The model of bidirectional synaptic modification through reversible changes in the phosphorylation of postsynaptic substrates (Lisman, 1989; Bear and Malenka, 1994) begged the question of which synaptic phosphoproteins are involved. Now there is direct evidence that LTD is associated with dephosphorylation of the GluR1 subunit of the AMPAR, the consequence of which is known to be depression of glutamate-evoked currents. In addition there are converging lines of evidence that LTD is associated with the removal of glutamate receptors from the postsynaptic membrane.

AMPARs are heteromeric complexes assembled from four homologous subunits (GluR1–4) in various combinations (Seeburg, 1993; Hollmann

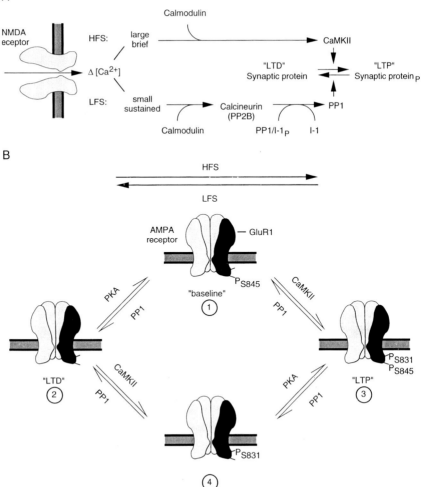

Figure 10.5. Molecular models for bidirectional plasticity of the Schaffer collateral synapse in CA1.

(A) Induction. Strong NMDAR activation resulting from high-frequency stimulation (HFS) causes a large, brief rise in intracellular Ca^{2+}. This Ca^{2+} signal triggers activation of CaMKII and the induction of LTP by phosphorylation of synaptic proteins. Weak NMDAR activation resulting from low-frequency stimulation (LFS) produces a smaller but more sustained rise in intracellular Ca^{2+}. This Ca^{2+} signal selectively activates calcineurin, which dephosphorylates inhibitor-1 (I-1). Dephosphorylation of I-1 relieves PP1 from inhibition. LTD results from dephosphorylation of the same synaptic proteins involved in LTP.

(B) Expression. (1) Under baseline conditions, the GluR1 subunit of AMPARs is highly phosphorylated at Ser-845, a PKA substrate. (2) LFS causes dephosphorylation of Ser-845 and LTD. (3) From the baseline state, HFS causes phosphorylation of Ser-831, a CaMKII substrate, and LTP. According to this model, LTD and LTP result from bidirectional modifications of AMPAR phosphorylation, but at different sites. Thus dedepression and LTP are not formally equivalent, nor are depotentiation and LTD. (4) The fourth possible state, in which Ser-831 is phosphorylated and Ser-845 is not, has not been observed experimentally.

Adapted from Kameyama et al. (1998).

and Heinemann, 1994). The large majority of AMPARs in the hippocampus contain both GluR1 and GluR2 subunits (Wenthold et al., 1996). The GluR1 subunit is highly regulated by protein phosphorylation and contains two identified phosphorylation sites on the intracellular carboxy-terminal domain. Serine 831 is phosphorylated by CaMKII and PKC, while Ser-845 is phosphorylated by PKA (Roche et al., 1996; Barria et al., 1997a). Phosphorylation of either of these sites has been shown to potentiate AMPAR function through distinct biophysical mechanisms (Derkach et al., 1999). Phosphorylation site–specific antibodies have been used to measure the phosphorylation status of receptors in situ (Mammen et al., 1997). The PKA site shows higher basal phosphorylation than the CaMKII-PKC (Lee et al., 1998).

To investigate the changes in AMPAR phosphorylation that occur following synaptic plasticity, Lee et al. (1998) devised a method to induce LTD chemically in hippocampus slices. The rationale behind this approach was to increase the probability of detecting biochemical changes by maximizing the number of affected synapses in the slice. They showed that brief bath application of NMDA induces synaptic depression (called chem-LTD) that shares a common expression mechanism with LFS-induced LTD. Biochemical analysis showed a selective dephosphorylation of Ser-845 (the PKA site) following induction of chem-LTD. Phosphorylation of Ser-831 (the CaMKII-PKC site), in contrast, was not altered by the treatment. This result was unexpected, as Ser-831 is the site phosphorylated during LTP (Barria et al., 1997b). Thus these findings contradict the simple notion that LTD and LTP reflect bidirectional changes in the phosphorylation of the same site.

By refining the biochemical detection method, these findings have now been confirmed using synaptically induced LTD and LTP (Lee et al., 2000). LFS delivered to naïve synapses causes dephosphorylation of the PKA site and LTD. Conversely theta-burst stimulation (TBS) delivered to naïve synapses causes phosphorylation of the CaMKII-PKC site and LTP. Both types of synaptic change are reversible, however. Thus LFS delivered after prior induction of LTP causes dephosphorylation of the CaMKII-PKC site and depotentiation of the synaptic response. TBS delivered after prior induction of LTD causes phosphorylation of the PKA site and de-depression of the synaptic response. The surprising implication of these findings is that the precise mechanisms of synaptic depression and potentiation caused by LFS and TBS, respectively, depend on the previous stimulation history of the synapse. The model suggested by these results appears in Fig. 10.5B.

Although the AMPAR phosphorylation changes could account for early expression of LTD and LTP, it must be conceded that they are merely biochemical correlates of the synaptic modifications. Nonetheless the model has withstood several experimental challenges. Experiments have now confirmed that dephosphorylation of postsynaptic PKA substrates (by postsynaptic injection of a PKA inhibitor) causes synaptic depression that occludes LFS-induced LTD. Moreover, activation of postsynaptic PKA reverses previously established LTD without affecting baseline synaptic transmission (Kameyama et al., 1998). Finally, recent experiments confirm that LTP and de-depression are differentially sensitive to inhibition of CaMKII (Lee et al., 2000). Taken together, the data strongly suggest that dephosphorylation of AMPARs is one expression mechanism for LTD in CA1.

Besides their regulation by phosphorylation, recent data have shown that AMPAR expression in the postsynaptic membrane is subject to rapid regulation. In the search for postsynaptic density proteins that interact with AMPARs, it was discovered that the carboxy terminus of GluR2 binds to the N-ethylmaleimide-sensitive factor (NSF), which plays an essential role in membrane fusion events (Nishimune et al., 1998; Osten et al., 1998; Song et al., 1998; Lüscher et al., 1999; Noel et al., 1999). Surprisingly, blocking this NSF-GluR2 interaction with specific peptide inhibitors produces the rapid (within minutes) internalization of receptors and the rundown of synaptic AMPAR currents. Although the mechanism by which the NSF-GluR2 interaction maintains AMPARs in the postsynaptic membrane remains to be determined, available evidence suggests that AMPAR internalization occurs via dynamin-dependent endocytosis (Lüscher et al., 1999). The pool of NSF-regulated AMPARs appears to be required for expression of LTD, because LFS has no effect after these receptors are internalized (Lüscher et al., 1999; Luthi et al., 1999).

Three lines of evidence suggest that AMPAR internalization is an expression mechanism for LTD. First, in hippocampal slices, prior saturation of LTD renders the AMPARs at the depressed synapses (but not at other synapses on the same neuron) insensitive to inhibitors of the NSF-GluR2 interaction (Luthi et al., 1999). Second, in hippocampal cell cultures, field stimulation at 5 Hz causes an NMDAR-dependent depression of spontaneous miniature EPSC amplitudes and the loss of surface-expressed GluR1 (Carroll et al., 1999). Third, in the adult hippocampus in vivo, there is an NMDAR-dependent loss of GluR1 and GluR2 from the synaptoneurosomal biochemical fraction following induction of LTD (Heynen et al., 2000).

In keeping with this evidence, the magnitude of the postsynaptic response to quantal release of glutamate (the quantal size) is decreased after LTD (Oliet et al., 1996). However, in addition, a robust finding is that the number of quantal responses to synaptic stimulation (the quantal content) is also decreased (Stevens and Wang, 1994; Goda and Stevens, 1996; Oliet et al., 1996; Carroll et al., 1999). According to traditional assumptions, decreased quantal content reflects the failure of neurotransmitter release in response to a presynaptic action potential. It is interesting, therefore, that a decrease in quantal content is also a consequence of disrupting the NSF-GluR2 interaction (Luthi et al., 1999). Presumably this results from the total loss of AMPARs from some synapses. Thus, although presynaptic changes may also occur following LFS (Ramakers et al., 1999), there is apparently no need to invoke a presynaptic mechanism to account for the key properties of NMDAR-LTD of AMPAR-mediated responses.

Another robust finding is that responses to activation of NMDARs are also depressed following LTD (Xiao et al., 1994; Selig et al., 1995a; Xiao et al., 1995). Although these observations are consistent with the parallel loss of AMPA and NMDA receptors from synaptoneurosomes following LTD in vivo (Heynen et al., 2000), they could also reflect a component of LTD expression that is presynaptic. The glutamate receptor internalization in response to intracellular manipulations of the NSF-GluR2 interaction (Luthi et al., 1999) or of dynamin-mediated endocytosis (Lüscher et al., 1999; Man et al., 2000; Wang and Linden, 2000) has been restricted to AMPARs. Thus the mechanism for NMDAR regulation is apparently distinct from that for AMPAR regulation.

Despite remarkable progress, many questions remain. How can activation of NMDARs lead to AMPAR internalization? What is the relationship of AMPAR dephosphorylation to receptor internalization? How are NMDARs regulated? If the current pace of research on LTD is sustained, we can expect answers (no doubt unexpected ones) in the near future.

Modulation of LTD

As mentioned, induction of LTD depends on postsynaptic phosphatase activation by Ca^{2+} passing through NMDAR channels. Thus it comes as no surprise that LTD is subject to modulation by factors that alter the Ca^{2+} flux in response to synaptic stimulation and by factors that alter the intracellular enzymatic response to a change in Ca^{2+} concentration.

An example of the first type of modulation is the effect of altered inhibition. Under some experimental conditions, reduced inhibition may be required to allow the NMDAR activation that is necessary to induce LTD (Wagner and Alger, 1995). However, under other conditions reduced inhibition may suppress LTD in response to LFS at certain frequencies by facilitating induction of LTP instead (Steele and Mauk, 1999). Similarly, the conditions required for LTD induction depend on the properties of postsynaptic NMDARs. For example, overexpression of the NR2B subunit, which leads to a prolongation of NMDAR-mediated synaptic currents, changes the frequency-response function to promote LTP and suppress LTD across a range of stimulation frequencies (Tang et al., 1999). Modulation of inhibition and NMDAR subunit composition are physiologically relevant, as these parameters change during development and are regulated by activity.

The intracellular response to a change in calcium depends on the availability of Ca^{2+}-binding proteins, such as calmodulin, and on the location, concentration, and activity of the kinases and phosphatases that regulate synaptic strength. Mutations that alter these parameters enhance (Mayford et al., 1995) or disrupt (Brandon et al., 1995; Qi et al., 1996; Migaud et al., 1998) LTD. PKA seems to play a pivotal role in the intracellular regulation of LTD. According to the current model for LTD induction, activation of PKA would be expected to inhibit LTD by preventing the activation of PP1 (via I-1 phosphorylation) and by maintaining AMPAR phosphorylation at a high level. Regulation of LTD via PKA is also physiologically relevant. For example, there is evidence that PKA activation in response to stimulation of noradrenergic β-receptors shifts the frequency response function to favor LTP over LTD (Blitzer et al., 1995; Thomas et al., 1996; Katsuki et al., 1997; Blitzer et al., 1998). Conversely, activation of muscarinic acetylcholine receptors facilitates LTD (Kirkwood et al., 1999), possibly by PKC-mediated inhibition of adenylyl cyclase (Stanton, 1995; Nouranifar et al., 1998).

Other variables that affect LTD are postnatal age and the behavioral state of the animal. Although the mechanism remains to be determined, it is well established that the magnitude and reliability of LTD decline with age (Dudek and Bear, 1993; Errington et al., 1995; Wagner and Alger, 1995; Kamal et al., 1998), supporting the idea that this phenomenon plays an important role in the refinement of circuits during critical periods of development (Rittenhouse et al., 1999). However, the existence of LTD in the adult hippocampus has been the subject of some controversy, with some groups reporting success (Thiels et al., 1994; Heynen et al., 1996; Manahan-Vaughan, 1997) and others reporting failure (Errington et al., 1995; Doyle et al., 1997; Staubli and Scafidi, 1997) to observe LTD in vivo.

Resolution of this controversy, however, may now be at hand. First, it has been established that there are rat strain differences in the expression of LTD (Manahan-Vaughan and Braunewell, 1999). Second, and most importantly, it has been shown that LTD is powerfully modulated by the behavioral state of the animal. When LTD-resistant animals are exposed to mild stress (Kim et al., 1994; Xu et al., 1997), or even simply to a novel environment (Manahan-Vaughan and Braunewell, 1999), there is a striking facilitation of LTD. The stress effect may be mediated by glucocorticoids (Coussens et al., 1997; Xu et al., 1998b); the mechanism for the novelty effect is unknown, although modulation by acetylcholine has been suggested (Bear, 1999). Whatever the mechanism, the results show that brain state is a crucial variable that must be controlled during studies of LTD in adults.

The effect of novelty exposure on LTD is particularly interesting. When LFS is delivered as animals explore a novel environment, the resulting LTD lasts for weeks regardless of the strain of rat. However, the facilitation of synaptic plasticity is so marked that the usual 1-Hz tetanus is no longer required to induce LTD. The electrical LFS normally used to monitor synaptic transmission (5 pulses given at 0.1 Hz every 5 min for 15 min) is enough to depress synaptic transmission significantly for several hours if it is delivered during novelty exposure. An exciting possibility is that the pattern of electrical stimulation imposed on the brain during the novel experience is incorporated into the memory of that experience and that this memory is stored as LTD of the synapses that were active at that time. Indeed recordings from neurons in the temporal lobes have consistently revealed that a cellular correlate of recognition memory is a diminished response to the learned stimulus (Xiang and Brown, 1998). Perhaps this reduced response and the memory trace are accounted for by the mechanisms of homosynaptic LTD.

A final type of LTD modulation is suggested by the BCM theory. The idea is that the value of θ_m (the LTD-LTP crossover point) should vary depending on the history of the integrated postsynaptic response (Bear et al., 1987; Bear, 1995). After periods of strong postsynaptic activity, the modification threshold slides to promote LTD over LTP; after periods of postsynaptic inactivity, the threshold adjusts to promote LTP over LTD. In this way the properties of synaptic plasticity adjust to keep the network of modifiable synapses within a useful dynamic range. There is now compelling evidence from a number of systems that the stimulation requirements for induction of LTD are indeed altered by prior post-synaptic activity (Kirkwood et al., 1996; Holland and Wagner, 1998; Wang and Wagner, 1999). The mechanisms for this plasticity of synaptic plasticity, or "metaplasticity" (Abraham and Bear, 1996), remain to be determined, but the obvious candidates are clear from this discussion of LTD modulation. Changes in inhibition (Huang et al., 1999b; Steele and Mauk, 1999), NMDAR properties (Quinlan et al., 1999), and the balance of post-synaptic kinases and phosphatases (Mayford et al., 1995; Migaud et al., 1998) have all been proposed as mechanisms for the sliding modification threshold of the BCM theory.

mGluR-Dependent LTD

Induction

In addition to activating ionotropic receptors, glutamate stimulates mGluRs. Three classes of mGluR are defined by their pharmacology and coupling to second messenger pathways (Pin and Bockaert, 1995). Group 1 mGluRs (designated mGluR1 and mGluR5) stimulate phospho-inositide (PI) turnover via activation of PLC. It is of historical interest to note that, based on theoretical considerations, the proposal was made that PI-coupled mGluRs play a role in triggering synaptic depression in the cerebral cortex (Dudek and Bear, 1989) and that the protocol of using LFS to induce homosynaptic LTD was originally designed with the aim of testing this hypothesis (Dudek and Bear, 1992). Although this early work implicated NMDARs instead, it was not long before a role for mGluRs was suggested for LTD. In particular, Bolshakov and Siegelbaum (1994) found that an mGluR antagonist α-methyl-4-carboxyphenyl-glycine (MCPG) prevents homosynaptic LTD in response to LFS in slices from very young rats (postnatal days 3–7). LTD in this preparation also required a rise in intracellular $[Ca^{2+}]$ and activation of voltage-gated Ca^{2+} channels during LFS, but not activation of NMDARs.

Confusion about mGluR involvement in LTD persisted for years owing to some failures to replicate these findings (Selig et al., 1995b), exacerbated by the finding that MCPG is actually a very weak antagonist of the action of glutamate at mGluR5 (Brabet et al., 1995; Huber et al., 1998). Fortunately the smoke has now cleared. It is now obvious that the activation of mGluRs necessary to induce LTD is often not achieved using the usual 1- to 5-Hz stimulus trains. Protocols that work reliably are those that enhance glutamate release during conditioning stimulation, such as by delivering prolonged trains of paired pulses (Kemp and Bashir, 1997a) or antagonizing the adenosine inhibition of glutamate release (de Mendonca et al., 1997; Kemp and Bashir, 1997b). These protocols produce LTD of large magnitude with a component that cannot be blocked with NMDAR antagonists. Moreover, the use of new, potent mGluR antagonists and genetically altered mice has established that the NMDAR-independent LTD requires activation of mGluR5 and, conversely, that induction of NMDAR-LTD does not (Bortolotto et al., 1999; Sawtell et al., 1999; Huber et al., 2000).

It has now been established that activation of mGluR5 induces LTD by a mechanism that is entirely distinct from that engaged by NMDAR activation (Oliet et al., 1997). This mGluR-dependent LTD (mGluR-LTD) can be induced by synaptic stimulation in the presence of NMDAR antagonists or by simple pharmacological activation of mGluR5 using the group 1 mGluR-selective agonist (RS)-3,5-dihydroxyphenylglycine (DHPG) (Fitzjohn et al., 1999; Huber et al., 2000). The requirement of voltage-gated Ca^{2+} entry for induction of mGluR-LTD by synaptic stimulation has been confirmed, although the type of channel (L- or T-type) apparently varies depending on the circumstances (Bolshakov and Siegelbaum, 1994; Oliet et al., 1997; Otani and Connor, 1998). In addition, synaptically induced mGluR-LTD requires activation of postsynaptic PLC (Reyes and Stanton, 1998) and PKC (Bolshakov and Siegelbaum, 1994; Oliet et al., 1997; Otani and Connor, 1998). Unlike the NMDAR-LTD, mGluR-LTD is not affected by inhibition of postsynaptic PP1 (Oliet et al., 1997).

Biochemical experiments suggest that activation of group 1 mGluRs in synaptoneurosomes stimulates, in a PKC-dependent manner, the aggregation of ribosomes and mRNA and the synthesis of the fragile X mental retardation protein (Weiler and Greenough, 1993; Weiler et al., 1997). Thus it is of considerable interest that mGluR-LTD is prevented by manipulations that interfere with protein synthesis. Huber et al. (2000) have shown that induction of mGluR-LTD is prevented by the postsynaptic inhibition of mRNA translation during conditioning stimulation. Because the LTD is homosynaptic and occurs even when the dendrites are isolated from their cell bodies, a requirement for rapid, synapse-specific synthesis of proteins from preexisting mRNA is strongly suggested. These findings are consistent with a number of converging lines of evidence pointing to a major role for mRNA translation in the mechanisms of mGluR5 action (Merlin et al., 1998; Raymond et al., 2000).

The discovery of polyribosomes at the base of dendritic spines has long invited speculation that synaptic activity regulates the protein composition, and therefore function, of synapses in the brain (Steward et al., 1988). Available data now indicate that mGluR5 activation triggers synapse-specific mRNA translation, and that one functional consequence is LTD (Fig. 10.6). The obvious questions to be examined next concern the mechanism of translation regulation, the identity of the essential transcripts, and the mechanism that couples new protein synthesis to a change in synaptic function. The mGluR-LTD model should prove extremely valuable for answering these questions.

Expression

At the present time more is known about how mGluR-LTD is *not* expressed than about how it is expressed. Specifically, mGluR-LTD is not expressed via the same mechanism as NMDAR-LTD. This conclusion is supported by the finding that the two forms of LTD are additive and do not mutually occlude one another. Moreover, whereas NMDAR-LTD is

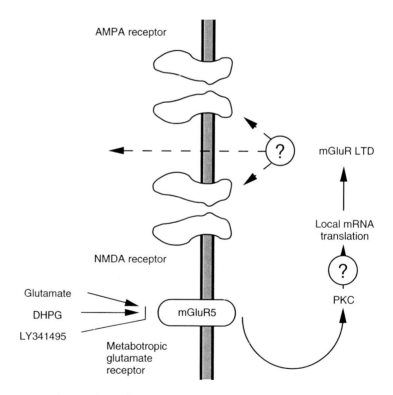

Figure 10.6. Model for mGluR-LTD in CA1. Activating mGluR5 during synaptic stimulation, or by means of the selective agonist DHPG, triggers LTD that can be prevented by the selective mGluR antagonist LY341495. Synaptically evoked mGluR-LTD requires activation of PKC in the postsynaptic neuron and local translation of preexisting mRNA, although it is not known if the PKC is involved in the protein synthesis regulation. The expression mechanism for mGluR LTD is unknown, but available evidence suggests a presynaptic modification.

reversed by induction of LTP (and vice versa), mGluR-LTD is not (Oliet et al., 1997; Fitzjohn et al., 1999; Huber et al., 2000). Finally, unlike NMDAR-LTD, mGluR-LTD is not associated with a dephosphorylation of AMPARs (Huber et al., 2000).

Two independent groups have reported that the magnitude of the postsynaptic response to the quantal synaptic release of glutamate (the quantal size) is unaffected after mGluR-LTD (Bolshakov and Siegelbaum, 1994; Oliet et al., 1997). Instead there is a large decrease in the quantal content. This observation is entirely consistent with a presynaptic mechanism, in which mGluR-LTD is the consequence of a decreased probability of glutamate release in response to a presynaptic action potential. An alternative explanation is the all-or-none loss of postsynaptic AMPARs at individual synapses (all-or-none because a graded decrease in receptors would be reflected in a decrease in quantal size). The surprisingly rapid regulation of AMPAR surface expression revealed by disruption of the NSF-GluR2 interaction makes such a mechanism plausible. However, if this mechanism does exist, the population of synapses susceptible to mGluR-LTD must be distinct from those participating in NMDAR-LTD, since the two forms of LTD are additive and not mutually occluding.

Depotentiation

The term *depotentiation* refers to the reversal of previously established LTP, which can be elicited by variations of the LFS protocol. Confusion has arisen because the same term has been used to describe two phenomena. One type of depotentiation, of course, is homosynaptic LTD of synapses from a potentiated baseline, which can be induced at any time following LTP induction and utilizes the same mechanisms previously discussed. The second type of depotentiation refers to the disruption of LTP that occurs when LFS is delivered within a relatively brief time window immediately following LTP induction. This time-sensitive depotentiation apparently is caused by interference with the transient intracellular biochemical reactions that are required to "fix" LTP in the period that follows strong NMDAR activation.

Time-Sensitive Depotentiation

High-frequency synaptic stimulation (HFS) typically induces LTP in CA1. However, establishment of stable LTP is prevented if the HFS is followed by certain types of synaptic stimulation, including (but not restricted to) LFS (Hesse and Teyler, 1976; Barrionuevo et al., 1980; Arai et al., 1990; Staubli and Lynch, 1990; Fujii et al., 1991; Barr et al., 1995; Holscher et al., 1997). This retrograde disruption of LTP is time-dependent. LFS within 5 min of HFS can completely prevent LTP; however, the same stimulation may have no effect when delivered 1 h after HFS.

A clear picture has finally emerged of the mechanism for time-sensitive depotentiation (TS-DP) (Fig. 10.7). Although the upstream regulation

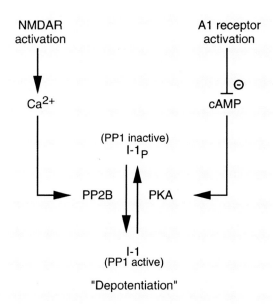

Figure 10.7. Converging pathways to depotentiation. Depotentiation occurs if synapses are given LFS in a narrow time window immediately following induction of LTP. Available evidence suggests that depotentiation results from dephosphorylation of postsynaptic proteins by PP1. PP1 is activated when I-1 is dephosphorylated by calcineurin (PP2B) at a PKA site. Stimuli that cause depotentiation include those that activate PP2B (e.g., NMDAR activation) and those that inhibit PKA (e.g., adenosine A1 receptor activation).

varies depending on the type of stimulation used, the critical downstream requirement for TS-DP is activation of postsynaptic PP1 during the sensitive period (O'Dell and Kandel, 1994; Staubli and Chun, 1996; Huang et al., 1999a). This period coincides with the time when LTP can be disrupted by inhibition of protein kinases (Huber et al., 1995). Thus the data indicate that stable establishment of LTP requires the active serine-threonine phosphorylation of synaptic substrates for a defined time period. If this phosphorylation is prevented or reversed, so is LTP.

Upstream regulation of TS-DP can occur in different ways. One route for TS-DP induction appears to be the now-familiar pathway involving NMDAR stimulation (Fujii et al., 1991; O'Dell and Kandel, 1994; Barr et al., 1995; Xiao et al., 1996) followed by activation of calcineurin and dephosphorylation of I-1 (O'Dell and Kandel, 1994; Zhuo et al., 1999). A second route to TS-DP induction, also leading to dephosphorylation of I-1, is the activation of A1 adenosine receptors (Larson et al., 1993; Staubli and Chun, 1996; Fujii et al., 1997; Huang et al., 1999a). A1 receptor activation inhibits adenylyl cyclase and, as a consequence, PKA (Dunwiddie and Fredholm, 1989). Injection of PKA activators into the postsynaptic neuron can prevent depotentiation caused by A1 receptor activation (Huang et al., 1999a).

Despite the apparent mechanistic similarities of TS-DP and NMDAR-LTD, the two phenomena differ. For example, TS-DP shows much less developmental regulation than LTD, and the patterns of activity that are optimal for induction are different (O'Dell and Kandel, 1994). In addition, TS-DP shows greater sensitivity to PP1 inhibitors and to deletion of the Aα isoform of calcineurin than does LTD (O'Dell and Kandel, 1994; Zhuo et al., 1999). Interestingly, however, like LTD, TS-DP in vivo is dramatically facilitated by exposure of animals to a novel environment (Xu et al., 1998a).

A simple way to reconcile the findings is to assume that LTP induction results in the transient exposure of postsynaptic phosphorylation sites to both protein kinases and phosphatases. Normally the kinase activation that follows strong NMDAR stimulation leads to a net phosphorylation of these sites and LTP. However, the sites are also vulnerable to dephosphorylation if PP1 is activated. Because the phosphorylation sites are exposed, the threshold level of PP1 activation for TS-DP is much lower than that for NMDAR-LTD. Thus, whereas stimulation that induces LTD also can always produce TS-DP, the converse is not true. However, the common requirement for PP1 activation makes both forms of synaptic modification subject to very similar types of modulation.

Time-Insensitive Depotentiation

Under conditions in which de novo LTD is induced by LFS, the same stimulation can also reverse LTP that was induced hours before. The precise mechanism of LTD and LTP reversal may not be identical, however. For example, LTD de novo is associated with the dephosphorylation of GluR1 at a PKA site. However, the same induction protocol given 1 h after induction of LTP causes dephosphorylation of the CaMKII-PKC site instead (Lee et al., 2000; see Fig. 10.5B). On the other hand, both LTD and LTP reversal in vivo are associated with a parallel decrease in AMPAR and NMDAR protein in the synaptoneurosomal fraction (Heynen et al., 2000).

Conclusion

The cerebellum and the hippocampus are distant, both physically and evolutionarily, and at high resolution the mechanisms of LTD are also clearly different. However, stepping back reveals many common themes. In the cerebellum the consequence of activating postsynaptic group 1 mGluRs in the appropriate context is a decrease in AMPAR function. Very similar changes in CA1 are caused by the appropriate level of NMDAR activation. Although apparently caused by a distinct expression mechanism, activation of group 1 mGluRs can also stimulate LTD in CA1.

It is possible that there will prove to be a limited number of pre- and postsynaptic LTD expression mechanisms that are widely expressed at glutamatergic synapses in the brain. Regional differences might be explained solely by variations in the signal transduction mechanisms that must be activated to induce a change. One line of evidence in support of this idea is the recent finding that both NMDAR-dependent LTD in hippocampal area CA1 (Lüscher et al., 1999; Man et al., 2000) and mGluR1-dependent cerebellar LTD appear to require clathrin-mediated internalization of postsynaptic AMPA receptors (Wang and Linden, 2000).

We hasten to add, however, that the list of possible LTD mechanisms is not exhausted by the study of the two synapses highlighted in this chapter. Inhibitory synapses have also been shown to exhibit LTD (Marty and Llano, 1995; Aizenman et al., 1998). Other novel forms of LTD have been described in excitatory synapses at other locations, such as the striatum (Lovinger et al., 1993; Calabresi et al., 1999), amygdala (Li et al., 1998; Wang and Gean, 1999), neocortex (Artola et al., 1996; Egger et al., 1999), and other parts of the hippocampal formation (Kobayashi et al., 1996; Manahan-Vaughan, 1998). For example, there is now good evidence from a number of different systems that activation of group 2 mGluRs, negatively coupled to adenylyl cyclase, triggers LTD. Time will tell if these and other forms of LTD converge on expression mechanisms that overlap with those we have described here.

An enormous amount of progress has been made in the investigation of LTD during the past decade. Ten years ago, the LTD club was very small. Indeed it is probably safe to say that at the end of the 1980s there was little general interest in LTD, or even any strong conviction that homosynaptic depression existed at all outside the cerebellum. This situation has clearly changed. The major challenge for the next ten years will be to see if the role of LTD in the brain lives up to its lofty theoretical promise.

References

Abraham, W., Bear, M. (1996). Metaplasticity: the plasticity of synaptic plasticity. *Trends Neurosci.* 19:126–130.

Abraham, W. C., Otani, S. (1991). Macromolecules and the maintenance of long-term potentiation. In *Kindling and Synaptic Plasticity: The Legacy of Graham Goddard,* Morrell, F., ed. Boston: Birkhauser. pp. 92–109.

Ahn, S., Ginty, D. D., Linden, D. J. (1999). A late phase of cerebellar long-term depression requires activation of CaMKIV and CREB. *Neuron* 23:559–568.

Aiba, A., Kano, M., Chen, C., Stanton, M. E., Fox, G. D., Herrup, K., Zwingman, T. A., Tonegawa, S. (1994). Deficient cerebellar long-term depression and impaired motor learning in mGluR1 mutant mice. *Cell* 79:377–388.

Aizenman, C., Manis, P. B., Linden, D. J. (1998). Polarity of long-term synaptic gain change is related to postsynaptic spike firing at a cerebellar inhibitory synapse. *Neuron* 21:827–835.

Albus, J. S. (1971). A theory of cerebellar function. *Math. Biosci.* 10:25–61.

Arai, A., Larson, J., Lynch, G. (1990). Anoxia reveals a vulnerable period in the development of long-term potentiation. *Brain Res.* 511:353–357.

Artola, A., Singer, W. (1993). Long-term depression of excitatory synaptic transmission and its relationship to long-term potentiation. *Trends Neurosci.* 16: 480–487.

Artola, A., Hensch, T., Singer, W. (1996). Calcium-induced long-term depression in the visual cortex of the rat in vitro. *J. Neurophysiol.* 76:984–994.

Barr, D. S., Lambert, N. A., Hoyt, K. L., Moore, S. D., Wilson, W. A. (1995). Induction and reversal of long-term potentiation by low- and high-intensity theta pattern stimulation. *J. Neurosci.* 15:5402–5410.

Barria, A., Derkach, V., Soderling, T. (1997a). Identification of the Ca^{2+}/ calmodulin-dependent protein kinase II regulatory phosphorylation site in the alpha-amino-3-hydroxyl-5-methyl-4-isoxazole-propionate-type glutamate receptor. *J. Biol. Chem.* 272:32,727–32,730.

Barria, A., Muller, D., Derkach, V., Griffith, L. C., Soderling, T. R. (1997b). Regulatory phosphorylation of AMPA-type glutamate receptors by CaM-KII during long-term potentiation. *Science* 276:2042–2045.

Barrionuevo, G., Schottler, F., Lynch, G. (1980). The effects of low frequency stimulation on control and "potentiated" synaptic responses in the hippocampus. *Life Sci.* 27:2385–2391.

Batini, C., Compoint, C., Buissert-Delmas, C., Daniel, H., Guegan, M. (1992). Cerebellar nuclei and the nucleocortical projections in the rat: retrograde tracing coupled to GABA and glutamate immunohistochemistry. *J. Comp. Neurol.* 315:74–84.

Bear, M. (1995). Mechanism for a sliding synaptic modification threshold. *Neuron* 15:1–4.

Bear, M. F. (1996). A synaptic basis for memory storage in the cerebral cortex. *Proc. Natl. Acad. Sci. USA* 93:13,453–13,459.

Bear, M. F. (1999). Homosynaptic long-term depression: a mechanism for memory? *Proc. Natl. Acad. Sci. USA* 96:9457–9458.

Bear, M. F., Abraham, W. C. (1996). Long-term depression in hippocampus. *Annu. Rev. Neurosci.* 19:437–462.

Bear, M. F., Malenka, R. C. (1994). Synaptic plasticity: LTP and LTD. *Curr. Opin. Neurobiol.* 4:389–399.

Bear, M. F., Cooper, L. N., Ebner, F. F. (1987). A physiological basis for a theory of synaptic modification. *Science* 237:42–48.

Berretta, N., Cherubini, E. (1998). A novel form of long-term depression in the CA1 area of the adult rat hippocampus independent of glutamate receptor activation. *Eur. J. Neurosci.* 10:2957–2963.

Berthier, N. E., Moore, J. W. (1986). Cerebellar Purkinje cell activity related to the classically conditioned nictitating membrane response. *Exp. Brain Res.* 63:341–350.

Berthier, N. E., Moore, J. W. (1990). Activity of deep cerebellar nuclear cells during classical conditioning of nictitating membrane extension in rabbits. *Exp. Brain Res.* 83:44–54.

Bezprozvanny, I. B., Ondrias, K., Kattan, E., Stoyanovsky, D. A., Erlich, B. A. (1993). Activation of the calcium release channel ryanodine receptor by heparin and other polyanions is calcium dependent. *Mol. Biol. Cell* 4:347–352.

Bi, G. Q., Poo, M. M. (1998). Synaptic modifications in cultured hippocampal neurons: dependence on spike timing, synaptic strength, and postsynaptic cell type. *J. Neurosci.* 18:10,464–10,472.

Bienenstock, E. L., Cooper, L. N., Munro, P. W. (1982). Theory for the development of neuron selectivity: orientation specificity and binocular interaction in visual cortex. *J. Neurosci.* 2:32–48.

Bito, H., Deisseroth, K., Tsien, R. W. (1996). CREB phosphorylation and dephosphorylation: a Ca^{2+}- and stimulus duration–dependent switch for hippocampal gene expression. *Cell* 87:1203–1214.

Blendy, J. A., Kaestner, K. H., Schmid, W., Schutz, G. (1996). Targeting of the CREB gene leads to up-regulation of a novel CREB mRNA isoform. *EMBO J.* 15:1098–1106.

Bliss, T. V. P., Lømo, T. (1973). Long-lasting potentiation of synaptic transmission in the dentate area of the anaesthetized rabbit following stimulation of the perforant path. *J. Physiol.* 232:331–356.

Blitzer, R. D., Wong, T., Nouranifar, R., Iyengar, R., Landau, E. M. (1995). Postsynaptic cAMP pathway gates early LTP in hippocampal CA1 region. *Neuron* 15:1403–1414.

Blitzer, R. D., Connor, J. H., Brown, G. P., Wong, T., Shenolikar, S., Iyengar, R., Landau, E. M. (1998). Gating of CaMKII by cAMP-regulated protein phosphatase activity during LTP. *Science* 280:1940–1942.

Blond, O., Daniel, H., Otani, S., Jaillard, D., Crepel, F. (1997). Presynaptic and postsynaptic effects of nitric oxide donors at synapses between parallel fibres and Purkinje cells: involvement in cerebellar long-term depression. *Neuroscience* 77:945–954.

Bolshakov, V. Y., Siegelbaum, S. A. (1994). Postsynaptic induction and presynaptic expression of hippocampal long-term depression. *Science* 264:1148–1152.

Bortolotto, Z. A., Fitzjohn, S. M., Collingridge, G. L. (1999). Roles of metabotropic glutamate receptors in LTP and LTD in the hippocampus. *Curr. Opin. Neurobiol.* 9:299–304.

Boxall, A. R., Garthwaite, J. (1996). Long-term depression in rat cerebellum requires both NO synthase and NO-sensitive guanylyl cyclase. *Eur. J. Neurosci.* 8:2209–2212.

Brabet, I., Mary, S., Bockaert, J., Pin, J.-P. (1995). Phenylglycine derivatives discriminate between mGluR1- and mGluR5-mediated responses. *Neuropharmacology* 34:895–903.

Bracha, V., Webster, M. L., Winters, N. K., Irwin, K. B., Bloedel, J. R. (1994). Effects of muscimol inactivation of the cerebellar interposed-dentate nuclear complex on the performance of the nictitating membrane response of the rabbit. *Exp. Brain Res.* 100:453–468.

Brandon, E. P., Zhuo, M., Huang, Y. Y., Qi, M., Gerhold, K. A., Burton, K. A., Kandel, E. R., McKnight, G. S., Idzerda, R. L. (1995). Hippocampal long-term depression and depotentiation are defective in mice carrying a targeted disruption of the gene encoding the RI beta subunit of cAMP-dependent protein kinase. *Proc. Natl. Acad. Sci. USA* 92:8851–8852.

Bredt, D. S., Hwang, P. M., Snyder, S. H. (1990). Localization of nitric oxide synthase indicating a neural role for nitric oxide. *Nature* 347:768–770.

Breese, C., Hampson, R., Deadwyler, S. (1989). Hippocampal place cells: stereotypy and plasticity. *J. Neurosci.* 9:1097–1111.

Calabresi, P., Centonze, D., Gubellini, P., Marfia, G. A., Bernardi, G. (1999). Glutamate-triggered events inducing corticostriatal long-term depression. *J. Neurosci.* 19:6102–6110.

Carroll, R. C., Lissin, D. V., von Zastrow, M., Nicoll, R. A., Malenka, R. C. (1999). Rapid redistribution of glutamate receptors contributes to long-term depression in hippocampal cultures. *Nature Neurosci.* 2:454–460.

Chavis, P., Mollard, P., Bockaert, J., Manzoni, O. (1998). Visualization of cyclic AMP–regulated presynaptic activity at cerebellar granule cells. *Neuron* 20:773–781.

Chawla, S., Hardingham, G. E., Quinn, D. R., Bading, H. (1998). CBP: a signal-regulated transcriptional coactivator controlled by nuclear calcium and CaM kinase IV. *Science* 281:1505–1509.

Chen, C., Regehr, W. G. (1997). The mechanism of cAMP-mediated enhancement at a cerebellar synapse. *J. Neurosci.* 17:8687–8694.

Chen, C., Thompson, R. F. (1995). Temporal specificity of long-term depression in parallel fiber–Purkinje synapses in rat cerebellar slice. *Learn. Mem.* 2:185–198.

Chen, C., Tonegawa, S. (1997). Molecular genetic analysis of synaptic plasticity, activity-dependent neural development, learning and memory in the mammalian brain. *Annu. Rev. Neurosci.* 20:157–184.

Chen, C., Kano, M., Abeliovich, A., Chen, L., Bao, S., Kim, J. J., Hashimoto, K., Thompson, R. F., Tonegawa, T. (1995). Impaired motor coordination correlates with persistent multiple climbing fiber innervation in PKC-mutant mice. *Cell* 83:1233–1242.

Christie, B. R., Magee, J. C., Johnston, D. (1996). The role of dendritic action potentials and Ca^{2+} influx in the induction of homosynaptic long-term depression in hippocampal CA1 pyramidal neurons. *Learn. Mem.* 3:160–169.

Cohen, P. (1989). The structure and regulation of protein phosphatases. *Annu. Rev. Biochem.* 58:453–508.

Conquet, F., Bashir, Z. I., Davies, C. H., Daniel, H., Ferraguti, F., Bordi, F., Franz-Bacon, K., Reggiani, A., Matarese, V., Conde, F., Collingridge, G. L., Crepel, F. (1994). Motor deficit and impairment of synaptic plasticity in mice lacking mGluR1. *Nature* 327:237–243.

Coussens, C. M., Kerr, D. S., Abraham, W. C. (1997). Glucocorticoid receptor activation lowers the threshold for NMDA-receptor-dependent homosynaptic long-term depression in the hippocampus through activation of voltage-dependent calcium channels. *J. Neurophysiol.* 78:1–9.

Crepel, F., Jaillard, D. (1990). Protein kinases, nitric oxide and long-term depression of synapses in the cerebellum. *NeuroReport* 1:133–136.

Crepel, F., Jaillard, D. (1991). Pairing of pre- and postsynaptic activities in cerebellar Purkinje cells induces long-term changes in synaptic efficacy in vitro. *J. Physiol.* 432:123–141.

Crepel F., Krupa, M. (1988). Activation of protein kinase C induces a long-term depression of glutamate sensitivity of cerebellar Purkinje cells: an in vitro study. *Brain Res.* 458:397–401.

Crepel, F., Audinat, E., Daniel, H., Hemart, N., Jaillard, D., Rossier, J., Lambolez, B. (1994). Cellular locus of the nitric oxide-synthase involved in cerebellar long-term depression induced by high external potassium concentration. *Neuropharmacology* 33:1399–1405.

Cummings, J. A., Nicola, S. M., Malenka, R. C. (1994). Induction in the rat hippocampus of long-term potentiation (LTP) and long-term depression (LTD) in the presence of a nitric oxide synthase inhibitor. *Neurosci. Lett.* 176:110–114.

Cummings, J. A., Mulkey, R. M., Nicoll, R. A., Malenka, R. C. (1996). Ca^{2+} signalling requirements for long-term depression in the hippocampus. *Neuron* 16:825–833.

Daniel, H., Hemart, N., Jaillard, D., Crepel, F. (1993). Long-term depression requires nitric oxide and guanosine $3'$-$5'$ cyclic monophosphate production in cerebellar Purkinje cells. *Eur. J. Neurosci.* 5:1079–1082.

Daniel, H., Levenes, C., Crepel, F. (1998). Cellular mechanisms of cerebellar LTD. *Trends Neurosci.* 21:401–407.

Daniel, H., Levenes, C., Fagni, L., Conquet, F., Bockaert, J., Crepel, F. (1999). Inositol-1,4,5-trisphosphate-mediated rescue of cerebellar long-term depression in subtype 1 metabotropic glutamate receptor mutant mouse. *Neuroscience* 92:1–6.

Debanne, D., Gahwiler, B. H., Thompson, S. M. (1994). Asynchronous pre- and postsynaptic activity induces associative long-term depression in area CA1 of the rat hippocampus in vitro. *Proc. Natl. Acad. Sci. USA* 91:1148–1152.

Debanne, D., Gahwiler, B. H., Thompson, S. M. (1997). Bidirectional associative plasticity of unitary CA3-CA1 EPSPs in the rat hippocampus in vitro. *J. Neurophysiol.* 77:2851–2855.

Deisseroth, K., Heist, E. K., Tsien, R. W. (1998). Translocation of calmodulin to the nucleus supports CREB phosphorylation in hippocampal neurons. *Nature* 392:198–202.

De Mendonca, A., Almeida, T., Bashir, Z. I., Ribeiro, J. A. (1997). Endogenous adenosine attenuates long-term depression and depotentiation in the CA1 region of the rat hippocampus. *Neuropharmacology* 36:161–167.

Derkach, V., Barria, A., Soderling, T. R. (1999). Ca^{2+}/calmodulin-kinase II enhances channel conductance of alpha-amino-3-hydroxy-5-methyl-4-isoxazolepropionate type glutamate receptors. *Proc. Natl. Acad. Sci. USA* 96:3269–3274.

De Zeeuw, C. I., Berrebi, A. S. (1995). Postsynaptic targets of Purkinje cell terminals in the cerebellar and vestibular nuclei of the rat. *Eur. J. Neurosci.* 7:2322–2333.

De Zeeuw, C. I., Hansel, C., Bian, F., Koekkoek, S. K. E., van Alphen, A. M., Linden, D. J., Oberdick, J. (1998). Expression of a protein kinase C inhibitor in Purkinje cells blocks cerebellar long-term depression and adaptation of the vestibulo-ocular reflex. *Neuron* 20:495–508.

Dodt, H., Eder, M., Frick, A., Zieglgansberger, W. (1999). Precisely localized LTD in the neocortex revealed by infrared-guided laser stimulation. *Science* 286:110–113.

Doyle, C. A., Cullen, W. K., Rowan, M. J., Anwyl, R. (1997). Low-frequency stimulation induces homosynaptic depotentiation but not long-term depression of synaptic transmission in the adult anaesthetized and awake rat hippocampus in vivo. *Neuroscience* 77:75–85.

Dudek, S. M., Bear, M. F. (1989). A biochemical correlate of the critical period for synaptic modification in the visual cortex. *Science* 246:673–675.

Dudek, S. M., Bear, M. F. (1992). Homosynaptic long-term depression in area CA1 of hippocampus and effects of N-methyl-D-aspartate receptor blockade. *Proc. Natl. Acad. Sci. USA* 89:4363–4367.

Dudek, S. M., Bear, M. F. (1993). Bidirectional long-term modification of synaptic effectiveness in the adult and immature hippocampus. *J. Neurosci.* 13:2910–2918.

Du Lac, S., Raymond, J. L., Sejnowski, T. J., Lisberger, S. G. (1995). Learning and memory in the vestibulo-ocular reflex. *Annu. Rev. Neurosci.* 18:409–441.

Dunwiddie, T. V., Fredholm, B. B. (1989). Adenosine A1 receptors inhibit adenylate cyclase activity and neurotransmitter release and hyperpolarize pyramidal neurons in rat hippocampus. *J. Pharmacol. Exp. Ther.* 249:31–37.

Egger, V., Feldmeyer, D., Sakmann, B. (1999). Coincidence detection and changes of synaptic efficacy in spiny stellate neurons in rat barrel cortex. *Nature Neurosci.* 2:1098–1105.

Ekerot, C.-F., Kano, M. (1985). Long-term depression of parallel fibre synapses following stimulation of climbing fibres. *Brain Res.* 342:357–360.

Ekerot, C.-F., Kano, M. (1989). Stimulation parameters influencing climbing fibre induced long-term depression of parallel fibre synapses. *Neurosci. Res.* 6:264–268.

Errington, M. L., Bliss, T. V. P., Richter-Levin, G., Yenk, K., Doyere, V., Laroche, S. (1995). Stimulation at 1–5 Hz does not produce long-term depression or depotentiation in the hippocampus of the adult rat in vivo. *J. Neurophysiol.* 74:1793–1799.

Fitzjohn, S. M., Kingston, A. E., Lodge, D., Collingridge, G. L. (1999). DHPG-induced LTD in area CA1 of juvenile rat hippocampus: characterisation and sensitivity to novel mGlu receptor antagonists. *Neuropharmacology* 38:1577–1583.

Fredette, B. J., Mugnaini, E. (1991). The GABAergic cerebello-olivary projection in the rat. *Anat. Embryol.* 184:225–243.

Freeman, J. H., Shi, T., Schreurs, B. G. (1998). Pairing-specific long-term depression prevented by blockade of PKC or intracellular Ca^{2+}. *NeuroReport* 9: 2237–2241.

Fujii, S., Saito, K., Miyakawa, H., Ito, K., Kato, H. (1991). Reversal of long-term potentiation (depotentiation) induced by tetanus stimulation of the input to CA1 neurons of guinea pig hippocampal slices. *Brain Res.* 555:112–122.

Fujii, S., Sekino, Y., Kuroda, Y., Sasaki, H., Ito, K., Kato, H. (1997). 8-Cyclopentyltheophylline, an adenosine A1 receptor antagonist, inhibits the reversal of long-term potentiation in hippocampal CA1 neurons. *Eur. J. Pharmacol.* 331:9–14.

Funabiki, K., Mishina, M., Hirano, T. (1995). Retarded vestibular compensation in mutant mice deficient in δ2 glutamate receptor subunit. *NeuroReport* 7: 189–192.

Gage, A. T., Reyes, M., Stanton, P. K. (1997). Nitric-oxide-guanylyl-cyclase-dependent and -independent components of multiple forms of long-term synaptic depression. *Hippocampus* 7:286–295.

Garcia, K. S., Mauk, M. D. (1998). Pharmacological analysis of cerebellar contributions to the timing and expression of conditioned eyelid responses. *Neuropharmacology* 37:471–480.

Ginty, D. D. (1997). Calcium regulation of gene expression: isn't that spatial? *Neuron* 18:183–186.

Glaum, S. R., Slater, N. T., Rossi, D. J., Miller, R. J. (1992). The role of metabotropic glutamate receptors at the parallel fiber–Purkinje cell synapse. *J. Neurophysiol.* 68:1453–1462.

Goda, Y., Stevens, C. F. (1996). Long-term depression properties in a simple system. *Neuron* 16:103–111.

Gusovsky, F., Hollingsworth, E. B., Daly, J. W. (1986). Regulation of phosphatidylinositol turnover in brain synaptoneurosomes: stimulatory effects of agents that enhance influx of sodium ions. *Proc. Natl. Acad. Sci. USA* 83:3003–3007.

Hansel, C., Linden, D. J. (2000). Long-term depression of the cerebellar climbing fiber–Purkinje neuron synapse. *Neuron* 26:473–482.

Hardiman, M. J., Ramnani, N., Yeo, C. H. (1996). Reversible inactivations of the cerebellum with muscimol prevent the acquisition and extinction of conditioned nictitating membrane responses in the rabbit. *Exp. Brain Res.* 110: 235–247.

Hartell, N. A. (1994). cGMP acts within cerebellar Purkinje cells to produce long-term depression via mechanisms involving PKC and PKG. *NeuroReport* 5:833–836.

Hartell, N. A. (1996). Strong activation of parallel fibers produces localized calcium transients and a form of LTD which spreads to distant synapses. *Neuron* 16:601–610.

Hebb, D. O. (1949). *The Organization of Behavior.* New York: John Wiley & Sons.

Hemart, N., Daniel, H., Jaillard, D., Crepel, F. (1994). Properties of glutamate receptors are modified during long-term depression in rat cerebellar Purkinje cells. *Neurosci. Res.* 19:213–221.

Hemart, N., Daniel, H., Jaillard, D., Crepel, F. (1995). Receptors and second messengers involved in long-term depression in rat cerebellar slices in vitro: a reappraisal. *Eur. J. Neurosci.* 7:45–53.

Herbert, J.-M., Maffrand, J.-P. (1991). Effect of pentosan polysulphate, standard heparin and related compounds on protein kinase C activity. *Biochim. Biophys. Acta* 1091:432–441.

Hesse, G. W., Teyler, T. J. (1976). Reversible loss of hippocampal long term potentiation following electronconvulsive seizures. *Nature* 264:562–564.

Heynen, A. J., Abraham, W. C., Bear, M. F. (1996). Bidirectional modification of CA1 synapses in the adult hippocampus *in vivo*. *Nature* 381:163–166.

Heynen, A., Quinlan, E. M., Bae, D., Bear, M. F. (2000). Bidirectional, activity-dependent regulation of glutamate receptors in the adult hippocampus *in vivo*. *Neuron* (submitted).

Hirano, T. (1990a). Depression and potentiation of the synaptic transmission between a granule cell and a Purkinje cell in rat cerebellar culture. *Neurosci. Lett.* 119:141–144.

Hirano, T. (1990b). Effects of postsynaptic depolarization in the induction of synaptic depression between a granule cell and a Purkinje cell in rat cerebellar culture. *Neurosci. Lett.* 119:145–147.

Hirano, T. (1991). Differential pre- and postsynaptic mechanisms for synaptic potentiation and depression between a granule cell and a Purkinje cell in rat cerebellar culture. *Synapse* 7:321–323.

Hirano, T., Kasono, K., Araki, K., Mishina, M. (1994a). Suppression of LTD in cultured Purkinje cells deficient in the glutamate receptor-2 subunit. *NeuroReport* 6:524–526.

Hirano, T., Kasono, K., Araki, K., Shiozuka, K., Mishina, M. (1994b). Involvement of the glutamate reception δ2 subunit in the long-term depression of glutamate responsiveness in cultured Purkinje cells. *Neurosci. Lett.* 182:172–176.

Holland, L. L., Wagner, J. J. (1998). Primed facilitation of homosynaptic long-term depression and depotentiation in rat hippocampus. *J. Neurosci.* 18:887–894.

Hollmann, M., Heinemann, S. (1994). Cloned glutamate receptors. *Annu. Rev. Neurosci.* 17:31–108.

Holscher, C., Anwyl, R., Rowan, M. J. (1997). Stimulation on the positive phase of hippocampal theta rhythm induces long-term potentiation that can be depotentiated by stimulation on the negative phase in area CA1 in vivo. *J. Neurosci.* 17:6470–6477.

House, C., Kemp, B. E. (1987). Protein kinase C contains a pseudosubstrate prototype in its regulatory domain. *Science* 238:1726–1728.

Hu, S.-C., Chrivia, J., Ghosh, A. (1999). Regulation of CBP-mediated transcription by neuronal calcium signaling. *Neuron* 22:799–808.

Huang, C. C., Liang, Y. C., Hsu, K. S. (1999a). A role for extracellular adenosine in time-dependent reversal of long-term potentiation by low-frequency stimulation at hippocampal CA1 synapses. *J. Neurosci.* 19:9728–9738.

Huang, J. Z., Kirkwood, A., Pizzorusso, T., Porciatti, V., Bear, M. F., Maffei, L., Tonegawa, S. (1999b). BDNF regulates the maturation of inhibition and the critical period of plasticity in mouse visual cortex. *Cell* 98:739–755.

Huber, K. M., Mauk, M. D., Thompson, C., Kelly, P. T. (1995). A critical period of protein kinase activity after tetanic stimulation is required for the induction of long-term potentiation. *Learn. Mem.* 2:81–100.

Huber, K. M., Sawtell, N. B., Bear, M. F. (1998). Effects of the metabotropic glutamate receptor antagonist MCPG on phosphoinositide turnover and synaptic plasticity in visual cortex. *J. Neurosci.* 18:1–9.

Huber, K. M., Kayser, M. S., Kameyama, K., Huganir, R. L., Roder, J. C., Bear, M. F. (2000). Activation of mGluR5 induces a distinct form of long-term depression in hippocampal area CA1 (in preparation).

Hummler, E., Cole, T. J., Blendy, J. A., Ganss, R., Aguzzi, A., Schmid, W., Beermann, F., Schutz, G. (1994). Targeted mutation of the CREB gene: compensation within the CREB/ATF family of transcription factors. *Proc. Natl. Acad. Sci. USA* 91:5647–5651.

Ikeda, M., Morita, I., Murota, S., Sekiguchi, F., Yuasa, T., Miyatake, T. (1993). Cerebellar nitric oxide synthase activity is reduced in nervous and Purkinje cell degeneration mutants but not in climbing fiber–lesioned mice. *Neurosci. Lett.* 155:148–150.

Impey, S., Obrietan, K., Wong, S. T., Poser, S., Yano, S., Wayman, G., Deloulme, J. C., Chan, G., Storm, D. R. (1998). Cross talk between ERK and PKA is required for Ca^{2+} stimulation of CREB-dependent transcription and ERK nuclear translocation. *Neuron* 21:869–883.

Inoue, T., Kato, K., Kohda, K., Mikoshiba, K. (1998). Type 1 inositol 1,4,5-trisphosphate receptor is required for induction of long-term depression in cerebellar Purkinje neurons. *J. Neurosci.* 15:5366–5373.

Ito, M. (1984). *The Cerebellum and Neural Control.* New York: Raven Press.

Ito, M. (1990). Long-term depression in the cerebellum. *Sem. Neurosci.* 2:381–390.

Ito, M., Karachot, L. (1992). Protein kinases and phosphatase inhibitors mediating long-term desensitization of glutamate receptors in cerebellar Purkinje cells. *Neurosci. Res.* 14:27–38.

Ito, M., Sakurai, M., Tongroach, P. (1982). Climbing fibre induced depression of both mossy fiber responsiveness and glutamate sensitivity of cerebellar Purkinje cells. *J. Physiol.* 324:113–134.

Izumi, Y., Zorumski, C. F. (1993). Nitric oxide and long-term synaptic depression in the rat hippocampus. *NeuroReport* 4:1131–1134.

Jeromin, A., Huganir, R., Linden, D. J. (1996). Suppression of the glutamate receptor δ2 subunit produces a specific impairment in cerebellar long-term depression. *J. Neurophysiol.* 76:3578–3583.

Kamal, A., Biessels, G. J., Gispen, W. H., Urban, I. J. (1998). Increasing age reduces expression of long-term depression and dynamic range of transmission plasticity in CA1 field of the rat hippocampus. *Neuroscience* 83:707–715.

Kamal, A., Ramakers, G. M., Urban, I. J., De, G. P., Gispen, W. H. (1999). Chemical LTD in the CA1 field of the hippocampus from young and mature rats. *Eur. J. Neurosci.* 11:3512–3516.

Kameyama, K., Lee, H. K., Bear, M. F., Huganir, R. L. (1998). Involvement of a postsynaptic protein kinase A substrate in the expression of homosynaptic long-term depression. *Neuron* 21:1163–1175.

Kandler, K., Katz, L. C., Kauer, J. A. (1998). Focal photolysis of caged glutamate produces long-term depression of hippocampal glutamate receptors. *Nature Neurosci.* 1:119–123.

Kano, M., Kato, M. (1987). Quisqualate receptors are specifically involved in cerebellar synaptic plasticity. *Nature* 325:276–279.

Kano, M., Kato, M. (1988). Mode of induction of long-term depression at parallel fibre–Purkinje cell synapses in rabbit cerebellar cortex. *Neurosci. Res.* 5:544–556.

Kano, M., Hashimoto, K., Chen, C., Abeliovich, A., Aiba, A., Kurihara, H., Watanabe, M., Inoue, Y., Tonegawa, S. (1995). Impaired synapse elimination during cerebellar development in PKC-mutant mice. *Cell* 83:1223–1231.

Kashiwabuchi, N., Ikeda, K., Araki, K., Hirano, T., Shibuki, K., Takayama, C., Inoue, Y., Kutsuwada, T., Yagi, T., Kang, Y., Aizawa, S., Mishina, M. (1995).

Impairment of motor coordination, Purkinje cell synapse formation, and cerebellar long-term depression in GluRδ2 mutant mice. *Cell* 81:245–252.

Kasono, K., Hirano T. (1995). Involvement of inositol trisphosphate in cerebellar long-term depression. *NeuroReport* 6:569–572.

Katsuki, H., Izumi, Y., Zorumski, C. F. (1997). Noradrenergic regulation of synaptic plasticity in the hippocampal CA1 region. *J. Neurophysiol.* 77:3013–3020.

Kelso, S. R., Ganong, A. H., Brown, T. (1986). Hebbian synapses in the hippocampus. *Proc. Natl. Acad. Sci. USA* 83:5326–5330.

Kemp, N., Bashir, Z. I. (1997a). NMDA receptor–dependent and –independent long-term depression in the CA1 region of the adult rat hippocampus in vitro. *Neuropharmacolgy* 36:397–399.

Kemp, N., Bashir, Z. I. (1997b). A role for adenosine in the regulation of long-term depression in the adult rat hippocampus in vitro. *Neurosci Lett.* 225:189–192.

Khodakhah, K., Armstrong, C. M. (1997). Induction of long-term depression and rebound potentiation by inositol trisphosphate in cerebellar Purkinje neurons. *Proc. Natl. Acad. Sci. USA* 94:14,009–14,014.

Khodakhah, K., Ogden, D. (1993). Functional heterogeneity of calcium release by inositol trisphosphate in single Purkinje neurones, cultured cerebellar astrocytes, and peripheral tissues. *Proc. Natl. Acad. Sci. USA* 90:4976–4980.

Kim, J. J., Thompson, R. F. (1997). Cerebellar circuits and synaptic mechanisms involved in classical eyeblink conditioning. *Trends Neurosci.* 20:177–181.

Kim, J. J., Foy, M. R., Thompson, R. F. (1994). Behavioral stress enhances LTD in rat hippocampus. *Soc. Neurosci. Abstr.* 20:1769.

Kirkwood, A., Bear, M. F. (1994). Homosynaptic long-term depression in the visual cortex. *J. Neurosci.* 14:3404–3412.

Kirkwood, A., Dudek, S. M., Gold, J. T., Aizenman, C. D., Bear, M. F. (1993). Common forms of synaptic plasticity in the hippocampus and neocortex in vitro. *Science* 260:1518–1521.

Kirkwood, A., Rioult, M. G., Bear, M. F. (1996). Experience-dependent modification of synaptic plasticity in visual cortex. *Nature* 381:526–528.

Kirkwood, A., Rozas, C., Kirkwood, J., Perez, F., Bear, M. F. (1999). Modulation of long-term synaptic depression in visual cortex by acetylcholine and norepinephrine. *J. Neurosci.* 19:1599–1609.

Knopfel, T., Vranesic, I., Staub, C., Gahwiler, B. H. (1990). Climbing fibre responses in olive-cerebellar slice cultures. II. Dynamics of cytosolic calcium in Purkinje cells. *Eur. J. Neurosci.* 3:343–348.

Kobayashi, K., Manabe, T., Takahashi, T. (1996). Presynaptic long-term depression at the hippocampal mossy fiber–CA3 synapse. *Science* 273:648–650.

Kohda, K., Inoue, T., Mikoshiba, K. (1995). Ca^{2+} release from Ca^{2+} stores, particularly from ryanodine-sensitive Ca^{2+} stores, is required for the induction of LTD in cultured cerebellar Purkinje cells. *J. Neurophysiol.* 74:2184–2188.

Konnerth, A., Dreessen, J., Augustine, G. J. (1992). Brief dendritic calcium signals initiate long-lasting synaptic depression in cerebellar Purkinje cells. *Proc. Natl. Acad. Sci. USA* 89:7051–7055.

Krupa, D. J., Thompson, R. F. (1995). Inactivation of the superior cerebellar peduncle blocks expression but not acquisition of the rabbit's classically conditioned eye-blink response. *Proc. Natl. Acad. Sci. USA* 92:5097–5101.

Krupa, D. J., Thompson, R. F. (1997). Reversible inactivation of the cerebellar interpositus nucleus completely prevents acquisition of the classically conditioned eye-blink response. *Learn. Mem.* 3:545–556.

Krupa, D. J., Thompson, J. K., Thompson, R. F. (1993). Localization of a memory trace in the mammalian brain. *Science* 260:989–991.

Kumoi, K., Saito, N., Kuno, T., Tanaka, C. (1988). Immunohistochemical localization of gamma-amino butyric acid– and aspartate-containing neurons in the rat deep cerebellar nuclei. *Brain Res.* 439:302–310.

Kurihara, H., Hashimoto, K., Kano, M,, Takayama, C., Sakimura, K., Mishina, M., Inoue, Y., Watanabe, M. (1997). Impaired parallel fiber–Purkinje cell synapse stabilization during cerebellar development of mutant mice lacking the glutamate receptor delta2 subunit. *J. Neurosci.* 17:9613–9623.

Larson, J., Xiao, P., Lynch, G. (1993). Reversal of LTP by theta frequency stimulation. *Brain Res.* 600:97–102.

Lavond, D. G., Steinmetz, J. E. (1989). Acquisition of classical conditioning without cerebellar cortex. *Behav. Brain Res.* 33:113–164.

Lee, H.-K., Kameyama, K., Huganir, R. L., Bear, M. F. (1998). NMDA induces long-term synaptic depression and dephosphorylation of the GluR1 subunit of AMPA receptors in hippocampus. *Neuron* 21:1151–1162.

Lee, H.-K., Barbarosie, M., Kameyama, K., Bear, M. F., Huganir, R. L. (2000). Regulation of distinct AMPA receptor phosphorylation sites during bidirectional synaptic plasticity. *Nature* 405:955–958.

Lev-Ram, V., Makings, L. R., Keitz, P. F., Kao, J. P. Y., Tsien, R. Y. (1995). Long-term depression in cerebellar Purkinje neurons results from coincidence of nitric oxide and depolarization-induced Ca^{2+} transients. *Neuron* 15:407–415.

Lev-Ram, V., Jiang, T., Wood, J., Lawrence, D. S., Tsien, R. Y. (1997a). Synergies and coincidence requirements between NO, cGMP, and Ca^{2+} in the induction of cerebellar long-term depression. *Neuron* 18:1025–1038.

Lev-Ram, V., Nebyelul, Z., Ellisman, M. H., Huang, P. L., Tsien, R. Y. (1997b). Absence of cerebellar long-term depression in mice lacking neuronal nitric oxide synthase. *Learn. Mem.* 4:169–177.

Li, H., Weiss, S. R., Chuang, D. M., Post, R. M., Rogawski, M. A. (1998). Bidirectional synaptic plasticity in the rat basolateral amygdala: characterization of an activity-dependent switch sensitive to the presynaptic metabotropic glutamate receptor antagonist 2S-alpha-ethylglutamic acid. *J. Neurosci.* 18: 1662–1670.

Li, J., Smith, S. S., McElligott, J. G. (1995). Cerebellar nitric oxide is necessary for vestibulo-ocular reflex adaptation, a sensorimotor model of learning. *J. Neurophysiol.* 74:489–494.

Linden, D. J. (1994). Input-specific induction of cerebellar long-term depression does not require presynaptic alteration. *Learn. Mem.* 1:121–128.

Linden, D. J. (1996). A protein synthesis–dependent late phase of cerebellar long-term depression. *Neuron* 17:483–490.

Linden, D. J. (1997). Long-term potentiation of glial synaptic currents in cerebellar culture. *Neuron* 18:983–994.

Linden, D. J. (1998). Synaptically evoked glutamate transport currents may be used to detect the expression of long-term potentiation in cerebellar culture. *J. Neurophysiol.* 79:3151–3156.

Linden, D. J. (1999). The return of the spike: postsynaptic action potentials and the induction of LTP and LTD. *Neuron* 22:661–666.

Linden, D. J., Ahn, S. (1999). Activation of presynaptic cAMP-dependent protein kinase is required for induction of cerebellar long-term potentiation. *J. Neurosci.* 19:10,221–10,227.

Linden, D. J., Connor, J. A. (1991). Participation of postsynaptic PKC in cerebellar long-term depression in culture. *Science* 254:1656–1659.

Linden, D. J., Connor, J. A. (1992). Long-term depression of glutamate currents in cultured cerebellar Purkinje neurons does not require nitric oxide signalling. *Eur. J. Neurosci.* 4:10–15.

Linden, D. J., Connor, J. A. (1995). Long-term synaptic depression. *Annu. Rev. Neurosci.* 18:319–357.

Linden, D. J., Dickinson, M. H., Smeyne, M., Connor, J. A. (1991). A long-term depression of AMPA currents in cultured cerebellar Purkinje neurons. *Neuron* 7:81–89.

Linden, D. J., Smeyne, M., Connor, J. A. (1993). Induction of cerebellar long-term depression in culture requires postsynaptic action of sodium ions. *Neuron* 11:1093–1100.

Linden, D. J., Dawson, T. M., Dawson, V. L. (1995). An evaluation of the nitric oxide/cGMP/cGMP-dependent protein kinase cascade in the induction of cerebellar long-term depression in culture. *J. Neurosci.* 15:5098–5105.

Lisman, J. (1989). A mechanism for the Hebb and the anti-Hebb processes underlying learning and memory. *Proc. Natl. Acad. Sci. USA* 86:9574–9578.

Lovinger, D. M., Tyler, E. C., Merritt, A. (1993). Short- and long-term synaptic depression in rat neostriatum. *J. Neurophysiol.* 70:1937–1949.

Lüscher, C., Xia, H., Beattie, E. C., Carrol, R. C., von Zastrow, M., Malenka, R. C., Nicoll, R. A. (1999). Role of AMPA receptor cycling in synaptic transmission and plasticity. *Neuron* 24:649–658.

Luthi, A., Chittajallu, R., Duprat, F., Palmer, M. J., Benke, T. A., Kidd, F. L., Henley, J. M., Isaac, J. T., Collingridge, G. L. (1999). Hippocampal LTD expression involves a pool of AMPARs regulated by the NSF-GluR2 interaction. *Neuron* 24:389–399.

McCormick, D. A., Thompson, R. F. (1984a). Cerebellum: essential involvement in the classically conditioned eyelid response. *Science* 223:296–299.

McCormick, D. A., Thompson, R. F. (1984b). Neuronal responses of the rabbit cerebellum during acquisition and performance of a classically conditioned nictitating membrane–eyelid response. *J. Neurosci.* 4:2811–2822.

McCormick, D. A., Clark, G. A., Lavond, D. G., Thompson, R. F. (1982). Initial localization of the memory trace for a basic form of learning. *Proc. Natl. Acad. Sci. USA* 79:2731–2735.

Malenka, R., Kauer, J., Perkel, D., Mauk, M., Kelly, P., Nicoll, R., Waxham, M. (1989). An essential role for postsynaptic calmodulin and protein kinase activity in long-term potentiation. *Nature* 340:554–557.

Malinow, R., Miller, J. P. (1986). Postsynaptic hyperpolarization during conditioning reversibly blocks induction of long-term potentiation. *Nature* 320:529–530.

Malinow, R., Schulman, H., Tsien, R. W. (1989). Inhibition of postsynaptic PKC or CaMKII blocks induction but not expression of LTP. *Nature* 245:862–865.

Mammen, A. L., Kameyama, K., Roche, K. W., Huganir, R. L. (1997). Phosphorylation of the alpha-amino-3-hydroxy-5-methylisoxazole-4-propionic acid receptor GluR1 subunit by calcium/calmodulin-dependent kinase II. *J. Biol. Chem.* 272:32,528–32,533.

Man, H. Y., Lin, J., Ju, W., Ahmadian, G., Liu, L. D., Becker, L. E., Sheng, M., Wang, Y. T. (2000). Regulation of AMPA receptor–mediated synaptic transmission by clathrin-dependent receptor internalization. *Neuron* 25:649–662.

Manahan-Vaughan, D. (1997). Group 1 and 2 metabotropic glutamate receptors play differential roles in hippocampal long-term depression and long-term potentiation in freely moving rats. *J. Neurosci.* 17:3303–3311.

Manahan-Vaughan, D. (1998). Priming of group 2 metabotropic glutamate receptors facilitates induction of long-term depression in the dentate gyrus of freely moving rats. *Neuropharmacology* 37:1459–1464.

Manahan-Vaughan, D., Braunewell, K. H. (1999). Novelty acquisition is associated with induction of hippocampal long-term depression. *Proc. Natl. Acad. Sci. USA* 96:8739–8744.

Markram, H., Lubke, J., Frotscher, M., Sakmann, B. (1997). Regulation of synaptic efficacy by coincidence of postsynaptic APs and EPSPs. *Science* 275:213–215.

Marr, D. (1969). A theory of cerebellar cortex. *J. Physiol.* 202:437–470.

Martin, L. J., Blackstone, C. D., Huganir, R. L., Price, D. L. (1992). Cellular localization of a metabotropic glutamate receptor in rat brain. *Neuron* 9:259–270.

Marty, A., Llano, I. (1995). Modulation of inhibitory synapses in the mammalian brain. *Curr. Opin. Neurobiol.* 5:335–341.

Mauk, M. D. (1997). Roles of cerebellar cortex and nuclei in motor learning: contradictions or clues? *Neuron* 18:343–346.

Mauk, M. D., Donegan, N. H. (1997). A model of Pavlovian eyelid conditioning based on the synaptic organization of the cerebellum. *Learn. Mem.* 3:130–158.

Mauk, M. D., Steinmetz, J. E., Thompson, R. F. (1986). Classical conditioning using stimulation of the inferior olive as the unconditioned stimulus. *Proc. Natl. Acad. Sci. USA* 83:5349–5353.

Mayford, M., Wang, J., Kandel, E., O'Dell, T. (1995). CaMKII regulates the frequency-response function of hippocampal synapses for the production of both LTD and LTP. *Cell* 81:1–20.

Medina, J. F., Mauk, M. D. (1999). Simulations of cerebellar motor learning: computational analysis of plasticity at the mossy fiber to deep nucleus synapse. *J. Neurosci.* 19:7140–7151.

Merlin, L. R., Bergold, P. J., Wong, R. K. S. (1998). Requirement of protein synthesis for group 1 mGluR-mediated induction of eplileptiform discharges. *J. Neurophys.* 80:989–993.

Migaud, M., Charlesworth, P., Dempster, M., Webster, L. C., Watabe, A. M., Makhinson, M., He, Y., Ramsay, M. F., Morris, R. G., Morrison, J. H., O'Dell, T. J., Grant, S. G. (1998). Enhanced long-term potentiation and impaired learning in mice with mutant postsynaptic density-95 protein. *Nature* 396:433–439.

Miyata, M., Okada, D., Hashimoto, K., Kano, M., Ito, M. (1999). Corticotropin-releasing factor plays a permissive role in cerebellar long-term depression. *Neuron* 22:763–775.

Mulkey, R. M., Malenka, R. C. (1992). Mechanisms underlying induction of homosynaptic long-term depression in area CA1 of the hippocampus. *Neuron* 9:967–975.

Mulkey, R. M., Herron, C. E., Malenka, R. C. (1993). An essential role for protein phosphatases in hippocampal long-term depression. *Science* 261:1051–1055.

Mulkey, R. M., Endo, S., Shenolikar, S., Malenka, R. C. (1994). Calcineurin and inhibitor-1 are components of a protein-phosphatase cascade mediating hippocampal LTD. *Nature* 369:486–488.

Murashima, M., Hirano, T. (1999). Entire course and distinct phases of day-lasting depression of miniature EPSC amplitudes in cultured Purkinje neurons. *J. Neurosci.* 19:7326–7333.

Nagao, S., Ito, M. (1991). Subdural application of hemoglobin to the cerebellum blocks vestibulo-ocular reflex adaptation. *NeuroReport* 2:193–196.

Nairn, A. C., Shenolikar, S. (1992). The role of protein phosphatases in synaptic transmission, plasticity and neuronal development. *Curr. Opin. Neurobiol.* 2:296–301.

Nakanishi, S., Maeda, N., Mikoshiba, K. (1991). Immunohistochemical localization of an inositol 1,4,5-trisphosphate receptor, P400, in neural tissue: studies in developing and adult mouse brain. *J. Neurosci.* 11:2075–2086.

Nakazawa, K., Mikawa, S., Hashikawa, T., Ito, M. (1995). Transient and persistent phosphorylation of AMPA-type glutamate receptor subunits in cerebellar Purkinje cells. *Neuron* 15:697–709.

Narasimhan, K., Linden, D. J. (1996). Defining a minimal computational unit for cerebellar long-term depression. *Neuron* 17:333–341.

Narasimhan, K., Pessah, I. N., Linden D. J. (1998). Inositol-1,4,5-trisphosphate receptor–mediated Ca mobilization is not required for the induction of cerebellar long-term depression in reduced preparations. *J. Neurophysiol.* 80: 2963–2974.

Neveu, D., Zucker, R. S. (1996). Postsynaptic levels of $[Ca^{2+}]_i$ needed to trigger LTD and LTP. *Neuron* 16:619–629.

Nishimune, A., Isaac, J. T., Molnar, E., Noel, J., Nash, S. R., Tagaya, M., Collingridge, G. L., Nakanishi, S., Henley, J. M. (1998). NSF binding to GluR2 regulates synaptic transmission. *Neuron* 21:87–97.

Noel, J., Ralph, G. S., Pickard, L., Williams, J., Molnar, E., Uney, J. B., Collingridge, G. L., Henley, J. M. (1999). Surface expression of AMPA receptors in hippocampal neurons is regulated by an NSF-dependent mechanism. *Neuron* 23:365–376.

Nouranifar, R., Blitzer, R. D., Wong, T., Landau, E. (1998). Metabotropic glutamate receptors limit adenylyl cyclase–mediated effects in rat hippocampus via protein kinase C. *Neurosci. Lett.* 244:101–105.

Oberdick, J., Smeyne, R. J., Mann, J. R., Zackson, S., Morgan, J. I. (1990). A promoter that drives transgene expression in cerebellar Purkinje and retinal bipolar neurons. *Science* 248:223–226.

O'Dell, T. J., Kandel, E. R. (1994). Low-frequency stimulation erases LTP through an NMDA receptor-mediated activation of protein phosphatases. *Learn. Mem.* 1:129–139.

Oliet, S. H., Malenka, R. C., Nicoll, R. A. (1996). Bidirectional control of quantal size by synaptic activity in the hippocampus. *Science* 271:1294–1297.

Oliet, S. H., Malenka, R. C., Nicoll, R. A. (1997). Two distinct forms of long-term depression coexist in CA1 hippocampal pyramidal cells. *Neuron* 18: 969–982.

Osten, P., Srivastava, S., Inman, G. J., Vilim, F. S., Khatri, L., Lee, L. M., States, B. A., Einheber, S., Milner, T. A., Hanson, P. I., Ziff, E. B. (1998). The AMPA receptor GluR2 C terminus can mediate a reversible, ATP-dependent interaction with NSF and alpha- and beta-SNAPs. *Neuron* 21:99–110.

Otani, S., Connor, J. A. (1998). Requirement of rapid Ca^{2+} entry and synaptic activation of metabotropic glutamate receptors for the induction of long-term depression in adult rat hippocampus. *J. Physiol.* 511:761–770.

Perrett, S., Mauk, M. (1995). Extinction of conditioned eyelid response requires the anterior lobe of the cerebellar cortex. *J. Neurosci.* 15:2074–2080.

Perrett, S., Ruiz, B., Mauk, M. (1993). Cerebellar cortex lesions disrupt learning-dependent timing of conditioned eyelid responses. *J. Neurosci.* 13:1708–1718.

Pin, J. P., Bockaert, J. (1995). Get receptive to metabotropic glutamate receptors. *Curr. Opin. Neurobiol.* 5:342–349.

Qi, M., Zhuo, M., Skalhegg, B. S., Brandon, E. P., Kandel, E. R., McKnight, G. S., Idzerda, R. L. (1996). Impaired hippocampal plasticity in mice lacking the Cbeta1 catalytic subunit of cAMP-dependent protein kinase. *Proc. Natl. Acad. Sci. USA* 93:1571–1576.

Quinlan, E. M., Olstein, D. H., Bear, M. F. (1999). Bidirectional, experience-dependent regulation of NMDA subunit composition in rat visual cortex during postnatal development. *Proc. Natl. Acad. Sci. USA* 96:12,876–12,880.

Ramakers, G. M., McNamara, R. K., Lenox, R. H., De, G. P. (1999). Differential changes in the phosphorylation of the protein kinase C substrates myristoylated alanine-rich C kinase substrate and growth-associated protein-43/B-50 following Schaffer collateral long-term potentiation and long-term depression. *J. Neurochem.* 73:2175–2183.

Ramnani, N., Yeo, C. H. (1996). Reversible inactivations of the cerebellum prevent the extinction of conditioned nictitating membrane responses in rabbits. *J. Physiol.* 495:159–168.

Raymond, C. R., Thompson, V. L., Tate, W. P., Abraham, W. C. (2000). Metabotropic glutamate receptors trigger homosynaptic protein synthesis to prolong LTP. *J. Neurosci.* 20:969–976.

Raymond, J. L., Lisberger, S. G., Mauk, M. D. (1996). The cerebellum: a neuronal learning machine? *Science* 272:1126–1131.

Reyes, H. M., Stanton, P. K. (1996). Induction of hippocampal long-term depression requires release of Ca^{2+} from separate presynaptic and postsynaptic intracellular stores. *J. Neurosci.* 16:5951–5960.

Reyes, H. M., Stanton, P. K. (1998). Postsynaptic phospholipase C activity is required for the induction of homosynaptic long-term depression in rat hippocampus. *Neurosci. Lett.* 252:155–158.

Reyes, H. M., Potter, B. V., Galione, A., Stanton, P. K. (1999). Induction of hippocampal LTD requires nitric-oxide-stimulated PKG activity and Ca^{2+} release from cyclic ADP-ribose-sensitive stores. *J. Neurophysiol.* 82:1569–1576.

Riedel, G., Micheau, J., Lam, A. G., Roloff, E. V., Martin, S. J., Bridge, H., Hoz, L. D., Poeschel, B., McCulloch, J., Morris, R. G. (1999). Reversible neural inactivation reveals hippocampal participation in several memory processes. *Nature Neurosci.* 2:898–905.

Rittenhouse, C. D., Shouval, H. Z., Paradiso, M. A., Bear, M. F. (1999). Monocular deprivation induces homosynaptic long-term depression in visual cortex. *Nature* 397:347–350.

Roche, K. W., O'Brien, R. J., Mammen, A. L., Bernhardt, J., Huganir, R. L. (1996). Characterization of multiple phosphorylation sites on the AMPA receptor GluR1 subunit. *Neuron* 16:1179–1188.

Rosenmund, C., Legendre, P., Westbrook, G. L. (1992). Expression of NMDA channels on cerebellar Purkinje cells acutely dissociated from newborn rats. *J. Neurophysiol.* 68:1901–1905.

Ross, W. N., Werman, R. (1987). Mapping calcium transients in the dendrites of Purkinje cells from the guinea-pig cerebellum in vitro. *J. Physiol.* 389:319–336.

Sakagami, H., Kondo, H. (1993). Cloning and sequencing of a gene encoding the beta polypeptide of Ca^{2+}/calmodulin-dependent protein kinase IV and its expression confined to the mature cerebellar granule cells. *Mol. Brain Res.* 19:215–218.

Sakagami, H., Watanabe, M., Kondo, H. (1992). Gene expression of Ca^{2+}/calmodulin-dependent protein kinase of the cerebellar granule cell type or type IV in the mature and developing rat brain. *Mol. Brain Res.* 16:20–28.

Sakurai, M. (1987). Synaptic modification of parallel fiber–Purkinje cell transmission in *in vitro* guinea-pig cerebellar slices. *J. Physiol.* 394:463–480.

Sakurai, M. (1990). Calcium is an intracellular mediator of the climbing fiber in induction of cerebellar long-term depression. *Proc. Natl. Acad. Sci. USA* 87:3383–3385.

Salin, P. A., Malenka, R. C., Nicoll, R. A. (1996). Cyclic AMP mediates a pre-synaptic form of LTP at cerebellar parallel fiber synapses. *Neuron* 16:797–806.

Sastry, R. B., Goh, J. W., Auyeung, A. (1986). Associative induction of posttetanic and long-term potentiation in CA1 neurons of rat hippocampus. *Science* 232: 988–990.

Sawtell, N. B., Huber, K. M., Roder, J. C., Bear, M. F. (1999). Induction of NMDA receptor–dependent long-term depression in visual cortex does not require metabotropic glutamate receptors. *J. Neurophysiol.* 82:3594–3597.

Schmahmann, J. D. (1997). The cerebellum and cognition. *Int. Rev. Neurobiol.* 41.

Schreurs, B. G., Alkon, D. L. (1993). Rabbit cerebellar slice analysis of long-term depression and its role in classical conditioning. *Brain Res.* 631:235–240.

Schreurs, B. G., Oh, M. M., Alkon, D. L. (1996). Pairing-specific long-term de-pression of Purkinje cell excitatory postsynaptic potentials results from a classical conditioning procedure in the rabbit cerebellar slice. *J. Neurophysiol.* 75:1051–1060.

Seeburg, P. H. (1993). The TINS/TiPS Lecture. The molecular biology of mam-malian glutamate receptor channels. *Trends Neurosci.* 16:359–365.

Selig, D. K., Hjelmstad, G. O., Herron, C., Nicoll, R. A., Malenka, R. C. (1995a). Independent mechanisms for long-term depression of AMPA and NMDA re-sponses. *Neuron* 15:417–426.

Selig, D. K., Lee, H.-K., Bear, M. F., Malenka, R. C. (1995b). Reexamination of the effects of MCPG on hippocampal LTP, LTD, and depotentiation. *J. Neurophys.* 74:1075–1082.

Shibuki, K., Okada, D. (1991). Endogenous nitric oxide release required for long-term synaptic depression in the cerebellum. *Nature* 349:326–328.

Shibuki, K., Okada, D. (1992). Cerebellar long-term potentiation under sup-pressed postsynaptic Ca^{2+} activity. *NeuroReport* 3:231–234.

Shibuki, K., Gomi, H., Chen, L., Bao, S., Kim, J. J., Wakatsuki, H., Fujisaki, T., Fujimoto, K., Ikeda, T., Chen, C., Thompson, R. F., Itohara, S. (1996). Defi-cient cerebellar long-term depression, impaired eyeblink conditioning and normal motor coordination in GFAP mutant mice. *Neuron* 16:587–599.

Shigemoto, R., Nakanishi, S., Hirano, T. (1994). Antibodies inactivating mGluR1 metabotropic glutamate receptor block long-term depression in cultured Purkinje cells. *Neuron* 12:1245–1255.

Silva, A. J., Kogan, J. H., Frankland, P. W., Kida, S. (1998). CREB and memory. *Annu. Rev. Neurosci.* 21:127–148.

Song, I., Kamboj, S., Xia, J., Dong, H., Liao, D., Huganir, R. L. (1998). Interaction of the *N*-ethylmaleimide-sensitive factor with AMPA receptors. *Neuron* 21: 393–400.

Stanton, P. K. (1995). Transient protein kinase C activation primes long-term depression and suppresses long-term potentiation of synaptic transmission in hippocampus. *Proc. Natl. Acad. Sci. USA* 92:1724–1728.

Staubli, U., Chun, D. (1996). Proactive and retrograde effects on LTP produced by theta pulse stimulation: mechanisms and characteristics of LTP reversal in vitro. *Learn. Mem.* 3:96–105.

Staubli, U., Lynch, G. (1990). Stable depression of potentiated synaptic responses in the hippocampus with 1–5 Hz stimulation. *Brain Res.* 513:113–118.

Staubli, U., Scafidi, J. (1997). Studies on long-term depression in area CA1 of the anesthetized and freely moving rat. *J. Neurosci.* 17:4820–4828.

Steele, P. M., Mauk, M. D. (1999). Inhibitory control of LTP and LTD: stability of synapse strength. *J. Neurophysiol.* 81:1559–1566.

Steinmetz, J. E., Rosen, D. J., Chapman, P. F., Lavond, D. G., Thompson, R. F. (1986). Classical conditioning of the rabbit eyelid response with a mossy-fiber stimulation CS: I. Pontine nuclei and middle cerebellar peduncle stimulation. *Behav. Neurosci.* 100:878–887.

Steinmetz, J. E., Lavond, D. G., Thompson, R. F. (1989). Classical conditioning in rabbits using pontine nucleus stimulation as a conditioned stimulus and inferior olive stimulation as an unconditioned stimulus. *Synapse* 3:225–233.

Steinmetz, J. E., Lavond, D. G., Ivkovich, D., Logan, C. G., Thompson, R. F. (1992). Disruption of classical eyelid conditioning after cerebellar lesions: damage to a memory trace system or a simple performance deficit? *J. Neurosci.* 12:4403–4426.

Stevens, C. F., Wang, Y. (1994). Changes in reliability of synaptic function as a mechanism for plasticity. *Nature* 371:704–707.

Steward, O., Davis, L., Dotti, C., Phillips, L. L., Rao, A., Banker, G. (1988). Protein synthesis and processing in cytoplasmic microdomains beneath postsynaptic sites on CNS neurons: a mechanism for establishing and maintaining a mosaic postsynaptic receptive surface. *Mol. Neurobiol.* 2:227–261.

Storm, D. R., Hansel, C., Hacker, B., Parent, A., Linden, D. J. (1998). Impaired cerebellar long-term potentiation in type I adenylyl cyclase mutant mice. *Neuron* 20:1199–1210.

Strata, P., Rossi, F. (1998). Plasticity of the olivocerebellar pathway. *Trends Neurosci.* 21:407–413.

Tang, Y. P., Shimizu, E., Dube, G. R., Rampon, C., Kerchner, G. A., Zhuo, M., Liu, G., Tsien, J. Z. (1999). Genetic enhancement of learning and memory in mice. *Nature* 401:63–69.

Tempia, F., Kano, M., Schneggenburger, R., Schirra, C., Garaschuk, O., Plant, T., Konnerth, A. (1996). Fractional calcium current through neuronal AMPA-receptor channels with a low calcium permeability. *J. Neurosci.* 16:456–466.

Teune, T. M., van der Burg, J., De Zeeuw, C. I., Voogd, J., Ruigrok, T. J. H. (1998). Single Purkinje cells can innervate multiple classes of projection neurons in the cerebellar nuclei of the rat: a light microscopic and ultrastructural triple-tracer study in the rat. *J. Comp. Neurol.* 392:164–178.

Thiels, E., Barrionuevo, G., Berger, T. W. (1994). Excitatory stimulation during postsynaptic inhibition induces long-term depression in hippocampus in vivo. *J. Neurophysiol.* 71:3009–3016.

Thiels, E., Norman, E. D., Barrionuevo, G., Klann, E. (1998). Transient and persistent increases in protein phosphatase activity during long-term depression in the adult hippocampus in vivo. *Neuroscience* 86:1023–1029.

Thomas, M. J., Moody, T. D., Makhinson, M., O'Dell, T. J. (1996). Activity-dependent beta-adrenergic modulation of low frequency stimulation induced LTP in the hippocampal CA1 region. *Neuron* 17:475–482.

Tsumoto, T. (1993). Long-term depression in cerebral cortex: a possible substrate of "forgetting" that should not be forgotten. *Neurosci. Res.* 16:263–270.

Vanhoutte, P., Barnier, J. V., Guibert, B., Pages, C., Besson, J. J., Hipskind, R. A., Caboche, J. (1999). Glutamate induces phosphorylation of Elk-1 and CREB, along with c-fos activation, via an extracellular signal-regulated kinase-dependent pathway in brain slices. *Mol. Cell. Biol.* 19:136–146.

Vincent, S. R., Kimura, H. (1992). Histochemical mapping of nitric oxide synthase in the rat brain. *Neuroscience* 46:755–784.

Wagner, J. J., Alger, B. E. (1995). GABAergic and developmental influences on homosynaptic LTD and depotentiation in rat hippocampus. *J. Neurosci.* 15:1577–1586.

Wang, H., Wagner, J. J. (1999). Priming-induced shift in synaptic plasticity in the rat hippocampus. *J. Neurophysiol.* 82:2024–2028.

Wang, S. J., Gean, P. W. (1999). Long-term depression of excitatory synaptic transmission in the rat amygdala. *J. Neurosci.* 19:10,656–10,663.

Wang, S. S., Augustine, G. J. (1995). Confocal imaging and local photolysis of caged compounds; dual probes of synaptic function. *Neuron* 15:755–760.

Wang, Y.-T., Linden, D. J. (2000). Expression of cerebellar long-term depression requires clathrin-mediated internalization of postsynaptic AMPA receptors. *Neuron* 25:635–647.

Weiler, I. J., Greenough, W. T. (1993). Metabotropic glutamate receptors trigger postsynaptic protein synthesis. *Proc. Natl. Acad. Sci. USA* 90: 7168–7171.

Weiler, I. J., Irwin, S. A., Klintsova, A. Y., Spencer, C. M., Brazelton, A. D., Miyashiro, K., Comery, T. A., Patel, B., Eberwine, J., Greenough, W. T. (1997). Fragile X mental retardation protein is translated near synapses in response to neurotransmitter activation. *Proc. Natl. Acad. Sci. USA* 94:5395–5400.

Welsh, J. P., Harvey, J. A. (1989). Cerebellar lesions and the nictitating membrane reflex: performance deficits of the conditioned and unconditioned response. *J. Neurosci.* 9:299–311.

Wenthold, R. J., Petralia, R. S., Blahos, J., II, Niedzielski, A. S. (1996). Evidence for multiple AMPA receptor complexes in hippocampal CA1/CA2 neurons. *J. Neurosci.* 16:1982–1989.

Wigstrom, M., Gustafsson, B. (1985). On long-lasting potentiation in the hippocampus: a proposed mechanism for its dependence on coincident pre- and postsynaptic activity. *Acta Physiol. Scand.* 123:519–522.

Wilson, M., McNaughton, B. (1993). Dynamics of the hippocampal ensemble that codes for space. *Science* 261:1055–1058.

Xiang, J. Z., Brown, M. W. (1998). Differential neuronal encoding of novelty, familiarity and recency in regions of the anterior temporal lobe. *Neuropharmacology* 37:657–676.

Xiao, M. Y., Wigstrom, H., Gustafsson, B. (1994). Long-term depression in the hippocampal CA1 region is associated with equal changes in AMPA and NMDA receptor-mediated synaptic potentials. *Eur. J. Neurosci.* 6:1055–1057.

Xiao, M. Y., Karpefors, M., Gustafsson, B., Wigstrom, H. (1995). On the linkage between AMPA and NMDA receptor-mediated EPSPs in homosynaptic long-term depression in the hippocampal CA1 region of young rats. *J. Neurosci.* 15:4496–4506.

Xiao, M. Y., Niu, Y. P., Wigstrom, H. (1996). Activity-dependent decay of early LTP revealed by dual EPSP recording in hippocampal slices from young rats. *Eur. J. Neurosci.* 8:1916–1923.

Xu, L., Anwyl, R., Rowan, M. J. (1997). Behavioural stress facilitates the induction of long-term depression in the hippocampus. *Nature* 387:497–500.

Xu, L., Anwyl, R., Rowan, M. J. (1998a). Spatial exploration induces a persistent reversal of long-term potentiation in rat hippocampus. *Nature* 394:891–894.

Xu, L., Holscher, C., Anwyl, R., Rowan, M. J. (1998b). Glucocorticoid receptor and protein/RNA synthesis-dependent mechanisms underlie the control of synaptic plasticity by stress. *Proc. Natl. Acad. Sci. USA* 95:3204–3208.

Yang, S. N., Tang, Y. G., Zucker, R. S. (1999). Selective induction of LTP and LTD by postsynaptic $[Ca^{2+}]_i$ elevation. *J. Neurophysiol.* 81:781–787.

Yeo, C. H., Hardiman, M. J. (1992). Cerebellar cortex and eyeblink conditioning: a reexamination. *Exp. Brain Res.* 88:623–638.

Yeo, C. H., Hardiman, M. J., Glickstein, M. (1985a). Classical conditioning of the nictitating membrane response of the rabbit. I. Lesions of the cerebellar nuclei. *Exp. Brain Res.* 60:87–98.

Yeo, C. H., Hardiman, M. J., Glickstein, M. (1985b). Classical conditioning of the nictitating membrane response of the rabbit. II. Lesions of the cerebellar cortex. *Exp. Brain Res.* 60:99–113.

Zhuo, M., Zhang, W., Son, H., Mansuy, I., Sobel, R. A., Seidman, J., Kandel, E. R. (1999). A selective role of calcineurin alpha in synaptic depotentiation in hippocampus. *Proc. Natl. Acad. Sci. USA* 96:4650–4655.

Zucker, R. S. (1999). Calcium- and activity-dependent synaptic plasticity. *Curr. Opin. Neurobiol.* 9:305–313.

Zuo, J., De Jager, P. L., Takahashi, K., Jiang, W., Linden, D. J., Heintz, N. (1997). Neurodegeneration in Lurcher mice results from a mutation in the δ2 receptor gene. *Nature* 388:769–772.

11

Synaptic Plasticity and Memory

Paul D. Grimwood, Stephen J. Martin,
and Richard G. M. Morris

519

t is widely assumed that patterns of neural activity encoding experience can elicit enduring changes in the strength of neural connectivity that, when appropriately reactivated, enable a memory of that experience. The discovery of long-term potentiation (LTP), whereby brief high-frequency stimulation of a neural pathway can induce long-lasting changes in synaptic efficacy (Bliss and Lømo, 1973), provided a reproducible phenomenon, a tool with which to investigate the assumption. Studies of the possible role of LTP or LTP-like changes in learning and memory have led to the general conclusion that we will call the synaptic plasticity and memory (SPM) hypothesis. This hypothesis states that "Activity-dependent synaptic plasticity is induced at appropriate synapses during memory formation and is both necessary and sufficient for the information storage underlying the type of memory mediated by the brain area in which that plasticity is observed."

Stated in this way, the hypothesis is intended to capture many of the properties and mechanisms of LTP, long-term depression (LTD), and other activity-dependent changes that have been discovered over the past 30 years (see Martin et al., 2000, for a more complete discussion). In order to create a framework for describing tests of this hypothesis, we outline a set of formal criteria and describe a range of experimental strategies that have been used to investigate it.

Investigating the Synaptic Plasticity and Memory Hypothesis

Logical Criteria

The SPM hypothesis must fulfill four logical criteria:

1. *Detectability:* If an animal displays memory of some previous experience, a change in synaptic efficacy should be detectable somewhere in its nervous system.
2. *Mimicry:* Conversely, if it were possible to induce the same spatial pattern of synaptic weight changes artificially, the animal should display "apparent" memory for some past experience that did not in practice occur.
3. *Anterograde alteration:* Interventions that prevent the induction of synaptic weight changes during a learning experience should impair the animal's memory of that experience.
4. *Retrograde alteration:* Interventions that alter the spatial distribution of synaptic weights induced by a prior learning experience should alter the animal's memory of that experience.

This chapter is adapted from Martin et al. (2000). It is published with permission of Annual Reviews, Inc. (www.AnnualReviews.org).

The paucity of synapses that change with any one learning experience, and their spatial distribution, may make the detectability criterion difficult to meet experimentally in certain neuronal circuits but pose less of a problem in others. Induction of LTP (or LTD) at an appropriate and distributed subset of hippocampal synapses to achieve an "apparent" memory of an event that never occurred (mimicry criterion) is unlikely to be feasible in the near future. It may be easier in brain areas such as the amygdala or in simpler vertebrate (or invertebrate) systems. For example, in the case of the goldfish escape reflex, it is known that repeated acoustic stimuli induce LTP at inhibitory synapses onto the Mauthner cell and also cause behavioral desensitization of this reflex (Oda et al., 1998). Satisfying the mimicry criterion is an enticing goal, for it would establish that a change in synaptic weight is indeed a sufficient condition for memory. On the other hand, satisfying both the anterograde and retrograde criteria is vital to the idea that changes in synaptic efficacy are necessary for the establishment of memory. An asymmetry about the anterograde criterion should be recognized. Treatments that definitively block synaptic plasticity in a brain area are predicted to have deleterious effects on learning mediated by that brain area; but treatments shown to affect learning need not impair synaptic plasticity. This asymmetry arises because there are likely to be many additional aspects of CNS function that influence learning and memory beyond synaptic plasticity.

Experimental Strategies

Five basic strategies have been used to investigate equivalents of the SPM hypothesis.

1. *Correlation:* The behavioral parameters of learning should be correlated with some but not necessarily all of the properties of synaptic plasticity.
2. *Induction:* Learning should be associated with the induction of measurable changes in synaptic efficacy at synapses in appropriate networks of the brain, and the induction of such changes at relevant synapses (were this to be feasible) should result in apparent memories.
3. *Occlusion:* Saturation of synaptic plasticity in a network should destroy the pattern of trace strengths corresponding to established memories and occlude new memory encoding.
4. *Intervention:* Blockade or enhancement of synaptic plasticity—achieved by pharmacological, genetic, or other manipulations—should have commensurate effects on learning or memory.
5. *Erasure:* Erasure of synaptic plasticity should, at least shortly after learning, induce forgetting.

The correlation (weakly) and induction (more strongly) strategies are relevant to the detectability criterion, that synaptic plasticity must occur during learning, and the mimicry criterion, that it must be sufficient for a memory trace to be established. The occlusion, intervention, and erasure strategies are relevant to the claim that changes in synaptic efficacy are necessary for memory trace formation or maintenance, and so relate to the anterograde and retrograde alteration criteria. Attempting to saturate LTP (occlusion) is relevant to both the anterograde and retrograde alteration criteria. Other strategies go beyond the formal criteria in interesting and important ways. For example, experiments establishing that pharmacological or genetic interventions that enhance LTP also improve learning may lead to discoveries about how to improve memory. The hypothesis, however, is not required to make this prediction, as synaptic plasticity may ordinarily be optimally tuned such that any disturbance of the balance between LTP and LTD would be deleterious.

The SPM hypothesis should be distinguished from others about LTP or LTD. These include the plasticity-pathology continuum hypothesis (McEachern and Shaw, 1996) and the notions that synaptic plasticity plays a role in attentional rather than memory processes (Shors and Matzel, 1997), or indeed that it has nothing to do with memory whatsoever (which we will call the null hypothesis). Distinguishing among these and alternative hypotheses about the functions of LTP is not always easy. The SPM and null hypotheses are easy to contrast, but much of the evidence thought to support the SPM hypothesis could also be said to support the view that an LTP-like mechanism underlies cognitive processes, such as attention, which are essential prerequisites for learning rather than integral to encoding or storage processes per se. The thrust of Shors and Matzel's (1997) critique, with which we have a measure of sympathy, is that few experiments have yet been conducted that can unambiguously distinguish these rival hypotheses.

Synaptic Plasticity: A Realistic Mnemonic Device?

Before discussing the experimental data that directly address the SPM hypothesis, it is worth considering the following question: do the properties of LTP, LTD, or both make them realistic representatives for the cellular mechanisms underlying learning and memory? We briefly discuss the relationship between the properties of LTP and the properties of learning, emphasizing the importance of the neuronal circuit. We also describe naturalistic patterns of induction of LTP and LTD and consider whether plasticity is important in each of the various phases of learning and memory (encoding, consolidation, storage, and retrieval).

The Properties of LTP and LTD: A Role in Learning?

It has often been pointed out that synaptic plasticity displays physiological properties that are highly suggestive of an information storage device (McNaughton, 1983; Lynch and Baudry, 1984; Goelet et al., 1986; Morris et al., 1990; Bliss and Collingridge, 1993; Barnes, 1995; Jeffery, 1997; Shors and Matzel, 1997). These classical properties include, at least for N-methyl-D-aspartate (NMDA) receptor–dependent LTP, that its induction is associative and that its expression is both input-specific and persistent over time. These may be relevant to associative or relational features of learning and memory (because associative induction implies the capacity to relate two arbitrary patterns of pre- and postsynaptic neural activity); to storage capacity (because a synapse-specific mechanism endows greater storage capacity than would changes in cell excitability); and to the permanence of memory (because the synaptic enhancement must last as long as the memory).

These assertions beg the question of whether all the properties of synaptic plasticity are likely to be homologous to the characteristics of learning at the behavioral level. Some properties, such as persistence over time, will be directly reflected in memory, while others are less likely to be. This is because the overt manifestations of memory are not solely due to synaptic properties—they also depend on the properties of the network in which that plasticity is embedded. LTP may serve a universal function in the encoding and storage of memory traces, but what gets encoded and how it is encoded will be an emergent property of the circuit, rather than of the mechanisms operating at the synapse in isolation. For example, the character of information processing in the hippocampus is different from that in the amygdala, and it would remain so even if the mechanisms of plasticity utilized in each brain area were conserved.

Furthermore, it may be noted that the temporal contiguity requirements for presynaptic glutamate release and postsynaptic depolarization in the induction of NMDA receptor–dependent LTP are much tighter than those governing various forms of associative conditioning. This has sometimes been presented as a weakness of the SPM hypothesis (Diamond and Rose, 1994). However, until more is known about how information is represented as spatiotemporal patterns of activity on pathways in the higher nervous system, and about the relative time scales of neural events within the brain and those of overt sensory stimuli and motor output, it is difficult to do more than speculate about the degree of isomorphism to be expected.

One problem with which the SPM hypothesis must contend is that we still do not understand enough about the information-processing and memory functions of different brain areas. The existence of multiple types of memory is now widely recognized (O'Keefe and Nadel, 1978; Squire, 1992; Cohen and Eichenbaum, 1993), but there are numerous unanswered questions about the representation of information, the algorithms computed by different kinds of neural architecture, and so on. For the present purpose, the key implication is that, with distributed representations and a variety of synaptic learning rules, there is unlikely to be any simple isomorphism between the pattern, extent, or direction of synaptic changes and the behavioral output observed as a result of conditioning (or other forms of learning). An adequate "circuit-level" description of information processing within a specific brain area will be essential to bridge the gap between synapse and behavior. Without this level of description, a satisfactory test of the mimicry criterion is impossible.

Naturalistic Patterns of Activity Can Induce Synaptic Plasticity

The traditional methods of inducing LTP and LTD—involving long bursts of presynaptic stimuli at high frequencies or prolonged periods of continuous low-frequency stimulation, respectively—almost certainly do not emulate natural patterns of neuronal activity. Previous variations of the SPM hypothesis have often been strongly criticized for this reason. However, at least in the hippocampus, LTP and LTD can be induced using stimulation that mimics firing patterns associated with the hippocampal theta rhythm that occurs as animals move around and explore the world. LTP lasting for several weeks in vivo occurs following the delivery of short bursts of 100-Hz stimulation at intervals of 200 msec (Larson et al., 1986; Rose and Dunwiddie, 1986; Stäubli and Lynch, 1987). It has also been reported that LTP is preferentially induced by burst stimulation on the positive phase of the theta rhythm in urethane-anesthetized rats (Pavlides et al., 1988). Similar findings have been reported in CA1 slices bathed in carbachol to elicit a theta rhythm: delivery of trains of single pulses, each locked to a positive theta peak, was sufficient to induce LTP, whereas stimulation on the negative phase had no effect or, occasionally, induced LTD (Huerta and Lisman, 1993). Furthermore, one of the key parameters for the induction of LTP, at least in the CA1 region of the hippocampus, appears to be "burst firing" in the postsynaptic cell (Thomas et al., 1998; Pike et al., 1999), a pattern of activity characteristic of principal cells within this region.

Role of Synaptic Plasticity in Various Phases of Learning

Learning and memory are generally divided into a set of constituent processes—encoding, consolidation, storage, and retrieval—that occur at different phases of learning and recall, that may involve different brain areas, and that are very likely to involve distinct activity patterns. When considering a role of synaptic plasticity in learning and memory, we must recognize that this role might be different at these different phases. It seems likely that synaptic plasticity is involved in encoding, storage, and the initial stages of consolidation of information, but not in retrieval of that information.

It is worth mentioning that memories should not be confused with the traces that subserve them. Trace encoding can be thought of as the momentary collective activity of large numbers of neurons whose patterns of firing give rise to increases and decreases of synaptic strength that then outlast these very patterns. Memory retrieval is the process of passing neural activity through the network to create patterns of firing that constitute a "memory." The SPM hypothesis asserts that activity-dependent synaptic plasticity is the fundamental mechanism responsible for creating and storing traces. In this sense, LTP enables memory; it is not equivalent to it.

Synaptic Plasticity and Hippocampus-, Cortex-, and Amygdala-Dependent Memories

We have chosen to focus on certain types of memory mediated by the hippocampus, cortex, and amygdala in adult animals, and we have structured our highly selective discussion around the five experimental strategies previously described. The role of the hippocampus in spatial learning has particularly dominated attempts to address the SPM hypothesis. The hippocampus is probably involved in the initial encoding and storage of spatial memory, somehow guiding the eventual consolidation of such information in the cortex to a point at which the participation of the hippocampal formation is no longer required. In addition to its role in explicit memory formation, the cortex is widely assumed to store traces underlying implicit or procedural learning. For this type of memory the cortex is thought to learn on its own, laying down a trace that later enables information to be called to mind but in a manner that disallows the capacity for constructive recollection. The involvement of synaptic plasticity in cortex-dependent memories is discussed particularly in relation to motor skill learning, odor discrimination learning, and conditioned taste aversion. The amygdala has been implicated in many forms of learning, but the link between LTP and learning has been most thoroughly investigated in relation to the acquisition of conditioned fear, and for this reason we focus on this particular form of amygdala-mediated learning. Numerous reviews concerning the types of learning mediated by the hippocampus, cortex, and amygdala have been published (O'Keefe and Nadel, 1978; Aggleton, 1992; Squire, 1992; Cohen and Eichenbaum, 1993; Bures et al., 1998; Holland and Gallagher, 1999).

The hippocampal formation, cortex, and amygdala (at least for the types of learning discussed) are all capable of supporting LTP in the specific pathways thought to be involved in the types of memory mediated by these areas. Of potential relevance to the acquisition of spatial and contextual information, LTP can be induced in all excitatory pathways of the hippocampus (Bliss and Collingridge, 1993). Relevant to the various forms of cortex-dependent learning discussed later, LTP can be induced in the motor cortex (Baranyi and Szente, 1987; Sakamoto et al., 1987), the insular cortex (Escobar et al., 1998b), and the piriform cortex (Stripling et al., 1988). With respect to the acquisition of conditioned fear, LTP can be induced at thalamic and cortical connections to the lateral amygdala (Clugnet and LeDoux, 1990; Chapman and Bellavance, 1992). The cellular mechanisms of LTP and LTD are covered elsewhere in this book (see Malenka and Siegelbaum, and Bear and Linden, respectively).

Strategy 1: Correlation

The first studies implicating LTP in memory were correlational. These can be thought of as loosely linked to the detectability and anterograde alteration criteria. Barnes (1979) and Barnes and McNaughton (1985) observed, in the course of work on the impact of aging, that the persistence of LTP was statistically correlated with the rate of learning, the degree of retention of spatial memories over time, or both. Similar correlations have been observed many times over the past 20 years. A recent example is the report that overexpression of mutant amyloid precursor protein (APP) in a murine model of Alzheimer's disease (Hsiao et al., 1996) is associated with an age-related decline in performance in a delayed spatial alternation task (Chapman et al., 1999). Task performance was positively correlated with the magnitude of LTP induced in both CA1 and dentate regions of the hippocampus.

Such correlations are hardly to be expected on the null hypothesis, but they clearly represent only a first step in understanding because the link reflects a statistical correlation rather than a mechanistic connection. Work on APP transgenic mice is a case in point because not enough is yet known about the normal function of beta amyloid for the link to mechanisms of induction or expression of LTP to be completely clear (Seabrook and Rosahl, 1998). Interestingly, one study of APP-transgenic mice (expressing a truncated form of APP at low concentration) revealed an impairment in water maze performance but no overall difference between groups with respect to LTP induction (see Fig. 11.1). Three main factors accounted for the variability in performance of the mutants: differences in swim speed (passivity), persistent swimming near the side walls (thigmotaxis), and variation in spatial memory. Once the contributions of the first two factors were removed, the magnitude of LTP in both mutants and controls correlated with the third factor: spatial memory (Wolfer et al., 1998). This work illustrates the importance of thorough behavioral analysis.

Strategy 2: Induction

The SPM hypothesis requires that synaptic changes must occur during learning (detectability criterion). As Morris and Davis (1994:368) put it, "no amount of research studying whether LTP is necessary for learning will ever be persuasive in the absence of studies definitively establishing that LTP occurs naturally during learning." The experimental design is

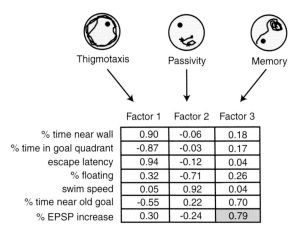

	Factor 1	Factor 2	Factor 3
% time near wall	0.90	-0.06	0.18
% time in goal quadrant	-0.87	-0.03	0.17
escape latency	0.94	-0.12	0.04
% floating	0.32	-0.71	0.26
swim speed	0.05	0.92	0.04
% time near old goal	-0.55	0.22	0.70
% EPSP increase	0.30	-0.24	0.79

Figure 11.1. Strategy 1: Correlation. An example of a statistical relationship between LTP and spatial memory. Mutant mice (*n* = 115) expressing low concentrations of a truncated form of βAPP were trained in a reference memory version of the water maze task, followed by a limited period of reversal learning. Dentate LTP was investigated in 45 of these mice. A factor analysis was carried out using data from all swim trials. The table shows correlation coefficients between the three principal factors extracted and a range of behavioral parameters. The names given to these factors—thigmotaxis, passivity, and memory—reflect the behavioral parameters with which the factors were most strongly correlated. Most of the variability in performance was attributable to the noncognitive factors of thigmotaxis, and to a lesser extent passivity. However, a small proportion of the variability was explained by a third factor, memory, which was strongly correlated with percentage time spent close to the original goal location, but not with any other attribute of performance. Significantly, the magnitude of LTP obtained was strongly correlated with the memory factor, but not with thigmotaxis or passivity. Note that such an association might easily be missed unless the contributions of noncognitive factors such as passivity and thigmotaxis are first eliminated from the analysis. Adapted from Wolfer et al. (1998).

ostensibly straightforward: synaptic efficacy is compared before and after a learning experience, the prediction being that a persistent increase in synaptic transmission should occur at appropriate synapses following certain types of learning. What is less clear is whether such changes would be readily detectable. The key problem is to monitor the appropriate synapses.

Hippocampus-Dependent Learning and the Induction of LTP

Considerable excitement surrounded early studies reporting that exploration of an unfamiliar environment was associated with increases in the perforant path–evoked dentate field excitatory postsynaptic potential (fEPSP) and decreases in the amplitude and latency of the population spike (Sharp et al., 1989; Green et al., 1990). However, Moser et al.

(1993a) highlighted potential problems with such observations—namely that changes in brain temperature (resulting from increased muscular activity) cause similar effects, potentially masking putative learning-induced enhancements of the fEPSP. Moser et al. (1993b) established calibration functions relating brain temperature to fEPSP magnitude, such that temperature effects could effectively be subtracted. They found that exploration of a novel environment resulted in a temperature-independent increase in the fEPSP. This occurred rapidly at the start of exploration and declined gradually to baseline over approximately 15 min. With so much focus on LTP in this field, the observation of a short-duration change is a timely reminder that other forms of plasticity occur in hippocampus that may be functionally important. Indeed genetic studies have shown that mice with mutations of presynaptic proteins display impaired short-term potentiation (STP) and profound learning difficulties in the absence of any known deficit in LTP (Silva et al., 1996).

The induction of more enduring LTP-like changes during learning has, however, sometimes been observed in both the extrinsic and intrinsic connections of the hippocampal formation. In the first instance, mice trained in two different tasks in a radial arm maze exhibited task-dependent potentiation of the connection of the fimbria to the lateral septum (Jaffard et al., 1996). Three observations reduce the likelihood that these changes were temperature-induced: the potentiation developed gradually over several days of training; it was positively correlated with learning as assessed by the probe trial; and it was not seen with a control task involving comparable muscular activity on a treadmill. Second, Ishihara et al. (1997) found that population spiking in the CA3 region induced by mossy fiber stimulation became potentiated during learning in a radial arm maze. It is interesting to note that a correlation between changes in population spike amplitude and performance was observed. Again several observations suggest that this phenomenon is unlikely to be due to temperature.

Cortex-Dependent Learning and the Induction of LTP

Roman et al. (1993) found that learning of an odor discrimination task, possibly a form of explicit learning, was associated with synaptic plasticity within the piriform cortex. The pathway concerned was the monosynaptic connection of olfactory bulb neurons within the lateral olfactory tract (LOT) onto cells in the piriform cortex. Rats were trained to discriminate between a true odor and olfactomimetic stimulation of the LOT (electrical odors). The key finding was that LTP occurred during learning but not control pseudoconditioning. A roughly parallel potentiation of the polysynaptic field potential occurs in the dentate gyrus during a similar odor discrimination protocol (Chaillan et al., 1996).

In addition to afferent inputs, intrinsic connections within the piriform cortex also appear to be upregulated following odor discrimination learning. Saar et al. (1999) trained rats to discriminate between two or three pairs of odors in a four-arm maze and subsequently (3–6 days later) prepared cortical slices from these animals. It was found that training was associated with an enhancement of synaptic transmission at intrinsic connections in the piriform cortex, relative to naïve and pseudoconditioned animals. It is interesting to note that a reduction of paired-pulse facilitation was also observed in trained rats. This effect only appeared on the third day after training and was transient. Whether synaptic enhancement in the piriform cortex reflects the storage of odor information or, as suggested by the authors, facilitates further learning is a matter for speculation. Nevertheless, these data, together with the study by Roman et al. (1993), provide demonstrations of learning being associated with measurable synaptic plasticity in both a neo- and an allocortical structure. The question remaining is what function it serves.

A further example of learning-induced cortical plasticity has been observed within the motor cortex: learning a motor skill is associated with strengthening of the horizontal connections within layers II and III of the motor cortex (Rioult-Pedotti et al., 1998). Rats were required to reach for food using one forelimb over a period of 3–5 days. Electrophysiological recordings were made 1–2 days later from slices of motor cortex taken from these and untrained animals. That this strengthening might be due to an LTP-like mechanism at existing connections, rather than other potential mechanisms such as synaptogenesis, is suggested by the observation that this increase is associated with an attenuation of LTP in this region. Furthermore, both the increase in fEPSP and a persistent improvement in motor skill learning are both NMDA-dependent (Margolis et al., 1999). That LTP was occluded begs the question of whether the learning of a new skill would also be occluded. Their observation that, following more extensive training (23–32 days), EPSPs are similarly enhanced but LTP is no longer occluded suggests that any potential saturation of learning would only be temporary (Rioult-Pedotti and Donoghue, 1999).

Amygdala-Dependent Learning and the Induction of LTP

Animals rapidly learn to associate various stimuli with painful situations: a previously innocuous stimulus such as a light or a tone (the conditioned stimulus; CS) can when paired with a footshock (the unconditioned stimulus; US) lead to a "fear" response to the CS alone (Davis et al., 1993; LeDoux, 1995). LeDoux and colleagues have sought to relate LTP to learning of conditioned fear, an attractive feature of this research being that the CS and US can be clearly identified, as can the anatomical pathways along which this information is projected (LeDoux, 1995). The role of LTP in amygdala-dependent learning is thus potentially easier to investigate than that in hippocampus-dependent learning. Evoked potentials elicited by a natural auditory CS (rather than artificial electrical stimulation) were monitored in the lateral nucleus of the amygdala before and after fear conditioning (Rogan et al., 1997b). As can be seen in Fig. 11.2, paired presentations of the auditory CS and footshock resulted in an increase in freezing behavior in response to the tone and a parallel potentiation of the CS-evoked potential. Furthermore, presentation of the CS in the absence of footshock led to the extinction of conditioned fear and the fall of the CS-evoked response back to baseline levels. As a control, animals trained with unpaired CS-US stimuli showed neither fear conditioning nor an increase in auditory evoked potentials. There is as yet no evidence that the increase in the CS-evoked response is mechanistically equivalent to electrically induced LTP or is selective for the specific CS used, although such experiments will no doubt be forthcoming.

A complementary study lends support to the idea that fear conditioning can induce a phenomenon resembling LTP. McKernan and Shinnick-Gallagher (1997) prepared brain slices from fear-conditioned

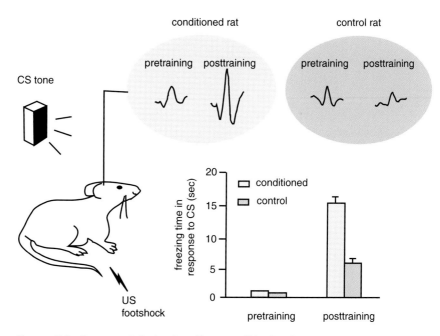

Figure 11.2. Strategy 2: Induction. Fear conditioning induces an LTP-like effect in the amygdala. Auditory-evoked potentials were recorded in the lateral amygdala before and after fear conditioning in freely moving rats. Fear-conditioned animals received paired presentations of CS (tone) and US (footshock), whereas controls received unpaired presentations. Presentation of the CS alone prior to training resulted in a complex extracellular field potential, of which the initial short-latency negative deflection is believed to reflect transmission from the auditory thalamus to the amygdala. After training, presentation of the CS alone revealed a potentiation of this short-latency component in conditioned rats but not in controls, a result that parallels the greater freezing in response to the CS observed in conditioned animals. Adapted from Rogan et al. (1997b).

rats 24 h after behavioral testing. (Control groups either received un-paired CS-US presentations or were experimentally naïve.) Excitatory postsynaptic currents (EPSCs) were recorded from neurons in the lateral amygdala in response to stimulation of afferents from the auditory thal-amus. Those from fear-conditioned rats were potentiated relative to those recorded from controls. On the other hand, EPSCs elicited by stimula-tion of the endopiriform nucleus projection to the lateral amygdala, a pathway not believed to be involved in fear conditioning, did not differ between fear-conditioned and control groups.

Together these amygdalar and cortical studies suggest that fear con-ditioning, and motor skill and odor discrimination learning, induce a form of LTP in the brain areas mediating these forms of learning, thus fulfilling the detectability criterion. However, it remains to be seen whether the artificial induction of LTP can induce behavioral responses analogous to conditioned fear or skilled motor control and thus fulfill the mimicry criterion (Stevens, 1998). For example, would pairing of stimulation in specific CS and US pathways to the amygdala result in potentiation of the CS pathway, and would behavioral testing reveal that this LTP constitutes the "engineering" of an emotional memory? It is unlikely such a mimicry experiment could be carried out in the hippo-campus in the near future, but in the amygdala at least we may not have so long to wait.

Difficulties Surrounding the Induction Strategy

Despite these positive findings, not all attempts to find persistent learn-ing-induced enhancements in synaptic efficacy have been successful. In addition to the purely short-term changes induced by exploratory be-havior (Moser et al., 1993b), an absence of any detectable changes in the parietal cortex following spatial learning has been reported (Beiko and Cain, 1998). Why might it be so difficult to see learning-associated synaptic changes, particularly in the hippocampus? Should we view the negative results as favoring alternative hypotheses?

One key point concerns whether we should expect to see major changes in synaptic efficacy. Long-term increases in synaptic efficacy should occur, at near-optimal signal-to-noise efficiency (Willshaw and Dayan, 1990), in proportion to the product of the probability of activity in a population of afferent fibers (p_{pre}) and the probability of sufficient depolarization in this same population (p_{post}). If it is assumed that dis-crete events are represented as spatiotemporal patterns of activity with a relatively sparse code to maximize storage capacity (i.e., that p_{pre} is small), the proportion of synapses potentiated following an individual learning experience will be a very small fraction of the whole (it will be proportional to the product $p_{pre} \times p_{post}$). Such small and possibly dis-persed changes are likely to be difficult to detect.

Second, if LTP-like changes were associated with heterosynaptic de-pression, putative learning-induced increases in fEPSP would be difficult to detect. Heterosynaptic LTD might serve a normalizing function by ensuring that the sum of the synaptic weights on any given neuron remains roughly constant; fEPSP amplitude would thus remain unchanged.

Having the capacity to modify synaptic efficacy bidirectionally improves the potential fidelity of memory recall in associative memory matrix models (Willshaw and Dayan, 1990). This theoretical argument is not based on assigning different functions to LTP and LTD (such as learning and forgetting, respectively); rather, they complement each other with respect to signal-to-noise ratio and hence storage capacity. A complication in accepting an important role for homosynaptic LTD in memory processing is that the phenomenon has, with rare exceptions (Doyère et al., 1996; Manahan-Vaughan, 1997), proved remarkably elusive in freely moving animals (Errington et al., 1995).

Alternative Assays of Induction

In many brain regions, the use of fEPSP measurements may be too crude an assay of synaptic change to test the detectability criterion. One way to get around this problem might be to use multiple single-unit recording. Wilson and McNaughton (1994) succeeded in finding nonrandom cross-correlation functions between the firing patterns of place cells with overlapping firing fields, and an increase in the cross-correlation when episodes of sleep followed repeated running through the relevant place fields. Such changes may be due to the consolidation of synaptic associations established during the earlier running.

An extension of this kind of experiment might be to examine whether the learning of a task is also associated with changes in cross-correlation functions. For example, short-term changes in the functional connectivity between two neurons have been observed in the monkey auditory cortex using cross-correlation histograms (Ahissar et al., 1992). Such changes are a function not only of the degree of contingency between neuronal pairs, but also of the behavioral relevance of the activity.

Cross-correlation approaches rely on the finding, if not of directly connected neurons, then of functionally connected pairs. However, the chances of finding connected pairs of CA3 and CA1 cells are very remote, even at the border of these areas. A potential second complication is that pairs of cells connected prior to learning may be the very cells at whose connections LTP has already occurred, as implied by findings of digital synaptic change (Petersen et al., 1998). If so, finding increases in the cross-correlation function would only be possible between cells that, before learning, are anatomically connected by nonfunctional synapses. Ironically, observing LTD in cell pairs might be easier because the nonrandom cross-correlation function might decrease. In our view, multiple single-unit recording is an extremely important new approach to the problem, but the design of analytically informative experiments is formidably difficult.

Markers of synaptic plasticity such as alterations in transmitter release, receptor sensitivity, or gene upregulation could also provide relevant information. For instance, an increase in glutamate release has been reported following both LTP (e.g., Dolphin et al., 1982; Bliss et al., 1986; but see Aniksztejn et al., 1989; Diamond et al., 1998; Lüscher et al., 1998) and water maze learning (McGahon et al., 1996; Richter-Levin et al., 1997). Richter-Levin et al. (1997) trained rats in a spatial water maze task for varying numbers of trials, then induced LTP in vivo on one side of the brain, and finally examined veratridine-induced glutamate release in synaptosomes prepared from the hippocampus of the trained animals. Learning was associated with an increase in glutamate release. Tissue prepared from the hemisphere in which LTP had been induced showed a greater increase in glutamate release when taken from rats at an early stage of training than when taken from rats at a later stage. Thus not only were both LTP and learning associated with an increase in glutamate release, but the learning-associated increase occluded the increase normally seen after LTP. This striking result would accord with the SPM hypothesis, were it not for the further finding that the amount of perforant path–induced LTP in vivo was the same in both the undertrained and extensively trained groups. This is puzzling. That the magnitude of electrophysiologically induced LTP is unaffected by prior spatial learning is consistent with the storage capacity argument outlined previously; learning may have enhanced only a small proportion of synapses in the dentate gyrus. However, if this is the case, why was the LTP seen after extensive learning not associated with an increase in glutamate release? Perhaps learning is associated with a shift in the relative expression of presynaptically mediated and postsynaptically mediated LTP.

Regional differences in the behavioral modulation of neural gene expression provide additional clues. Hess et al. (1995a,b) and Gall et al. (1998) have found that the immediate early gene c-*fos* and the dendritically localized mRNA Arc (or Arg3.1; Link et al., 1995; Lyford et al., 1995) are both upregulated during exploration and odor discrimination learning, with, interestingly, the regional pattern of activation reflecting different stages of learning. Activation of c-*fos* in the CA3 region was seen during the earliest stages of odor learning, whereas both exploratory activity and overtraining in odor discrimination resulted in higher CA1 activation. Similarly, Wan et al. (1999) report differential regional patterns of c-*fos* expression in animals observing novel stimuli and novel spatial arrangements of stimuli. These studies imply that the type of task and stage of learning are likely determinants of when and where LTP-like changes may occur. However, c-*fos* is not an unambiguous marker of activity-dependent synaptic plasticity, being regulated by various patterns of neural activity (Kaczmarek, 1992). The discovery of genes more tightly coupled to the induction of late LTP would be helpful.

Strategy 3: Occlusion

Saturation of LTP or LTD and Prevention of Encoding of New Memory Traces

The concept of saturation of LTP is poorly defined. We regard saturation of LTP as a neural state in which, at least for a period of time, no further LTP is possible. This need not mean that all synapses are maximally potentiated (cf. Bliss, 1998; Fig. 11.3C). All that is implied by the term (and all that is required for the experimental strategy to work) is that the target cells (the granule cells) are no longer capable of supporting any further potentiation. According to the SPM hypothesis, LTP saturation prior to behavioral training should prevent new learning. Similar considerations might also apply to saturation of LTD (Thiels et al., 1998), but preliminary experiments by Doyère (personal communication) indicate inconsistent behavioral effects of LTD-inducing stimulation in a radial maze task. Research on the behavioral effects of LTP saturation reached an impasse in 1993 when a series of papers (Cain et al., 1993; Jeffery and Morris, 1993; Korol et al., 1993; Sutherland et al., 1993) reported an inability to replicate earlier findings indicating that saturation induced a reversible occlusion of subsequent spatial learning (McNaughton et al., 1986; Castro et al., 1989).

Bliss and Richter-Levin (1993) pointed to several reasons why—quite apart from the SPM hypothesis being wrong—negative results were obtained: (1) Cumulative LTP of perforant path terminals may not have reached a true state of saturation. (2) Perforant path terminals may have been sufficiently saturated, but not those of other extrinsic or intrinsic hippocampal pathways that may also be critical for learning (e.g., CA3-CA1 terminals). (3) Appropriate saturation of the full septotemporal axis of the hippocampus may not have been achieved with stimulation at a single site within the angular bundle.

The study by Moser et al. (1998) was designed with these issues in mind (Fig. 11.3). There were three key features: (1) the use of an array of cross-bundle stimulation electrodes designed to activate the perforant path maximally, with the cathode switched frequently between active electrodes; (2) the use of a separate midbundle stimulating electrode to test whether the asymptotic LTP induced by the electrode array was a true saturation of LTP on that pathway; and (3) the use of animals given unilateral hippocampal lesions (Mumby et al., 1993). Subsequent to multiple high-frequency (HF) trains or control low-frequency (LF) stimulation, the rats were trained in a standard water maze task. Although the controls learned normally, the HF group showed a bimodal distribution with some animals learning where the platform was located and others failing to learn. When all animals were subsequently tested for the induction of LTP from the probe site within the perforant path, those HF animals in which it was impossible to induce further LTP (the

A. The saturation approach

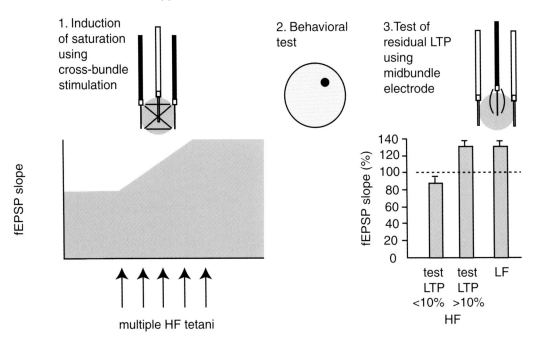

B. Saturation of dentate LTP prevents learning of spatial information

C. What does saturation involve?

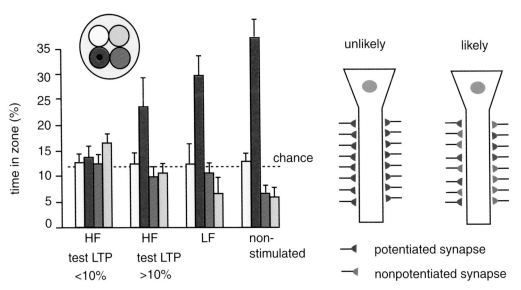

"saturated" subgroup) were the ones that failed to learn the water maze. By contrast, those in whom LTP could still be induced did learn a little about where the platform was located. Thus a true saturation of LTP in the perforant path does appear to impair water maze performance, and these findings therefore vindicate the earlier claims of McNaughton et al. (1986) and Castro et al. (1989).

Despite these positive findings, there remains skepticism about the analytic potential of saturation experiments. One concern is that repeated tetanization might result in acute pathological phenomena, such as seizurelike afterdischarges, that would cause learning deficits (McEachern and Shaw, 1996). However, Moser et al. (1998) found no such afterdischarges. A second concern arises because LTP saturation results in a global increase in the efficacy of synaptic transmission that might disrupt normal hippocampal information processing. However, the number of studies reporting normal learning despite the induction of substantial LTP suggests that an increase in synaptic weights does not in itself disrupt the encoding of new information. In fact, Moser et al. (1998) found no correlation between the magnitude of LTP induced by

Figure 11.3. (opposite) Strategy 3: Occlusion. Saturation of dentate LTP impairs spatial learning.

(A) An array of concentric bipolar stimulating electrodes was positioned in the angular bundle of the perforant path. Two electrodes were placed on either side of the perforant path, and a third was placed in the center of the angular bundle. Five episodes of cross-bundle tetanic stimulation were delivered, using all possible combinations of tip and shaft as cathode and anode, respectively, resulting in cumulative LTP (high-frequency group, HF). Controls received either low-frequency stimulation that did not induce LTP (low-frequency group, LF) or no stimulation at all. After the last tetanization episode, rats were trained in a spatial reference memory version of the water maze task. Finally, to check whether LTP was truly saturated, an attempt was made to induce LTP using the midbundle stimulating electrode. Low-frequency controls showed good LTP, as expected. The high-frequency group could be divided into two subgroups: those rats showing >10% residual LTP and those showing <10% LTP.

(B) Following spatial training, LF and nonstimulated controls searched accurately in the correct zone during a probe trial with the platform absent. However, those animals with a true saturation of LTP (i.e., <10% LTP following tetanization of the midbundle electrode) completely failed to learn the task, as indicated by chance performance on the probe trial. Animals with >10% residual LTP performed as well as LF controls.

(C) It is unlikely that LTP saturation constitutes a state in which all synapses are maximally potentiated. An alternative definition is that saturation is a neural state in which, at least for a period of time, no further LTP is possible. Whether 10%, 20%, or a very high proportion of the synapses have been subject to maximal LTP is irrelevant; the point is that saturation is defined as occurring when no further LTP can be induced even though the Hebbian induction criteria are met.

Adapted from Moser et al. (1998).

cross-bundle tetanization and subsequent learning. A third potential difficulty concerns homeostatic compensatory changes—such as alterations in inhibitory transmission, alterations in synapse formation, and reductions in postsynaptic sensitivity—that may accompany LTP saturation. It might be argued that such changes, rather than saturation itself, are responsible for the learning impairment.

Two recent studies go some way toward addressing concerns about nonspecific effects of LTP saturation. Otnaess et al. (1999) found that spatial pretraining (5 days of eight trials per day to a fixed location) in a second water maze a week before high-frequency stimulation can eliminate the deficits in spatial learning induced by LTP saturation. This result is analogous to the findings of Bannerman et al. (1995) using an NMDA antagonist to block LTP during training. Second, a preliminary study by Molden et al. (1999) using a delayed matching-to-place (DMP) task observed that saturation induced a small learning impairment with a 2-h interval between trials, but not with a 15-sec interval. One month later these animals performed as well as the low-frequency control animals, suggesting that high-frequency stimulation did not cause permanent electrolytic damage. This time-dependent component to the effect of saturation is similar to Steele and Morris's (1999) observation concerning AP5-induced learning impairments in the DMP task. Together these LTP saturation studies provide good evidence that repeated stimulation at high frequency in order to ensure saturation of LTP does not induce significant nonspecific effects or sensorimotor impairments. However, an important question emerges from the fact that pretraining can eliminate learning deficits induced by both saturation and NMDA receptor antagonism: what aspects of water maze performance require dentate LTP?

The study by Moser et al. (1998) is unlikely to be the last word. Saturation might also be realized by bilateral stimulation of the ventral hippocampal commissure, to potentiate the commissural-associational pathway in CA3 and CA1 (Bliss and Richter-Levin, 1993). A pharmacological rather than an electrophysiological approach should also be considered, using drugs such as agonists of adenylyl cyclase (AC), protein kinase A, or mitogen-activated protein kinase (MAPK) to induce a slow-onset but asymptotic synaptic potentiation.

Saturation of LTP or LTD and Prevention of Retrieval of Old Memory Traces

The second type of occlusion experiment involves the induction of saturating levels of LTP after learning has occurred. Such a procedure should in theory scramble the encoded information. This strategy is different from the erasure strategy, but both are relevant to the retrograde alteration criterion. It has been reported that intense induction of LTP in a way similar to that used in the study by Moser et al. (1998) will, when performed shortly or a week after spatial information has been learned, prevent the retrieval of that information (Brun et al., 1999). This finding is in agreement with earlier observations that brief high-frequency stimulation of the perforant path, when performed shortly after training, will disrupt recently acquired spatial information (McNaughton et al., 1986). The added touch in the more recent study is the attempt to control for the nonspecific consequences of high-frequency stimulation, essentially by repeating the procedure in the presence of the NMDA antagonist CPP. Under these conditions high-frequency stimulation did not affect retrieval. The implication of the saturation-induced deficit in retrieval is that alteration of a precise distribution of synaptic weights results in the loss of spatial information.

Strategy 4: Intervention

The earliest intervention studies were pharmacological: many behavioral studies have been conducted with NMDA antagonists that block both LTP and LTD. Other drugs acting downstream of the NMDA receptor or on alternative pathways, such as mGluR antagonists, have also been used. More recently pharmacological studies have been complemented by targeted molecular engineering approaches. The first of these to be applied systematically to examine synaptic plasticity and learning was the use of gene knockouts (Grant et al., 1992; Silva et al., 1992a,b; Chen and Tonegawa, 1997). Drugs and genetic manipulations that enhance plasticity have also been examined. These strategies are relevant to the anterograde alteration criterion.

Considerations of Pharmacological and Genetic Intervention Studies

Before discussing how various pharmacological and genetic intervention studies have contributed to the SPM hypothesis, it is worth considering the experimental approaches themselves. The first point concerns regional selectivity, and the problems associated with affecting structures beyond those of interest. For example, the intracerebroventricular (i.c.v.) and particularly the intraperitoneal (i.p.) method of drug infusion result in diffusion to many regions of the brain, which, in the case of AP5, is likely to affect sensorimotor, cognitive, and NMDA receptor–dependent learning processes in all these structures. In a similar way, genetic manipulations that affect large areas of the central nervous system are subject to the same problems: what function does the target protein have in other processes and brain areas besides, for example, LTP in area CA1 of the hippocampus?

Considering the pharmacological approach, greater regional selectivity can be achieved by local acute infusions. For example, acute infusions of nanomolar quantities of AP5 into the dorsal hippocampus (at a dose that blocks LTP) are sufficient to impair spatial learning in the water maze (Morris et al., 1989). Regional selectivity is also possible with genetic approaches, with either forebrain-specific promoters, antisense techniques, or the use of viral transfection. The ability to perform cell-type-specific interventions using the cre/loxP system (Tsien et al., 1996a) is unique to the targeted molecular engineering approach. This technique offers the potential to investigate gene function in pre- or post-synaptic sites independently. Because it is impossible to target a drug to area CA1 (along its full axis) without also invading the dentate gyrus, another example concerns the role of LTP in subregions of the hippocampal formation.

A second concern, and one not entirely divorced from the first, is that a number of sensorimotor disturbances can arise from pharmacological and genetic interventions. For example, sensorimotor impairments are sometimes (although not always) observed during water maze training with diffuse NMDA receptor blockade. These include falling off the platform during a "wet-dog" shake, thigmotaxis, and failure to climb onto the platform (Cain et al., 1996; Saucier et al., 1996; Morris et al., 2000). Such abnormalities could be due to diffusion of drug to the thalamus, disrupting the normal transmission of somatosensory and visual information (Sillito, 1985; Salt, 1986; Salt and Eaton, 1989) or to the striatum causing motor disturbances such as akinesia (Turski et al., 1990). Disturbances are also seen in other tasks and with other interventions, and these represent a problem for both pharmacological and genetic approaches. Clearly learning cannot proceed when animals can neither see, feel, nor move properly.

A third point concerns the temporal control of manipulation. Precise inactivation (for instance the blockade of a given receptor) followed by a return to normal function was once the exclusive realm of the pharmacological approach. However, the use of inducible promoters to put gene activation or inactivation under experimental control (such as Bujard's tetracycline transactivator systems, rtTA and tTA; Furth et al., 1994; Kistner et al., 1996) or the use of viral transfection (Harding et al., 1997) is narrowing this advantage. (Incidentally, these techniques should also finesse the complications of altered neuronal development that can occur with standard knockouts [Lathe and Morris, 1994; Mayford et al., 1997].) The issue of temporal control is, among other things, relevant to studies addressing the role of plasticity in the various phases of learning and recall (encoding, consolidation, storage, and retrieval).

The last point concerns three potential problems with genetic studies. First, for certain critically important proteins there can be a catastrophic outcome, such as embryonic or perinatal lethality. Second, other mutants display the opposite problem, a null phenotype, but it is hard to believe that key brain enzymes have no function. Either inappropriate tests have been used or there has been biochemical compensation by other closely related genes (Grant et al., 1995). Third, the issue of genetic background is very important. The use of embryonic stem cells from one strain when crossbred with another can introduce a number of flanking genes derived from the original strain that will still be expressed alongside the mutated gene for several generations. Aspects of the resulting phenotype may reflect these flanking genes (Gerlai, 1996), an outcome that can be problematic as certain strains are notoriously poor at learning. Recommendations about the desirability of backcrosses into more suitable strains have now been discussed in the gene-targeting community, and guidelines have been published (Anonymous, 1997).

Intervention Strategies and Hippocampus-Dependent Learning

Following the original observations of Morris et al. (1986) that the NMDA antagonist AP5 blocks spatial but not visual discrimination learning, numerous studies have found that competitive NMDA antagonists impair hippocampus-dependent learning. Learning paradigms used include spatial learning, T-maze alternation, certain types of olfactory learning, contextual fear conditioning, delayed reinforcement of low rates of response, and other operant tasks (Danysz et al., 1988; Tonkiss et al., 1988; Stäubli et al., 1989; Shapiro and Caramanos, 1990; Tonkiss and Rawlins, 1991; Bolhuis and Reid, 1992; Cole et al. 1993; Lyford et al. 1993; Caramanos and Shapiro, 1994; Fanselow et al., 1994; Li et al., 1997; see Danysz et al., 1995, for review). The impairment is dose-related and occurs over a range of intrahippocampal drug concentrations comparable to those that impair hippocampal LTP in vivo and in vitro (e.g., Davis et al., 1992). These data strongly support the SPM hypothesis, but there are problems associated with drug diffusion, sensorimotor side effects, and the fact that NMDA antagonists may affect neuronal processes other than LTP.

Cain et al. (1996) showed that the impairment of spatial learning in a water maze is correlated with the degree of sensorimotor impairment. It is therefore tempting to look upon the sensorimotor deficit as primary and the learning deficit as secondary. However, Morris et al. (2000) have recently used the massed trial training protocol of Cain et al. (1996) and found that AP5-induced sensorimotor disturbances were modest on trial 1 but gradually built up across trials. The correlation that Cain et al. observed between sensorimotor disturbances and learning could, therefore, have arisen because the AP5-induced failure to learn resulted in fatigue, which then exacerbated the situation. That is, the direction of causality could, at least in part, be the opposite of what Cain et al. surmise.

Saucier and Cain (1995) also observed that the impairment in water maze learning that normally occurs following i.p. administration of a competitive NMDA antagonist disappears if the animals are given sufficient pretraining to prevent the drug-induced sensorimotor disturbances seen in experimentally naïve animals. Their pretrained animals showed a clear block of LTP, no sensorimotor impairment, and normal rates of spatial learning. Other data indicate, however, that a deficit in spatial learning can still be observed, relative to appropriate control groups, following nonspatial pretraining (Morris, 1989; Bannerman et al., 1995). Bannerman et al. (1995) discovered that the usual AP5-induced learning deficit all but disappeared in animals trained first as normal animals in one water maze ("downstairs") before being trained in a second water maze ("upstairs") under the influence of the drug. However, if training in the downstairs water maze was nonspatial in character, with sight of extramaze cues occluded, an AP5-induced deficit in spatial learning was again seen in the second task. Bannerman et al. (1995) suggested that blocking NMDA receptors dissociated different components of spatial learning: it may impair an animal's ability to learn the required strategy rather than the map of landmarks in the room in which the water maze is situated. This explanation is apparently refuted by the report by Hoh et al. (1999) that water maze strategy learning is unaffected by i.p. administration of the NMDA antagonist CGS19755 at a dose that successfully blocks LTP in freely moving animals in both CA1 and the dentate gyrus. Drug-treated rats learned nonspatial strategies adequately and showed performance equivalent to that of controls in subsequent spatial learning and spatial reversal. Hoh et al. (1999) suggest that relative task difficulty (their platform size was smaller) may explain the different outcome of their study compared to those of Bannerman et al. (1995) and Steele and Morris (1999). However, their use of i.p. drug administration is problematic, as many of the 10 trials per session (with a 5-min intertrial interval) would probably have been completed before adequate penetration of CGS19755 into the brain.

Results similar to those reported by Bannerman et al. (1995) have recently been obtained using saturation of dentate LTP rather than AP5 administration: spatially pretrained rats no longer exhibited a learning deficit in a separate pool following LTP saturation (Otnaess et al., 1999). The attenuation of sensorimotor side effects by pretraining cannot account for these results because saturated animals exhibit no obvious sensorimotor impairments (see the section on occlusion studies). Resolution of these issues will not be easy. The procedural simplicity of the water maze task belies an underlying complexity that is inadequately captured by the notion of "task difficulty"; rats learn several qualitatively different things in the task and dissociable components of spatial learning can be revealed with different protocols.

A study by Steele and Morris (1999) also argues against a purely sensorimotor explanation for the observed NMDA receptor antagonist–induced impairments in water maze performance (see Fig. 11.4). They trained rats in a variant of the DMP task in the water maze, in which the platform is hidden in a different location each day and stays there for four trials. Normal rats show quite long escape latencies on trial 1 (when they do not know where the platform is hidden) but much shorter latencies on subsequent trials (when they do). Most of the "savings" in escape latency occur between trials 1 and 2, indicative of one-trial learning.

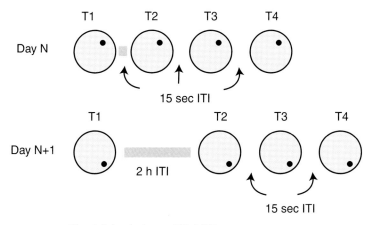

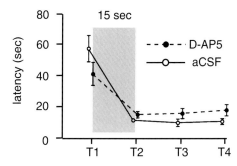

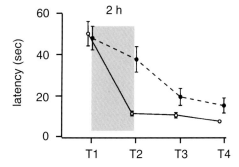

Figure 11.4. Strategy 4: Intervention. NMDAR blockade causes a delay-dependent memory deficit in a DMP task. In this variant of the DMP task, animals are given four trials per day with the platform remaining in the same location. The platform location is changed between days. On any given day the intertrial interval (ITI) between trials 1 and 2 is either 15 sec or 2 h; all subsequent intertrial intervals remain at 15 sec. Note that aCSF-treated controls show a dramatic improvement in escape latencies between trials 1 and 2 at both delays. However, animals given intrahippocampal infusions of D-AP5 30 min prior to trial 1 perform as well as controls at the 15-sec delay but show a marked deficit at the 2-h delay. The delay-dependent nature of this deficit suggests that NMDA receptor blockade causes a genuine spatial memory impairment that can be dissociated from any sensorimotor side effects of drug infusion. Adapted from Steele and Morris (1999).

Infusion of AP5 had no effect on performance at a short memory delay (15-sec intertrial interval) but caused a pronounced impairment at 20 min and 2 h. This delay-dependent deficit occurred irrespective of whether the animals stayed in the training context throughout the memory delay or were returned to the room where they lived, and irrespective of whether the drug was infused chronically i.c.v. or acutely into the hippocampus. If the AP5-induced impairment of matching-to-place performance were sensorimotor or attentional in nature, a deficit would be expected at all delays.

There are other complications. Several studies indicate that low doses of NMDA receptor antagonists can, paradoxically, enhance the learning of certain tasks, such as step-down inhibitory avoidance (Mondadori et al., 1989) and social learning (Lederer et al., 1993). These findings embarrass but do not really challenge the SPM hypothesis because the effects are observed at doses too low to block LTP in vivo, and different mechanisms are likely to be involved in the antagonist-induced facilitation of learning. For instance, the facilitation of inhibitory avoidance by low doses of NMDA antagonists is sensitive to pretreatment by steroids such as aldosterone or corticosterone, whereas the impairment of inhibitory avoidance caused by high doses is steroid-insensitive (Mondadori and Weiskrantz, 1993; Mondadori et al., 1996). Work with the noncompetitive antagonist memantine has led to counterintuitive findings by virtue of its rapid on- and off-channel blocking kinetics. At therapeutic doses, memantine impairs neither learning nor LTP, but it does limit neurotoxicity and so prevents impairment in cognitive function (Parsons et al., 1999).

These pharmacological studies are complemented, and in certain respects taken further, by studies in which the NMDA receptor is targeted genetically. In particular, Tsien et al. (1996a,b) created mice in which the NMDAR1 knockout was restricted to area CA1 of the hippocampus. They showed no LTP in area CA1, normal LTP in the dentate gyrus and neocortex, and a modest learning impairment in the water maze. Using multiple single-unit recording, McHugh et al. (1996) discovered that these mice had abnormal place fields and a reduction in the correlated firing of cells with overlapping place fields.

In summary, there is now an overwhelming body of data indicating that blockade of hippocampal NMDA receptors during learning disrupts the acquisition of hippocampus-dependent memory tasks, and this is supported by NMDAR knockout studies. Before accepting that such studies provide evidence for the SPM hypothesis, it is worth considering two questions. First, are the effects of an intervention on learning dissociable from its effects on sensorimotor processes? Although not desirable, a degree of sensorimotor impairment can be accommodated: one of the advantages of the DMP study by Steele and Morris (1999) is that the within-subject design allows isolation of the learning variable. Second, do NMDA receptors in the targeted region serve functions besides enabling the induction of LTP? Leung and Desborough (1988) showed that acute i.c.v. infusions of AP5 disrupt the hippocampal theta rhythm. Effects on the firing of complex spike cells (Abraham and Kairiss, 1988), place cell stability (Kentros et al., 1998), and a decrease in population spike amplitude have been reported (Errington et al., 1987; Abraham and Mason, 1988). NMDA currents contribute to normal synaptic transmission at somatosensory and visual relays in the thalamus (Salt, 1986; Salt and Eaton, 1989; Sillito et al., 1990), and NMDA currents may similarly contribute to more than just plasticity in the hippocampus. One potential way to circumvent these problems might be to target plasticity at sites independent of the NMDA receptor (such as mGluRs) or at sites downstream of the NMDA receptor.

The Effect on Learning of Drugs and Genetic Manipulations Acting at Sites Other Than NMDA Receptors

Experiments using the mGluR antagonist MCPG are interesting because some data suggest that it blocks LTP while leaving STP unaffected both in vitro (Bashir et al., 1993) and in freely moving rats (Riedel et al., 1995). It should, therefore, cause a different sensitivity to memory delay than that induced by AP5 (Steele and Morris, 1999). Unfortunately its reliability in blocking LTP has been called into question, as it appears to block it under some circumstances (Bashir et al., 1993; Richter-Levin et al., 1994; Riedel et al., 1995; Breakwell et al., 1996) but not others, either in hippocampal slices or in vivo (Chinestra et al., 1993; Manzoni et al., 1994; Bordi and Ugolini, 1995; Martin and Morris, 1997; Anwyl, 1999). The mGluR subtype involved in LTP also remains unknown (Breakwell et al., 1998; Fitzjohn et al., 1998). Nonetheless MCPG and the group I–selective drug AIDA have been reported to impair spatial learning and contextual fear conditioning in rats (Richter-Levin et al., 1994; Riedel et al., 1994; Bordi et al., 1996; Nielsen et al., 1997), but effects on tasks with varying memory delay are yet to be reported.

There have been several reports that interfering with the synthesis of the putative intercellular messenger nitric oxide (NO) can cause impairment of spatial learning and olfactory recognition (Chapman et al., 1992; Böhme et al., 1993; Hölscher et al., 1996; Kendrick et al., 1997). However, these findings are controversial because the precise role of NO in LTP is unresolved (Hawkins et al., 1998). Nor is it clear whether the alterations in behavioral performance that occur with broad-spectrum nitric oxide synthase (NOS) inhibitors can ever be fully independent of the cerebrovascular consequences of inhibiting endothelial NOS (e.g., high blood pressure; Bannerman et al., 1994). Studies using an inhibitor of neuronal rather than endothelial NO, such as the compound 7-NI, could illuminate this issue. Hölscher et al. (1996) report that 7-NI does impair learning in a radial maze at a dose that has been shown to block LTP in area CA1 in vivo, but the effect obtained is not a robust one.

Of those studies that address the SPM hypothesis by intervening at sites downstream of the NMDA receptor, many have used a targeted molecular engineering approach. Mayford et al. (1996) described a transgenic mouse in which the overexpression of constitutive Ca^{2+}/calmodulin-dependent protein kinase II (CaMKII) was under the control of tetracycline. This study built upon previous work using standard transgenic mice that overexpress the autophosphorylated form of CaMKII through a point mutation of Thr-286 (Mayford et al., 1995). These animals exhibited normal CA1 LTP in response to high-frequency stimulation at 100 Hz, but stimulation in the 5- to 10-Hz range (encompassing theta) preferentially resulted in LTD rather than LTP. Hippocampus-dependent learning was impaired: the mice showed impaired spatial learning in a Barnes maze but normal contextual fear conditioning (Bach et al., 1995), a finding clarified by later lesion work establishing that contextual fear conditioning is not always hippocampus-dependent in mice (Frankland et al., 1998). Mayford et al. (1996) replicated the learning impairments and deficits in LTP found by Bach et al. (1995) and Mayford et al. (1995), with suppression of the transgene by administration of doxycycline relieving the impairment of both learning and synaptic plasticity.

Work by Giese et al. (1998) complements these studies. They introduced a point mutation into the gene encoding CaMKII to block autophosphorylation at Thr-286, thereby preventing the transition of this kinase into a CaM-independent state without disrupting its CaM-dependent activity. CA1 LTP could not be elicited in the mutant mice across a range of stimulation frequencies, a slightly different profile from that shown by Mayford's transgenic mouse. The mice in the Giese et al. study also exhibited profound deficits in spatial learning in the water maze and showed an altered dependence on extra- versus intramaze cues in single-unit recording studies of place cells (Cho et al., 1998). However, Giese and his colleagues have not yet used inducible techniques to control this site-directed mutation.

In addition to CaMKII, further studies highlight the requirement for MAPK and certain ACs for both LTP and hippocampus-dependent memory. These results also support a conclusion that LTP-like phenomena may play a role in contextual fear conditioning. The specific MAPK inhibitor SL327 was found to impair both LTP at CA3-CA1 connections in vitro (without affecting baseline transmission) and contextual fear conditioning in rats (Atkins et al., 1998). Note, however, that cued fear conditioning was also impaired, suggesting that the deficits in both forms of conditioning may have arisen from the inhibition of amygdalar rather than hippocampal MAPK. Knockout mice lacking both AC1 and AC8 do not exhibit the late phase of LTP at CA3-CA1 connections and show a significant impairment in long-term memory for contextual (but not cued) fear conditioning and passive avoidance learning (Wong et al., 1999). That administration of forskolin to area CA1 is able to restore both LTP and long-term memory for passive avoidance learning not only confirms that the impairment is due to a deficit in cAMP signaling but also suggests that these double knockout mice have the capacity to perform the task.

The first studies to use both the tTA and rtTA techniques are those of Mansuy et al. (1998a,b) and Winder et al. (1998). They report that transgenic mice overexpressing a truncated form of the phosphatase calcineurin display normal early LTP and short-term memory, but defective late LTP and long-term memory. However, evidence that the latter deficit was secondary to some other problem came from behavioral work showing that a change in training protocol could "rescue" the impairment in long-term memory. This suggests that the deficit in these animals is more likely in the transition from short- to long-term memory than in the mechanisms underlying either on their own. Regulation of calcineurin overexpression using the rtTA technique was examined in animals tested in the water maze (Mansuy et al., 1998b). In the latter study, animals completed training and were first tested with the transgene "off." They learned the platform location, as indexed by good performance in a posttraining probe test. When the transgene was then turned on, performance in a second probe test fell to chance, and, astonishingly, when it was later turned off again performance recovered. These results suggest that calcineurin contributes to retrieval mechanisms, performance, or both.

Although there are many studies reporting parallel deficits in CA3-CA1 LTP and spatial learning, this is not always the case (Meiri et al., 1998; Okabe et al., 1998; Zamanillo et al., 1999). Similarly, a number of reports have found no detectable deficits in spatial learning despite an impairment in dentate LTP (Nosten-Bertrand et al., 1996; but see Schurmans et al., 1997) or mossy fiber LTP (Huang et al., 1995). In the study by Okabe et al. (1998), it is possible that the residual CA1 LTP present in the NR2D mutants was sufficient to support normal learning. However, Zamanillo et al. (1999) generated mice lacking the GluR-A α-amino-3-hydroxyl-5-methyl-4-isoxazolepropionic acid (AMPA) receptor subunit and found that these animals completely lacked LTP in area CA1 but showed no deficits in the rate of learning a standard reference memory task in the water maze. It is difficult to reconcile these results with reports that deficits in spatial learning accompany deficits in plasticity. It may, however be worth drawing attention to the extensive 13-day pretraining period in this study. It has been noted that pretraining can eliminate both NMDA antagonist- and saturation-induced deficits in water maze performance (Bannerman et al., 1995; Saucier and Cain, 1995; Otnaess et al., 1999).

A further dissociation between LTP and learning has been reported following antisense disruption of the presynaptic A-type potassium channel, Kv 1.4. This treatment completely eliminated both early and late phases of CA1 LTP without affecting spatial learning (Meiri et al., 1998). Information about the regional spread of the antisense oligonucleotide is lacking. This latter point is important, as LTP may have been monitored in an area along the longitudinal axis of the hippocampus that was affected by the oligonucleotide, while learning may have utilized neurons along the full length of this axis. It should be recognized, however, that antisense disruption of Kv 1.1—a different potassium channel that is highly localized within dendrites of CA3 neurons—had no effect on LTP in either the CA1 subfield or the dentate gyrus but did cause profound deficits in spatial learning (Meiri et al., 1997). Clearly the longitudinal axis objection cannot apply here.

A separate consideration concerns the value of looking at LTP in vivo. Using standard knockout techniques, Nosten-Bertrand et al. (1996) found that *Thy-1* mutants had normal LTP in area CA1 but no LTP in the dentate gyrus when measured in anesthetized animals in vivo. Further studies revealed, however, that normal LTP could be obtained in dentate slices when bicuculline was added to reduce inhibition, implying that the machinery for inducing LTP must still be present in *Thy-1* mutants. When bicuculline was infused locally in vivo in small quantities (to avoid seizure activity), the now-disinhibited area of the dentate showed normal LTP. Errington et al. (1997) also found that LTP in freely moving *Thy-1* mutants was compromised but not totally abolished. Zamanillo et al. (1999) did not investigate LTP in vivo, but in fairness they did investigate the effects of bicuculline and found no effect. Still the important general message of the *Thy-1* study is that electrophysiological results in brain slices are not infallible predictors of what might happen to synaptic plasticity in the whole animal.

In summary, there are many reports of parallel deficits in LTP and spatial learning and a few reports of apparent dissociations. With respect to those data that reveal a dichotomy between LTP and learning, interpretation must take into account the possibility that an apparent blockade of LTP need not mean that the capacity for LTP is completely eliminated. There are a number of reasons why this might be so:

1. The intervention itself might shift the frequency response curve for the induction of synaptic plasticity (Migaud et al., 1998). For example, Mayford et al. (1996) reported that LTP induced with theta-frequency, but not high-frequency, stimuli was impaired in their animals. Alternatively it is possible that LTP induced with a 1-sec, 100-Hz tetanus may be blocked, but not that induced by other patterns of activity.

2. The intensity of stimulation needed to induce LTP may have increased. For example, Kiyama et al. (1998) noted that the LTP deficit observed in mice lacking the $\varepsilon 1$ subunit of the NMDA receptor could be alleviated by increasing the intensity of tetanization (by an amount that produced a twofold change in the baseline slope value).

3. LTP or LTD in other untested hippocampal pathways may be unaffected, and plasticity in these pathways may be important for learning.

4. Plasticity under in vitro conditions may have been affected, whereas it is possible that LTP in vivo was less affected or unimpaired, or vice versa (Nosten-Bertrand et al., 1996). Consequently there is a need to assess synaptic plasticity using a range of induction parameters, in a number of pathways, and to perform such studies in vitro and in vivo.

A separate point concerns the fact that the mechanisms of experimentally induced LTP and the mechanisms of learning may be dissociated to some extent. Despite much evidence in support of a link between LTP and learning, there remains a possibility that certain factors may be important for tetanically induced LTP and less so for the putative changes in synaptic efficacy that may underlie learning. It follows that intervening with these factors may affect the magnitude of LTP but not alter learning—a hypothetical argument that is still in accordance with the SPM hypothesis. Nevertheless, a significant amount of data supports a conclusion that the mechanisms underlying tetanically induced LTP also underlie hippocampus-dependent learning.

The Role of Synaptic Plasticity in Encoding and Retrieval

Conditional on the assumption that in the hippocampus NMDA receptor antagonists leave AMPA receptor–mediated fast synaptic transmission relatively unaffected, such drugs should freeze the spatial distribution of synaptic weights throughout the intrinsic circuitry of the hippocampal formation. Neurons in the network should still be able to fire and transmit information. NMDA antagonists may therefore impair the encoding of memory "traces" but have no effect on retrieval.

Consistent with this idea, administration of AP5 to animals after they have been trained in odor discrimination learning has no effect on retention, although the drug does impair new learning (Stäubli et al., 1989). Entorhinal cortex lesions, on the other hand, cause rapid forgetting of olfactory information (Stäubli et al., 1984). Likewise Morris (1989) and Morris et al. (1990) found that AP5 had no effect on the retention of a previously trained water maze task, whereas lesions of the hippocampal formation were disruptive when made shortly after the end of training. Similar deficits in encoding but not retrieval were noted in the DMP task of Steele and Morris (1999).

Prevention of LTP and Cortex-Dependent Learning

Perhaps the best example of the use of the intervention strategy to investigate the role of LTP in cortex-dependent learning has been in the study of conditioned taste aversion (CTA), in which a rat learns to avoid a novel taste when it is followed by digestive malaise (usually through LiCl injections). This form of learning involves multiple brain regions, including the insular cortex (Bures et al., 1998). Acquisition but not retention of CTA is impaired by local infusion of NMDA receptor antagonists into the insular cortex, without significant impairment of sensory, motivational, or motor abilities necessary to acquire or express the behavior (Rosenblum et al., 1997; Escobar et al., 1998a; Gutierrez et al., 1999). However, CTA is sensitive to AP5 injections up to at least 2 h after the acquisition trial (Gutierrez et al., 1999). This time delay is intriguing as it suggests that the critical associative events can occur long after ingestion. It is thought that upon feeling ill a few hours postingestion, the animal forms an association between the retrieved memory of the taste and the then-current state of malaise. This would be an instance of what Holland (1990) calls mediated associative learning, and it is consistent with NMDA receptors serving to detect stimulus conjunctions. But one of the stimuli is a memory.

In a separate series of experiments Escobar et al. (1998b) found that intraperitoneal injections of CPP or MK801 blocked LTP in the insular cortex following high-frequency stimulation of the basolateral amygdala. Thus an NMDAR-triggered mechanism appears to be necessary for both the induction of LTP in the insular cortex and the encoding (but not retrieval) of CTA. However, it may be premature to conclude that an LTP-like mechanism underlies the learning-related functions of the insular cortex during CTA because local infusion of AP5 dramatically blocks taste responses in cortical taste areas (Otawa et al., 1995). The lack of effect of AP5 on retention of CTA, although not conclusive, argues against a major sensory effect of the drug. Additional support for a role of LTP-like mechanisms in CTA is, however, provided by observations that blocking molecules downstream of the NMDA receptor also impairs CTA (Yasoshima and Yamamoto, 1997; Berman et al., 1998). Further support for a role of LTP in CTA is correlational in nature. LTP in the dentate gyrus (Rosenblum et al., 1996) and the learning of a novel taste (Rosenblum et al., 1997) are both associated with phosphorylation of the NMDA receptor 2B subunit.

Nonetheless, even if NMDAR-dependent LTP must occur in the insular cortex during CTA, it is unlikely to be the only mechanism of trace formation. Changes occur elsewhere in the brain. Novel tastes evoke long-lasting bursting activity in the nucleus of the solitary tract (McCaughey et al., 1997) and larger responses in the parabrachial nucleus (Shimura et al., 1997). NMDA receptor blockade (Yamamoto and Fujimoto, 1991; Tucci et al., 1998), PKC inhibition (Yasoshima and Yamamoto, 1997), and CREB disruption (Lamprecht et al., 1997) in the amygdala also all impair CTA. Trace formation in CTA is clearly more complex than a simple upregulation of synaptic strength in a single brain region. This will make satisfying our mimicry criterion difficult to meet in subsequent research.

Intervention Strategies and Amygdala-Dependent Learning

If NMDA receptor–dependent LTP is the induction mechanism underlying the association of CS and US information within the amygdala, then infusion of AP5 into the amygdala ought to block the acquisition, but not the expression, of fear conditioning. Exactly this result was obtained using fear-potentiated startle to visual and auditory CSs (Miserendino et al., 1990; Campeau et al., 1992). Similar results have been obtained in several other amygdala-dependent conditioning tasks, including second-order fear conditioning (Gewirtz and Davis, 1997) and discriminated approach to an appetitive CS (Burns et al., 1994). However, in view of recent electrophysiological data, studies of conditioned fear involving intra-amygdala application of NMDA receptor antagonists should be interpreted with some caution.

The first concern is that induction of LTP in the lateral amygdala by high-frequency stimulation of thalamic afferents is NMDA receptor–independent and is instead mediated via L-type voltage-gated Ca^{2+} channels (Weisskopf et al., 1999). Second, application of AP5 reveals an NMDA receptor–mediated component of normal, low-frequency synaptic transmission (Li et al., 1995), suggesting that NMDA receptors are involved in the routine transmission of sensory information from the thalamus. Consistent with this observation are findings that, in addition to impairing the acquisition of conditioning in naïve and pretrained rats, AP5 can also impair the expression of conditioned fear to both auditory and visual stimuli (Lee and Kim, 1998; cf. Bannerman et al., 1995). These results suggest that amygdalar NMDA receptors may have a role in memory retrieval or non-mnemonic processes related to task performance (e.g., attention; Shors and Matzel, 1997).

A second route for the transmission of information to the lateral amygdala is via the external capsule, which relays information from the cortex. LTP induced by high-frequency stimulation of this input has also been reported to be NMDA receptor–independent (Chapman and Bellavance, 1992), although a recent study has provided evidence that LTP in this pathway may depend on NMDA receptor activation under certain stimulus conditions (Huang and Kandel, 1998). However, low-frequency transmission in this pathway, unlike that in the thalamic input, is mediated solely by AMPA receptors (Li et al., 1996). Based on the reports of Li et al. (1996) and Huang and Kandel (1998), the attenuation of LTP in the cortical input following AP5 infusion should impair those components of fear conditioning mediated by this pathway, without affecting the expression of fear responses conveyed by this route. Thus the use of NMDA receptor blockade coupled with selective lesioning of the thalamic pathways (Romanski and LeDoux, 1992) might resolve this issue. The implication of the present findings, however, is that the use of NMDA antagonists during learning may not control all forms of plasticity in this network and may also have additional performance effects.

The potential for genetic intervention in amygdalar synaptic plasticity is becoming a focus of interest. Mayford et al. (1996) created a number of strains of transgenic mice in which the autophosphorylated form of CaMKII was under the control of tTA. In one of these strains, expression was moderate in the hippocampus, subiculum, striatum, and amygdala; but in another strain, there was little expression in the hippocampus and neocortex but prominent expression in the striatum and the lateral and anterior nuclei of the amygdala. The former strain was impaired in a hippocampus-dependent spatial learning task but unimpaired in fear conditioning, while the latter strain was unimpaired in spatial learning but showed a severe impairment in fear conditioning.

Acquisition of conditioned fear was normal after suppression of the transgene by doxycycline. In a further experiment, mice were trained in the presence of doxycycline, then tested for retention of fear conditioning after doxycycline withdrawal and the resumption of transgene expression. A retention deficit was obtained, which, after control experiments, could not be attributed to differences in the perception of the US or changes in the performance of the conditioned response. This ingenious use of genetic engineering techniques suggests that the CaMKII signaling pathway in the amygdala is also involved in the consolidation or retrieval of conditioned fear.

Other signaling pathways have also been implicated in amygdala-dependent synaptic plasticity and fear conditioning. Brambilla et al. (1997) created mice deficient in Ras-GRF, one member of a protein subfamily normally associated with the control of cell proliferation and differentiation. Ras-GRF is specific to cells of the central nervous system and is involved in the activation of the Ras-MAPK pathway in response to postsynaptic calcium influx. Knockout mice showed normal retention of fear conditioning when tested 30 min after acquisition but were impaired relative to controls when tested 24 h later, suggesting that Ras signaling may be involved in memory consolidation. Subsequent experiments revealed that slices from knockout mice exhibited deficient LTP in the basolateral amygdala in response to tetanization of the external capsule. Despite the high degree of expression of Ras-GRF in wild-type CA1, in addition to the amygdala, hippocampus-dependent spatial learning was normal in knockout animals, as was CA1 LTP in vitro. The apparent differential dependence of hippocampal and amygdalar synaptic plasticity and memory on Ras signaling could reflect either the compensatory actions of other regulatory molecules within the hippocampus or a difference in LTP induction mechanisms between the two structures (Orban et al., 1999). Indeed muscarinic receptors, which are known to activate Ras-GRF, are highly expressed in the amygdala, and the antagonism of these receptors blocks the induction of LTP.

Enhancement of LTP and Memory

Before concluding this section, note should be made of studies in which pharmacological and genetic techniques have achieved an enhancement of LTP, learning, or both. As discussed earlier, the SPM hypothesis is not required to predict that enhancement of LTP must also enhance learning. However, such a finding would be not only icing on the cake, but also of potential clinical significance in its own right.

Perhaps the best-known drugs that enhance learning are the "ampakines," which decrease the rate of AMPA receptor desensitization and slow the deactivation of receptor currents after agonist application (Arai et al., 1994, 1996). Ampakines facilitate the induction of hippocampal LTP (Stäubli et al., 1994), and there is now a considerable body of evidence that they can enhance the encoding of memory in a variety of tasks (Lynch, 1998). For example, infusion of the ampakine BDP-12 results in faster acquisition of fear conditioning but does not affect the final level of conditioned fear attained (Rogan et al., 1997a). This result parallels the effect of the drug on LTP induction, in which the rate of potentiation with successive tetani is increased but the asymptotic level remains unchanged. A number of other compounds have also been reported to enhance both learning and hippocampal LTP, including benzodiazepine inverse antagonists (Fontana et al., 1997; Letty et al., 1997; Marchetti-Gauthier et al., 1997; Seabrook et al., 1997).

Mice lacking the nociceptin-orphanin FQ receptor show enhanced LTP in area CA1 (possibly also due to a change in K^+ channel function) and both a modest but significant decrease in escape latency in the water maze and enhanced memory consolidation in step-through avoidance learning (Manabe et al., 1998). In contrast, a PSD-95 mutant that shows enhanced hippocampal LTP and decreased LTD across a range of induction frequencies displays a profound impairment of water maze performance (Migaud et al., 1998). How can this discrepancy be explained? Migaud et al. (1998) suggest that deletion of PSD-95 has shifted θ_m of the BCM function (Bienenstock et al., 1982) well to the left of its optimal position for bidirectional plasticity. (LTD was not investigated by Manabe et al., 1998.) Finally, a thorough behavioral analysis by Tang et al. (1999) revealed that overexpression of the "juvenile" 2B subunit of the NMDA receptor facilitates LTP across a range of induction frequencies and enhances memory in a novel object recognition task, cue and context fear conditioning, and probe test performance in the earliest stages of learning a water maze. The multiple determinants of LTP and LTD offer numerous sites at which genetic mutations to enhance function will be explored in coming years.

Strategy 5: Erasure

Does reversal of LTP cause forgetting? According to the retrograde alteration criterion, if traces related to a recent learning experience are stored within a given brain region, procedures that successfully reverse LTP in that structure should cause forgetting. Erasure might be achieved using (1) trains of suitable depotentiating (e.g., low-frequency) stimulation or (2) drugs or enzyme inhibitors that interrupt the expression of LTP when administered shortly after its induction (e.g., kinase inhibitors).

Depotentiation can be induced using continuous trains of single pulses at 1–5 Hz (Barrionuevo et al., 1980; Stäubli and Lynch, 1990; Bashir and Collingridge, 1994). Stäubli and colleagues report that 5-Hz stimulation can depotentiate recently induced LTP in area CA1 in vitro (Stäubli and Chun, 1996) and in vivo (Stäubli and Scafidi, 1999). The efficacy of depotentiation declines rapidly as the interval between tetanus and 5-Hz stimulation is increased, with little effect being obtained 30 min after LTP induction. Dentate LTP in vivo can also be reversed by 5-Hz stimulation when delivered up to 2 min after tetanization, but such stimulation has little effect after 10 min or 30 min (Martin, 1998). The ability to erase recently induced LTP preferentially, while sparing established LTP, might provide a novel tool for future behavioral studies of the SPM hypothesis.

However, none of the protocols for inducing depotentiation have yet been tested for their ability to cause forgetting in behaving animals. A proposed experiment is illustrated in Fig. 11.5. One problem with such an experiment is that it may be difficult to induce depotentiation on all relevant pathways of the hippocampal formation. The same practical difficulties that beset the saturation approach also lurk menacingly here. Arguably it might not be necessary to depotentiate very many hippocam-

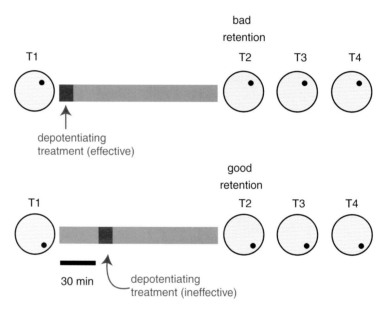

Figure 11.5. Strategy 5: Erasure. Treatments that reverse LTP in a time-dependent manner might selectively erase recently encoded spatial memories while sparing more remote ones. A number of treatments have been reported to reverse LTP in a time-dependent fashion, typically reversing LTP when applied within minutes of tetanization, but becoming progressively less effective as the interval between tetanus and treatment is increased. Such treatments may offer a means of selectively targeting synapses that have recently undergone potentiation during memory encoding. For instance, a depotentiating treatment might be applied soon after trial 1 of a DMP task. If synaptic potentiation underlies memory encoding and storage, then this potentiation—and hence the memory—should be erased by the treatment, resulting in poor retention performance on trial 2. If, however, a longer period (e.g., 30 min) is allowed to intervene between trial 1 and the depotentiating treatment, learning-induced potentiation should be spared, and performance on trial 2 should be normal. To our knowledge, no experiments of this kind have yet been conducted.

pal synapses after an individual learning experience, because if LTP-like changes are sparsely distributed at specific synapses, only some terminals may have to be depotentiated to disrupt a stored trace. However, there are added difficulties, such as the finding that long trains of 5-Hz stimulation can cause seizures (although this problem might be avoided by the use of more naturalistic patterns of stimulation).

A pharmacological approach would have the advantage that it is easier to target all the relevant synapses with a drug, albeit at the cost of unknown side effects. For example, bath application of zinc protoporphyrin IX, an inhibitor of heme-oxygenase (the enzyme that makes carbon monoxide in the brain), has been reported to bring a recently potentiated pathway back to its pre-LTP baseline without affecting an independent nonpotentiated pathway (Stevens and Wang, 1993). Unfortunately the reliability of these results has been questioned (Meffert et al., 1994), and it may therefore be necessary to await newer drugs before this approach can be satisfactorily explored. A promising compound is the integrin antagonist Gly-Arg-Gly-Asp-Ser-Pro (GRGDSP), which has been reported to reverse LTP in a pathway-specific manner within a time window of up to 10–15 min after LTP induction (Stäubli et al., 1998).

Interestingly, Xu et al. (1998b) have reported that exposing freely moving animals to a novel but nonstressful recording chamber can reverse recently induced LTP without affecting a control pathway. They speculate that exposure to novelty has the effect of erasing hitherto unconsolidated information. Support for this interpretation is offered by the work of Izquierdo et al. (1999), in which exposure to novelty limited the ability of an animal to remember a one-trial inhibitory avoidance task carried out up to 1 h previously. Exploration of novelty shortly before or long after the training trial was without effect. The effect appears to be NMDA receptor- and CaMKII-dependent.

Conclusion

It has been three decades since Kandel first discovered that synaptic plasticity occurs during nonassociative conditioning of reflex behavior. Fueled by Bliss and Lømo's report about the discovery of LTP in 1973, the elucidation of NMDA receptor physiology and pharmacology, new insights into the biochemical pathways of LTP, and advances in genetic engineering, much progress has been made in understanding the cellular basis of learning and memory. Advances in our understanding of multiple memory systems have occurred in parallel. Throughout this time, variants of what we have termed the SPM hypothesis have proved useful in probing the cellular mechanisms of memory. Varying degrees of enthusiasm and skepticism surround such hypotheses, and clearly we are among the enthusiasts.

We began by asserting that there are multiple types of memory and that the exact role of synaptic plasticity in trace storage would depend very much on the neural network in which it is embedded. It is difficult to build upon this bald assertion because serious testing of neural network models of memory is likely to require the simultaneous recording of hundreds (perhaps thousands) of single cells. One would look for alterations in neuronal connectivity and establish whether these were mediated by an LTP-like change and/or could be blocked by suitable pharmacological and genetic manipulations. Experimental research would, in addition, be guided by the very specific predictions of particular models. The field has not yet reached this state of sophistication.

What we have been able to do, however, is outline a set of formal criteria by which to judge a more generalized SPM hypothesis: detectability, mimicry, and both anterograde and retrograde intervention. These criteria have been met to varying degrees with respect to different types of memory formation.

The detectability criterion has proved difficult to meet in studies of spatial learning. One such study found only short-term temperature-independent changes in dentate field potentials during exploration. However, persistent "behavioral" LTP induced by learning has been found in projections to the lateral septum. The detectability criterion has also been met in the amygdala during fear conditioning, in the motor cortex during motor skill acquisition, and in the piriform cortex during odor discrimination learning, all arguably in a much more convincing way than in the hippocampus. These data constitute some of the strongest evidence in support of the SPM hypothesis.

With respect to hippocampus-, amygdala-, and cortex-dependent memory (and, as far as we know, all other forms of memory), the mimicry criterion has not yet been met. Indeed until we know more about how information is represented in distributed networks it is unlikely to be met in the near future, at least for hippocampus- and cortex-dependent forms of learning. As we asserted, the relationship between the induction of synaptic plasticity and a particular behavioral change will be a complex function of the neuronal network in which that plasticity is embedded. However, satisfying the mimicry criterion with respect to fear conditioning may be a more realistic goal, owing to the fact that the amygdala receives identifiable inputs conveying information about conditioned and unconditioned stimuli.

The anterograde alteration criterion has been the main focus in studies of hippocampus-dependent learning, and the strategies relevant to this criterion—saturation and intervention—have each provided strong support for the SPM hypothesis. With respect to amygdala-dependent learning, genetic intervention approaches that target the biochemical pathways of LTP also satisfy the anterograde intervention criterion. However, pharmacological intervention studies involving the blockade of amygdalar or cortical NMDA receptors are subject to the problem that NMDA receptors may play a role in routine synaptic transmission in these structures.

Finally, the retrograde intervention criterion has been met with respect to hippocampus-dependent learning, in that saturation of LTP after learning impairs spatial memory. To our knowledge similar studies with respect to amygdala- and cortex-dependent learning have not been reported.

A thorough evaluation of the SPM hypothesis requires experiments addressing both necessity and sufficiency. The current shortfall is that sufficiency has barely been tested. To do so requires the artificial induction of synaptic changes to create what would constitute an "apparent" memory of an event that did not occur (the mimicry criterion). This apart, evidence in favor of the SPM hypothesis is growing, and exploring whether synaptic plasticity is sufficient for memory remains an enticing if somewhat intangible goal to add to the already large body of supportive evidence.

What of the future? We doubt there is any single definitive experiment yet to be done. At present, progress is hampered by our lack of knowledge about what information is represented as spike trains across extrinsic and intrinsic pathways of memory-processing areas and how it is represented. We anticipate ever more interest in the technology of multiple single-unit recording and the possibility of combining this with pharmacological or genetic intervention. The sophisticated nature of the field means that few laboratories can marshal within their walls the myriad of multidisciplinary techniques necessary to advance our understanding. Such diverse technological requirements dictate a collaborative approach. There is also growing interest in the role of the neocortex in memory storage, and in its interactions with allocortical areas such as the hippocampus and amygdala during memory consolidation. We anticipate that tackling this issue will be a particular focus of future research.

References

Abraham, W. C., Kairiss, E. W. (1988). Effects of the NMDA antagonist 2AP5 on complex spike discharge by hippocampal pyramidal cells. *Neurosci. Lett.* 89: 36–42.

Abraham, W. C., Mason, S. E. (1988). Effects of the NMDA receptor/channel antagonists CPP and MK801 on hippocampal field potentials and long-term potentiation in anesthetized rats. *Brain Res.* 462:40–46.

Aggleton, J. P. (1992). *The Amygdala.* New York: Wiley-Liss.

Ahissar, M., Ahissar, E., Bergman, H., Vaadia, E. (1992). Encoding of sound-source location and movement: activity of single neurons and interactions between adjacent neurons in the monkey auditory cortex. *J. Neurophysiol.* 67:203–215.

Aniksztejn, L., Roisin, M. P., Amsellem, R., Ben-Ari, Y. (1989). Long-term potentiation in the hippocampus of the anesthetized rat is not associated with a sustained enhanced release of endogenous excitatory amino acids. *Neuroscience* 28:387–392.

Anonymous. (1997). Mutant mice and neuroscience: recommendations concerning genetic background. *Neuron* 19:755–759.

Anwyl, R. (1999). Metabotropic glutamate receptors: electrophysiological properties and role in plasticity. *Brain Res. Brain Res. Rev.* 29:83–120.

Arai, A., Kessler, M., Xiao, P., Ambros-Ingerson, J., Rogers, G., Lynch, G. (1994). A centrally active drug that modulates AMPA receptor gated currents. *Brain Res.* 638:343–346.

Arai, A., Kessler, M., Ambros-Ingerson, J., Quan, A., Yigiter, E., Rogers, G., Lynch, G. (1996). Effects of a centrally active benzoylpyrrolidine drug on AMPA receptor kinetics. *Neuroscience* 75:573–585.

Atkins, C. M., Selcher, J. C., Petraitis, J. J., Trzaskos, J. M., Sweatt, J. D. (1998). The MAPK cascade is required for mammalian associative learning. *Nature Neurosci.* 1:602–609.

Bach, M. E., Hawkins, R. D., Osman, M., Kandel, E. R., Mayford, M. (1995). Impairment of spatial but not contextual memory in CaMKII mutant mice with a selective loss of hippocampal LTP in the range of the theta frequency. *Cell* 81:905–915.

Bannerman, D. M., Chapman, P. F., Kelly, P. A. T., Butcher, S. P., Morris, R. G. M. (1994). Inhibition of nitric oxide synthase does not prevent the induction of long-term potentiation *in vivo*. *J. Neurosci.* 14:7415–7425.

Bannerman, D. M., Good, M. A., Butcher, S. P., Ramsay, M., Morris, R. G. M. (1995). Distinct components of spatial learning revealed by prior training and NMDA receptor blockade. *Nature* 378:182–186.

Baranyi, A., Szente, M. B. (1987). Long lasting potentiation of synaptic transmission requires postsynaptic modifications in the neocortex. *Brain Res.* 423:378–384.

Barnes, C. A. (1979). Memory deficits associated with senescence: a neurophysiological and behavioral study in the rat. *J. Comp. Physiol. Psychol.* 93:74–104.

Barnes, C. A. (1995). Involvement of LTP in memory: are we searching under the street light? *Neuron* 15:751–754.

Barnes, C. A., McNaughton, B. L. (1985). An age comparison of the rates of acquisition and forgetting of spatial information in relation to long-term enhancement of hippocampal synapses. *Behav. Neurosci.* 99:1040–1048.

Barrionuevo, G., Schottler, F., Lynch, G. (1980). The effects of repetitive low frequency stimulation on control and "potentiated" synaptic responses in the hippocampus. *Life Sci.* 27:2385–2391.

Bashir, Z. I., Collingridge, G. L. (1994). An investigation of depotentiation of long-term potentiation in the CA1 region of the hippocampus. *Exp. Brain Res.* 100:437–443.

Bashir, Z. I., Bortolotto, Z. A., Davies, C. H., Berretta, N., Irving, A. J., Seal, A. J., Henley, J. M., Jane, D. E., Watkins, J. C., Collingridge, G. L. (1993). Induction of LTP in the hippocampus needs synaptic activation of glutamate metabotropic receptors. *Nature* 363:347–350.

Beiko, J., Cain, D. P. (1998). The effect of water maze spatial training on posterior parietal cortex transcallosal evoked field potentials in the rat. *Cereb. Cortex* 8:407–414.

Berman, D. E., Hazvi, S., Rosenblum, K., Seger, R., Dudai, Y. (1998). Specific and differential activation of mitogen-activated protein kinase cascades by unfamiliar taste in the insular cortex of the behaving rat. *J. Neurosci.* 18: 10,037–10,044.

Bienenstock, E. L., Cooper, L. N., Munro, P. W. (1982). Theory for the development of neuron selectivity: orientation specificity and binocular interaction in visual cortex. *J. Neurosci.* 2:32–48.

Bliss, T. V. P. (1998). The saturation debate. *Science* 281:1975–1977.

Bliss, T. V. P., Collingridge, G. L. (1993). A synaptic model of memory: long-term potentiation in the hippocampus. *Nature* 361:31–39.

Bliss, T. V. P., Lømo, T. (1973). Long-lasting potentiation of synaptic transmission in the dentate area of the anaesthetized rabbit following stimulation of the perforant path. *J. Physiol.* 232:331–356.

Bliss, T. V. P., Richter-Levin, G. (1993). Spatial learning and the saturation of long-term potentiation. *Hippocampus* 3:123–126.

Bliss, T. V. P., Douglas, R. M., Errington, M. L., Lynch, M. A. (1986). Correlation between long-term potentiation and release of endogenous amino acids from dentate gyrus of anaesthetized rats. *J. Physiol.* 377:391–408.

Böhme, G. A., Bon, C., Lemaire, M., Reibaud, M., Piot, O., Stutzmann, J. M., Doble, A., Blanchard, J. C. (1993). Altered synaptic plasticity and memory formation in nitric oxide synthase inhibitor–treated rats. *Proc. Natl. Acad. Sci. USA* 90:9191–9194.

Bolhuis, J. J., Reid, I. C. (1992). Effects of intraventricular infusion of the *N*-methyl-D-aspartate (NMDA) receptor antagonist AP5 on spatial memory of rats in a radial arm maze. *Behav. Brain Res.* 47:151–157.

Bordi, F., Ugolini, A. (1995). Antagonists of the metabotropic glutamate receptor do not prevent induction of long-term potentiation in the dentate gyrus of rats. *Eur. J. Pharmacol.* 273:291–294.

Bordi, F., Marcon, C., Chiamulera, C., Reggiani, A. (1996). Effects of the metabotropic glutamate receptor antagonist MCPG on spatial and context-specific learning. *Neuropharmacology* 35:1557–1565.

Brambilla, R., Gnesutta, N., Minichiello, L., White, G., Roylance, A., Herron, C. E., Ramsey, M., Wolfer, D. P., Cestari, V., Rossi-Arnaud, C., Grant, S. G., Chapman, P. F., Lipp, H. P., Sturani, E., Klein, R. (1997). A role for the Ras signalling pathway in synaptic transmission and long-term memory. *Nature* 390:281–286.

Breakwell, N. A., Rowan, M. J., Anwyl, R. (1996). Metabotropic glutamate receptor dependent EPSP and EPSP-spike potentiation in area CA1 of the submerged rat CA1 slice. *J. Neurophysiol.* 76:3126–3135.

Breakwell, N. A., Rowan, M. J., Anwyl, R. (1998). (+)-MCPG blocks induction of LTP in CA1 of rat hippocampus via agonist action at an mGluR group II receptor. *J. Neurophysiol.* 79:1270–1276.

Brun, V. H., Ytterbø, K., Otnaess, M. K., Moser, M.-B., Moser, E. I. (1999). Blockade of retrieval following saturation of LTP is reversed by NMDA receptor antagonism. *Soc. Neurosci. Abstr.* 25:1622.

Bures, J., Bermudez-Rattoni, F., Yamamoto, T. (1998). *Conditioned Taste Aversion: Memory of a Special Kind.* London: Oxford University Press.

Burns, L. H., Everitt, B. J., Robbins, T. W. (1994). Intra-amygdala infusion of the *N*-methyl-D-aspartate receptor antagonist AP5 impairs acquisition but not performance of discriminated approach to an appetitive CS. *Behav. Neural Biol.* 61:242–250.

Cain, D. P., Hargreaves, E. L., Boon, F., Dennison, Z. (1993). An examination of the relations between hippocampal long-term potentiation, kindling, after-discharge, and place learning in the water maze. *Hippocampus* 3:153–163.

Cain, D. P., Saucier, D., Hall, J., Hargreaves, E. L., Boon, F. (1996). Detailed behavioral analysis of water maze acquisition under APV or CNQX: contribution of sensorimotor disturbances to drug-induced acquisition deficits. *Behav. Neurosci.* 110:86–102.

Campeau, S., Miserendino, M. J. D., Davis, M. (1992). Intra-amygdala infusion of the *N*-methyl-D-aspartate receptor antagonist AP5 blocks acquisition but not expression of fear-potentiated startle to an auditory conditioned stimulus. *Behav. Neurosci.* 106:569–574.

Caramanos, Z., Shapiro, M. L. (1994). Spatial memory and *N*-methyl-D-aspartate receptor antagonists APV and MK-801: memory impairments depend on familiarity with the environment, drug dose and training duration. *Behav. Neurosci.* 108:30–43.

Castro, C. A., Silbert, L. H., McNaughton, B. L., Barnes, C. A. (1989). Recovery of spatial learning deficits after decay of electrically induced synaptic enhancement in the hippocampus. *Nature* 342:545–548.

Chaillan, F. A., Roman, F. S., Soumireu-Mourat, B. (1996). Modulation of synaptic plasticity in the hippocampus and piriform cortex by physiologically meaningful olfactory cues in an olfactory association task. *J. Physiol. (Paris)* 90:343–347.

Chapman, P. F., Bellavance, L. L. (1992). Induction of long-term potentiation in the basolateral amygdala does not depend on NMDA receptor activation. *Synapse* 11:310–318.

Chapman, P. F., Atkins, C. M., Allen, M. T., Haley, J. E., Steinmetz, J. E. (1992). Inhibition of nitric oxide synthesis impairs two different forms of learning. *NeuroReport* 3:567–570.

Chapman, P. F., White, G. L., Jones, M. W., Cooper-Blacketer, D., Marshall, V. J., Irizarry, M., Younkin, L., Good, M. A., Bliss, T. V. P., Hyman, B. T., Younkin, S. G., Hsiao, K. K. (1999). Impaired synaptic plasticity and learning in aged amyloid precursor protein transgenic mice. *Nature Neurosci.* 2:271–276.

Chen, C., Tonegawa, S. (1997). Molecular genetic analysis of synaptic plasticity, activity-dependent neural development, learning, and memory in the mammalian brain. *Annu. Rev. Neurosci.* 20:157–184.

Chinestra, P., Aniksztejn, L., Diabira, D., Ben-Ari, Y. (1993). (RS)-α-methyl-4-carboxyphenylglycine neither prevents induction of LTP nor antagonizes metabotropic glutamate receptors in CA1 hippocampal neurons. *J. Neurophysiol.* 70:2684–2689.

Cho, Y. H., Giese, K. P., Tanila, H., Silva, A. J., Eichenbaum, H. (1998). Abnormal hippocampal spatial representations in alpha CaMKII T286A and CREB alpha delta-mice. *Science* 279:867–869.

Clugnet, M.-C., LeDoux, J. E. (1990). Synaptic plasticity in fear conditioning circuits: induction of LTP in the lateral nucleus of the amygdala by stimulation of the medial geniculate body. *J. Neurosci.* 10:2818–2824.

Cohen, N. J., Eichenbaum, H. E. (1993). *Memory, Amnesia and the Hippocampal System.* Cambridge, Mass.: MIT Press.

Cole, B. J., Klewer, M., Jones, G. H., Stephens, D. N. (1993). Contrasting effects of the competitive NMDA antagonist CPP and non-competitive NMDA antagonist MK801 on performance of an operant delayed matching to position task in rats. *Psychopharmacology* 111:465–471.

Danysz, W., Wroblewski, J. T., Costa, E. (1988). Learning impairment in rats by *N*-methyl-D-aspartate receptor antagonists. *Neuropharmacology* 27:653–656.

Danysz, W., Zajaczkowski, W., Parsons, C. G. (1995). Modulation of learning processes by ionotropic glutamate receptor ligands. *Behav. Pharmacol.* 6:455–474.

Davis, M., Falls, W. A., Campeau, S., Kim, M. (1993). Fear-potentiated startle: a neural and pharmacological analysis. *Behav. Brain Res.* 58:175–198.

Davis, S., Butcher, S. P., Morris, R. G. M. (1992). The NMDA receptor antagonist D-2-amino-5-phosphonopentanoate (D-AP5) impairs spatial-learning and LTP in vivo at intracerebral concentrations comparable to those that block LTP in vitro. *J. Neurosci.* 12:21–34.

Diamond, D. M., Rose, G. M. (1994). Does associative LTP underlie classical conditioning? *Psychobiology* 22:263–269.

Diamond, J. S., Bergles, D. E., Jahr, C. E. (1998). Glutamate release monitored with astrocyte transporter currents during LTP. *Neuron* 21:425–433.

Dolphin, A. C., Errington, M. L., Bliss, T. V. P. (1982). Long-term potentiation of the perforant path in vivo is associated with increased glutamate release. *Nature* 297:496–498.

Doyère, V., Errington, M. L., Laroche, S., Bliss, T. V. P. (1996). Low-frequency trains of paired stimuli induce long-term depression in area CA1 but not in dentate gyrus of the intact rat. *Hippocampus* 6:52–57.

Errington, M. L., Lynch, M. A., Bliss, T. V. P. (1987). Long-term potentiation in the dentate gyrus: induction and increased glutamate release are blocked by D(–) aminophosphonovalerate. *Neuroscience* 20:279–284.

Errington, M. L., Bliss, T. V. P., Richter-Levin, G., Yenk, K., Doyère, V., Laroche, S. (1995). Stimulation at 1–5 Hz does not produce long-term depression or depotentiation in the hippocampus of the adult rat in vivo. *J. Neurophysiol.* 74:1793–1799.

Errington, M. L., Bliss, T. V. P., Morris, R. J., Laroche, S., Davis, S. (1997). Long-term potentiation in awake mutant mice. *Nature* 387:666–667.

Escobar, M. L., Alcocer, I., Chao, V. (1998a). The NMDA receptor antagonist CPP impairs conditioned taste aversion and insular cortex long-term potentiation in vivo. *Brain Res.* 812:246–251.

Escobar, M. L., Chao, V., Bermúdez-Rattoni, F. (1998b). In vivo long term potentiation in the insular cortex: NMDA receptor dependence. *Brain Res.* 779: 314–319.

Fanselow, M. S., Kim, J. J., Yipp, J., De Oca, B. (1994). Differential effects of the *N*-methyl-D-aspartate antagonist DL-2-amino-5-phosphonovalerate on acquisition of fear of auditory and contextual cues. *Behav. Neurosci.* 108:235–240.

Fitzjohn, S. M., Bortolotto, Z. A., Palmer, M. J., Doherty, A. J., Ornstein, P. L., Schoepp, D. D., Kingston, A. E., Lodge, D., Collingridge, G. L. (1998). The potent mGlu receptor antagonist LY341495 identifies roles for both cloned and novel mGlu receptors in hippocampal synaptic plasticity. *Neuropharmacology* 37:1445–1458.

Fontana, D. J., Daniels, S. E., Wong, E. H., Clark, R. D., Eglen, R. M. (1997). The effects of novel, selective 5-hydroxytryptamine (5-HT)4 receptor ligands in rat spatial navigation. *Neuropharmacology* 36:689–696.

Frankland, P. W., Cestari, V., Filipkowski, R. K., McDonald, R. J., Silva, A. J. (1998). The dorsal hippocampus is essential for context discrimination but not for contextual conditioning. *Behav. Neurosci.* 112:863–874.

Furth, P. A., St. Onge, L., Böger, H., Gruss, P., Gossen, M., Kistner, A., Bujard, H., Hennighausen, L. (1994). Temporal control of gene expression in transgenic mice by a tetracycline-responsive promoter. *Proc. Natl. Acad. Sci. USA* 91:9302–9306.

Gall, C. M., Hess, U. S., Lynch, G. (1998). Mapping brain networks engaged by, and changed by, learning. *Neurobiol. Learn. Mem.* 70:14–36.

Gerlai, R. (1996). Gene-targeting studies of mammalian behaviour: is it the mutation or the background genotype? *Trends Neurosci.* 19:177–181.

Gewirtz, J. C., Davis, M. (1997). Second-order fear conditioning prevented by blocking NMDA receptors in amygdala. *Nature* 388:471–474.

Giese, K. P., Fedorov, N. B., Filipkowski, R. K., Silva, A. J. (1998). Autophosphorylation at Thr^{286} of the alpha calcium-calmodulin kinase II in LTP and learning. *Science* 279:870–873.

Goelet, P., Castellucci, V. F., Schacher, S., Kandel, E. R. (1986). The long and the short of long-term memory: a molecular framework. *Nature* 322:419–422.

Grant, S. G. N., O'Dell, T. J., Karl, K. A., Stein, P. L., Sorian, O. P., Kandel, E. R. (1992). Impaired long-term potentiation, spatial learning, and hippocampal development in *fyn* mutant mice. *Science* 258:1903–1910.

Grant, S. G. N., Karl, K. A., Kiebler, M. A., Kandel, E. R. (1995). Focal adhesion kinase in the brain: novel subcellular localization and specific regulation by *Fyn* tyrosine kinase in mutant mice. *Genes Dev.* 9:1909–1921.

Green, E. J., McNaughton, B. L., Barnes, C. A. (1990). Exploration-dependent modulation of evoked responses in fascia dentata: dissociation of motor, EEG, and sensory factors and evidence for a synaptic efficacy change. *J. Neurosci.* 10:1455–1471.

Gutierrez, H., Hernandez-Esheagaray, E., Ramirez-Amaya, V., Bermudez-Rattoni, F. (1999). Blockade of *N*-methyl-D-aspartate receptors in the insular cortex disrupts taste aversion and spatial memory formation. *Neuroscience* 89:751–758.

Harding, T. C., Geddes, B. J., Noel, J. D., Murphy, D., Uney, J. B. (1997). Tetracycline-regulated transgene expression in hippocampal neurons following transfection with adenoviral vectors. *J. Neurochem.* 69:2620–2623.

Hawkins, R. D., Son, H., Arancio, O. (1998). Nitric oxide as a retrograde messenger during long-term potentiation in hippocampus. *Prog. Brain Res.* 118: 155–172.

Hess, U. S., Lynch, G., Gall, C. M. (1995a). Changes in c-*fos* mRNA expression in rat brain during odor discrimination learning: differential involvement of hippocampal subfields CA1 and CA3. *J. Neurosci.* 15:4786–4795.

Hess, U. S., Lynch, G., Gall, C. M. (1995b). Regional patterns of c-*fos* mRNA expression in rat hippocampus following exploration of a novel environment versus performance of a well-learned discrimination. *J. Neurosci.* 15:7796–7809.

Hoh, T., Beiko, J., Boon, F., Weiss, S., Cain, D. P. (1999). Complex behavioral strategy and reversal learning in the water maze without NMDA receptor dependent long-term potentiation. *J. Neurosci.* 19(RC2):1–5.

Holland, P. C. (1990). Forms of memory in Pavlovian conditioning. In *Brain Organization and Memory: Cells, Systems, and Circuits*, McGaugh, J. L., Weinberger, N. M., Lynch, G., eds. New York: Oxford University Press. pp. 78–105.

Holland, P. C., Gallagher, M. (1999). Amygdala circuitry in attentional and representational processes. *Trends Cog. Sci.* 3:65–73.

Hölscher, C., McGlinchey, L., Anwyl, R., Rowan, M. J. (1996). 7-Nitro indazole, a selective neuronal nitric oxide synthase inhibitor in vivo, impairs spatial learning in the rat. *Learn. Mem.* 2:267–278.

Hsiao, K., Chapman, P., Nilsen, S., Eckman, C., Harigaya, Y., Younkin, S., Yang, F., Cole, G. (1996). Correlative memory deficits, Aβ elevation, and amyloid plaques in transgenic mice. *Science* 274:99–102.

Huang, Y.-Y., Kandel, E. (1998). Postsynaptic induction and PKA-dependent expression of LTP in the lateral amygdala. *Neuron* 21:169–178.

Huang, Y.-Y., Kandel, E. R., Varshavsky, L., Brandon, E. P., Qi, M., Idzerda, R. L., McKnight, G. S., Bourtchouladze, R. (1995). A genetic test of the effects of mutations in PKA on mossy fiber LTP and its relation to spatial and contextual learning. *Cell* 83:1211–1222.

Huerta, P. T., Lisman, J. E. (1993). Heightened synaptic plasticity of hippocampal CA1 neurons during a cholinergically induced rhythmic state. *Nature* 364:723–725.

Ishihara, K., Mitsuno, K., Ishikawa, M., Sasa, M. (1997). Behavioral LTP during learning in rat hippocampal CA3. *Behav. Brain Res.* 83:235–238.

Izquierdo, I., Schroder, N., Netto, C. A., Medina, J. H. (1999). Novelty causes time-dependent retrograde amnesia for one-trial avoidance in rats through NMDA receptor– and CaMKII-dependent mechanisms in hippocampus. *Eur. J. Neurosci.* 11:3323–3328.

Jaffard, R., Vouimba, R. M., Marighetto, A., Garcia, R. (1996). Long-term potentiation and long-term depression in the lateral septum in spatial working and reference memory. *J. Physiol. (Paris)* 90:339–341.

Jeffery, K. J. (1997). LTP and spatial learning: where to next? *Hippocampus* 7: 95–110.

Jeffery, K. J., Morris, R. G. M. (1993). Cumulative long-term potentiation in the rat dentate gyrus correlates with, but does not modify, performance in the water maze. *Hippocampus* 3:133–140.

Kaczmarek, L. (1992). Expression of c-*fos* and other genes encoding transcription factors in long-term potentiation. *Behav. Neural Biol.* 57:263–266.

Kendrick, K. M., Guevara-Guzman, R., Zorrilla, J., Hinton, M. R., Broad, K. D., Mimmack, M., Ohkura, S. (1997). Formation of olfactory memories mediated by nitric oxide. *Nature* 388:670–674.

Kentros, C., Hargreaves, E., Hawkins, R. D., Kandel, E. R., Shapiro, M., Muller, R. V. (1998). Abolition of long-term stability of new hippocampal place cell maps by NMDA receptor blockade. *Science* 280:2121–2126.

Kistner, A., Gossen, M., Zimmermann, F., Jerecic, J., Ullmer, C., Lübbert, H., Bujard, H. (1996). Doxycycline-mediated quantitative and tissue-specific control of gene expression in transgenic mice. *Proc. Natl. Acad. Sci. USA* 93: 10,933–10,938.

Kiyama, Y., Manabe, T., Sakimura, K., Kawakami, F., Mori, H., Mishina, M. (1998). Increased thresholds for long-term potentiation and contextual learning in mice lacking the NMDA-type glutamate receptor epsilon1 subunit. *J. Neurosci.* 18:6704–6712.

Korol, D. L., Abel, T. W., Church, L. T., Barnes, C. A., McNaughton, B. L. (1993). Hippocampal synaptic enhancement and spatial learning in the Morris swim task. *Hippocampus* 3:127–132.

Lamprecht, R., Hazvi, S., Dudai, Y. (1997). cAMP response element–binding protein in the amygdala is required for long- but not short-term conditioned taste aversion memory. *J. Neurosci.* 17:8443–8450.

Larson, J., Wong, D., Lynch, G. (1986). Patterned stimulation at the theta frequency is optimal for the induction of hippocampal long-term potentiation. *Brain Res.* 368:347–350.

Lathe, R., Morris, R. G. M. (1994). Analyzing brain function and dysfunction in transgenic animals. *Neuropathol. Appl. Neurobiol.* 20:350–358.

Lederer, R., Radeke, E., Mondadori, C. (1993). Facilitation of social learning by treatment with an NMDA receptor antagonist. *Behav. Neural Biol.* 60:220–224.

LeDoux, J. E. (1995). Emotion: clues from the brain. *Annu. Rev. Psychol.* 46: 209–235.

Lee, H., Kim, J. J. (1998). Amygdalar NMDA receptors are critical for new fear learning in previously fear conditioned rats. *J. Neurosci.* 18:8444–8454.

Letty, S., Child, R., Dumuis, A., Pantaloni, A., Bockaert, J., Rondouin, G. (1997). 5-HT4 receptors improve social olfactory memory in the rat. *Neuropharmacology* 36:681–687.

Leung, L. W., Desborough, K. A. (1988). APV, an *N*-methyl-D-aspartate receptor antagonist, blocks the hippocampal theta rhythm in behaving rats. *Brain Res.* 463:148–152.

Li, H., Matsumoto, K., Yamamoto, M., Watanbe, H. (1997). NMDA but not AMPA receptor antagonists impair the delay-interposed radial maze performance of rats. *Pharmacol. Biochem. Behav.* 58:249–253.

Li, X. F., Phillips, R., LeDoux, J. E. (1995). NMDA and non-NMDA receptors contribute to synaptic transmission between the medial geniculate body and the lateral nucleus of the amygdala. *Exp. Brain Res.* 105:87–100.

Li, X. F., Stutzmann, G. E., LeDoux, J. E. (1996). Convergent but temporally separated inputs to lateral amygdala neurons from the auditory thalamus and auditory cortex use different postsynaptic receptors: in vivo intracellular and extracellular recordings in fear conditioning pathways. *Learn. Mem.* 3:229–242.

Link, W., Konietzko, U., Kauselmann, G., Krug, M., Schwanke, B., Frey, U., Kuhl, D. (1995). Somatodendritic expression of an immediate early gene is regulated by synaptic activity. *Proc. Natl. Acad. Sci. USA* 92:5734–5738.

Lüscher, C., Malenka, R. C., Nicoll, R. A. (1998). Monitoring glutamate release during LTP with glial transporter currents. *Neuron* 21:443–453.

Lyford, G. L., Gutnikov, S. A., Clark, A. M., Rawlins, J. N. (1993). Determinants of non-spatial working memory deficits in rats given intraventricular infusions of the NMDA antagonist AP5. *Neuropsychologia* 31:1079–1098.

Lyford, G. L., Yamagata, K., Kaufmann, W. E., Barnes, C. A., Sanders, L. K., Copeland, N. G., Gilbert, D. J., Jenkins, N. A., Lanahan, A. A., Worley, P. F. (1995). Arc, a growth factor and activity-regulated gene, encodes a novel cytoskeleton-associated protein that is enriched in neuronal dendrites. *Neuron* 14:433–445.

Lynch, G. (1998). Memory and the brain: unexpected chemistries and a new pharmacology. *Neurobiol. Learn. Mem.* 70:82–100.

Lynch, G., Baudry, M. (1984). The biochemistry of memory: a new and specific hypothesis. *Science* 224:1057–1063.

McCaughey, S. A., Giza, B. K., Nolan, L. J., Scott, T. R. (1997). Extinction of a conditioned taste aversion in rats. II. Neural effects in the nucleus of the solitary tract. *Physiol. Behav.* 61:373–379.

McEachern, J. C., Shaw, C. A. (1996). An alternative to the LTP orthodoxy: a plasticity-pathology continuum model. *Brain Res. Brain Res. Rev.* 22:51–92.

McGahon, B., Hölscher, C., McGlinchey, L., Rowan, M. J., Lynch, M. A. (1996). Training in the Morris water maze occludes the synergism between ACPD and arachidonic acid on glutamate release in synaptosomes prepared from rat hippocampus. *Learn. Mem.* 3:296–304.

McHugh, T. J., Blum, K. I., Tsien, J. Z., Tonegawa, S., Wilson, M. A. (1996). Impaired hippocampal representation of space in CA1-specific NMDAR1 knock-out mice. *Cell* 87:1339–1349.

McKernan, M. G., Shinnick-Gallagher, P. (1997). Fear conditioning induces a lasting potentiation of synaptic currents *in vitro*. *Nature* 390:607–611.

McNaughton, B. L. (1983). Activity-dependent modulation of hippocampal synaptic efficacy: some implications for memory processes. In *Neurobiology of the Hippocampus*, Siefert, W., ed. London: Academic Press. pp. 233–252.

McNaughton, B. L., Barnes, C. A., Rao, G., Baldwin, J., Rasmussen, M. (1986). Long-term enhancement of hippocampal synaptic transmission and the acquisition of spatial information. *J. Neurosci.* 6:563–571.

Manabe, T., Noda, Y., Mamiya, T., Katagiri, H., Houtani, T., Nishi, M., Noda, T., Takahashi, T., Sugimoto, T., Nabeshima, T., Takeshima, H. (1998). Facilitation of long-term potentiation and memory in mice lacking nociceptin receptors. *Nature* 394:577–581.

Manahan-Vaughan, D. (1997). Group 1 and 2 metabotropic glutamate receptors play differential roles in hippocampal long-term depression and long-term potentiation in freely moving rats. *J. Neurosci.* 17:3303–3311.

Mansuy, I. M., Mayford, M., Jacob, B., Kandel, E. R., Bach, M. E. (1998a). Restricted and regulated overexpression reveals calcineurin as a key component in the transition from short-term to long-term memory. *Cell* 92:39–49.

Mansuy, I. M., Winder, D. G., Moallem, T. M., Osman, M., Mayford, M., Hawkins, R. D., Kandel, E. R. (1998b). Inducible and reversible gene expression with the rtTA system for the study of memory. *Neuron* 21:257–265.

Manzoni, O. J., Weisskopf, M. G., Nicoll, R. A. (1994). MCPG antagonizes metabotropic glutamate receptors but not long-term potentiation in the hippocampus. *Eur. J. Neurosci.* 6:1050–1054.

Marchetti-Gauthier, E., Roman, F. S., Dumuis, A., Bockaert, J., Soumireu-Mourat, B. (1997). BIMU1 increases associative memory in rats by activating 5-HT4 receptors. *Neuropharmacology* 36:697–706.

Margolis, D. J., Donoghue, J. P., Rioult, M.-G., Rioult-Pedotti, M.-S. (1999). Role of NMDA receptors in skill learning and learning-induced synaptic strengthening. *Soc. Neurosci. Abstr.* 25:888.

Martin, S. J. (1998). Time-dependent reversal of dentate LTP by 5 Hz stimulation. *NeuroReport* 9:3775–3781.

Martin, S. J., Morris, R. G. M. (1997). (R,S)-α-methyl-4-carboxyphenylgline (MCPG) fails to block LTP under urethane anaesthesia *in vivo*. *Neuropharmacology* 36:1339–1354.

Martin, S. J., Grimwood, P. D., Morris, R. G. M. (2000). Synaptic plasticity and memory: an evaluation of the hypothesis. *Annu. Rev. Neurosci.* 23:649–711.

Mayford, M., Wang, J., Kandel, E. R., O'Dell, T. J. (1995). CaMKII regulates the frequency-response function of hippocampal synapses for the production of both LTD and LTP. *Cell* 81:891–904.

Mayford, M., Bach, M. E., Huang, Y.-Y., Wang, L., Hawkins, R., Kandel, E. R. (1996). Control of memory formation through regulated expression of a CaMKII transgene. *Science* 274:1678–1683.

Mayford, M., Mansuy, I. M., Muller, R. U., Kandel, E. R. (1997). Memory and behavior: a second generation of genetically modified mice. *Curr. Biol.* 7:R580–R589.

Meffert, M. K., Haley, J. E., Schuman, E. M., Schulman, H., Madison, D. V. (1994). Inhibition of hippocampal heme oxygenase, nitric oxide synthase, and long-term potentiation by metalloporphyrins. *Neuron* 13:1225–1233.

Meiri, N., Ghelardini, C., Tesco, G., Galeotti, N., Dahl, D., Tomsic, D., Cavallaro, S., Quattrone, A., Capaccioli, S., Bartolini, A., Alkon, D. L. (1997). Reversible antisense inhibition of Shaker-like Kv1.1 potassium channel expression impairs associative memory in mouse and rat. *Proc. Natl. Acad. Sci. USA* 94:4430–4434.

Meiri, N., Sun, M. K., Segal, Z., Alkon, D. L. (1998). Memory and long-term potentiation (LTP) dissociated: normal spatial memory despite CA1 LTP elimination with Kv1.4 antisense. *Proc. Natl. Acad. Sci. USA* 95:15,037–15,042.

Migaud, M., Charlesworth, P., Dempster, M., Webster, L. C., Watabe, A. M., Makhinson, M., He, Y., Ramsay, M. F., Morris, R. G. M., Morrison, J. H., O'Dell, T. J., Grant, S. G. N. (1998). Enhanced long-term potentiation and impaired learning in mice with mutant postsynaptic density-95 protein. *Nature* 396:433–439.

Miserendino, M. J. D., Sananes, C. B., Melia, K. R., Davis, M. (1990). Blocking of acquisition but not expression of conditioned fear-potentiated startle by NMDA antagonists in the amygdala. *Nature* 345:716–718.

Molden, S., Morris, R. G. M., Moser, M.-B., Moser, E. I. (1999). Preservation of short term working memory after massive tetanization of the perforant path. *Soc. Neurosci. Abstr.* 25:1622.

Mondadori, C., Weiskrantz, L. (1993). NMDA receptor blockers facilitate and impair learning via different mechanisms. *Behav. Neural Biol.* 60:205–210.

Mondadori, C., Weiskrantz, L., Buerki, H., Petschke, F., Fagg, G. E. (1989). NMDA receptor antagonists can enhance or impair learning performance in animals. *Exp. Brain Res.* 75:449–456.

Mondadori, C., Borkowski, J., Gentsch, C. (1996). The memory-facilitating effects of the competitive NMDA-receptor antagonist CGP 37849 are steroid-sensitive, whereas its memory-impairing effects are not. *Psychopharmacology* 124:380–383.

Morris, R. G. M. (1989). Synaptic plasticity and learning: selective impairment of learning in rats and blockade of long-term potentiation in vivo by the *N*-methyl-D-aspartate receptor antagonist AP5. *J. Neurosci.* 9:3040–3057.

Morris, R. G. M., Davis, M. (1994). The role of NMDA receptors in learning and memory. In *The NMDA Receptor,* Collingridge, G. L., Watkins, J. C., eds. Oxford: Oxford University Press. pp. 340–375.

Morris, R. G. M., Anderson, E., Lynch, G. S., Baudry, M. (1986). Selective impairment of learning and blockade of long-term potentiation by an *N*-methyl-D-aspartate receptor antagonist, AP5. *Nature* 319:774–776.

Morris, R. G. M., Halliwell, R. F., Bowery, N. (1989). Synaptic plasticity and learning. II. Do different kinds of plasticity underlie different kinds of learning? *Neuropsychologica* 27:41–59.

Morris, R. G. M., Davis, S., Butcher, S. P. (1990). Hippocampal synaptic plasticity and NMDA receptors: a role in information storage? *Philos. Trans. R. Soc. London Ser. B* 329:187–204.

Morris, R. G. M., Steele, R. J., Martin, S. J., Bell, J. E. (2000). *N*-methyl-D-aspartate receptors, learning and memory: chronic intraventricular infusion of the NMDA antagonist D-AP5 interacts directly with neural mechanisms of learning. *Behav. Neurosci.* (submitted).

Moser, E. I., Mathiesen, I., Andersen, P. (1993a). Association between brain temperature and dentate field-potentials in exploring and swimming rats. *Science* 259:1324–1326.

Moser, E. I., Moser, M.-B., Andersen, P. (1993b). Synaptic potentiation in the rat dentate gyrus during exploratory learning. *NeuroReport* 5:317–320.

Moser, E. I., Krobert, K. A., Moser, M.-B., Morris, R. G. M. (1998). Impaired spatial learning after saturation of long-term potentiation. *Science* 281:2038–2042.

Mumby, D. G., Weisand, M. P., Barela, P. B., Sutherland, R. J. (1993). LTP saturation contralateral to a hippocampal lesion impairs place learning in rats. *Soc. Neurosci. Abstr.* 19:437.

Nielsen, K. S., Macphail, E. M., Riedel, G. (1997). Class I mGlu receptor antagonist 1-aminoindan-1,5-dicarboxylic acid blocks contextual but not cue conditioning in rats. *Eur. J. Pharmacol.* 36:105–108.

Nosten-Bertrand, M., Errington, M. L., Murphy, K. P., Tokugawa, Y., Barboni, E., Kozlova, E., Michalovich, D., Morris, R. G. M., Silver, J., Stewart, C. L., Bliss, T. V. P., Morris, R. J. (1996). Normal spatial learning despite regional inhibition of LTP in mice lacking Thy-1. *Nature* 379:826–829.

Oda, Y., Kawasaki, K., Morita, M., Korn, H., Matsui, H. (1998). Inhibitory long-term potentiation underlies auditory conditioning of goldfish escape behavior. *Nature* 394:182–185.

Okabe, S., Collin, C., Auerbach, J. M., Meiri, N., Bengzon, J., Kennedy, M. B., Segal, M., McKay, R. D. (1998). Hippocampal synaptic plasticity in mice overexpressing an embryonic subunit of the NMDA receptor. *J. Neurosci.* 18: 4177–4188.

O'Keefe, J., Nadel, L. (1978). *The Hippocampus as a Cognitive Map.* Oxford: Clarendon Press.

Orban, P. C., Chapman, P. F., Brambilla, R. (1999). Is the Ras-MAPK signalling pathway necessary for long-term memory formation? *Trends Neurosci.* 22: 38–44.

Otawa, S., Takagi, K., Ogawa, H. (1995). NMDA and non-NMDA receptors mediate taste afferent inputs to cortical taste neurons in rats. *Exp. Brain Res.* 106:391–402.

Otnaess, M. K., Moser, M.-B., Moser, E. I. (1999). Pretraining prevents spatial learning impairment following saturation of LTP. *Soc. Neurosci. Abstr.* 25:1623.

Parsons, C. G., Danysz, W., Quack, G. (1999). Memantine is a clinically well tolerated *N*-methyl-D-aspartate (NMDA) receptor antagonist—a review of preclinical data. *Neuropharmacology* 38:735–767.

Pavlides, C., Greenstein, Y. J., Grudman, M., Winson, J. (1988). Long-term potentiation in the dentate gyrus is induced preferentially on the positive phase of the theta rhythm. *Brain Res.* 439:383–387.

Petersen, C. C., Malenka, R. C., Nicoll, R. A., Hopfield, J. J. (1998). All-or-none potentiation at CA3-CA1 synapses. *Proc. Natl. Acad. Sci. USA* 95:4732–4737.

Pike, F. G., Meredith, R. M., Olding, A. W. A., Paulsen, O. (1999). Postsynaptic bursting is essential for "Hebbian" induction of LTP at excitatory synapses in rat hippocampus. *J. Physiol.* 518:571–576.

Richter-Levin, G., Errington, M. L., Maegawa, H., Bliss, T. V. P. (1994). Activation of metabotropic glutamate receptors is necessary for long-term potentiation in the dentate gyrus and for spatial learning. *Neuropharmacology* 33:853–857.

Richter-Levin, G., Canevari, L., Bliss, T. V. P. (1997). Spatial training and high-frequency stimulation engage a common pathway to enhance glutamate release in the hippocampus. *Learn. Mem.* 4:445–450.

Riedel, G., Wetzel, W., Reymann, K. G. (1994). (R,S)-α-methyl-4-carboxyphenylglycine (MCPG) blocks spatial learning in rats and long-term potentiation in the dentate gyrus *in vivo*. *Neurosci. Lett.* 167:141–144.

Riedel, G., Casabona, G., Reymann, K. G. (1995). Inhibition of long-term potentiation in the dentate gyrus of freely moving rats by the metabotropic glutamate receptor antagonist MCPG. *J. Neurosci.* 15:87–98.

Rioult-Pedotti, M.-S., Donoghue, J. P. (1999). Persistent synaptic modification and reappearance of LTP after long term motor skill training. *Soc. Neurosci. Abstr.* 25:888.

Rioult-Pedotti, M.-S., Friedman, D., Hess, G., Donoghue, J. P. (1998). Strengthening of horizontal cortical connections following skill learning. *Nature Neurosci.* 1:230–234.

Rogan, M. T., Stäubli, U. V., LeDoux, J. E. (1997a). AMPA receptor facilitation accelerates fear learning without altering the level of conditioned fear acquired. *J. Neurosci.* 17:5928–5935.

Rogan, M. T., Stäubli, U. V., LeDoux, J. E. (1997b). Fear conditioning induces associative long-term potentiation in the amygdala. *Nature* 390:604–607.

Roman, F. S., Simonetto, I., Soumireu-Mourat, B. (1993). Learning and memory of odor-reward association: selective impairment following horizontal diagonal band lesions. *Behav. Neurosci.* 107:72–81.

Romanski, L. M., LeDoux, J. E. (1992). Equipotentiality of thalamo-amygdala and thalamo-cortico-amygdala circuits in auditory fear conditioning. *J. Neurosci.* 12:4501–4509.

Rose, G. M., Dunwiddie, T. V. (1986). Induction of hippocampal long-term potentiation using physiologically patterned stimulation. *Neurosci. Lett.* 69:244–248.

Rosenblum, K., Dudai, Y., Richter-Levin, G. (1996). Long-term potentiation increases tyrosine phosphorylation of the *N*-methyl-D-aspartate receptor subunit 2B in rat dentate gyrus in vivo. *Proc. Natl. Acad. Sci. USA* 93:10,457–10,460.

Rosenblum, K., Berman, D. E., Hazvi, S., Lamprecht, R., Dudai, Y. (1997). NMDA receptor and the tyrosine phosphorylation of its 2B subunit in taste learning in the rat insular cortex. *J. Neurosci.* 17:5129–5135.

Saar, D., Grossman, Y., Barkai, E. (1999). Reduced synaptic facilitation between pyramidal neurons in the piriform cortex after odor learning. *J. Neurosci.* 19:8616–8622.

Sakamoto, T., Porter, L. L., Asanuma, H. (1987). Long lasting potentiation of synaptic potentials in the motor cortex produced by stimulation of the sensory cortex in the cat: a basis for motor learning. *Brain Res.* 413:360–364.

Salt, T. E. (1986). Mediation of thalamic sensory input by both NMDA receptors and non-NMDA receptors. *Nature* 322:263–265.

Salt, T. E., Eaton, S. A. (1989). Function of non-NMDA receptors and NMDA receptors in synaptic responses to natural somatosensory stimulation in the ventrobasal thalamus. *Exp. Brain Res.* 77:646–652.

Saucier, D., Cain, D. P. (1995). Spatial learning without NMDA receptor dependent long-term potentiation. *Nature* 378:186–189.

Saucier, D., Hargreaves, E. L., Boon, F., Vanderwolf, C. H., Cain, D. P. (1996). Detailed behavioral analysis of water maze acquisition under systemic NMDA or muscarinic antagonism: nonspatial pretraining eliminates spatial learning deficits. *Behav. Neurosci.* 110:103–116.

Schurmans, S., Schiffmann, S. N., Gurden, H., Lemaire, M., Lipp, H. P., Schwam, V., Pochet, R., Imperato, A., Böhme, G. A., Parmentier, M. (1997). Impaired long-term potentiation induction in dentate gyrus of calretinin-deficient mice. *Proc. Natl. Acad. Sci. USA* 94:10,415–10,420.

Seabrook, G. R., Rosahl, T. W. (1998). Transgenic animals relevant to Alzheimer's disease. *Neuropharmacology* 38:1–77.

Seabrook, G. R., Easter, A., Dawson, G. R., Bowery, B. J. (1997). Modulation of long-term potentiation in CA1 region of mouse hippocampal brain slices by $GABA_A$ receptor benzodiazepine site ligands. *Neuropharmacology* 36:823–830.

Shapiro, M. L., Caramanos, Z. (1990). NMDA antagonist MK-801 impairs acquisition but not performance of spatial working and reference memory. *Psychobiology* 18:231–243.

Sharp, P. E., McNaughton, B. L., Barnes, C. A. (1989). Exploration-dependent modulation of evoked response in fascia dentata: fundamental observations and time-course. *Psychobiology* 17:257–269.

Shimura, T., Tanaka, H., Yamamoto, T. (1997). Salient responsiveness of parabrachial neurons to the conditioned stimulus after the acquisition of taste aversion learning in rats. *Neuroscience* 81:239–247.

Shors, T. J., Matzel, L. D. (1997). Long-term potentiation: what's learning got to do with it? *Behav. Brain Sci.* 20:597–655.

Sillito, A. M. (1985). Inhibitory circuits and orientation selectivity in the visual cortex. In *Models of the Visual Cortex*, Rose, D., Dobson, V. G., eds. New York: John Wiley and Sons. pp. 396–407.

Sillito, A. M., Murphy, P. C., Salt, T. E., Moody, C. I. (1990). Dependence of retinogeniculate transmission in cat on NMDA receptors. *J. Neurophysiol.* 63:347–355.

Silva, A. J., Stevens, C. F., Tonegawa, S., Wang, Y. (1992a). Deficient hippocampal long-term potentiation in α-calcium-calmodulin kinase II mutant mice. *Science* 257:201–206.

Silva, A. J., Paylor, R., Wehner, J. M., Tonegawa, S. (1992b). Impaired spatial learning in α-calcium-calmodulin kinase II mutant mice. *Science* 257:206–211.

Silva, A. J., Rosahl, T. W., Chapman, P. F., Marowitz, Z., Friedman, E., Frankland, P. W., Cestari, V., Cioffi, D., Südhof, T. C., Bourtchuladze, R. (1996). Impaired learning in mice with abnormal short-lived plasticity. *Curr. Biol.* 6: 1509–1518.

Squire, L. R. (1992). Memory and the hippocampus: a synthesis from findings with rats, monkeys, and humans. *Psychol. Rev.* 99:195–231.

Stäubli, U., Chun, D. (1996). Factors regulating the reversibility of long-term potentiation. *J. Neurosci.* 16:853–860.

Stäubli, U., Lynch, G. (1987). Stable hippocampal long-term potentiation elicited by theta pattern stimulation. *Brain Res.* 435:227–234.

Stäubli, U., Lynch, G. (1990). Stable depression of potentiated synaptic responses in the hippocampus with 1–5 Hz stimulation. *Brain Res.* 513:113–118.

Stäubli, U., Scafidi, J. (1999). Time-dependent reversal of long-term potentiation in area CA1 of the freely moving rat induced by theta pulse stimulation. *J. Neurosci.* 19:8712–8719.

Stäubli, U., Ivy, G., Lynch, G. (1984). Hippocampal denervation causes rapid forgetting of olfactory information in rats. *Proc. Natl. Acad. Sci. USA* 81:5885–5887.

Stäubli, U., Thibault, O., DiLorenzo, M., Lynch, G. (1989). Antagonism of NMDA receptors impairs acquisition but not retention of olfactory memory. *Behav. Neurosci.* 103:54–60.

Stäubli, U., Perez, Y., Xu, F. B., Rogers, G., Ingvar, M., Stone-Elander, S., Lynch, G. (1994). Centrally active modulators of glutamate receptors facilitate the induction of long-term potentiation in vivo. *Proc. Natl. Acad. Sci. USA* 91:11,158–11,162.

Stäubli, U., Chun, D., Lynch, G. (1998). Time-dependent reversal of long-term potentiation by an integrin antagonist. *J. Neurosci.* 18:3460–3469.

Steele, R. J., Morris, R. G. M. (1999). Delay-dependant impairment of a matching to place task with chronic and intrahippocampal infusion of the NMDA antagonist D-AP5. *Hippocampus* 9:118–136.

Stevens, C. F. (1998). A million dollar question: does LTP equal memory? *Neuron* 20:1–2.

Stevens, C. F., Wang, Y. (1993). Reversal of long-term potentiation by inhibitors of haem oxygenase. *Nature* 364:147–149.

Stripling, J. S., Patneau, D. K., Gramlich, C. A. (1988). Selective long term potentiation in the pyriform cortex. *Brain Res.* 441:281–291.

Sutherland, R. J., Dringenberg, H. C., Hoesing, J. M. (1993). Induction of long-term potentiation at perforant path dentate synapses does not affect place learning or memory. *Hippocampus* 3:141–147.

Tang, Y. P., Shimizu, E., Dube, G. R., Rampon, C., Kerchner, G. A., Zhuo, M., Liu, G., Tsien, J. Z. (1999). Genetic enhancement of learning and memory in mice. *Nature* 401:63–69.

Thiels, E., Norman, E. D., Barrionuevo, G., Klann, E. (1998). Transient and persistent increases in protein phosphatase activity during long-term depression in the adult hippocampus *in vivo*. *Neuroscience* 86:1023–1029.

Thomas, M. J., Watabe, A. M., Moody, T. D., Makhinson, M., O'Dell, T. J. (1998). Postsynaptic complex spike bursting enables the induction of LTP by theta frequency synaptic stimulation. *J. Neurosci.* 18:7118–7126.

Tonkiss, J., Rawlins, J. N. P. (1991). The competitive NMDA antagonist AP5, but not the noncompetitive antagonist MK801, induces a delay-related impairment in spatial working memory in rats. *Exp. Brain Res.* 85:349–358.

Tonkiss, J., Morris, R. G. M., Rawlins, J. N. P. (1988). Intra-ventricular infusion of the NMDA antagonist AP5 impairs performance on a non-spatial operant DRL task in the rat. *Exp. Brain Res.* 73:181–188.

Tsien, J. Z., Chen, D. F., Gerber, D., Tom, C., Mercer, E. H., Anderson, D. J., Mayford, M., Kandel, E. R., Tonegawa, S. (1996a). Subregion and cell type-restricted gene knockout in mouse brain. *Cell* 87:1317–1326.

Tsien, J. Z., Huerta, P. A., Tonegawa, S. (1996b). The essential role of hippocampal CA1 NMDA receptor–dependant synaptic plasticity in spatial memory. *Cell* 87:1327–1338.

Tucci, S., Rada, P., Hernandez, L. (1998). Role of glutamate in the amygdala and lateral hypothalamus in conditioned taste aversion. *Brain Res.* 813:44–49.

Turski, L., Klockgether, T., Turski, W. A., Schwarz, M., Sontag, K. H. (1990). Blockade of excitatory neurotransmission in the globus pallidus induces rigidity and akinesia in the rat: implications for excitatory neurotransmission in pathogenesis of Parkinson's disease. *Brain Res.* 512:125–131.

Wan, H., Aggleton, J. P., Brown, M. W. (1999). Different contributions of the hippocampus and perirhinal cortex to recognition memory. *J. Neurosci.* 19:1142–1148.

Weisskopf, M. G., Bauer, E. P., LeDoux, J. E. (1999). L-type voltage-gated calcium channels mediate NMDA-independent associative long-term potentiation at thalamic input synapses to the amygdala. *J. Neurosci.* 19:10,512–10,519.

Willshaw, D., Dayan, P. (1990). Optimal plasticity from matrix memories: what goes up must come down. *Neural Computation* 2:85–93.

Wilson, M. A., McNaughton, B. L. (1994). Reactivation of hippocampal ensemble memories during sleep. *Science* 265:676–682.

Winder, D. G., Mansuy, I. M., Osman, M., Moallem, T. M., Kandel, E. R. (1998). Genetic and pharmacological evidence for a novel, intermediate phase of long-term potentiation suppressed by calcineurin. *Cell* 92:25–37.

Wolfer, D. P., Stagljar-Bozicevic, M., Errington, M. L., Lipp, H.-P. (1998). Spatial memory and learning in transgenic mice: fact or artifact? *News Physiol. Sci.* 13:118–123.

Wong, S. T., Athos, J., Figueroa, X. A., Pineda, V. V., Schafer, M. L., Chavkin, C. C., Muglia, L. J., Storm, D. R. (1999). Calcium-stimulated adenylyl cyclase activity is critical for hippocampus-dependent long-term memory and late phase LTP. *Neuron* 23:787–798.

Xu, L., Anwyl, R., Rowan, M. J. (1998b). Spatial exploration induces a persistent reversal of long-term potentiation in rat hippocampus. *Nature* 394:891–894.

Yamamoto, T., Fujimoto, Y. (1991). Brain mechanisms of taste aversion learning in the rat. *Brain Res. Bulletin* 27:403–406.

Yasoshima, Y., Yamamoto, T. (1997). Rat gustatory memory requires protein kinase C activity in the amygdala and cortical gustatory area. *NeuroReport* 8:1363–1367.

Zamanillo, D., Sprengel, R., Hvalby, O., Jensen, V., Burnashev, N., Rozov, A., Kaiser, K. M., Köster, H. J., Borchardt, T., Worley, P., Lübke, J., Frotscher, M., Kelly, P. H., Sommer, B., Andersen, P., Seeburg, P. H., Sakmann, B. (1999). Importance of AMPA receptors for hippocampal synaptic plasticity but not for spatial learning. *Science* 284:1805–1811.

12 Synapse Formation and Maturation

Ann Marie Craig and Jeff W. Lichtman

One of the most amazing feats of development is the elaboration of the innumerable synaptic connections that functionally link the cells of the nervous system in highly specific ways. Synapse formation can be considered the culminating stage of neural development, which requires earlier events, including the production and migration of neurons and the extension of axons and dendrites, all to occur correctly. Synaptogenesis, however, is itself a discrete phenomenon. Throughout phylogeny, axons and their postsynaptic partners undergo a multistep, bidirectional signaling cascade, which ensures that synapses form between appropriate cells and, even more remarkably, that the precise location of neurotransmitter release by the presynaptic cell aligns with aggregated receptors of the appropriate kind in the postsynaptic cell membrane. After axons and their postsynaptic partners make functional contact, further enhancements in structure and function improve the strength and reliability of the synapses. While some synapses are maturing, others wither and ultimately disappear. The synapses that survive this period of retrenchment must be either maintained for long periods or constantly replaced so as to preserve the storage of information that presumably resides in synaptic circuitry.

In this chapter we highlight the main steps in the formation, maturation, and elimination or maintenance of synapses. We compare and contrast what is known about the accessible but highly specialized vertebrate NMJ with central synapses, focusing primarily on excitatory synapses in the mammalian cerebral cortex. We examine the principles of development of central and peripheral nervous system synapses to determine if common themes emerge and to place the rapidly expanding molecular information into a cellular perspective.

Synapse Structure and Composition

Although chemical synapses come in a wide variety of structural designs, all have certain common features. In the words of an earlier review of neuromuscular synapse structure and development, "pre- and postsynaptic specializations form a superstructure that spans two cells and links their interiors" (Hall and Sanes, 1993:104). Fast-regulated focal release and detection of transmitter are accomplished by alignment of synaptic vesicles, calcium channels, fusion apparatus, neurotransmitter receptors, and signal transduction machinery. Thus the synaptic superstructure includes not just cleft and plasma membrane components but also specialized cytoskeletal elements and organelles of both pre- and postsynaptic partners.

The most thoroughly studied synapse types, for which many of the molecular components have been identified, are the vertebrate neuromuscular junction (NMJ), a cholinergic synapse, and the excitatory and inhibitory synapses of the mammalian cortex (Fig. 12.1). The excitatory synapses in the central nervous system (CNS) release glutamate and

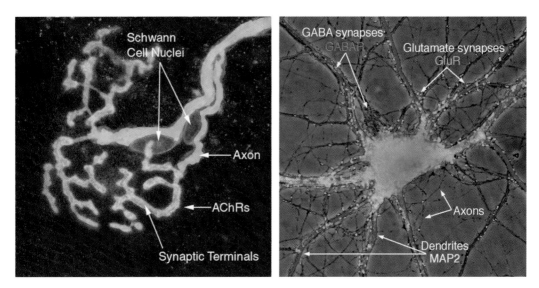

Figure 12.1. Immunofluorescence images of synaptic contacts between moto-neuron axons and muscle (left) and between hippocampal axons and dendrites in culture (right).

(Left) The NMJ is a site of confluence of three cell types: neurons, skeletal muscle fibers, and Schwann cells. Each adult skeletal muscle fiber (in mammals) is innervated at a single site (constituting less than 0.1% of the muscle's surface area) by one motor axon. The axon in this image was fluorescently stained using antibodies to neurofilaments and synaptic vesicle proteins (green) and can be seen to ramify over a small circumscribed area. Each terminal axonal branch is precisely juxtaposed with a high density of AChRs in the muscle fiber membrane, labeled red with a rhodamine-tagged snake toxin (alpha bungarotoxin). Enveloping the axon are Schwann cell processes, labeled blue with antibodies to a glial-specific marker, S100. The Schwann cell nuclei are closely apposed to the axon branches. The locations of some of the 6–10 muscle fiber nuclei that congregate at the NMJ are also shown (dark yellow ovals).

(Right) Synapses between hippocampal neurons in culture as visualized by immunofluorescence signals overlaid onto a phase-contrast image showing the neuronal cell soma, dendrites, and axons. Immunoreactivity against microtubule-associated protein MAP2 marks the soma and major dendrites (blue). Excitatory AMPA-type glutamate receptors (GluR1, green) are concentrated at synaptic contacts between axons and dendrites, mainly on dendritic spines. Inhibitory GABA receptors ($GABA_A R\beta2/3$ subunit, red) are concentrated at a different subset of synapses, mainly on dendrite shafts and on the axon initial segment. Additional studies have shown that these receptor clusters in dendrites are apposed to clusters of synaptic vesicles carrying the corresponding neurotransmitters in axons.

Image courtesy of A. Rao.

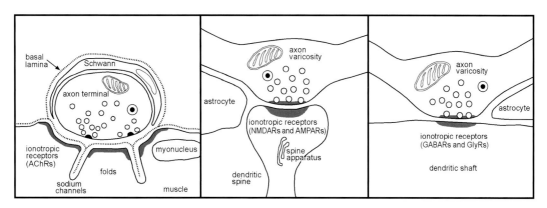

Figure 12.2. Common structural and molecular features of synaptic junctions. (Left) Neuromuscular junction. *(Middle)* Central excitatory synapse. *(Right)* Central inhibitory synapse.

correspond to the classic asymmetric or Gray's type I synapse (Gray, 1959). Symmetric or type II synapses (Gray, 1959) are generally inhibitory and include γ-aminobutyric acid (GABA)–releasing synapses in the cortex and glycinergic synapses in the spinal cord. All of these types of synapses share key structural features and functionally homologous or analogous, if not identical, molecular components (Fig. 12.2).

Presynaptic specializations often occur at nerve endings, although in the CNS they are found more frequently at en passant boutons or varicosities along the length of the axon. These specializations contain numerous small (circa 50 nm diameter), clear synaptic vesicles, which release fast-acting neurotransmitters; fewer large (circa 200 nm diameter), dense core vesicles, which mediate the release of amines and peptides; mitochondria to meet local energy needs; and membranous organelles, which are probably endosomes. With the application of many different research methodologies—including purification of synaptic vesicle protein components, biochemical screens for proteins that mediate vesicle fusion (SNAPs and SNAREs), studies of the mode of action of toxins that perturb synaptic function (e.g., tetanus toxin), and genetic studies of organisms from yeast through mammals—much progress has been made in understanding the synaptic vesicle cycle that leads to neurotransmitter release via exocytosis and membrane recycling (Südhof, 1995).

Synaptic vesicles are highly organized within the terminal. Small synaptic vesicle fusion occurs at active zones, at regions of electron-dense material opposite postsynaptic regions of high receptor content associated with the postsynaptic density of central synapses or with the junctional folds of neuromuscular synapses. Whereas the extended neuromuscular presynaptic elements contain hundreds of active zones, central synapses generally contain only one or a few active zones. Since release probability is less than 1 per active zone, this arrangement leads to fail-safe excitation at neuromuscular synapses but variable subthreshold responses at most central synapses (see Burns and Augustine, 1995, for review). On the order of 5–50 vesicles are docked per active zone, and these may constitute a readily releasable pool (Schikorski and Stevens, 1997). At the neuromuscular synapse, vesicles are docked in two rows on either side of the presynaptic density; the vesicle docking sites are thought to be within a few tens of nanometers of the rows of voltage-gated Ca^{2+} channels, allowing fast and reliable coupling between Ca^{2+} entry and vesicle fusion. Docked vesicles appear to be arranged in a gridlike array within most CNS presynaptic active zones, but the precise arrangement of Ca^{2+} channels is not known. In all terminals, the undocked vesicles seem to be tethered near the active zone.

Compared with presynaptic specializations, postsynaptic specializations appear to be more diverse with respect to form and molecular composition. There are, however, a number of common elements. One such element is the postsynaptic density, named for its dark appearance in stained transmission electron microscope sections. The postsynaptic density contains neurotransmitter receptors along with associated signal-transducing and scaffolding proteins. Morphologically, at least, inhibitory synapses have no additional apparent specializations, whereas excitatory synapses often have particular morphological characteristics, such as the junctional folds at the neuromuscular synapse and the dendritic spines associated with the majority of central glutamatergic synapses. Spines are narrow protrusions (0.2–3 µm long) with a bulbous head, which is the site of excitatory neurotransmission (see Harris and Kater, 1994, and Harris, 1999, for review). They are actin-rich motile structures containing smooth endoplasmic reticulum, which can be specialized to form a spine apparatus, and often polyribosomes. The function of the dendritic spines is not understood. One view is that they serve as local signaling compartments; indeed, upon synaptic stimulation calcium elevations can be restricted to single dendritic spines (Denk et al., 1996). Folds at the NMJ also isolate release sites. At the junction, the voltage-gated Na^+ channels are located at the bottom of the postsynaptic folds (Flucher and Daniels, 1989; Le Treut et al., 1990; Boudier et al., 1992). This arrangement may help ensure an extremely high density of neurotransmitter receptor channels near the presynaptic release sites as well as prevent shunting of the postsynaptic current.

Neurotransmitter receptors that transduce signals via ion flow across the membrane (ionotropic receptors) are concentrated opposite the active zones of all fast synapses. Another class of neurotransmitter receptor (metabotropic receptors) is found at some synapses but also extra-synaptically. Ionotropic receptors generally cluster opposite terminals releasing the corresponding transmitter (e.g., Craig et al., 1994). At the mature NMJ, acetylcholine receptor (AChR) number and density are quite high (circa 10,000 per square micron) opposite the active zones and much lower elsewhere (circa 10 per square micron) (Salpeter and Harris 1983; Salpeter et al., 1988). At the frog NMJ the receptive area containing highly infolded membrane has been estimated to be 1500 square microns (Matthews-Bellinger and Salpeter, 1978). At a typical NMJ about 1000 AChR channels open in response to fusion of a single vesicle (Katz and Miledi, 1972), and therefore an endplate potential resulting from hundreds of quanta simultaneously released from an activated nerve will lead to activation of less than 10% of the total receptive area, meaning that the receptors are in extreme excess. Despite this mismatch, at sites of vesicle fusion it is thought that the released ACh saturates the adjacent receptors in the postsynaptic membrane (Kuffler and Yoshikami, 1975; Land et al., 1980). At central synapses fewer than 200 or even fewer than 20 receptor channels often open in response to fusion of a single vesicle, and it is not yet clear whether receptors are in excess or limiting (Frerking and Wilson, 1996; Nusser et al., 1997; Liu et al., 1999; Mainen et al., 1999).

Recent data suggest that receptors may be saturated following quantal release at some central synapses but not at others, owing partly to a high variability in receptor content. Based on physiological studies and quantitative immunoelectron microscopy, Nusser et al. (1997) reported that a quantum of GABA saturates receptors at smaller but not larger synapses. Synaptic receptor density was fairly constant at about 1250 functional GABA receptors per square micron, corresponding to an estimated 15–460 receptors per synapse. Comparisons of synaptic with extrasynaptic GABA receptor labeling indicated a 180- to 230-fold enrichment at synapses (Nusser et al., 1995). For glutamate receptors, the synaptic to extrasynaptic ratio has not been reported, although extrasynaptic immunolabeling is usually low. Generally all mature glutamate synapses contain concentrations of some type of ionotropic glutamate receptor, but the synaptic density of different glutamate receptor subtypes appears to be pathway-specific, dependent on both pre- and postsynaptic partners. For example, in the hippocampus, α-amino-3-hydroxyl-5-methyl-4-isoxazolepropionic acid (AMPA)–type glutamate receptors are concentrated at all mossy fiber–CA3 synapses, with receptor content related to synapse size (ranging from an estimated 11 to 297 receptors per synapse) (Nusser et al., 1998b). However, a significant fraction (15%) of Schaffer collateral–CA1 synapses lack AMPA receptors, but all contain N-methyl-D-aspartate (NMDA) receptors (Nusser et al., 1998b; Takumi et al., 1999).

Another common component of postsynaptic specializations is receptor anchoring or scaffolding proteins. At the NMJ, rapsyn, a 63-kDa peripheral membrane protein, is tightly associated with the AChR with 1:1 stoichiometry and is required for receptor clustering (Gautam et al., 1995). Although it bears no sequence homology with rapsyn, gephyrin is a peripheral membrane protein associated with the glycine receptor (Kirsch et al., 1993) and is thought to be a functional homologue of rapsyn for central inhibitory synapses. For glutamate synapses it is less clear if there is a single receptor-anchoring protein like rapsyn or gephyrin. Numerous postsynaptic proteins bind to glutamate receptors, including the PSD-95 (Postsynaptic Density Protein of 95 kDa) and S-SCAM families of PDZ domain proteins, the actin-binding proteins α-actinin and spectrin, and yotiao for NMDA receptors and the PDZ domain proteins GRIP, ABP, and PICK1 for AMPA receptors (see Ziff, 1997; and Hata et al., 1998, for review).

PDZ domains are regions of some 90 amino acids that form binding pockets for specific C-terminal sequences of interacting proteins. Some of these glutamate-binding proteins have multiple PDZ or other protein interaction domains and form scaffolds for cell adhesion proteins and downstream signal-transducing proteins as well as receptors. For example, PSD-95 binds to neuroligin (Irie et al., 1997), neuronal nitric oxide synthase (Brenman et al., 1996), and a synaptic Ras-GTPase activating protein (Chen et al., 1998; Kim et al., 1998) as well as to NMDA receptors. At the NMJ there is also a complex molecular specialization associated with clustered receptor and Na^+ channels in the postsynaptic membrane. Utrophin, a protein with strong homology to dystrophin, co-localizes with rapsyn and AChRs, although its absence apparently has only subtle effects on AChR clustering (Deconinck et al., 1997; Grady et al., 1997). The clustering of Na^+ channels at the NMJ may be related to their binding to PDZ domains of syntrophins, which are also concentrated at the NMJ. Syntrophins bind to dystrophin, which is linked both to cortical actin and to the extracellular matrix via the transmembrane dystroglycan complex (see Michalak and Opas, 1997, and Colledge and Froehner, 1998, for review). The dystroglycan complex, by virtue of its binding to so many intracellular and extracellular molecules, may be a central organizing molecule for synaptic structure.

A general feature of chemical synapses is the tight adhesion between pre- and postsynaptic components. At the NMJ separation of the nerve from muscle requires proteolytic treatment (Kuffler and Yoshikami, 1975; Wilkinson et al., 1996). In the CNS purified presynaptic terminals (known as synaptosomes) usually contain a piece of postsynaptic membrane and density attached (De Robertis, 1967). The space between nerve and postsynaptic cell, however, may serve other functions as well. Classic denervation and muscle irradiation experiments showed that when muscle fibers or nerves were removed key synaptogenic signals remain in the basal lamina of the synaptic cleft at the NMJ (see Sanes and Lichtman, 1999a, for review). These results suggest that the synaptic basal lamina is the repository of signaling components. The synaptic cleft of central synapses is narrower (10–15 nm versus about 50 nm for the NMJ) and contains electron-dense material but lacks a basal lamina. These ultrastructural differences raise the possibility that some of the key synaptogenic signals for central synapses may be mediated by direct transmembrane protein interactions.

Besides the pre- and postsynaptic cells, a third cell that is increasingly acknowledged to be a potential key component of the synapse is the glial cell. Schwann cells cover the nerve terminal at the NMJ, and over half of hippocampal CA1 synapses are in direct contact with astrocytes

(Ventura and Harris, 1999). At the NMJ, glia have been shown to respond to neurotransmitter release via a G protein cascade that affects subsequent presynaptic release (Robitaille, 1998). In the CNS, the major plasma membrane glutamate transporters are present on astrocytes, where they function to limit the time course of glutamate in the synaptic cleft (Bergles et al., 1997; Mennerick et al., 1999) and perhaps to limit spillover to nearby synapses. The classic view that glial cells simply provide support or are at most modulatory is being challenged by new evidence that glia may act directly as either a presynaptic or a postsynaptic partner (Araque et al., 1999).

Interestingly, although central synapses come in many forms, there is conservation in the relative sizes of elements within a synapse. For example, for hippocampal spine synapses, the spine head volume, smooth endoplasmic reticulum volume, area of the postsynaptic density, area of the presynaptic active zone, bouton volume, number of synaptic vesicles, and number of docked vesicles all correlate (Harris and Kater, 1994; Schikorski and Stevens, 1997). Thus some kind of feedback mechanism must adjust the various elements to make them match.

Synapse Formation

Synapse development involves a series of very gradual structural, functional, and molecular changes. We attempt to differentiate here between synapse formation, the earliest events that generate a functional synapse, and subsequent maturation. There is some uncertainty about the earliest initiating events that lead to synaptic contact in both the CNS and muscle. In muscle, for example, AChR clusters are first observed on developing muscle fibers before nerves have made contact, although they are in the general vicinity of the nerve and depend on its presence (Dahm and Landmesser, 1991). It is unclear whether any of these early AChR clusters serve as preferential sites for synaptic contact. At the time uninnervated muscle fibers are beginning to differentiate sites capable of neurotransmitter reception, the axon is capable of transmitter release near the muscle. Growing cholinergic axons are capable of quantal ACh release even in the absence of postsynaptic targets (Hume et al., 1983; Young and Poo, 1983). This release is based on molecular machinery similar to that found at synapses and becomes progressively restricted toward the growth cones as axons approach their targets (Antonov et al., 1999; Zakharenko et al., 1999). Thus it would seem that both the release machinery in presynaptic terminals and neurotransmitter receptor clusters in postsynaptic cells are being established in advance of the first connections. The presence of these precocious synaptic characteristics in both pre- and postsynaptic cells makes determining which partner is the initiator of synaptogenesis difficult. Given the synaptic characteristics of nerve and muscle prior to synaptic contact, it is perhaps not surprising that synaptic transmission begins within minutes of actual contact (Xie and Poo, 1986; Sun and Poo, 1987; Evers et al., 1989).

Similarly, it is not known in the case of CNS neurons which partner—the axon or the postsynaptic process—initiates synaptogenesis. Both "free" postsynaptic densities (Blue and Parnavelas, 1983a) and "free" presynaptic densities with associated synaptic vesicles (Newman-Gage et al., 1987) have been observed ultrastructurally in developing cortex. However, from such static observations it is difficult to define a sequence of synaptogenic events in a heterogeneous developing tissue. By following a wave of synaptogenesis as spinal cord growth cones contacted superior cervical ganglion cells in culture, Rees et al. (1976) observed close contact, the appearance of coated vesicles in the postsynaptic cell and the postsynaptic density with enlargement of the cleft, followed by clustering of synaptic vesicles in the axon and the formation of the presynaptic density.

More recent studies have used molecular markers and live imaging, mostly in neuronal culture systems, to probe the early events in synapse development. Due to their relative homogeneity and the advantages of low-density culture systems (Banker and Cowan, 1977; Goslin et al., 1998), hippocampal neurons have been used fairly extensively as a model system for excitatory synapse development. Prior to cell-cell contact, functional NMDA- and non-NMDA-type glutamate receptors are present on the cell surface (Verderio et al., 1994). AMPA-type glutamate receptors are diffusely distributed in developing dendrites, but NMDA receptors and PSD-95 cluster in the absence of axon contact (Craig et al., 1993; Rao et al., 1998). It is not yet known whether these clusters are intracellular or whether they are on the cell surface, perhaps as precursors of postsynaptic specializations. Early in development there are small mobile clusters of synaptic vesicles in isolated axons; these may represent precursors to presynaptic specializations (Kraszewski et al., 1995). Synaptic vesicles in the isolated axons release glutamate via calcium- and depolarization-dependent exocytosis even before synapse formation. Thus, as with developing motoneuron axons and muscle, hippocampal axons and dendrites develop precocious synaptic characteristics prior to contact. A similar situation applies to central inhibitory synaptic components. Prior to contact and synaptic maturation, the inhibitory synaptic scaffolding protein gephyrin forms spontaneous clusters (Colin et al., 1996, 1998). Furthermore, quantitative data suggest that unapposed clusters of synaptic vesicles in axons, or of glycine or GABA receptors in dendrites, occur separately very early in the development of spinal neurons in culture (Levi et al., 1999).

Much recent interest has focused on the cellular substrates involved in initial synaptogenic contact. Ziv and Smith (1996) performed live imaging of postsynaptic structures labeled with a membrane dye followed by identification of sites of recycling synaptic vesicles in hippocampal cultures. They showed that the numbers of labile dendritic filopodia declined and the numbers of stable dendritic spines increased during development, a finding that is consistent with previous studies of fixed material (Papa et al., 1995). In one instance Ziv and Smith (1996) observed a dendritic filopodium contact an axon and remain in contact for 90 min, whereupon this contact site was found to label for recycling synaptic vesicles. This study suggested the interesting model that dendritic filopodia may initiate synaptogenic contact (Fig. 12.3A; Smith, 1999). Synapses on the tips and along the lengths of dendritic filopodia have been observed by electron microscopy in hippocampal tissue

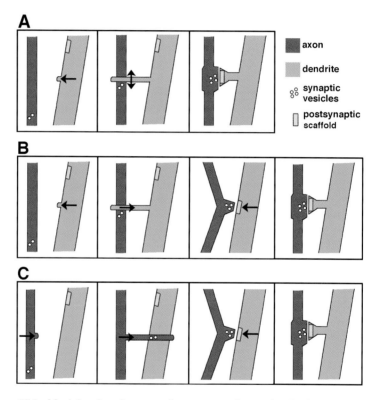

Figure 12.3. Models of excitatory spine synapse formation initiated by filopodial contact.

(A) Dendritic filopodia may contact an axon, persist, and transform into a spine shape (see also Smith, 1999).

(B) Dendritic filopodia may contact an axon and then retract while maintaining contact, thus pulling the axon in to form a shaft synapse. Gradual outgrowth from the dendrite may then convert the synapse to a spine shape (see also Fiala et al., 1998).

(C) Axonal filopodia may contact a dendrite and then contract while maintaining contact to pull the axon in to form a shaft synapse. Gradual outgrowth from the dendrite may then convert the synapse to a spine shape.

In all models, the structural alterations would be accompanied by local aggregation of synaptic components, probably synaptic vesicles first, followed by postsynaptic scaffolds and receptors.

preparations. Early in development over 20% of the total synapses can be found associated with dendritic filopodia (Fiala et al., 1998). However, the density of filopodial synapses is always less than half that of shaft synapses. The implications of this study (Fiala et al., 1998) depend on the half-life of the filopodia bearing synapses: if they are labile and retract rapidly, as do most filopodia in culture, then the numbers of synapses generated by such a mechanism would be too high to match the actual rate of synapse generation, indicating that some filopodial synapses would not be retained while others would be retained and converted to shaft synapses (Fig. 12.3B). Conversely, if the filopodia that bear synapses are very stable they may not be converted to shaft synapses but remain as filopodia or be converted directly to dendritic spines (Fig. 12.3A); this would leave open the possibility of a large percentage of synapses also forming directly on shafts (Fig. 12.3C). Perhaps such issues will soon be resolved by further imaging of developing synapses in culture or slice systems.

An interesting related question concerns the mode of genesis of excitatory synapses on nonspiny neurons. Such neurons generally do have filopodia-like dendritic appendages during development. Although synapses on such appendages are rare in fixed tissue (Wong et al., 1992; Linke et al., 1994), more dynamic analyses may reveal a transient role of filopodia in synaptogenesis. This model for the initiation of synaptogenesis by dendritic filopodia is in contrast to the classic model of neuromuscular synapse initiation by the axonal growth cone. However, neuromuscular synapses in invertebrates such as *Caenorhabditis elegans* occur on "muscle dendrites" and synaptogenesis at *Drosophila* NMJs is preceded by the transient outgrowth of postsynaptic filopodia from the muscle (Chiba, personal communication), perhaps indicating that the apparently passive role of the muscle fiber at the mammalian NMJ may be the exception. The en passant nature of most mammalian central synapses is in keeping with a model for synapse initiation between dendritic filopodia and axon shafts, although it seems equally possible that contacts between axonal filopodia and dendrite shafts may play a role.

The mechanisms for aligning the developing presynaptic and postsynaptic specializations are not well understood, but they clearly involve two-way signaling. Evidence for bidirectional signaling comes from noting that disruptions of either the pre- or postsynaptic element during development can alter the differentiation of the remaining partner. For example, postsynaptic receptor clustering depends on agrin release from nerve terminals (Gautam et al., 1996). However, even over months, axons will not differentiate normal nerve terminals if muscle fibers lack rapsyn or muscle-specific tyrosine kinase (MuSK) (Nguyen et al., submitted). Heterochronic co-culture experiments also indicate bidirectional signaling in the development of synapses between hippocampal neurons (Fletcher et al., 1994). Presynaptic specializations, defined as clusters of synaptic vesicles in axons at sites of contact with dendrites, can form in the axons of neurons that have been in culture for 1 day, but only at contact points with neuron dendrites that have been in culture for at least 3 days. Thus axons are competent to participate in synapse formation immediately upon outgrowth, but dendrites require several days of maturation before they become competent to participate in synapse formation. Moreover, both partners play an active role: vesicle clustering in the axons requires some signal that can only be generated by more mature dendrites. Immature dendrites may be incapable of responding to an initial synaptogenic signal from the axon, incapable of generating a retrograde synaptogenic signal, or both. Identification of the molecular nature of the signals will be required to resolve the cascade of cellular signaling events. An intriguing finding from Hazuka et al. (1999) is the periodic distribution of sec6/8 along axons and the correlation of subsets of sec6/8 concentrations with developing synapses. Given the demonstrated role of the sec6/8 complex in regulating specific vesicle fusion in nonneuronal cells, this study suggests that axons may be delimited into synapse-competent and -incompetent zones by the distribution of specific vesicle fusion machinery. Insight into how synapse-competent zones are established may be gained through imaging (Ahmari et al., 2000).

Based on immunocytochemical analyses of hippocampal neurons at different stages of development, it seems most likely that presynaptic vesicle clusters stabilize first at contact sites, followed by the PSD-95 postsynaptic scaffold, and then neurotransmitter receptors (Craig et al., 1993; Rao et al., 1998). However, the early presence of some clusters of PSD-95 (Rao et al., 1998) and even glutamate receptors in spinal cultures at axon-dendrite contact sites but not apposed to synaptic vesicles (O'Brien et al., 1997) leaves open the possibility that postsynaptic elements may stabilize first at some sites. Although relatively few studies have directly compared electrophysiological and immunocytochemical measures of synapse development between central neurons, the emergence of spontaneous synaptic transmission appears to correlate with the presence of both synaptic vesicle clusters and receptor clusters at contact points (e.g., O'Brien et al., 1997). Glutamate release properties are different at synapses than in isolated axons, even though both are dependent on calcium and depolarization. Basal exocytosis rate is lower, calcium sensitivity is lower, and sensitivity to tetanus toxin is higher at synapses than along isolated axons (Coco et al., 1998; Verderio et al., 1999). It is not yet known how quickly this change occurs, but it appears to be locally regulated by target contact, as vesicle exocytosis from synaptic versus extrasynaptic regions of a single axon can exhibit differential sensitivity to tetanus toxin (Verderio et al., 1999).

One way to probe the requirements for specific axon-dendrite contact in synapse development is to grow neurons under conditions in which only one type of presynaptic terminal exists, either as a purified population of a single cell type or as a single neuron in isolation. By analysis of purified motoneuron cultures, Levi et al. (1999) showed that signals associated with GABA or glycinergic terminals are specifically required to induce synaptic clustering of the corresponding receptors. In purified cholinergic motoneuron cultures, GABA receptors and glycine receptors formed clusters at nonsynaptic sites but did not cluster opposite the cholinergic terminals. When GABAergic neurons were added to these cholinergic cultures, GABA receptors clustered exclusively opposite the GABAergic terminals. The further addition of glycinergic innervation resulted in synaptic clustering of glycine receptors. Similarly, signals associated with glutamatergic terminals are required to induce synaptic clustering of AMPA glutamate receptors (O'Brien et al., 1997; Rao et al., submitted). In this case, in the absence of appropriate innervation AMPA receptors did not form any clusters. However, the principles governing the matching of pre- and postsynaptic elements are not always so straightforward. Rao et al. (submitted) have observed "mismatched appositions" in single-neuron cultures but not in multi-innervated cultures from hippocampus. Isolated GABA cells cluster PSD-95 and NMDA receptors specifically opposite GABA terminals, and isolated glutamatergic cells cluster GABA receptors and gephyrin specifically opposite glutamate terminals. Furthermore, on an isolated pyramidal cell some glutamate terminals are apposed to normal glutamate receptor clusters, whereas others are apposed to GABA receptors but not glutamate receptor clusters. These results suggest that multiple signals may be involved in aligning pre- and postsynaptic components, with perhaps one signal in common between glutamate and GABA synapses that allows these mismatches to occur, and a second specificity signal that ensures appropriate matching in most situations. They further suggest the existence of some type of exclusionary signal that prevents mixing of excitatory and inhibitory receptor clusters.

In muscle one of the most impressive aspects of the alignment of pre- and postsynaptic elements is the juxtaposition of presynaptic active zones with the postsynaptic folds. Thus not only are receptors clustered opposite release sites, but the postsynaptic membrane has a specialized topography that aligns with the release sites. Generating such alignment requires a kind of signaling between the two synaptic partners that is not yet understood. In a few situations, such as ribbon synapses in the retina, contacts between neurons also show complex and stereotyped membrane bulges and folding. At the NMJ one proposal for getting postsynaptic folds to align with active zones is based on different rates of membrane addition during growth of attached pre- and postsynaptic elements (Marques et al., 2000).

As for the early cellular events, there is also uncertainty regarding the nature of the molecular signals mediating synapse formation. At the NMJ nerve release of agrin seems to be an indispensable early step in the molecular cascade that initiates postsynaptic differentiation (see Meier and Wallace, 1998, and Sanes and Lichtman, 1999a, for review). Agrin synthesized by motoneurons is transported to synaptic terminals, where it is released and adheres to the synaptic basal lamina. Agrin mediates postsynaptic differentiation through activation of MuSK, which is clustered by agrin. This kinase, through steps that are not well worked out, causes the clustering of a membrane-associated protein, rapsyn, which binds tightly to AChRs and is essential for AChR clustering to occur. Less clear is the initiating signal necessary for the specific transcription of AChR-subunit genes by myonuclei that are located at the synapse. Neuregulin, which is synthesized by both motoneurons and muscle fibers, seems to induce receptor synthesis via activation of members of the epidermal growth factor–related receptor kinase (erbB) family. Three different erbB proteins are concentrated at the NMJ. Because agrin can induce AChR subunit transcription via the erbB pathway (Jones et al., 1996; Meier et al., 1998a,b), it is possible that neuregulins act downstream of agrin, suggesting agrin's primacy in the signaling cascade. In addition to the nerve effects that cause receptor clustering and synthesis in skeletal muscle, the nerve also plays a role in inhibiting receptor expression at nonsynaptic sites. The restriction of receptors to synaptic sites is due to the ability of muscle activity (usually induced by nerve activity) to downregulate receptor subunit transcription by non-synaptic myonuclei (see Sanes and Lichtman, 1999a, for review). This effect of the nerve on receptor expression is thought to be mediated by a direct effect of activity on intracellular Ca^{2+} levels in muscle fibers (e.g., Huang and Schmidt, 1994). This seems to be an agrin-independent pathway.

For most central synapses, axon-dendrite contact probably initiates synaptogenesis via transmembrane or secreted factors, although these are yet to be characterized. Although agrin is widely expressed in the CNS, it appears not to be a key synaptogenic factor as it is at the NMJ. Cortical cultures from agrin-deficient mice develop glutamate and GABA synapses that are molecularly and functionally indistinguishable from the wild type (Z. Li et al., 1999; Serpinskaya et al., 1999; but see also Ferreira, 1999). Synaptic activity appears to be dispensable for the initial formation of glutamate or GABA synapses (Craig et al., 1994; Verderio et al., 1994; O'Brien et al., 1997), although it is required for the formation of glycinergic synapses. Strychnine, a glycine receptor antagonist, prevents clustering of glycine receptors at synapses in spinal cord cultures (Kirsch and Betz, 1998; Levi et al., 1998). Inhibition of voltage-dependent Ca^{2+} channels has a similar effect, suggesting that the key event required for receptor stabilization is depolarization-induced Ca^{2+} influx (Kirsch and Betz, 1998). Activity presumably works in concert with additional molecular cues to mediate glycinergic synapse formation.

One of the intracellular mediators of inhibitory receptor clustering has been identified. Gephyrin, a peripheral membrane protein isolated by co-purification with the glycine receptor, is essential for synaptic clustering of glycine receptors and of many GABA receptors (Kirsch et al., 1993; Essrich et al., 1998; Feng et al., 1998b; Kneussel et al., 1999). Gephyrin binds directly to the intracellular loop of the glycine receptor β subunit (Meyer et al., 1995). Direct binding of gephyrin to GABA receptors has not been reported, but co-dependent clustering of gephyrin and the GABA receptor γ2 subunit suggests that these proteins form part of a complex (Essrich et al., 1998). At neuronal cholinergic synapses, although rapsyn is present it does not seem to be necessary for clustering neuronal AChRs (Feng et al., 1998a). At glutamate synapses, it is not yet clear which of the many glutamate receptor–binding proteins mediate clustering for any of the receptor subtypes. For example, although PSD-95 has been suggested as a mediator of NMDA receptor clustering, PSD-95 mutant mice still have normal NMDA receptor clusters (Migaud et al., 1998). It may be that the highly related and co-localized proteins PSD-93/chapsyn-110 and SAP102 mediate NMDA receptor clustering in the PSD-95 mutant, or it may be that other NMDA receptor–binding proteins are involved. Presumably there must exist different receptor-clustering factors for each receptor type and subtype to allow for the observed differences in their distribution patterns. Furthermore, multiple pathways may act in parallel to control the localization of a single receptor type.

Several other proteins have been suggested to be involved in early synapse formation, including cadherins (Fannon and Colman, 1996; Uchida et al., 1996; Benson and Tanaka, 1998), cadherin-related neuronal receptors (Kohmura et al., 1998), ephrins and Eph receptors (Torres et al., 1998; Buchert et al., 1999), integrins (Einheber et al., 1996; Nishimura et al., 1998), neurexins and neuroligins (Irie et al., 1997; Butz et al., 1998; Song et al., 1999), densin-180 (Apperson et al., 1996), and neuronal activity–regulated pentraxin (Narp) (O'Brien et al., 1999). Narp secreted onto the cell surface has the interesting ability to induce aggregation of AMPA receptors on spinal cord neurons, and Narp is localized to aspiny excitatory synapses in spinal and hippocampal cultures. However, Narp is not found at spiny excitatory synapses between pyramidal cells and thus cannot be a generally required glutamatergic synaptogenic factor. Evidence for the involvement of any of these other proteins in synapse initiation is tentative, based mainly on their transmembrane nature and localization to synapses, their biochemical association with other synaptic proteins, or both. Considering the roles of the cadherin, ephrin, and integrin families in earlier steps of development, such as neuronal migration and axon outgrowth and guidance, a specific role in synaptogenesis would be surprising but not inconceivable. Many of these protein families consist of multiple members generated from different genes or by alternative splicing, and so they have the capacity to mediate signaling at diverse synapse types. Considering the heterogeneity in molecular composition of individual glutamate and GABA postsynaptic elements, it is likely that different combinations of molecules function in synapse initiation at different subsets of synapses.

Synapse Maturation

At nerve-muscle contacts there is a period of nearly a month between the first synaptic contacts and the attainment of a mature NMJ. From a functional standpoint several important alterations occur. First the reliability and amount of neurotransmitter release increases. This alteration is probably associated with the severalfold increase in the area of the presynaptic terminals, which stretch as they remain adherent to a rapidly growing muscle fiber. As the synaptic efficacy is increasing all but one of the axons initially converging at the synaptic site are eliminated. Second, the postsynaptic membrane undergoes dramatic changes in topography as an initial flat, plaque-shaped region containing a moderate cluster of AChRs transforms into a spoon-shaped plaque and finally becomes pretzel-shaped. The postsynaptic membrane is thrown into shallow gutters containing secondary infoldings. These secondary folds possess a very high density of AChRs at their crests, whereas nearby regions lose nearly all receptors.

The earliest synapses recognizable ultrastructurally contain only a few synaptic vesicles, and it has been found in nearly all systems studied that synaptic vesicle numbers increase with development (Jones, 1983; Vaughn, 1989). Over the first month of postnatal development of the rat visual cortex, Blue and Parnavelas (1983b) observed a gradual fourfold increase in the number of vesicles per terminal. However, most studies suggest that there is little developmental change in the size of several other CNS synaptic parameters, including synaptic contact length, postsynaptic density length, and terminal size (Hinds and Hinds, 1976; Blue and Parnavelas, 1983b; Jones, 1983; Weber and Kalil, 1987; Vaughn, 1989). Synapse-associated organelles may change; for example, the frequency of polyribosomes in the vicinity of synapses peaks during the first postnatal week in rat hippocampus (Steward and Falk, 1991).

The other major structural change in developing central synapses is the form of the postsynaptic element at excitatory synapses. Perhaps analogous to the changes in postsynaptic membrane topography at the NMJ, dendritic spines develop gradually following initial synapse formation. Early models suggesting that spine synapses develop from shaft synapses by outgrowth through a stage of stubby spines (Cotman et al., 1973; Miller and Peters, 1981) are supported by quantitative data comparing P15 versus adult hippocampi (Harris et al., 1992). Recent experiments, discussed previously, have resulted in modified models according to which postsynaptic form may change from filopodia to shaft to spine (Fiala et al., 1998; Fig. 12.3B) or directly from filopodia to spine (Ziv and Smith, 1996; Fig. 12.3A). These models are not mutually exclusive, and the exact sequence may depend on developmental stage. For example, conversion of a synapse from dendritic filopodium to shaft to spine would require considerable movement of the axon and dendrite, and this may occur most readily in young tissue. Even once formed, dendritic spines are not static but highly motile structures driven by actin dynamics (Fischer et al., 1998), although the degree of motility declines with postnatal age (Dunaevsky et al., 1999).

A common maturational event for many synapse types is a change in transmitter receptor subunit composition. At the NMJ, for example, the γ subunits of the AChR that are expressed throughout fetal life are replaced by ε subunits during the first postnatal week. This change of the AChR from fetal to adult form causes a decrease in the duration of miniature endplate currents but also, perhaps more importantly, causes a threefold increase in calcium permeability (Villarroel and Sakmann, 1996). In mutant mice that have had the ε gene deleted, the fetal form of the AChR persists but with deleterious consequences, including impaired neuromuscular transmission with a gradual reduction in AChR density, abnormal structure of the NMJ, muscle fiber atrophy, and finally the animal's death within several months (Witzemann et al., 1996; Missias et al., 1997). Similar to the γ-to-ε subunit switch for AChRs at the NMJ, spinal cord glycine receptors switch postnatally from the α2 to the α1 subunit, and embryonic forms may be α2 homomers compared with the adult α1/β heteromers (see Vannier and Triller, 1997, and Betz et al., 1999, for review). The neonatal receptor has a higher conductance, a longer channel open time, and a broader agonist range than the adult form. Some hereditary motor disorders in human and mouse are due to mutations in the gene encoding the adult glycine receptor α1 subunit. Most such disorders result from dominant missense mutations leading to reduced glycine sensitivity of the encoded receptors and exaggerated startle response (Andrew and Owen, 1997). Complete loss of the glycine receptor α1 subunit leads to progressive neurological symptoms and juvenile lethality in mice (Buckwalter et al., 1994) but may be compensated in humans (Brune et al., 1996).

NMDA-type glutamate receptors also undergo a subunit switch in early postnatal life. NMDA receptors are either tetramers or pentamers containing the essential NR1 subunit and some combination of NR2A–D. Cortical neurons express NR1/NR2B NMDA-type glutamate receptors early in development, and postnatal maturation involves expression of NR2A and formation of NR1/NR2B/NR2A and NR1/NR2A receptors (Monyer et al., 1994; Sheng et al., 1994). The embryonic NR2B-containing receptors have a higher glutamate affinity and longer channel open time (see Feldmeyer and Cull-Candy, 1996, for review). Mice lacking the NR2A subunit are viable and show no gross morphological abnormalities but exhibit reduced hippocampal long-term potentiation and spatial learning (Sakimura et al., 1995). The developmental switch in NMDA receptor subunit composition in neocortex correlates with the end of the critical period for synaptic reorganization and may both regulate activity-dependent structural changes and be regulated by activity (Scheetz and Constantine-Paton, 1994). For example, eye opening induces a rapid insertion of NR2A at synapses in dark-reared rats (Quinlan et al., 1999). Experiments in primary culture systems indicate that although NR1/NR2B receptors can function at synapses, NR2A-containing receptors may be preferentially targeted to synapses (Li et al., 1998; Tovar and Westbrook, 1999; but see also Rao et al., 1998).

AMPA-type glutamate receptors do not show a concerted developmental subunit switch, but in fact the appearance of synaptic AMPA receptors may be a late maturational event. At early postnatal hippocampal CA1 and thalamocortical synapses, recordings at hyperpolarized versus depolarized potentials suggest that many synapses exhibit NMDA receptor–mediated currents but lack AMPA receptor–mediated currents (Durand et al., 1996; Liao and Malinow, 1996; Isaac et al., 1997). Such silent synapses appear to be physically lacking AMPA receptors (Rao and Craig, 1997; Nusser et al., 1998b; Liao et al., 1999; Petralia et al., 1999; Takumi et al., 1999). A current model of synapse development is that AMPA receptor insertion at synapses is a late event regulated by activation of NMDA receptors (Constantine-Paton and Cline, 1998; Shi et al., 1999). However, this is not an obligatory sequence of events since, at least in culture, synaptic AMPA receptors can develop prior to synaptic NMDA receptors and in the absence of NMDA receptor function (Rao and Craig, 1997).

GABAergic and glycinergic synapses also show an important developmental maturation due to a change in the Cl⁻ reversal potential. In fact, although inhibitory in the adult, these synapses are excitatory in neonatal life (Cherubini et al., 1991; Boehm et al., 1997). The switch may be due to the onset of expression of the Cl⁻ transporter KCC2, which changes the Cl⁻ equilibrium potential to a more hyperpolarized value (Rivera et al., 1999). Excitatory GABAergic transmission in the neonatal hippocampus may contribute to activation of the early NMDA-only-type glutamate synapses (Leinekugel et al., 1997), and, as mentioned earlier, depolarization mediated by synaptic transmission is required for receptor stabilization at developing glycinergic synapses (Kirsch and Betz, 1998; Levi et al., 1998).

Developmental mechanisms must account for the high degree of heterogeneity in molecular composition of individual glutamatergic and GABAergic postsynaptic elements. Some differences in composition, such as variability in AMPA and NMDA receptor density, may be influenced by synaptic activity, but it is likely that additional activity-independent, input-specific signals are involved. Receptor composition often varies at different sites on a single postsynaptic cell depending on the presynaptic partner. For example, the NR2B subunit of the NMDA receptor is selectively excluded from hippocampal CA3 synapses postsynaptic to mossy fiber inputs (Fritschy et al., 1998; Watanabe et al., 1998); the GluR2 subunit of the AMPA receptor is present at hippocampal interneuron synapses postsynaptic to CA3 collaterals but not mossy fibers (Toth and McBain, 1998); in the cerebellum, the δ2 glutamate receptor is present at parallel fiber but not climbing fiber synapses on adult Purkinje cells, but it is present at both synapses early in development (Landsend et al., 1997; Zhao et al., 1998); the GABA$_A$ receptor α2 subunit is selectively concentrated at synapses on the axon initial segment compared with a lower density at somatodendritic synapses on hippocampal pyramidal cells (Nusser et al., 1996); and many GABA$_A$ receptor subunits show selective subcellular targeting in cerebellar granule cells and in the retina (Koulen et al., 1996; Nusser et al., 1998c). Explant experiments have confirmed that differential presynaptic innervation directly regulates NMDA receptor subunit composition in hippocampal pyramidal neurons (Gottmann et al., 1997).

In addition to changes in ligand-gated ion channels, at the NMJ there are also maturational changes in the basal lamina as laminin subtypes change and become differentiated from the extrasynaptic basal lamina (Patton et al., 1997). At the same time, and in a phenomenon that is perhaps causally related, the laminin receptors (integrins) in the postsynaptic membrane are also undergoing alterations in distribution (Martin et al., 1996; Anderson et al., 1997). Maturational changes in cell adhesion proteins also occur at CNS synapses. N-cadherin is concentrated at both glutamate and GABA synapses early in development in hippocampal cultures, but it is lost from GABA synapses with maturation (Benson and Tanaka, 1998). Maturational changes in structural proteins occur at CNS synapses. Whereas putative scaffolding proteins of the PSD-95, GKAP/SAPAP, and Shank families localize to synapses early, many spine-associated components—such as α-actinin-2, drebrin, Ca^{2+}/calmodulin-dependent protein kinase II α (CaMKIIα), and syndecan-2—cluster at synapses late in development, concurrent with the outgrowth or maturation of spines (Rao et al., 1998; Ethell and Yamaguchi, 1999; Hsueh and Sheng, 1999; Lim et al., 1999). Indeed overexpression of either drebrin or syndecan-2 regulates the morphology of dendritic protrusions (Hayashi and Shirao, 1999; Ethell and Yamaguchi, 1999). Some of the maturational events at glutamate synapses are summarized in Fig. 12.4. Future challenges will require relating molecular changes to structural and functional maturation of the synapse.

A few recent experiments have shown that maturation of presynaptic terminals can also be regulated. Retinal ganglion neurons cultured in the absence of glia form synapses that are morphologically normal but functionally deficient in neurotransmitter release (Pfrieger and Barres, 1997). These synapses are presumably lacking in some molecular component mediating regulated vesicle fusion. Functional synapses can be induced by addition of soluble glial-derived factors. Similarly, hippocampal neurons cultured from immature E16 rats are deficient in synaptic transmission unless neurotrophins are added to stimulate presynaptic development (Vicario-Abejon et al., 1998). In contrast, neurons from E18 or older rats (as in the previous experiments discussed) can form functional synapses in the absence of added neurotrophins. Further evidence for regulated maturation of presynaptic function is the ability of cAMP analogues to enhance synaptic vesicle recycling (Ma et al., 1999). It is not yet clear whether most presynaptic specializations in vivo undergo a maturational delay, beyond the increase in vesicle content.

Activity can regulate the postsynaptic structure and the molecular composition of existing synapses and may also regulate synapse formation. However, there is not yet a consensus on how activity influences synaptic parameters. Part of the difficulty is that the results of individual studies differ widely, for example, from activity enhancement to activity

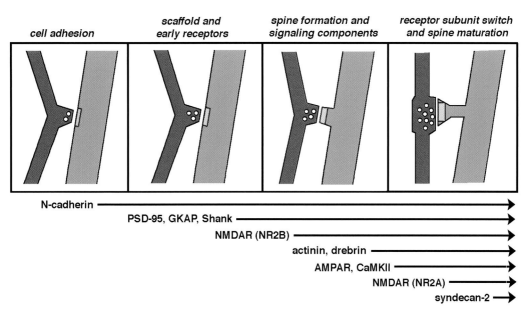

Figure 12.4. Maturational events at central excitatory spine synapses. An approximate sequence of assembly of postsynaptic components is related to maturational changes in synapse structure. Many other molecular components have been found at mature glutamate synapses and are probably involved in these processes but have not yet been studied developmentally. The figure is based on data from Benson and Tanaka (1998), Li et al. (1998), Rao et al. (1998), Ethell and Yamaguchi (1999), Hsueh and Sheng (1999), Liao et al. (1999), Lim et al. (1999), Tovar and Westbrook (1999), and Crump, Allison, Fong, and Craig (unpublished). Note that in some cell culture systems NR2A appears early along with NR2B (Rao et al., 1998).

inhibition of dendritic spine formation. Such variations may reflect differences in cell type, developmental stage, experimental preparation, time course, and mode of activity manipulation. Some of the key studies on the relation between activity and synapse formation include local activity induction of new spine or filopodia-like protrusions in developing hippocampal slices (Engert and Bonhoeffer, 1999; Maletic-Savatic et al., 1999), enhanced spine formation caused by activity blockade in mature hippocampus (Kirov and Harris, 1999; Kirov et al., 1999), loss of spines due to blockade of spontaneous vesicle fusion in hippocampal slice cultures (McKinney et al., 1999), and enhanced spine formation due to tetrodotoxin blockade along proximal dendrites of mature Purkinje cells (Bravin et al., 1999). Activity can also regulate spine shape (e.g., Korkotian and Segal, 1999) and motility (Kaech et al., 1999). Interpreting changes in the number of spines is complicated by the fact that these are dynamic structures. Alterations in initiation, average lifetime, or active elimination processes could all result in net changes in spine number. The challenge will be to determine which of these processes lead to changes in spine number. In addition, it will be necessary to understand better the relation between filopodia and spine dynamics and the presence and function of synapses. For example, changes in the number and dynamics of dendritic protrusions may occur without any synaptic change.

Synaptic composition, particularly the density of neurotransmitter receptors, can also be modulated by activity. It appears that each receptor subtype is regulated independently by activity, such that different synaptic receptors can be either up- or downregulated over different time courses. Some of these differences may reflect the ability of activity initially to induce synapse-specific changes according to Hebbian mechanisms and later to induce more global homeostatic changes to rebalance the system (Turrigiano, 1999). Recent studies have focused on rapid regulation of synaptic targeting of AMPA receptors and its role in long-term depression and potentiation (Carroll et al., 1999; P. Li et al., 1999; Shi et al., 1999). Some AMPA receptors may be continually undergoing cycles of exocytosis and endocytosis regulated by interactions with N-ethylmaleimide-sensitive fusion protein (Noel et al., 1999). Long-term activity manipulations can regulate synaptic densities of AMPA- and NMDA-type glutamate receptors and glycine receptors (see Craig, 1998, for review). Chronic activity blockade increases synaptic levels of AMPA receptors, a form of homeostasis that may be mediated through brain-derived neurotrophic factor (Rutherford et al., 1998). Chronic NMDA-receptor blockade also enhances synaptic targeting of NMDA receptors, perhaps a form of metaplasticity (Rao and Craig, 1997). Synaptic targeting of CaMKII can be rapidly enhanced by glutamate stimulation (Strack et al., 1997; Shen and Meyer, 1999). Excitatory transmission at the NMJ affects AChR density. Spontaneous release of neurotransmitter is sufficient to prevent receptor loss from the junction, perhaps via intracellular calcium (Caroni et al., 1993; Akaaboune et al., 1999). Thus while Ca^{2+} downregulates extrajunctional AChR expression, it plays the opposite role of maintaining the high density of AChRs at the synapse. Cortical inhibitory transmission can be regulated by activity—both presynaptically by downregulation of glutamic acid decarboxylase, the GABA synthetic enzyme (Rutherford et al., 1997), and postsynaptically by changing levels of synaptic GABA receptors (Nusser et al., 1998a).

Synapse Elimination

In both the CNS and peripheral nervous system (PNS), synapse formation generates many connections that exist only transiently in development. Proof that synapses are being lost comes from the finding that in many parts of the nervous system axons are synaptically connected during development to partners to which they are no longer connected later in life. At the NMJ this is quite obvious because at each adult NMJ (one per muscle fiber) there is only one motor axon, whereas at birth almost all junctions receive convergent innervation from two or sometimes as many as six motor axons (Redfern, 1970). In autonomic ganglia, paraganglionic axons also transiently hyperinnervate ganglion cells in early neonatal life (Lichtman, 1977; Lichtman and Purves, 1980). In several regions of the developing CNS, early axonal inputs are also removed. For example, in rodents multiple climbing fibers innervate each Purkinje cell at birth, and all but one are removed in early postnatal life (Lohof et al., 1996). In the avian auditory system, there is also a loss of axonal input to individual neurons (Jackson and Parks, 1982).

Perhaps the best-known examples of synapse elimination occur in the visual system. In both the thalamus (lateral geniculate) and the primary visual cortex, inputs associated with the two eyes initially overlap extensively but later segregate into eye-specific regions (Hubel et al., 1977; Sretavan and Shatz, 1986). In most of the developing CNS, however, it is unknown whether synapse elimination occurs. The difficulty is that assaying for a change in the number of innervating axons requires being able to count the number of axons, or at least having a situation in which distinct but overlapping pathways can be separately labeled or stimulated (such as the two eyes).

Interestingly, in most cases in which synapse elimination occurs, it does not lead to a net reduction in synapses, because the remaining axons elaborate more synapses than are lost. For example, in the submandibular ganglion of rats there is an approximately twofold increase in the number of synapses while the number of innervating axons is reduced from five to one (Lichtman, 1977). Thus the elimination process reduces the amount of axonal convergence on postsynaptic cells while the strength of the maintained connections increases. The concurrence of synaptogenesis and synapse elimination makes the interpretation of net changes in the number of synapses difficult.

The purpose of synapse elimination seems not to be some kind of gross error correction, because the lost connections are from the same presynaptic populations as the connections that are maintained. One possibility is that the loss is a consequence of the extreme degree to which mammalian (and other vertebrate) nervous systems are composed of duplicated neurons (Lichtman and Colman, 2000). For example, pools of motoneurons that may number in the hundreds innervate individual skeletal muscles containing thousands of muscle fibers. Rather than having a single identified motoneuron innervating a single identifiable muscle fiber, as can occur in invertebrate nervous systems, in vertebrates many duplicated neurons seem to have nearly identical roles, as does the large population of duplicated postsynaptic cells that constitutes a muscle. Two consequences of this redundancy may be increased axonal convergence and divergence. Multiple neurons seem to be the appropriate presynaptic partner for each muscle fiber, so multiple axons converge. Because multiple muscle fibers are the appropriate target for each motoneuron, each motoneuron diverges to innervate many muscle fibers. Thus at the outset there are overlapping converging and diverging pathways. In developing muscle, as in other locations, this state of affairs is short lived. Axonal branches are pruned, causing each axon's projection to become nonoverlapping and thus distinct. From a functional standpoint the removal of overlap in motor axon projections eliminates redundancy and allows the firing of each motor axon to give rise to a substantive increase in tension of the muscle. In like manner, depth perception (stereopsis) may require removal of overlap in the visual streams originating from the two eyes. Therefore synapse elimination tends to parse a highly redundant circuitry into multiple, functionally distinct circuits.

A number of lines of evidence suggest that the loss of synapses is the consequence of competition based on activity between inputs to the same target cells. The classic studies of Hubel and Wiesel showed segregation of the geniculocortical projections driven by the two eyes during a critical period in early postnatal life. Competition between the eyes for cortical space was suggested by their finding that imbalances between the eyes in terms of visual experience tipped the outcome of the segregation in favor of one eye over the other. The skewing that resulted from depriving one eye of vision was due both to additional losses in the connections driven by the inactive eye and to additional maintenance of the connections from the normally active nondeprived eye. Ordinarily each eye's afferents relinquish their connections with approximately half of the postsynaptic cells initially shared in the visual cortex. But when one eye is deprived of vision, its projection to the cortex loses connections with nearly all cells, while the nondeprived eye's inputs are not eliminated from any of the territory they initially contacted. Furthermore, brief binocular eye closure during the critical period appears to have far less serious effects than monocular occlusion (Hubel et al., 1977). These results support the idea that the loss is due to an activity-mediated competitive interaction between the connections driven by the two eyes.

How might activity mediate such competition? At the NMJ, it is likely that the activity of the postsynaptic cell is the critical intermediary in the competition, and perhaps the same is true in the visual system. Based on an experiment somewhat analogous to unilateral eye closure, postsynaptic silencing at one site in an NMJ causes local synapse elimination at that site (Balice-Gordon and Lichtman, 1994). Postsynaptic silencing of an entire NMJ, however, has no such effect. These results have been used to argue that synaptic activity is necessary for synapse elimination at sites that are not active at the same time. Thus synaptic activity might destabilize inactive synapses.

At the NMJ, synapse elimination is a gradual process (Fig. 12.5), and it has both a pre- and a postsynaptic component because at the time nerve terminals are withdrawing from postsynaptic sites by a process of lift-off and atrophy (Gan and Lichtman, 1998; Bernstein and Lichtman, 1999) the underlying postsynaptic site is losing its density of AChRs, rapsyn, and other postsynaptic markers (Balice-Gordon and Lichtman,

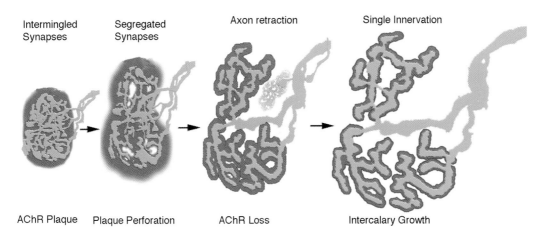

Intermingled Synapses   Segregated Synapses   Axon retraction   Single Innervation

AChR Plaque   Plaque Perforation   AChR Loss   Intercalary Growth

Figure 12.5. Steps in the synapse elimination process at the developing neuro-muscular junction. Between birth and the second and third postnatal week the vast majority of NMJs in rodents undergo the transition from multiple to single innervation. Based on a number of vital imaging and histological studies, it appears that the transition is gradual. At birth, each input to an NMJ is intermingled, as shown for a green- and a blue-colored axon in the diagram. The two inputs gradually segregate from each other by retracting branches that undergo atrophy and lift off the surface of the muscle fiber. During this process one axon is typically losing more territory than the other, so that the relatively even proportions of junctional area occupied by the two axons become progressively skewed in favor of one competitor. Finally the last site occupied by the losing axon is relinquished as that axon becomes atrophic and lifts off the muscle fiber. A bulb-shaped ending is often seen for both the final withdrawing branch and the receding small branches during the segregation phase. Once a junction is singly innervated it continues to grow in proportion to the growth of the muscle fiber by intercalary expansion. While the nerve terminals are segregating the underlying receptor plaque is transformed into a pretzel shape. This sculpting occurs by the progressive loss of AChR density within the original oval-shaped plaque. In some cases at least, the changes to the receptor sites begin before the nerve has vacated the site, suggesting that the postsynaptic cell may be instigating nerve terminal withdrawal by dismantling the postsynaptic sites.

1993; Culican et al., 1998). The postsynaptic changes begin before the nerve has been removed, suggesting that the postsynaptic cell might instigate the presynaptic removal. It remains to be determined whether muscle fibers initiate nerve terminal removal by action (e.g., expression of a synaptolytic factor, such as a protease; Liu et al., 1994; Zoubine et al., 1996) or inaction (e.g., downregulation of a synaptic maintenance factor; Snider and Lichtman, 1996; Nguyen et al., 1998). One way the postsynaptic cell might act as an arbiter of the competition is if synaptic transmission elicited intersynaptic signals that destabilized sites not active at the same time. Evidence supporting the idea of signals passing between competing synapses comes from the finding that during synaptic competition the two inputs gradually segregate from each other (Fig. 12.5; Gan and Lichtman, 1998). This suggests that the activity-mediated "punishment signals" that destabilize synaptic sites first eliminate nearby competing sites and only later eliminate more distant ones. Thus the destabilizing signals decrease in potency with distance and may be diffusible agents.

Mechanisms related to synapse elimination have also been studied in the cerebellum, where all but one of the three to four climbing fiber inputs to each Purkinje cell is eliminated in early postnatal life. As in muscle, the input removal is due to the retraction of branches of incoming axons, in this case from the inferior olivary nucleus (Delhaye-Bouchaud et al., 1985). At the time inputs are being eliminated, there is also a translocation of climbing fiber synapses from the soma of the Purkinje cell to its dendrites. It should be noted that the removal of climbing fiber inputs is probably not the only synapse elimination occurring in the developing cerebellum, as many mossy fiber synapses to granule cells may also be removed during the first postnatal month (Hamori and Somogyi, 1983); in this case, however, it is not known if mossy fiber inputs are completely removed.

A number of manipulations can prevent the elimination of climbing fiber inputs. It has been known for several years that intact parallel fiber innervation of Purkinje cells from granule cells is necessary for synapse elimination to occur. Thus mouse mutants such as *weaver* and *reeler* (which lack granule cells) and *staggerer* (which has disrupted parallel fiber synapses on Purkinje cells) all have persistent multiple climbing fiber innervation (Crepel et al., 1980; Mariani and Changeux, 1980). Furthermore, viral infections or X irradiation, which prevents proliferation of granule cells in early postnatal life, also allows multiple climbing fiber innervation to persist (see Lohof et al., 1996, for review). A case has also been made for the involvement of NMDA receptor activation; thus half the Purkinje cells remained multiply innervated by climbing fibers beyond the normal period of elimination when the developing cerebellum was chronically perfused with D,L-APV during early postnatal life (Rabacchi et al., 1992). A similar degree of inhibition of synapse elimination was noted in mutant mice deficient in protein kinase C (γ isoform) (Kano et al., 1995). These mice have impaired motor coordination that may be related to the extra climbing fiber innervation (Chen et al., 1995). Because in the cerebellum this kinase is expressed selectively in Purkinje cells and peaks at about the time climbing fiber input is decreasing, it is possible that it acts as a downstream element in a postsynaptic cascade that instigates climbing fiber removal. Also involved downstream may be isoforms of phospholipase C, as mutants lacking an isoform ($\beta4$) that is the only major isoform expressed in the cerebellar vermis have deficient climbing fiber removal in the vermis, but not elsewhere in the cerebellum where other isoforms are present (Kano et al., 1998). As with studies of other synaptic phenomena (e.g., Sanes and Lichtman, 1999b), it remains a challenge to gain insight into this cellular-structural phenomenon by perturbations of molecules whose effects can be indirect.

Unfortunately there are few detailed studies of the cell biology of synapse elimination in the CNS. The degree to which synaptic segregation, postsynaptic receptor density losses, or axon retraction reactions occur in the CNS remains speculative. It is, however, important to realize that the retraction bulbs and thinning observed in motor axon branches that withdraw from muscle fibers during synapse elimination are similar to the changes seen in central axons when they are deprived of their targets (Bernstein and Lichtman, 1999). This similarity may mean that central axons are susceptible to the absence of maintaining signals from their targets and respond during development in the same way.

As most central neurons maintain connections with hundreds of axons or more, it is reasonable to ask whether synapse elimination (and the associated axon branch removal) plays a general role in the development of the CNS, or if it only occurs in a few specialized situations (e.g., the competition between the two eyes for space in the visual system or climbing fiber competition for Purkinje cells in the cerebellum). As previously mentioned, for technical reasons it is difficult to prove that a single axon is removing its complete complement of synapses from a postsynaptic cell, but it is possible that synapse elimination is commonplace in the CNS. Axon branch withdrawal from the vicinity of targets is found throughout the developing CNS just as in the developing peripheral nervous system. For example, in the developing corpus callosum large numbers of axons are present in developing animals that do not project there later (Elberger, 1994a,b; Innocenti, 1995), and branching by axons within the callosum itself is also reduced (Kadhim et al., 1993). Some of this exuberance is known to connect visual areas that do not remain connected later (Bressoud and Innocenti, 1999). Axon collateral elimination also removes large numbers of inappropriate projections into the pyramidal tract (Stanfield et al., 1982; O'Leary, 1987) and removes cerebrocerebellar projections (Tolbert and Panneton, 1984). Although some of these collaterals appear to have varicosities and thus are potentially presynaptic specializations (e.g., Elberger, 1994a), it remains unknown whether the removal of these projections is secondary to the kind of competitive synapse elimination previously described. There are, however, also examples of unambiguous synapse loss, of unknown cause, in the developing CNS. One particularly well-documented case concerns the stripping of synapses from the somas of motoneurons during early postnatal life (Conradi and Ronnevi, 1977; Ronnevi, 1979; Gibson and Clowry, 1999). Whether this relates to some kind of synaptic segregation or full-blown input elimination is unclear. In sum, there is a good possibility that synapse elimination and axon retraction—as occurs in the visual system, cerebellum, autonomic ganglia, and muscle—is a general feature of CNS development. Newer labeling techniques using green fluorescent protein (GFP) or lipophilic dyes may help to resolve this issue.

Synapse Maintenance

Essentially nothing is known about the lifetime of a typical central synapse in vivo—is it measured in hours, days, or years? This question is of more than passing interest, as it will reveal the degree to which the developmental principles of synapse formation must continue to operate in the adult. The recent evidence that neurons may be generated in the adult brain and integrated into the relevant connectivity patterns means that synapse formation must continue into adulthood. If synapses have a limited lifetime compared with that of the organism, then synapse turnover may be commonplace in adult animals. One clear instance of central synapse turnover is the natural cycle of excitatory synapse and dendritic spine formation and elimination that occurs in adult rat hippocampal CA1 pyramidal cells across the estrous cycle (see Woolley, 1999a,b, for review). Spine density fluctuates by 22–45% over the course of a few days. Similar transient increases in spine density can be induced by estradiol and progesterone in ovariectomized animals. The new spines are selectively coupled to multiple synapse boutons, indicating formation of new postsynaptic elements on existing presynaptic elements. In one of the first studies to visualize central synapse turnover directly, at least in cell culture, Okabe et al. (1999) observed clusters of the PSD-95 tagged with GFP in individual cultured hippocampal neurons for periods of up to 1 day. Most of the PSD-95 clusters were stable, but more than 20% of the original number were newly formed and about 20% were eliminated within 1 day. Considering these clusters as markers of glutamatergic postsynaptic specializations (an assumption that is supported by appropriate antibody labeling), then it would seem that synapse turnover does occur, at least in this developing culture system. Activity regulation of dendritic spines, as described previously, further supports the possibility of central synapse turnover within a time frame of hours.

At the NMJ accessibility of individual muscle fibers in living animals has allowed study of junctions over long periods. In certain muscles, individual motor nerve terminals monitored over months were maintained with a recognizable branching pattern for most of the adult life of a mouse with only minor changes (Lichtman et al., 1987; Balice-Gordon and Lichtman, 1990; Balice-Gordon et al., 1990; Wigston, 1990). In other mammalian and in amphibian muscles, there is more evidence for remodeling over time (Wigston, 1989; Herrera et al., 1990; Ko and Chen, 1996). The ability of axons to remain connected to the same postsynaptic receptor-rich site for very long periods suggests that there is no obligatory requirement for synapses to undergo turnover. Of course the molecules in synapses are undergoing constant turnover. For example, the AChRs at the NMJ normally reside in the membrane for only about 2 weeks (Akaaboune et al., 1999), and it is possible that other synaptic molecules are tuning over even faster.

In the PNS, synapse stability has also been studied at neuronal-neuronal synapses in autonomic ganglia. Purves and his colleagues used fluorescent dye injections into sympathetic ganglion cells of the superior cervical ganglion in young adult mice to reveal that the dendritic tree of these cells was undergoing continual remodeling (Purves et al., 1986), which suggests that synapses made on these dendrites were transient. These workers also used vital dyes to monitor presynaptic terminals of preganglionic axons. When the same cells were viewed before and several weeks after reinnervation, the location of synapses was not the same (Purves and Lichtman, 1987), suggesting that—unlike the NMJ, where axons faithfully reinnervate the same postsynaptic sites (Rich and Lichtman, 1989)—the sites of synaptic contact are less stable on autonomic neurons. A direct test of the stability of neuronal synapses was accomplished using a mitochondrial membrane label that highlights the terminal synaptic boutons on submandibular ganglion cells in the mouse (Purves et al., 1987). As these cells have few dendrites (Snider, 1987), axosomatic synapses are the most numerous. Views of the same ganglion cells over 1–3 weeks showed that the distribution of the synapses on the surface of these cells changed gradually, suggesting ongoing remodeling. The advent of mice that express GFP and its variants in axons has provided an opportunity to study the details of synaptic plasticity with greater resolution. These recent experiments show evidence of rapid (i.e., over minutes) growth and retraction of fine processes from preganglionic axon terminals, and at the same time a surprising stability of the general location of synapses on particular regions of the postsynaptic cell soma, even over many months (Gan et al., in preparation). Thus it seems likely that some synapses between neurons can be stably maintained for long periods.

Conclusion

One of the foremost challenges in neuroscience in the coming years will be to find ways to relate molecular and cellular aspects of neuronal development. Synapse development is likely to be at the forefront of this effort as both cellular and molecular information is in ample supply. Progress toward this unification will be hastened by an improved sense of the cellular behavior of synapses, which will provide answers to questions regarding their life span, whether ongoing signaling is required for synapse maintenance, and, if so, whether the mechanisms are the same as those that caused synapses to form in the first place. Determining the degree of plasticity and regulation of the structure and composition of synapses over different time courses has become a priority. These questions will require better techniques for monitoring synapses over time and better ways of relating light microscopic, ultrastructural, and molecular phenomena. Finally, the molecular analysis of synapse development will need to accommodate better the dynamic nature of molecular associations and the multitude of intersecting pathways operating in parallel.

References

Ahmari, S. E., Buchanan, J., Smith, S. J. (2000). Assembly of presynaptic active zones from cytoplasmic transport packets. *Nature Neurosci.* 3:445–451.

Akaaboune, M., Culican, S. M., Turney, S. G., Lichtman, J. W. (1999). Rapid and reversible effects of activity on acetylcholine receptor density at the neuromuscular junction in vivo. *Science* 286:503–507.

Anderson, M. J., Shi, Z. Q., Zackson, S. L. (1997). Nerve-induced disruption and reformation of beta1-integrin aggregates during development of the neuromuscular junction. *Mech. Dev.* 67:125–139.

Andrew, M., Owen, M. J. (1997). Hyperekplexia: abnormal startle response due to glycine receptor mutations. *Br. J. Psychiatry* 170:106–108.

Antonov, I., Chang, S., Zakharenko, S., Popov, S. V. (1999). Distribution of neurotransmitter secretion in growing axons. *Neuroscience* 90:975–984.

Apperson, M. L., Moon, I. S., Kennedy, M. B. (1996). Characterization of densin-180, a new brain-specific synaptic protein of the O-sialoglycoprotein family. *J. Neurosci.* 16:6839–6852.

Araque, A., Parpura, V., Sanzgiri, R. P., Haydon, P. G. (1999). Tripartite synapses: glia, the unacknowledged partner. *Trends Neurosci.* 22:208–215.

Balice-Gordon, R. J., Lichtman, J. W. (1990). In vivo visualization of the growth of pre- and postsynaptic elements of neuromuscular junctions in the mouse. *J. Neurosci.* 10:894–908.

Balice-Gordon, R. J., Lichtman, J. W. (1993). In vivo observations of pre- and postsynaptic changes during the transition from multiple to single innervation at developing neuromuscular junctions. *J. Neurosci.* 13:834–855.

Balice-Gordon, R. J., Lichtman, J. W. (1994). Long-term synapse loss induced by focal blockade of postsynaptic receptors. *Nature* 372:519–524.

Balice-Gordon, R. J., Breedlove, S. M., Bernstein, S., Lichtman, J. W. (1990). Neuromuscular junctions shrink and expand as muscle fiber size is manipulated: in vivo observations in the androgen-sensitive bulbocavernosus muscle of mice. *J. Neurosci.* 10:2660–2671.

Banker, G. A., Cowan, W. M. (1977). Rat hippocampal neurons in dispersed cell culture. *Brain Res.* 126:397–342.

Benson, D. L., Tanaka, H. (1998). N-cadherin redistribution during synaptogenesis in hippocampal neurons. *J. Neurosci.* 18:6892–6904.

Bergles, D. E., Dzubay, J. A., Jahr, C. E. (1997). Glutamate transporter currents in Bergmann glial cells follow the time course of extrasynaptic glutamate. *Proc. Natl. Acad. Sci. USA* 94:14,821–14,825.

Bernstein, M., Lichtman, J. W. (1999). Axonal atrophy: the retraction reaction. *Curr. Opin. Neurobiol.* 9:364–370.

Betz, H., Kuhse, J., Schmieden, V., Laube, B., Kirsch, J., Harvey, R. J. (1999). Structure and functions of inhibitory and excitatory glycine receptors. *Ann. N.Y. Acad. Sci.* 868:667–676.

Blue, M. E., Parnavelas, J. G. (1983a). The formation and maturation of synapses in the visual cortex of the rat. I. Qualitative analysis. *J. Neurocytol.* 12:599–616.

Blue, M. E., Parnavelas, J. G. (1983b). The formation and maturation of synapses in the visual cortex of the rat. II. Quantitative analysis. *J. Neurocytol.* 12:697–712.

Boehm, S., Harvey, R. J., von Holst, A., Rohrer, H., Betz, H. (1997). Glycine receptors in cultured chick sympathetic neurons are excitatory and trigger neurotransmitter release. *J. Physiol.* 504:683–694.

Boudier, J. L., Le Treut, T., Jover, E. (1992). Autoradiographic localization of voltage-dependent sodium channels on the mouse neuromuscular junction using ^{125}I-alpha scorpion toxin. II. Sodium distribution on postsynaptic membranes. *J. Neurosci.* 12:454–466.

Bravin, M., Morando, L., Vercelli, A., Rossi, F., Strata, P. (1999). Control of spine formation by electrical activity in the adult rat cerebellum. *Proc. Natl. Acad. Sci. USA* 96:1704–1709.

Brenman, J. E., Chao, D. S., Gee, S. H., McGee, A. W., Craven, S. E., Santillano, D. R., Wu, Z., Huang, F., Xia, H., Peters, M. F., Froehner, S. C., Bredt, D. S. (1996). Interaction of nitric oxide synthase with the postsynaptic density protein PSD-95 and alpha1-syntrophin mediated by PDZ domains. *Cell* 84: 757–767.

Bressoud, R., Innocenti, G. M. (1999). Typology, early differentiation, and exuberant growth of a set of cortical axons. *J. Comp. Neurol.* 406:87–108.

Brune, W., Weber, R. G., Saul, B., von Knebel Doeberitz, M., Grond-Ginsbach, C., Kellerman, K., Meinck, H. M., Becker, C. M. (1996). A GLRA1 null mutation in recessive hyperekplexia challenges the functional role of glycine receptors. *Am. J. Hum. Genet.* 58:989–997.

Buchert, M., Schneider, S., Meskenaite, V., Adams, M. T., Canaani, E., Baechi, T., Moelling, K., Hovens, C. M. (1999). The junction-associated protein AF-6 interacts and clusters with specific Eph receptor tyrosine kinases at specialized sites of cell-cell contact in the brain. *J. Cell Biol.* 144:361–371.

Buckwalter, M. S., Cook, S. A., Davisson, M. T., White, W. F., Camper, S. A. (1994). A frameshift mutation in the mouse alpha 1 glycine receptor gene (Glra1) results in progressive neurological symptoms and juvenile death. *Hum. Mol. Genet.* 3:2025–2030.

Burns, M. E., Augustine, G. J. (1995). Synaptic structure and function: dynamic organization yields architectural precision. *Cell* 83:187–194.

Butz, S., Okamoto, M., Südhof, T. C. (1998). A tripartite protein complex with the potential to couple synaptic vesicle exocytosis to cell adhesion in brain. *Cell* 94:773–782.

Caroni, P., Rotzler, S., Britt, J. C., Brenner, H. R. (1993). Calcium influx and protein phosphorylation mediate the metabolic stabilization of synaptic acetylcholine receptors in muscle. *J. Neurosci.* 13:1315–1325.

Carroll, R. C., Lissin, D. V., von Zastrow, M., Nicoll, R. A., Malenka, R. C. (1999). Rapid redistribution of glutamate receptors contributes to long-term depression in hippocampal cultures. *Nature Neurosci.* 2:454–460.

Chen, C., Kano, M., Abeliovich, A., Chen, L., Bao, S., Kim, J. J., Hashimoto, K., Thompson, R. F., Tonegawa, S. (1995). Impaired motor coordination correlates with persistent multiple climbing fiber innervation in PKC gamma mutant mice. *Cell* 83:1233–1242.

Chen, H. J., Rojas-Soto, M., Oguni, A., Kennedy, M. B. (1998). A synaptic Ras-GTPase activating protein (p135 SynGAP) inhibited by CaM kinase II. *Neuron* 20:895–904.

Cherubini, E., Gaiarsa, J. L., Ben-Ari, Y. (1991). GABA: an excitatory transmitter in early postnatal life. *Trends Neurosci.* 14:515–519.

Coco, S., Verderio, C., De Camilli, P., Matteoli, M. (1998). Calcium dependence of synaptic vesicle recycling before and after synaptogenesis. *J. Neurochem.* 71:1987–1992.

Colin, I., Rostaing, P., Triller, A. (1996). Gephyrin accumulates at specific plasmalemma loci during neuronal maturation in vitro. *J. Comp. Neurol.* 374: 467–479.

Colin, I., Rostaing, P., Augustin, A., Triller, A. (1998). Localization of components of glycinergic synapses during rat spinal cord development. *J. Comp. Neurol.* 398:359–372.

Colledge, M., Froehner, S. C. (1998). Signals mediating ion channel clustering at the neuromuscular junction. *Curr. Opin. Neurobiol.* 8:357–363.

Conradi, S., Ronnevi, L. O. (1977). Ultrastructure and synaptology of the initial axon segment of cat spinal motoneurons during early postnatal development. *J. Neurocytol.* 6:195–210.

Constantine-Paton, M., Cline, H. T. (1998). LTP and activity-dependent synaptogenesis: the more alike they are, the more different they become. *Curr. Opin. Neurobiol.* 8:139–148.

Cotman, C., Taylor, D., Lynch, G. (1973). Ultrastructural changes in synapses in the dentate gyrus of the rat during development. *Brain Res.* 63:205–213.

Craig, A. M. (1998). Activity and synaptic receptor targeting: the long view. *Neuron* 21:459–462.

Craig, A. M., Blackstone, C. D., Huganir, R. L., Banker, G. (1993). The distribution of glutamate receptors in cultured rat hippocampal neurons: postsynaptic clustering of AMPA-selective subunits. *Neuron* 10:1055–1068.

Craig, A. M., Blackstone, C. D., Huganir, R. L., Banker, G. (1994). Selective clustering of glutamate and gamma-aminobutyric acid receptors opposite terminals releasing the corresponding neurotransmitters. *Proc. Natl. Acad. Sci. USA* 91:12,373–12,377.

Crepel, F., Delhaye-Bouchaud, N., Guastavino, J. M., Sampaio, I. (1980). Multiple innervation of cerebellar Purkinje cells by climbing fibres in staggerer mutant mouse. *Nature* 283:483–484.

Culican, S. M., Nelson, C. C., Lichtman, J. W. (1998). Axon withdrawal during synapse elimination at the neuromuscular junction is accompanied by disassembly of the postsynaptic specialization and withdrawal of Schwann cell processes. *J. Neurosci.* 18:4953–4965.

Dahm, L. M., Landmesser, L. T. (1991). The regulation of synaptogenesis during normal development and following activity blockade. *J. Neurosci.* 11:238–255.

Deconinck, A. E., Potter, A. C., Tinsley, J. M., Wood, S. J., Vater, R., Young, C., Metzinger, L., Vincent, A., Slater, C. R., Davies, K. E. (1997). Postsynaptic abnormalities at the neuromuscular junctions of utrophin-deficient mice. *J. Cell Biol.* 136:883–894.

Delhaye-Bouchaud, N., Geoffroy, B., Mariani, J. (1985). Neuronal death and synapse elimination in the olivocerebellar system. I. Cell counts in the inferior olive of developing rats. *J. Comp. Neurol.* 232:299–308.

Denk, W., Yuste, R., Svoboda, K., Tank, D. W. (1996). Imaging calcium dynamics in dendritic spines. *Curr. Opin. Neurobiol.* 6:372–378.

De Robertis, E. (1967). Ultrastructure and cytochemistry of the synaptic region: the macromolecular components involved in nerve transmission are being studied. *Science* 156:907–914.

Dunaevsky, A., Tashiro, A., Majewska, A., Mason, C., Yuste, R. (1999). Developmental regulation of spine motility in the mammalian central nervous system. *Proc. Natl. Acad. Sci. USA* 96:13,438–13,443.

Durand, G. M., Kovalchuk, Y., Konnerth, A. (1996). Long-term potentiation and functional synapse induction in developing hippocampus. *Nature* 381:71–75.

Einheber, S., Schnapp, L. M., Salzer, J. L., Cappiello, Z. B., Milner, T. A. (1996). Regional and ultrastructural distribution of the alpha 8 integrin subunit in developing and adult rat brain suggests a role in synaptic function. *J. Comp. Neurol.* 370:105–134.

Elberger, A. J. (1994a). Transitory corpus callosum axons projecting throughout developing rat visual cortex revealed by DiI. *Cerebral Cortex* 4:279–299.

Elberger, A. J. (1994b). The corpus callosum provides a massive transitory input to the visual cortex of cat and rat during early postnatal development. *Behav. Brain Res.* 64:15–33.

Engert, F., Bonhoeffer, T. (1999). Dendritic spine changes associated with hippocampal long-term synaptic plasticity. *Nature* 399:66–70.

Essrich, C., Lorez, M., Benson, J. A., Fritschy, J. M., Lüscher, B. (1998). Postsynaptic clustering of major $GABA_A$ receptor subtypes requires the gamma 2 subunit and gephyrin. *Nature Neurosci.* 1:563–571.

Ethell, I. M., Yamaguchi, Y. (1999). Cell surface heparan sulfate proteoglycan syndecan-2 induces the maturation of dendritic spines in rat hippocampal neurons. *J. Cell Biol.* 144:575–586.

Evers, J., Laser, M., Sun, Y. A., Xie, Z. P., Poo, M. M. (1989). Studies of nerve-muscle interactions in *Xenopus* cell culture: analysis of early synaptic currents. *J. Neurosci.* 9:1523–1539.

Fannon, A. M., Colman, D. R. (1996). A model for central synaptic junctional complex formation based on the differential adhesive specificities of the cadherins. *Neuron* 17:423–434.

Feldmeyer, D., Cull-Candy, S. (1996). Functional consequences of changes in NMDA receptor subunit expression during development. *J. Neurocytol.* 25:857–867.

Feng, G., Steinbach, J. H., Sanes, J. R. (1998a). Rapsyn clusters neuronal acetylcholine receptors but is inessential for formation of an interneuronal cholinergic synapse. *J. Neurosci.* 18:4166–4176.

Feng, G., Tintrup, H., Kirsch, J., Nichol, M. C., Kuhse, J., Betz, H., Sanes, J. R. (1998b). Dual requirement for gephyrin in glycine receptor clustering and molybdoenzyme activity. *Science* 282:1321–1324.

Ferreira, A. (1999). Abnormal synapse formation in agrin-depleted hippocampal neurons. *J. Cell Sci.* 112:4729–4738.

Fiala, J. C., Feinberg, M., Popov, V., Harris, K. M. (1998). Synaptogenesis via dendritic filopodia in developing hippocampal area CA1. *J. Neurosci.* 18:8900–8911.

Fischer, M., Kaech, S., Knutti, D., Matus, A. (1998). Rapid actin-based plasticity in dendritic spines. *Neuron* 20:847–854.

Fletcher, T. L., De Camilli, P., Banker, G. (1994). Synaptogenesis in hippocampal cultures: evidence indicating that axons and dendrites become competent to form synapses at different stages of neuronal development. *J. Neurosci.* 14:6695–6706.

Flucher, B. E., Daniels, M. P. (1989). Distribution of Na^+ channels and ankyrin in neuromuscular junctions is complementary to that of acetylcholine receptors and the 43 kd protein. *Neuron* 3:163–175.

Frerking, M., Wilson, M. (1996). Saturation of postsynaptic receptors at central synapses? *Curr. Opin. Neurobiol.* 6:395–403.

Fritschy, J. M., Weinmann, O., Wenzel, A., Benke, D. (1998). Synapse-specific localization of NMDA and GABA(A) receptor subunits revealed by antigen-retrieval immunohistochemistry. *J. Comp. Neurol.* 390:194–210.

Gan, W. B., Lichtman, J. W. (1998). Synaptic segregation at the developing neuromuscular junction. *Science* 282:1508–1511.

Gautam, M., Noakes, P. G., Mudd, J., Nichol, M., Chu, G. C., Sanes, J. R., Merlie, J. P. (1995). Failure of postsynaptic specialization to develop at neuromuscular junctions of rapsyn-deficient mice. *Nature* 377:232–236.

Gautam, M., Noakes, P. G., Moscoso, L., Rupp, F., Scheller, R. H., Merlie, J. P., Sanes, J. R. (1996). Defective neuromuscular synaptogenesis in agrin-deficient mutant mice. *Cell* 85:525–535.

Gibson, C. L., Clowry, G. J. (1999). Retraction of muscle afferents from the rat ventral horn during development. *NeuroReport* 10:231–235.

Goslin, K., Asmussen, H., Banker, G. (1998). Hippocampal neurons in low density culture. In *Culturing Nerve Cells*, Banker, G., Goslin, K., eds. Cambridge, Mass.: MIT Press. pp. 339–370.

Gottmann, K., Mehrle, A., Gisselmann, G., Hatt, H. (1997). Presynaptic control of subunit composition of NMDA receptors mediating synaptic plasticity. *J. Neurosci.* 17:2766–2774.

Grady, R. M., Merlie, J. P., Sanes, J. R. (1997). Subtle neuromuscular defects in utrophin-deficient mice. *J. Cell Biol.* 136:871–882.

Gray, E. G. (1959). Axo-somatic and axo-dendritic synapses of the cerebral cortex: an electron microscopic study. *J. Anat.* 93:420–433.

Hall, Z. W., Sanes, J. R. (1993). Synaptic structure and development: the neuromuscular junction. *Cell* 72:99–121.

Hamori, J., Somogyi, J. (1983). Differentiation of cerebellar mossy fiber synapses in the rat: a quantitative electron microscope study. *J. Comp. Neurol.* 220: 365–377.

Harris, K. M. (1999). Structure, development, and plasticity of dendritic spines. *Curr. Opin. Neurobiol.* 9:343–348.

Harris, K. M., Kater, S. B. (1994). Dendritic spines: cellular specializations imparting both stability and flexibility to synaptic function. *Annu. Rev. Neurosci.* 17:341–371.

Harris, K. M., Jensen, F. E., Tsao, B. (1992). Three-dimensional structure of dendritic spines and synapses in rat hippocampus (CA1) at postnatal day 15 and adult ages: implications for the maturation of synaptic physiology and long-term potentiation. *J. Neurosci.* 12:2685–2705. [Erratum appears in *J. Neurosci.* 12(8) (1992) following table of contents.]

Hata, Y., Nakanishi, H., Takai, Y. (1998). Synaptic PDZ domain–containing proteins. *Neurosci. Res.* 32:1–7.

Hayashi, K., Shirao, T. (1999). Change in the shape of dendritic spines caused by overexpression of drebrin in cultured cortical neurons. *J. Neurosci.* 19: 3918–3925.

Hazuka, C. D., Foletti, D. L., Hsu, S. C., Kee, Y., Hopf, F. W., Scheller, R. H. (1999). The sec6/8 complex is located at neurite outgrowth and axonal synapse-assembly domains. *J. Neurosci.* 19:1324–1334.

Herrera, A. A., Banner, L. R., Nagaya, N. (1990). Repeated in vivo observation of frog neuromuscular junctions: remodelling involves concurrent growth and retraction. *J. Neurocytol.* 19:85–99.

Hinds, J. W., Hinds, P. L. (1976). Synapse formation in the mouse olfactory bulb. II. Morphogenesis. *J. Comp. Neurol.* 169:41–61.

Hsueh, Y. P., Sheng, M. (1999). Regulated expression and subcellular localization of syndecan heparan sulfate proteoglycans and the syndecan-binding protein CASK/LIN-2 during rat brain development. *J. Neurosci.* 19:7415–7425.

Huang, C. F., Schmidt, J. (1994). Calcium influx blocks the skeletal muscle acetylcholine receptor alpha-subunit gene *in vivo*. *FEBS Lett.* 338:277–280.

Hubel, D. H., Wiesel, T. N., LeVay, S. (1977). Plasticity of ocular dominance columns in monkey striate cortex. *Philos. Trans. R. Soc. London Ser. B* 278: 377–409.

Hume, R. I., Role, L. W., Fischbach, G. D. (1983). Acetylcholine release from growth cones detected with patches of acetylcholine receptor–rich membranes. *Nature* 305:632–634.

Innocenti, G. M. (1995). Exuberant development of connections, and its possible permissive role in cortical evolution. *Trends Neurosci.* 18:397–402.

Irie, M., Hata, Y., Takeuchi, M., Ichtchenko, K., Toyoda, A., Hirao, K., Takai, Y., Rosahl, T. W., Südhof, T. C. (1997). Binding of neuroligins to PSD-95. *Science* 277:1511–1515.

Isaac, J. T., Crair, M. C., Nicoll, R. A., Malenka, R. C. (1997). Silent synapses during development of thalamocortical inputs. *Neuron* 18:269–280.

Jackson, H., Parks, T. N. (1982). Functional synapse elimination in the developing avian cochlear nucleus with simultaneous reduction in cochlear nerve axon branching. *J. Neurosci.* 2:1736–1743.

Jones, D. G. (1983). Recent perspectives on the organization of central synapses. *Anesth. Analg.* 62:1100–1112.

Jones, G., Herczeg, A., Ruegg, M. A., Lichtsteiner, M., Kroger, S., Brenner, H. R. (1996). Substrate-bound agrin induces expression of acetylcholine receptor epsilon-subunit gene in cultured mammalian muscle cells. *Proc. Natl. Acad. Sci. USA* 93:5985–5990.

Kadhim, H. J., Bhide, P. G., Frost, D. O. (1993). Transient axonal branching in the developing corpus callosum. *Cerebral Cortex* 3:551–566.

Kaech, S., Brinkhaus, H., Matus, A. (1999). Volatile anesthetics block actin-based motility in dendritic spines. *Proc. Natl. Acad. USA* 96:10,433–10,437.

Kano, M., Hashimoto, K., Chen, C., Abeliovich, A., Aiba, A., Kurihara, H., Watanabe, M., Inoue, Y., Tonegawa, S. (1995). Impaired synapse elimination during cerebellar development in PKC gamma mutant mice. *Cell* 83:1223–1231.

Kano, M., Hashimoto, K., Watanabe, M., Kurihara, H., Offermanns, S., Jiang, H., Wu, Y., Jun, K., Shin, H. S., Inoue, Y., Simon, M. I., Wu, D. (1998). Phospholipase C beta4 is specifically involved in climbing fiber synapse elimination in the developing cerebellum. *Proc. Natl. Acad. Sci. USA* 95:15,724–15,729.

Katz, B., Miledi, R. (1972). The statistical nature of the acetylcholine potential and its molecular components. *J. Physiol.* 224:665–699.

Kim, J. H., Liao, D., Lau, L. F., Huganir, R. L. (1998). SynGAP: a synaptic Ras-GAP that associates with the PSD-95/SAP90 protein family. *Neuron* 20:683–691.

Kirov, S. A., Harris, K. M. (1999). Dendrites are more spiny on mature hippocampal neurons when synapses are inactivated. *Nat. Neurosci.* 2:878–883.

Kirov, S. A., Sorra, K. E., Harris, K. M. (1999). Slices have more synapses than perfusion-fixed hippocampus from both young and mature rats. *J. Neurosci.* 19:2876–2886.

Kirsch, J., Betz, H. (1998). Glycine-receptor activation is required for receptor clustering in spinal neurons. *Nature* 392:717–720.

Kirsch, J., Wolters, I., Triller, A., Betz, H. (1993). Gephyrin antisense oligonucleotides prevent glycine receptor clustering in spinal neurons. *Nature* 366:745–748.

Kneussel, M., Brandstatter, J. H., Laube, B., Stahl, S., Muller, U., Betz, H. (1999). Loss of postsynaptic GABA(A) receptor clustering in gephyrin-deficient mice. *J. Neurosci.* 19:9289–9297.

Ko, C. P., Chen, L. (1996). Synaptic remodeling revealed by repeated *in vivo* observations and electron microscopy of identified frog neuromuscular junctions. *J. Neurosci.* 16:1780–1790.

Kohmura, N., Senzaki, K., Hamada, S., Kai, N., Yasuda, R., Watanabe, M., Ishii, H., Yasuda, M., Mishina, M., Yagi, T. (1998). Diversity revealed by a novel family of cadherins expressed in neurons at a synaptic complex. *Neuron* 20:1137–1151.

Korkotian, E., Segal, M. (1999). Bidirectional regulation of dendritic spine dimensions by glutamate receptors. *NeuroReport* 10:2875–2877.

Koulen, P., Sassoe-Pognetto, M., Grunert, U., Wassle, H. (1996). Selective clustering of GABA(A) and glycine receptors in the mammalian retina. *J. Neurosci.* 16:2127–2140.

Kraszewski, K., Mundigl, O., Daniell, L., Verderio, C., Matteoli, M., De Camilli, P. (1995). Synaptic vesicle dynamics in living cultured hippocampal neurons visualized with CY3-conjugated antibodies directed against the lumenal domain of synaptotagmin. *J. Neurosci.* 15:4328–4342.

Kuffler, S. W., Yoshikami, D. (1975). The distribution of acetylcholine sensitivity at the post-synaptic membrane of vertebrate skeletal twitch muscles: iontophoretic mapping in the micron range. *J. Physiol.* 244:703–730.

Land, B. R., Salpeter, E. E., Salpeter, M. M. (1980). Acetylcholine receptor site density affects the rising phase of miniature endplate currents. *Proc. Natl. Acad. Sci. USA* 77:3736–3740.

Landsend, A. S., Amiry-Moghaddam, M., Matsubara, A., Bergersen, L., Usami, S., Wenthold, R. J., Ottersen, O. P. (1997). Differential localization of delta glutamate receptors in the rat cerebellum: coexpression with AMPA receptors in parallel fiber–spine synapses and absence from climbing fiber–spine synapses. *J. Neurosci.* 17:834–842.

Leinekugel, X., Medina, I., Khalilov, I., Ben-Ari, Y., Khazipov, R. (1997). Ca^{2+} oscillations mediated by the synergistic excitatory actions of GABA(A) and NMDA receptors in the neonatal hippocampus. *Neuron* 18:243–255.

Le Treut, T., Boudier, J. L., Jover, E., Cau, P. (1990). Localization of voltage-sensitive sodium channels on the extrasynaptic membrane surface of mouse skeletal muscle by autoradiography of scorpion toxin binding sites. *J. Neurocytol.* 19:408–420.

Levi, S., Vannier, C., Triller, A. (1998). Strychnine-sensitive stabilization of postsynaptic glycine receptor clusters. *J. Cell Sci.* 111:335–345.

Levi, S., Chesnoy-Marchais, D., Sieghart, W., Triller, A. (1999). Synaptic control of glycine and GABA(A) receptors and gephyrin expression in cultured motoneurons. *J. Neurosci.* 19:7434–7449.

Li, J. H., Wang, Y. H., Wolfe, B. B., Krueger, K. E., Corsi, L., Stocca, G., Vicini, S. (1998). Developmental changes in localization of NMDA receptor subunits in primary cultures of cortical neurons. *Eur. J. Neurosci.* 10:1704–1715.

Li, P., Kerchner, G. A., Sala, C., Wei, F., Huettner, J. E., Sheng, M., Zhuo, M. (1999). AMPA receptor–PDZ interactions in facilitation of spinal sensory synapses. *Nature Neurosci.* 2:972–977.

Li, Z., Hilgenberg, L. G., O'Dowd, D. K., Smith, M. A. (1999). Formation of functional synaptic connections between cultured cortical neurons from agrin-deficient mice. *J. Neurobiol.* 39:547–557.

Liao, D., Malinow, R. (1996). Deficiency in induction but not expression of LTP in hippocampal slices from young rats. *Learn. Mem.* 3:138–149.

Liao, D., Zhang, X., O'Brien, R., Ehlers, M. D., Huganir, R. L. (1999). Regulation of morphological postsynaptic silent synapses in developing hippocampal neurons. *Nature Neurosci.* 2:37–43.

Lichtman, J. W. (1977). The reorganization of synaptic connexions in the rat submandibular ganglion during post-natal development. *J. Physiol.* 273:155–177.

Lichtman, J. W., Colman, H. (2000). Synapse elimination and indelible memory. *Neuron* 25:269–278.

Lichtman, J. W., Purves, D. (1980). The elimination of redundant preganglionic innervation to hamster sympathetic ganglion cells in early post-natal life. *J. Physiol.* 301:213–228.

Lichtman, J. W., Magrassi, L., Purves, D. (1987). Visualization of neuromuscular junctions over periods of several months in living mice. *J. Neurosci.* 7:1215–1222.

Lim, S., Naisbitt, S., Yoon, J., Hwang, J. I., Suh, P. G., Sheng, M., Kim, E. (1999). Characterization of the Shank family of synaptic proteins: multiple genes, alternative splicing, and differential expression in brain and development. *J. Biol. Chem.* 274:29,510–29,518.

Linke, R., Soriano, E., Frotscher, M. (1994). Transient dendritic appendages on differentiating septohippocampal neurons are not the sites of synaptogenesis. *Brain Res. Dev. Brain Res.* 83:67–78.

Liu, G., Choi, S., Tsien, R. W. (1999). Variability of neurotransmitter concentration and nonsaturation of postsynaptic AMPA receptors at synapses in hippocampal cultures and slices. *Neuron* 22:395–409.

Liu, Y., Fields, R. D., Festoff, B. W., Nelson, P. G. (1994). Proteolytic action of thrombin is required for electrical activity–dependent synapse reduction. *Proc. Natl. Acad. Sci. USA* 91:10,300–10,304.

Lohof, A. M., Delhaye-Bouchaud, N., Mariani, J. (1996). Synapse elimination in the central nervous system: functional significance and cellular mechanisms. *Rev. Neurosci.* 7:85–101.

Ma, L., Zablow, L., Kandel, E. R., Siegelbaum, S. A. (1999). Cyclic AMP induces functional presynaptic boutons in hippocampal CA3-CA1 neuronal cultures. *Nature Neurosci.* 2:24–30.

McKinney, R. A., Capogna, M., Durr, R., Gahwiler, B. H., Thompson, S. M. (1999). Miniature synaptic events maintain dendritic spines via AMPA receptor activation. *Nature Neurosci.* 2:44–49.

Mainen, Z. F., Malinow, R., Svoboda, K. (1999). Synaptic calcium transients in single spines indicate that NMDA receptors are not saturated. *Nature* 399:151–155.

Maletic-Savatic, M., Malinow, R., Svoboda, K. (1999). Rapid dendritic morphogenesis in CA1 hippocampal dendrites induced by synaptic activity. *Science* 283:1923–1927.

Mariani, J., Changeux, J. P. (1980). Multiple innervation of Purkinje cells by climbing fibers in the cerebellum of the adult staggerer mutant mouse. *J. Neurobiol.* 11:41–50.

Marques, M. J., Conchello, J.-A., Lichtman, J. W. (2000). From plaque to pretzel: fold formation and acetylcholine receptor loss at the developing neuromuscular junction. *J. Neurosci.* 20:3663–3675.

Martin, P. T., Kaufman, S. J., Kramer, R. H., Sanes, J. R. (1996). Synaptic integrins in developing, adult, and mutant muscle: selective association of alpha1, alpha7A, and alpha7B integrins with the neuromuscular junction. *Dev. Biol.* 174:125–139.

Matthews-Bellinger, J., Salpeter, M. M. (1978). Distribution of acetylcholine receptors at frog neuromuscular junctions with a discussion of some physiological implications. *J. Physiol.* 279:197–213.

Meier, T., Wallace, B. G. (1998). Formation of the neuromuscular junction: molecules and mechanisms. *BioEssays* 20:819–829.

Meier, T., Marangi, P. A., Moll, J., Hauser, D. M., Brenner, H. R., Ruegg, M. A. (1998a). A minigene of neural agrin encoding the laminin-binding and acetylcholine receptor–aggregating domains is sufficient to induce postsynaptic differentiation in muscle fibres. *Eur. J. Neurosci.* 10:3141–3152.

Meier, T., Masciulli, F., Moore, C., Schoumacher, F., Eppenberger, U., Denzer, A. J., Jones, G., Brenner, H. R. (1998b). Agrin can mediate acetylcholine receptor gene expression in muscle by aggregation of muscle-derived neuregulins. *J. Cell Biol.* 141:715–726.

Mennerick, S., Shen, W., Xu, W., Benz, A., Tanaka, K., Shimamoto, K., Isenberg, K. E., Krause, J. E., Zorumski, C. F. (1999). Substrate turnover by transporters curtails synaptic glutamate transients. *J. Neurosci.* 19:9242–9251.

Meyer, G., Kirsch, J., Betz, H., Langosch, D. (1995). Identification of a gephyrin binding motif on the glycine receptor beta subunit. *Neuron* 15:563–572.

Michalak, M., Opas, M. (1997). Functions of dystrophin and dystrophin associated proteins. *Curr. Opin. Neurol.* 10:436–442.

Migaud, M., Charlesworth, P., Dempster, M., Webster, L. C., Watabe, A. M., Makhinson, M., He, Y., Ramsay, M. F., Morris, R. G., Morrison, J. H., O'Dell, T. J., Grant, S. G. (1998). Enhanced long-term potentiation and impaired learning in mice with mutant postsynaptic density-95 protein. *Nature* 396:433–439.

Miller, M., Peters, A. (1981). Maturation of rat visual cortex. II. A combined Golgi–electron microscope study of pyramidal neurons. *J. Comp. Neurol.* 203:555–573.

Missias, A. C., Mudd, J., Cunningham, J. M., Steinbach, J. H., Merlie, J. P., Sanes, J. R. (1997). Deficient development and maintenance of postsynaptic specializations in mutant mice lacking an "adult" acetylcholine receptor subunit. *Development* 124:5075–5086.

Monyer, H., Burnashev, N., Laurie, D. J., Sakmann, B., Seeburg, P. H. (1994). Developmental and regional expression in the rat brain and functional properties of four NMDA receptors. *Neuron* 12:529–540.

Newman-Gage, H., Westrum, L. E., Bertram, J. F. (1987). Stereological analysis of synaptogenesis in the molecular layer of piriform cortex in the prenatal rat. *J. Comp. Neurol.* 261:295–305.

Nguyen, Q. T., Parsadanian, A. S., Snider, W. D., Lichtman, J. W. (1998). Hyperinnervation of neuromuscular junctions caused by GDNF overexpression in muscle. *Science* 279:1725–1729.

Nishimura, S. L., Boylen, K. P., Einheber, S., Milner, T. A., Ramos, D. M., Pytela, R. (1998). Synaptic and glial localization of the integrin alphavbeta8 in mouse and rat brain. *Brain Res.* 791:271–282.

Noel, J., Ralph, G. S., Pickard, L., Williams, J., Molnar, E., Uney, J. B., Collingridge, G. L., Henley, J. M. (1999). Surface expression of AMPA receptors in hippocampal neurons is regulated by an NSF-dependent mechanism. *Neuron* 23:365–376.

Nusser, Z., Roberts, J. D., Baude, A., Richards, J. G., Somogyi, P. (1995). Relative densities of synaptic and extrasynaptic GABA$_A$ receptors on cerebellar granule cells as determined by a quantitative immunogold method. *J. Neurosci.* 15:2948–2960.

Nusser, Z., Sieghart, W., Benke, D., Fritschy, J. M., Somogyi, P. (1996). Differential synaptic localization of two major gamma-aminobutyric acid type A receptor alpha subunits on hippocampal pyramidal cells. *Proc. Natl. Acad. Sci. USA* 93:11,939–11,944.

Nusser, Z., Cull-Candy, S., Farrant, M. (1997). Differences in synaptic GABA(A) receptor number underlie variation in GABA mini amplitude. *Neuron* 19:697–709.

Nusser, Z., Hajos, N., Somogyi, P., Mody, I. (1998a). Increased number of synaptic GABA(A) receptors underlies potentiation at hippocampal inhibitory synapses. *Nature* 395:172–177.

Nusser, Z., Lujan, R., Laube, G., Roberts, J. D., Molnar, E., Somogyi, P. (1998b). Cell type and pathway dependence of synaptic AMPA receptor number and variability in the hippocampus. *Neuron* 21:545–559.

Nusser, Z., Sieghart, W., Somogyi, P. (1998c). Segregation of different GABA$_A$ receptors to synaptic and extrasynaptic membranes of cerebellar granule cells. *J. Neurosci.* 18:1693–1703.

O'Brien, R. J., Mammen, A. L., Blackshaw, S., Ehlers, M. D., Rothstein, J. D., Huganir, R. L. (1997). The development of excitatory synapses in cultured spinal neurons. *J. Neurosci.* 17:7339–7350.

O'Brien, R. J., Xu, D., Petralia, R. S., Steward, O., Huganir, R. L., Worley, P. (1999). Synaptic clustering of AMPA receptors by the extracellular immediate-early gene product Narp. *Neuron* 23:309–323.

Okabe, S., Kim, H. D., Miwa, A., Kuriu, T., Okado, H. (1999). Continual remodeling of postsynaptic density and its regulation by synaptic activity. *Nature Neurosci.* 2:804–811.

O'Leary, D. D. (1987). Remodelling of early axonal projections through the selective elimination of neurons and long axon collaterals. *Ciba Found. Symp.* 126:113–142.

Papa, M., Bundman, M. C., Greenberger, V., Segal, M. (1995). Morphological analysis of dendritic spine development in primary cultures of hippocampal neurons. *J. Neurosci.* 15:1–11.

Patton, B. L., Miner, J. H., Chiu, A. Y., Sanes, J. R. (1997). Distribution and function of laminins in the neuromuscular system of developing, adult, and mutant mice. *J. Cell Biol.* 139:1507–1521.

Petralia, R. S., Esteban, J. A., Wang, Y. X., Partridge, J. G., Zhao, H. M., Wenthold, R. J., Malinow, R. (1999). Selective acquisition of AMPA receptors over postnatal development suggests a molecular basis for silent synapses. *Nature Neurosci.* 2:31–36.

Pfrieger, F. W., Barres, B. A. (1997). Synaptic efficacy enhanced by glial cells in vitro. *Science* 277:1684–1687.

Purves, D., Lichtman, J. W. (1987). Synaptic sites on reinnervated nerve cells visualized at two different times in living mice. *J. Neurosci.* 7:1492–1497.

Purves, D., Hadley, R. D., Voyvodic, J. T. (1986). Dynamic changes in the dendritic geometry of individual neurons visualized over periods of up to three months in the superior cervical ganglion of living mice. *J. Neurosci.* 6:1051–1060.

Purves, D., Voyvodic, J. T., Magrassi, L., Yawo, H. (1987). Nerve terminal remodeling visualized in living mice by repeated examination of the same neuron. *Science* 238:1122–1126.

Quinlan, E. M., Philpot, B. D., Huganir, R. L., Bear, M. F. (1999). Rapid, experience-dependent expression of synaptic NMDA receptors in visual cortex in vivo. *Nature Neurosci.* 2:352–357.

Rabacchi, S., Bailly, Y., Delhaye-Bouchaud, N., Mariani, J. (1992). Involvement of the *N*-methyl-D-aspartate (NMDA) receptor in synapse elimination during cerebellar development. *Science* 256:1823–1825.

Rao, A., Craig, A. M. (1997). Activity regulates the synaptic localization of the NMDA receptor in hippocampal neurons. *Neuron* 19:801–812.

Rao, A., Kim, E., Sheng, M., Craig, A. M. (1998). Heterogeneity in the molecular composition of excitatory postsynaptic sites during development of hippocampal neurons in culture. *J. Neurosci.* 18:1217–1229.

Redfern, P. A. (1970). Neuromuscular transmission in new-born rats. *J. Physiol.* 209:701–709.

Rees, R. P., Bunge, M. B., Bunge, R. P. (1976). Morphological changes in the neuritic growth cone and target neuron during synaptic junction development in culture. *J. Cell Biol.* 68:240–263.

Rich, M. M., Lichtman, J. W. (1989). In vivo visualization of pre- and post-synaptic changes during synapse elimination in reinnervated mouse muscle. *J. Neurosci.* 9:1781–1805.

Rivera, C., Voipio, J., Payne, J. A., Ruusuvuori, E., Lahtinen, H., Lamsa, K., Pirvola, U., Saarma, M., Kaila, K. (1999). The K^+/Cl^- co-transporter KCC2 renders GABA hyperpolarizing during neuronal maturation. *Nature* 397:251–255.

Robitaille, R. (1998). Modulation of synaptic efficacy and synaptic depression by glial cells at the frog neuromuscular junction. *Neuron* 21:847–855.

Ronnevi, L. O. (1979). Spontaneous phagocytosis of C-type synaptic terminals by spinal alpha-motoneurons in newborn kittens. An electron microscopic study. *Brain Res.* 162:189–199.

Rutherford, L. C., DeWan, A., Lauer, H. M., Turrigiano, G. G. (1997). Brain-derived neurotrophic factor mediates the activity-dependent regulation of inhibition in neocortical cultures. *J. Neurosci.* 17:4527–4535.

Rutherford, L. C., Nelson, S. B., Turrigiano, G. G. (1998). BDNF has opposite effects on the quantal amplitude of pyramidal neuron and interneuron excitatory synapses. *Neuron* 21:521–530.

Sakimura, K., Kutsuwada, T., Ito, I., Manabe, T., Takayama, C., Kushiya, E., Yagi, T., Aizawa, S., Inoue, Y., Sugiyama, H., Mishima, M. (1995). Reduced hippocampal LTP and spatial learning in mice lacking NMDA receptor epsilon 1 subunit. *Nature* 373:151–155.

Salpeter, M. M., Harris, R. (1983). Distribution and turnover rate of acetylcholine receptors throughout the junction folds at a vertebrate neuromuscular junction. *J. Cell Biol.* 96:1781–1785.

Salpeter, M. M., Marchaterre, M., Harris, R. (1988). Distribution of extrajunctional acetylcholine receptors on a vertebrate muscle: evaluated by using a scanning electron microscope autoradiographic procedure. *J. Cell Biol.* 106: 2087–2093.

Sanes, J. R., Lichtman, J. W. (1999a). Development of the vertebrate neuromuscular junction. *Annu. Rev. Neurosci.* 22:389–442.

Sanes, J. R., Lichtman, J. W. (1999b). Can molecules explain long-term potentiation? *Nature Neurosci.* 2:597–604.

Scheetz, A. J., Constantine-Paton, M. (1994). Modulation of NMDA receptor function: implications for vertebrate neural development. *FASEB J.* 8:745–752.

Schikorski, T., Stevens, C. F. (1997). Quantitative ultrastructural analysis of hippocampal excitatory synapses. *J. Neurosci.* 17:5858–5867.

Serpinskaya, A. S., Feng, G., Sanes, J. R., Craig, A. M. (1999). Synapse formation by hippocampal neurons from agrin-deficient mice. *Dev. Biol.* 205:65–78.

Shen, K., Meyer, T. (1999). Dynamic control of CaMKII translocation and localization in hippocampal neurons by NMDA receptor stimulation. *Science* 284: 162–166.

Sheng, M., Cummings, J., Roldan, L. A., Jan, Y. N., Jan, L. Y. (1994). Changing subunit composition of heteromeric NMDA receptors during development of rat cortex. *Nature* 368:144–147.

Shi, S. H., Hayashi, Y., Petralia, R. S., Zaman, S. H., Wenthold, R. J., Svoboda, K., Malinow, R. (1999). Rapid spine delivery and redistribution of AMPA receptors after synaptic NMDA receptor activation. *Science* 284:1811–1816.

Smith, S. J. (1999). Dissecting dendrite dynamics. *Science* 283:1860–1861.

Snider, W. D. (1987). The dendritic complexity and innervation of submandibular neurons in five species of mammals. *J. Neurosci.* 7:1760–1768.

Snider, W. D., Lichtman, J. W. (1996). Are neurotrophins synaptotrophins? *Mol. Cell. Neurosci.* 7:433–442.

Song, J. Y., Ichtchenko, K., Südhof, T. C., Brose, N. (1999). Neuroligin 1 is a postsynaptic cell-adhesion molecule of excitatory synapses. *Proc. Natl. Acad. Sci. USA* 96:1100–1105.

Sretavan, D. W., Shatz, C. J. (1986). Prenatal development of retinal ganglion cell axons: segregation into eye-specific layers within the cat's lateral geniculate nucleus. *J. Neurosci.* 6:234–251.

Stanfield, B. B., O'Leary, D. D., Fricks, C. (1982). Selective collateral elimination in early postnatal development restricts cortical distribution of rat pyramidal tract neurones. *Nature* 298:371–373.

Steward, O., Falk, P. M. (1991). Selective localization of polyribosomes beneath developing synapses: a quantitative analysis of the relationships between polyribosomes and developing synapses in the hippocampus and dentate gyrus. *J. Comp. Neurol.* 314:545–557.

Strack, S., Choi, S., Lovinger, D. M., Colbran, R. J. (1997). Translocation of autophosphorylated calcium/calmodulin-dependent protein kinase II to the postsynaptic density. *J. Biol. Chem.* 272:13,467–13,470.

Südhof, T. C. (1995). The synaptic vesicle cycle: a cascade of protein-protein interactions. *Nature* 375:645–653.

Sun, Y. A., Poo, M. M. (1987). Evoked release of acetylcholine from the growing embryonic neuron. *Proc. Natl. Acad. Sci. USA* 84:2540–2544.

Takumi, Y., Ramirez-Leon, V., Laake, P., Rinvik, E., Ottersen, O. P. (1999). Different modes of expression of AMPA and NMDA receptors in hippocampal synapses. *Nature Neurosci.* 2:618–624.

Tolbert, D. L., Panneton, W. M. (1984). The transience of cerebrocerebellar projections is due to selective elimination of axon collaterals and not neuronal death. *Brain Res.* 318:301–306.

Torres, R., Firestein, B. L., Dong, H., Staudinger, J., Olson, E. N., Huganir, R. L., Bredt, D. S., Gale, N. W., Yancopoulos, G. D. (1998). PDZ proteins bind, cluster, and synaptically colocalize with Eph receptors and their ephrin ligands. *Neuron* 21:1453–1463.

Toth, K., McBain, C. J. (1998). Afferent-specific innervation of two distinct AMPA receptor subtypes on single hippocampal interneurons. *Nature Neurosci.* 1:572–578.

Tovar, K. R., Westbrook, G. L. (1999). The incorporation of NMDA receptors with a distinct subunit composition at nascent hippocampal synapses in vitro. *J. Neurosci.* 19:4180–4188.

Turrigiano, G. G. (1999). Homeostatic plasticity in neuronal networks: the more things change, the more they stay the same. *Trends Neurosci.* 22:221–227.

Uchida, N., Honjo, Y., Johnson, K. R., Wheelock, M. J., Takeichi, M. (1996). The catenin/cadherin adhesion system is localized in synaptic junctions bordering transmitter release zones. *J. Cell Biol.* 135:767–779.

Vannier, C., Triller, A. (1997). Biology of the postsynaptic glycine receptor. *Int. Rev. Cytol.* 176:201–244.

Vaughn, J. E. (1989). Fine structure of synaptogenesis in the vertebrate central nervous system. *Synapse* 3:255–285.

Ventura, R., Harris, K. M. (1999). Three-dimensional relationships between hippocampal synapses and astrocytes. *J. Neurosci.* 19:6897–6906.

Verderio, C., Coco, S., Fumagalli, G., Matteoli, M. (1994). Spatial changes in calcium signaling during the establishment of neuronal polarity and synaptogenesis. *J. Cell Biol.* 126:1527–1536.

Verderio, C., Coco, S., Bacci, A., Rossetto, O., De Camilli, P., Montecucco, C., Matteoli, M. (1999). Tetanus toxin blocks the exocytosis of synaptic vesicles clustered at synapses but not of synaptic vesicles in isolated axons. *J. Neurosci.* 19:6723–6732.

Vicario-Abejon, C., Collin, C., McKay, R. D., Segal, M. (1998). Neurotrophins induce formation of functional excitatory and inhibitory synapses between cultured hippocampal neurons. *J. Neurosci.* 18:7256–7271.

Villarroel, A., Sakmann, B. (1996). Calcium permeability increase of endplate channels in rat muscle during postnatal development. *J. Physiol.* 496:331–338.

Watanabe, M., Fukaya, M., Sakimura, K., Manabe, T., Mishina, M., Inoue, Y. (1998). Selective scarcity of NMDA receptor channel subunits in the stratum lucidum (mossy fibre–recipient layer) of the mouse hippocampal CA3 subfield. *Eur. J. Neurosci.* 10:478–487.

Weber, A. J., Kalil, R. E. (1987). Development of corticogeniculate synapses in the cat. *J. Comp. Neurol.* 264:171–192.

Wigston, D. J. (1989). Remodeling of neuromuscular junctions in adult mouse soleus. *J. Neurosci.* 9:639–647.

Wigston, D. J. (1990). Repeated in vivo visualization of neuromuscular junctions in adult mouse lateral gastrocnemius. *J. Neurosci.* 10:1753–1761.

Wilkinson, R. S., Son, Y. J., Lunin, S. D. (1996). Release properties of isolated neuromuscular boutons of the garter snake. *J. Physiol.* 495:503–514.

Witzemann, V., Schwarz, H., Koenen, M., Berberich, C., Villarroel, A., Wernig, A., Brenner, H. R., Sakmann, B. (1996). Acetylcholine receptor epsilon-subunit deletion causes muscle weakness and atrophy in juvenile and adult mice. *Proc. Natl. Acad. Sci. USA* 93:13,286–13,291.

Wong, R. O. L., Yamawaki, R. M., Shatz, C. (1992). Synaptic contacts and the transient dendritic spines of developing retinal ganglion cells. *Eur. J. Neurosci.* 4:1387–1397.

Woolley, C. S. (1999a). Electrophysiological and cellular effects of estrogen on neuronal function. *Crit. Rev. Neurobiol.* 13:1–20.

Woolley, C. S. (1999b). Effects of estrogen in the CNS. *Curr. Opin. Neurobiol.* 9:349–354.

Xie, Z. P., Poo, M. M. (1986). Initial events in the formation of neuromuscular synapse: rapid induction of acetylcholine release from embryonic neuron. *Proc. Natl. Acad. Sci. USA* 83:7069–7073.

Young, S. H., Poo, M. M. (1983). Spontaneous release of transmitter from growth cones of embryonic neurones. *Nature* 305:634–637.

Zakharenko, S., Chang, S., O'Donoghue, M., Popov, S. V. (1999). Neurotransmitter secretion along growing nerve processes: comparison with synaptic vesicle exocytosis. *J. Cell Biol.* 144:507–518. [Erratum appears in *J. Cell Biol.* 144(4) (1999) following p. 801.]

Zhao, H. M., Wenthold, R. J., Petralia, R. S. (1998). Glutamate receptor targeting to synaptic populations on Purkinje cells is developmentally regulated. *J. Neurosci.* 18:5517–5528.

Ziff, E. B. (1997). Enlightening the postsynaptic density. *Neuron* 19:1163–1174.

Ziv, N. E., Smith, S. J. (1996). Evidence for a role of dendritic filopodia in synaptogenesis and spine formation. *Neuron* 17:91–102.

Zoubine, M. N., Ma, J. Y., Smirnova, I. V., Citron, B. A., Festoff, B. W. (1996). A molecular mechanism for synapse elimination: novel inhibition of locally generated thrombin delays synapse loss in neonatal mouse muscle. *Dev. Biol.* 179:447–457.

13 Neurotrophins and Refinement of Visual Circuitry

Edward S. Lein and Carla J. Shatz

The refinement of connections during nervous system development is thought to involve mechanisms to detect coincident activation of presynaptic and postsynaptic cells. A favorite model system for studying activity-dependent competitive interactions is the formation of ocular dominance columns in higher mammals, in which thalamic axons carrying information from the two eyes segregate from one another into eye-specific patches within the visual cortex. Although the phenomenology of ocular dominance column formation has been extensively described, the underlying molecular mechanisms and the molecular objects of competition (if such exist) are not well understood.

Recent research has suggested that members of the family of neurotrophins are important for ocular dominance column formation and plasticity. Several neurotrophins have profound effects on axonal and dendritic branching within the visual cortex and can also modulate cellular mechanisms of activity-dependent synaptic modification, such as long-term potentiation (LTP). Furthermore, a number of aspects of neurotrophin expression and activity-dependent regulation make neurotrophins particularly attractive candidate molecular signals for activity-dependent synaptic modification. Finally, neurotrophins have profound effects on inhibitory circuitry utilizing γ-aminobutyric acid (GABA), which can also affect ocular dominance plasticity. This review examines critically the available evidence regarding neurotrophins within the visual system in the context of a number of proposed actions for neurotrophins in the formation of visual circuitry.

The concept that patterns and levels of neuronal activity shape the functional properties of neuronal circuits is prevalent in current thinking about mechanisms underlying processes such as learning and memory. Work in a variety of in vivo and in vitro systems has demonstrated that changes in connectivity can result from alterations of neural activity, yet the molecular mechanisms underlying such phenomena are largely uncharacterized. Within the visual system, many studies have demonstrated that alterations of visual experience during a relatively short period of postnatal visual system development, termed the critical period, dramatically alter the final connectivity and functionality of visual circuitry (Hubel and Wiesel, 1970; Blakemore and Van Sluyters, 1974; Shatz, 1990; Daw et al., 1992; Katz and Shatz, 1996). Following the critical period, alterations of visual experience are not able to influence connectivity within visual pathways, and conversely are not able to reverse the effects on these pathways caused by prior manipulation during the critical period (Blakemore and Van Sluyters, 1974; Daw et al., 1992). The highly stereotyped circuitry, the robust nature of the changes induced by alterations in the sensory periphery, and the limited period during which plasticity within visual pathways can occur make the developing visual system an excellent model system for examining mechanisms of neuronal plasticity and the molecules involved in these processes. We first consider some of the expected properties of molecular signals involved in the development of visual circuitry and then explore the suitability of neurotrophins as such signaling molecules.

Visual Pathways

The retina has direct projections to a variety of structures in the brain, the largest of which terminates in the dorsal lateral geniculate nucleus (LGN) of the thalamus (Rodieck, 1979). Retinal ganglion cells from each eye representing the same region of visual space terminate on morphologically distinct layers in the LGN in an eye-specific fashion, such that the A lamina receives only contralateral input, lamina A1 receives only ipsilateral input, and so on for the other laminae (Hickey and Guillery, 1974). Thalamic relay neurons in the A laminae of the LGN project through the optic radiations to areas 17 (primary visual cortex) and 18, where they arborize largely in layer 4, which consists mainly of small stellate neurons (Humphrey et al., 1985a,b). In the adult, transneuronal transport of tritiated proline or wheat germ agglutinin–conjugated horseradish peroxidase (WGA-HRP) demonstrates that the projections from each eye are strictly segregated from one another into eye-specific laminae in the LGN and "patches" (or "stripes" depending on the plane of section used) roughly 500 µm across in layer 4 of the visual cortex (LeVay et al., 1980; Anderson et al., 1988). Single-axon studies on LGN neurons projecting to visual cortex complement these findings, demonstrating that individual LGN neurons have terminal arbors that form dense clusters of the same width as these patches (Humphrey et al., 1985a,b; Antonini and Stryker, 1993a). This anatomical segregation of eye-specific inputs into visual cortex underlies the physiologically defined ocular dominance columns spanning the entire radial dimension of the primary visual cortex, such that cells in a given column are more responsive to input from one eye than the other (Shatz and Stryker, 1978; LeVay et al., 1980). Different subtypes of neurons in the LGN project either to different sublaminae of layer 4 (4ab, 4c) or to all sublaminae of layer 4, based on single-axon reconstructions following HRP injections into physiologically defined LGN neurons (Friedlander et al., 1981; Friedlander, 1982; Humphrey et al., 1985a,b). This laminar specificity of axonal arborization suggests that molecular positional cues exist within the cortex that help to target specific populations of afferent axons to their correct laminar targets (Bolz et al., 1990).

Development of Eye-Specific Connections

Although eye-specific connections are strictly segregated in the mature geniculocortical pathway, this segregation is not present during the early formation of connections, but rather emerges during development from an initially unsegregated state (Rakic, 1977; LeVay et al., 1978, 1980;

Antonini and Stryker, 1993a). Within the cat visual system, transneuronal transport experiments demonstrate that the projection from one eye forms a band of continuous label within layer 4 of visual cortex at 2 weeks after birth (LeVay et al., 1978) and becomes increasing periodic over the next two months. Tracer studies using *Phaseolus* lectin injected into the LGN illustrate that single axons initially have sparse branches over a wider area than that of an adult column, whereas later in development single axon terminal arbors are more compact and dense (Antonini and Stryker, 1993a). This observation has been interpreted to mean that LGN afferents are sculpted during development through a process of pruning of inappropriate axonal branches and elaboration of appropriate branches. These changes in the morphology of LGN axons are accompanied by functional changes in the synaptic response properties of cortical cells. Initially most cortical neurons in layer 4 can be binocularly driven but come to respond only to stimulation of one eye during the same developmental period (Shatz and Stryker, 1978; LeVay et al., 1980). These observations suggest the involvement of molecules regulating the elaboration and maintenance of axonal branches; furthermore, it might be expected that these molecules would only be expressed during this period of axonal rearrangement.

Role for Neuronal Activity in the Formation of Eye-Specific Connections

A large body of experimental evidence has accumulated that suggests an essential role for neuronal activity in the formation of ocular dominance columns. Complete blockade of retinal activity via binocular injections of tetrodotoxin (TTX) prevents the segregation of geniculocortical axons into ocular dominance columns within layer 4 of primary visual cortex (Stryker and Harris, 1986). On the other hand, depriving animals of patterned visual experience, by either dark-rearing or binocular eyelid suture, prolongs but does not prevent the formation of ocular dominance columns (Wiesel and Hubel, 1965; Mower et al., 1985). The eventual formation of ocular dominance columns if retinal activity is not actually silenced suggests that remaining spontaneous retinal activity (or unpatterned activity in the case of eyelid suture) is sufficient to direct the segregation of geniculocortical axons into eye-specific circuits, as is thought to be the case also in the developing retinogeniculate pathway (Shatz and Stryker, 1988; Penn et al., 1998). Indeed recent experiments have indicated that ocular dominance columns are detectable even prior to visual experience in cats (Crair et al., 1998) and monkeys (Horton and Hocking, 1996).

Several other experiments add further evidence that the overall level of cortical activity is important for ocular dominance plasticity. During early postnatal development, the visual cortex receives excitatory input from glutamatergic subplate neurons underlying the cortical plate

(Finney et al., 1998), which in turn receive input from LGN axons (Friauf et al., 1990; Friauf and Shatz, 1991). Selective ablation of subplate neurons in cats early during the first postnatal week with the excitotoxin kainic acid (which removes excitatory drive onto cortical neurons) prevents the segregation of LGN axons into anatomical ocular dominance columns in the overlying cortical plate (Ghosh and Shatz, 1992, 1994). The neocortex also receives modulatory inputs from cholinergic and noradrenergic nuclei in the basal forebrain and brainstem, respectively. Depletion of both cortical acetylcholine and noradrenaline substantially reduces the physiological shift induced by monocular eyelid suture (Bear and Singer, 1986). Since these inputs provide excitatory input to the cortex (tonic in the case of acetylcholine and noradrenaline), these experiments show that the overall level of excitatory input to the cortex is important for experience-dependent synaptic plasticity. Collectively the activity dependence of ocular dominance column formation and plasticity suggests that molecules intimately involved in these processes should be in some fashion regulated by neuronal activity.

Importance of the Pattern of Presynaptic Activity for Ocular Dominance Column Formation

Not only is the level of neuronal activity important for the formation of eye-specific circuitry, the pattern of that activity is critical to the final connectivity. The first indication of this requirement came from the pioneering studies of Hubel and Wiesel, who demonstrated that the balance of activity driven by the two eyes during development determines the amount of cortical circuitry controlled by each eye in adulthood (Wiesel and Hubel, 1963a,b; Hubel and Wiesel, 1970; Hubel et al., 1977). If one eyelid is sutured shut during the critical period of postnatal development, the proportion of cells in visual cortex responding principally to the open eye increases dramatically, while the number of cells responding mainly to the closed eye diminishes. Anatomically geniculocortical afferents subserving the nondeprived eye end up occupying a much greater territory within layer 4 of the primary visual cortex than those subserving the deprived eye, although the total width of an open eye–deprived eye column remains the same as in normal animals (Hubel et al., 1977; Shatz and Stryker, 1978). Experimental modulation of retinal ganglion cell activity by blocking activity in the retina with TTX and stimulating the optic nerves directly reemphasizes the importance of the pattern of presynaptic input for ocular dominance refinement. If the two optic nerves are stimulated synchronously, the majority of cortical cells remain binocularly driven, but if the stimulation is asynchronous, the responses of most cortical neurons become dominated by one eye or the other, as in normal development (Stryker and Strickland, 1984). These results suggest that co-active presynaptic inputs form synapses on the same postsynaptic cells, while asynchronously active inputs segregate their connections from one another. Furthermore, there must be a mechanism for detecting co-active inputs on postsynaptic cortical neurons.

Importance of Postsynaptic Activity in the Formation of Ocular Dominance Columns: Cellular Mechanisms and Implications for Retrograde Signaling

There is strong evidence that the activity levels of postsynaptic cortical cells are also important for ocular dominance column plasticity. If muscimol (a $GABA_A$ receptor agonist) is infused into visual cortex during the critical period in conjunction with monocular deprivation, the physiological and anatomical shift in ocular dominance toward the nondeprived eye is prevented, and there is even a shift in favor of the deprived eye (Reiter and Stryker, 1988; Hata and Stryker, 1994). Since this treatment probably prevents large-scale depolarization of cortical neurons, this result implies that the ability of presynaptic inputs to excite postsynaptic cells is critical for normal plasticity. Indeed if stimulation of one eye is paired with the depolarization of postsynaptic cortical cells by "puffing on" potassium, there is a long-lasting shift in ocular dominance toward the stimulated eye in those cells (Fregnac et al., 1988). These studies show that simultaneous activation of presynaptic LGN axons and postsynaptic layer 4 cortical neurons can lead to a strengthening of the connections between them. If there is a great mismatch between those patterns of activation, as in the case of the muscimol infusions (Reiter and Stryker, 1988), the active presynaptic inputs are actually weakened and lost (Hata and Stryker, 1994).

The necessity for co-activation of pre- and postsynaptic activity for ocular dominance plasticity is characteristic of so-called Hebbian synapses (Hebb, 1949) and is thought to be a common mechanism for experience-dependent synaptic strengthening. This form of synaptic plasticity has been studied most thoroughly in the hippocampus, where high-frequency stimulation of afferent pathways leads to LTP (Bliss and Gardner-Medwin, 1973; Madison et al., 1991; Malenka and Nicoll, 1993; Malenka, 1994). At the synapse between the Schaffer collaterals of CA3 neurons upon the pyramidal cells of field CA1, a particular type of glutamate-gated ion channel, the N-methyl-D-aspartate (NMDA) receptor, has been proposed as a molecular coincidence detector that could underlie LTP (Malenka and Nicoll, 1993). In order for the NMDA channel to open, two things must occur simultaneously: the receptor must bind glutamate and the postsynaptic membrane must be depolarized to relieve a postsynaptic blockade of the channel by magnesium ions (Bliss and Collingridge, 1993). As glutamate is released from the presynaptic axon, and depolarization of the postsynaptic membrane is a measure of the overall stimulation of the neuron, this receptor opens only when pre- and postsynaptic activity occurs at the same time. Once open, the NMDA channel fluxes calcium, a potent intracellular messenger that effects the potentiation of that synapse. As described previously for the visual system (Fregnac et al., 1988), pairing of low-frequency stimulation of presynaptic fibers with depolarization of the postsynaptic cell also leads to LTP of those synapses (Madison et al., 1991; Malenka, 1994).

LTP can be induced within the visual cortex as well as the hippocampus (Artola and Singer, 1987; Komatsu et al., 1988; Kirkwood and Bear, 1995; Kirkwood et al., 1995). For example, white matter stimulation in slices of visual cortex from young rats produces robust LTP in layer 3 (Kirkwood et al., 1995). Furthermore, the ability to induce LTP using this paradigm declines with age in parallel with the ability of monocular deprivation to influence the physiological ocular dominance of cortical neurons (Kirkwood et al., 1995; Gordon and Stryker, 1996), implying a link between LTP and ocular dominance plasticity. In addition, infusion of the NMDA receptor antagonist 2-amino-5-phosphonovalerate (APV) during monocular deprivation prevents the physiological shift in ocular dominance (Kleinschmidt et al., 1987; Bear et al., 1990). (This result must be interpreted carefully, however, because in the developing cortex ion flow through the NMDA channel makes up a large part of the overall current [Miller et al., 1989]. Thus blockade of these channels may be similar to complete blockade of activity with TTX rather than to a selective requirement for NMDA receptor activation.) Despite these correlations, it remains to be shown definitively that LTP in visual cortex in vivo is necessary for ocular dominance column formation and plasticity. This distinction is emphasized by recent studies in transgenic mice that either lack ocular dominance plasticity following monocular deprivation yet display normal cortical LTP (Hensch et al., 1998a) or exhibit the reverse characteristics (Hensch et al., 1998b).

In addition to synaptic strengthening, ocular dominance column formation requires some form of synaptic weakening and eventual axon withdrawal. Studies in the hippocampus have demonstrated another form of synaptic plasticity called long-term depression (LTD), which acts to weaken synapses (Stanton and Sejnowski, 1989; Dudek and Bear, 1993; Malenka, 1994; Bear and Abraham, 1996). One form of synaptic depression, termed homosynaptic depression, occurs if the presynaptic fibers are stimulated at very low frequencies. In another form, heterosynaptic depression, stimulation of one pathway leads to depression of other, unstimulated pathways (Bear and Malenka, 1994). This type of plasticity could be viewed as a mechanism to stabilize the overall level of excitatory input to a cell—if some inputs are strengthened, then others must be weakened in order to avoid overstimulation of the cell. How does synaptic weakening apply to actual loss of axonal branches? Studies at the neuromuscular junction have indicated that, prior to the physical loss of a synapse, that synapse becomes weaker, which implies that as a synapse is progressively weakened it eventually disappears (Lo and Poo,

1991, 1994; Colman et al., 1997). LTD can also be induced in slices of developing visual cortex (Aroniadou and Teyler, 1991; Haruta et al., 1994; Kirkwood and Bear, 1995), although, as in the case of LTP, a requirement for LTD in the normal development or plasticity of ocular dominance columns has not yet been demonstrated.

The site of synaptic strengthening during LTP seems to be at least partly presynaptic (Bliss et al., 1990; Stevens, 1993). Since both pre- and postsynaptic activity are necessary for LTP, some kind of retrograde messenger must be transmitted back to the presynaptic cell to bring about this effect. Although the locus for hippocampal LTP is still a matter of debate (Huang and Stevens, 1997; Malenka and Nicoll, 1997), there is clearly a presynaptic component to ocular dominance column formation, as the overall anatomical pattern of the LGN axon terminal arborizations is dramatically remodeled. Based on these arguments, it seems highly likely that a retrograde signal passes from target neurons in layer 4 of the primary visual cortex to the incoming LGN axons, modulating synaptic strengths and ultimately the axonal arborization of LGN axons.

Evidence for Competitive and Trophic Interactions during Ocular Dominance Column Formation

Monocular deprivation during the critical period for ocular dominance column formation leads not only to a shrinkage of the terminal arbors of LGN axons subserving the deprived eye, but also to a shrinkage of the cell bodies of those LGN neurons (Wiesel and Hubel, 1963a). This observation is important because cell body shrinkage is characteristic of neurons whose trophic support is removed (Levi-Montalcini, 1987) and suggests that LGN neurons receive trophic support from their cortical target neurons in an activity-dependent manner. As discussed subsequently, trophic factors are known to be retrograde messengers in the peripheral nervous system (PNS), involved in the maintenance of neuronal survival and branching, and are the basis for certain competitive mechanisms during development. A number of experiments that have led to the shrinkage of LGN cell bodies have clearly demonstrated that LGN neurons carrying information from the two eyes compete with one another in order to maintain their terminal arbors (and consequently their soma size) during ocular dominance column formation and plasticity. First, binocular deprivation fails to cause LGN cell shrinkage (Guillery, 1973). Second, while monocular deprivation leads to shrinkage of LGN cells that project to binocularly innervated visual cortex, LGN cells that project to the monocular segment do not shrink (Guillery and Stelzner, 1970). Third, if a small lesion is made in the nondeprived retina, and the cell bodies of LGN axons that serve the deprived eye (which project to the region of visual cortex corresponding to the retinal lesion) are examined, no atrophy is seen (Guillery, 1972).

Thus atrophy of LGN neurons following visual deprivation occurs only if the deprived axons are co-innervating cortical neurons along with active LGN axons, a finding that in turn suggests a competitive mechanism rather than simple atrophy from lack of use.

To summarize, a number of predictions can be made regarding the expected properties of molecules that may be involved in the refinement of visual circuitry. Such molecules should be expressed in appropriate cell populations, and one might expect them to be differentially regulated during the period of peak plasticity (i.e., the critical period). Furthermore, regulation of these molecules should reflect the activity levels of pre- and postsynaptic neurons. Segregation of thalamocortical axons involves the maintenance and elaboration of axonal branches, and this segregation involves retrograde signaling from the target cortical neurons onto presynaptic LGN neurons, presumably based on their co-activation. Finally, compromising the activity levels of LGN axons leads to a loss of axonal branches and atrophy of cell bodies through some form of competitive mechanism. In short, the process of ocular dominance column formation has many features that suggest it involves trophic interactions. Members of the family of neurotrophins have emerged as attractive candidate retrograde trophic messengers for these processes. In the following section we review some of the established actions of neurotrophins on neuronal survival and aspects of neuronal plasticity; in the final section we discuss potential roles for neurotrophins in visual system plasticity.

Neurotrophins and Neuronal Survival

The neurotrophins are a family of small proteins that display trophic, or growth-supporting, activities for specific subsets of neurons that express the cognate receptors (Barde, 1989; Korsching, 1993). Nerve growth factor (NGF), the prototypic member of this family, was first discovered on the basis of its ability to promote the survival of sympathetic and sensory neurons in the PNS (see Levi-Montalcini, 1987, for review). Since the discovery of NGF, several homologous proteins have been discovered that, along with NGF, are collectively termed neurotrophins, although a number of other molecules have also been shown to have similar "neurotrophic" properties. Like NGF, brain-derived neurotrophic factor (BDNF) was isolated based on its ability to promote neuronal survival (Hofer and Barde, 1988; Leibrock et al., 1989). The others—neurotrophin-3 (NT-3), neurotrophin-4/5 (NT-4/5), and neurotrophin-6 (NT-6)—were cloned using the polymerase chain reaction based on regions of homology between the first two members (Ernfors et al., 1990; Hohn et al., 1990; Berkemeier et al., 1991; Ip et al., 1992; Gotz et al., 1994). These neurotrophins exert their cellular actions through members of the trk family of transmembrane protein tyrosine kinase receptors, such that NGF acts through trkA (Martin-Zanca et al., 1986; Cordon-Cardo et al., 1991; Klein et al., 1991a), BDNF and NT-4 act through trkB (Klein et al., 1991b, 1992), and NT-3 acts through trkC (Lamballe et al., 1991). Like NGF, BDNF, NT-3, and NT-4/5 have similarly been shown to have survival-promoting effects on various subpopulations of PNS neurons (Lindsay et al., 1985; Hohn et al., 1990; Davies et al., 1993).

Each of the neurotrophins and their high-affinity trk receptors have been knocked out in transgenic mice. For the most part, each of these knockouts demonstrates clear losses of peripheral neurons that express the relevant trk receptors, confirming the essential role of neurotrophins in neuronal survival in the PNS (see Snider, 1994, and Conover and Yancopoulos, 1997, for review). Trophic dependency of neurons in the central nervous system (CNS) appears to be much more complicated than that in the PNS. Although neurotrophins have survival-promoting activities on distinct populations of cultured neurons from the CNS (see Korsching, 1993, for review), many of these populations are not depleted in transgenic mice that lack a particular neurotrophin or receptor, suggesting that there is a great deal of redundancy in CNS trophic interactions (Snider, 1994).

The observed effects of NGF on sensory and sympathetic neurons in the PNS led to the formulation of the target-derived neurotrophic hypothesis and to the concept that neurotrophins are involved as the basis of competition (see Levi-Montalcini, 1987; Purves, 1988; Barde, 1989; and Oppenheim, 1991, for review). According to this model, the size of an initially overabundant population of trkA-expressing afferent neurons is matched to the size of its NGF-expressing target through a competition for limiting amounts of the target-derived neurotrophin. In support of this hypothesis, the addition of extra target tissue during development lessens the extent of normal developmental cell death; conversely, removal of target tissue leads to greater neuronal death. These effects can be mimicked by adding exogenous neurotrophin (Hamburger et al., 1981) or by blocking neurotrophin action (Aloe et al., 1981), respectively. The neurotrophins and their appropriate high-affinity receptors are expressed by the target tissues and the afferent neuronal populations, respectively; moreover, neurotrophins are taken up at the nerve terminals and retrogradely transported back to the cell body. This retrograde mode of action influences neuronal survival by altering or maintaining gene expression and protein synthesis (Deshmukh and Johnson, 1997). The neurons that are best able to obtain the relevant trophic factor (presumably those with the most functionally robust connections) survive, while those that are unable to receive an adequate supply of the neurotrophin atrophy and die through a program of apoptotic cell death (Pettmann and Henderson, 1998). The relevance of such a model to synaptic remodeling rather than to neuronal survival is considered in greater detail later in this chapter.

Effects of Neurotrophins on Neurite Outgrowth and Maintenance

In addition to their well-known ability to promote the survival of specific neuronal populations, neurotrophins have clear effects on neurite outgrowth and maintenance. The first demonstration of this point came from the so-called halo assay used to identify NGF (Levi-Montalcini, 1987). Not only does NGF promote the survival of cultured sympathetic and spinal sensory neurons, it leads to a massive proliferation of processes extending radially away from the neuron cell bodies. Refinements of these experiments by Campenot (1982, 1994) demonstrated that NGF can induce directed sympathetic axonogenesis and that NGF at the nerve terminal is sufficient for the elaboration and maintenance of neurites independent of neuronal survival. Thus NGF has a dual action—on neuronal survival and neurite extension.

Several of the neurotrophins have effects on both axonal and dendritic morphology of CNS neurons (see McAllister et al., 1999, for review). One clear example of this phenomenon was obtained in the *Xenopus* visual system (Cohen-Cory and Fraser, 1995). If the optic nerve is crushed in *Xenopus* embryos, the retinal ganglion cells regenerate axons back into their major target, the optic tectum. When the retinal ganglion cells are then labeled with the lipophilic dye 1,1'-dioctadecyl-3,3,3',3'-tetramethylindocarbocyanine perchlorate (DiI), this regrowth can be visualized in the living embryo. If BDNF is perfused onto the optic tectum during regrowth, many more ganglion cell axonal branches are seen within the tectum (it is known that ganglion cells express the trkB receptor). Conversely, if a blocking antibody to BDNF is applied during retinal axon ingrowth, fewer branches are seen in the tectum, indicating that endogenous BDNF plays a role in regulating axonal branch extension as the ganglion cell axons navigate to their correct locations in the optic tectum. BDNF and NT-3 also have effects on the dendritic architecture of cortical neurons (McAllister et al., 1995, 1997). In slices of ferret visual cortex cultured in the presence of BDNF or NT-3, total dendritic length and the complexity of pyramidal neuron dendritic arbors increase dramatically in specific cortical layers. Thus, as in the PNS, neurotrophins can affect connectivity in the CNS by regulating the sprouting and maintenance of neuronal processes.

Regulation of Neurotrophins by Neuronal Activity

A large body of literature has demonstrated that the expression of neurotrophins, and to a lesser extent their release, can be regulated by neuronal activity. The first evidence for this came from studies of changes in gene expression following seizure induction in the hippocampus. Induction of seizures by the systemic administration of kainic acid leads to a rapid and dramatic upregulation of NGF and BDNF mRNA levels in a variety of CNS structures, including the hippocampus and neo-cortex (Dugich-Djordjevic et al., 1992). These neurotrophins can also be regulated by less dramatic alterations in activity levels, such as stimuli used to induce LTP (Patterson et al., 1992; Castren et al., 1993). NT-4 levels can be upregulated by activity in muscle (Funakoshi et al., 1995), and NT-3 levels in the hippocampus are downregulated following seizures or LTP-inducing increases in neuronal activity (Castren et al., 1993; Garcia et al., 1997). Modulation of neurotrophin protein levels has proven to be much more difficult to demonstrate experimentally, because neuro-trophins are normally present at very low levels and because they are secreted molecules. However, several studies using antibodies directed against small peptide fragments of BDNF have shown that the adminis-tration of kainic acid leads to an increase in BDNF protein levels in the hippocampus (Yan et al., 1997). Such treatment also causes a relocal-ization of the protein from the cell body to the dendrites, from which it may be released (Wetmore et al., 1994). In order to deal with the de-tection problem posed by the extremely low levels of endogenous neuro-trophins in vivo, the release properties of neurotrophins in cultured neurons overexpressing NGF and BDNF have been examined (Blochl and Thoenen, 1995, 1996). These studies demonstrate that there is both a constitutive pathway of release and a pathway for regulated release fol-lowing depolarization of the neuron. The depolarization-induced release of the neurotrophin and the rapid induction of neurotrophin mRNA production suggest that the neurotrophins are in an excellent position to play a role in activity-induced changes in synaptic strength and struc-ture. The remainder of this review assesses the potential roles played by neurotrophins in visual system development and plasticity.

Neurotrophins and Development of Intrinsic Cortical Circuitry

Several observations have suggested that neurotrophins play a role in the formation of layer-specific intrinsic circuitry within the visual cortex (see McAllister et al., 1999, for review). The first indication that neuro-trophins played such a role was the demonstration that locally applied NGF could influence the direction of cultured sympathetic neurite out-growth toward the source of neurotrophin (Campenot, 1982, 1994). Furthermore, these neurites are maintained in the continued presence of NGF (Campenot, 1982, 1994). Although such tropic actions of neuro-trophins have not been directly assessed on cultured cortical neurons, correlations between patterns of cortical connectivity, neurotrophin ex-pression, and neurotrophin action on axonal and dendritic morphology are all highly suggestive that neurotrophins play similar roles in cortical development.

Laminar axonal and dendritic arborization patterns are highly stereo-typed in the visual cortex (Gilbert and Wiesel, 1983). As described ear-lier, different subtypes of LGN axons arborize in different sublaminae within layer 4 (Friedlander et al., 1981; Friedlander, 1982; Humphrey et al., 1985a,b), and neurons in the visual cortex send their axons into specific layers and have their dendritic arbors in others (Gilbert and Wiesel, 1983; Katz and Callaway, 1992). In culture experiments, NT-3 has been shown to influence axon outgrowth in a layer-specific fashion. NT-3 is repulsive for axon outgrowth from layer 2/3 explants of rat visual cortex and permissive for axon outgrowth from layer 5/6 explants (Bolz et al., 1997; Castellani and Bolz, 1998). In most cortical areas NT-3 is expressed only in the superficial layers (2/3 and 4) (Lein et al., 2000), which, based on the culture experiments discussed previously, would make them attractive for deep cortical axons while preventing layer 2/3 neurons from making connections in these layers. These observations are consistent with the known anatomy of the cortex, as neurons in su-perficial cortical layers (2/3) send their axons to deeper cortical layers and to the white matter en route to other cortical areas, whereas the neurons in the deeper cortical layers (5/6) project subcortically and send axon collaterals into superficial layers (Gilbert and Wiesel, 1983). (Inconsistent with these correlations, however, is the fact that neurons in layers 2/3 also project axon collaterals horizontally within the same layers [Gilbert and Wiesel, 1983; Katz and Callaway, 1992].) In primary visual cortex, NT-3 expression is even more restricted than elsewhere, being confined mainly to a subset of neurons in layer 4 (4c; Lein et al., 2000). This specificity of expression is particularly tantalizing because the axons of certain LGN neurons (e.g., Y cells; Friedlander et al., 1981; Fried-lander, 1982; Humphrey et al., 1985a,b) and GABAergic interneurons (e.g., basket cells; Naegele and Katz, 1990) specifically avoid arborizing in this sublamina of layer 4. Furthermore, NT-3 is only transiently ex-pressed in neocortex during the first few postnatal weeks (Lein et al., 2000), during the period in which many intrinsic cortical connections are forming.

BDNF and NT-3 also have effects on the dendritic branching of cor-tical neurons (see McAllister et al., 1999, for review). Addition of these neurotrophins (or agents that prevent endogenous neurotrophin signal-ing) to slices of visual cortex leads to profound changes in the dendritic morphology of the neurons in a manner that is neurotrophin-specific, layer-specific, and even different with respect to basal versus apical den-drites (McAllister et al., 1995, 1997). Furthermore, in some cases the ac-tions of BDNF and NT-3 on dendritic architecture are opposite or even mutually antagonistic (McAllister et al., 1997) (Fig. 13.1). In this context it is interesting to note that during early postnatal development in the cat, BDNF and NT-3 mRNAs are expressed in a highly complementary pattern, such that the layers that express BDNF (2/3 and 5/6) do not express NT-3, whereas NT-3 is nearly exclusively expressed in layer 4 (Lein et al., 2000) (Fig. 13.2). Although purely correlative at this point,

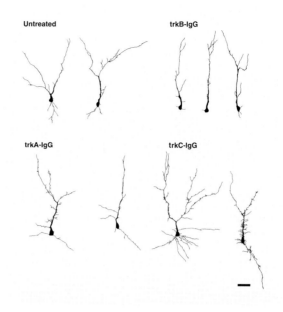

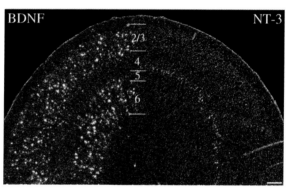

Figure 13.1. (top) Opposite effects of endogenous BDNF and NT-3 on dendritic growth of layer 4 neurons. Camera lucida reconstructions of layer 4 pyramidal neurons from slices of P14 ferret visual cortex, visualized by particle-mediated gene transfer of β-galactosidase. Neutralizing endogenous BDNF and NT-4/5 with trkB-IgG (upper right) causes a pronounced decrease in the growth of basal dendrites, whereas neutralizing NT-3 with trkC-IgG (lower right) dramatically enhances dendritic arbors. Neutralizing NGF with trkA-IgG (lower left) has no significant effect on dendritic growth relative to untreated neurons (upper left). Scale bar = 30 μm. Reprinted with permission from McAllister et al. (1997).

Figure 13.2. (bottom) Reciprocal expression of BDNF and NT-3 during the critical period for ocular dominance formation. In situ hybridization for BDNF mRNA (left) or NT-3 mRNA (right) in the primary visual cortex of a postnatal day 28 (P28) cat. BDNF mRNA is localized primarily to neurons in superficial (2/3) and deep (5/6) cortical layers, while NT-3 is nearly exclusively expressed in neurons at the base of layer 4 (4c). Sections are counterstained with the fluorescent dye bisbenzimide (blue). Scale bar = 100 μm. Modified from Lein et al. (2000).

these observations are highly suggestive that BDNF and NT-3 have a role in establishing layer-specific patterns of connectivity within the visual cortex. Studies of the axonal and dendritic morphology of single neurons in neurotrophin knockout mice could help to confirm the exact nature of these actions.

Neurotrophins as Mediators of Competitive Interactions between LGN Axons and Their Target Neurons

As previously noted, a number of aspects of ocular dominance column development and plasticity, including the sprouting and maintenance of axonal branches and the atrophy of inputs following visual deprivation, suggest the involvement of trophic factors. The formation of ocular dominance columns in the primary visual cortex of carnivores and primates involves an activity-dependent competition between LGN axons carrying information from the two eyes for synaptic territory within layer 4 of primary visual cortex (see Shatz, 1990, for review). Similarities between this process and extensively studied competitive mechanisms thought to regulate neuronal survival or death in the PNS (Levi-Montalcini, 1987) have suggested that common cellular mechanisms may underlie both phenomena (see Thoenen, 1995; Bonhoeffer, 1996; Shatz, 1997; and McAllister et al., 1999, for review). In the PNS, initially overabundant populations of neurons are thought to be matched to the sizes of their target tissues through a competition for limiting quantities of trophic factors secreted by the target cells (Levi-Montalcini, 1987). Those neurons best able to obtain these factors survive, whereas those unable to do so die. In the developing visual system, thalamic axons subserving each eye initially overlap with one another, branching widely within layer 4 of primary visual cortex (LeVay et al., 1978; Antonini and Stryker, 1993a). Through a process of selective elimination and elaboration of axonal branches, the terminal arbors of LGN axons are refined such that axons carrying information from each eye become segregated from one another into eye-specific stripes or patches within layer 4. If the projection from one eye is put at a competitive disadvantage by decreasing the activity levels within that eye, the territory devoted to the nondeprived eye expands at the expense of that devoted to the deprived eye (Hubel et al., 1977; Shatz and Stryker, 1978; Antonini and Stryker, 1993b). In the visual cortex (unlike the PNS), the competition between eye-specific inputs to layer 4 does not involve the survival of LGN neurons, but rather the "survival" and elaboration of individual axonal branches and synapses (the period of naturally occurring cell death for LGN neurons occurs much earlier; Williams and Rakic, 1988). Thus, by analogy, the branches of LGN axons may compete for limiting amounts of trophic factors available in the cortex, which they are able to obtain or use in an activity-dependent manner; those branches able to obtain trophins are maintained and elaborated, whereas those unable to obtain these factors are weakened and finally lost. This model has been dubbed the synapto-trophic hypothesis (Snider and Lichtman, 1996).

If this mechanism underlies the segregation of LGN axons in the visual cortex, a number of minimal postulates should be true for the developing thalamocortical pathway. First, the addition of excess neuro-trophin should remove the basis for competition: if exogenous neuro-trophin is provided to the LGN axon terminals during the critical period, the axonal branches of LGN neurons subserving both eyes should be maintained. Second, decreasing the levels of neurotrophins in the cortex might be expected to prevent the formation of ocular dominance columns by preventing the elaboration of LGN axonal branches. Third, the appropriate neurotrophins and trks should be expressed by target neurons (layer 4 neurons) and afferent inputs (LGN neurons), respectively. Finally, neurotrophin levels might be expected to be regulated by neuronal activity in a manner that closely mirrors activity levels.

The first experimental evidence for an involvement of neurotrophins in ocular dominance plasticity came from studies on the effects of intra-ventricular infusions of NGF in rodents (Maffei et al., 1992). If NGF is applied during the critical period in rats, the physiological shift in ocular dominance toward the nondeprived eye following monocular deprivation is prevented. Furthermore, the shrinkage of LGN cell bodies correspon-ding to the deprived eye that normally occurs following this manipula-tion is prevented (Domenici et al., 1993). Administration of antibodies to NGF, on the other hand, extends the period during which visual ma-nipulation can lead to changes in the eye-specific physiological properties of cortical neurons, an effect similar to that produced by dark-rearing (Domenici et al., 1994). These experiments have been interpreted to mean that cortically produced NGF, whose availability is regulated by neural activity, regulates the branching of LGN axons.

However, examination of the expression pattern of the high-affinity NGF receptor trkA throws doubt upon this interpretation. Virtually the only trkA-expressing neurons in or projecting to the visual cortex are cholinergic neurons from the basal forebrain (Venero and Hefti, 1993). NGF is known to have both survival-promoting effects on these neurons in culture and potentiating effects on the expression of choline acetyl-transferase, the key enzyme in the pathway leading to acetylcholine syn-thesis (Johnston et al., 1987). As discussed previously, manipulations of cholinergic synaptic transmission are known to affect ocular dominance column plasticity (Bear and Singer, 1986). Since infusion of NGF was directed into the ventricles, the cholinergic afferents from the basal fore-brain certainly had access to this NGF. Thus the most likely interpreta-tion of the experiments of Maffei and others is that NGF, acting through trkA on basal forebrain neurons, potentiates cholinergic transmission in the visual cortex and consequently affects ocular dominance plasticity only indirectly.

Recent experiments have provided much more compelling evidence for a direct role in thalamocortical competition for the ligands of the trkB receptor, BDNF and NT-4. If BDNF or NT-4, but not NGF or NT-3, is infused into kitten visual cortex via an osmotic minipump during the peak of the critical period, LGN axons fail to segregate into ocular dominance columns near the site of neurotrophin infusion (Cabelli et al., 1995) (Fig. 13.3). Similar administration of NT-4 directly into layer 4 of the ferret visual cortex, in conjunction with monocular deprivation, prevents the shrinkage of LGN neurons related to the deprived eye (Riddle et al., 1995). Furthermore, ablation of subplate neurons, which is known to prevent the formation of ocular dominance columns (Ghosh and Shatz, 1992, 1994), leads to a dramatic upregulation of BDNF mRNA in the same area of cortex in which ocular dominance columns are absent (Lein et al., 1999) (Fig. 13.4). In terms of the synaptotrophic hypothesis, the excess levels of BDNF can be viewed as swamping out competitive mechanisms, stabilizing all axonal branches, and thus preventing segregation. Blockade of neurotrophin signaling during the critical period similarly affects ocular dominance column formation. Infusion of antagonists of trkB signaling (trkB-IgG receptor bodies), but not of trkA or trkC signaling, prevents the segregation of LGN axons near the site of infusion (Cabelli et al., 1997). These experiments clearly demonstrate the involvement of an endogenous ligand of trkB in ocular dominance column formation.

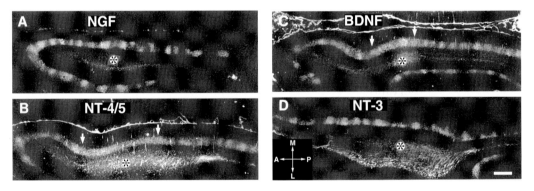

Figure 13.3. Infusion of ligands of trkB prevents the formation of ocular dominance columns. Dark-field photomicrographs demonstrating transneuronal transport of ^{3}H-proline in the visual cortex of animals that received minipump infusions of neurotrophins for 2 weeks during the critical period for ocular dominance column formation. The presence of periodic patches of label in layer 4 of NGF- or NT-3-treated animals indicates the formation of normal ocular dominance columns. In contrast, ocular dominance columns fail to form following infusion of BDNF or NT-4, as demonstrated by uniform labeling overlying infusion sites (area between arrows in *B* and *C*). Asterisks indicate infusion sites. Scale bar = 1 mm. Reprinted with permission from Cabelli et al. (1995); copyright 1995 American Association for the Advancement of Science.

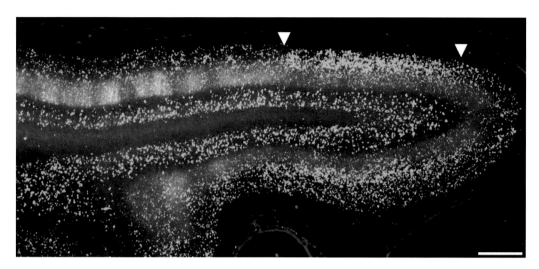

Figure 13.4. Subplate neuron ablation leads to a long-lasting upregulation of BDNF mRNA in visual cortex in the same region in which ocular dominance columns are disrupted. Pseudocolor overlay of adjacent horizontal sections of primary visual cortex labeled for transneuronally transported ^{3}H-proline to visualize ocular dominance columns (red patches) or BDNF mRNA (green) in a P56 animal following ablation of subplate neurons with kainic acid at P8. The extent of increased BDNF mRNA expression correlates closely with the region in which ocular dominance columns are absent (area between arrowheads). Scale bar = 2 mm. Modified from Lein et al. (1999).

Expression data for trks and neurotrophins are also consistent with the view that cortically produced ligands of trkB have a direct action on thalamic axons. LGN neurons express trkB and trkC during early postnatal life, as assayed by immunohistochemistry (Cabelli et al., 1996), in situ hybridization (Schoups et al., 1995; Lein and Shatz, unpublished observations), and chemical crosslinking of iodinated neurotrophins to extracts from the LGN during the critical period (Allendoerfer et al., 1994). BDNF mRNA is present in visual cortex and is developmentally regulated during the critical period (Castren et al., 1992; Lein et al., 2000). Within layer 4, BDNF mRNA is detectable at high levels late in the critical period, although its levels are very low earlier in ocular dominance column formation (Lein et al., 2000). Thus, in contrast to trkA and NGF, trkB and BDNF are expressed in appropriate structures during the segregation of thalamocortical afferents, a finding consistent with a target-derived mode of action.

There is also considerable evidence for activity-dependent regulation of BDNF during visual system development. Levels of BDNF mRNA in visual cortex increase shortly after eye opening (Castren et al., 1992; Lein et al., 2000). BDNF expression can be downregulated by total blockade of retinal activity or by eyelid suture, and the normal developmental upregulation can be delayed by dark-rearing (Castren et al., 1992; Bozzi et al., 1995; Schoups et al., 1995; Lein and Shatz, 2000). Furthermore,

BDNF mRNA levels are rapidly modulated within eye-specific neuronal circuits by alterations in the levels of retinal activity (Lein and Shatz, 2000). Short-term monocular TTX blockade during the critical period downregulates BNDF mRNA specifically within the ocular dominance columns serving the treated eye (Lein and Shatz, 2000). Taken together, these observations indicate that BDNF mRNA levels provide a fairly accurate measure of the activity levels of postsynaptic cortical neurons.

Thus, based on the available evidence, BDNF has emerged as the most likely candidate neurotrophin to play a role as a molecular object of competition underlying the segregation of thalamocortical axons that carry information from the two eyes into ocular dominance columns within layer 4 of the primary visual cortex. However, a number of important questions remain unanswered. First, in order to understand the effects that neurotrophins have on thalamocortical axonal morphology, it is essential to know in detail the effects of adding exogenous neurotrophin or blocking neurotrophin signaling on the morphology of single axons. Thus far, transneuronal autoradiographic techniques have been used to examine the consequences of neurotrophin infusions on LGN axons (Cabelli et al., 1995, 1997). Consequently the branching patterns of individual LGN axons have been inferred indirectly from an examination of the distribution of silver grains over layer 4. Ideally single-axon labeling techniques such as those used by Antonini (Antonini and Stryker, 1993a,b) should be combined with neurotrophin infusion studies like those of Cabelli (Cabelli et al., 1995, 1997). In the frog visual system, exogenous BDNF is known to cause an increase in axonal branching of regenerating retinal ganglion cells in the optic tectum (Cohen-Cory and Fraser, 1995). By analogy, BDNF in cat visual cortex may act to enhance sprouting or stabilize thalamocortical axon branches, as suggested by an increased density of ^{3}H-proline labeling following BDNF infusion (Cabelli et al., 1995; see Fig. 13.3). Conversely, the decreased density of labeling following infusion of antagonists of trkB signaling suggests that the branching density of LGN axons may have been diminished under these circumstances (Cabelli et al., 1997).

An even more general question is simply whether BDNF in the cortex acts directly on thalamic axons at all, particularly given the large number of neurons that express trkB and BDNF in the LGN and visual cortex (Cabelli et al., 1996; Lein et al., 2000). The reason for assuming that it does is based squarely on the fact that LGN neurons express trkB, the receptor for BDNF (Allendoerfer et al., 1994; Schoups et al., 1995; Cabelli et al., 1996). However, it remains to be shown that neurons in the LGN indeed bind and respond to BDNF by altering their axonal morphology. It should be possible to address these questions by studying LGN neurons in vitro and by examining whether exogenous labeled BDNF can be transported retrogradely from the visual cortex to the LGN in vivo. These studies are critical, because it is known that the morphology of LGN axons (and the size of the cell bodies) can be affected through indirect mechanisms (Domenici et al., 1993; Hata and Stryker, 1994).

BDNF mRNA levels in layer 4 of primary visual cortex are extremely low early in the critical period and increase toward its end (Lein et al., 2000), consistent with a role for BDNF in the elaboration of synapses that occurs late in thalamocortical segregation (Antonini and Stryker, 1993a). Near the end of the critical period, once segregation is largely complete, thalamic axons within an eye-specific column drive the post-synaptic neurons in layer 4 almost exclusively. At this point, something of a positive feedback loop may be in effect, such that, as the input from the relevant eye becomes stronger, more BDNF is released; this in turn induces more synaptic growth, which further increases the excitatory drive from that eye, and so on. Such a role in synapse formation and growth is consistent with findings from several other studies. For example, BDNF induces excitatory synapse formation in cultured hippocampal neurons (Vicario-Abejón et al., 1998), whereas fewer axon collaterals, varicosities, and synaptic contacts are observed in the hippocampus of trkB knock-out mice (Martínez et al., 1998). However, the fact that blockade of trkB function prior to the completion of LGN axon segregation prevents ocular dominance column formation (Cabelli et al., 1997) implies that early low levels of BDNF are functionally relevant to the competitive mechanisms that lead to the loss of synaptic connections as well. It remains unclear what functionally relevant levels of neurotrophins in visual cortex are, given currently available detection methods and the poorly understood relationship between neurotrophin mRNA levels and protein levels (Wetmore et al., 1994; Yan et al., 1997).

Although it is an appealing concept, the idea that neurotrophins underlie the competition between thalamocortical axons should be viewed with caution. Even in simpler systems, such as the neuromuscular junction, it is still unclear exactly what role trophic factors play in the activity-dependent elaboration or elimination of synapses (Snider and Lichtman, 1996). At the neuromuscular junction, each muscle fiber is initially innervated by several motoneurons, and through an activity-dependent process synapses are progressively eliminated, until each muscle fiber is innervated by only one axon (Lichtman and Balice-Gordon, 1990). If muscle cells are transfected with glial-derived neurotrophic factor, a potent trophic factor for motoneurons, the early polyinnervation persists, suggesting that an excess of trophic factor allows competing axon terminals to be maintained (Nguyen et al., 1998). However, eventually the transfected muscle fibers do become singly innervated, although the process of synapse elimination takes much longer. These observations suggest that trophic factor levels may not directly underlie the mechanism responsible for competitive elimination of axonal branches, but rather may act to maintain a state of plasticity in the motoneurons such that they continue to try to innervate multiple targets.

The segregation of LGN axons into eye-specific patches in layer 4 of visual cortex is exquisitely sensitive to the pattern of neuronal activity from the two eyes (see Katz and Shatz, 1996, for review). This process is thought to be Hebbian (Hebb, 1949), in the classical sense that coincident pre- and postsynaptic activation leads to synaptic strengthening, whereas asynchronous activity levels leads to destabilization. How could a competition for neurotrophins selectively stabilize inputs that are synchronously active while destabilizing adjacent inactive or asynchronous inputs? Such a mechanism would seem to require that the trophic factor release be extremely localized or highly efficacious. There is evidence that BDNF is released through both constitutive and regulated pathways (Thoenen, 1995). For example, the depolarization of cultured neurons overexpressing BDNF leads to a rapid release of BDNF; however, it is not clear how local this release can be (can it occur at individual synapses?) and how far the released neurotrophin can diffuse in vivo. Some recent experiments have indicated that the action of neurotrophins may be limited to synapses. Several forms of the trkB receptor exist, including truncated forms lacking the signal-transducing tyrosine kinase domain (Bothwell, 1995) that may be dominant negative with regard to trkB signaling (Eide et al., 1996). In adult animals at least, this truncated form of the receptor is localized to glia (Frisén et al., 1993) and the total neuronal cell membrane, but it is absent from membranes derived from synaptosomal preparations (Wu et al., 1996). In contrast, full-length functional trkB is enriched in synaptosomal fractions (as well as other fractions) (Wu et al., 1996). This finding suggests that neurotrophin efficacy can be restricted to synapses and argues for a local action, if not local release. Furthermore, there is evidence that the ability of neurons to utilize neurotrophins may also depend on neuronal activity. For example, the survival of cultured cortical neurons or retinal ganglion cells is enhanced by BDNF only if the cells are also depolarized or have increased intracellular calcium levels (Ghosh et al., 1994; Meyer-Franke et al., 1995).

These findings suggest that neurotrophins, even if not released locally, may act locally through a synergy with activity levels. Finally, there is evidence that low levels of trkB activation in neurons dramatically decrease the ability of the neurons to respond to subsequent BDNF application (Carter et al., 1995). Taken together, these observations suggest that depolarization of a neuron may cause local BDNF release, which may act to strengthen co-active synapses and perhaps also weaken asynchronous inputs through low-level trkB signaling. This "instructive" mechanism of neurotrophin regulation of thalamocortical connectivity is of course highly speculative, and it will require many additional experiments to verify the hypothesis.

Neurotrophins and Enhancement of Excitatory Synaptic Transmission

The discussion thus far has focused on the ability of neurotrophins to maintain and induce axonal branching, that is, long-term changes in axonal structure. In addition, neurotrophins can have potent and rapid effects on synaptic transmission. This was first demonstrated in frog motoneuron-myocyte cultures, where it was found that exogenous BDNF (or NT-3) causes a rapid and reversible increase in the rate of spontaneous motoneuron firing (Lohof et al., 1993). It has since been shown that BDNF can facilitate synaptic transmission in a number of systems, with both short- (seconds to minutes) and longer-latency (hours) effects (see Schuman, 1999, for review). In the cortex, BDNF has been shown to facilitate LTP at layer 4–layer 2/3 synapses in slice cultures (Akaneya et al., 1997) and also to block the induction of LTD (Akaneya et al., 1996). Furthermore, LTP induction paradigms lead to increases in BDNF mRNA levels (Patterson et al., 1992; Castren et al., 1993). Thus BDNF levels are increased by stimuli that induce changes in synaptic strength, and these increased BDNF levels can lead in turn to synaptic strengthening.

As we pointed out earlier, studies of hippocampal LTP have suggested the existence of a retrograde signal that affects presynaptic enhancement. A number of studies have indicated that, at least in certain circumstances, BDNF could act in such a fashion to trigger a cascade of intracellular signaling during LTP. First, LTP is impaired in mice that lack one or both copies of the BDNF gene (Korte et al., 1995; Patterson et al., 1996). Conversely, the addition of BDNF to hippocampal slice cultures leads to a potentiation of synapses at the CA3-CA1 synapse (Kang and Schuman, 1995), suggesting that BDNF could act as a retrograde signal downstream of the initial electrical events. However, tetanus-induced LTP can be induced over and above this neurotrophin-induced potentiation and vice versa (Kang and Schuman, 1995), implying that the two forms of potentiation occur through at least partially different signaling pathways. Consistent with this idea, bath application of trkB-IgG molecules to hippocampal slices prevents LTP induced by theta-burst stimulation but not by tetanic stimulation (Kang et al., 1997). Thus BDNF signaling may be important for some but not all forms of LTP induction.

Changes in synaptic efficacy observed following the induction of LTP occur on a time scale of tens to hundreds of milliseconds. Can the release of neurotrophin, and its action through the trkB receptor and subsequent signaling cascades, occur on this time scale? That this may be possible has been elegantly demonstrated by Boulanger and Poo, who have shown using co-cultures of neurons and myocytes that potentiation of synaptic transmission by BDNF only occurs in those presynaptic terminals in which the elevation of Ca^{2+} coincides with BDNF application (Boulanger and Poo, 1999). Thus BDNF released locally could act in an instructive fashion to alter synaptic strength only at co-active synapses. On the other hand, the action of BDNF could be permissive, affecting the ability of other signaling mechanisms to modulate synaptic efficacy. Bear and colleagues (Kirkwood et al., 1996) have proposed that there is a sliding threshold for synaptic modification, such that the ability of a given synapse to be modified depends upon the past history of the cell.

Experiments by Bear's group using slices of visual cortex have shown that the addition of BDNF slides this threshold toward potentiation, so that in the presence of BDNF tetanic stimulation at frequencies lower than normal can induce LTP (Huber et al., 1998) (Fig. 13.5). Furthermore, the impaired LTP in BDNF knockout animals can be rescued either by transfecting hippocampal neurons with BDNF (Korte et al., 1996) or by bath application of recombinant BDNF (Patterson et al., 1996), findings that suggest an overall regulatory function for this neurotrophin rather than its acting as an "instructive" signal. Thus it seems most likely that BDNF acts as a modulator of synaptic plasticity by affecting the ability of other mechanisms, such as LTP, to alter synaptic strengths based on patterns of neuronal activity.

To summarize, BDNF can have potent effects on specific aspects of synaptic plasticity. This implies that, in addition to their long-term control of the structural consequences of competitive interactions, neurotrophins may also rapidly and directly affect the physiological mechanisms underlying synaptic competition. It is unclear at this point how the physical addition or loss of synapses relates to changes in synaptic strength. Studies of the neuromuscular junction have shown that synaptic weakening occurs prior to axon withdrawal (Colman et al., 1997). This observation suggests that the causative mechanisms for these two phenomena overlap and that neurotrophin signaling could affect not only rapid modulation of synaptic strength but also longer-latency morphological alterations in the architecture of axonal branches.

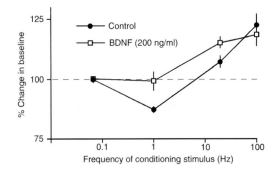

Figure 13.5. *BDNF treatment shifts the function relating stimulation frequency and synaptic plasticity in the visual cortex.* Average field potentials (percent change from baseline ± SEM) recorded in layer 2/3 of rat visual cortical slices 30 min after tetanic stimulation in layer 4, either in control slices (filled circles) or in the presence of 200 ng/ml BDNF (open squares). BDNF application attenuates LTD induced at low frequencies (e.g., 1 Hz) and enhances LTP induced at higher frequencies (e.g., 20 Hz). Reprinted with permission from Huber et al. (1998).

Neurotrophin Regulation of Inhibitory GABAergic Transmission and Visual System Plasticity

Strict Hebbian models of synaptic competition, such as those considered earlier, cannot explain all the experimental data concerning ocular dominance column plasticity and the role of neurotrophins. First, it is known that early in the critical period there is a large physiological bias in favor of the contralateral eye (Crair et al., 1998). A simple model of synaptic strengthening based on coincident activity of pre- and postsynaptic cells would predict that the contralateral eye should take over the entire cortex; this obviously does not occur. Furthermore, if postsynaptic activity levels in the primary visual cortex are decreased by infusion of muscimol (a $GABA_A$ receptor agonist) following monocular eyelid suture, there is an anatomical and physiological shift in favor of the deprived eye rather than the nondeprived eye (Reiter and Stryker, 1988; Hata and Stryker, 1994). These experiments suggest that active synapses that are unable to stimulate the related postsynaptic cells are actually weakened, whereas inactive synapses are strengthened or at least maintained. A similar shift in favor of the deprived eye, as assessed physiologically, occurs following cortical infusion of BDNF in conjunction with monocular deprivation (Galuske et al., 1996), a finding that hints at a common mechanism underlying both of these phenomena.

A possible explanation for these observations is that BDNF may act to enhance inhibitory transmission through an action on GABAergic interneurons. There is, in fact, substantial evidence in favor of this idea (see Marty et al., 1997, for review). First, BDNF has been shown to enhance the phenotype of GABAergic interneurons in the cerebral cortex in vivo (Nawa et al., 1994). BDNF also increases the formation of inhibitory synapses in cultures of immature hippocampal neurons (Vicario-Abejón et al., 1998). Transgenic mice that overexpress BDNF in forebrain neurons have been found to contain increased numbers of neurons that express GABAergic markers (Huang et al., 1999), whereas BDNF knock-out mice display reduced numbers of interneurons that are immuno-positive for GABAergic markers (Jones et al., 1994). Finally, subplate neuron ablation results in both an increase in BDNF mRNA expression in primary visual cortex and an increase in immunostaining for GAD-67, the major synthetic enzyme for GABA (Lein et al., 1999). As the ablation of subplate neurons also prevents LGN axons from segregating to form ocular dominance columns (Ghosh and Shatz, 1992, 1994), these observations provide a tantalizing link between neurotrophins, cortical inhibition, and visual system synaptic plasticity.

It has been known for some time that inhibitory neurotransmission can affect synaptic plasticity. For example, depressing inhibitory transmission facilitates the induction of LTP in the hippocampus (Bliss and Collingridge, 1993) and visual cortex (Kirkwood and Bear, 1995). However, it is surprising, given the effects of BDNF on phenotypic markers of inhibitory interneurons, that in the hippocampus BDNF has been reported to attenuate GABAergic inhibitory transmission onto pyramidal neurons (Tanaka et al., 1997). The details of BDNF's action on excitatory and inhibitory transmission within the visual cortex remain to be elucidated, but since pharmacological treatments that either increase GABAergic transmission (Reiter and Stryker, 1988) or elevate BDNF levels (Galuske et al., 1996; Huang et al., 1999) give similar results, it would appear that BDNF can enhance inhibitory transmission in developing visual cortex.

Data regarding the development of GABAergic circuitry in visual cortex also lend correlative support to the idea that BDNF facilitates the maturation of GABAergic synaptic transmission. The developmental upregulation of BDNF mRNA in cat visual cortex (Lein et al., 2000) correlates closely with developmental changes of levels of glutamic acid decarboxylase (GAD), an enzyme in the synthetic pathway for GABA (Fosse et al., 1989). Furthermore, BDNF mRNA levels are decreased within deprived-eye ocular dominance columns by monocular retinal activity blockade (Lein and Shatz, 2000); similar manipulations also lead to a downregulation of levels of GAD in deprived-eye columns (Hendry and Jones, 1986). These correlations suggest that BDNF within the visual cortex can locally modulate levels of inhibition, which in turn could indirectly influence the competitive mechanisms that determine correlations between LGN axons and cortical neurons.

Several experiments also bear on the interrelationship between maturation of cortical inhibitory circuitry, visual system synaptic plasticity, and neurotrophins. Bear and his colleagues (Kirkwood et al., 1995) have shown that early in development it is possible to induce LTP in layer 2/3 cortical neurons by delivering a tetanus to the underlying white matter, but following the critical period for monocular deprivation such stimulation no longer evokes LTP in layers 2/3. However, blockade of GABAergic transmission in adult cortical slices can restore white matter–evoked LTP in layers 2/3 (Kirkwood and Bear, 1995), suggesting that the critical period is brought to a close by the maturation of cortical inhibition. In an extension of this observation, transgenic mice that overexpress BDNF have been found not only to undergo premature maturation of GABAergic neurons in the visual cortex but also to have an earlier and shorter critical period for the effects of monocular deprivation (Huang et al., 1999). Similarly, premature elevation of BDNF mRNA (and enhanced GABAergic phenotype) in the visual cortex of cats following ablation of the underlying subplate prevents the remodeling of LGN axon branches into ocular dominance columns (Lein et al., 1999; see Fig. 13.4), again implying that enhanced intracortical inhibition can diminish activity-dependent synaptic plasticity and could thus bring the critical period to a close.

Although this hypothesis is attractive, it is clearly oversimplified. In an elegant experiment that more directly assesses the role of inhibition, Hensch et al. (1998b) examined visual cortical plasticity in mice deficient for GAD-65, which is required for most of GABA synthesis. Fast synaptic inhibition in these mice was absent, yet monocular deprivation failed to cause a shift in ocular dominance in favor of the open eye. Furthermore, plasticity could be restored by administering diazepam (a GABAergic use-dependent agonist), implying that a lack of cortical inhibition, rather than premature or enhanced inhibition, prevents visual system plasticity.

How can these seemingly contradictory observations be reconciled? One possibility is that a delicate balance between excitation and inhibition within the visual cortex is required for synaptic plasticity, and that too much of either activity prevents plasticity. In this context, perhaps BDNF functions to regulate the total amount of synaptic input onto neurons by modulating the levels of both excitatory and inhibitory synaptic input—acting directly on geniculocortical synapses and also on GABAergic interneurons—as it will be recalled that the geniculate and cortical GABAergic neurons are trkB-positive (Cabelli et al., 1996). Consistent with this idea, Turrigiano and colleagues have shown, in visual cortex cultures that contain both excitatory and inhibitory neurons, that quantal amplitudes at excitatory and inhibitory synapses change as a function of the total level of neuronal activity (Turrigiano et al., 1998). Blocking activity with TTX leads to an increase in excitatory quantal amplitudes and a decrease in inhibitory quantal amplitudes. Conversely, increasing activity (by blocking GABAergic transmission) reduces excitatory quantal amplitudes. These changes, which can be viewed as a cellular homeostatic mechanism that operates to maintain a steady level of excitation, can be mimicked by changes in BDNF levels (Rutherford et al., 1997, 1998). The exogenous addition of BDNF can prevent the decrease in excitatory quantal amplitudes under TTX blockade and can increase inhibitory quantal amplitudes. It has long been thought that there must be mechanisms to stabilize the overall levels of excitatory input during intense periods of synaptic reorganization (Miller and Stryker, 1990). It may be that neurotrophins that are produced and released as a function of neuronal activity, and that act simultaneously on both excitatory and inhibitory synapses, could maintain such a homeostatic process. The fact that BDNF mRNA levels mirror postsynaptic activity levels (Lein and Shatz, 2000) supports this model.

Conclusion

Neurotrophins display a broad spectrum of actions during the development of the visual system, and many features of their regulation make them attractive candidate molecules for activity-dependent circuit formation. Several neurotrophins are developmentally regulated during the critical period, and BDNF expression in particular is tightly regulated by afferent neuronal activity during the relevant period. There is evidence for trophic action in many aspects of visual cortical development, including the formation of specific patterns of axonal and dendritic branching, the maintenance of axonal branching, and the regulation of synaptic competition between LGN axons during ocular dominance column formation. Furthermore, neurotrophins may act to influence the physiological properties of cortical circuits, both by modulating mechanisms of synaptic plasticity, such as LTP, and by influencing the development and function of cortical inhibitory circuitry. It is exciting finally to have compelling molecular candidates for mechanisms of visual system plasticity, but much more research is needed to elucidate the many possible roles of neurotrophins.

References

Akaneya, Y., Tsumoto, T., Hatanaka, H. (1996). Brain-derived neurotrophic factor blocks long-term depression in rat visual cortex. *J. Neurophysiol.* 76: 4198–4201.

Akaneya, Y., Tsumoto, T., Kinoshita, S., Hatanaka, H. (1997). Brain-derived neurotrophic factor enhances long-term potentiation in rat visual cortex. *J. Neurosci.* 17:6707–6716.

Allendoerfer, K. L., Cabelli, R. J., Escandon, E., Kaplan, D. R., Nikolics, K., Shatz, C. J. (1994). Regulation of neurotrophin receptors during the maturation of the mammalian visual system. *J. Neurosci.* 14:1795–1811.

Aloe, L., Cozzari, C., Calissano, P., Levi-Montalcini, R. (1981). Somatic and behavioral postnatal effects of fetal injections of nerve growth factor antibodies in the rat. *Nature* 291:413–415.

Anderson, P. A., Olavarria, J., Van Sluyters, R. C. (1988). The overall pattern of ocular dominance bands in cat visual cortex. *J. Neurosci.* 8:2183–2200.

Antonini, A., Stryker, M. P. (1993a). Development of individual geniculocortical arbors in cat striate cortex and effects of binocular impulse blockade. *J. Neurosci.* 13:3549–3573.

Antonini, A., Stryker, M. P. (1993b). Rapid remodeling of axonal arbors in the visual cortex. *Science* 260:1819–1821.

Aroniadou, V. A., Teyler, T. J. (1991). The role of NMDA receptors in long-term potentiation (LTP) and depression (LTD) in rat visual cortex. *Brain Res.* 562: 136–143.

Artola, A., Singer, W. (1987). Long-term potentiation and NMDA receptors in rat visual cortex. *Nature* 330:649–652.

Barde, Y. A. (1989). Trophic factors and neuronal survival. *Neuron* 2:1525–1534.

Bear, M. F., Abraham, W. C. (1996). Long-term depression in hippocampus. *Annu. Rev. Neurosci.* 19:437–462.

Bear, M. F., Malenka, R. C. (1994). Synaptic plasticity: LTP and LTD. *Curr. Opin. Neurobiol.* 4:389–399.

Bear, M. F., Singer, W. (1986). Modulation of visual cortical plasticity by acetylcholine and noradrenaline. *Nature* 320:172–176.

Bear, M. F., Kleinschmidt, A., Gu, Q. A., Singer, W. (1990). Disruption of experience-dependent synaptic modifications in striate cortex by infusion of an NMDA receptor antagonist. *J. Neurosci.* 10:909–925.

Berkemeier, L. R., Winslow, J. W., Kaplan, D. R., Nikolics, K., Goeddel, D. V., Rosenthal, A. (1991). Neurotrophin-5: a novel neurotrophic factor that activates trk and trkB. *Neuron* 4:189–201.

Blakemore, C., Van Sluyters, R. C. (1974). Reversal of the physiological effects of monocular deprivation in kittens: further evidence for a sensitive period. *J. Physiol.* 237:195–216.

Bliss, T. V., Collingridge, G. L. (1993). A synaptic model of memory: long-term potentiation in the hippocampus. *Nature* 361:31–39.

Bliss, T. V., Gardner-Medwin, A. R. (1973). Long-lasting potentiation of synaptic transmission in the dentate area of the unanaesthetized rabbit following stimulation of the perforant path. *J. Physiol.* 232:357–374.

Bliss, T. V., Errington, M. L., Lynch, M. A., Williams, J. H. (1990). Presynaptic mechanisms in hippocampal long-term potentiation. *Cold Spring Harbor Symp. Quant. Biol.* 55:119–129.

Blochl, A., Thoenen, H. (1995). Characterization of nerve growth factor (NGF) release from hippocampal neurons: evidence for a constitutive and an unconventional sodium-dependent regulated pathway. *Eur. J. Neurosci.* 7:1220–1228.

Blochl, A., Thoenen, H. (1996). Localization of cellular storage compartments and sites of constitutive and activity-dependent release of nerve growth factor (NGF) in primary cultures of hippocampal neurons. *Mol. Cell. Neurosci.* 7:173–190.

Bolz, J., Novak, N., Gotz, M., Bonhoeffer, T. (1990). Formation of target-specific neuronal projections in organotypic slice cultures from rat visual cortex. *Nature* 346:359–362.

Bolz, J., Castellani, V., Batardiere, A. (1997). Neurotrophic factors play a role in the elaboration of local cortical circuits. *Soc. Neurosci. Abstr.* 23:1433.

Bonhoeffer, T. (1996). Neurotrophins and activity-dependent development of the neocortex. *Curr. Opin. Neurobiol.* 6:119–126.

Bothwell, M. (1995). Functional interactions of neurotrophins and neurotrophin receptors. *Annu. Rev. Neurosci.* 18:223–253.

Boulanger, L., Poo, M.-M. (1999). Presynaptic depolarization facilitates neurotrophin-induced synaptic potentiation. *Nature Neurosci.* 2:346–351.

Bozzi, Y., Pizzorusso, T., Cremisi, F., Rossi, F. M., Barsacchi, G., Maffei, L. (1995). Monocular deprivation decreases the expression of messenger RNA for brain-derived neurotrophic factor in the rat visual cortex. *Neuroscience* 69:1133–1144.

Cabelli, R. J., Hohn, A., Shatz, C. J. (1995). Inhibition of ocular dominance column formation by infusion of NT-4/5 or BDNF. *Science* 267:1662–1666.

Cabelli, R. J., Allendoerfer, K. L., Radeke, M. J., Welcher, A. A., Feinstein, S. C., Shatz, C. J. (1996). Changing patterns of expression and subcellular localization of TrkB in the developing visual system. *J. Neurosci.* 16:7965–7980.

Cabelli, R. J., Shelton, D. L., Segal, R. A., Shatz, C. J. (1997). Blockade of endogenous ligands of trkB inhibits formation of ocular dominance columns. *Neuron* 19:63–76.

Campenot, R. B. (1982). Development of sympathetic neurons in compartmentalized cultures. II. Local control of neurite survival by nerve growth factor. *Dev. Biol.* 93:13–21.

Campenot, R. B. (1994). NGF and the local control of nerve terminal growth. *J. Neurobiol.* 25:599–611.

Carter, B. D., Zirrgiebel, U., Barde, Y. A. (1995). Differential regulation of p21ras activation in neurons by nerve growth factor and brain-derived neurotrophic factor. *J. Biol. Chem.* 270:21,751–21,757.

Castellani, V., Bolz, J. (1998). NT-3 is a repellent and an attractive guidance cue for subpopulations of cortical neurons. *Soc. Neurosci. Abstr.* 24:789.

Castren, E., Zafra, F., Thoenen, H., Lindholm, D. (1992). Light regulates expression of brain-derived neurotrophic factor mRNA in rat visual cortex. *Proc. Natl. Acad. Sci. USA* 89:9444–9448.

Castren, E., Pitkanen, M., Sirvio, J., Parsadanian, A., Lindholm, D., Thoenen, H., Riekkinen, P. J. (1993). The induction of LTP increases BDNF and

NGF mRNA but decreases NT-3 mRNA in the dentate gyrus. *NeuroReport* 4: 895–898.

Cohen-Cory, S., Fraser, S. E. (1995). Effects of brain-derived neurotrophic factor on optic axon branching and remodelling in vivo. *Nature* 378:192–196.

Colman, H., Nabekura, J., Lichtman, J. W. (1997). Alterations in synaptic strength preceding axon withdrawal. *Science* 275:356–361.

Conover, J. C., Yancopoulos, G. D. (1997). Neurotrophin regulation of the developing nervous system: analyses of knockout mice. *Rev. Neurosci.* 8:13–27.

Cordon-Cardo, C., Tapley, P., Jing, S. Q., Nanduri, V., O'Rourke, E., Lamballe, F., Kovary, K., Klein, R., Jones, K. R., Reichardt, L. F. (1991). The trk tyrosine protein kinase mediates the mitogenic properties of nerve growth factor and neurotrophin-3. *Cell* 66:173–183.

Crair, M. C., Gillespie, D. C., Stryker, M. P. (1998). The role of visual experience in the development of columns in cat visual cortex. *Science* 279:566–570.

Davies, A. M., Horton, A., Burton, L. E., Schmelzer, C., Vandlen, R., Rosenthal, A. (1993). Neurotrophin-4/5 is a mammalian-specific survival factor for distinct populations of sensory neurons. *J. Neurosci.* 13:4961–4967.

Daw, N. W., Fox, K., Sato, H., Czepita, D. (1992). Critical period for monocular deprivation in the cat visual cortex. *J. Neurophysiol.* 67:197–202.

Deshmukh, M., Johnson, E. M., Jr. (1997). Programmed cell death in neurons: focus on the pathway of nerve growth factor deprivation–induced death of sympathetic neurons. *Mol. Pharmacol.* 51:897–906.

Domenici, L., Cellerino, A., Maffei, L. (1993). Monocular deprivation effects in the rat visual cortex and lateral geniculate nucleus are prevented by nerve growth factor (NGF). II. Lateral geniculate nucleus. *Proc. R. Soc. London Ser. B* 251:25–31.

Domenici, L., Cellerino, A., Berardi, N., Cattaneo, A., Maffei, L. (1994). Antibodies to nerve growth factor (NGF) prolong the sensitive period for monocular deprivation in the rat. *NeuroReport* 5:2041–2044.

Dudek, S. M., Bear, M. F. (1993). Bidirectional long-term modification of synaptic effectiveness in the adult and immature hippocampus. *J. Neurosci.* 13: 2910–2918.

Dugich-Djordjevic, M. M., Tocco, G., Lapchak, P. A., Pasinetti, G. M., Najm, I., Baudry, M., Hefti, F. (1992). Regionally specific and rapid increases in brain-derived neurotrophic factor messenger RNA in the adult rat brain following seizures induced by systemic administration of kainic acid. *Neuroscience* 47: 303–315.

Eide, F. F., Vining, E. R., Eide, B. L., Zang, K., Wang, X. Y., Reichardt, L. F. (1996). Naturally occurring truncated trkB receptors have dominant inhibitory effects on brain-derived neurotrophic factor signaling. *J. Neurosci.* 16:3123–3129.

Ernfors, P., Ibanez, C. F., Ebendal, T., Olson, L., Persson, H. (1990). Molecular cloning and neurotrophic activities of a protein with structural similarities to nerve growth factor: developmental and topographical expression in brain. *Proc. Natl. Acad. Sci. USA* 87:5454–5458.

Finney, E. M., Stone, J. R., Shatz, C. J. (1998). A major glutamatergic projection from subplate into visual cortex during development. *J. Comp. Neurol.* 398: 105–118.

Fosse, V. M., Heggelund, P., Fonnum, F. (1989). Postnatal development of glutamatergic, GABAergic, and cholinergic neurotransmitter phenotypes in the visual cortex, lateral geniculate nucleus, pulvinar, and superior colliculus in cats. *J. Neurosci.* 9:426–435.

Fregnac, Y., Schulz, D., Thorpe, S., Bienenstock, E. (1988). A cellular analogue of visual cortical plasticity. *Nature* 333:367–370.

Friauf, E., Shatz, C. J. (1991). Changing patterns of synaptic input to subplate and cortical plate during development of visual cortex. *J. Neurophysiol.* 66: 2059–2071.

Friauf, E., McConnell, S. K., Shatz, C. J. (1990). Functional synaptic circuits in the subplate during fetal and early postnatal development of cat visual cortex. *J. Neurosci.* 10:2601–2613.

Friedlander, M. J. (1982). Structure of physiologically classified neurons in the kitten dorsal lateral geniculate nucleus. *Nature* 300:180–183.

Friedlander, M. J., Lin, C.-S., Stanford, L. R., Sherman, S. M. (1981). Morphology of functionally identified neurons in lateral geniculate nucleus of the cat. *J. Neurophysiol.* 46:80–129.

Frisén, J., Verge, V. M., Fried, K., Risling, M., Persson, H., Trotter, J., Hökfelt, T., Lindholm, D. (1993). Characterization of glial trkB receptors: differential response to injury in the central and peripheral nervous systems. *Proc. Natl. Acad. Sci. USA* 90:4971–4975.

Funakoshi, H., Belluardo, N., Arenas, E., Yamamoto, Y., Casabona, A., Persson, H., Ibanez, C. F. (1995). Muscle-derived neurotrophin-4 as an activity-dependent trophic signal for adult motor neurons. *Science* 268:1495–1499.

Galuske, R. A. W., Kim, D. S., Castren, E., Thoenen, H., Singer, W. (1996). Brain-derived neurotrophic factor reverses experience-dependent synaptic modifications in kitten visual cortex. *Eur. J. Neurosci.* 8:1554–1559.

Garcia, M. L., Garcia, V. B., Isackson, P. J., Windebank, A. J. (1997). Long-term alterations in growth factor mRNA expression following seizures. *NeuroReport* 8:1445–1449.

Ghosh, A., Shatz, C. J. (1992). Involvement of subplate neurons in the formation of ocular dominance columns. *Science* 255:1441–1443.

Ghosh, A., Shatz, C. J. (1994). Segregation of geniculocortical afferents during the critical period: a role for subplate neurons. *J. Neurosci.* 14:3862–3880.

Ghosh, A., Carnahan J., Greenberg M. E. (1994). Requirement for BDNF in activity-dependent survival of cortical neurons. *Science* 263:1618–1623.

Gilbert, C. D., Wiesel, T. N. (1983). Laminar specialization and intracortical connections in cat primary visual cortex. In *The Organization of the Cerebral Cortex,* Schmitt, F. O., Worden, F. G., Adelman, G., Dennis, S., eds. Cambridge, Mass.: MIT Press. pp. 163–191.

Gordon, J. A., Stryker, M. P. (1996). Experience-dependent plasticity of binocular responses in the primary visual cortex of the mouse. *J. Neurosci.* 16:3274–3286.

Gotz, R., Koster, R., Winkler, C., Raulf, F., Lottspeich, F., Schartl, M., Thoenen, H. (1994). Neurotrophin-6 is a new member of the nerve growth factor family. *Nature* 372:266–269.

Guillery, R. W. (1972). Binocular competition in the control of geniculate cell growth. *J. Comp. Neurol.* 144:117–129.

Guillery, R. W. (1973). The effect of lid suture upon the growth of cells in the dorsal lateral geniculate nucleus of kittens. *J. Comp. Neurol.* 148:417–422.

Guillery, R. W., Stelzner, D. J. (1970). The differential effects of unilateral lid closure upon the monocular and binocular segments of the dorsal lateral geniculate nucleus in the cat. *J. Comp. Neurol.* 139:413–422.

Hamburger, V., Brunso-Bechtold, J. K., Yip, J. W. (1981). Neuronal death in the spinal ganglia of the chick embryo and its reduction by nerve growth factor. *J. Neurosci.* 1:60–71.

Haruta, H., Kamishita, T., Hicks, T. P., Takahashi, M. P., Tsumoto, T. (1994). Induction of LTD but not LTP through metabotropic glutamate receptors in visual cortex. *NeuroReport* 5:1829–1832.

Hata, Y., Stryker, M. P. (1994). Control of thalamocortical afferent rearrangement by postsynaptic activity in developing visual cortex. *Science* 265:1732–1735.

Hebb, D. O. (1949). *The Organization of Behavior.* New York: John Wiley and Sons.

Hendry, S. H., Jones, E. G. (1986). Reduction in number of immunostained GABAergic neurones in deprived-eye dominance columns of monkey area 17. *Nature* 320:750–753.

Hensch, T. K., Fagiolini, M., Mataga, N., Stryker, M. P., Baekkeskov, S., Kash, S. F. (1998a). Local GABA circuit control of experience-dependent plasticity in developing visual cortex. *Science* 282:1504–1508.

Hensch, T. K., Gordon, J. A., Brandon, E. P., McKnight, G. S., Idzerda, R. L., Stryker, M. P. (1998b). Comparison of plasticity in vivo and in vitro in the developing visual cortex of normal and protein kinase A RIbeta–deficient mice. *J. Neurosci.* 18:2108–2117.

Hickey, T. L., Guillery, R. W. (1974). An autoradiographic study of retinogeniculate pathways in the cat and the fox. *J. Comp. Neurol.* 156:239–254.

Hofer, M. M., Barde, Y. A. (1988). Brain-derived neurotrophic factor prevents neuronal death in vivo. *Nature* 331:261–262.

Hohn, A., Leibrock, J., Bailey, K., Barde, Y. A. (1990). Identification and characterization of a novel member of the nerve growth factor/brain-derived neurotrophic factor family. *Nature* 344:339–341.

Horton, J. C., Hocking, D. R. (1996). An adult-like pattern of ocular dominance columns in striate cortex of newborn monkeys prior to visual experience. *J. Neurosci.* 16:1791–1807.

Huang, E. P., Stevens, C. F. (1997). Estimating the distribution of synaptic reliabilities. *J. Neurophysiol.* 78:2870–2880.

Huang, Z. J., Kirkwood, A., Pizzorusso, T., Porciatti, V., Morales, B., Bear, M. F., Maffei, L., Tonegawa, S. (1999). BDNF regulates the maturation of inhibition and the critical period of plasticity in mouse visual cortex. *Cell* 98:739–755.

Hubel, D. H., Wiesel, T. N. (1970). The period of susceptibility to the physiological effects of unilateral eye closure in kittens. *J. Physiol.* 206:419–436.

Hubel, D. H., Wiesel, T. N., LeVay, S. (1977). Plasticity of ocular dominance columns in monkey striate cortex. *Philos. Trans. R. Soc. London Ser. B* 278:377–409.

Huber, K. M., Sawtell, N. B., Bear, M. F. (1998). Brain-derived neurotrophic factor alters the synaptic modification threshold in visual cortex. *Neuropharmacology* 37:571–579.

Humphrey, A. L., Sur, M., Uhlrich, D. J., Sherman, S. M. (1985a). Projection patterns of individual X- and Y-cell axons from the lateral geniculate nucleus to cortical area 17 in the cat. *J. Comp. Neurol.* 233:159–189.

Humphrey, A. L., Sur, M., Uhlrich, D. J., Sherman, S. M. (1985b). Termination patterns of individual X- and Y-cell axons in the visual cortex of the cat: projections to area 18, to the 17/18 border region, and to both areas 17 and 18. *J. Comp. Neurol.* 233:190–212.

Ip, N. Y., Ibáñez, C. F., Nye, S. H., McClain, J., Jones, P. F., Gies, D. R., Belluscio, L., Le Beau, M. M., Espinosa, R., 3d, Squinto, S. P., Persson, H., Yancopoulos, G. D. (1992). Mammalian neurotrophin-4: structure, chromosomal localization, tissue distribution, and receptor specificity. *Proc. Natl. Acad. Sci. USA* 89:3060–3064.

Johnston, M. V., Rutkowski, J. L., Wainer, B. H., Long, J. B., Mobley, W. C. (1987). NGF effects on developing forebrain cholinergic neurons are regionally specific. *Neurochem. Res.* 12:985–994.

Jones, K. R., Farinas, I., Backus, C., Reichardt, L. F. (1994). Targeted disruption of the BDNF gene perturbs brain and sensory neuron development but not motor neuron development. *Cell* 76:989–999.

Kang, H., Schuman, E. M. (1995). Long-lasting neurotrophin-induced enhancement of synaptic transmission in the adult hippocampus. *Science* 267:1658–1662.

Kang, H., Welcher, A. A., Shelton, D., Schuman, E. M. (1997). Neurotrophins and time: different roles for TrkB signaling in hippocampal long-term potentiation. *Neuron* 19:653–664.

Katz, L. C., Callaway, E. M. (1992). Development of local circuits in mammalian visual cortex. *Annu. Rev. Neurosci.* 15:31–56.

Katz, L. C., Shatz, C. J. (1996). Synaptic activity and the construction of cortical circuits. *Science* 274:1133–1138.

Kirkwood, A., Bear, M. F. (1995). Elementary forms of synaptic plasticity in the visual cortex. *Biol. Res.* 28:73–80.

Kirkwood, A., Lee, H. K., Bear, M. F. (1995). Co-regulation of long-term potentiation and experience-dependent synaptic plasticity in visual cortex by age and experience. *Nature* 375:328–331.

Kirkwood, A., Rioult, M. C., Bear, M. F. (1996). Experience-dependent modification of synaptic plasticity in visual cortex. *Nature* 381:526–528.

Klein, R., Jing, S., Nanduri, V., O'Rourke, E., Barbacid, M. (1991a). The trk proto-oncogene encodes a receptor for nerve growth factor. *Cell* 65:189–197.

Klein, R., Nanduri, V., Jing, S. A., Lamballe, F., Tapley, P., Bryant, S., Cordon-Cardo, C., Jones, K. R., Reichardt, L. F., Barbacid, M. (1991b). The trkB tyrosine protein kinase is a receptor for brain-derived neurotrophic factor and neurotrophin-3. *Cell* 66:395–403.

Klein, R., Lamballe, F., Bryant, S., Barbacid, M. (1992). The trkB tyrosine protein kinase is a receptor for neurotrophin-4. *Neuron* 8:947–956.

Kleinschmidt, A., Bear, M. F., Singer, W. (1987). Blockade of NMDA receptors disrupts experience-dependent plasticity of kitten striate cortex. *Science* 238:355–358.

Komatsu, Y., Fujii, K., Maeda, J., Sakaguchi, H., Toyama, K. (1988). Long term potentiation of synaptic transmission in kitten visual cortex. *J. Neurophysiol.* 59:124–141.

Korsching, S. (1993). The neurotrophic factor concept: a reexamination. *J. Neurosci.* 13:2739–2748.

Korte, M., Carroll, P., Wolf, E., Brem, G., Thoenen, H., Bonhoeffer, T. (1995). Hippocampal long-term potentiation is impaired in mice lacking brain-derived neurotrophic factor. *Proc. Natl. Acad. Sci. USA* 92:8856–8860.

Korte, M., Griesbeck, O., Gravel, C., Carroll, P., Staiger, V., Thoenen, H., Bonhoeffer, T. (1996). Virus-mediated gene transfer into hippocampal CA1 region restores long-term potentiation in brain-derived neurotrophic factor mutant mice. *Proc. Natl. Acad. Sci. USA* 93:12,547–12,552.

Lamballe, F., Klein, R., Barbacid, M. (1991). trkC, a new member of the trk family of tyrosine protein kinases, is a receptor for neurotrophin-3. *Cell* 66:967–979.

Leibrock, J., Lottspeich, F., Hohn, A., Hofer, M., Hengerer, B., Masiakowski, P., Thoenen, H., Barde, Y. A. (1989). Molecular cloning and expression of brain-derived neurotrophic factor. *Nature* 341:149–152.

Lein, E. S., Shatz, C. J. (2000). Rapid regulation of brain-derived neurotrophic factor mRNA within eye-specific circuits during ocular dominance column formation. *J. Neurosci.* 20:1470–1483.

Lein, E. S., Finney, E. M., McQuillen, P. S., Shatz, C. J. (1999). Subplate neuron ablation alters neurotrophin expression and ocular dominance column formation. *Proc. Natl. Acad. Sci. USA* 96:13,491–13,495.

Lein, E. S., Hohn, A., Shatz, C. J. (2000). Dynamic regulation of BDNF and NT-3 expression during visual system development. *J. Comp. Neurol.* 420:1–18.

LeVay, S., Stryker, M. P., Shatz, C. J. (1978). Ocular dominance columns and their development in layer IV of the cat's visual cortex: a quantitative study. *J. Comp. Neurol.* 179:223–244.

LeVay, S., Wiesel, T. N., Hubel, D. H. (1980). The development of ocular dominance columns in normal and visually deprived monkeys. *J. Comp. Neurol.* 191:1–51.

Levi-Montalcini, R. (1987). The nerve growth factor 35 years later. *Science* 237:1154–1162.

Lichtman, J. W., Balice-Gordon, R. J. (1990). Understanding synaptic competition in theory and in practice. *J. Neurobiol.* 21:99–106.

Lindsay, R. M., Thoenen, H., Barde, Y. A. (1985). Placode and neural crest–derived sensory neurons are responsive at early developmental stages to brain-derived neurotrophic factor. *Dev. Biol.* 112:319–328.

Lo, Y.-J., Poo, M.-M. (1991). Activity-dependent synaptic competition in vitro: heterosynaptic suppression of developing synapses. *Science* 254:1019–1022.

Lo, Y.-J., Poo, M.-M. (1994). Heterosynaptic suppression of developing neuromuscular synapses in culture. *J. Neurosci.* 14:4684–4693.

Lohof, A. M., Ip, N. Y., Poo, M.-M. (1993). Potentiation of developing neuromuscular synapses by the neurotrophins NT-3 and BDNF. *Nature* 363:350–353.

McAllister, A. K., Lo, D. C., Katz, L. C. (1995). Neurotrophins regulate dendritic growth in developing visual cortex. *Neuron* 15:791–803.

McAllister, A. K., Katz, L. C., Lo, D. C. (1997). Opposing roles for endogenous BDNF and NT-3 in regulating cortical dendritic growth. *Neuron* 18:767–778.

McAllister, A. K., Katz, L. C., Lo, D. C. (1999). Neurotrophins and synaptic plasticity. *Annu. Rev. Neurosci.* 22:295–318.

Madison, D. V., Malenka, R. C., Nicoll, R. A. (1991). Mechanisms underlying long-term potentiation of synaptic transmission. *Annu. Rev. Neurosci.* 14:379–397.

Maffei, L., Berardi, N., Domenici, L., Parisi, V., Pizzorusso, T. (1992). Nerve growth factor (NGF) prevents the shift in ocular dominance distribution of visual cortical neurons in monocularly deprived rats. *J. Neurosci.* 12:4651–4662.

Malenka, R. C. (1994). Synaptic plasticity in the hippocampus: LTP and LTD. *Cell* 78:535–538.

Malenka, R. C., Nicoll, R. A. (1993). NMDA-receptor-dependent synaptic plasticity: multiple forms and mechanisms. *Trends Neurosci.* 16:521–527.

Malenka, R. C., Nicoll, R. A. (1997). Silent synapses speak up. *Neuron* 19:473–476.

Martin-Zanca, D., Mitra, G., Long, L. K., Barbacid, M. (1986). Molecular characterization of the human trk oncogene. *Cold Spring Harbor Symp. Quant. Biol.* 51:983–992.

Martínez, A., Alcántara, S., Borrell, V., Del Río, J. A., Blasi, J., Otal, R., Campos, N., Boronat, A., Barbacid, M., Silos-Santiago, I., Soriano, E. (1998). TrkB and TrkC signaling are required for maturation and synaptogenesis of hippocampal connections. *J. Neurosci.* 18:7336–7350.

Marty, S., Berzaghi Mda, P., Berninger, B. (1997). Neurotrophins and activity-dependent plasticity of cortical interneurons. *Trends Neurosci.* 20:198–202.

Meyer-Franke, A., Kaplan, M. R., Pfrieger, F. W., Barres, B. A. (1995). Characterization of the signaling interactions that promote the survival and growth of developing retinal ganglion cells in culture. *Neuron* 15:805–819.

Miller, K. D., Stryker, M. P. (1990). Development of ocular dominance columns: mechanisms and models. In *Connectionist Modeling and Brain Function: The Developing Interface,* Olson, C. R., Hanson, S. J., eds. Cambridge, Mass.: MIT Press. pp. 255–305.

Miller, K. D., Keller, J. B., Stryker, M. P. (1989). Ocular dominance column development: analysis and simulation. *Science* 245:605–615.

Mower, G. D., Caplan, C. J., Christen, W. G., Duffy, F. H. (1985). Dark rearing prolongs physiological but not anatomical plasticity of the cat visual cortex. *J. Comp. Neurol.* 235:448–466.

Naegele, J. R., Katz, L. C. (1990). Cell surface molecules containing *N*-acetyl-galactosamine are associated with basket cells and neurogliaform cells in cat visual cortex. *J. Neurosci.* 10:540–557.

Nawa, H., Pelleymounter, M. A., Carnahan, J. (1994). Intraventricular administration of BDNF increases neuropeptide expression in newborn rat brain. *J. Neurosci.* 14:3751–3765.

Nguyen, Q. T., Parsadanian, A. S., Snider, W. D., Lichtman, J. W. (1998). Hyperinnervation of neuromuscular junctions caused by GDNF overexpression in muscle. *Science* 279:1725–1729.

Oppenheim, R. W. (1991). Cell death during development of the nervous system. *Annu. Rev. Neurosci.* 14:453–501.

Patterson, S. L., Grover, L. M., Schwartzkroin, P. A., Bothwell, M. (1992). Neurotrophin expression in rat hippocampal slices: a stimulus paradigm inducing LTP in CA1 evokes increases in BDNF and NT-3 mRNAs. *Neuron* 9:1081–1088.

Patterson, S. L., Abel, T., Deuel, T. A., Martin, K. C., Rose, J. C., Kandel, E. R. (1996). Recombinant BDNF rescues deficits in basal synaptic transmission and hippocampal LTP in BDNF knockout mice. *Neuron* 16:1137–1145.

Penn, A. A., Riquelme, P. A., Feller, M. B., Shatz, C. J. (1998). Competition in retinogeniculate patterning driven by spontaneous activity. *Science* 279:2108–2112.

Pettmann, B., Henderson, C. E. (1998). Neuronal cell death. *Neuron* 20:633–647.

Purves, D. (1988). *Body and Brain: A Trophic Theory of Neural Connections.* Cambridge, Mass.: Harvard University Press.

Rakic, P. (1977). Prenatal development of the visual system in rhesus monkey. *Philos. Trans. R. Soc. London Ser. B.* 278:245–260.

Reiter, H. O., Stryker, M. P. (1988). Neural plasticity without postsynaptic action potentials: less-active inputs become dominant when kitten visual cortical cells are pharmacologically inhibited. *Proc. Natl. Acad. Sci. USA* 85:3623–3627.

Riddle, D. R., Lo, D. C., Katz, L. C. (1995). NT-4-mediated rescue of lateral geniculate neurons from effects of monocular deprivation. *Nature* 378:189–191.

Rodieck, R. W. (1979). Visual pathways. *Annu. Rev. Neurosci.* 2:193–255.

Rutherford, L. C., DeWan, A., Lauer, H. M., Turrigiano, G. G. (1997). Brain-derived neurotrophic factor mediates the activity-dependent regulation of inhibition in neocortical cultures. *J. Neurosci.* 17:4527–4535.

Rutherford, L. C., Nelson, S. B., Turrigiano, G. G. (1998). BDNF has opposite effects on the quantal amplitude of pyramidal neuron and interneuron excitatory synapses. *Neuron* 21:521–530.

Schoups, A. A., Elliott, R. C., Friedman, W. J., Black, I. B. (1995). NGF and BDNF are differentially modulated by visual experience in the developing geniculocortical pathway. *Dev. Brain Res.* 86:326–334.

Schuman, E. M. (1999). Neurotrophin regulation of synaptic transmission. *Curr. Opin. Neurobiol.* 9:105–109.

Shatz, C. J. (1990). Impulse activity and the patterning of connections during CNS development. *Neuron* 5:745–756.

Shatz, C. J. (1997). Neurotrophins and visual system plasticity. In *Molecular and Cellular Approaches to Neural Development,* Cowan, W. M., Jessell, T. M., Zipursky, S. L., eds. New York: Oxford University Press. pp. 509–524.

Shatz, C. J., Stryker, M. P. (1978). Ocular dominance in layer IV of the cat's visual cortex and the effects of monocular deprivation. *J. Physiol.* 281:267–283.

Shatz, C. J., Stryker, M. P. (1988). Prenatal tetrodotoxin infusion blocks segregation of retinogeniculate afferents. *Science* 242:87–89.

Snider, W. D. (1994). Functions of the neurotrophins during nervous system development: what the knockouts are teaching us. *Cell* 77:627–638.

Snider, W. D., Lichtman, J. W. (1996). Are neurotrophins synaptotrophins? *Mol. Cell. Neurosci.* 7:433–442.

Stanton, P. K., Sejnowski, T. J. (1989). Associative long term depression in the hippocampus: induction of synaptic plasticity by Hebbian covariance. *Nature* 339:215–218.

Stevens, C. F. (1993). Quantal release of neurotransmitter and long-term potentiation. *Cell* 72(Suppl.):55–63.

Stryker, M. P., Harris, W. A. (1986). Binocular impulse blockade prevents the formation of ocular dominance columns in cat visual cortex. *J. Neurosci.* 6: 2117–2133.

Stryker, M. P., Strickland, S. L. (1984). Physiological segregation of ocular dominance columns depends on the pattern of afferent electrical activity. *Invest. Opthalmol. Vis. Sci.* 25(Suppl.):278.

Tanaka, T., Saito, H., Matsuki, N. (1997). Inhibition of GABA$_A$ synaptic responses by brain-derived neurotrophic factor (BDNF) in rat hippocampus. *J. Neurosci.* 17:2959–2966.

Thoenen, H. (1995). Neurotrophins and neuronal plasticity. *Science* 270:593–598.

Turrigiano, G. G., Leslie, K. R., Desai, N. S., Rutherford, L. C., Nelson, S. B. (1998). Activity-dependent scaling of quantal amplitude in neocortical neurons. *Nature* 391:892–896.

Venero, J. L., Hefti, F. (1993). TrkA NGF receptor expression by non-cholinergic thalamic neurons. *NeuroReport* 4:959–962.

Vicario-Abejón, C., Collin, C., McKay, R. D., Segal, M. (1998). Neurotrophins induce formation of functional excitatory and inhibitory synapses between cultured hippocampal neurons. *J. Neurosci.* 18:7256–7271.

Wetmore, C., Olson, L., Bean, A. J. (1994). Regulation of brain-derived neurotrophic factor (BDNF) expression and release from hippocampal neurons is mediated by non-NMDA type glutamate receptors. *J. Neurosci.* 14:1688–1700.

Wiesel, T. N., Hubel, D. H. (1963a). Effects of visual deprivation on morphology and physiology of cells in the cat's lateral geniculate body. *J. Neurophysiol.* 26:978–993.

Wiesel, T. N., Hubel, D. H. (1963b). Single cell responses in striate cortex of kittens deprived of vision in one eye. *J. Neurophysiol.* 26:1003–1017.

Wiesel, T. N., Hubel, D. H. (1965). Comparison of the effects of unilateral and bilateral eye closure on cortical unit responses in kittens. *J. Neurophysiol.* 28: 1029–1040.

Williams, R. W., Rakic, P. (1988). Elimination of neurons from the rhesus monkey's lateral geniculate nucleus during development. *J. Comp. Neurol.* 272: 424–436.

Wu, K., Xu, J. L., Suen, P. C., Levine, E., Huang, Y. Y., Mount, H. T., Lin, S. Y., Black, I. B. (1996). Functional trkB neurotrophin receptors are intrinsic components of the adult brain postsynaptic density. *Mol. Brain Res.* 43: 286–290.

Yan, Q., Rosenfeld, R. D., Matheson, C. R., Hawkins, N., Lopez, O. T., Bennett, L., Welcher, A. A. (1997). Expression of brain-derived neurotrophic factor protein in the adult rat central nervous system. *Neuroscience* 78:431–448.

14 Novel Neurotransmitters and Their Neuropsychiatric Relevance

Solomon H. Snyder and Christopher D. Ferris

Neurotransmitters are the key information molecules of the brain and mediate the actions of all known psychoactive drugs. Presently we know of at least 50–100 possible neurotransmitters representing diverse chemical classes, including the biogenic amines, amino acids, peptides, and gases. As recently as the late 1950s, we knew of only two neurotransmitters, acetylcholine and norepinephrine. Acetylcholine was well established as the neurotransmitter of the neuromuscular junction and in the autonomic nervous system, whereas norepinephrine was appreciated as the neurotransmitter of postganglionic sympathetic neurons. It was assumed that acetylcholine was important in the brain, but its central nervous system (CNS) functions were unclear. Norepinephrine had not yet been identified in CNS neurons. During these years, seminal observations were spawning the field of psychopharmacology. Newly developed fluorometric techniques permitted the simple and specific measurement of biogenic amines. Soon it was discovered that reserpine, which causes depression in many patients, depleted the brain of serotonin and norepinephrine, and that monoamine oxidase inhibitors, which had antidepressant effects, elevated the levels of these biogenic amines. There emerged a simplistic but reasonably accurate notion that norepinephrine and serotonin were important determinants of mood, with low levels causing and elevated levels reversing depression.

What was extraordinary about this early work, besides its prescience, was that this functional-psychiatric insight into biogenic amines preceded any evidence that these chemicals were neurotransmitters or even that they were contained within neurons. Establishing the presence of a substance in neurons and visualizing the specific neuronal populations containing it provides a quantum leap in insight. For biogenic amines, this information did not emerge until the mid 1960s, when a histochemical technique to visualize the amines, developed by the Swedish investigator Nils-Erik Hillarp, was applied to the brain by his students Kjell Fuxe and Annica Dahlstrom. These investigators successfully mapped out the biogenic amine neuronal pathways in the brain (Hillarp et al., 1966).

These historical reflections highlight a long-standing question in the neurosciences that has not yet been fully resolved: what is a neurotransmitter? Over the years, criteria for defining a neurotransmitter have become so problematic that many neuroscientists despairingly gloss over the problem with vague words such as *neuromodulator*. For purposes of classifying neurotransmitters, neurophysiologists have discriminated two principal types of physiological actions, fast and slow. Fast effects usually involve opening or closing of ion channels that are often part of a receptor protein, such as the sodium channel that is contained within the

This article is adapted from, and reproduced with the permission of, the *American Journal of Psychiatry*.

nicotinic acetylcholine receptor. Slow actions typically arise from metabolic alterations in target cells whose receptor proteins are termed metabotropic. Classic examples include the formation of cyclic GMP, cyclic AMP, or inositol (1,4,5) trisphosphate by enzymes that are linked to receptors via GTP-binding G proteins (e.g., the muscarinic receptors for acetylcholine). Neurotransmitters are believed to signal through both fast and slow mechanisms. What else defines a neurotransmitter?

Although most neuroscientists have a vague idea of what to expect in a neurotransmitter, some researchers have enunciated concrete criteria for "transmitterhood." Since acetylcholine was the first neurotransmitter to be identified and characterized, the early criteria assumed that candidate neurotransmitters should be "just like" acetylcholine. For example, it was known that acetylcholine is synthesized by choline acetyltransferase, stored in synaptic vesicles, and released into the synapse following fusion of the vesicles with the plasma membrane in response to elevated calcium through exocytosis. Thus a candidate transmitter should be localized to neurons, have a special biosynthetic enzyme, and be released in a calcium-dependent fashion when the nerve is depolarized. Of course, the candidate transmitter should mimic the actions of the endogenous transmitter when applied to postsynaptic cells. Acetylcholine, when acting postsynaptically, binds to specialized receptor proteins like the nicotinic and muscarinic acetylcholine receptors. Accordingly, dogma expected new candidate transmitters likewise to bind to a receptor protein localized to the external surface of the target cell's plasma membrane. Finally, to mimic acetylcholine, an enzyme near the receptor (i.e., like acetylcholinesterase) should degrade and thereby inactivate the transmitter at the synapse.

Thus a neurotransmitter has been thought of as a chemical stored in a nerve terminal that is released when the nerve fires to act on adjacent cells, altering their level of excitation. However, there are many challenges to this simpleminded conceptualization. For example, neurophysiologists have now identified many unconventional ways in which neurons can signal to each other via their cell bodies, their dendrites, and portions of their axons other than the nerve terminals. These findings have generated new questions regarding neurotransmitter functions. Might a neurotransmitter be released from a dendrite rather than a nerve terminal? Can a chemical that acts on the same neuron that released it be a neurotransmitter? Neurons account for only a minority of cells in the brain, with glia constituting about 85% of the total. Though originally characterized as "supporting cells," glia are now known to display a number of interesting electrical and chemical features. Might a substance be a "neurotransmitter" if it acts on glial cells rather than neurons? Can glia release neurotransmitters?

By the early 1960s, difficulties with the acetylcholine-based definition of a neurotransmitter became evident as research moved beyond acetylcholine. In the early characterization of norepinephrine, monoamine oxidase, the only known degrading enzyme, was presumed to provide synaptic inactivation. However, monoamine oxidase inhibitors did not potentiate sympathetic neurotransmission. When Julius Axelrod discovered catechol-*O*-methyltransferase (COMT), he thought that this enzyme might constitute the synaptic inactivating system. However, COMT inhibitors also failed to potentiate synaptic transmission. Axelrod and his associates then made a key discovery: synaptic inactivation of norepinephrine did not involve enzymes at all, but rather a novel reuptake mechanism whereby the presynaptic neuron removed the neurotransmitter from the synapse by transporting it back into the nerve that had released it (Axelrod, 1971). If one applied the original, rigid, criteria for a transmitter based upon acetylcholine, norepinephrine would fail to qualify. Over time, it became evident that reuptake is common. Almost all biogenic amines, including serotonin and dopamine, are inactivated by reuptake. Interestingly, histamine provides an exception to this generalization. Although histamine is formed by a specific enzyme, histidine decarboxylase, and is localized to discrete neuronal populations in the brain, there is no evidence that reuptake accounts for its synaptic inactivation. In addition, the histamine-metabolizing enzymes, histamine methyltransferase and diamine oxidase, do not account for histamine's inactivation. Thus whether specific synaptic inactivation terminates histamine signaling remains an open question.

In addition to biogenic amines, amino acids were found to have neurotransmitter functions. Signaling by γ-aminobutyric acid (GABA), glutamate, glycine, and other amino acid transmitters is terminated by specific reuptake transport proteins like the biogenic amines. Gradually neuroscientists accepted neurotransmitter reuptake as a valid means of providing synaptic inactivation, like enzymatic degradation. In fact, reuptake eventually became appreciated as the rule rather than the exception.

The discovery and characterization of amino acid transmitters in the brain challenged other dogma about what constitutes a neurotransmitter. GABA was readily accepted as a transmitter because it is primarily, if not exclusively, devoted to a neurotransmitter role, and it is synthesized by a specific enzyme, glutamic acid decarboxylase. Moreover, the enzyme is localized to discrete neurons, GABA neurons. Glutamate and glycine were far more difficult for the scientific community to accept, because these amino acids are involved in protein synthesis and other metabolic pathways, with only a minor fraction of the total neuronal pool involved in a transmitter role. Glutamate posed a particular challenge, because its total concentration in the brain is extraordinarily high (circa 20 mM), and glutamate is crucial in numerous pathways of intermediary metabolism. Even today, it is difficult to identify histochemically exactly which neurons in the brain employ glutamate as a neurotransmitter. Since one cannot identify specific biosynthetic enzymes for glutamate and glycine, one could argue that these molecules do not satisfy classic criteria for a neurotransmitter.

Despite such initial concerns, neuroscientists now accept amino acids as neurotransmitters with widespread functions in the brain. Although exact percentages are hard to pin down, glutamate is likely to be the neurotransmitter for 50% or more of synapses in the brain and is unquestionably the principal excitatory neurotransmitter. Aspartate chemically resembles glutamate, having just one less methylene group, so it is difficult to distinguish these two in terms of transmitter function and localization. However, aspartate is thought to be an important excitatory neurotransmitter in the spinal cord. GABA is the principal inhibitory neurotransmitter in the brain and occupies 25–40% of synapses depending on the brain region. In the spinal cord glycine may be the major inhibitory transmitter, involving 25–30% of synapses. By contrast, dopamine, norepinephrine, and serotonin each are thought to account for only about 1% of brain synapses, while acetylcholine might occupy up to 5%. Thus quantitatively amino acids are the principal neurotransmitters in the brain.

Like the amino acids, adenosine is involved in multiple metabolic functions and occurs in high concentrations in all tissues. Adenosine is neither a biogenic amine nor an amino acid, though it does possess aminelike properties. Discrete neuronal populations, with uniquely high densities of adenosine, have been identified by immunohistochemistry, suggesting a transmitter role (Snyder, 1985). Adenosine is neuroactive, acting upon nerve terminals to inhibit the release of most neurotransmitters. Moreover, a robust reuptake system for adenosine exists, and potent, selective adenosine reuptake inhibitors potentiate adenosine effects on neural activity.

By the end of the 1960s, acetylcholine, biogenic amines, and amino acids had been generally accepted as neurotransmitters. The 1970s became the decade of the peptide transmitters. Substance P was the first appreciated neuropeptide. It was discovered in the 1930s as an unidentified factor from tissue extracts that caused smooth muscle contraction. Substance P was isolated and sequenced by Susan Leeman, culminating her efforts to identify a tissue component that stimulates salivation in rats (Chang and Leeman, 1970). Widespread interest in neuropeptides followed the identification of opiate receptors and the discovery of enkephalins as their endogenous ligands (Hughes et al., 1975; Snyder, 1975). Enkephalins are small peptides containing five amino acids. They are highly localized to the same discrete sites as opiate receptors. Actions of the enkephalins closely mimic those of morphine, which led to rapid acceptance of the enkephalins as neuroactive substances with physiological roles in the brain. The localization of the enkephalins made use of immunohistochemistry, a technology pioneered through the elegant studies of Hökfelt and associates (Hökfelt et al., 1984) as a tool to characterize neuropeptides. In the late 1970s and early 1980s immunohistochemistry contributed to a veritable explosion in the number of neuropeptides. During this time, substance P was found to be enriched in intestinal neurons as well as brain neurons. Indeed many neuropeptides were first identified in the intestine and related organs such as the pancreas. Examples include vasoactive intestinal polypeptide, cholecystokinin, gastrin, and even insulin (Krieger 1983).

During the time neuropeptides were being discovered in the brain and intestine, one might have argued that the amines and amino acids had "filled up the brain" so that there were no neuronal populations remaining without transmitters. What neurons would use neuropeptides? This dilemma was addressed by Hökfelt and others establishing evidence for co-transmitters (Hökfelt et al., 1984). We now know that most neurons contain an amino acid transmitter stored together with a biogenic amine or a peptide. While any given neuropeptide probably accounts for only 1% or fewer of synapses, the 50–100 known neuropeptides may collectively occupy a substantial portion of CNS neurons. Now it is generally accepted that few if any brain neurons contain a single transmitter.

As transmitters, neuropeptides also challenge the dogmatic criteria for neurotransmitters. In terms of synaptic inactivation, peptides are presumably hydrolyzed by various peptidases. However, no one has rigorously demonstrated that a specific peptidase accounts for synaptic inactivation of individual neuropeptides. Most peptides do not elicit neuronal excitation or inhibition like acetylcholine and GABA. Rather, neuropeptides appear to have a modulatory action that has been difficult for neurophysiologists to characterize clearly. In some instances, certain actions of serotonin, norepinephrine, and dopamine also appear modulatory. This has led some to deny these molecules "transmitter status" and relegate them to the muddled role of neuromodulators. Perhaps, if the discovery of neuropeptides had preceded that of acetylcholine, the latter would have been challenged to fulfill transmitter criteria modeled on neuropeptide function.

By the mid-1980s, so many neuropeptides had been identified that most neuroscientists were ready to close the book on chemical classes of neurotransmitters, with completed chapters on biogenic amines, amino acids, and neuropeptides. The 1990s have brought us two new major chemical classes of novel neurotransmitters, which we will now explore in detail: gases and D-amino acids.

Gases

Nitric Oxide

Nitric oxide (NO) was first appreciated in the 1980s for its role in blood vessel and macrophage physiology. In 1980 Robert Furchgott discovered that the ability of acetylcholine to relax blood vessels did not involve direct actions through cholinergic receptors on smooth muscle. Instead acetylcholine's action required the endothelium to elaborate a small vasoactive substance that would enter the smooth muscle to relax it

(Furchgott and Zawadzki, 1980). This endothelial-derived relaxing factor (EDRF) was later shown to be NO (Ignarro et al., 1981, 1987; Palmer et al., 1987). Independently other workers were trying to explain how activated macrophages kill tumor cells and bacteria. Arginine, later found to be the precursor of NO, was known to be crucial, and NO was identified as the key active molecule (Hibbs et al., 1987; Ignarro et al., 1987; Marletta et al., 1988; Stuehr et al., 1989). Garthwaite and colleagues (Garthwaite et al., 1988) reported evidence for a substance with EDRF activity in the cerebellum. We became curious about a possible role for NO in the brain. Since it was virtually impossible at that time to measure

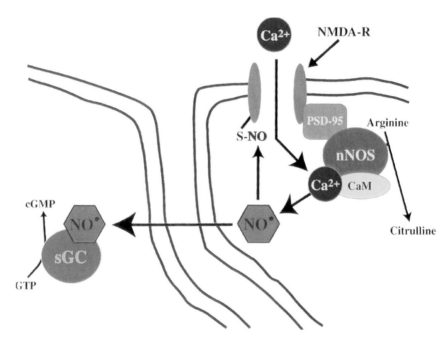

Figure 14.1. Generation of nitric oxide from arginine following glutamate activation of NMDA receptors and calcium influx. In the brain NO production is linked to glutamate receptor activation (see text). The binding of glutamate to NMDA-type glutamate receptors (NMDA-R) results in the opening of an ion channel that allows Ca^{2+} to enter the neuron. nNOS is physically linked to NMDA receptors through postsynaptic density protein-95 (PSD-95). Following activation by Ca^{2+}/calmodulin (CaM), arginine is converted to NO and citrulline. In this way Ca^{2+} coming through the NMDA receptor efficiently activates NO production through CaM. NO, in turn, regulates NMDA receptor function through direct modification of a cysteine residue contained within a subunit of the NMDA receptor. Covalent modification of the sulfur atom (S) in cysteine by NO is known as S-nitrosylation and, like phosphorylation, may represent a general mechanism of posttranslational protein modification. In addition NO functions as a neurotransmitter by diffusing through the membranes of postsynaptic cells, where it binds the heme in soluble guanylate cyclase (sGC), activating this enzyme to make the second messenger, cGMP.

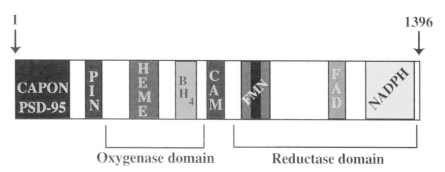

Figure 14.2. nNOS contains oxygenase and reductase domains and has binding sites for multiple regulatory molecules. nNOS has 1396 amino acids. The most N-terminal region contains binding sites for interacting proteins such as PSD-95, CAPON, and PIN (see text). nNOS also contain an oxygenase domain that binds heme and tetrohydrobiopterin (BH_4). A calmodulin-binding site links the oxygenase domain to the reductase domain in the C-terminal half of nNOS. The oxygenase domain displays homology to cytochrome P450 reductase and contains binding sites for FMN, FAD, and NADPH, as shown. As indicated by a stripe in the center of the FMN domain, nNOS contains a 42-amino-acid insert that confers sensitivity to calmodulin (Daff et al., 1999). This insert is absent in iNOS, which is not dependent upon calmodulin.

NO gas directly, we decided to focus upon the enzyme that makes NO. NO is generated in a single step from the amino acid arginine via an enzyme designated NO synthase (NOS; Figs. 14.1 and 14.2). Utilizing NADPH as an electron donor, NOS oxidizes one of the guanidino nitrogens of arginine to form NO, with citrulline as an amino acid co-product. Despite extensive investigations, a neuroactive role for citrulline has not been found.

To establish whether NO plays a role in neurotransmission in the brain, we took advantage of what was already known about how NO relaxes smooth muscle in blood vessels. By stimulating the activity of soluble guanylyl cyclase, the enzyme that makes cGMP, NO initiates a signaling cascade that involves protein phosphorylation and causes smooth muscle relaxation. The cerebellum contains the highest levels of cGMP in the brain. In addition, it was known that glutamate could trigger a 10-fold augmentation of cGMP levels via activation of the N-methyl-D-aspartate (NMDA) subtype of glutamate receptors. As an index of NOS activity, we monitored the conversion of radiolabeled arginine to citrulline in ex vivo preparations of rat cerebellum. Following NMDA receptor activation, NOS activity rapidly tripled in parallel with increased cGMP levels (Bredt and Snyder, 1989; Garthwaite et al., 1989). We utilized arginine derivatives that inhibit NOS activity by competing with arginine for the active site in NOS to explore a link between NOS and cGMP. NOS inhibitors specifically blocked the rise in cGMP. Therefore, we could attribute increased cGMP production to NO, establishing a role for NO in mediating actions of glutamate in the brain.

Clearly NO dynamics are determined by NOS activity. Thus it was crucial to characterize this enzyme. Several groups had tried to purify the NO-generating protein without success, because with most purification steps enzyme activity was lost. We guessed that the enzyme itself might not be particularly labile, but that a co-factor might be dissociated during the purification. Knowing that calcium augmented enzyme activity, we hypothesized that the calcium-binding protein, calmodulin, might be the co-factor. When added to partially purified preparations, calmodulin completely restored NOS activity (Bredt and Snyder, 1990). Besides permitting purification of the enzyme, this finding explained how glutamate, through NMDA receptor activation, could almost immediately stimulate NO formation. After opening in response to binding glutamate, the NMDA receptor ion channel allows calcium entry into neurons, where it binds calmodulin, forming a calcium-calmodulin complex that binds and activates NOS.

Subsequently we biochemically purified NOS (Bredt and Snyder, 1990) and developed specific antibodies allowing immunohistochemical visualization of NO neurons (Bredt et al., 1990). NOS-containing neurons have very discrete localizations and represent only about 1% of neuronal cells in the brain. However, their axons ramify so extensively that virtually every cell in the brain may encounter an NOS nerve terminal (Bredt et al., 1991a,b). Presumably glutamate neurons synapse upon NOS neurons, enabling NMDA receptor activation to trigger NO formation. As a diatomic gas, NO is freely diffusible and thus can readily enter adjacent neuronal or other cells (Fig. 14.1). Once inside target cells, NO binds the iron in heme contained within the active site of soluble guanylyl cyclase, activating the enzyme to form cGMP.

Given what we know about the synthesis and action of NO, does NO fulfill classic criteria for a neurotransmitter? In the brain NO is formed in neurons in response to calcium influx, reminiscent of calcium-dependent exocytic release of other neurotransmitters. However, being a gas, NO cannot be stored in synaptic vesicles or released by exocytosis. Moreover, there are no "receptors" for NO on the postsynaptic membrane of adjacent cells. Instead NO diffuses from NOS neurons into neighboring cells, where it binds guanylyl cyclase, an "enzyme receptor."

For most neurotransmitters, only a small percentage of the total store is released with each nerve stimulation, as there is a large storage pool of neurotransmitter in synaptic vesicles. Since NO cannot be stored, NOS must be activated every time a neuron needs to release it. Accordingly one might expect NOS enzymatic activity to be exquisitely regulated. Biochemical purification of NOS allowed us to clone the gene encoding it, obtain the full amino acid sequence, and express the recombinant protein in vitro for detailed characterization (Bredt et al., 1991b). NOS contains multiple sites for regulation (Bredt et al., 1992). Though most oxidative enzymes utilize one electron donor, NOS is more complex. Besides NADPH, NOS possesses tightly bound flavin adenine mononucleotide (FMN) and flavin adenine dinucleotide (FAD).

It also utilizes heme and tetrahydrobiopterin as electron donors. NOS possesses sites for phosphorylation by the major phosphorylating enzymes cAMP-dependent protein kinase, protein kinase C (PKC), calcium-calmodulin-dependent protein kinase, and cGMP-dependent protein kinase (Brune and Lapetina, 1991; Nakane et al., 1991; Bredt et al., 1992). Recently NOS, in endothelial cells, has been shown to be phosphorylated by protein kinase B, also known as Akt, which participates in signaling cascades that affect nuclear function (Dimmeler et al., 1999; Fulton et al., 1999; Michell et al., 1999).

NOS can be regulated by interactions with other proteins. In the brain NOS is physically linked to the postsynaptic membrane near NMDA receptors through its interaction with postsynaptic density protein-95 (Brenman et al., 1996). Using the yeast two-hybrid technique we identified two other NOS binding partners: a protein inhibitor of NOS (PIN) (Jaffrey and Snyder, 1996) and carboxyl-terminal PDZ ligand of NOS (CAPON) (Jaffrey et al., 1998). PIN is a phylogenetically conserved small protein that inhibits neuronal NOS by preventing the formation of NOS dimers, the form required for enzyme activity. CAPON appears to be a chaperone or scaffolding protein that links neuronal NOS to other proteins.

Molecular cloning of NOS from brain permitted the identification of genes for distinct NOS isozymes from endothelial cells in blood vessels (Janssens et al., 1992; Lamas et al., 1992; Sessa et al., 1992) and macrophages (Lowenstein et al., 1992; Lyons et al., 1992; Xie et al., 1992). The form of NOS that we initially purified and cloned has been designated neuronal NOS (nNOS), while the macrophage form is termed inducible NOS (iNOS) and the endothelial form is called endothelial NOS (eNOS). iNOS is so designated because, under resting conditions, macrophages and other cells display negligible enzyme activity. However, in response to physiological stimuli, such as exposure to lipopolysaccharide from gram-negative bacteria, these cells are induced to express iNOS and generate large amounts of NO sufficient to kill bacteria or tumor cells. Such stimuli provoke new synthesis of iNOS enzyme protein in just 1–2 h. By contrast, nNOS and eNOS proteins are constitutively present and, as described previously for nNOS, are activated by calcium-calmodulin.

Functions of NO

Experimentally establishing that a substance is a neurotransmitter in the brain can be exceedingly difficult. The intestine provides an excellent model system for the study of neurotransmitters since smooth muscle function can be used to monitor neurotransmitter release from the enteric nervous system (ENS), and the ENS uses nearly all transmitters found in the brain. Indeed the most direct evidence for NO as a neurotransmitter comes from studies of the intestine, in which many neurons express nNOS. The excitatory transmitters in the ENS—

including acetylcholine, norepinephrine, and some neuropeptides—
have been known for some time. In contrast, the identity of the major
inhibitory transmitters has been controversial. Since inhibitory transmission persists in the presence of selective blockade of adrenergic
and cholinergic receptors, inhibitory transmission was designated non-
adrenergic, noncholinergic (NANC) transmission. Stimulation of intestinal neurons causes the elaboration of a vasorelaxant factor that is
indistinguishable from NO (Bult et al., 1990). NOS inhibitors block
NANC transmission (Boeckxstaens et al., 1990; Stark et al., 1991; Stark
and Szurszewski, 1992). To clarify further the role of NO as a neurotransmitter, mice with targeted genomic deletion of the nNOS gene
(nNOS knockout mice) were established (Huang et al., 1993). Using
small intestinal smooth muscle preparations from these mice, we found
that NANC transmission was reduced by about 50% (Zakhary et al.,
1997). More recently we have found that smooth muscle cells from
nNOS knockout mice have abnormal resting membrane potentials,
implying that basal NO production determines the excitability of intestinal smooth muscle (Xue et al., 2000). Taken together with the ability
of NO donors to mimic NANC transmission, these findings establish a
role for NO as a neurotransmitter.

   We also found nNOS localized to neurons that innervate the corpora
cavernosae and blood vessels of the penis (Burnett et al., 1992). Electrical stimulation of the cavernous nerves provokes erection, which is
blocked by NOS inhibitors, establishing NO as a neurotransmitter mediating penile erection. As in other organs, NO mediates erection
through the stimulation of cGMP formation. Sildenafil (Viagra^{TM}) inhibits phosphodiesterase type 5, an enzyme that selectively degrades
cGMP (Corbin and Francis, 1999). Thus the therapeutic effect of sildenafil for patients with erectile dysfunction involves potentiation of NO
neurotransmission.

   Mice lacking nNOS and therefore NO production afford an opportunity to ascertain physiological roles for neuronally derived NO. Gross
anatomical observations revealed only a greatly dilated stomach with hypertrophy of the pylorus (Huang et al., 1993). Recently we found that
NANC relaxation of the pyloric sphincter is abolished in nNOS knockout mice and restored by NO donors (Watkins et al., 2000). As the
pylorus contains a plexus of nNOS neurons, NO appears to be the neurotransmitter mediating NANC relaxation of the pylorus. This finding
has clinical relevance, as nNOS knockout mice appear to be an animal
model of infantile hypertrophic pyloric stenosis. In patients with this
condition, nNOS protein cannot be detected in the pylorus, and the
pyloric muscle is hypertrophied as in the nNOS knockout mice (Vanderwinden et al., 1992; Chung et al., 1996). As the gastric dilation and
pyloric dysfunction of the nNOS knockout mice resemble the gastric
dysfunction observed in many diabetic patients, we reasoned that the
nNOS knockout mice might also represent an animal model of this

condition, diabetic gastropathy (Kassander, 1958; Mearin et al., 1986; Koch, 1999). In diabetic mice, we found that NO-dependent NANC transmission is lost in the pyloric muscle (Watkins et al., 2000). At the same time, nNOS protein and mRNA are absent from the pyloric neurons, though the nerves themselves remain intact. Insulin treatment restores nNOS protein and mRNA expression in conjunction with a return of NO-mediated NANC pyloric relaxation. These results are consistent with a regulatory effect of insulin or glucose on nNOS expression, perhaps through regulatory sites in the promoter of the nNOS gene (Hall et al., 1994; Wang and Marsden, 1995; Wang et al., 1999).

Insight into a physiological role for NO in the brain comes from behavioral studies of nNOS knockout mice. These mice are extraordinarily aggressive, more so than virtually any other form of genetically mutant mice previously described (Nelson et al., 1995). This behavior is dependent upon testosterone, as it is abolished by castration, is reproduced by testosterone administration, and is absent in females (Kriegsfeld et al., 1997). nNOS knockout male mice also display abnormal, excessive sexual activity. When a normal male mouse is placed together with a female that is not in estrus (the sexually receptive phase of the estrous cycle), the male will begin to mount the female but will cease when he perceives that the female is not responsive. Male nNOS knockout mice fail to heed such clues and will repeatedly mount the female (Nelson et al., 1995). These studies imply that in males nNOS neurons normally restrain aggressive and sexual behaviors. In contrast, in females maternal aggression is reduced in nNOS-deficient mice (Gammie and Nelson, 1999).

NO may also play a role in learning and memory. Long-term potentiation (LTP) is a model of synaptic plasticity in which powerful stimulation of a synaptic input to a neuronal system potentiates subsequent synaptic transmission in the system for long time periods. In intact animals, LTP can persist for weeks. Chemicals that release NO facilitate LTP. There has been controversy about possible decreased LTP in nNOS-deficient mice, perhaps because of compensation by related genes like eNOS. Elegant studies by Kandel and associates show a clear decrease of LTP in mice with deletion of both eNOS and nNOS (Son et al., 1996). Whether eNOS is located in neurons or only blood vessels in the brain remains controversial. Perhaps NO produced in blood vessels can influence neural transmission. This hypothesis fits with evidence for behavioral changes in eNOS knockout mice. These mice display a pronounced decrease in aggressive behavior, unlike the nNOS knockout mice (Demas et al., 1999).

NO Dysfunction in Stroke

Medical students used to be taught that stroke causes permanent damage in infarcted tissue where cerebral arteries have been occluded. However, evidence accumulated over the past decade indicates that a major fraction of the neural damage in stroke is due to oxygen-derived free radicals that are produced following reperfusion of the ischemic area or by hypoxic mitochondria in surviving neurons. Now we also know that hypoxia in the brain triggers a massive release of excitatory transmitters, especially glutamate. Glutamate levels may reach 50 times their normal levels, literally "exciting to death" partially hypoxic cells (Choi, 1988). Evidence for a role of glutamate in stroke includes the ability of glutamate antagonists, especially NMDA receptor antagonists, to reduce stroke damage by about 50–60% (Choi, 1988). Glutamate toxicity associated with stroke can be mimicked in cerebral cortical cultures, where NMDA receptor activation can kill up to 90% of neurons, whereas NMDA receptor antagonists prevent this damage (Meldrum and Garthwaite, 1990). Glutamate neurotoxicity is diminished in cultures from nNOS knockout mice or following treatment with NOS inhibitors (Dawson et al., 1996). Stroke damage is also markedly reduced following treatment with NOS inhibitors (Nowicki et al., 1991; Buisson et al., 1992; Nagafuji et al., 1992; Trifiletti, 1992; Nishikawa et al., 1993) and in nNOS knockout mice (Huang et al., 1994).

If NO mediates neurotoxicity in stroke, how exactly does NO kill cells? Though NO is a free radical, it is not a particularly toxic one. When NO combines with superoxide (O_2^-) (e.g., from hypoxic mitochondria) it forms peroxynitrite ($ONOO^-$), which degenerates into the extremely toxic hydroxyl radical ($OH^\bullet$). Peroxynitrite and hydroxyl radicals can damage all major biomolecules, including lipids (through peroxidation of cell membranes, leading to calcium entry), proteins (both directly and through calcium-dependent proteases), and DNA. Though all of these may play some role, recent evidence suggests that DNA damage is the major mechanism. $OH^\bullet$ causes DNA strand breaks that activate the enzyme poly(ADP-ribose)polymerase (PARP). PARP is a nuclear enzyme that facilitates the DNA repair process. PARP's substrate is NAD, and when activated it transfers 50–200 ADP-ribose groups in branched chains to several nuclear proteins. A major substrate is PARP itself. When it is overactivated by massive amounts of DNA damage, NAD levels are depleted and, in efforts to resynthesize NAD, ATP is also depleted from cells (Berger, 1985; Pieper et al., 1999). Accordingly DNA damage leading to PARP overactivation can kill cells by energy loss and cellular starvation. Evidence that such a mechanism mediates NO killing comes from studies showing that PARP inhibitors block neuronal death in cultures elicited by NO donors or NMDA receptor activation (Zhang et al., 1994). Even more strikingly, brain cultures from PARP knockout mice are completely protected from such neurotoxicity (Eliasson et al., 1997). PARP-mediated neuronal cell death may have clinical relevance. In PARP knockout mice, there is an 80% reduction in stroke damage following reversible occlusion of the middle cerebral artery (Eliasson et al., 1997; Endres et al., 1997), and postischemic administration of a PARP inhibitor can ameliorate stroke damage (Takahashi et al., 1999).

Carbon Monoxide

Neurotransmitters are grouped by chemical classes, such as the biogenic amines, amino acids, and neuropeptides. Thus we wondered whether there might exist gaseous transmitters besides NO. Carbon monoxide (CO) is formed physiologically by heme oxygenase (HO). HO cleaves the porphyrin ring of heme to form biliverdin, which is rapidly reduced by biliverdin reductase to bilirubin. In the process ferrous iron is released and a one-carbon fragment is released as CO (Fig. 14.3). Might CO be the second member of a new group of neurotransmitters, the gases?

HO was first identified as an inducible enzyme activated by many cellular stresses. This inducible form of HO was found to be identical

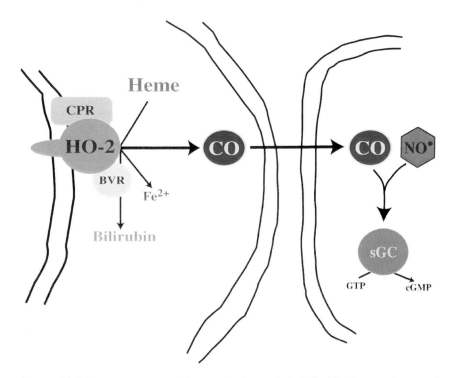

Figure 14.3. *Heme oxygenase-2 converts heme into bilirubin, ferrous iron, and carbon monoxide.* Heme catabolism involves the collaboration of three enzymes: cytochrome P450 reductase (CPR), biliverdin reductase (BVR), and either heme oxygenase-1 (HO-1) or -2 (HO-2). HO-2 is enriched in neurons and its activity results in the production of carbon monoxide (CO), which functions similarly to NO. Thus CO diffuses through membranes to target cells, where it can bind the heme of soluble guanylate cyclase (sGC) to regulate cGMP production. CO may modulate the effect of NO by competing for binding to the heme in sGC (Ingi et al., 1996a). The co-localization of nNOS and HO-2 in myenteric neurons suggests that NO and CO function as co-neurotransmitters (Zakhary et al., 1997; Xue et al., 2000). Bilirubin production may be neuroprotective (Dore et al., 1999a,b), and the iron (Fe^{2+}) released from heme is extruded from cells to prevent cell death (Ferris et al., 1999).

to a heat shock protein (HSP-32) and is now designated HO1. HO1 is most concentrated in the spleen, the repository of aged red blood cells. In the spleen, HO degrades heme from red blood cell hemoglobin, which otherwise would be quite toxic. In other cells, HO destroys heme from mitochondrial and other enzymes. In the course of purifying HO1, Maines and associates discovered a second HO protein, which they designated HO2 (Maines et al., 1986; Trakshel et al., 1986). HO2 is not inducible and is most concentrated in the brain and testes. We reasoned that if CO were to be a neurotransmitter, its biosynthetic enzyme should be discretely localized to certain neurons. In situ hybridization studies revealed that HO2 mRNA is highly localized to specific neurons in the brain, with localizations similar to those of soluble guanylyl cyclase (Verma et al., 1993). Like NO, CO can activate this enzyme, triggering cGMP formation (Fig. 14.3). In cultures of olfactory neurons, which have very high concentrations of HO2 and lack nNOS, cGMP levels are depleted by HO inhibitors (Ingi and Ronnett, 1995; Ingi et al., 1996b).

As in the case of NO, direct evidence for CO as a neurotransmitter emerged from studies of the myenteric plexus of the intestine. Immunohistochemical staining revealed significant co-localization of HO2 and nNOS in myenteric neuronal cells, suggesting that CO and NO might be co-transmitters (Zakhary et al., 1996, 1997). As discussed earlier, NANC transmission is reduced by half in intestinal preparations from nNOS knockout mice. In similar experiments we employed mice with a targeted genomic deletion of HO2 (HO2 knockout mice) and found that NANC transmission is also reduced by 50% in the intestine (Zakhary et al., 1997). Recently we have extended these findings by demonstrating depolarization of the resting membrane potential in jejunal smooth muscle cells from HO2 knockout mice (Xue et al., 2000). Intestinal preparations from these mice also have decreased NANC inhibitory junctional potentials measured electrophysiologically. Using mice deficient in both nNOS and HO2, we observed additive effects on the resting membrane potential and the inhibitory junctional potentials (Xue et al., 2000). Application of exogenous CO mimics NANC neurotransmission. Thus CO, like NO, appears to be an inhibitory NANC neurotransmitter.

HO2 is also concentrated in nerves that innervate the vas deferens, a smooth muscle whose contractions underlie ejaculation. Reflex activity of the bulbospongiosus muscle that underlies ejaculation is profoundly reduced in preparations from HO2 knockout mice, and the mice display a marked reduction in ejaculatory function (Burnett et al., 1998). Bulbospongiosus nerves do not contain nNOS. Conversely the NO nerves mediating penile erection do not contain HO2. Although NO is critical for penile erections, CO appears to be the neurotransmitter of nerves mediating ejaculation. Thus in the myenteric plexus NO and CO may be co-transmitters, but in other peripheral autonomic nerves they have distinct localizations and functions.

As described previously, CO appears to regulate cGMP physiologically in olfactory neurons. However, its functional link to cGMP in other parts of the brain is not clear despite the apparent co-localization of HO2 and soluble guanylate cyclase. Though some relatively nonspecific inhibitors of HO diminish LTP, no major deficit in LTP is evident in HO2 knockout mice (Meffert et al., 1994; Shinomura et al., 1994; Poss et al., 1995). In contrast to the current paucity of evidence regarding a role for CO in normal brain function, other products of HO enzymatic activity do display physiological roles.

Ames and co-workers identified an antioxidant function for bilirubin many years ago (Stocker et al., 1987a,b). Recently we showed a neuroprotectant role for bilirubin (Dore et al., 1999b). This work stemmed from our efforts to identify a means for activating HO2 analogous to the calcium-calmodulin activation of nNOS. Phorbol esters, which stimulate PKC, protect cortical neurons from death following treatment with hydrogen peroxide. The neuroprotection of phorbol esters is greatly reduced by inhibitors of HO and abolished in neuronal cultures from HO2 knockout mice. Bilirubin, in very low concentrations, mimics the neuroprotective effects of phorbol esters. Consistent with a neuroprotective role for bilirubin, stroke damage is substantially worsened in HO2 knockout mice (Dore et al., 1999a). Sometimes mice with a specific genetic deficiency are unhealthy and therefore susceptible to injury or illness in general. HO1 knockout mice are generally debilitated, with weight loss, anemia, and signs of chronic inflammation developing after 20 weeks of age. However, HO1 knockout mice do not display accentuated stroke damage. In contrast, HO2 knockout mice appear normal and healthy, though they suffer enlarged strokes in ischemic models.

The neuroprotective effects of bilirubin are surprising considering the well-known neurotoxicity of bilirubin in jaundiced babies with very high blood levels of bilirubin. Kernicterus (Gourley, 1997) requires high micromolar levels of bilirubin, a thousand times the concentrations of bilirubin that are neuroprotective in vitro. Interestingly, there is evidence that the moderately "elevated" plasma levels of bilirubin in most normal babies may be protective, as babies with moderately elevated bilirubin levels are less susceptible to oxygen radical–mediated injury (Belanger et al., 1997).

Iron, the third product of HO, is highly toxic through the Fenton reaction that produces OH$^\bullet$ (Meneghini, 1997). Accordingly there ought to exist some mechanism to stimulate the efflux from the cell of iron formed by HO. HO activity appears to be linked to cellular iron efflux. Thus transfection of HO1 into mammalian cells stimulates iron efflux, and iron efflux is greatly diminished in fibroblasts from HO1 knockout mice (Ferris et al., 1999). These findings fit with evidence of low serum iron and accumulation of iron in the tissues of HO1 knockout mice (Poss and Tonegawa, 1997a,b). Whether HO2 similarly regulates iron efflux is not yet established, although HO2 knockout mice develop iron accumulation in their lungs in response to hyperoxia (Dennery et al., 1998).

Other gases might also function as neurotransmitters. Hydrogen sulfide (H_2S) exerts specific effects on the electrical properties of serotonin neurons in the dorsal raphe nucleus (Kombian et al., 1993). Kimura and Abe demonstrated that H_2S is produced in the brain. In addition, exogenous H_2S enhances NMDA neurotransmission and facilitates the induction of LTP in the hippocampus at concentrations comparable to those that occur physiologically (Abe and Kimura, 1996). Cystathionine beta-synthase is also expressed in smooth muscle and where physiologically relevant concentrations of H_2S enhance NO-mediated muscle relaxation (Hosoki et al., 1997).

D-Amino Acids: Focus on D-Serine

Organic molecules, and therefore biological molecules, are based upon the chemistry of the carbon atom. Carbon atoms can have up to four bonded groups attached to them in three-dimensional space, forming a tetrahedron. Because of this structure, carbon-containing molecules can have the same four constituents, yet differ in their structure by the location of the four groups in space. This property of carbon-based molecules is known as chirality. As an example, one's hands are mirror images of each other, but they cannot be superimposed one upon the other. Similarly carbon atoms with four different groups attached occur in two forms that are nonsuperimposable mirror images of each other, known as enantiomers. The elegant precision of biological reactions requires the recognition sites in proteins to be exceedingly specific. Just as a right-handed glove will not fit a left hand, enzymes and receptors can discriminate between the enantiomers of ligands and substrates.

Biology students are taught that organisms exclusively employ the D-enantiomers for sugars, while amino acids for proteins are always in the L-form. D-amino acids had been observed in bacteria and invertebrates (Corrigan, 1969), but a role for D-amino acids in higher species was deemed unlikely. However, recent observations point to the existence of substantial quantities of some D-amino acids in higher species, including humans (Chouinard et al., 1993; Hashimoto et al., 1993a–c; Hashimoto and Oka, 1997). In the brain, levels of D-serine are up to a third those of L-serine, and, in a variety of tissues including the brain and certain glands, D-aspartate levels are 20–30% those of L-aspartate (Hashimoto and Oka, 1997).

The existence of significant amounts of D-serine and D-aspartate in the brain suggested that some D-amino acids serve specific neuroactive roles. There are marked regional variations in the levels of D-serine, with highest concentrations in the forebrain, where NMDA-type glutamate receptors are enriched (Hashimoto and Oka, 1997). This finding meshed with certain known features of the NMDA receptor. For instance, Ascher and colleagues (Johnson and Ascher, 1987) had observed that the loss of NMDA receptor activation following rapid perfusion of neural preparations can be reversed by glycine. Subsequent studies confirmed the existence of a glycine recognition site on the NMDA receptor and established that this receptor requires co-activation by glycine as well as glutamate (Kleckner and Dingledine, 1988). These conclusions were somewhat puzzling, as glycine concentrations in the CNS are lowest in the forebrain, areas that are enriched in NMDA receptors, and highest in the spinal cord and hindbrain, where glycine is known to function as an inhibitory neurotransmitter. We suspected that D-serine might act as an endogenous ligand for the "glycine site" of the NMDA receptor, because numerous studies had established that D-serine was at least as potent as glycine at the "glycine site."

Based on these hints, we developed antisera to D-serine and mapped its localization in the CNS. D-serine is indeed highly concentrated in areas of the brain enriched in NMDA receptors, where glycine levels are lowest (Schell et al., 1995). In contrast, in some parts of the brain, such as the adult cerebellum and the accessory olfactory bulb, glycine, rather than D-serine, may be associated with NMDA receptors (Schell et al., 1997a). Most striking was our observation that D-serine occurs in glia, not neurons. D-serine is exclusively localized to protoplasmic astrocytes that are enriched in gray matter together with NMDA receptors (Fig. 14.4). One reason that neurotransmitters have been presumed to derive exclusively from neurons is that only neurons were thought to release transmitters upon appropriate excitation. In cultures of protoplasmic type II astrocytes, activation of non-NMDA glutamate receptors stimulates release of D-serine (Schell et al., 1995). In the brain, these astrocytes ensheath the synapse so that any D-serine released from astrocytes would have close proximity to NMDA receptors. Thus we have proposed that synaptic release of glutamate from a presynaptic neuron triggers the release of D-serine from adjacent astrocytes to co-activate the NMDA receptors on nearby postsynaptic neurons (Fig. 14.4).

The notion that D-serine possesses a physiological role in the brain helped to clarify a biochemical oddity dating back to the 1930s. At that time Krebs discovered an enzyme that selectively deaminates D-amino acids and designated it "D-amino acid oxidase" (DAAOX) (Krebs, 1935). Because D-amino acids were unknown in mammalian tissues, it was assumed that the enzyme was an evolutionary vestige from bacteria, a mistake of nature, or served to oxidize glycine, which is not chiral and therefore does not have enantiomeric forms. Our histochemical investigations showed marked regional variations in DAAOX localization, with concentrations exactly reciprocal to those of D-serine (Schell et al., 1995). These data suggest that DAAOX degrades D-serine physiologically.

How might one definitively determine whether D-serine is an endogenous modulator of NMDA receptor function? We utilized DAAOX as a tool, first establishing that D-serine is the only substance in brain preparations degraded by DAAOX (Mothet et al., 2000). Then we showed that DAAOX treatment markedly reduces NMDA neurotransmission in cerebellar preparations by monitoring both NOS activity and cGMP levels

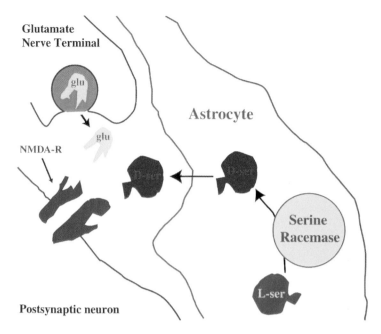

Figure 14.4. D-serine is synthesized in astrocytes and released near NMDA receptors, where it regulates glutaminergic neurotransmission. The NMDA receptor
(NMDA-R) requires ligand binding at two sites to function as an ion channel. Thus
glutamate only opens the NMDA receptor if the "glycine" site is occupied by the
agonist. D-serine is synthesized in astrocytes by serine racemase (Wolosker et al.,
1999a,b), is released by glutamate, and binds to the "glycine" site of the NMDA
receptor.

as well as by electrophysiological studies in slice and culture preparations
of the cerebellum and the hippocampus (Parent et al., 1999; Mothet et
al., 2000).

How might the brain synthesize D-serine? After developing a specific
assay to monitor D-serine levels, we isolated an enzyme from mammalian
brain that converts L-serine to D-serine, designating it serine racemase
(Wolosker et al., 1999b), and then cloned its cDNA (Wolosker et al.,
1999a). Serine racemase is localized to the D-serine-containing protoplasmic astrocytes in areas of the brain enriched in NMDA receptors
(Wolosker et al., 1999a). It is a novel protein, though it possesses some
modest sequence similarity to other enzymes that use serine as a substrate.
These enzymes, like serine racemase, require pyridoxal phosphate (vitamin B_6) as a co-factor (Wolosker et al., 1999a,b).

D-serine challenges various dogmas regarding neurotransmitters,
perhaps to a greater extent even than NO and CO. The notion of a
function for D-amino acids in mammals, especially as a neurotransmitter, goes against long-standing biochemical and physiological teachings.
Does D-serine satisfy criteria for a neurotransmitter (Table 14.1)? It is
localized to the sites of its receptors and possesses a dedicated biosyn-

Table 14.1. Neurotransmitter-like Properties of D-Serine

1. Specific localization: D-serine is enriched in astrocytes and specifically localized near glutamate–NMDA receptor synapses.

2. Biosynthesis: D-serine is synthesized in astrocytes from L-serine by serine racemase.

3. Synaptic release: D-serine is released from astrocytes by glutamate through activation of non-NMDA receptors.

4. Mimicry of normal physiology: Exogenous D-serine binds the "glycine" site of NMDA receptors and restores normal NMDA receptor function in vitro.

5. Requirement for normal physiology: Selective depletion of endogenous D-serine with DAAOX blocks NMDA receptor–dependent neurotransmission.

thetic enzyme. D-serine is released upon appropriate stimulation and mimics the actions of the physiological transmitter. Thus in many ways D-serine fulfills more criteria than many neuropeptides that are well accepted as transmitters. However, the concept of a neurotransmitter arising from glia may discomfort some neuroscientists. Glia do not have synaptic vesicles, so D-serine release cannot occur through classical exocytosis. Instead its release may result from reversing the directionality of an amino acid transporter that might otherwise remove substances from the synapse. Thus, as for NO and CO, aberrant release processes would confound those who insist that neurotransmitters adhere to traditional criteria.

The cloning of serine racemase provides a potentially powerful approach to learning about D-serine in the brain and to exploring therapeutic ramifications. Future studies will allow the monitoring of NMDA neurotransmission, LTP, and overall behavior in serine racemase gene knockout mice. Inhibitors of serine racemase would be expected to diminish NMDA neurotransmission, and so, like NMDA receptor antagonists, serine racemase inhibitors might be beneficial in treating stroke and other conditions associated with excess excitation. Glutamate neurotoxicity may be relevant for the therapy of neurodegenerative diseases, whether or not excitotoxicity is directly involved in their pathophysiology. We do not know the precise etiology of conditions such as Parkinson's and Alzheimer's diseases. However, regardless of the initiating event, neurons degenerate in patients with these disorders. Because glutamate release is augmented in the presence of hypoxic neural damage, it is likely that excessive levels of glutamate are released as the disease progresses and that glutamate may work together with other insults to permit cell death. Conceivably, diminishing glutamate receptor activation may block the progression of neurodegenerative conditions, such as Alzheimer's disease, Parkinson's disease, Huntington's disease, and amyotrophic lateral sclerosis (ALS). Although reliable animal models of Alzheimer's and Huntington's diseases have been difficult to establish, there are model systems for ALS and Parkinson's disease. Glutamate receptor antagonists are therapeutic in these models, so that serine racemase inhibitors merit examination as well.

D-serine and NMDA transmission may be relevant to schizophrenia. The psychotic state observed following administration of NMDA antagonists such as phencyclidine closely resembles certain features of schizophrenia, more than most drug psychoses. According to the NMDA receptor model of schizophrenia, one would expect glutamate agonists to be therapeutic. Since glutamate itself does not pass the blood-brain barrier, researchers have administered glycine, D-serine, or cycloserine (which mimics D-serine at NMDA receptors) to assess their potential efficacy in schizophrenic patients. In several studies beneficial effects have been reported (Krystal and D'Souza, 1998; Tsai et al., 1998).

D-serine may not be the only D-amino acid neurotransmitter candidate. D-aspartate was first discovered in invertebrates (D'Aniello and Guiditta, 1977, 1978). Several groups identified D-aspartate in mammalian tissues, with notable concentrations in endocrine glands and the brain (Dunlop et al., 1986; Neidle and Dunlop, 1990; Lee et al., 1997; Sakai et al., 1997). D'Aniello and associates (D'Aniello et al., 1996, 1998a) observed that D-aspartate is released from the testes and may stimulate testosterone synthesis, and some studies have found changes in D-aspartate levels in the brains of Alzheimer's disease patients (D'Aniello et al., 1998b; Fisher et al., 1998). With antisera to D-aspartate, we localized the amino acid to selected neuronal populations in the brain as well as to epinephrine-containing cells in the adrenal medulla, to the vasopressin-releasing hypothalamic neurons innervating the posterior pituitary gland, and to pinealocytes in the pineal gland (Schell et al., 1997b). Though D-aspartate can activate NMDA receptors, the localizations of D-aspartate do not match those of NMDA receptors, and D-aspartate levels are far lower than glutamate levels. Thus, at the present time, too little is known about D-aspartate to draw any firm conclusions as to its function in endocrine glands or the brain.

Conclusion

This chapter has focused on NO, CO, and D-serine as recently identified candidate neurotransmitters. A historical review reveals that every neurotransmitter candidate after acetylcholine has altered one or more preconceptions regarding the defining characteristics of a neurotransmitter. With so much controversy, one may be tempted to discard the term *neurotransmitter* altogether. Alternatively, bearing in mind these historical lessons, we might try to be reasonably liberal in our definition of a neurotransmitter in the following way: *A transmitter is a molecule, released by neurons or glia, that physiologically influences the electrochemical state of adjacent cells.*

Outside the CNS, those adjacent target cells need not be neurons and, in most instances, would be smooth muscle or glandular cells. Within the CNS, researchers usually think in terms of neurotransmitters influencing adjacent neurons, but one need not exclude influences upon glia or blood vessels. Advocating changes in classical definitions of neurotransmitters may seem heretical. However, we should remember that previous assumptions about neurotransmission have been repeatedly challenged over the past 50 years. In all of these instances, we have benefited from new insights, some of which have important therapeutic implications for neuropsychiatry.

Acknowledgments

This work was supported by USPHS grants MH-18501 and DA-00266, and a Research Scientist Award (DA-00074) to S.H.S. C.D.F. is the recipient of a Howard Hughes Fellowship for Physicians. The authors thank Bertil Hille and David Linden for helpful discussions.

References

Abe, K., Kimura, H. (1996). The possible role of hydrogen sulfide as an endogenous neuromodulator. *J. Neurosci.* 16:1066–1071.

Axelrod, J. (1971). Noradrenaline: fate and control of its biosynthesis. *Science* 173:598–606.

Belanger, S., Lavoie, J. C., Chessex, P. (1997). Influence of bilirubin on the antioxidant capacity of plasma in newborn infants. *Biol. Neonate* 71:233–238.

Berger, N. A. (1985). Poly(ADP-ribose) in the cellular response to DNA damage. *Radiat. Res.* 101:4–15.

Boeckxstaens, G. E., Pelckmans, P. A., Bult, H., De Man, J. G., Herman, A. G., Van Maercke, Y. M. (1990). Non-adrenergic non-cholinergic relaxation mediated by nitric oxide in the canine ileocolonic junction. *Eur. J. Pharmacol.* 190:239–246.

Bredt, D. S., Snyder, S. H. (1989). Nitric oxide mediates glutamate-linked enhancement of cGMP levels in the cerebellum. *Proc. Natl. Acad. Sci. USA* 86: 9030–9033.

Bredt, D. S., Snyder, S. H. (1990). Isolation of nitric oxide synthetase, a calmodulin-requiring enzyme. *Proc. Natl. Acad. Sci. USA* 87:682–685.

Bredt, D. S., Hwang, P. M., Snyder, S. H. (1990). Localization of nitric oxide synthase indicating a neural role for nitric oxide. *Nature* 347:768–770.

Bredt, D. S., Glatt, C. E., Hwang, P. M., Fotuhi, M., Dawson, T. M., Snyder, S. H. (1991a). Nitric oxide synthase protein and mRNA are discretely localized in neuronal populations of the mammalian CNS together with NADPH diaphorase. *Neuron* 7:615–624.

Bredt, D. S., Hwang, P. M., Glatt, C. E., Lowenstein, C., Reed, R. R., Snyder, S. H. (1991b). Cloned and expressed nitric oxide synthase structurally resembles cytochrome P-450 reductase. *Nature* 351:714–718.

Bredt, D. S., Ferris, C. D., Snyder, S. H. (1992). Nitric oxide synthase regulatory sites. Phosphorylation by cyclic AMP—dependent protein kinase, protein kinase C, and calcium/calmodulin protein kinase; identification of flavin and calmodulin binding sites. *J. Biol. Chem.* 267:10,976–10,981

Brenman, J. E., Chao, D. S., Gee, S. H., McGee, A. W., Craven, S. E., Santillano, D. R., Wu, Z., Huang, F., Xia, H., Peters, M. F., Froehner, S. C., Bredt, D. S. (1996). Interaction of nitric oxide synthase with the postsynaptic density protein PSD-95 and alpha1-syntrophin mediated by PDZ domains. *Cell* 84: 757–767.

Brune, B., Lapetina, E. G. (1991). Phosphorylation of nitric oxide synthase by protein kinase A. *Biochem. Biophys. Res. Commun.* 181:921–926.

Buisson, A., Plotkine, M., Boulu, R. G. (1992). The neuroprotective effect of a nitric oxide inhibitor in a rat model of focal cerebral ischaemia. *Br. J. Pharmacol.* 106:766–767.

Bult, H., Boeckxstaens, G. E., Pelckmans, P. A., Jordaens, F. H., Van Maercke, Y. M., Herman, A. G. (1990). Nitric oxide as an inhibitory non-adrenergic non-cholinergic neurotransmitter. *Nature* 345:346–347.

Burnett, A. L., Lowenstein, C. J., Bredt, D. S., Chang, T. S., Snyder, S. H. (1992). Nitric oxide: a physiologic mediator of penile erection. *Science* 257:401–403.

Burnett, A. L., Johns, D. G., Kriegsfeld, L. J., Klein, S. L., Calvin, D. C., Demas, G. E., Schramm, L. P., Tonegawa, S., Nelson, R. J., Snyder, S. H., Poss, K. D. (1998). Ejaculatory abnormalities in mice with targeted disruption of the gene for heme oxygenase-2. *Nature Med.* 4:84–87.

Chang, M. M., Leeman, S. E. (1970). Isolation of a sialogogic peptide from bovine hypothalamic tissue and its characterization as substance P. *J. Biol. Chem.* 245: 4784–4790.

Choi, D. W. (1988). Glutamate neurotoxicity and diseases of the nervous system. *Neuron* 1:623–634.

Chouinard, M. L., Gaitan, D., Wood, P. L. (1993). Presence of the N-methyl-D-aspartate-associated glycine receptor agonist, D-serine, in human temporal cortex: comparison of normal, Parkinson, and Alzheimer tissues. *J. Neurochem.* 61:1561–1564.

Chung, E., Curtis, D., Chen, G., Marsden, P. A., Twells, R., Xu, W., Gardiner, M. (1996). Genetic evidence for the neuronal nitric oxide synthase gene (NOS1) as a susceptibility locus for infantile pyloric stenosis. *Am. J. Hum. Genet.* 58: 363–370.

Corbin, J. D., Francis, S. H. (1999). Cyclic GMP phosphodiesterase-5: target of sildenafil. *J. Biol. Chem.* 274:13,729–13,732.

Corrigan, J. J. (1969). D-amino acids in animals. *Science* 164:142–149.

Daff, S., Sagami, I., Shimizu, T. (1999). The 42-amino acid insert in the FMN domain of neuronal nitric-oxide synthase exerts control over Ca^{2+}/calmodulin-dependent electron transfer. *J. Biol. Chem.* 274:30,589–30,595.

D'Aniello, A., Guiditta, A. (1977). Identification of D-aspartic acid in the brain of *Octopus vulgaris* Lam. *J. Neurochem* 29:1053–1057.

D'Aniello, A., Guiditta, A. (1978). Presence of D-aspartate in squid axoplasm and in other regions of the cephalopod nervous system. *J. Neurochem.* 31: 1107–1108.

D'Aniello, A., Di Cosmo, A., Di Cristo, C., Annunziato, L., Petrucelli, L., Fisher, G. (1996). Involvement of D-aspartic acid in the synthesis of testosterone in rat testes. *Life Sci.* 59:97–104.

D'Aniello, A., Di Fiore, M. M., D'Aniello, G., Colin, F. E., Lewis, G., Setchell, B. P. (1998a). Secretion of D-aspartic acid by the rat testis and its role in endocrinology of the testis and spermatogenesis. *FEBS Lett.* 436:23–27.

D'Aniello, A., Lee, J. M., Petrucelli, L., Di Fiore, M. M. (1998b). Regional decreases of free D-aspartate levels in Alzheimer's disease. *Neurosci. Lett.* 250: 131–134.

Dawson, V. L., Kizushi, V. M., Huang, P. L., Snyder, S. H., Dawson, T. M. (1996). Resistance to neurotoxicity in cortical cultures from neuronal nitric oxide synthase–deficient mice. *J. Neurosci.* 16:2479–2487.

Demas, G. E., Kriegsfeld, L. J., Blackshaw, S., Huang, P., Gammie, S. C., Nelson, R. J., Snyder, S. H. (1999). Elimination of aggressive behavior in male mice lacking endothelial nitric oxide synthase. *J. Neurosci.* 19(RC30):1–5.

Dennery, P. A., Spitz, D. R., Yang, G., Tatarov, A., Lee, C. S., Shegog, M. L., Poss, K. D. (1998). Oxygen toxicity and iron accumulation in the lungs of mice lacking heme oxygenase-2. *J. Clin. Invest.* 101:1001–1011.

Dimmeler, S., Fleming, I., Fisslthaler, B., Hermann, C., Busse, R., Zeiher, A. M. (1999). Activation of nitric oxide synthase in endothelial cells by Akt-dependent phosphorylation. *Nature* 399:601–605.

Dore, S., Sampei, K., Goto, S., Alkayed, N. J., Guastella, D., Blackshaw, S., Gallagher, M., Traystman, R. J., Hurn, P. D., Koehler, R. C., Snyder, S. H. (1999a). Heme oxygenase-2 is neuroprotective in cerebral ischemia. *Mol. Med.* 5:656–663.

Dore, S., Takahashi, M., Ferris, C. D., Hester, L. D., Guastella, D., Snyder, S. H. (1999b). Bilirubin, formed by activation of heme oxygenase-2, protects neurons against oxidative stress injury. *Proc. Natl. Acad. Sci. USA* 96:2445–2450.

Dunlop, D. S., Neidle, A., McHale, D., Dunlop, D. M., Lajtha, A. (1986). The presence of free D-aspartic acid in rodents and man. *Biochem. Biophys. Res. Commun.* 141:27–32.

Eliasson, M. J., Sampei, K., Mandir, A. S., Hurn, P. D., Traystman, R. J., Bao, J., Pieper, A., Wang, Z. Q., Dawson, T. M., Snyder, S. H., Dawson, V. L. (1997). Poly(ADP-ribose) polymerase gene disruption renders mice resistant to cerebral ischemia. *Nature Med.* 3:1089–1095.

Endres, M., Wang, Z. Q., Namura, S., Waeber, C., Moskowitz, M. A. (1997). Ischemic brain injury is mediated by the activation of poly(ADP-ribose)polymerase. *J. Cereb. Blood Flow Metab.* 17:1143–1151.

Ferris, C. D., Jaffrey, S. R., Sawa, A., Takahashi, M., Brady, S. D., Barrow, R. K., Tysoe, S. A., Wolosker, H., Baranano, D. E., Dore, S., Poss, K. D., Snyder, S. H. (1999). Heme oxygenase-1 prevents cell death by regulating cellular iron. *Nature Cell Biol.* 1:152–157.

Fisher, G., Lorenzo, N., Abe, H., Fujita, E., Frey, W. H., Emory, C., Di Fiore, M. M., Aniello, D. A. (1998). Free D- and L-amino acids in ventricular cerebrospinal fluid from Alzheimer and normal subjects. *Amino Acids* 15:263–269.

Fulton, D., Gratton, J. P., McCabe, T. J., Fontana, J., Fujio, Y., Walsh, K., Franke, T. F., Papapetropoulos, A., Sessa, W. C. (1999). Regulation of endothelium-derived nitric oxide production by the protein kinase Akt. *Nature* 399:597–601.

Furchgott, R. F., Zawadzki, J. V. (1980). The obligatory role of endothelial cells in the relaxation of arterial smooth muscle by acetylcholine. *Nature* 288:373–376.

Gammie, S. C., Nelson, R. J. (1999). Maternal aggression is reduced in neuronal nitric oxide synthase–deficient mice. *J. Neurosci.* 19:8027–8035.

Garthwaite, J., Charles, S. L., Chess-Williams, R. (1988). Endothelium-derived relaxing factor release on activation of NMDA receptors suggests role as intercellular messenger in the brain. *Nature* 336:385–388.

Garthwaite, J., Garthwaite, G., Palmer, R. M., Moncada, S. (1989). NMDA receptor activation induces nitric oxide synthesis from arginine in rat brain slices. *Eur. J. Pharmacol.* 172:413–416.

Gourley, G. R. (1997). Bilirubin metabolism and kernicterus. *Adv. Pediatr.* 44:173–229.

Hall, A. V., Antoniou, H., Wang, Y., Cheung, A. H., Arbus, A. M., Olson, S. L., Lu, W. C., Kau, C. L., Marsden, P. A. (1994). Structural organization of the human neuronal nitric oxide synthase gene (NOS1). *J. Biol. Chem.* 269:33,082–33,090.

Hashimoto, A., Oka, T. (1997). Free D-aspartate and D-serine in the mammalian brain and periphery. *Prog. Neurobiol.* 52:325–353.

Hashimoto, A., Kumashiro, S., Nishikawa, T., Oka, T., Takahashi, K., Mito, T., Takashima, S., Doi, N., Mizutani, Y., Yamazaki, T. (1993a). Embryonic development and postnatal changes in free D-aspartate and D-serine in the human prefrontal cortex. *J. Neurochem.* 61:348–351.

Hashimoto, A., Nishikawa, T., Konno, R., Niwa, A., Yasumura, Y., Oka, T., Takahashi, K. (1993b). Free D-serine, D-aspartate and D-alanine in central nervous system and serum in mutant mice lacking D-amino acid oxidase. *Neurosci. Lett.* 152:33–36.

Hashimoto, A., Nishikawa, T., Oka, T., Takahashi, K. (1993c). Endogenous D-serine in rat brain: *N*-methyl-D-aspartate receptor–related distribution and aging. *J. Neurochem.* 60:783–786.

Hibbs, J. B., Jr., Vavrin, Z., Taintor, R. R. (1987). L-arginine is required for expression of the activated macrophage effector mechanism causing selective metabolic inhibition in target cells. *J. Immunol.* 138:550–565.

Hillarp, N. A., Fuxe, K., Dahlstrom, A. (1966). Demonstration and mapping of central neurons containing dopamine, noradrenaline, and 5-hydroxytryptamine and their reactions to psychopharmaca. *Pharmacol. Rev.* 18:727–741.

Hökfelt, T., Johansson, O., Goldstein, M. (1984). Chemical anatomy of the brain. *Science* 225:1326–1334.

Hosoki, R., Matsuki, N., Kimura, H. (1997). The possible role of hydrogen sulfide as an endogenous smooth muscle relaxant in synergy with nitric oxide. *Biochem. Biophys. Res. Commun.* 237:527–531.

Huang, P. L., Dawson, T. M., Bredt, D. S., Snyder, S. H., Fishman, M. C. (1993). Targeted disruption of the neuronal nitric oxide synthase gene. *Cell* 75:1273–1286.

Huang, Z., Huang, P. L., Panahian, N., Dalkara, T., Fishman, M. C., Moskowitz, M. A. (1994). Effects of cerebral ischemia in mice deficient in neuronal nitric oxide synthase. *Science* 265:1883–1885.

Hughes, J., Smith, T. W., Kosterlitz, H. W., Fothergill, L. A., Morgan, B. A., Morris, H. R. (1975). Identification of two related pentapeptides from the brain with potent opiate agonist activity. *Nature* 258:577–580.

Ignarro, L. J., Buga, G. M., Wood, K. S., Byrns, R. E., Chaudhuri, G. (1987). Endothelium-derived relaxing factor produced and released from artery and vein is nitric oxide. *Proc. Natl. Acad. Sci. USA* 84:9265–9269.

Ignarro, L. J., Lippton, H., Edwards, J. C., Baricos, W. H., Hyman, A. L., Kadowitz, P. J., Gruetter, C. A. (1981). Mechanism of vascular smooth muscle relaxation by organic nitrates, nitrites, nitroprusside and nitric oxide: evidence for the involvement of S-nitrosothiols as active intermediates. *J. Pharmacol. Exp. Ther.* 218:739–749.

Ingi, T., Cheng, J., Ronnett, G. V. (1996a). Carbon monoxide: an endogenous modulator of the nitric oxide–cyclic GMP signaling system. *Neuron* 16:835–842.

Ingi, T., Chiang, G., Ronnett, G. V. (1996b). The regulation of heme turnover and carbon monoxide biosynthesis in cultured primary rat olfactory receptor neurons. *J. Neurosci.* 16:5621–5628.

Jaffrey, S. R., Snyder, S. H. (1996). PIN: an associated protein inhibitor of neuronal nitric oxide synthase. *Science* 274:774–777.

Jaffrey, S. R., Snowman, A. M., Eliasson, M. J., Cohen, N. A., Snyder, S. H. (1998). CAPON: a protein associated with neuronal nitric oxide synthase that regulates its interactions with PSD95. *Neuron* 20:115–124.

Janssens, S. P., Shimouchi, A., Quertermous, T., Bloch, D. B., Bloch, K. D. (1992). Cloning and expression of a cDNA encoding human endothelium–derived relaxing factor/nitric oxide synthase. *J. Biol. Chem.* 267:14,519–14,522.

Johnson, J. W., Ascher, P. (1987). Glycine potentiates the NMDA response in cultured mouse brain neurons. *Nature* 325:529–531.

Kassander, P. (1958). Asymptomatic gastric retention in diabetes (gastroparesis diabeticorum). *Ann. Int. Med.* 48:797–812.

Kleckner, N. W., Dingledine, R. (1988). Requirement for glycine in activation of NMDA-receptors expressed in Xenopus oocytes. *Science* 241:835–837.

Koch, K. L. (1999). Diabetic gastropathy: gastric neuromuscular dysfunction in diabetes mellitus: a review of symptoms, pathophysiology, and treatment. *Dig. Dis. Sci.* 44:1061–1075.

Kombian, S. B., Reiffenstein, R. J., Colmers, W. F. (1993). The actions of hydrogen sulfide on dorsal raphe serotonergic neurons in vitro. *J. Neurophysiol.* 70: 81–96.

Krebs, H. A. (1935). CXCVII. Metabolism of amino acids. III. Deamination of amino acids. *Biochem. J.* 29:1620–1644.

Krieger, D. T. (1983). Brain peptides: what, where, and why? *Science* 222:975–985.

Kriegsfeld, L. J., Dawson, T. M., Dawson, V. L., Nelson, R. J., Snyder, S. H. (1997). Aggressive behavior in male mice lacking the gene for neuronal nitric oxide synthase requires testosterone. *Brain Res.* 769:66–70.

Krystal, J. H., D'Souza, D. C. (1998). D-serine and the therapeutic challenge posed by the *N*-methyl-D-aspartate antagonist model of schizophrenia. *Biol. Psychiatry* 44:1075–1076.

Lamas, S., Marsden, P. A., Li, G. K., Tempst, P., Michel, T. (1992). Endothelial nitric oxide synthase: molecular cloning and characterization of a distinct constitutive enzyme isoform. *Proc. Natl. Acad. Sci. USA* 89:6348–6352.

Lee, J. A., Homma, H., Sakai, K., Fukushima, T., Santa, T., Tashiro, K., Iwatsubo, T., Yoshikawa, M., Imai, K. (1997). Immunohistochemical localization of D-aspartate in the rat pineal gland. *Biochem. Biophys. Res. Commun.* 231:505–508.

Lowenstein, C. J., Glatt, C. S., Bredt, D. S., Snyder, S. H. (1992). Cloned and expressed macrophage nitric oxide synthase contrasts with the brain enzyme. *Proc. Natl. Acad. Sci. USA* 89:6711–6715.

Lyons, C. R., Orloff, G. J., Cunningham, J. M. (1992). Molecular cloning and functional expression of an inducible nitric oxide synthase from a murine macrophage cell line. *J. Biol. Chem.* 267:6370–6374.

Maines, M. D., Trakshel, G. M., Kutty, R. K. (1986). Characterization of two constitutive forms of rat liver microsomal heme oxygenase. Only one molecular species of the enzyme is inducible. *J. Biol. Chem.* 261:411–419.

Marletta, M. A., Yoon, P. S., Iyengar, R., Leaf, C. D., Wishnok, J. S. (1988). Macrophage oxidation of L-arginine to nitrite and nitrate: nitric oxide is an intermediate. *Biochemistry* 27:8706–8711.

Mearin, F., Camilleri, M., Malagelada, J. R. (1986). Pyloric dysfunction in diabetics with recurrent nausea and vomiting. *Gastroenterology* 90:1919–1925.

Meffert, M. K., Haley, J. E., Schuman, E. M., Schulman, H., Madison, D. V. (1994). Inhibition of hippocampal heme oxygenase, nitric oxide synthase, and long-term potentiation by metalloporphyrins. *Neuron* 13:1225–1233.

Meldrum, B., Garthwaite, J. (1990). Excitatory amino acid neurotoxicity and neurodegenerative disease. *Trends Pharmacol. Sci.* 11:379–387.

Meneghini, R. (1997). Iron homeostasis, oxidative stress, and DNA damage. *Free Radical Biol. Med.* 23:783–792.

Michell, B. J., Griffiths, J. E., Mitchelhill, K. I., Rodriguez-Crespo, I., Tiganis, T., Bozinovski, S., de Montellano, P. R., Kemp, B. E., Pearson, R. B. (1999). The Akt kinase signals directly to endothelial nitric oxide synthase. *Curr. Biol.* 12: 845–848.

Mothet, J. P., Parent, A. T., Wolosker, H., Brady, R. O., Jr., Linden, D. J., Ferris, C. D., Rogawski, M. A., Snyder, S. H. (2000). D-Serine is an endogenous ligand for the glycine site of the N-methyl-D-aspartate receptor. *Proc. Natl. Acad. Sci. USA* 97:4926–4931.

Nagafuji, T., Matsui, T., Koide, T., Asano, T. (1992). Blockade of nitric oxide formation by N-omega-nitro-L-arginine mitigates ischemic brain edema and subsequent cerebral infarction in rats. *Neurosci. Lett.* 147:159–162.

Nakane, M., Mitchell, J., Forstermann, U., Murad, F. (1991). Phosphorylation by calcium calmodulin–dependent protein kinase II and protein kinase C modulates the activity of nitric oxide synthase. *Biochem. Biophys. Res. Commun.* 180: 1396–1402.

Neidle, A., Dunlop, D. S. (1990). Developmental changes in free D-aspartic acid in the chicken embryo and in the neonatal rat. *Life Sci.* 46:1517–1522.

Nelson, R. J., Demas, G. E., Huang, P. L., Fishman, M. C., Dawson, V. L., Dawson, T. M., Snyder, S. H. (1995). Behavioural abnormalities in male mice lacking neuronal nitric oxide synthase. *Nature* 378:383–386.

Nishikawa, T., Kirsch, J. R., Koehler, R. C., Bredt, D. S., Snyder, S. H., Traystman, R. J. (1993). Effect of nitric oxide synthase inhibition on cerebral blood flow and injury volume during focal ischemia in cats. *Stroke* 24:1717–1724.

Nowicki, J. P., Duval, D., Poignet, H., Scatton, B. (1991). Nitric oxide mediates neuronal death after focal cerebral ischemia in the mouse. *Eur. J. Pharmacol.* 204:339–340.

Palmer, R. M., Ferrige, A. G., Moncada, S. (1987). Nitric oxide release accounts for the biological activity of endothelium-derived relaxing factor. *Nature* 327: 524–526.

Parent, A., Mothet, J. P., Snyder, S. H., Rogawski, M. A. (1999). D-serine is an endogenous ligand for the glycine site of the NMDA receptor: electrophysiological characterization in cultured hippocampal neurons. *Soc. Neurosci. Abstr.* 25:1714.

Pieper, A. A., Verma, A., Zhang, J., Snyder, S. H. (1999). Poly (ADP-ribose) polymerase, nitric oxide and cell death. *Trends Pharmacol. Sci.* 20:171–181.

Poss, K. D., Tonegawa, S. (1997a). Heme oxygenase 1 is required for mammalian iron reutilization. *Proc. Natl. Acad. Sci. USA* 94:10,919–10,924.

Poss, K. D., Tonegawa, S. (1997b). Reduced stress defense in heme oxygenase 1-deficient cells. *Proc. Natl. Acad. Sci. USA* 94:10,925–10,930.

Poss, K. D., Thomas, M. J., Ebralidze, A. K., O'Dell, T. J., Tonegawa, S. (1995). Hippocampal long-term potentiation is normal in heme oxygenase-2 mutant mice. *Neuron* 15:867–873.

Sakai, K., Homma, H., Lee, J. A., Fukushima, T., Santa, T., Tashiro, K., Iwatsubo, T., Imai, K. (1997). D-aspartic acid localization during postnatal development of rat adrenal gland. *Biochem. Biophys. Res. Commun.* 235:433–436.

Schell, M. J., Molliver, M. E., Snyder, S. H. (1995). D-serine, an endogenous synaptic modulator: localization to astrocytes and glutamate-stimulated release. *Proc. Natl. Acad. Sci. USA* 92:3948–3952.

Schell, M. J., Brady, R. O., Jr., Molliver, M. E., Snyder, S. H. (1997a). D-serine as a neuromodulator: regional and developmental localizations in rat brain glia resemble NMDA receptors. *J. Neurosci.* 17:1604–1615.

Schell, M. J., Cooper, O. B., Snyder, S. H. (1997b). D-aspartate localizations imply neuronal and neuroendocrine roles. *Proc. Natl. Acad. Sci. USA* 94:2013–2018.

Sessa, W. C., Harrison, J. K., Barber, C. M., Zeng, D., Durieux, M. E., D'Angelo, D. D., Lynch, K. R., Peach, M. J. (1992). Molecular cloning and expression of a cDNA encoding endothelial cell nitric oxide synthase. *J. Biol. Chem.* 267: 15,274–15,276.

Shinomura, T., Nakao, S., Mori, K. (1994). Reduction of depolarization-induced glutamate release by heme oxygenase inhibitor: possible role of carbon monoxide in synaptic transmission. *Neurosci. Lett.* 166:131–134.

Snyder, S. H. (1975). Opiate receptor in normal and drug altered brain function. *Nature* 257:185–189.

Snyder, S. H. (1985). Adenosine as a neuromodulator. *Annu. Rev. Neurosci.* 8:103–124.

Son, H., Hawkins, R. D., Martin, K., Kiebler, M., Huang, P. L., Fishman, M. C., Kandel, E. R. (1996). Long-term potentiation is reduced in mice that are doubly mutant in endothelial and neuronal nitric oxide synthase. *Cell* 87:1015–1023.

Stark, M. E., Szurszewski, J. H. (1992). Role of nitric oxide in gastrointestinal and hepatic function and disease. *Gastroenterology* 103:1928–1949.

Stark, M. E., Bauer, A. J., Szurszewski, J. H. (1991). Effect of nitric oxide on circular muscle of the canine small intestine. *J. Physiol.* 444:743–761.

Stocker, R., Glazer, A. N., Ames, B. N. (1987a). Antioxidant activity of albumin-bound bilirubin. *Proc. Natl. Acad. Sci. USA* 84:5918–5922.

Stocker, R., Yamamoto, Y., McDonagh, A. F., Glazer, A. N., Ames, B. N. (1987b). Bilirubin is an antioxidant of possible physiological importance. *Science* 235:1043–1046.

Stuehr, D. J., Gross, S. S., Sakuma, I., Levi, R., Nathan, C. F. (1989). Activated murine macrophages secrete a metabolite of arginine with the bioactivity of endothelium-derived relaxing factor and the chemical reactivity of nitric oxide. *J. Exp. Med.* 169:1011–1020.

Takahashi, K., Pieper, A. A., Croul, S. E., Zhang, J., Snyder, S. H., Greenberg, J. H. (1999). Post-treatment with an inhibitor of poly(ADP-ribose) polymerase attenuates cerebral damage in focal ischemia. *Brain Res.* 829:46–54.

Trakshel, G. M., Kutty, R. K., Maines, M. D. (1986). Purification and characterization of the major constitutive form of testicular heme oxygenase: the noninducible isoform. *J. Biol. Chem.* 261:11,131–11,137.

Trifiletti, R. R. (1992). Neuroprotective effects of NG-nitro-L-arginine in focal stroke in the 7-day-old rat. *Eur. J. Pharmacol.* 218:197–198.

Tsai, G., Yang, P., Chung, L. C., Lange, N., Coyle, J. T. (1998). D-serine added to antipsychotics for the treatment of schizophrenia. *Biol. Psychiatry* 44:1081–1089.

Vanderwinden, J. M., Mailleux, P., Schiffmann, S. N., Vanderhaeghen, J. J., De Laet, M. H. (1992). Nitric oxide synthase activity in infantile hypertrophic pyloric stenosis. *N. Engl. J. Med.* 327:511–515. [Erratum appears in *N. Engl. J. Med.* 327:1252.]

Verma, A., Hirsch, D. J., Glatt, C. E., Ronnett, G. V., Snyder, S. H. (1993). Carbon monoxide: a putative neural messenger. *Science* 259:381–384.

Wang, Y., Marsden, P. A. (1995). Nitric oxide synthases: gene structure and regulation. *Adv. Pharmacol.* 34:71–90.

Wang, Y., Newton, D. C., Robb, G. B., Kau, C. L., Miller, T. L., Cheung, A. H., Hall, A. V., VanDamme, S., Wilcox, J. N., Marsden, P. A. (1999). RNA diversity has profound effects on the translation of neuronal nitric oxide synthase. *Proc. Natl. Acad. Sci. USA* 96:12,150–12,155.

Watkins, C., Blackshaw, S., Sawa, A., Snyder, S., Ferris, C. (2000). Diabetic gastropathy in mice reflects loss of neuronal nitric oxide synthase that is restored by insulin. *J. Clin. Invest.* (in press).

Wolosker, H., Blackshaw, S., Snyder, S. H. (1999a). Serine racemase: A glial enzyme synthesizing D-serine to regulate glutamate-N-methyl-D-aspartate neurotransmission. *Proc. Natl. Acad. Sci. USA* 96:13,409–13,414.

Wolosker, H., Sheth, K. N., Takahashi, M., Mothet, J. P., Brady, R. O., Jr., Ferris, C. D., Snyder, S. H. (1999). Purification of serine racemase: biosynthesis of the neuromodulator D-serine. *Proc. Natl. Acad. Sci. USA* 96:721–725.

Xie, Q. W., Cho, H. J., Calaycay, J., Mumford, R. A., Swiderek, K. M., Lee, T. D., Ding, A., Troso, T., Nathan, C. (1992). Cloning and characterization of inducible nitric oxide synthase from mouse macrophages. *Science* 256:225–228.

Xue, L., Farrugia, G., Miller, S. M., Ferris, C. D., Snyder, S. H., Szurszewski, J. H. (2000). Carbon monoxide and nitric oxide as coneurotransmitters in the enteric nervous system: evidence from genomic deletion of biosynthetic enzymes. *Proc. Natl. Acad. Sci. USA* 97:1851–1855.

Zakhary, R., Gaine, S. P., Dinerman, J. L., Ruat, M., Flavahan, N. A., Snyder, S. H. (1996). Heme oxygenase 2: endothelial and neuronal localization and role in endothelium-dependent relaxation. *Proc. Natl. Acad. Sci. USA* 93:795–798.

Zakhary, R., Poss, K. D., Jaffrey, S. R., Ferris, C. D., Tonegawa, S., Snyder, S. H. (1997). Targeted gene deletion of heme oxygenase 2 reveals neural role for carbon monoxide. *Proc. Natl. Acad. Sci. USA* 94:14,848–14,853.

Zhang, J., Dawson, V. L., Dawson, T. M., Snyder, S. H. (1994). Nitric oxide activation of poly(ADP-ribose) synthetase in neurotoxicity. *Science* 263:687–689.

Synaptic Variability
New Insights from Reconstructions and Monte Carlo Simulations with MCell

Joel R. Stiles, Thomas M. Bartol, Miriam M. Salpeter,
Edwin E. Salpeter, and Terrence J. Sejnowski

B iological structures show tremendous complexity and diversity at the subcellular level. For example, a single cubic millimeter of the cerebral cortex may contain on the order of five billion interdigitated synapses of different shapes and sizes.[1] Subcellular communication is based on a wide variety of chemical signaling pathways, and for synaptic transmission these include neurotransmitter and neuromodulator molecules, proteins involved with exo- and endocytosis, receptor proteins, transport proteins, and oxidative and hydrolytic enzymes. Synaptic crosstalk may result when ligand molecules released from one synapse diffuse to another (Clements, 1996; Barbour and Hausser, 1997; Rusakov and Kullmann, 1998; Rusakov et al., 1999), so the range over which ligands can act likely extends from nanometers to microns. In addition, chemical and structural plasticity at synapses undoubtedly contributes to information storage and processing, and it is widely discussed in relation to high-level cognitive functions such as learning and memory (Edwards, 1995a,b; see Grimwood et al., this volume).

Theoretical studies of ligand diffusion and chemical reaction have been used to investigate synaptic structure-function relationships since the late 1950s (Eccles and Jaeger, 1958), but until recently computer hardware and software limitations have precluded highly realistic three-dimensional (3-D) simulations of reconstructed synapses. As a result the contribution of actual ultrastructure to synaptic current variability (or other signaling phenomena) has gone largely unexplored, and quantitative modeling of synaptic physiology has been severely hampered. Computer hardware limitations have now been significantly overcome by massively parallel systems and, on a smaller scale, by affordable multiprocessor workstations with large amounts of memory and fast graphics-handling capabilities. The major remaining bottleneck is the development of programs that not only provide the requisite simulation capabilities but also interface smoothly with interactive 3-D reconstruction and animation programs. We describe here a program that opens up the study of synapses with realistic geometries at the subcellular level.

Our simulation program, MCell,[2] is based on Monte Carlo (MC) algorithms and incorporates highly realistic 3-D reconstructions into models of ligand diffusion (e.g., neurotransmitter exocytosis) and signaling (e.g., synaptic currents). As shown in Fig. 15.1, MCell simulations are positioned at a biological scale above molecular dynamics but below whole-cell and network modeling studies. Its structural realism lies between the space-filled atomic resolution of molecular dynamics (e.g., AMBER or CHARMM simulations; Cornell et al., 1995; MacKerell et al., 1998) and that of less structure-dependent approaches, such as proteomics (e.g., E-CELL simulations; Tomita et al., 1999) and compartmental models of "Hodgkin-Huxley"–style neurons (e.g., NEURON or GENESIS simulations; Hines, 1993; Bower and Beeman, 1995). In a compartmental model, the complex geometry of an individual cell is subdivided into approximately isopotential parts, each of which becomes a resistive-capacitive element in a branched electrical circuit representation. The actual 3-D

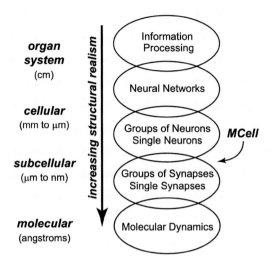

Figure 15.1. *Physical scales versus structural realism in computational neuro-science models.* MCell simulations encompass diffusion and chemical reaction of molecules in 3-D reconstructions and hence are situated between compartmental models of single- or multiple-neuron excitation and molecular dynamics models of single-molecule structure.

configuration of the neuron(s) and surrounding tissue volume is not explicitly considered, nor is the molecular nature of various conductances and currents that can be incorporated into different compartments. MCell and these other modeling approaches are mostly complementary, and integration of the various approaches is an important direction for future research.

In MCell simulations, the diffusion of individual ligand molecules within a reconstructed 3-D environment is simulated using a Brownian dynamics random walk algorithm, and bulk solution rate constants are converted into MC probabilities so that the diffusing ligands can inter-act stochastically with individual binding sites, such as receptor proteins, enzymes, and transporters. Such methods can be applied to many questions related to synaptic transmission, such as:

1. How does the architecture of the synaptic cleft affect quantal and multiquantal current amplitudes and time courses at central and peripheral synapses?
2. How do the kinetics and precise localization of transmitter release, receptors, and reuptake sites influence current variability and cross-talk between synapses?
3. How might intra- and extracellular ion fluxes from repeated synaptic activation in realistic synaptic architectures, together with discrete transmembrane conductances and pumps, influence particular regions of neuropil?

4. How do the many calcium-binding proteins in 3-D subcellular micro-environments trigger and regulate a multitude of signaling pathways?

We begin this chapter with a short historical look at the vertebrate neuromuscular junction (NMJ), which has long served as a model synapse for investigations of structure and function. We then review our modeling approach, validate the latest algorithms, and discuss factors that affect numerical accuracy. MCell is used to simulate acetylcholine (ACh) exocytosis at the NMJ and to show how realistic synaptic architecture may account for a significant fraction of miniature endplate current (MEPC) variability. We conclude by describing future directions for continued synaptic reconstruction, simulation, and combined theoretical and experimental studies.

Historical Overview of Structure and Function at the Vertebrate Neuromuscular Junction

The vertebrate NMJ has long served as a model synapse owing to its easy accessibility, large size, and singular distribution. A wealth of physiological and morphological data is available, and, in spite of its unique characteristics, the NMJ has provided many insights into our understanding of central synapses.

Basic features of the NMJ on vertebrate twitch fibers were first described in the 1840s (Doyere, 1840), yet the modern view of NMJ organization began with Couteaux in the 1940s. In a series of illuminating articles, he identified the postjunctional muscle surface as a unique palisade-like specialization rich in acetylcholinesterase (AChE), which is now known as the junctional folds (JFs). Couteaux also helped clarify the relationship between Schwann cells and the NMJ. From him came some of the best early descriptions of the morphology of the vertebrate NMJ, as well as a fascinating view of some early controversies (e.g., Couteaux, 1946, 1955, 1958). However, the detailed morphology of the NMJ was not revealed until the advent of transmission electron microscopy (TEM) in the 1950s and 1960s.

It is now clear that chemically transmitting synapses like the NMJ consist of three distinct compartments: (1) the presynaptic nerve terminal, containing large numbers of synaptic vesicles, numerous mitochondria, and one or more active zones (thickened presynaptic membrane); (2) the postsynaptic apparatus, characterized in part by a dense membrane with an underlying filamentous cytoskeleton; and (3) an intervening synaptic cleft that at the NMJ contains basal lamina material.

A major difference between neuroneuronal synapses and the NMJ is the organization of the second and third compartments. The postsynaptic membrane of the NMJ often is extensively folded, forming JFs of variable depth (generally 0.5–1 μm). These JFs create secondary extracellular cleft spaces connected to the primary cleft, which lies between the nerve terminal membrane and the tops of the folds. For neuro-neuronal synapses the width of the synaptic cleft is about 10–20 nm, while for the NMJ it is about 50 nm in both the primary and secondary cleft spaces. The overall organization of NMJs on typical vertebrate twitch fibers is shown in Fig. 15.2. The primary and secondary clefts contain AChE active sites and basal lamina proteins, and the latter are part of a complex scaffold that connects neural and musculocytoskeletal elements (Froehner, 1986; Hall and Sanes, 1993; Matsumura and Campbell, 1994; Apel and Merlie, 1995; Sanes and Lichtman, 1999).

Small round vesicles (circa 50 nm in diameter; Fig. 15.2B) were seen in the presynaptic terminals in the first TEM images (Palade and Palay, 1954; Reger, 1955; Robertson, 1956). The anatomical description of synaptic vesicles coincided with physiological studies showing that excitation at the NMJ occurs in multiples of quantal depolarizing events (del Castillo and Katz, 1954). This led to the suggestion that synaptic vesicles contain neurotransmitter molecules that are released by fusion of vesicles with the presynaptic membrane, and that there is a one-to-one correspondence between the release of a single vesicle's contents and a quantal physiological signal. Numerous studies were undertaken in the 1950s and 1960s using biochemical (DeRobertis et al., 1961; Whitaker, 1965), physiological, and electron microscopic (EM; Ceccarelli et al., 1973; Heuser and Reese, 1973) techniques, which supported the claim that vesicles contain the releasable pool of ACh. In addition, the TEM visualization of "omega figures" in fast-frozen tissue (Ceccarelli et al., 1973; Heuser and Reese, 1973)—vesicles fused to the presynaptic membrane by means of an open pore—finally convinced many investigators that ACh is released by exocytosis of quantal packets from synaptic vesicles.

The JFs were initially thought necessary to increase the muscle's postsynaptic receptive surface area. This idea was supported by the proposed mosaic model of acetylcholine receptor (AChR) and AChE distribution (Barnard et al., 1971; Porter et al., 1973). However, the mosaic model was suspect on physiological and theoretical grounds, because it seemed unlikely that many ACh molecules would reach AChRs at the bottom of the JFs during the rising phase of an MEPC. EM autoradiography finally provided direct evidence that AChRs are, in fact, highly concentrated on the crests (i.e., the top 200–300 nm) of the JFs and are present at decreasing average density along the depth of the JF, with essentially none at the bottom (Salpeter et al., 1984). High AChR density therefore is spatially correlated with the presence of the postsynaptic membrane density, which also is mostly observed at the crests of JFs (Fig. 15.2B; Albuquerque et al., 1974; Fertuck and Salpeter, 1974, 1976). Another important result provided by EM autoradiography is that AChE active sites are distributed throughout the primary and secondary cleft spaces, but at average densities much lower than that for the crest AChRs (Salpeter, 1967, 1969; Rogers et al., 1969; Anglister et al., 1994). The high ratio of AChR- to AChE-binding sites at the receptive surface gave rise to the "saturated disc" model of MEPC generation (Matthews-Bellinger and

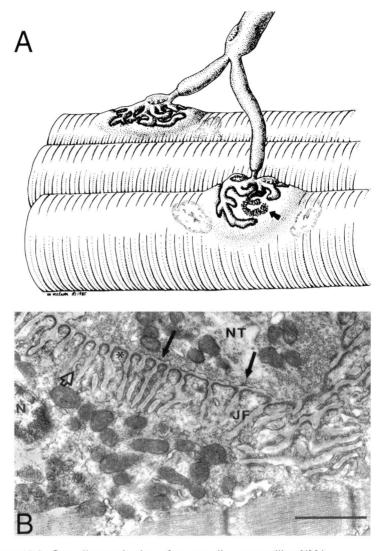

Figure 15.2. Overall organization of mammalian or reptilian NMJ.

(A) Schematic diagram of a branched, myelinated axon innervating two muscle fibers. The myelin sheath ends above the nerve terminals, which are encapsulated by Schwann cells and are twisted to form compact endplates. If the nerve terminal is removed, the openings into JFs are visible (arrow). Schwann cell and postsynaptic sole plate nuclei are also indicated.

(B) TEM view of mouse sternomastoid NMJ. Synaptic vesicles and mitochondria are visible throughout the nerve terminal (NT), and active zones (filled arrows) are localized across from the openings of JFs. Dense postsynaptic membrane is present mostly at the crests of JFs (asterisk), while nondense membrane is mostly lower and sometimes includes fused vesicles (open arrowhead). One sole plate nucleus (N) and numerous mitochondria are visible. Note also the relative regularity of JF contours at the left and center, and the extreme structural complexity at the right (cf. Fig. 15.18A). Scale bar = 1 μm.

From Salpeter (1987), reprinted by permission of Wiley-Liss, Inc., a subsidiary of John Wiley & Sons, Inc.

Salpeter, 1978; Land et al., 1980, 1981, 1984; Salpeter, 1987), which illustrated how quantal packets of ACh would interact with discrete, almost saturated, nonoverlapping areas of postsynaptic receptive surface. The saturated disc model agreed well with contemporary physiological studies that indicated nonoverlapping areas of AChR activation at NMJs with normally active AChE, but suggested the potential for overlap if AChE was inhibited (Hartzell et al., 1975).

Other molecules, such as Na^+ channels, have also been identified at the bottom of JFs (Flucher and Daniels, 1989; Boudier et al., 1992). As the extensive complexity of JFs is incorporated into high-resolution 3-D reconstructions and MCell simulations, the role of JFs and other aspects of neuromuscular function (normal and pathological) are increasingly open to detailed biophysical investigations. Similarly, as the site densities and distributions of the many molecules that participate in central synaptic transmission are defined and incorporated into realistic simulations, complex aspects of synaptic communication and plasticity will become increasingly accessible.

Rationale for Monte Carlo Approach to Simulation of 3-D Diffusion and Chemical Reaction

The simultaneous diffusion of molecules and their chemical reactions in three dimensions can be simulated using one of two fundamentally different methods, based either on finite element (FE) or MC algorithms. Both require significant computer programming and computational resources, and each approximates reality in a different way. The MC approach is more realistic and is generally more expensive computationally, but as 3-D models become more detailed the difference in computer requirements for MC versus FE simulations is not so great as it once was. Prior to the recent explosion in computing power, limited resources dictated that 3-D problems be simplified dramatically, for example by replacing a real in situ synaptic cleft with a flat, isolated, radially symmetrical model. In this way 3-D problems could be reduced to one dimension, and the mathematics of diffusion and chemical reaction could be handled with analytic approximations or sets of ordinary differential equations that could be evaluated with simple finite difference methods. Such *equation-based approaches* disregard the discrete nature of ligand and receptor molecules and instead use hypothetical concentrations that vary smoothly as a function of position and time. The strengths of such simplified approaches are rapid and precise predictions for the average behavior of the model (assuming a small enough time step). However, the drawbacks of simplification may in some circumstances lead to significant errors, and stochastic variability is ignored.

For 3-D FE simulations, the molecular nature of the real system is disregarded, and differential equations are used to compute fluxes and reaction rates between and within spatial subdivisions or voxels (the finite elements). Concentration gradients are ignored within each voxel (i.e., the contents are assumed to be well mixed), and flow occurs across the interfaces between adjacent voxels (except, of course, where a voxel wall coincides with a diffusion boundary). Thus, as for the simplified one-dimensional (1-D) approach, reactant concentrations vary smoothly throughout space, albeit with small stepwise changes from one voxel to the next. Whereas 1-D approaches subdivide only time, the 3-D FE approach subdivides both time and space, and overly large spatial subdivisions (coarse granularity) can markedly degrade numerical accuracy. Fine granularity is easy to achieve for simple overall spatial configurations (e.g., a cell represented as a subdivided box), and under such conditions this method can be extremely efficient. For complex realistic structures, however, the spatial subdivisions become correspondingly complex to plan and implement, and the number of voxels can grow to be very large. Under such conditions the computational expense increases tremendously, and in no case do FE simulations provide direct information about stochastic variability arising from the spatial configuration and actual finite numbers of participating molecules.

In contrast to the equation-based simulation methods, the MC approach to 3-D diffusion and chemical reaction begins with an arbitrary set of surfaces that represent a subcellular environment (e.g., cell and organelle membranes), and then the surfaces and surrounding space are populated with individual ligand and ligand-binding molecules. With the present version of MCell, the shape of the surfaces can be as realistically complex as desired, as can the biochemical pathways followed by ligand and ligand-binding (effector) sites. Ligand movements approximate Brownian motion by means of random walk displacements, and collisions with surfaces and effector sites can be detected without the use of voxels by tracing random walk trajectories through space with algorithms similar to those used to trace light rays in photorealistic computer graphics. The average radial distance ($\bar{l}_r$) traveled in a random walk step depends on both the specified ligand mobility (diffusion coefficient, D) and the simulation time step (Δt). High numerical accuracy can generally be obtained under conditions in which $\bar{l}_r$ and Δt are orders of magnitude larger than the mean free path and time between collisions, respectively, for true Brownian motion. Another relationship of importance to numerical accuracy is the ratio of Δt to the average lifetimes of the chemical reactant states in the simulation. (Such issues underlying the choice of input parameter values and resulting accuracy are illustrated later in the chapter.)

During each MC time step, Δt, decisions about distance and direction of motion, binding, unbinding, conformational changes, and all other possible events are made by comparing the values of random numbers to MC probabilities that are precalculated for each type of event. The MC probabilities depend on input values for Δt, D, reaction rate constants, and the surface area of effector sites, and they ensure that results for equilibrium or steady-state conditions match analytic expectations based on bulk solution rate equations. The use of random numbers to make decisions during the simulation is reminiscent of throwing dice, and the term *Monte Carlo* was originally coined in this context by Ulam and Von Neumann during the Manhattan Project (Rubenstein, 1981). Since events occur on a molecule-by-molecule basis, the simulation results include realistic stochastic noise arising from the spatial arrangement and finite number of participating molecules. Averaging over a number, n, of different simulations (run with different random numbers) decreases the noise, but generally only in proportion to $1/\sqrt{n}$. This noise may seem to be a disadvantage of MC simulations, but it is in fact a blessing in disguise because its spectral properties can be compared with the observed properties of noise present in experimental data.

Although the potential realism and generality of MC methods were clear many years ago, large increases in computer speed and memory were required before many applications became feasible. Because of such increases, as well as algorithm optimizations developed for MCell, it is now possible to simulate complex cellular reconstructions on workstations, and very large projects can be ported effectively to massively parallel computer architectures. As mentioned earlier, this level of realism is not simple to implement using FE methods, nor is it clear that an FE approach would be appreciably more efficient. Most importantly, MC simulations can now provide insights into the stochastic variability and nonintuitive behavior of complex 3-D systems containing small numbers of reactant molecules.

MCell Overview

History and Scope

A novel Monte Carlo approach to MEPC simulation was initiated more than a decade ago (see Figs. 21–23 in Salpeter, 1987) in order to advance saturated disc modeling at the NMJ beyond a flat cleft configuration and earlier equation-based simulation methods (Land et al., 1980, 1981, 1984). The new approach led to a computer program that could simulate planar (simplified) JFs (see Fig. 15.13) and was run on an IBM supercomputer (Bartol et al., 1991; Bartol, 1992). A similar program written independently at about the same time focused primarily on delineating and tracking stochastic molecular events that occur during MC simulations of MEPCs (Stiles, 1990) and was initially run on large clusters of VAX superminicomputers. These two programs were merged, and the resulting code again was run on a supercomputer to simulate the effect of AChE site density and hydrolytic rate at the NMJ (Anglister et al., 1994). Other investigators began using the underlying random walk and MC binding-unbinding algorithms (Bartol et al., 1991) to simulate other simplified synaptic systems (Faber et al., 1992; Bennett et al., 1995, 1997, 1998; Wahl et al., 1996; Kruk et al., 1997).

These early MC programs were designed to simulate one type of simplified structure and could not be applied to more realistic problems. Such limits were removed by generalizing and optimizing the MC methods, and the earliest versions of MCell emerged in successive stages (Stiles et al., 1996, 1998). A Model Description Language (MDL) was created to design and control large-scale simulations, as well as to integrate models with 3-D imaging software, and it has evolved into a standardized interface and archiving system.

MCell has been used worldwide at research laboratories since 1997 (see Web sites in note 2), and it now includes new features that dramatically increase its range of application, memory efficiency, and execution speed. For example, many simulations can be run on workstations rather than supercomputers, the realism and speed of the Brownian dynamics random walk algorithms have been greatly enhanced, and the addition of *spatial partitions* renders execution speed essentially independent of the model's geometric complexity. Thus simulations based on large-scale tissue reconstructions can now be run in about the same time previously required for highly simplified structures, in effect reducing the required computer time from many months to hours or even minutes.

Typical events that occur during an MCell simulation include the release of ligand molecules from a structure (e.g., a vesicle), de novo creation or destruction of ligand molecules (e.g., synthesis, hydrolysis, or redox reactions), ligand diffusion within spaces defined by arbitrary surfaces (e.g., pre- and postsynaptic membranes or a cell membrane with attached patch clamp micropipette), and chemical reactions undergone by diffusing ligand and fixed effector (e.g., receptor or enzyme) molecules. Ligands, effectors, reaction mechanisms, 3-D surfaces, and other simulation components are specified using the MDL (Fig. 15.3),[3] a simple programming language that was designed with biologists in mind (Stiles and Bartol, 2000). When a simulation is run, one or more MDL input files are interpreted (parsed) to create the simulation objects, and then execution begins for a specified number of iterations. Each iteration corresponds to one MC time step, Δt, which typically is on the order of 1 μsec for synapses. Simulations can be stopped and subsequently restarted at user-specified *checkpoints* (Fig. 15.3), and when a simulation restarts updated information can be read from the input MDL file(s). Checkpointing is thus a powerful and general way to change run-time parameters such as Δt, reaction rate constants, and surface positions; it can also be used to split long simulations into segments that are run sequentially.

Brownian Dynamics Random Walk

MCell simulates diffusion using a novel Brownian dynamics random walk algorithm that has been highly optimized for speed, numerical accuracy, and use with complex structures. In essence, extensive sets of equally probable radial distances and directions are stored in two look-up

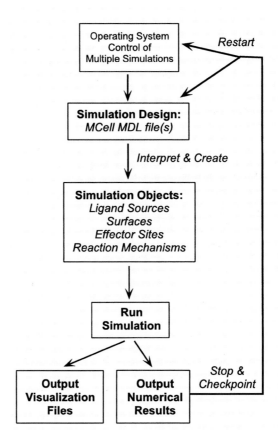

Figure 15.3. Overview of MCell simulation design, execution, and output. Simulations are run in UNIX or Windows environments and are designed using Model Description Language (MDL) text files created by the user or from 3-D surface reconstruction data. When the simulation is initialized, the MDL files are interpreted to create the objects used during execution. Execution continues for a specified number of time steps, and the amount and type of output are under the user's control. Output files fall into three general categories: (1) visualization, used with a variety of 3-D imaging and animation software; (2) numerical results, used to tally the number of reaction intermediates and transitions as a function of time and space; and (3) checkpointing results, used as all or part of the initial conditions for subsequent simulations. In addition, sets of simulations may be controlled using command files that are specific to the computer's operating system (e.g., UNIX shell scripts or DOS batch files).

tables, and one value is chosen from each table to generate each random walk movement. The use of such tables increases execution speed dramatically, and any reduction in accuracy is effectively immeasurable.

The values for the first table are based on diffusion theory and are calculated from a standardized probability distribution function (p_s) for a dimensionless parameter *s:*

$$p_s = \frac{4}{\sqrt{\pi}} \, s^2 e^{-s^2} \, ds. \qquad (15.1)$$

At the beginning of each simulation, equation (15.1) is integrated numerically to obtain the cumulative probability of s, which then is finely subdivided into bins (generally 1024) of equal area. The mean value of s for each bin is stored in the look-up table, and then, while the simulation runs, values can be chosen from the table as needed using uniformly distributed random numbers. For a ligand molecule with diffusion constant D_L, a chosen value of s is converted to a radial step length (l_r) using a multiplicative scaling factor given by $\sqrt{4D_L(\Delta t)}$, which has units of distance.

For a given value of D_L, the distribution of random walk step lengths changes according to the value specified for Δt. For accurate simulation of diffusion, Δt must be chosen so that the mean radial step length ($\bar{l}_r$) is smaller than the dimensions of restrictive structures in the model. The value of $\bar{l}_r$ is obtained from the expectation value of s and the scaling factor $\sqrt{4D_L(\Delta t)}$ and is given by

$$\bar{l}_r = 2\sqrt{\frac{4D_L(\Delta t)}{\pi}}. \qquad (15.2)$$

In order to calculate MC binding probabilities for effector sites located on a surface, it is necessary to know the average random walk displacement with respect to a linear direction oriented perpendicular to the surface ($\bar{l}_\perp$). From a form of equation (15.1) for linear rather than radial displacements,

$$\bar{l}_\perp = \sqrt{\frac{4D_L \Delta t}{\pi}} \qquad (15.3)$$

or $\bar{l}_r/2$.

To generate a ligand movement, a randomly chosen radial step length must be paired with an unbiased choice of radial direction. Complete elimination of bias is critical because even a tiny asymmetry can accumulate over thousands of iterations to produce substantial drift. The second look-up table is used to store a set of equally probable radial directions, that is, unit vectors that originate from a point and radiate out in all directions with equal probability. These vectors are calculated numerically when the simulation begins (there are generally more than 130,000), using methods that guarantee the absence of directional bias. Once a value of s and a direction have been chosen for a particular movement,[4] the x, y, and z components of the direction vector are multiplied by the product of s and the scaling factor $\sqrt{4D_L(\Delta t)}$, and the results are added to the present (x, y, z) coordinates of the molecule.

Over any interval of time, Δt, longer than the time between actual Brownian collisions (sub-picosecond scale at room temperature; e.g., Barrow, 1981), a real diffusing molecule follows some tortuous path between a starting position (P1) and an ending position (P2). The molecule's thermal velocity and the duration of Δt determine the *total* distance traveled, but the *net radial* distance traveled (the length of the vector $\vec{l}_r$ between P1 and P2) is proportional to $\sqrt{\Delta t}$. This proportionality is shown directly for the average distances $\bar{l}_r$ and $\bar{l}_\perp$ in equations (15.2) and (15.3).

In an MCell simulation, the molecule moves along the shorter, straight-line path $\vec{l}_r$ rather than the actual tortuous path, and hence the apparent velocity of motion (1) is less than the true thermal velocity, (2) is not constant for different chosen values of Δt, and (3) decreases as Δt increases. When ligand molecules must diffuse through a restriction (e.g., a vesicle fusion pore) or can bind to effector sites, the apparent velocity of motion directly influences the apparent net flux and also the MC binding probability and hence directly influences numerical accuracy.

Surfaces and Effector Sites

Each curved surface imported into an MCell simulation is actually composed of a polygon mesh (e.g., Figs 15.7, 15.11, 15.16B, and 15.18B), where each mesh element (ME) must be a convex planar polygon. Triangles are used for the most part, because they are guaranteed to be planar (three points in space define a plane) and are required for surfaces that include effector sites. Triangulated meshes are also the typical form of output from 3-D surface reconstruction software.

The MEs within a surface can be individually classified as reflective, transparent, or absorptive with respect to diffusing ligand molecules. Each time that a random walk trajectory (a ray) is generated, it must be traced to see if it intersects with an ME before the endpoint of motion is reached. If so, the final result depends on the properties of the ME at the point of intersection (see Stiles and Bartol, 2000, for a more complete discussion). If the ME contains an effector site at that point and binding occurs, the motion stops and the ligand molecule's fate in subsequent time steps depends on the reaction pathways defined for the effector site. If binding does not occur (whether or not an effector site is present), then (1) if the ME is absorptive, the motion stops and the ligand molecule is removed from the simulation; (2) if the ME is transparent, the intersection is detected and the ray continues through unchanged; or (3) if the ME is reflective, the ray undergoes specular reflection. In the case of (2) or (3), the random walk movement continues until either (a) binding or absorption occurs upon a subsequent intersection or (b) no further reflections occur and the ligand molecule is placed at the end of its remaining trajectory.[5]

Effector sites can be added to one or more MEs of any surface, whether the MEs are reflective, transparent, or absorptive. As illustrated subsequently, effector sites on a reflective surface are typically used to model various types of membrane-bound proteins, while effectors on transparent surfaces generally represent enzyme or other sites localized in an intra- or extracellular scaffold that does not specifically impede ligand movement. Effector sites on absorptive surfaces are a special case and could be used to sample the flux across the surface. The kinetic behavior of effector sites is defined by reaction mechanisms specified in an MDL file; it includes not only binding, unbinding, and conformational transitions but also the directionality of binding and unbinding with respect to the surface. For example, simple receptor sites would bind and unbind ligand molecules on only one side of a reflective surface, re-uptake sites would bind on the extracellular side and release on the intracellular side, and sites on transparent surfaces would bind and unbind from either side.

To use effector sites in a simulation, a global *effector tile grid density*, denoted here as σ_{EG}, must be specified in units of tiles/μm^2. This parameter determines the maximum density of effector sites that can be added to any surface, so if more than one type of effector is added to the same mesh (e.g., intermixed α-amino-3-hydroxyl-5-methyl-4-isoxazolepropionic acid and N-methyl-D-aspartate glutamate receptors), the sum of their densities should not exceed σ_{EG}. To add effector sites to MEs of arbitrary triangular shape, an individual grid is created for each ME, using *barycentric subdivision*. In short this method creates interdigitated triangular effector tiles that cover each ME exactly and have the same triangular aspect ratio as the ME on which they reside (Fig. 15.4). If the ME is large enough to accommodate more than just a few tiles (which is almost always the case), the aspect ratio is unimportant to numerical results. However, long and thin triangles are not optimal for imaging, because effector molecule glyphs placed at the center of mass of each tile (part of MCell's visualization output) appear in unrealistic linear arrays (Fig. 15.4B). With nearly equilateral triangles, on the other hand, the molecule glyphs occupy positions in a seemingly realistic hexagonal array (Fig. 15.4A).[6]

When a simulation begins, MCell makes one pass through each surface that contains effector sites, and for each ME it (1) calculates its area (A_{ME}); (2) performs barycentric subdivision (Fig. 15.4) to obtain an integer number (N_{ET}) of triangular effector *tiles* that cover its surface exactly; (3) calculates the expected number of effector *sites;* (4) calculates the probability (p) that each effector *tile* is occupied by an effector *site* (as opposed to remaining bare surface); and (5) compares the value of a random number, $\kappa(0 \leq \kappa \leq 1)$, with the value of p, to determine the identity of each effector tile.

Chemical Reactions: Monte Carlo Probabilities and Heuristics

Consider a simple reversible reaction between ligand L and effector E:

$$L + E \underset{k_-}{\overset{k_+}{\rightleftarrows}} LE. \qquad (15.4)$$

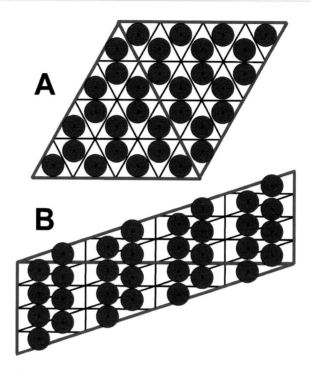

Figure 15.4. Simple examples of barycentric tiling. (A) Two equilateral MEs (heavy blue lines), each with area A_{ME}, subdivided according to the value of A_{ME} and σ_{EG}. In this case a 4×4 barycentric grid is obtained (narrow black lines), producing 16 effector tiles per ME. *(B)* Two nonequilateral MEs, each also with identical area A_{ME}, and therefore also subdivided using a 4×4 barycentric grid. When simulation results are visualized, color-coded glyphs are used to indicate the positions and chemical states of tiles occupied by effector sites. The glyphs typically are shaped and scaled to reflect the molecule being modeled (e.g., AChRs). In this 2-D illustration, red discs are used as glyphs, and every tile is occupied by an effector site. The glyphs are placed at the center of mass of each tile, and with equilateral MEs *(A)* the result is a hexagonal array. With long and thin MEs *(B)* the result is linear arrays. For realistic imaging results, optimized meshes containing nearly equilateral MEs are preferable (cf. Figs 15.13, 15.15, 15.16, and 15.18).

The rate constants k_{+} and k_{-} are phenomenological scaling factors that relate an observed rate of reaction to reactant concentrations in bulk solution (e.g., Hammes, 1978). At a given point in space, the rate equation is

$$-\partial(L) = -\partial(E) = \partial(LE) = [k_{+}(L)(E) - k_{-}(LE)]\partial t. \qquad \textbf{(15.5)}$$

Under well-mixed conditions, the concentration terms are independent of space at all times, and the partial differentials can be replaced by finite differences (e.g., $\partial(E)$ and ∂t become ΔE and Δt, respectively). If concentration gradients exist, then equation (15.5) can be simulated using the FE or MC methods discussed earlier. With the FE approach, equation (15.5) would be used in finite difference form within each voxel, and flux J between each voxel would be calculated with a 1-D finite difference simplification of the diffusion equation $J = -D_L \nabla (C_L)$, where C_L is ligand concentration. With the MC approach, net random walk motion determines ligand flux directly, and (as outlined below) the rate constants k_+ and k_- must be converted into dimensionless probabilities p_b and p_k, respectively.

If the binding step in equation (15.4) is taken in isolation, then equation (15.5) becomes

$$-\partial(L) = -\partial(E) = \partial(LE) = k_+(L)(E)\partial t, \qquad (15.6)$$

and the number of binding events ($N_B = \Delta(LE)$) expected per unit effector concentration (E), in bulk solution during time Δt, is given by $(k_+)(L)(\Delta t)$. The value of p_b for an effector site E is calculated from the ratio N_B/N_H, where N_H is the average number of times during Δt that ligand molecules "hit" (i.e., random walk trajectories intersect) the effector tile on which site E is located. (This use of N_B and N_H is a useful conceptual simplification; for a full derivation of p_b, N_H, and p_k, see Stiles and Bartol [2000].) The value of N_H depends on A_{ET}, the area of the tile, and since A_{ET} generally differs for each ME ($A_{ET} = A_{ME}/N_{ET}$), N_H and therefore p_b also differ for each ME. As shown in Fig. 15.5, N_H can be derived conceptually by first defining a direction vector ψ that is perpendicular to the ME. The average net distance traveled by ligand molecules along ψ is given by $\bar{l}_\perp$ (equation (15.3)). On average, half of the molecules within this distance will step away from the ME, while the other half will step toward it (and therefore will hit it). For an effector E in particular, $2N_H$ ligand molecules thus are contained within a volume V_{ET} that extends both "above" and "below" the tile for a distance ($\pm$) $\bar{l}_\perp$ along ψ. The value of V_{ET} therefore is $2(\bar{l}_\perp)(A_{ET})$ and N_H is given by $(L)(N_a)(V_{ET})/2$, where N_a is Avogadro's number. A final expression for p_b is

$$p_b = \frac{(f_s)(f_A)(k_+)(\sigma_{EG})}{2(N_a)} \left(\frac{\pi(\Delta t)}{D_L} \right)^{1/2}, \qquad (15.7)$$

where σ_{EG} is the effector tile grid density. The factor f_A (≥ 1) is given by the ratio of two areas, A_{ET} and $1/\sigma_{EG}$, i.e., $f_A = 1/(A_{ET}\sigma_{EG})$. The additional term ($f_s$) is unity if ligand molecules bind to either side of the ME (e.g., the transparent surfaces as illustrated in Fig. 15.7). For a reflective surface with binding allowed from only one side, V_{ET} would extend a distance $\bar{l}_\perp$ either "above" or "below" the ME and therefore would be halved. As a result N_H would also be halved, and $f_s = 2$ would be used in equation (15.7) to double the value of p_b.

Whenever a diffusing ligand molecule hits an effector site on an ME, the occurrence of binding is tested by comparing the value of a random number ($0 \leq \kappa \leq 1$) to the value of p_b for the ME. Because the apparent MC velocity of motion decreases if Δt is increased, p_b must increase to keep the average MC binding rate equal to the bulk solution binding rate. Equation (15.7) shows that p_b scales with $\sqrt{\Delta t}$ (because $\bar{l}_\perp$ scales with $\sqrt{\Delta t}$; equation (15.3)), so if the simulation time step is doubled and all

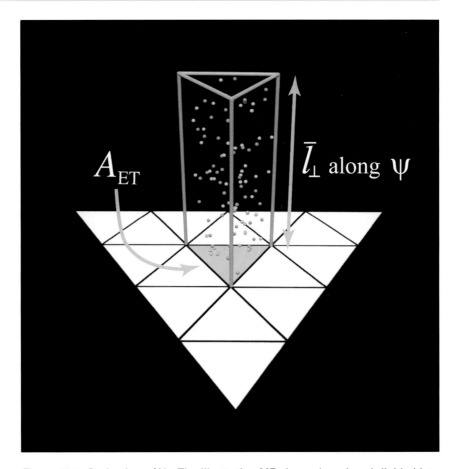

Figure 15.5. Derivation of N_H. The illustrative ME shown here is subdivided into 16 effector tiles, one of which is an effector site E (gray). During a time step, Δt, diffusing ligand molecules moving in random directions are displaced an average distance $\bar{l}_\perp$ along a direction ψ perpendicular to the ME. On average, half of the molecules (e.g., blue spheres) move away from the ME and half (e.g., yellow spheres) move toward it, so the value of N_H is one-half the number of molecules contained within the volume $V_{ET} = 2(\bar{l}_\perp)(A_{ET})$ indicated by the extruded gray prism for the portion of V_{ET} above the ME. Of course some of the yellow molecules actually move outside V_{ET} during Δt (and hence do not hit E), but on average they are replaced by an equal number of molecules moving into V_{ET} from the surrounding space.

other input parameters are held constant, the value of p_b increases by $\sqrt{2}$. The resulting impact on numerical accuracy depends on the new value of p_b and other factors. It is necessary that the value of p_b not exceed unity.

If the unbinding step in equation (15.4) is taken in isolation, the transition is a unimolecular Poisson process, and equation (15.5) becomes

$$d(L) = d(E) = -d(LE) = k\_(LE) \; dt. \qquad \textbf{(15.8)}$$

Other types of unimolecular transitions in MCell simulations include effector site isomerizations and ligand production, transformation, and destruction (Fig. 15.6). In any case, the generic first-order rate constant k has the units of inverse time, and the lifetime of the source state is exponentially distributed with a mean value (τ) of $1/k$. The MC probability (p_k) is the likelihood that the transition occurs during time Δt and is given by $1 - \exp(-k\Delta t)$. If the source state has a choice of n unimolecular transitions with rate constants $k_1, \ldots, k_n$ (e.g., the double-bound A_2R state in Fig. 15.6C has two unbinding paths and one isomerization path), τ is given by $1/\sum^n k$, the total probability (p_T) of any transition is

$$p_T = 1 - \exp\left[-\left(\sum_{}^{n} k \right) \cdot \Delta t \right], \qquad (15.9)$$

and the probability of the pathway with rate k_i is

$$p_i = p_T \cdot \frac{k_i}{\sum_{}^{n} k}. \qquad (15.10)$$

The decision between all possible events (including no transition) is made by comparing a single random number ($0 \leq \kappa \leq 1$) to the cumulative set of probabilities ($p_1, p_1 + p_2, \ldots, \Sigma p, 1$).

When an irreversible reaction is simulated using MC methods, the algorithm design and use of p_b and p_k are straightforward because each ligand and effector molecule can undergo a maximum of only one transition per time step. When reversible reactions are simulated, however, the algorithm design can take different forms. In a "first-order" approach, each molecule remains limited to a maximum of one transition per time step. "Higher-order" approaches allow individual molecules to undergo multiple "sub-Δt" transitions and thus allow some degree of "hidden" reversibility during each iteration. Using the reaction of equation (15.4) as a simple example, a particular effector site initially in the LE state might unbind at some point during the time step, and then sometime later during the same time step it could become bound again.

If a higher-order approach can be suitably balanced for complex cyclic reactions (e.g., Fig. 15.6B), its advantage is improved numerical accuracy for a given value of Δt. Testing different higher-order approaches involves simulating a set of simple and complex reactions at equilibrium and comparing the fractional amounts of each reactant to analytic predictions or to a finite difference simulation of the corresponding rate equations. MCell uses an extensive set of optimized rules that govern sub-Δt transitions, and the numerical accuracy for both simple and highly complex reactions is illustrated briefly in the following section.

Chemical Reactions: Numerical Accuracy

The use of equilibrium conditions to quantify the accuracy of MCell simulations is summarized in Tables 15.1 and 15.2, and Figs. 15.7 and 15.8.

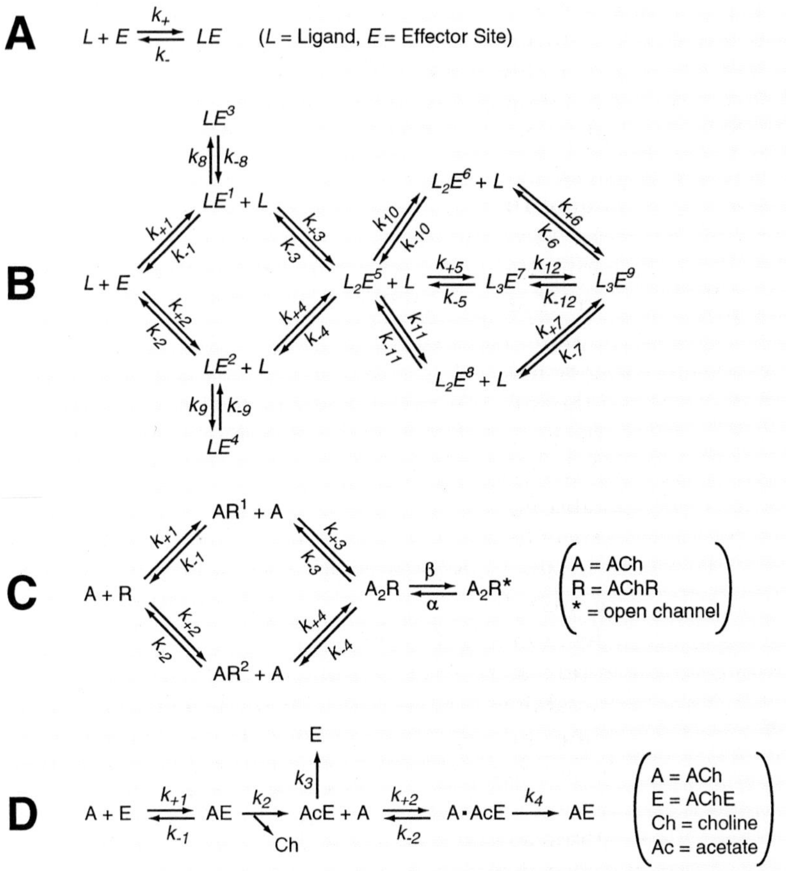

Figure 15.6. Examples of chemical reaction mechanisms used with MCell simulations. Rate constants for bimolecular associations (k_{+n}) are given in units of $M^{-1}sec^{-1}$, and for unimolecular transitions $(k_{-n}$ or $k_n)$ in units of sec^{-1}. Ligand diffusion constants are given as cm^2/sec.

(A) Simple reversible binding reaction, as used for the equilibrium simulations of Table 15.1 and the relaxation simulations of Fig. 15.8. Except as indicated otherwise in Table 15.1, the values of k_+ and k_- were 2×10^8 and 50,000, respectively. D_L was 2×10^{-6}.

(B) Complex reaction mechanism for an effector with three ligand-binding sites and 10 possible states (indicated by superscripts). This mechanism, with both parallel and sequential binding, unbinding, and isomerization transitions, was designed specifically to test MCell's rules governing sub-Δt transitions (see text). Numerical accuracy for this mechanism is illustrated in Table 15.2. D_L was 2×10^{-6}, and rate constant values were as follows: k_{+1}, 1×10^8; k_{-1}, 10,000; k_{+2}, 1.5×10^8; k_{-2}, 10,000; k_{+3}, 5×10^7; k_{-3}, 7000; k_{+4}, 7.5×10^7; k_{-4}, 7000; k_{+5}, 8.5×10^7; k_{-5}, 15,000; k_{+6}, 1×10^7; k_{-6}, 50,000; k_{+7}, 2×10^7; k_{-7}, 50,000; k_8, 20,000; k_{-8}, 35,000; k_9, 25,000; k_{-9}, 33,000; k_{10}, 40,000; k_{-10}, 45,000; k_{11}, 38,000; k_{-11}, 38,000; k_{12}, 50,000; k_{-12}, 2000.

(C) Mechanism used for AChR activation in MEPC simulations. D_{ACh}, 2.1×10^{-6}; k_{+1}, k_{+2}, k_{+3}, k_{+4}, 1.35×10^8; k_{-1}, k_{-2}, k_{-3}, k_{-4}, 64,286; β, 48750; α, 1250.

Effector sites were dispersed uniformly throughout space within a reflective spherical shell, on a series of inner concentric shells made of transparent polygons (Fig. 15.7). A reaction mechanism was chosen (e.g., Fig. 15.6A or 15.6B),[7] and the effector sites were initialized either to the unbound state or to an expected equilibrium distribution (to simulate either the approach to or maintenance of equilibrium, respectively). MCell's checkpointing feature was used to introduce ligand molecules at uniform concentration throughout the shell. Checkpointing was also used to create an instantaneous change in free ligand concentration during a simulation,[8] so that the relaxation to new equilibrium conditions could be quantified (Fig. 15.8).

Table 15.1 shows how simulation results for a simple reversible binding reaction (equation (15.4) and Fig. 15.6A) changed as a function of the MC time step, Δt. With all other input parameters held constant, a decrease in Δt (1) decreases the length of random walk movements; (2) decreases the values of the MC binding and unbinding probabilities, p_b and p_k; and (3) decreases the simulation's temporal granularity with respect to the theoretical average lifetimes of reaction intermediates (average lifetimes are constant at equilibrium and are shown in Table 15.1). Because the reaction space in this example is symmetrical, the scaling effect of Δt on random walk movements is unimportant, and the simulation accuracy can be assessed in terms of temporal granularity and MC probability values. When Δt is small (i.e., when the temporal granularity is fine and p_b and p_k are both much less than unity), the relative error in the simulation results (average fractional concentration of each intermediate) is far below 1%. If an inappropriately large value of Δt is chosen so that p_b exceeds unity (equation (15.7)), the results reflect the nonsensical value of p_b and will show an excess of unbound ligand and effector molecules.

Values of p_b that approach, but do not exceed, unity are perhaps surprisingly well tolerated, as long as the increase in Δt does not also degrade the temporal granularity by a large amount. This last point is clearly shown by another set of results in Table 15.1, obtained from simulations in which the effector sites were all placed on the inside of the outer shell, rather than on the inner concentric shells. Under these conditions, identical equilibrium results are expected because the same amounts of ligand and effector sites are present in the same total reaction volume.

Figure 15.6. (opposite) continued

(D) Mechanism for ACh hydrolysis by AChE in MEPC simulations. This reaction illustrates unimolecular transitions in which ligand molecules are produced (choline, Ch), transformed (ACh to acetate, Ac), or destroyed (Ac). It also includes ACh binding to the acetylated AChE intermediate (AcE) to simulate excess substrate inhibition (Rosenberry, 1979). The effective turnover number for ACh hydrolysis was 16,000/sec (Stiles, 1990). k_{+1}, 2×10^8; k_{-1}, 14,000; k_2, 112,000; k_3, 18,667; k_{+2}, 5×10^6; k_{-2}, 21,429; k_4, 1867.

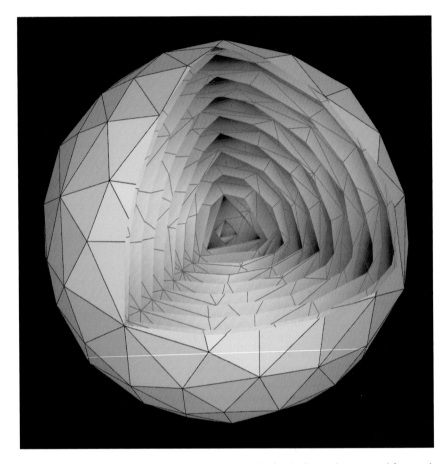

Figure 15.7. Cutaway view of radially symmetrical reaction volume used for equilibrium simulations. The outer shell (light blue mesh, actual volume 0.0628 μm³, effective radius 0.247 μm) is reflective to diffusing ligand molecules, while the 10 inner shells (gray) are transparent. Effector sites (not shown) were distributed either on the inner shells to simulate a well-mixed solution (nearest-neighbor distance was the same within a shell as between adjacent shells) or on the inside surface of the outer shell (see text and Table 15.1).

However, in this case the ligand molecules bind to a reflective (rather than transparent) surface, and this doubles the value of p_b for a given value of Δt (Table 15.1, and see discussion of the factor f_s in equation (15.7)). Comparison of these results (circa 2.5% error obtained with $\Delta t = 3$ μsec and $p_b = 0.72$) to the earlier results obtained with (1) the same time step and smaller binding probability (circa 2.5% error, $p_b = 0.36$) or (2) a longer time step but similar binding probability (6–13% error, $\Delta t = 10$ μsec and $p_b = 0.66$) shows clearly that simulation accuracy can be more sensitive to temporal granularity than to binding probability.

As shown in Table 15.2, equilibrium simulations of extremely complex reactions (Fig. 15.6B) also yield relative errors much less than 1%,

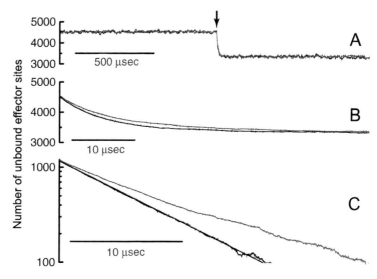

Figure 15.8. Example of equilibrium relaxation simulations for the reaction mechanism of Fig. 15.6A. Effector sites were present either on the inner shells shown in Fig. 15.7 (black data points) or on the inside surface of the outer shell (red data points).

(A) The initial equilibrium conditions were perturbed by doubling the number of free ligand molecules at the time indicated by the arrow; hence the number of unbound effector sites (ordinate values) relaxes to a new, lower equilibrium value. Results are shown for individual simulations ($\Delta t = 0.1$ μsec).

(B) The relaxation transition shown on an expanded time scale, with results averaged across five simulations for each distribution of effector sites.

(C) A semilog plot of the averaged relaxation transition (final average equilibrium number of unbound effector sites subtracted from ordinate values). With effector sites distributed on the inner shells, the transition is first order (linear fit superimposed on black data points). With effector sites on the outer shell (red data points), the curved transition reflects higher-order kinetics that include ligand diffusion time.

given reasonably small values of Δt. Thus simulation accuracy is largely independent of reaction complexity, and this independence arises largely from the optimized implementation of sub-Δt binding and unbinding transitions discussed earlier. Figure 15.8 shows results from equilibrium relaxation simulations for the simple reaction of Fig. 15.6A, obtained with effector sites located either on the concentric shells or on the outer shell alone. In the former case, the reaction space simulates a well-mixed solution, and hence the relaxation follows an expected exponential time course. With effectors on the outer shell (circa 0.25 μm radius), however, the diffusion time from the center of the sphere to the periphery becomes a component of the relaxation, which therefore displays complex higher-order kinetics. This result underscores the importance of accurate 3-D simulations for problems such as second messenger diffusion and signaling in subcellular locales.

Table 15.1. Numerical Accuracy for a Simple Reversible Binding Reaction

Δt(μsec)	p_b	p_k	Percentage error		
			$L(\tau = 42$ μsec)	$E(\tau = 28$ μsec)	$LE(\tau = 20$ μsec)
Effector sites on concentric shells					
0.01	0.021	0.00050	+0.10	+0.15	−0.21
0.10	0.066	0.0050	+0.13	+0.20	−0.49
1.00	0.21	0.49	−0.39	−0.59	+0.82
3.00	0.36	0.14	−1.7	−2.6	+3.6
10.0	0.66	0.39	−6.3	−9.4	+13
10.0*	1.32	0.39	+5.7	+4.3	−6.0
Effector sites on outer shell					
3.00	0.72	0.14	−1.6	−2.4	+3.4

Note: The reaction mechanism and rate constants are given in Fig. 15.6A, and simulations were run using the radially symmetrical structure shown in Fig. 15.7. For each set of conditions, 20 simulations were started at equilibrium and run for 20,000 iterations with a total of 7750 effector sites. The total number of ligand molecules was 10,000, except in one case (*) where k_+ was doubled and the amount of ligand was reduced to keep the expected mean number of bound (LE_{eq}) and unbound (E_{eq}) effector sites constant. LE_{eq}, E_{eq}, and L_{eq} (expected mean for unbound ligand molecules) were calculated from analytic equilibrium expressions, as were the expected mean lifetimes (τ) for each intermediate. During a run, the number of molecules in each state fluctuates around some average value (see Fig. 15.8A), which was calculated after the run had been completed. These values then were averaged across the set of 20 simulations to obtain grand averages for the MC results (L_{MC}, E_{MC}, and LE_{MC}). To calculate the percent error values shown in the table, the MC grand averages were compared with the analytically predicted means, e.g., ($E_{MC}/E_{eq} - 1) \times 100$ for unbound effector sites. With small values of Δt and correspondingly small values of p_b and p_k, the MC results are fully converged. As Δt, p_b, and p_k increase, the accuracy is reduced, with an excess of the bound state (LE) as long as p_b remains less than unity. If p_b exceeds unity (*), then there can never be enough binding events per unit time, and the direction of error reverses. The temporal granularity (value of Δt relative to the values of τ) is another important factor underlying accuracy, as can be seen by comparing the results obtained with effector sites on the outer shell to the results obtained with effectors on the inner concentric shells (see text).

Simulation of Acetylcholine Exocytosis

The earliest models of MEPC generation simplified NMJ architecture to a flat, coin-shaped diffusion space and used either an analog computer (Rosenberry, 1979) or differential equation–based simulation methods (Wathey et al., 1979; Land et al., 1980, 1981, 1984; Pennefather and Quastel, 1981; Madsen et al., 1984). Synaptic ACh appeared instantaneously as either a point source or another simple distribution centered within the space. With the introduction of early MC algorithms, planar

Table 15.2. Numerical Accuracy for a Complex Cyclic Reaction

	L	E	LE	LE^2	LE^3	LE^4
τ (μsec)	232	32.5	27.7	22.6	28.6	30.3
Percentage error	−0.25	−0.63	+0.021	−0.28	+0.34	−0.32

	L_2E^5	L_2E^6	L_2E^8	L_3E^7	L_3E^9
τ (μsec)	9.76	21.6	24.7	15.4	9.80
Percentage error	+0.12	+0.22	+0.28	+0.063	+0.0079

Note: The reaction mechanism and rate constants are given in Fig. 15.6B, and otherwise the simulations were run and results were analyzed as described in the note to Table 15.1. A single example is shown for fully converged conditions with effector sites on the concentric shells ($\Delta t = 0.1$ μsec).

JFs were added to a model NMJ (Bartol et al., 1991), and the impact of AChE site density and hydrolytic rate was investigated (Anglister et al., 1994). More recently early versions of MCell have been used to add a simplified vesicle and fusion pore to simulations of ACh exocytosis and MEPC generation (Stiles et al., 1996, 1998), and they have also been used for detailed modeling of MEPC temperature sensitivity (Stiles et al., 1999). In all of these studies, the model results and predictions were tested against *average* values of experimental measurements (e.g., amplitude, rise time, and fall time), and therefore the simplified models were adequate. However, more realistic models are required to investigate *distributions* of measured values (i.e., synaptic variability and plasticity).

The first step in developing more realistic models is to replace flat surfaces with membrane contours. In general this requires high-resolution 3-D reconstructions and polygon meshes that are optimized for use in simulations. To introduce key concepts and methods, in this section we illustrate the use of a complex polygon mesh to represent a synaptic vesicle that merges smoothly with an expanding exocytotic fusion pore. The diffusion of ACh into a synaptic cleft is simulated, and the predicted time course is shown to match earlier results that had been tested carefully to ensure high accuracy (Stiles et al., 1996, 1998). These earlier results were obtained with a much simpler shape for the vesicle and pore, so the agreement with the present findings both confirms the earlier conclusions and validates the accuracy of the new simulations. We show that the choice of $\bar{l}_r$ (and therefore Δt) is particularly important to numerical accuracy when ligand molecules diffuse through a constriction like the fusion pore, and we also briefly discuss the use of an optimization (spatial partitioning) that renders execution speed nearly independent of mesh size and complexity.

Figure 15.9 shows a same-scale comparison of the polygon mesh vesicle-pore model and the earlier simplified configuration (cube with cylindrical pore composed of rectangular facets). The cube and sphere have the same volume (2.7×10^4 nm^3) and represent the lumen of a

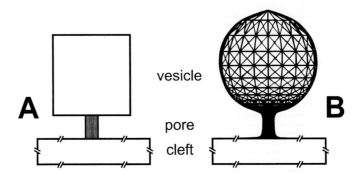

Figure 15.9. Simplified and realistic models of a synaptic vesicle and exocytotic fusion pore.

(A) Simplified, cube-shaped vesicle (30 nm side length) that communicates with the synaptic cleft through a cylindrical pore composed of 16 rectangular facets.

(B) Polygon mesh model of vesicle and pore, composed of 1580 triangles. See text for discussion of vesicle volume and pore dimensions.

synaptic vesicle at the NMJ, with actual (sphere) or equivalent (cube) radius of 18.6 nm. The pore height (h) is 9 nm in both cases (i.e., midway between a single and double thickness of plasma membrane). The average radius (r) of the mesh pore is matched to the exact radius of the cylindrical pore (2.5 nm in this illustration). Simulations were run either with a fixed value of r as shown or with r increasing at a constant rate of 25 nm/msec to simulate fusion pore expansion at the NMJ (Stiles et al., 1996).

For this model of ACh exocytosis, the steepest concentration gradient is across the pore height h. Therefore the rate of ACh efflux is mostly determined by the pore geometry, that is, its "resistance." Concentration gradients do form inside the vesicle and cleft as well, but their significance depends on the value of r. At very small r, the concentration drop across h is very steep (the pore "resistance" is very high) and almost completely limits ACh efflux. For larger values of r, the vesicle and cleft gradients become increasingly significant, but they are appreciably steep only within a distance several times r from the inner and outer pore openings (Stiles et al., 1996). The consequences are that (1) the bounding shape of the vesicle should not affect ACh efflux significantly, although the vesicle volume is an important factor (thus simulation results obtained with the polygon mesh model should match results obtained with the simplified vesicle and pore), and (2) for accurate simulation of ACh efflux, the random walk algorithm must correctly produce concentration gradients over distances comparable to r.

Because the MCell random walk algorithm is grid-free and selects from an extensive set of radial distances and directions for each movement, concentration gradients are simulated accurately within distances nearly as small as the average step length $\bar{l}_r$. Thus, for accurate simulation of ACh exocytosis, the time step Δt need only be chosen so that $\bar{l}_r$ (equation (15.2)) is somewhat smaller than r.[9] However, if the value specified for Δt is too long and $\bar{l}_r$ is appreciably larger than r, then the concentration gradients inside the vesicle, pore, and cleft will not be steep enough. As illustrated subsequently, the apparent rate of ACh efflux can then easily be slowed by an order of magnitude.

Each simulation of ACh efflux was begun using checkpointing to ensure uniform initial ACh concentration within the vesicle.[10] Figure 15.10 (inset) shows four examples of ACh efflux curves that were obtained using the simplified vesicle structure and fixed r (Fig. 15.9A) and illustrates the dramatically slowed emptying that occurs with increasing Δt. Each of these curves has the form of an exponential decay (Stiles et al., 1996, 1998), and the main panel of Fig. 15.10 shows the e-fold emptying time (τ) plotted as a function of the ratio $\bar{l}_r / (\bar{l}_r + r)$. Essentially identical

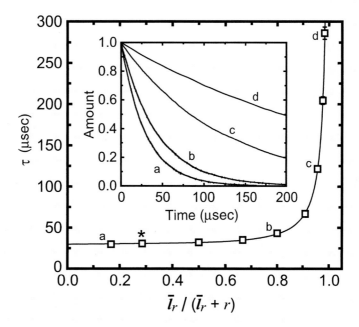

Figure 15.10. Simulation of ACh exocytosis through a fusion pore with constant radius (r = 2.5 nm, Fig. 15.9A). Inset shows the time course of vesicle emptying obtained with Δt values of approximately 82 psec, 33 nsec, 1.0 μsec, and 7.3 μsec for curves a–d, respectively ($D_{ACh} = 6 \times 10^{-6}$ cm²/sec). Ordinate values are normalized to the starting amount of ACh, and each curve is the averaged result of 10 simulations. Such exponential emptying curves were fitted to obtain e-fold times, and the resulting values of τ were plotted (open squares in main panel, mean ± SD for 10 simulations) as a function of the ratio $\bar{l}_r/(\bar{l}_r + r)$. As indicated by the fitted curve, τ converges to the correct value when $\bar{l}_r$ is less than r. The point marked with an asterisk was obtained with $\bar{l}_r = 0.4r$, and this condition was used for simulations of exocytosis through an expanding pore (Figs. 15.11 and 15.12).

results are obtained with the polygon mesh structure (not shown, but see subsequently), so r in this ratio is either the average (polygon mesh) or exact (simple model) pore radius. As the value of $\bar{l}_r / (\bar{l}_r + r)$ decreases (i.e., as $\bar{l}_r$ becomes small compared with r), the value of τ converges to circa 30 μsec. For the case with the longest time step (i.e., with $\bar{l}_r$ much greater than r), the apparent value of τ (circa 300 μsec) is almost an order of magnitude larger than the correct value, because the ACh concentration gradients do not form properly in the pore and immediate vicinity.

To illustrate and validate simulation of ACh exocytosis through a realistic expanding fusion pore, the polygon mesh structure was interpolated (morphed) 200 times, between limiting average radii of 0.93 nm and 5.0 nm (Fig. 15.11). This starting radius simulates instantaneous initial opening to a conductance of circa 300 pS, that is, gap junction dimensions. The ending radius corresponds to an omega figure at the NMJ and is reached by the time the vesicle empties if the pore expands at 25 nm/msec after the initial opening (see Stiles et al., 1996, for additional details). The sequence of morphed structures then was used in a series of checkpointed simulations, in which the elapsed time for each checkpoint (dictated by the expansion rate and number of morphs) was circa 0.8 μsec. For each run in the sequence, Δt was chosen so that $\bar{l}_r$ remained about 40% of r. This maintained high numerical accuracy throughout the sequence (see equivalent value of $\bar{l}_r / (\bar{l}_r + r)$ marked by an asterisk in

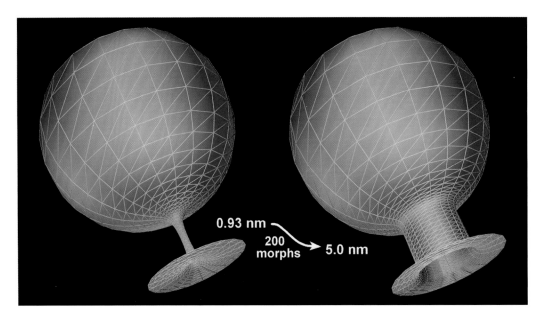

Figure 15.11. Realistic expanding exocytotic fusion pore. The structures for the beginning and ending average pore radii (0.93 and 5.0 nm, respectively) are shown. Using 3-D solid modeling software, smoothly interpolated structures (morphs) were created, so that a total of 200 could be used in a sequence of MCell checkpoints.

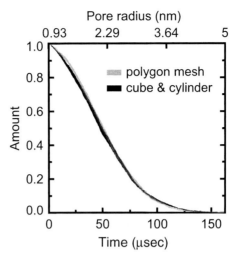

Figure 15.12. Simulation of ACh exocytosis through an expanding fusion pore. The time course of vesicle emptying is shown for the realistic (polygon mesh) and simplified (cube and faceted cylinder) models of a fusion pore expanding at 25 nm/msec ($D_{ACh} = 6 \times 10^{-6}$ cm^2/sec). The thickness of each curve represents the 95% confidence interval obtained from five simulations for each structure. Ordinate values normalized as for Fig. 15.10 (inset).

Fig. 15.10). The value of Δt therefore increased from about 45 psec for the first run, to about 1.3 nsec for the last run (D_{ACh} was 6×10^{-6} cm^2/sec).

Figure 15.12 shows the sigmoidal ACh efflux curve obtained with the expanding fusion pore, and for comparison it also shows the analogous curve obtained when r was increased at the same rate for the simplified structure. The two curves are essentially indistinguishable and therefore validate the accuracy of MCell's diffusion algorithms used with complex polygon meshes. As previously reported (Stiles et al., 1996), the vesicle empties within about 150 μsec, at which time r reaches about 5 nm. In addition the vesicle content is reduced to about 20% of the starting amount within about 80 μsec, a length of time comparable to the 20–80% rise time of MEPCs at room temperature (Stiles et al., 1996, 1999).

Each time that a diffusing ACh molecule takes a random walk step on its way out of the vesicle, its trajectory must be traced to find the nearest intersection with a polygon (if any). This search for intersections is one of the most time-consuming steps in an MCell simulation, and, unless the search algorithm is optimized using spatial partitions, computation time scales with the number of polygons that compose the surface(s) in the simulation. Spatial partitions are simply transparent planes (ligands pass through unhindered) that can be placed in arbitrary positions along the x, y, and/or z axes to subdivide the simulation space into smaller compartments (see Stiles and Bartol, 2000, for further details). When the simulation begins, the polygons contained (wholly or partly) within each compartment are identified. As the simulation runs, random walk trajectories that originate in a certain compartment need only be traced for

intersections with those polygons in the same compartment. If no intersections occur and the trajectory projects into the next compartment, the
process continues as required. If each compartment contains only a small
number (n_c) of polygons and the total number of polygons in the simulation is N_s, then the optimal increase in execution speed is of order
N_s/n_c, with no loss of numerical accuracy.

For example, the vesicle-pore mesh shown previously in Figs. 15.9 and
15.11 contains 1580 polygons, and if an exocytosis simulation is run without spatial partitions the computer time increases by a factor of nearly
300. Thus a simulation that ordinarily requires 1 min with partitions
would require 4.5 h without partitions, and a set of checkpoint simulations that ordinarily runs in 1 day would require about 9 months. Meshes
from synaptic reconstructions can easily contain 10^5–10^6 polygons
rather than several thousand, so even single simulations are not feasible
in the absence of partitions. With partitions, however, such large-scale
simulations are now routine.

Sources of Miniature Endplate Current Variability: Reconstruction and Simulation of Realistic Endplate Ultrastructure

Planar Junctional Fold Model

As outlined previously, MC simulations based on a simplified (planar)
model of JFs at the NMJ (Fig. 15.13) are sufficient when comparing predictions to average values of experimental measurements, such as MEPC
amplitude, rise time, and fall time. To address factors that underlie MEPC
variability, the model must be more realistic, and in this section we introduce high-resolution reconstructions and simulations of the synaptic
ultrastructure. To begin, we first establish a baseline for comparison by
examining the variability predicted by the planar JF model itself.

The planar JF model (Fig. 15.13) represents a "thick section" through
one nerve terminal and underlying muscle from a vertebrate NMJ (see
Fig. 15.2). The real membrane topology is simplified to regular and
constant dimensions, so the structure is characterized by the following
parameters (typical values used for reptile or mammalian endplates
are shown in parentheses, e.g., Matthews-Bellinger and Salpeter, 1978;
Salpeter et al., 1984):

1. Primary cleft length and width (2–4 μm).
2. Primary cleft height and secondary cleft width (50 nm).
3. Distance between JFs (0.2–0.4 μm).
4. Depth of the JFs (0.5–1.0 μm).
5. AChR site density on the crests of the JFs (7000–10,000/μm^2).
6. Depth to which crest AChRs extend within JFs (0.2–0.3 μm).
7. Additional depth to which AChRs extend at reduced density (0.2–
   0.3 μm at circa 30% of crest density).

8. Uniform AChE site density throughout basal lamina planes centered within the primary (1500–$2500/\mu m^2$) and secondary (effectively double the primary cleft value; see Fig. 15.13) clefts.

9. Position of ACh release site.

Based on the size of a single AChR (circa 8.5 nm diameter; Unwin, 1998), the maximum packing density in postsynaptic membrane would be some $15,000/\mu m^2$. This is considerably larger than quantitative values obtained with EM autoradiography, so even crest AChRs are unlikely to cover the entire membrane area. The actual microscopic distribution within a given area remains unknown owing to the limited resolution of experimental measurements. As a first approximation, in MCell simulations we set σ_{EG} (the global effector grid density) to a value near the maximum packing density, and then when AChR effector sites are

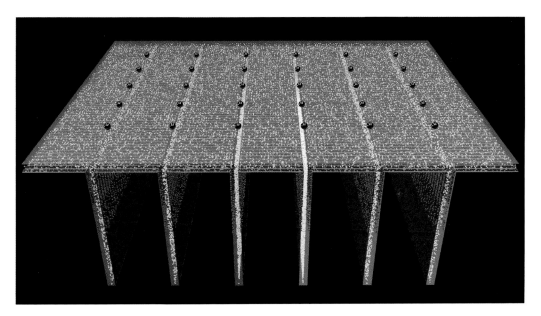

Figure 15.13. Simplified (planar) model of a vertebrate NMJ with JFs. Pre- and postsynaptic membranes (semitransparent and blue, respectively) are represented by rectangular planes (each rectangle is actually bisected into two triangles). The primary cleft is 2.0×2.7 μm in length and width and is 50 nm in height. The secondary cleft width is 50 nm; the JFs are 0.8 μm deep and are separated by 0.4 μm. AChRs (blue glyphs) are present at high density ($7250/\mu m^2$) on the crests of the JFs (i.e., to a depth of 0.22 μm) and at reduced density (70% less) for an additional 0.2 μm. Below that AChRs are absent. AChE active sites (white glyphs) are located in transparent planes that represent anchoring points in the basal lamina ($1800/\mu m^2$ in the primary cleft, and effectively $3600/\mu m^2$ in the secondary clefts where the basal lamina doubles back as it follows the folded postsynaptic membrane). Black vesicles overlying the JFs represent 30 ACh release sites. The apparent linear "stripes" of AChR glyphs result from barycentric tiling of long, thin triangles (the bisected rectangles), as illustrated schematically in Fig. 15.4, and as seen in closer view in Fig. 15.15.

added at a density less than σ_{EG} their actual placement in different membrane regions reflects random choices between different available effector tiles (Fig. 15.4). The same is true for AChE sites in the basal lamina. If a constant amount of ACh (vesicle content, N_{ACh}) is released during each simulation, and fixed values are chosen for all of the parameters listed previously, then the only remaining sources of stochastic variability are ACh random walk movements and individual AChR and AChE reaction transitions. If the ACh release site is moved to different locations, then the regional differences in AChR and AChE positions become an added component of variability.

Figure 15.13 shows a planar JF model with 30 different ACh release sites located in active zones that are centered above 6 JFs (active zones generally are observed above the opening of a secondary cleft; see Fig. 15.2). Ten different simulations were run at each release site, using the AChR and AChE reaction mechanisms and rate constants shown in Fig. 6C,D. Exocytosis of ACh (i.e., diffusion through an expanding fusion pore) was simulated using a timesaving method that omits the vesicle and pore but reproduces the time course of release shown in Fig. 15.12. This method allows the use of a constant time step on the microsecond scale rather than an adaptive time step on the picosecond-to-nanosecond scale and hence reduces the required computer time by several orders of magnitude (Stiles et al., 1998).

To summarize the MEPC variability predicted by the planar JF model, the 10 individual MEPCs simulated for each release site were averaged together, and the resulting 30 traces are superimposed in Fig. 15.14A. The traces are very nearly identical (e.g., the coefficient of variation [CV] for peak amplitude is less than 1%). The mean amplitude is circa 1000 open AChR channels, and the difference between the minimum and maximum amplitudes is less than 40 channels (essentially the same result is obtained for each set of 10 MEPCs per release site; not shown).

For each of the 300 simulated MEPCs, ACh diffuses away from the release site and becomes diluted as it binds to AChRs and AChEs. The efficiency of binding therefore decreases over time and radial distance. A snapshot at the time of peak amplitude (Fig. 15.15A) shows a central area of double-bound open AChRs (yellow) surrounded by a sparse fringe of single-bound states (red), that is, the saturated disc of postsynaptic activation. The size of the saturated disc is not large enough to reach the edge of the primary cleft, but it is large enough to sample many AChRs over a radial distance of circa 0.2 μm in the primary and secondary cleft. The size of the saturated disc and postsynaptic response overwhelms noise that arises from stochastic single-channel behavior and regional membrane differences in AChR and AChE placement. Thus, on the scale of the saturated disc, the structure of the planar JF model is essentially constant and MEPCs show little variability.

When analogous simulations are run with AChE completely inhibited, the mean amplitude is 35% larger and the falling phase is prolonged (Fig. 15.14B). However, the CV remains very small (2.6%), and the 30

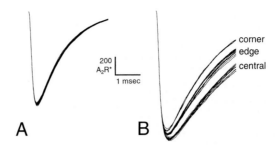

Figure 15.14. MEPC variability is minimal for the planar JF model and 30 ACh release sites shown in Fig.15.13. In both A and B, 30 MEPCs are superimposed (one per release site, each an average of 10 individual simulations in which 6200 ACh molecules were released, $\Delta t = 1.0$ µsec). In A, AChE was normally active (reaction mechanism shown in Fig. 15.6D), and in B, AChE activity was completely inhibited (k_{+1} in Fig. 15.6D was set to zero, so ACh could never bind). The reaction mechanism for AChR activation was as shown in Fig. 15.6C, and MEPC amplitude is expressed as the number of receptors in the double-bound, open conformation (A_2R^*). In B, the mean amplitude is circa 35% larger than in A, and the 30 MEPCs fall into three groups according to the position of the ACh release site (corner, edge, or central; see text and Fig. 15.15).

traces fall into three easily discernible categories based on their amplitudes and fall times. The smallest traces are also those with the shortest duration, and they correspond to the four corner release sites. The traces that are intermediate in size and decay time correspond to the 14 edge (noncorner) positions, and the largest, slowest-decaying traces correspond to the 12 central release sites.

With AChE inhibited, the full amount of released ACh (N_{ACh}) spreads radially and binds to additional AChRs during formation of the saturated disc. For release from a central location, the additional AChRs are readily available in all directions, so the radius of the disc is somewhat larger (Fig. 15.15B) and the increased peak amplitude is reached somewhat later. During the falling phase, ACh binds repeatedly and hence diffuses and dilutes slowly (buffered diffusion; Katz and Miledi, 1973). Although the efficiency of channel opening is low, additional openings are frequent enough to prolong the fall time circa threefold. With release from a corner or edge site, peak amplitude increases less because a significant fraction of spreading ACh escapes from the primary cleft even during the rising phase. More escapes during the falling phase, which reduces buffered diffusion and attenuates the prolongation of fall time.

The variability of quantal endplate signals has been an integral factor in classical quantal analysis for many years (del Castillo and Katz, 1954; Redman, 1990; Edwards, 1995a,b) and is markedly larger than the variability predicted by the planar JF model. Experimental MEPC amplitudes vary by a factor of two to three, and the distribution is approximately

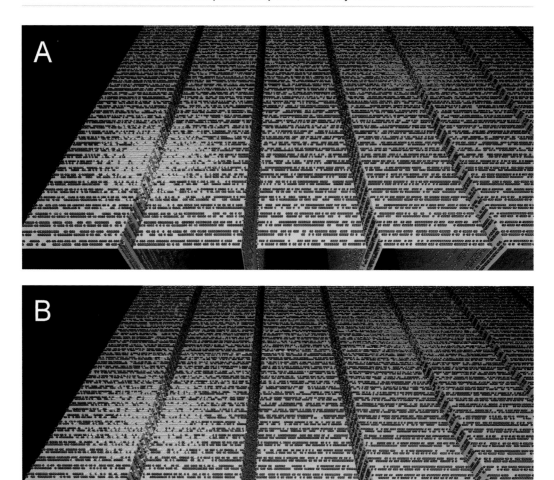

Figure 15.15. Appearance of the saturated disc in the planar JF model. For this illustration, ACh was released from two sites simultaneously (central and corner),
and the presynaptic membrane and AChE sites are omitted for clarity.

(A) Snapshot at peak amplitude with AChE active (circa 300 μsec after onset of
ACh release). AChR glyphs are color-coded according to their reaction state (AR[1]
and AR[2], red; A_2R, green; A_2R^*, yellow; see Fig. 15.6C). ACh molecules are tiny
dots. If bound, they have the same color as the corresponding AChR glyphs. If unbound, they are cyan. The radial extent of saturated disc formation is small enough
that the edge of the primary cleft space has no effect on MEPC variability (see text
and Fig. 15.14A).

(B) Snapshot at peak amplitude with AChE inhibited (circa 490 μsec). The saturated disc has spread farther than it does with active AChE, so edge effects are
apparent for peak amplitude and fall time (see text and Fig. 15.14B). Single-bound
AChRs arising from the two release sites show clear overlap.

Gaussian (broader and shifted slightly to larger values after AChE inhibition). It has mostly been assumed that this variability arises from differences in N_{ACh} from one vesicle to another (Edwards, 1995a), but the planar JF model can be used to suggest other alternatives. For example, small changes (much less than twofold) in primary cleft height, JF width, or global AChR density can all affect MEPC amplitude to an extent that greatly outweighs the stochastic variability illustrated previously. In addition, all of these factors introduce opposing changes in amplitude and rise time (e.g., an increase in primary cleft height would decrease amplitude but increase rise time). Changes in N_{ACh} or the distance between JFs, on the other hand, cause amplitude and rise time to change in the same direction. Experimental MEPCs show little or no correlation between amplitude and rise time (Land et al., 1980; Bartol, 1992; J. Stiles, unpublished data), so presynaptic factors (i.e., N_{ACh}, ACh exocytosis) and postsynaptic architecture must both contribute significantly to MEPC variability, with offsetting influences on amplitude and the time course of the rising phase. With AChE active, MEPC fall time is almost exclusively determined by the apparent open time of the AChR channel (Anderson and Stevens, 1973; Anglister et al., 1994; Stiles et al., 1999), so variability in single-channel kinetics from one AChR to another will also be required for a complete model of MEPC variability.

Curved and Branched Junctional Fold Model

As a first step toward modeling these complexities, we designed a more realistic "thick section" of NMJ that has curved and branched JFs, but that otherwise can easily be compared to the planar JF model. A portion of pre- and postsynaptic membrane contours was traced from a TEM image of rat diaphragm NMJ (Fig. 15.16A), and the length of primary cleft, number of JFs, and spacing between JFs were similar to the corresponding parameters in the planar model. The fold structure was morphed smoothly to obtain additional successive "sections," which then were reconstructed into optimized pre- and postsynaptic meshes for MCell simulations (circa 3400 and 25,000 triangles in the nerve and muscle membrane meshes, respectively; Fig. 15.16B). As shown in Fig. 15.16D, the muscle membrane was clipped into pieces representing crest, intermediate, and deep JF regions, and AChR effector sites were added to the top two regions (circa 91,000 total), as had been done for the planar JF model. A third polygon mesh containing AChE sites (circa 59,000) was created to model the basal lamina and followed the contour of the postsynaptic membrane (Fig. 15.16C). Thirty ACh release sites were placed in active zones above the folds (Fig. 15.16C,D), and ACh exocytosis was simulated using the timesaving method described previously.

Ten MEPCs were simulated and averaged for each release site of the branched JF model, and the resulting 30 traces are shown superimposed in Fig. 15.17A. Comparison of these results to those for the planar JF model (Fig. 15.14A) shows that the MEPC variability is markedly

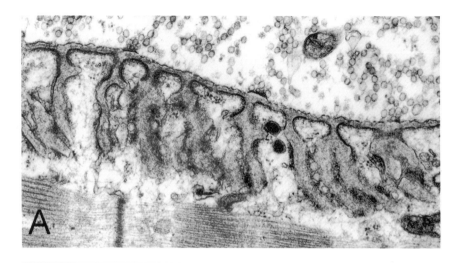

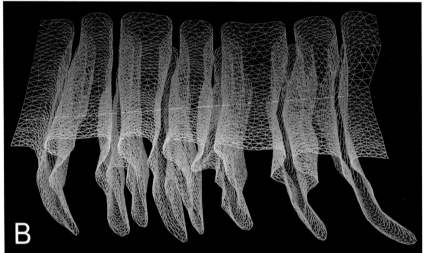

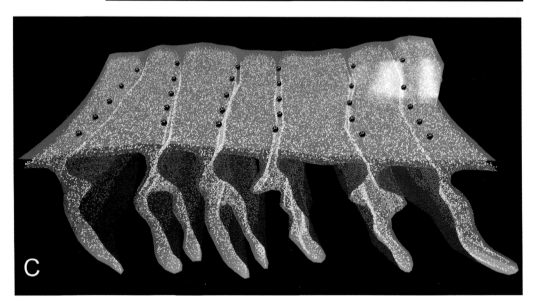

FRONT

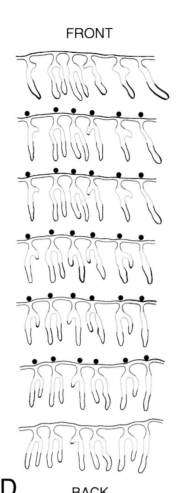

D   BACK

Figure 15.16. Model of a vertebrate NMJ with curved and branched JFs.

(A) Typical TEM of rat diaphragm NMJ. Three active zones are clearly seen above the openings of JFs. From Peters et al. (1991), used by permission of Oxford University Press, Inc.

(B) Wireframe image of the optimized mesh created after tracing and morphing a portion of the postsynaptic membrane shown in *A.*

(C) Pre- and postsynaptic membranes, AChR and AChE glyphs, and 30 ACh release sites visualized as described previously for the planar JF model (Fig. 15.13). The contour of the basal lamina (1800 AChE sites/μm^2) follows that of the postsynaptic membrane throughout the branched and variable JFs. The postsynaptic membrane was subdivided into crest, intermediate, and deep JF regions, as shown in *D,* and the density of AChRs in each region was the same as for the planar JF model. The apparently random positions of AChR glyphs reflect barycentric tiling of the optimized mesh, which contains nearly equilateral triangles (*B;* and see Fig. 15.4A).

(D) JF structure and ACh release sites. The pre- and postsynaptic membranes shown in *C* were cut in a manner analogous to TEM thin sections (40 nm). At the top is the leading edge (front) section, followed by those sections that included a row of ACh release sites, and the final (back) section. The postsynaptic membrane is color-coded according to JF region (and therefore AChR density): crest, red; intermediate, cyan; deep, black. The line thickness along each contour is determined by the orientation of the membrane relative to the plane of section.

increased. The CV is 6%, that is, more than sevenfold larger than the previous result, the range of amplitudes is larger by more than fivefold, and a relative independence of amplitude and rise time is qualitatively evident. The mean amplitude is reduced by some 25% and reflects increased diffusion and dilution space within the curved and branched secondary cleft contours. The variability for each set of 10 MEPCs per release site was no different than that for the planar JF model (not shown), so the results shown here originate entirely from geometric differences between release sites. The relationship between MEPC amplitude and the detailed structure underlying each release site (see contours of Fig. 15.16D) is not easy to predict, as some of the largest and smallest MEPCs occurred at corner and edge positions, over single and branched JFs. In essence the MEPC amplitude is very sensitive to the cleft volume within and around the region of saturated disc formation, and this 3-D parameter is very difficult to estimate by eye from membrane contours.

When AChE inhibition is simulated (Fig. 15.17B), the mean amplitude is 38% larger than with AChE active (Fig. 15.17A). This relative change is almost exactly the same as that described previously for the planar NMJ model (35%), but the range and variability of amplitudes are markedly increased (CV of 7.6%, about threefold larger; compare Fig. 15.17B to Fig. 15.14B). Neither the amplitudes nor the falling phases shown in Fig. 15.17B can be separated into nonoverlapping groups on the basis of corner, edge, or central ACh release position. In addition, the amplitude range nearly overlaps the range obtained with normal AChE activity (Fig. 15.17A). When AChE inhibition is simulated using the branched rather than planar JF model, diffusing ACh travels farther into the larger available cleft volume and therefore "samples" more geometric variability. This diminishes the influence of the primary cleft boundary and broadens and smoothes the distributions of MEPC size and time course to resemble experimental distributions more closely.

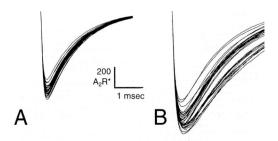

Figure 15.17. MEPC variability for the NMJ model with curved and branched JFs. As in Fig. 15.14, 30 MEPCs (one averaged trace per ACh release site) are superimposed in A (AChE active) and B (AChE inhibited). The mean amplitude is circa 38% larger in B than in A, that is, about the same relative increase as obtained with the planar JF model and AChE inhibition. However, the MEPC variability is many times larger than that obtained with the planar JF model (see text). $\Delta t = 1.0$ μsec.

Neuromuscular Junction Reconstruction and Simulation

Even with N_{ACh} held constant (i.e., assuming no variation in the amount of ACh released per quantal event), the variability introduced by the branched JF model represents an appreciable fraction of experimental MEPC variability (e.g., 30–50% based on range of amplitudes). Hence this model illustrates very well the importance of architectural realism in synaptic simulations. However, the branched JF model itself remains a simplification (surprisingly so), and therefore it also illustrates very well the need for high-resolution, quantitative reconstructions of synaptic architecture. It is considerably more difficult to create such reconstructions for use with simulations than it is to create them for imaging and morphometrics, and we will detail our methods and large-scale efforts elsewhere. We conclude here with an illustrative example and briefly compare preliminary results from MCell simulations to those obtained with the branched JF model.

Since the distance across primary and secondary cleft spaces is about 50 nm, the resolution for reconstructions must be about 10 nm or better in all three dimensions for use with simulations. On the other hand, the overall size of the reconstruction is on the order of microns in each dimension, so the scale of the problem is quite large and presents many technical challenges. In essence, creation of a reconstruction for use with MCell simulations entails three steps: (1) generation of a high-resolution EM data set for a volume of tissue, (2) extraction of membrane or other surfaces of interest from the volumetric data set, and (3) subdivision of the surfaces as needed, to add simulation objects such as ligand release sites and different populations of effector sites (e.g., subdivision of JFs according to AChR density).

The conventional approach to step 1 is serial TEM sections, but typically even the thinnest sections (say, 40 nm) are thicker than the desired resolution, and as section thickness decreases the section uniformity and number of serial sections can become problematic. EM tomography ultimately may be a better approach and is becoming more accessible, and the specific application of such methods to MC simulations is an important area for future research. Step 2 requires both the identification of the surfaces and their subsequent transformation into optimized polygon meshes, and such problems are presently the focus of much computer science and graphics research, particularly for highly convoluted structures like cell membranes. Step 3 is likewise a present-day computer science and graphics problem, and it poses particular difficulties because of the need for interactive control over very large-scale structures.

For the example illustrated here, we reconstructed a length of nerve terminal and postsynaptic membrane from a mouse sternomastoid NMJ. Approximately 20 sequential TEM sections were cut to a thickness of 100 nm to ensure high uniformity, and 7 were chosen for the reconstruction. The resulting mesh thus had an aggregate thickness of 0.6 μm, that is, several times the diameter of a saturated disc. Figure 15.18A (inset)

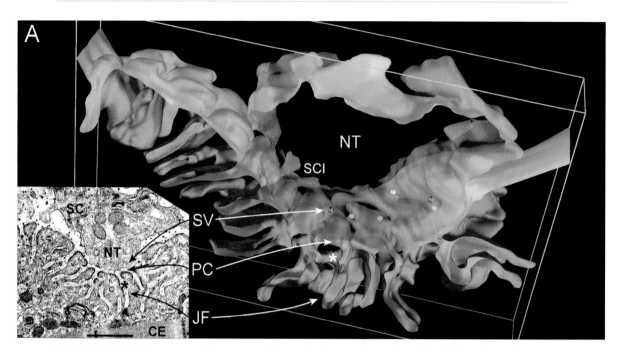

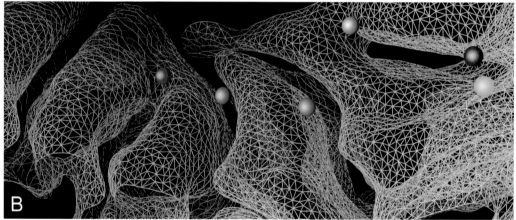

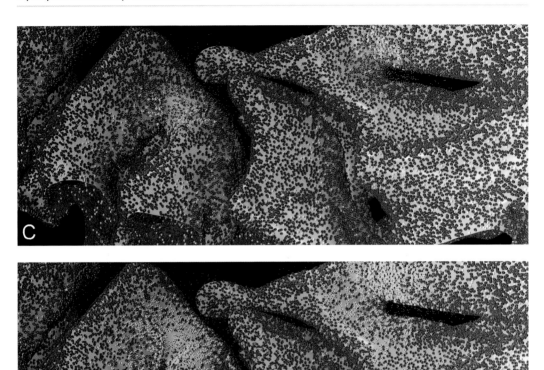

Figure 15.18. Simulation of MEPCs in a high-resolution, partial reconstruction of mouse sternomastoid NMJ.

(A) The inset shows a TEM image from one section used for the reconstruction. CE, contractile elements; JF, junctional fold; NT, nerve terminal; PC, primary cleft; SC, Schwann cell; SV, synaptic vesicle. Scale bar = 1 μm. The segment of reconstructed nerve terminal is shown in translucent gray, and six ACh release sites are indicated by color-coded synaptic vesicles that correspond to the six color-coded MEPCs shown in Fig. 15.19. Extra- and intracellular faces of the postsynaptic muscle membrane are light and dark blue, respectively, and the bounding box is 4.2 × 0.6 × 2.5 μm. SCI, Schwann cell process invaginations. Asterisks mark the position of a tunnel between JFs.

(B) Close-up image of the ACh release sites above the postsynaptic membrane, shown as a wireframe view of the optimized mesh.

(C–D) Illustrative snapshots of MEPC generation for the yellow and blue ACh release sites, 100 μsec *(C)* and 400 μsec *(D)* after onset of release. AChRs and AChE active sites were distributed uniformly over the postsynaptic membrane at 7250 and 1800/μm², respectively. The colors for AChR and ACh glyphs are the same as in Fig. 15.15. Unbound AChE sites are shown as white spherical glyphs, and bound AChE states (Fig. 15.6D) are black.

shows a micrograph from one of the sections, and Fig. 15.2B shows another representative TEM view of a mouse sternomastoid NMJ. In contrast to textbook depictions, the JF structure and disposition of thickened postsynaptic membrane are extremely complicated, and tunnels can even cross from one secondary cleft to another (asterisk in Fig. 15.18A).

The seven sections chosen for the reconstruction were digitized, and the pre- and postsynaptic membrane contours were traced as smooth curves in each. The positions of synaptic structures can change a great deal from one TEM section to the next, and such changes can markedly degrade the output of surface reconstruction algorithms. To overcome this problem, a method was devised to interpolate between the contours of adjacent sections and thus in effect create additional "sections" at finely spaced intervals between the originals.[11] After interpolation, a total of 121 "sections" were obtained at 5-nm intervals, containing smoothly curved pre- and postsynaptic membrane contours that changed very little from one section to the next. The sections were then used to create corresponding images of the contours with a pixel scale of 5 nm, and the resulting "stack" of images at $5 \times 5 \times 5$-nm final resolution was passed through a "marching cubes" surface reconstruction algorithm (Schroeder et al., 1998) that in our experience is the most reliable for highly convoluted structures. The initial meshes obtained for the pre- and postsynaptic membranes each contained several million triangles, and they were subsequently decimated (reduced to fewer triangles) and optimized for MCell simulations without appreciable degradation of topology. The final meshes for the nerve and muscle surfaces contained 20,000 and 80,000 triangles, respectively.

Figure 15.18A shows the reconstructed nerve and muscle membranes and their overall dimensions. To those preconditioned by typical cartoons of vertebrate NMJs, the structure is exceedingly complex. Because of numerous twists, pockets, and interconnections at various depths, the structure of the JFs is closer to that of a sponge than a pleated sheet. A close-up view of the optimized mesh for the muscle membrane is shown in Fig. 15.18B, together with six synaptic vesicles positioned at different possible ACh release sites above secondary cleft openings.

The next level of detail for this reconstruction would be realistic distributions and densities of AChR and AChE sites. However, to illustrate the isolated effect of synaptic topology on MEPC variability, we added AChR effector sites at uniform crest density over the entire muscle surface and then intermixed AChE sites onto the muscle surface as well. Exocytosis of ACh was simulated as described previously for the planar and branched JF models, and Fig. 15.18C,D shows different stages of saturated disc formation during release from two of the six vesicles indicated in Fig. 15.18A,B. Individual color-coded MEPCs that correspond to the six different release sites are illustrated in Fig. 15.19, and the range of amplitudes is almost as large as the range obtained with 30 different release sites and the branched JF model (Fig. 15.17A). Although

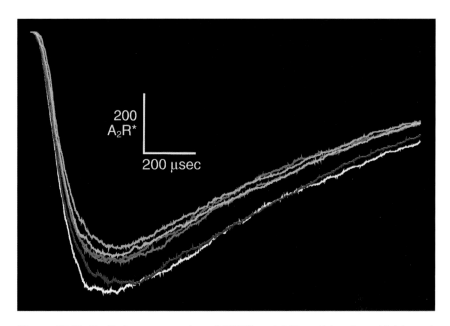

Figure 15.19. Preliminary examples of MEPC variability arising from highly realistic postsynaptic topology at the mouse sternomastoid NMJ. Individual color-coded MEPCs are shown, corresponding to the ACh release positions indicated in Fig. 15.18A,B ($\Delta t = 0.5$ µsec). Whereas classical assumptions have mostly attributed MEPC variability to differences in the amount of ACh released from each synaptic vesicle, these examples, based only on a thin strip of reconstructed tissue, show that significant differences in amplitude and rise time can arise just from local differences in cleft architecture.

the yellow release site is positioned near the edge of the primary cleft space, the corresponding MEPC is not the smallest of the group. The remaining release sites are positioned about midway across the thickness of the structure, and there is no easily discernible relationship between their positions, the surrounding JF structure, and the relative sizes of the MEPCs. In addition, there is no apparent correlation between MEPC amplitude and the time-to-peak. Therefore, preliminary results obtained with this high-resolution reconstruction substantiate the results described previously for the branched JF model.

Close examination of TEM sections for mouse sternomastoid NMJs (e.g., Figs. 15.18A [inset] and 15.2) shows that the depth and disposition of thick, ostensibly AChR-rich, postjunctional membrane is highly variable, and that vesicles and active zones can be positioned at the "edge" of the nerve terminal where the height of the primary cleft can increase markedly. The addition of these factors to reconstructions of extended thickness (at least 4–5 µm) will allow a quantitative description of MEPC variability (with and without AChE inhibition) arising from postsynaptic structural variations. Thereafter distributions for additional pre- and postsynaptic factors (e.g., N_{ACh}, vesicle volume [which would affect the rate of ACh release], and single channel rate constants) can be added to the model in order to deconvolve their contribution to the total observed variability of amplitude, rise time, and fall time, and explain correlations or the absence of correlations between the quantal size and time course.

A Brief Look to the Future

In this chapter we have reviewed and illustrated MC methods for simulation of molecular diffusion and chemical reaction in 3-D subcellular environments. Our focus was synaptic physiology and detailed structural realism. Large-scale reconstructions of complex endplate morphology presently underlie ongoing simulations of normal and pathological neuromuscular function.

Although 3-D reconstructions are becoming increasingly common in biological research, the combination of reconstruction and simulation is in its infancy. Quantitative simulations require reconstructions with very high resolution and accuracy, and thus the reconstruction and model design become significant computational challenges in and of themselves. The design and use of such detailed quantitative models drives the acquisition of more accurate experimental data and (particularly with highly realistic MC models) often changes the way one thinks about the structure and function of the modeled system. This in turn leads to different avenues of experimental investigation, the results of which then can be used to test or extend the model. An excellent case in point is detailed relationships between synaptic current size and time course, which can be used to identify and constrain pre- and postsynaptic parameters that contribute to synaptic variability and plasticity. Once this approach has been worked out in a relatively simple system (e.g., the NMJ), it may be applied in more complex settings (e.g., a reconstructed volume of neuropil).

Computational power continues to grow exponentially and is expected to do so throughout the foreseeable future. New algorithms will take advantage of such increased power and will continue to add new levels of realism to MC simulations. For example:

1. Bimolecular associations, as described here for diffusing ligand molecules and fixed effector sites, will be extended to include interactions between multiple diffusing species (e.g., Ca^{2+}, Ca^{2+}-sensitive dyes, and mobile Ca^{2+}-binding proteins).
2. Surface properties such as membrane potential will be added and combined with electric field and pressure gradient modifications to diffusion algorithms, so that classical transport physiology and cellular excitation can be simulated at the level of stochastic 3-D interactions.
3. Biomechanical properties will be added to components of models, to simulate cellular motility and the active intracellular trafficking of molecules and organelles.

Such new and existing modeling capabilities, together with a rapidly expanding experimental data base of 3-D structure, molecular constituents, biochemical pathways, and physiological measurements, will help provide quantitative answers to many existing questions centered on synaptic plasticity, crosstalk, modulation, and functional relationships between neurons and glia. Increasingly realistic models may also help to identify previously unforeseen principles of peripheral and central synaptic function. Perhaps in time functional analogies will be drawn between JFs at the NMJ, perforated spine synapses in the brain, and intersynaptic regions between dendritic spines—all structures that provide interspersed receptive and nonreceptive postsynaptic area and cleft space. At present it seems that all may somehow reflect or influence neurotransmitter receptor localization and metabolism (e.g., endplate AChR turnover; Akaaboune et al., 1999; Salpeter, 1999) or glutamate receptor mobilization and recruitment during long-term potentiation of excitatory hippocampal synapses (Shi et al., 1999). Another outcome of interleaved experimental and realistic modeling studies could be guided development of new clinical interventions for peripheral and central nervous system diseases.

Appendix: Symbols

A_{ET}	Area of effector tile
A_{ME}	Area of mesh element
C_L	Ligand concentration
D_L	Ligand diffusion constant
Δt	Monte Carlo time step
E	Effector site
h	Exocytotic pore height
κ	Random number
k	Rate constant for unimolecular transition
k_+	Rate constant for bimolecular association
L	Ligand
l_r	Random walk radial step length
$\bar{l}_r$	Random walk average radial step length
$\bar{l}_\perp$	Random walk average step length along direction perpendicular to a mesh element
N_{ACh}	Number of acetylcholine molecules in a synaptic vesicle
N_B	Average number of ligand molecules that bind to an effector site during Δt
N_{ET}	Number of effector tiles on a mesh element
N_H	Average number of ligand molecules that hit an effector tile during Δt
p	Probability
p_b	Monte Carlo binding probability, obtained conceptually from N_B/N_H
p_k	Monte Carlo probability of a unimolecular transition
r	Radial distance, or exocytotic pore radius
V_{ET}	Volume "above" and "below" effector tile, product of area A_{ET} and height $2(\bar{l}_\perp)$

Acknowledgments

We thank the following for excellent technical assistance: Maria Szabo for electron microscopy, Philip Davidson for 3-D reconstructions, and Krishna Juluri for MCell simulations. J.R.S. and T.M.B. also personally acknowledge the invaluable assistance of Mark Stiles and Bruce Land in the early years of Monte Carlo code development on VAX and IBM computer systems, respectively. This work is supported by NIH grants K08 NS01776 and RR-06009 (J.R.S.), NSF grant IBN-9603611 (T.M.B. and T.J.S.), and NIH grants NS09315 and GM10422 (M.M.S.).

Notes

1. Based on observed distances between synaptic densities in 3-D reconstructions (e.g., Ventura and Harris, 1999).

2. <http://www.mcell.cnl.salk.edu> and <www.mcell.psc.edu>.

3. MCell's MDL allows simulation models to be archived and exchanged in readable text. An MDL Reference Guide and tutorial examples are available at the MCell Web sites given in note 2.

4. Random numbers are used to choose step lengths and directions, and to make other choices, and the computer time required to calculate random numbers is a significant fraction of the time required to run a simulation. MCell includes a self-contained cryptographic-quality random number generator that ensures identical results across different computer platforms and is optimized for speed in various ways (e.g., the bits obtained from calculation of a single random number are subdivided so that two or more decisions can be made for the price of one).

5. In computer graphics, ray tracing entails following a light ray through successive surface intersections and reflections, and ray marching is a special case in which some property of the ray (e.g., intensity) is decremented after each intersection. By analogy, MCell employs highly optimized ray tracing and ray marching algorithms for ligand movements.

6. Software packages for surface reconstruction and mesh optimization (including adaptation to nearly equilateral triangles) are freely available in many forms, but their design, reliability, and usefulness are highly variable. Additional details can be found at the Web sites given in note 2.

7. If a reaction that consumes ligand is used, such as that shown for AChE hydrolysis of ACh in Fig. 15.6D, then the ligand consumption can be balanced by introducing additional effector sites that produce ligand at the necessary rate. The simulation thus is run under steady-state rather than equilibrium conditions.

8. Each simulation was begun with all rate constants in the reaction mechanism set to zero, so no transitions would occur and the initial state of the effector sites would be preserved. The ligand molecules were introduced at a point in the center of the sphere, and the first checkpoint was reached when the ligands had diffused to uniform average concentration. The run was then continued from this checkpoint with all rate constants changed from zero to their desired values, so that the reaction could either proceed to equilibrium or remain at equilibrium, depending on the initial state of the effector sites. To simulate a relaxation, a second checkpoint was used to stop the run, set all rate constants back to zero, and introduce additional free ligand molecules, which then diffused to uniform concentration together with the pre-existing ligand. A final checkpoint reset the rate constants so that the relaxation would proceed.

9. But not, for example, an order of magnitude smaller than r, which would likely be the case with a simpler random walk algorithm based on a fixed step

length and movements on a lattice. Equivalent accuracy with the simpler method (which would not execute any faster) thus would probably require a step length about 4-fold smaller than $\bar{l}_r$. Because step length scales with $\sqrt{\Delta t}$, the simulation would require about 16-fold more time step iterations and a corresponding increase in computer time.

10. A reflective plane was added to cut across the pore where it joins the vesicle. The ACh molecules inside the vesicle thus could not escape and were allowed to diffuse until they became uniformly dispersed. The simulation then was halted and restarted from this checkpoint without the reflective plane, so that ACh molecules could diffuse through the pore and into the cleft space.

11. Automated interpolation of highly irregular, convoluted contours is not generally possible because there is no simple way to keep corresponding portions, or segments, of the contours in register. When registration errors occur, the interpolated surface is likely to twist and pass through itself, which will ruin a reconstruction that is to be used with simulations. In brief, our method for interpolation entails (1) animating the sequence of original TEM images, because the correspondence between structures in different sections is much easier to grasp when the structures are seen "in motion"; (2) hand-subdivision of the complex contours of each section into shorter segments with simple shapes (e.g., a portion of a single JF); and (3) automated linear interpolation between corresponding segments of adjacent contours, to create entire new contours between those of the original sections.

References

Akaaboune, M., Culican, S. M., Turney, S. G., Lichtman, J. W. (1999). Rapid and reversible effects of activity on acetylcholine receptor density at the neuromuscular junction *in vivo*. *Science* 286:503–507.

Albuquerque, E. X., Barnard, E. A., Porter, C. W., Warnick, J. E. (1974). The density of acetylcholine receptors and their sensitivity in the postsynaptic membrane of muscle endplates. *Proc. Natl. Acad. Sci. USA* 71:2818–2822.

Anderson, C. R., Stevens, C. F. (1973). Voltage clamp analysis of acetylcholine produced end-plate current fluctuations at frog neuromuscular junction. *J. Physiol.* 235:655–691.

Anglister, L., Stiles, J. R., Salpeter, M. M. (1994). Acetylcholinesterase density and turnover number at frog neuromuscular junctions, with modeling of their role in synaptic function. *Neuron* 12:783–794.

Apel, E. D., Merlie, J. P. (1995). Assembly of the postsynaptic apparatus. *Curr. Opin. Neurobiol.* 5:62–67.

Barbour, B., Hausser, M. (1997). Intersynaptic diffusion of neurotransmitter. *Trends Neurosci.* 20:377–384.

Barnard, R. J., Wieckowski, J., Chiu, T. H. (1971). Cholinergic receptor molecules and cholinesterase molecules at mouse skeletal muscle junctions. *Nature* 234:207–209.

Barrow, G. M. (1981). *Physical Chemistry for the Life Sciences*. New York: McGraw-Hill.

Bartol, T. M., Jr. (1992). A study of miniature endplate current generation at the vertebrate neuromuscular junction using electrophysiology and Monte Carlo simulation. Ph.D. thesis, Department of Neurobiology and Behavior, Cornell University.

Bartol, T. M., Jr., Land, B. R., Salpeter, E. E., Salpeter, M. M. (1991). Monte Carlo simulation of MEPC generation in the vertebrate neuromuscular junction. *Biophys. J.* 59:1290–1307.

Bennett, M. R., Farnell, L., Gibson, W. G. (1995). Quantal transmission at purinergic synapses: stochastic interaction between ATP and its receptors. *J. Theor. Biol.* 175:397–404.

Bennett, M. R., Farnell, L., Gibson, W. G. Lavidis, N. A. (1997). Synaptic transmission at visualized sympathetic boutons: stochastic interaction between acetylcholine and its receptors. *Biophys. J.* 72:1595–1606.

Bennett, M. R., Farnell, L., Gibson, W. G. (1998). On the origin of skewed distributions of spontaneous synaptic potentials in autonomic ganglia. *Proc. R. Soc. London Ser. B* 265:271–277.

Boudier, J. L., Le Treut, T., Jover, E. (1992). Autoradiographic localization of voltage-dependent sodium channels on the mouse neuromuscular junction using ^{125}I-alpha scorpion toxin. II. Sodium distribution on postsynaptic membranes. *J. Neurosci.* 12:454–466.

Bower, J. M., Beeman, D. (1995). *The Book of GENESIS: Exploring Realistic Neural Models with the GEneral NEural SImulation System.* Santa Clara, Calif.: TELOS.

Ceccarelli, B., Hurlburt, W. P., Mauro, A. (1973). Turnover of transmitter and synaptic vesicles at the frog neuromuscular junction. *J. Cell Biol.* 57:499–524.

Clements, J. D. (1996). Transmitter timecourse in the synaptic cleft: its role in central synaptic function. *Trends Neurosci.* 19:163–171.

Cornell, W. D., Cieplak, P., Bayly, C. I., Gould, I. R., Merz, K. M., Jr., Ferguson, D. M., Spellmeyer, D. C., Fox, T., Caldwell, J. W., Kollman, P. A. (1995). A second generation force field for the simulation of proteins and nucleic acids. *J. Am. Chem. Soc.* 117:5179–5197.

Couteaux, R. (1946). Sur les gouttières synaptiques du muscle strie. *C. R. Soc. Biol. (Paris)* 140:270–271.

Couteaux, R. (1955). Localization of cholinesterases at neuromuscular junctions. *Int. Rev. Cytol.* 5:335–375.

Couteaux, R. (1958). Morphological and cytochemical observations on the postsynaptic membrane at motor end-plates and ganglion synapses. *Exp. Cell Res.* 5(Suppl.):294–322.

Del Castillo, J., Katz, B. (1954). Quantal components of the endplate potential. *J. Physiol.* 124:560–573.

DeRobertis, E., Pellegrino DeIraldi, A., Rodriguez, G., Gomez, C. J. (1961). On the isolation of nerve endings and synaptic vesicles. *J. Biophys. Biochem. Cytol.* 9:229–235.

Doyere, L. (1840). Mémoire sur des tardigrades. *Ann. Sci. Nat. Zool.* 14:269–361.

Eccles, J. C., Jaeger, J. C. (1958). The relationship between the mode of operation and the dimensions of the junctional regions at synapses and motor end-organs. *Proc. R. Soc. London Ser. B* 148:38–56.

Edwards, F. A. (1995a). LTP: a structural model to explain the inconsistencies. *Trends Neurosci.* 18:250–255.

Edwards, F. A. (1995b). Anatomy and electrophysiology of fast central synapses lead to a structural model for long-term potentiation. *Physiol. Rev.* 75:759–787.

Faber, D. S., Young, W. S., Legendre, P., Korn, H. (1992). Intrinsic quantal variability due to stochastic properties of receptor-transmitter interactions. *Science* 258:1494–1498.

Fertuck, H. C., Salpeter, M. M. (1974). Sensitivity in electron microscope autoradiography for I-125. *J. Histochem. Cytochem.* 22:80–87.

Fertuck, H. C., Salpeter, M. M. (1976). Quantitation of junctional and extrajunctional acetylcholine receptors by electron microscope autoradiography after ^{125}I-α-bungarotoxin binding at mouse neuromuscular junctions. *J. Cell Biol.* 69:144–158.

Flucher, B. E., Daniels, M. P. (1989). Distribution of Na$^+$ channels and ankyrin in neuromuscular junctions is complementary to that of acetylcholine receptors and the 43 kd protein. *Neuron* 3:163–175.

Froehner, S. C. (1986). The role of the postsynaptic cytoskeleton in acetylcholine receptor organization. *Trends Neurosci.* 9:37–41.

Hall, Z. W., Sanes, J. R. (1993). Synaptic structure and development: the neuromuscular junction. *Cell* 72/*Neuron* 10(Suppl.):99–121.

Hammes, G. G. (1978). *Principles of Chemical Kinetics.* New York: Academic Press.

Hartzell, H., Kuffler, S., Yoshikami, D. (1975). Post-synaptic potentiation: interaction between quanta of acetylcholine at the skeletal neuromuscular synapse. *J. Physiol.* 251:427–463.

Heuser, J. E., Reese, T. S. (1973). Evidence for recycling of synaptic vesicle membrane during transmitter release at the frog neuromuscular junction. *J. Cell Biol.* 57:315–344.

Hines, M. (1993). Neuron: a program for simulation of nerve equations with branching geometries. *Int. J. Biomed. Comput.* 24:55–68.

Katz, B., Miledi, R. (1973). The binding of acetylcholine to receptors and its removal from the synaptic cleft. *J. Physiol.* 231:549–574.

Kruk, P. J., Korn, H., Faber, D. S. (1997). The effects of geometrical parameters on synaptic transmission: a Monte Carlo simulation study. *Biophys. J.* 73:2874–2890.

Land, B. R., Salpeter, E. E., Salpeter, M. M. (1980). Acetylcholine receptor site density affects the rising phase of miniature endplate currents. *Proc. Natl. Acad. Sci. USA* 77:3736–3740.

Land, B. R., Salpeter, E. E., Salpeter, M. M. (1981). Kinetic parameters for acetylcholine interaction in intact neuromuscular junction. *Proc. Natl. Acad. Sci. USA* 78:7200–7204.

Land, B. R., Harris, W. V., Salpeter, E. E., Salpeter, M. M. (1984). Diffusion and binding constants for acetylcholine derived from the falling phase of miniature endplate currents. *Proc. Natl. Acad. Sci. USA* 81:1594–1598.

MacKerell, A. D., Jr., Brooks, B., Brooks, C. L., III, Nilsson, L., Roux, B., Won, Y., Karplus, M. (1998). CHARMM: the energy function and its parameterization with an overview of the program. In *The Encyclopedia of Computational Chemistry,* Schleyer, P. V. R., ed. Chichester, U.K.: John Wiley and Sons. pp. 271–277.

Madsen, B. W., Edeson, R. O., Lam, H. S., Milne, R. K. (1984). Numerical simulation of miniature endplate currents. *Neurosci. Lett.* 48:67–74.

Matsumura, K. M., Campbell, K. P. (1994). Dystrophin-glycoprotein complex: its role in the molecular pathogenesis of muscular dystrophies. *Muscle Nerve* 17:2–15.

Matthews-Bellinger, J., Salpeter, M. M. (1978). Distribution of acetylcholine receptors at frog neuromuscular junctions with a discussion of some physiological implications. *J. Physiol.* 279:197–213.

Palade, G. E., Palay, S. L. (1954). Electron microscope observation of interneuronal and neuromuscular synapses. *Anat. Rec.* 118:335–336.

Pennefather, P., Quastel, D. M. (1981). Relation between subsynaptic receptor blockade and response to quantal transmitter at the mouse neuromuscular junction. *J. Gen. Physiol.* 78:313–344.

Peters, A., Palay, S. L., Webster, H. de F. (1991). *The Fine Structure of the Nervous System: Neurons and Their Supporting Cells,* 3rd ed. New York: Oxford University Press.

Porter, C. W., Barnard, E. A., Chiu, T. H. (1973). The ultrastructural localization and quantitation of cholinergic receptors at the mouse motor endplate. *J. Membr. Biol.* 14:383–401.

Redman, S. (1990). Quantal analysis of synaptic potentials in neurons of the central nervous system. *Physiol. Rev.* 70:165–198.

Reger, J. F. (1955). Electron microscopy of the motor endplate in rat intercostal muscle. *Anat. Rec.* 122:1–16.

Robertson, J. D. (1956). The ultrastructure of a reptilian myoneural junction. *J. Biophys. Biochem. Cytol.* 2:381–394.

Rogers, A. W., Derzynkiewicz, Z., Salpeter, M. M., Ostrowski, K., Barnard, E. A. (1969). Quantitative studies on enzymes in structures in striated muscles by labeled inhibitor methods. I. The number of acetylcholinesterase molecules and of other DFP reactive sites at motor endplates, measured by radioautography. *J. Cell Biol.* 41:665–685.

Rosenberry, T. (1979). Quantitative simulation of endplate currents at neuromuscular junctions based on the reaction of acetylcholine with acetylcholine receptor and acetylcholinesterase. *Biophys. J.* 26:263–290.

Rubenstein, R. Y. (1981). *Simulation and the Monte Carlo Method.* New York: John Wiley and Sons.

Rusakov, D. A., Kullmann, D. M. (1998). Extrasynaptic glutamate diffusion in the hippocampus: ultrastructural constraints, uptake, and receptor activation. *J. Neurosci.* 18:158–170.

Rusakov, D. A., Kullmann, D. M., Stewart, M. G. (1999). Hippocampal synapses: do they talk to their neighbors? *Trends Neurosci.* 22:382–388.

Salpeter, M. M. (1967). Electron microscope autoradiography as a quantitative tool in enzyme cytochemistry: the distribution of acetylcholinesterase at motor endplates of a vertebrate twitch muscle. *J. Cell Biol.* 32:379–389.

Salpeter, M. M. (1969). Electron microscope radioautography as a quantitative tool in enzyme cytochemistry. II. The distribution of DFP-reactive sites at motor endplates of a vertebrate twitch muscle. *J. Cell Biol.* 42:122–134.

Salpeter, M. M. (1987). Vertebrate neuromuscular junctions: general morphology, molecular organization, and functional consequences. In *The Vertebrate Neuromuscular Junction,* Salpeter, M. M., ed. New York: Alan R. Liss. pp. 1–54.

Salpeter, M. M. (1999). Neurobiology: the constant junction. *Science* 286:424–425.

Salpeter, M. M., Smith, C. D., Matthews-Bellinger, J. A. (1984). Acetylcholine receptor at neuromuscular junctions by EM autoradiography using mask analysis and linear sources. *J. Electron Microsc. Tech.* 1:63–81.

Sanes, J. R., Lichtman, J. W. (1999). Development of the vertebrate neuromuscular junction. *Annu. Rev. Neurosci.* 22:389–442.

Schroeder, W., Martin, K., Lorensen, B. (1998). *The Visualization Toolkit,* 2nd ed. Upper Saddle River, N.J.: Prentice-Hall.

Shi, S. H., Hayashi, Y., Petralia, R. S., Zaman, S. H., Wenthold, R. J., Svoboda, K., Malinow, R. (1999). Rapid spine delivery and redistribution of AMPA receptors after synaptic NMDA receptor activation. *Science* 284:1811–1816.

Stiles, J. R. (1990). Acetylcholinesterase molecular forms: studies on their assay, hydrodynamic discrimination, subcellular distribution and response to neuromuscular perturbations. With: Monte Carlo computer simulations of miniature endplate currents based on the reaction of acetylcholine with acetylcholine receptor and acetylcholinesterase. Ph.D. thesis, Department of Physiology, University of Kansas.

Stiles, J. R., Bartol, T. M. (2000). Monte Carlo methods for simulating realistic synaptic microphysiology using MCell. In *Computational Neuroscience: Realistic Modeling for Experimentalists,* De Schutter, E., ed. New York: CRC Press. In press.

Stiles, J. R., Van Helden, D., Bartol, T. M., Salpeter, E. E., Salpeter, M. M. (1996). Miniature endplate current rise times <100 µs from improved dual recordings can be modeled with passive acetylcholine diffusion from a synaptic vesicle. *Proc. Natl. Acad. Sci. USA* 93:5747–5752.

Stiles, J. R., Bartol, T. M., Salpeter, E. E., Salpeter, M. M. (1998). Monte Carlo simulation of neurotransmitter release using MCell, a general simulator of cellular physiological processes. In *Computational Neuroscience,* Bower, J. M., ed. New York: Plenum Press. pp. 279–284.

Stiles, J. R., Kovyazina, I. V., Salpeter, E. E., Salpeter, M. M. (1999). The temperature sensitivity of miniature endplate currents is mostly governed by channel gating: evidence from optimized recordings and Monte Carlo simulations. *Biophys. J.* 77:1177–1187.

Tomita, M., Hashimoto, K., Takahashi, K., Shimizu, T., Matsuzaki, Y., Miyoshi, F., Saito, K., Tanida, S., Yugi, K., Venter, J. C., Hutchison, C. (1999). E-CELL: software environment for whole cell simulation. *Bioinformatics* 15:72–84.

Unwin, N. (1998). The nicotinic acetylcholine receptor of the Torpedo electric ray. *J. Struct. Biol.* 121:181–190.

Ventura, R., Harris, K. M. (1999). Three-dimensional relationship between hippocampal synapses and astrocytes. *J. Neurosci.* 19:6897–6906.

Wahl, L. M., Pouzat, C., Stratford, K. J. (1996). Monte Carlo simulation of fast excitatory synaptic transmission at a hippocampal synapse. *J. Neurophysiol.* 75:597–608.

Wathey, J., Nass, M. M., Lester, H. A. (1979). Numerical reconstruction of the quantal event at nicotinic synapses. *Biophys. J.* 27:145–164.

Whitaker, V. P. (1965). The application of subcellular fractionation techniques to the study of brain function. *Progr. Biophys.* 15:39–96.

Index

Page numbers followed by *f, t,* or *n* denote entries for figures, tables, or notes, respectively.